18th International Conference PhysicA.SPb 2015

Journal of Physics: Conference Series Volume 769

Saint Petersburg, Russia
26 – 29 October 2015

Editors:

Nikita S. Averkiev
Sergey A. Poniaev
Grigorii S. Sokolovskii

ISBN: 978-1-5108-3385-2
ISSN: 1742-6588

Printed from e-media with permission by:

Curran Associates, Inc.
57 Morehouse Lane
Red Hook, NY 12571

proceedings.com

Some format issues inherent in the e-media version may also appear in this print version.

Printed by Curran Associates, Inc. (2017)

For permission requests, please contact the Institute of Physics
at the address below.

Institute of Physics
Dirac House, Temple Back
Bristol BS1 6BE UK

Phone: 44 1 17 929 7481
Fax: 44 1 17 920 0979

techtracking@iop.org

Additional copies of this publication are available from:

Curran Associates, Inc.
57 Morehouse Lane
Red Hook, NY 12571 USA
Phone: 845-758-0400
Fax: 845-758-2633
Email: curran@proceedings.com
Web: www.proceedings.com

Table of contents

Volume 769

18th International Conference PhysicA.SPb
26–29 October 2015, Saint Petersburg, Russian Federation

Accepted papers received: 19 October 2016
Published online: 23 November 2016

Preface

Papers

Astronomy and astrophysics

Atomic and elementary particles physics

Biophysics

Nanostructured and thin-film materials

Optics and spectroscopy

Optoelectronic devices

Physics and technology of energy conversion

Physics of ferroics

Physics of quantum structures

Plasma physics, hydro- and aero-dynamics

Surface effects

THz and UHF materials and devices

Journal of Physics: Conference Series **769** (2016) 011001 doi:10.1088/1742-6596/769/1/011001

International Conference PhysicA.SPb

The eighteenth International Conference PhysicA.SPb was held 26-29 October 2015 in Saint-Petersburg, Russia. The Conference continues the tradition of Saint-Petersburg Seminars on Physics and Astronomy originating from mid-90s. Since then PhysicA.SPb maintains both the scientific and educational quality of contributions delivered to the young audience. This is the main feature of the Conference that makes it possible to combine the whole spectrum of modern Physics and Astronomy within one event.

PhysicA.SPb/2015 has brought together about 300 students, young scientists and their colleague professors from many universities and research institutes across the whole of Russia as well as from Belarus, Ukraine, Finland, Netherlands, Iceland, and France. Oral and poster presentations were combined into well-defined sections among which one should name Astronomy and Astrophysics, Optics and spectroscopy, Atomic and elementary particles physics, Physics of ferroics, Nanostructured and thin-film materials, Mathematical physics and numerical methods, THz and UHF materials and devices, Biophysics, Optoelectronic devices, Surface effects, Physics and technology of energy conversion, Plasma physics, hydro- and aero-dynamics, and Physics of quantum structures.

This issue of the *Journal of Physics: Conference Series* presents the extended contributions from participants of PhysicA.SPb/2015 that were peer-reviewed by expert referees through processes administered by the Presiders of the Organising and Programme Committees to the best professional and scientific standards. This was made possible by the efforts of the Sectional Editors of this Issue: Prof. Petr Arseev (Lebedev Physical Institute), Prof. Yuri Kusraev (Ioffe Institute), Prof. Alexander Ivanchik (Ioffe Institute), Prof. Polina Ryabochkina (Ogarev Mordova State University), Dr. Karina Litvinova (Aston University), Dr. Nikolay Bert (Ioffe Institute), Dr. Yana Kuznetsova (Ioffe Institute), Dr. Prokhor Alekseev (Ioffe Institute), Dr. Mikhail Dunaevskii (Ioffe Institute), Dr. Alexandra Kalashnikova (Ioffe Institute), Dr. Leonid Lutsev (Ioffe Institute).

The Editors: **Nikita S. Averkiev, Sergey A. Poniaev and Grigorii S. Sokolovskii**

Journal of Physics: Conference Series **769** (2016) 011002 doi:10.1088/1742-6596/769/1/011002

Peer review statement

All papers published in this volume of *Journal of Physics: Conference Series* have been peer reviewed through processes administered by the proceedings Editors. Reviews were conducted by expert referees to the professional and scientific standards expected of a proceedings journal published by IOP Publishing.

Diffusion in dense non-isothermal Coulomb plasma

M V Beznogov[1] **and D G Yakovlev**[2]

[1] St. Petersburg Academic University, 8/3 Khlopina Street, St. Petersburg 194021, Russia
[2] Ioffe Physical Technical Institute, 26 Politekhnicheskaya, St. Petersburg 194021, Russia

E-mail: mikavb89@gmail.com

Abstract. We generalize our previous work on diffusion of ions in isothermal plasmas to handle non-isothermal systems. This approach, combined with the method of effective potentials for computing diffusion coefficients, allows us to formulate the diffusion problem in Coulomb systems of arbitrary coupling. The results are applied to study heat-blanketing envelopes of neutron stars in diffusive equilibrium.

1. Introduction

Diffusion in ion mixtures is an important problem of stellar physics. Here we focus on the diffusion problem in dense stellar plasmas where the ions can be moderately or strongly coupled by Coulomb forces. Such plasmas are characteristic for compact stars, i.e. for neutron stars and white dwarfs.

Consider a strongly coupled and non-magnetized multicomponent plasma consisting of several ion species ($\alpha = j$, $j = 1, 2, \ldots$) and neutralizing electron background ($\alpha = e$). Let A_j and Z_j be the mass and charge numbers of ion species j, and n_α be the number density of particles α, with $n_e = \sum_j Z_j n_j$ due to electric neutrality.

Coulomb coupling parameter of ions is proportional to a dimensionless parameter $\Gamma_0 = e^2/(ak_{\rm B}T)$ (see, e.g., [1–4], for more details). Here $e > 0$ is an elementary charge, $a = (4\pi n/3)^{-1/3}$ is the average ion sphere radius, $n = \sum_j n_j$ is the total number density of the ions, $k_{\rm B}$ is the Boltzmann constant and T is the temperature. It is convenient to introduce the average Coulomb coupling parameter defined as $\overline{\Gamma} = \Gamma_0 \overline{Z^{5/3}}\,\overline{Z}^{1/3}$, where the average $\overline{f}$ of any quantity f is given by $\overline{f} = n^{-1}\sum_j n_j f_j$. If $\overline{\Gamma} \gg 1$ the ions are strongly coupled (i.e. highly non-ideal), whereas at $\overline{\Gamma} \ll 1$ they are weakly coupled; $\overline{\Gamma} \sim 1$ refers to the intermediate coupling.

2. General expressions for the diffusive flux

The general approach for deriving the diffusive flux is the same as for an isothermal plasma [5, 6]. Now we need to include properly a temperature gradient. To this aim, we introduce an additional term to the generalized thermodynamic force acting on particles α,

$$\widetilde{\boldsymbol{F}_\alpha} = \boldsymbol{F}_\alpha - \left(\boldsymbol{\nabla}\mu_\alpha - \left.\frac{\partial \mu_\alpha}{\partial T}\right|_P \boldsymbol{\nabla} T \right). \tag{1}$$

Here $\boldsymbol{F}_\alpha$ is a total force, acting on particles α, and μ_α is a chemical potential. In particular, we consider $\boldsymbol{F}_\alpha = e_\alpha \boldsymbol{E} + m_\alpha \boldsymbol{g}$, where e_α and m_α are charge and mass of particles α, respectively;

$\boldsymbol{E}$ is an electric field due to plasma polarization in an external gravitational field (to maintain plasma electric neutrality) and $\boldsymbol{g}$ is a gravitational acceleration. Note that $\partial\mu_\alpha/\partial T$ has to be calculated at constant pressure P.

The quantities $\boldsymbol{d}_\alpha$, which characterize deviations from diffusive equilibrium, are introduced in the same way as in [5, 6],

$$\boldsymbol{d}_\alpha = \frac{\rho_\alpha}{\rho}\sum_\beta n_\beta \widetilde{\boldsymbol{F}_\beta} - n_\alpha \widetilde{\boldsymbol{F}_\alpha}. \tag{2}$$

They have the same properties $\sum_\alpha \boldsymbol{d}_\alpha = 0$ ($\rho_\alpha = m_\alpha n_\alpha$ being a mass density of particles α and ρ a total mass density). Using the Gibbs-Duhem relation $\sum_\alpha n_\alpha \boldsymbol{\nabla}\mu_\alpha = \boldsymbol{\nabla}P - S\boldsymbol{\nabla}T$ (S being the entropy density) we obtain

$$\sum_\alpha n_\alpha \widetilde{\boldsymbol{F}_\alpha} = \rho\boldsymbol{g} - \boldsymbol{\nabla}P. \tag{3}$$

We are mainly interested in the outer heat-blanketing envelopes of neutron stars which are in hydrostatic equilibrium as a whole. Then $\sum_\alpha n_\alpha \widetilde{\boldsymbol{F}_\alpha} = \rho\boldsymbol{g} - \boldsymbol{\nabla}P = 0$ and we can rewrite (2) in the form

$$\boldsymbol{d}_\alpha = -\frac{\rho_\alpha}{\rho}\boldsymbol{\nabla}P - Z_\alpha n_\alpha e\boldsymbol{E} + n_\alpha\left(\boldsymbol{\nabla}\mu_\alpha - \left.\frac{\partial\mu_\alpha}{\partial T}\right|_P \boldsymbol{\nabla}T\right), \tag{4}$$

with $Z_e = -1$ for the electrons. The electric field can be calculated assuming the electron quasi-equilibrium (with respect to ion motion) which occurs mostly because the electrons are much lighter than the ions (e.g. [5]). Then $\boldsymbol{d}_e = 0$ and $m_e \to 0$, leading to $\widetilde{\boldsymbol{F}_e} = 0$, and to

$$e\boldsymbol{E} = -\left(\boldsymbol{\nabla}\mu_e - \left.\frac{\partial\mu_e}{\partial T}\right|_P \boldsymbol{\nabla}T\right). \tag{5}$$

This expression can be written in alternate forms using standard thermodynamic relations.

Chemical potentials are usually known as functions of temperature and number densities, but not the pressure. It is useful to rewrite $\partial\mu_\alpha/\partial T$ at constant P in terms of $\partial\mu_\alpha/\partial T$ at constant n_α. In the following relation, written for a binary ionic mixture (BIM), the pressure is assumed to be known as a function of T and n_α (in other words, the equation of state – EOS – of the system is known),

$$\left.\frac{\partial\mu}{\partial T}\right|_P = \left.\frac{\partial\mu}{\partial T}\right|_{n_1,n_2} - \frac{\partial P}{\partial T}\left(\frac{n_1}{n_2}\frac{\partial\mu}{\partial n_1} + \frac{\partial\mu}{\partial n_2}\right)\left(\frac{n_1}{n_2}\frac{\partial P}{\partial n_1} + \frac{\partial P}{\partial n_2}\right)^{-1}. \tag{6}$$

The general expression for the diffusive flux reads (also see [5])

$$\boldsymbol{J}_\alpha = \Phi\sum_{\beta\neq\alpha} m_\alpha m_\beta D_{\alpha\beta}\boldsymbol{d}_\beta + D_{\alpha,T}\frac{\boldsymbol{\nabla}T}{T}. \tag{7}$$

Here Φ is a normalization function, $\Phi = n/(\rho k_B T)$ for a BIM; $D_{\alpha\beta}$ is a diffusion coefficient for particles α with respect to particles β; $D_{\alpha,T}$ is a thermal diffusion coefficient of particles α. Writing the diffusive flux in such a form guaranties that the diffusion and thermal diffusion coefficients introduced here coincide with their standard definitions in physical kinetics of weakly coupled plasma [7, 8]. For the diffusive currents one always have $\sum_\alpha \boldsymbol{J}_\alpha = 0$.

3. Examples

As an example let us study diffusive equilibrium configurations of BIMs in heat-blanketing envelopes of neutron stars. Although the thermal diffusion term can affect an equilibrium configuration, it is usually small compared to ordinary diffusion. This allows us to neglect thermal diffusion. In addition, electrons weakly affect the ion transport [9] so that the ion subsystem can be studied (quasi-)independently. This simplifies the diffusive flux,

$$\boldsymbol{J}_2 = -\boldsymbol{J}_1 = \frac{n m_1 m_2}{\rho k_\mathrm{B} T} D_{12} \boldsymbol{d}_1, \tag{8}$$

where D_{12} is the interdiffusion coefficient. According to (8) the diffusion equilibrium $\boldsymbol{J}_2 = 0$ is equivalent to the condition $\boldsymbol{d}_1 = 0$. The latter equation can then be used to calculate the equilibrium configuration. In fact, it represents the chemical equilibrium approach of [10] modified for non-isothermal systems. Note that in our particular case we do not need an explicit expression for D_{12}. However, generally, one needs both the diffusion and thermal diffusion coefficients to find the equilibrium configuration.

Let us stress the importance of the last (temperature) term in (1). For strongly coupled ions and degenerate electrons (described in [5]) all temperature derivatives exactly cancel out and the resulting expressions for the diffusive currents are the same as in [5], with the equilibrium configuration equivalent to chemical equilibrium [10]. However, for high enough surface temperatures and light elements like hydrogen, helium or carbon in the heat-blanketing envelope (especially in the outermost parts of the envelope, at $\rho \lesssim 10^7$ g cm^{-3}) the ions can be coupled not strongly but moderately, and the temperature term becomes important.

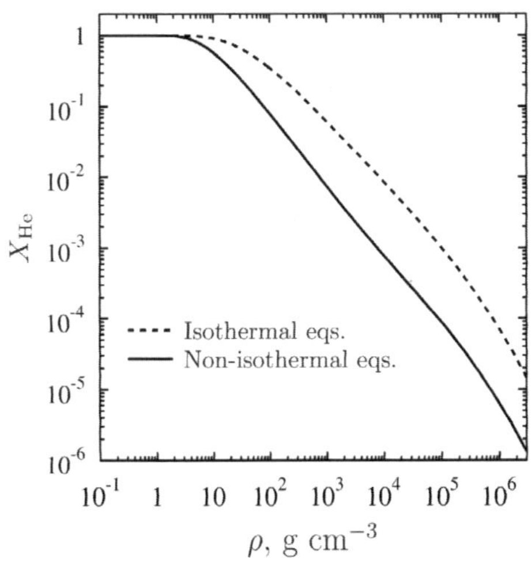

Figure 1: Comparison of isothermal (dashed curve) and non-isothermal (solid curve) approaches to diffusive equilibrium in a BIM in the heat-blanketing envelope of a neutron star. X_He is the number fraction of He in He – C mixture shown versus the total mass density. The neutron star has the mass $M = 1.4$ M$_\odot$, radius $R = 10$ km and the surface temperature $T_s = 0.755$ MK.

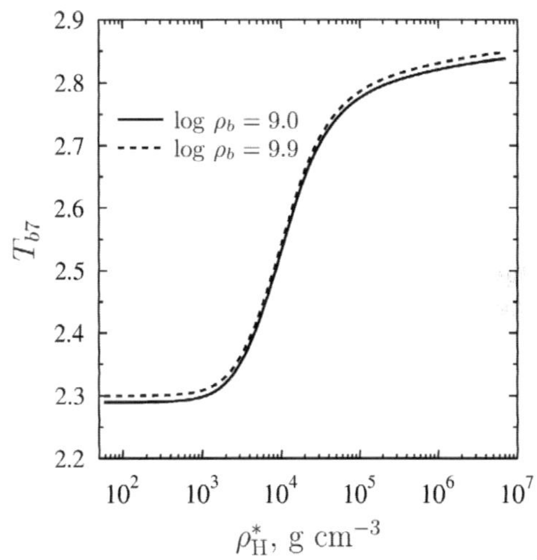

Figure 2: Internal temperature of a neutron star $T_{b7} = T_b/10^7$K in an H – He mixture as a function of ρ^*_H. One can see the transition from pure He (small amount of H, low ρ^*_H) to pure H (a lot of H, large ρ^*_H). Neutron star parameters are the same as in Fig. 1; ρ^*_H is an effective density of the transition from H to He. Solid and dashed curves correspond to different densities ρ_b at the bottom of the envelope.

Fig. 1 compares the fraction of helium in a helium – carbon mixture computed using the isothermal approach [5, 10] (the dashed curve) and the non-isothermal approach of this paper (the solid curve). The difference is quite visible. The helium fraction can be important in the heat-blanketing envelopes, in particular, for diffusive nuclear burning. Note that other quantities can be significantly less affected by the temperature term in (1). For example, the relation between internal and surface temperatures of the star, the pressure and total density profiles differ only slightly when computed via the isothermal and non-isothermal approaches.

Fig. 2 shows a transition from a pure hydrogen to a pure helium envelope as well as a weak dependence of internal stellar temperature on the density ρ_b at the bottom of the heat-blanketing envelope. The effective density $\rho^*_{\rm H}$ of the transition is determined by the total amount of hydrogen in the envelope. It is introduced by artificial replacement of a smooth transition by a step-like one. More details will be published elsewhere.

4. Conclusions

We have derived general expressions for the diffusive flux in multicomponent non-isothermal Coulomb systems with arbitrary Coulomb coupling. In the limit of weakly coupled plasma these expressions turn into the classical expressions for diffusion in ideal gas mixtures. Our new expressions are valid not only for Coulomb systems, but for any gaseous or liquid system, provided that one knows chemical potentials of its constituents and corresponding diffusion coefficients (although diffusion is generally available in solids, it is greatly suppressed there compared to gases and liquids).

The expressions for the diffusive flux combined with the diffusion coefficients (see, e.g., [4]) allow one to calculate not only diffusive equilibrium configurations of heat-blanketing envelopes of neutron stars, but also the equilibration of these configurations with time provided that the initial configurations are out of equilibrium.

Acknowledgments

The authors are grateful to A.Y. Potekhin for many useful advises. The work of MB was partly supported by the Dynasty Foundation, and the work of DY by the Russian Foundation for Basic Research (grants 14-02-00868-a and 13-02-12017-ofi-M).

References

[1] Hansen J P, Torrie G M and Vieillefosse P 1977 *Phys. Rev. A* **16** 2153–2168
[2] Hansen J P, Joly F and McDonald I R 1985 *Physica A* **132** 472–488
[3] Haensel P, Potekhin A Y and Yakovlev D G 2007 *Neutron Stars. 1. Equation of State and Structure* (*Astrophysics and Space Science Library* vol 326) (Springer, New York)
[4] Beznogov M V and Yakovlev D G 2014 *Phys. Rev. E* **90** 033102
[5] Beznogov M V and Yakovlev D G 2013 *Phys. Rev. Lett.* **111** 161101
[6] Beznogov M V and Yakovlev D G 2014 *J. Phys.: Conf. Ser.* **572** 012001
[7] Chapman S and Cowling T G 1952 *The Mathematical Theory of Non-Uniform Gases* (Cambridge Univ. Press, Cambridge)
[8] Hirschfelder J O, Curtiss C F and Bird R B 1954 *Molecular Theory of Gases and Liquids* (Wiley, New York)
[9] Paquette C, Pelletier C, Fontaine G and Michaud G 1986 *Astrophys. J. Suppl.* **61** 177–195
[10] Chang P, Bildsten L and Arras P 2010 *Astrophys. J.* **723** 719–728

X-ray Analysis of the Pulsar Wind Nebula DA 495 and its Central Object

A Karpova[1,2], D Zyuzin[1], A Danilenko[1], Yu Shibanov[1]

[1]Ioffe Institute, Politekhnicheskaya 26, St. Petersburg, 194021, Russia
[2]Peter the Great St.Petersburg Polytechnic University, Politekhnicheskaya 29, St. Petersburg, 195251, Russia

E-mail: annakarpova1989@gmail.com

Abstract. We report the results of a simultaneous analysis of the *Chandra* and *XMM-Newton* data on the pulsar wind nebula DA 495 and its central object, J1952.2+2925, which is presumably a pulsar. The J1952.2+2925 pure thermal spectrum can be equally well described either by the blackbody model with a temperature of 215 eV and an emitting area radius of 0.6 km or magnetized neutron star atmosphere models with temperatures of 80–90 eV. We also used the high temporal resolution *XMM-Newton*/EPIC-pn data to search for pulsations from J1952.2+2925. No pulsations were found so we set an upper limit for the pulsed fraction which is 40%. Using of the interstellar absorption-distance relation allowed us to estimate the distance to DA 495, which lies between 1 and 5 kpc.

1. Introduction

The pulsar wind nebula (PWN) DA 495 (G65.7+1.2) was discovered in the radio band[1]. The distance to DA 495 D was estimated as ≈ 2.1 kpc under the assumption of association with the open cluster NGC 6834 [2]. Another estimations of 1–1.5 kpc are based on H I absorption measurements [3, 4]. DA 495 age is suggested to be about 20 kyr [4]. A spectral break was found at 1.3 GHz [4], spectral indices $\alpha_\nu = 0.45 \pm 0.20$ and $\alpha_\nu = 0.87 \pm 0.10$ below and above the break, respectively[1]. A compact source, 1WGA J1952.2+2925 (hereafter J1952), surrounded by diffuse emission was discovered in X-rays by *ROSAT* and *ASCA* and later observed by *Chandra* [5]. It was supposed to be a pulsar powering the PWN, although the analysis of the *ASCA* data allowed only to set an upper limit on the pulsed fraction of 50% for periods > 30 ms [6]. The nebula spectrum is described by a power law with $\Gamma = 1.6 \pm 0.3$. The J1952 spectrum is pure thermal and well fitted either with the blackbody model with a temperature $T \approx 2.5$ MK and an emitting area radius $R \approx 0.3$ km or with the neutron star atmosphere (NSA) model [7] with an effective temperature of ≈ 1 MK and radius of 10 km [5]. Here we briefly report the results of *Chandra* and unpublished *XMM-Newton* X-ray data analysis clarifying the object parameters[2].

2. Timing analysis

We searched for pulsations from J1952 using the EPIC-pn data with the time resolution of 5.6 ms. J1952 light curve was extracted from a circular region of $20''$ around the source in the 0.5–1.4

[1] Assuming the spectral flux $F_\nu \propto \nu^{-\alpha_\nu}$, where ν is the frequency.
[2] See [8] for details.

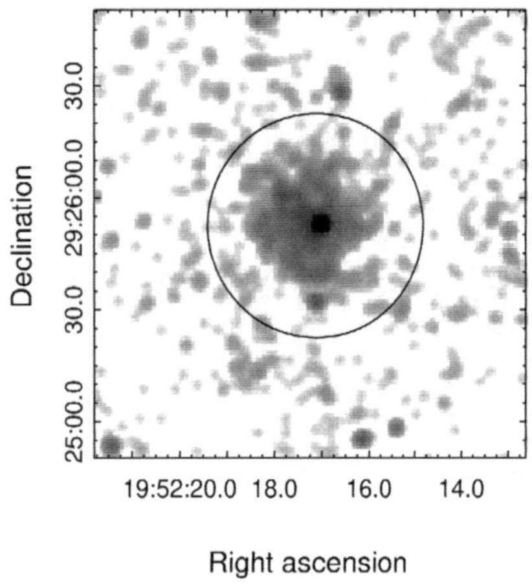

(*a*)

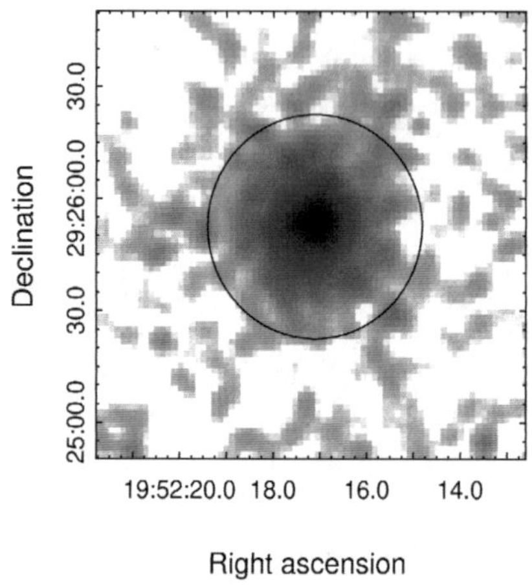

(*b*)

Figure 1. $2' \times 2'$ X-ray *Chandra*/ACIS-I (a) and *XMM-Newton*/MOS1 (b) images of DA 495 in 0.3–8 keV range. J1952 is seen in the centre of the PWN. The solid circle of $30''$ radius shows the extraction region for the PWN+J1952 emission.

keV energy band. A Fast Fourier Transform (FFT) analysis was used to search for pulsations. No pulsations were found, so we calculated an upper limit for the pulsed fraction (PF) of 40% (at the 99% confidence level) for periods ≥ 12 ms using the method presented in [9].

3. Spectral analysis

The *Chandra* and *XMM-Newton* images of DA 495 are shown in Fig. 1. We extracted individual spectra of J1952 and the PWN from the *Chandra* data, and the spectrum of combined PWN+J1952 emission from the *XMM-Newton* data.

The J1952 spectrum was fitted with the NS atmosphere models NSMAX [10] with magnetic fields $B = 10^{12}$ and 10^{13} G and the blackbody (BB) model. The PWN emission was modelled by a power law (PL). We fitted the spectra simultaneously in 0.3–10 keV energy range. We utilize the absorption column density $N_{\rm H}$–distance relation along its line of sight using the interstellar extinction distribution obtained in [11]. This allowed us to estimate the distance to DA 495 and the emitting area radius for J1952. The best-fit spectral parameters are presented in Table 1 and the example of the fit is shown in Fig. 2.

The values of the emitting area radius R show that the emission originates from a hot spot on the pulsar surface for the BB model and from the entire NS surface for the NSMAX models. In the former case, a second BB component was added to derive a 3σ upper limit on the entire NS surface temperature of ≈ 60 eV. R was fixed at 13 km.

A PL component was added to the thermal model for J1952 to estimate an upper limit on the pulsar non-thermal luminosity $L^{\rm pl}_{\rm J1952}$. We assumed the photon index Γ range of 0.5–2.0. The resulting $L^{\rm pl}_{\rm J1952} \approx 5 \times 10^{31}$ erg s^{-1}.

4. Distance to DA 495

The distance values for different models (see Table 1) within uncertainties are in agreement with one of the previous estimations, $D \approx 2.1$ kpc, which assumes association of DA 495 with the open cluster NGC 6834 [2]. Another distance estimation of 1.0 ± 0.4 kpc [4] is not compatible with the NSMAX models. It was based on H I absorption measurements indicating a systemic

Table 1. Best-fit spectral parameters. All errors correspond to 90% credible intervals. Emitting area radii R and temperatures T are given as seen by a distant observer. $N_{\rm pwn}$ is the normalisation constant for the PL model of the PWN.

Model/Parameters	BB	NSMAX ($B = 10^{12}$ G)	NSMAX ($B = 10^{13}$ G)
N_H, 10^{21} cm^{-2}	$2.6^{+0.5}_{-0.4}$	$3.5^{+0.7}_{-0.6}$	$3.4^{+0.7}_{-0.6}$
D, kpc	$2.4^{+1.3}_{-1.1}$	$3.3^{+1.7}_{-1.3}$	$3.2^{+1.5}_{-1.2}$
R, km	$0.6^{+0.6}_{-0.3}$	10^{+21}_{-7}	$6^{+9}_{-3.5}$
T, eV	215^{+23}_{-23}	76^{+19}_{-16}	91^{+17}_{-15}
$\Gamma_{\rm pwn}$	$1.71^{+0.12}_{-0.12}$	$1.83^{+0.14}_{-0.13}$	$1.82^{+0.13}_{-0.13}$
$N_{\rm pwn}$, 10^{-5} ph keV^{-1} cm^{-2}	$4.4^{+0.7}_{-0.6}$	$5.1^{+0.9}_{-0.7}$	$5.1^{+0.8}_{-0.6}$
χ^2/dof	455/461	455/461	454/461

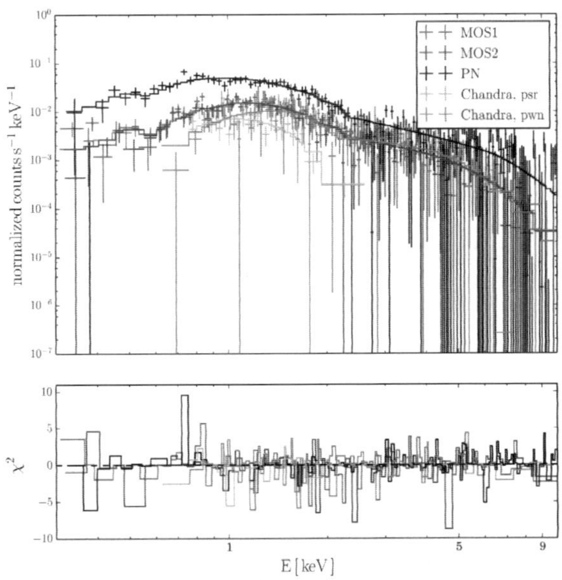

Figure 2. Spectra of J1952+PWN system obtained with *Chandra* and *XMM-Newton* MOS1, MOS2 and pn detectors. Lines represent the best-fit model (NSMAX with $B = 10^{12}$ G for J1952 and PL for the PWN).

velocity for DA 495 of about +12 km s^{-1}, which is likely associated with the Local Arm [4]. In fact, another distance of 5 kpc also corresponds to this velocity. However, it was rejected for two reasons. First, the low absorbing column density was obtained from *Chandra* data for the BB model [5] meaning that the distance to DA 495 cannot be large. Second, the distance of 5 kpc was assumed to be beyond the Local Arm tangent point but the H I absorption from the tangent point was absent [4]. We note, that the radio data have a low signal-to-noise ratio so the latter absorption cannot be completely excluded. Moreover, it was shown [12] that the Local Arm extends between $l \approx 72°$ and 52° toward the Perseus Arm and may be a separate branch of the Perseus. Thus, the gas within this arm can be a natural source of the larger absorption column density at the larger DA 495 distance suggested by the NSMAX models.

5. PWN emission

The PWN best-fit spectral indices $\alpha_\nu = \Gamma - 1$ (Table 1) are consistent within uncertainties with $\alpha_\nu = 0.87 \pm 0.10$ (68% confidence) obtained from the radio data for frequencies >1.3 GHz [4]. We investigated whether a second break between the radio and X-ray is needed to describe the PWN spectrum. We fitted the X-ray spectra and the radio data taken from [4] simultaneously using power law models with one and two spectral breaks. To compare models with single and double-knee breaks we used an F-test which showed that in the case of the NSMAX+BPL models the second break is not needed , but in the case of the BB+BPL the PWN spectrum in X-rays is flatter. The additional break position lies in a range of $10^{14} - 10^{17}$ Hz.

6. Summary

J1952 has a pure thermal spectrum. For the BB model, the emission originates from a hot spot and for the NSMAX models – from the entire NS surface. The upper limit for the pulsed fraction of J1952 is 40% (99% confidence) for periods ≥ 12 ms. The distance to DA 495 can be larger, up to 5 kpc, than it was supposed previously. The BB model for J1952 suggests that a second spectral break between radio and X-ray bands is needed to describe the PWN data while the atmosphere models do not require the additional break.

Acknowledgments

The scientific results reported in this article are based on data obtained from the *Chandra* Data Archive and observations obtained with *XMM-Newton*, an ESA science mission with instruments and contributions directly funded by ESA Member States and the USA (NASA). YS acknowledges the support from Russian Foundation for Basic Researches (grant 14-02-00868-a).

References

[1] Galt J A and Kennedy J E D 1968 *AJ* **73** 135–151
[2] Landecker T L and Caswell J L 1983 *AJ* **88** 1810–1815
[3] Kothes R, Landecker T L and Wolleben M 2004 *ApJ* **607** 855–864
[4] Kothes R, Landecker T L, Reich W, Safi-Harb S and Arzoumanian Z 2008 *ApJ* **687** 516–531
[5] Arzoumanian Z, Safi-Harb S, Landecker T L, Kothes R and Camilo F 2008 *ApJ* **687** 505–515
[6] Arzoumanian Z, Safi-Harb S, Landecker T L and Kothes R 2004 *ApJ* **610** L101–L104
[7] Zavlin V E, Pavlov G G and Shibanov Y A 1996 *A&A* **315** 141–152
[8] Karpova A, Zyuzin D, Danilenko A and Shibanov Y 2015 *MNRAS* **453** 2241–2249
[9] Vaughan B A, van der Klis M, Wood K S, Norris J P, Hertz P, Michelson P F, van Paradijs J, Lewin W H G, Mitsuda K and Penninx W 1994 *ApJ* **435** 362–371
[10] Ho W C G, Potekhin A Y and Chabrier G 2008 *ApJS* **178** 102–109
[11] Marshall D J, Robin A C, Reylé C, Schultheis M and Picaud S 2006 *A&A* **453** 635–651
[12] Xu Y, Li J J, Reid M J, Menten K M, Zheng X W, Brunthaler A, Moscadelli L, Dame T M and Zhang B 2013 *ApJ* **769** 15

X-ray Analysis of the Bright Source in the Supernova Remnant G350.0−2.0 Field

A Karpova[1,2], P Shternin[1], D Zyuzin[1], A Danilenko[1], Yu Shibanov[1]
[1]Ioffe Institute, Politekhnicheskaya 26, St. Petersburg, 194021, Russia
[2]Peter the Great St. Petersburg Polytechnic University, Politekhnicheskaya 29, St. Petersburg, 195251, Russia

E-mail: annakarpova1989@gmail.com

Abstract. We present results of the analysis of the *XMM-Newton* data on the bright source 1RXS J172653.4−382157 in the field of the supernova remnant G350.0−2.0. Its spectrum is well described by power law plus thermal component (blackbody or neutron star atmosphere) model. Therefore the source can be a rotation powered pulsar. Alternatively, it can be a cataclysmic variable star since its spectrum is equally well fitted by the two-temperature optically thin thermal plasma model. No periodic pulsations and flux time variability were found. The upper limit on the pulsed fraction of 27% cannot help to state whether the source is a pulsar or a cataclysmic variable. A faint source was detected in the *XMM-Newton* optical/UV monitor image and found in the GSC-II catalog and ESO Hα and optical/near infrared broadband sky survey images on the X-ray position of J172653.4−382157. Its spectral energy distribution favors the cataclysmic variable interpretation. Further optical and X-ray observations are needed to confirm this.

1. Introduction

The supernova remnant (SNR) G350.0−2.0 was discovered in the radio with Molonglo and Parkes telescopes at 408 MHz and 5 GHz, respectively [1]. The Very Large Array (VLA) observations [2] showed a complicated morphology of G350.0−2.0 that consists of three spatially distinct emission regions. In the optical, some Hα filaments and clumps spatially coincided with the radio structures were found [3]. G350.0−2.0 was observed in X-rays with *ROSAT* and *ASCA*. The multiwavelength data show that the SNR belongs to a mixed-morphology SNR type, which is characterized by a shell-like radio and center-filled X-ray morphology (see Fig. 1, left panel). The distance to G350.0−2.0 $D \approx 3.7$ kpc was estimated basing on the radio surface brightness–to–diameter relationship (Σ–D) [4]. A point-like X-ray source, 1RXS J172653.4−382157 (hereafter J1726), was detected inside the SNR with *ROSAT* and *ASCA*. Its nature is unclear. It may be an associated neutron star (NS) revealing itself either as a compact central object (CCO) or a rotation powered pulsar. However, no radio pulsar was detected within the SNR [5]. On the other hand, it may be an unrelated object, e.g. a cataclysmic variable (CV) star. X-ray spectral and timing analyses of the object using the *ROSAT* and *ASCA* data are problematic due to low count statistics. To understand J1726 nature and to better study the SNR properties we performed *XMM-Newton*[1] observations. Here we present some results of the observations focusing on the analysis of the J1726 emission.

[1] PI Zyuzin, *XMM-Newton*/EPIC, ObsID 0724220101

Journal of Physics: Conference Series **769** (2016) 012003 doi:10.1088/1742-6596/769/1/012003

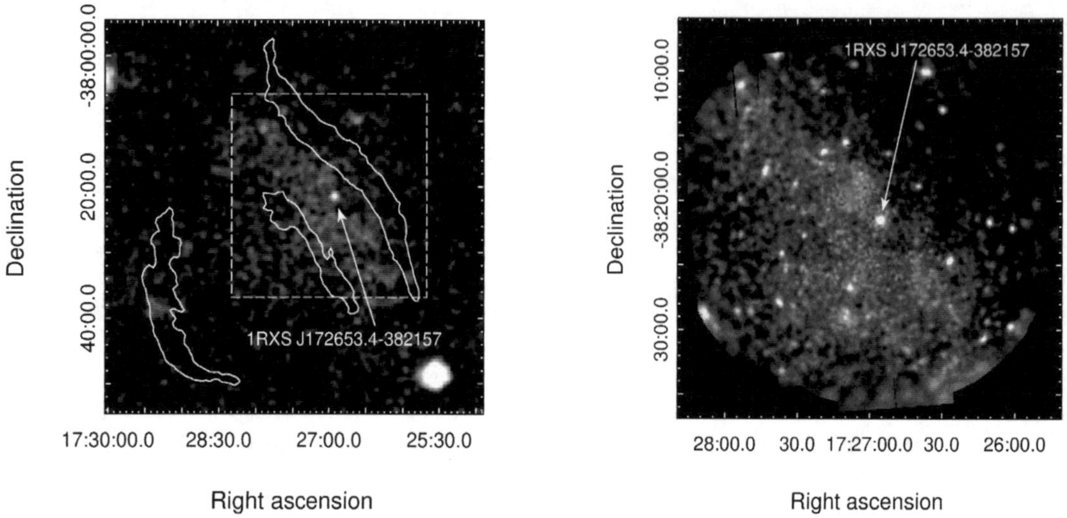

Figure 1. $60' \times 60'$ *ROSAT* image of the G350.0−2.0 field in 0.1–2.4 keV band (left panel) together with the VLA 1.4 GHz radio contours [2]. The dashed square corresponds to the $31' \times 31'$ *XMM-Newton* image in 0.4–7.2 keV band shown in the right panel (the data of MOS and pn detectors are combined). Square root brightness scale is used for both images. The X-ray source 1RXS J172653.4−382157 is marked by an arrow.

2. X-ray data

The *XMM-Newton* observations of G350.0−2.0 were carried out on 2013 September 21 with total exposure of 38 ks. All EPIC cameras were operated in the Full Frame Mode. The `XMM-SAS` v.13.5.0 software was used for data processing. Single and double pixel events (PATTERN $\leq$ 4) were selected for the EPIC-pn and single to quadruple-pixel events (PATTERN $\leq$ 12) – for the EPIC-MOS data. The periods of background flares were removed. The resulting effective exposures were 19.5, 21.3 and 15.8 ks for the MOS1, MOS2, and pn data, respectively. The *XMM-Newton* image of the G350.0−2.0 field is presented in the right panel of Fig. 1. The point-like source J1726 with coordinates RA = $261\overset{\circ}{.}7334$ and Dec = $-38\overset{\circ}{.}3595$ obtained with `edetect_chain` tool, is clearly detected in the centre of the image. The extended emission is associated with the SNR.

3. Spectral analysis

The J1726 spectra were extracted from MOS1, MOS2 and pn data within the $30''$ radius aperture. For the background, we chose the $45''$–$75''$ annulus region around J1726. The total number of the source counts was 410(MOS1)+625(MOS2)+1340(pn). Spectra were grouped to ensure > 15 counts per energy bin and fitted simultaneously in the 0.3–10 keV range using `XSPEC` package. We tried the absorbed (model `phabs` in `XSPEC`) power law (PL), power law plus blackbody (PL+BB) and power law plus the NS hydrogen atmosphere [6] (PL+NSA) models, assuming the NS origin of J1726. Another model we tried was the absorbed two-temperature optically thin thermal plasma (2MEKAL) model [7] which is frequently used to describe CVs spectra. Best-fit parameters are presented in Table 1. According to χ^2 values, all models are consistent with the data.

To check whether the thermal component is required if J1726 has a NS nature, we used F-test. The resulting null-hypothesis probability was $\approx 10^{-6}$, showing that the thermal component is needed to describe the source spectra. The fit for the PL+BB model is shown in Fig. 2. The derived photon index Γ and the BB temperature are typical for pulsar emission. At the same time, the thermal component can be equally well described by the hydrogen atmosphere model NSA with NS mass $M_{NS} = 1.4M_{\odot}$ and radius $R_{NS} = 13$ km. As always, the effective temperature, $T \approx 71$ eV, is smaller for the hydrogen atmosphere.

For the 2MEKAL model we obtained temperatures of 0.8 and 8.6 keV which are typical for cataclysmic variables. We also tried the single MEKAL model but it was rejected by the fit ($\chi^2_\nu = 1.16$ for dof=135) and the F-test probability of $\approx 6 \times 10^{-7}$. A simple single-temperature model does not always describe CV spectra well (see, e.g., [8, 9]). The 2MEKAL fit reflect the fact that just above the white dwarf photosphere the accretion flow is shocked and the X-ray emission originates from plasma with a temperature distributed from the hotter shock to the cooler photosphere and unshocked flow.

Table 1. J1726 best-fit spectral parameters for different models. All errors correspond to 90% confidence intervals. Unabsorbed fluxes f_X refer to the 0.3–10 keV energy range. For the NSA+PL model, magnetic field $B = 10^{12}$ G and distance to the object D was fixed at 3.7 kpc. Temperature T_{NSA} is given as seen by a distant observer.

Model/Parameters	PL	PL+BB	PL+NSA	2MEKAL
Column density $N_{\rm H}$, 10^{21} cm^{-2}	$2.37^{+0.35}_{-0.33}$	$5.05^{+2.03}_{-1.71}$	$4.29^{+0.40}_{-0.39}$	$1.97^{+0.37}_{-0.31}$
Photon index Γ	$1.81^{+0.12}_{-0.11}$	$1.71^{+0.20}_{-0.20}$	$1.60^{+0.14}_{-0.14}$	
PL normalization N_{PL}, 10^{-5} ph keV^{-1} cm^{-2} s^{-1}	$7.82^{+0.96}_{-0.85}$	$7.26^{+2.30}_{-1.79}$	$6.08^{+1.11}_{-1.03}$	
Temperature $T_{BB/NSA}$, eV		138^{+42}_{-25}	73^{+3}_{-3}	
Emitting area radius R, km		$1.29^{+2.82}_{-0.89}\ D_{\rm kpc}$	13 (fixed)	
Temperature T_1, keV				$0.82^{+0.30}_{-0.14}$
Temperature T_2, keV				$8.56^{+4.13}_{-2.13}$
Unabsorbed flux f_X, 10^{-13} erg s^{-1} cm^{-2}	4.98	10.2	8.3	4.6
χ^2/dof	146.44/135	119.38/133	122.30/134	126.42/133

We also extracted spectra of the whole SNR using the *XMM-Newton* Extended Source Analysis Software (`XMM-ESAS`) and fitted it with the collisionally-ionized plasma model VAPEC with solar abundances. The resulting absorption column density $N_{\rm H} = (6.0 \pm 0.3) \times 10^{21}$ cm^{-2}. This value is compatible within uncertainties with $N_{\rm H}$ obtained for J1726 with the PL+BB(NSA) model while it is considerably larger than $N_{\rm H}$ obtained for the single PL and 2MEKAL models.

4. Timing analysis

Detection of periodic pulsations in the J1726 X-ray emission would help to establish that it is indeed the NS or CV. For the timing analysis we used EPIC-pn data with time resolution of 73.4 ms. The event times were extracted from the 30″ radius aperture in 0.3–10 keV range and corrected to the solar system barycentre using the `SAS` task `barycen` and J1726 coordinates. We searched for pulsations in the 0.15–4000 s range using Z^2_n test [10]. The number of harmonics n was varied from 1 to 5. No pulsations were found. We estimated an upper limit for the pulsed fraction (PF) of 27% (at the 99% confidence level) assuming $n = 1$ (i.e. sinusoidal signal) and following the method described in [11].

5. Optical/UV data

At the X-ray position of J1726 we found a faint optical source in the *XMM-Newton* Optical/UV Monitor (OM) image in the U ($\lambda_{\rm eff} = 344$ nm) filter. The `SAS` task `omdetect` resulted in the

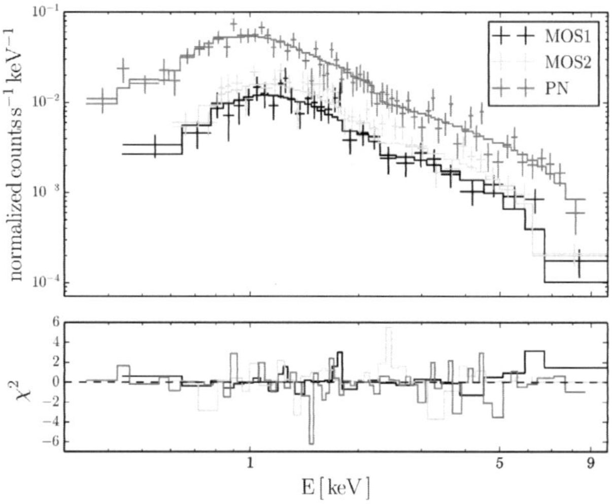

Figure 2. X-ray spectra of J1726, obtained with *XMM-Newton* MOS1, MOS2 and pn detectors (*top panel*). Lines demonstrate the best-fit BB+PL model. *Bottom panel* shows the fit residuals.

background subtracted count rate of 0.15 ± 0.03 cts s^{-1} that corresponds to the instrumental magnitude $U = 20.35 \pm 0.19$ ($U_{AB} = 21.28 \pm 0.19$)[2].

We also found this source in the Guide Star Catalog II (GSC-II)[3], in the SuperCOSMOS Hα survey (SHS) [12] and in archived images obtained during two ESO surveys: the VISTA Variables in the Via Lactea Survey (VVV) and the VST Photometric Hα Survey of the Southern Galactic Plane and Bulge (VPHAS+)[4]. Visual magnitudes and observed fluxes measured in different bands are presented in Table 2. The SHS Hα and OM U-band images where the putative counterpart is marked by an arrow are shown in Fig. 3. The source coordinates are RA = $17^h26^m55\overset{s}{.}9$, Dec = $-38°21'34\overset{''}{.}4$ with the uncertainty of $0\overset{''}{.}3$ for both coordinates, and fully consistent with the X-ray position of J1726 obtained with the nominal *XMM-Newton* pointing accuracy of $2''$(90% confidence)[5].

6. Discussion

The X-ray spectrum of the point source J1726 can be described by BB(NSA)+PL or by 2MEKAL models. In former case, parameters obtained from spectral fits are typical for a rotation powered pulsar. The corresponding column density for J1726 is consistent with that of the SNR supporting the NS origin of the source.

However, we found a faint optical source at the X-ray position of J1726. This source has a non-stellar spectral energy distribution with a strong excess in the narrow-band Hα filter (Table 2). These spectral properties are common for CVs (see, e.g., [14]). If we assume that the optical source is the J1726 counterpart, then J1726 is a CV. This is supported by the X-ray spectrum of J1726, which can be described by the two-temperature thermal plasma model typical for CVs [8, 9]. Assuming that the optical source is the counterpart of J1726, we calculated the X-ray-to-optical flux ratio f_X/f_{B_J}. The magnitude B_J was corrected for interstellar extinction

[2] This value should be considered with caution due to the source was on top of a straylight artefact, so it was difficult to measure its count rate accurately.

[3] http://gsss.stsci.edu/Catalogs/GSC/GSC2/GSC2.htm

[4] https://www.eso.org/sci/observing/PublicSurveys/sciencePublicSurveys.html

[5] See http://xmm2.esac.esa.int/docs/documents/CAL-TN-0018.pdf

Journal of Physics: Conference Series **769** (2016) 012003 doi:10.1088/1742-6596/769/1/012003

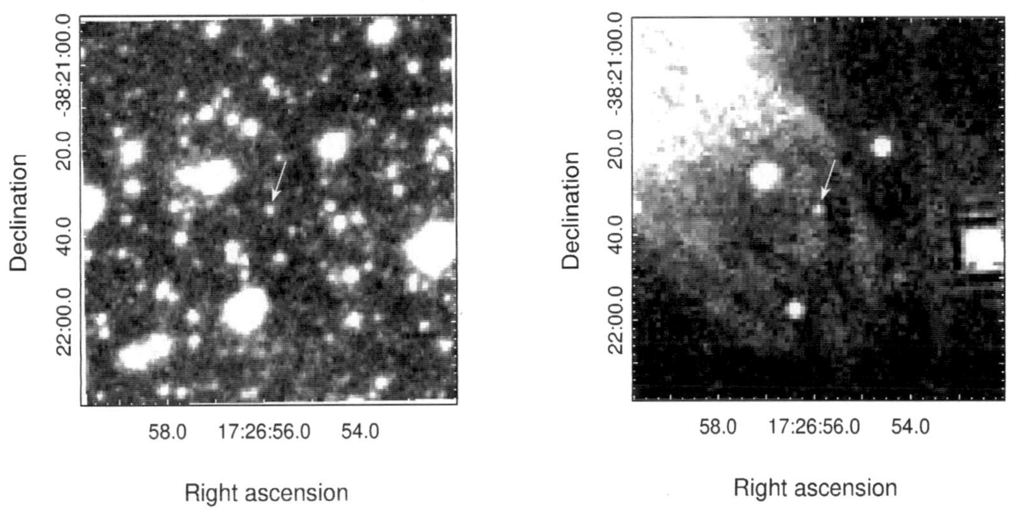

Figure 3. 1.′5×1.′5 Hα (*left*) and OM *U*-band (*right*) images of J1726 field. The J1726 possible counterpart is marked by an arrow.

Table 2. Visual magnitudes, observed and dereddened fluxes of the J1726 possible optical counterpart obtained from the VVV and VPHAS+ surveys. *u*, *g*, *r*, H_α and *i* magnitudes obtained using point spread function (PSF) fitting are taken from the VPHAS+ catalogue. *Z* magnitude was obtained from VVV catalogue and B_J – from GSC-II. Dereddening was done with the interstellar absorption $A_V = 0.89$, which was obtained from N_H–A_V relation [13] using N_H=1.97×10^{21} cm^{-2} obtained from the 2MEKAL fit.

Filter	MJD	Magnitude	Flux (μJy)	Dereddened flux (μJy)
U(OM)	56556.83866	20.35 ± 0.19	11 ± 2	41.7 ± 7.6
B_J	42270.43179	21.61 ± 0.49	9^{+6}_{-3}	$24.5^{+16.3}_{-8.2}$
u	56566.01030	19.81 ± 0.07	17.8 ± 1.2	65.8 ± 4.4
	56566.01254	19.74 ± 0.07	19.0 ± 1.2	70.3 ± 4.4
g	56566.02029	21.04 ± 0.05	15.7 ± 0.7	42.1 ± 1.9
	56566.02224	21.17 ± 0.05	13.8 ± 0.7	37.0 ± 1.9
r	56566.02705	20.33 ± 0.06	23.7 ± 1.3	48.6 ± 2.7
	56566.02785	20.51 ± 0.08	20.0 ± 1.5	41.0 ± 3.1
	56149.08459	20.58 ± 0.06	18.8 ± 1.0	38.6 ± 2.1
	56149.08535	20.66 ± 0.06	17.5 ± 1.0	35.9 ± 2.1
H_α	56149.07406	19.13 ± 0.04	55.4 ± 2.1	108.5 ± 4.3
i	56149.09070	20.02 ± 0.06	25.4 ± 1.4	43.6 ± 2.4
	56149.09146	19.76 ± 0.07	32.1 ± 2.0	55.2 ± 3.4
Z	55725.26322	18.97 ± 0.09	$58.2^{+5.2}_{-4.8}$	$87.6^{+7.9}_{-7.3}$

of A_V=0.89, which was obtained from N_H–A_V relation taken from [13] and using $N_H=1.97\times10^{21}$ cm^{-2} obtained from the 2MEKAL fit, using the extinction law of [15] and then the optical flux was calculated as in [16]. The unabsorbed X-ray flux was obtained in the 0.3–10 keV energy band. $f_X/f_{B_J} \approx 10$, which less than the values obtained for isolated NSs ($\sim 10^3$) or low-mass X-ray binaries ($\sim 10^2 - 10^3$) [17], but is typical for CVs [16], supporting the CV interpretation of J1726. However, a situation when a CV visible in the optical and a NS visible in the X-rays spatially coincide can not be ruled out. However, this situation appears to be implausible.

We did not find any time variability typical for CVs in the X-ray data. The obtained upper limit on PF of periodic flux variations 27% is non-informative since PFs of many CVs are lower than this value.

To conclude, it is most likely that J1726 is a CV. Detailed optical and X-ray multi-epoch studies can help to confirm that.

Acknowledgments

The scientific results reported in this article are based on observations obtained with *XMM-Newton*, an ESA science mission with instruments and contributions directly funded by ESA Member States and the USA (NASA). The work of DZ was supported by RF Presidential Programme MK-2837.2014.2.

References

[1] Caswell J L, Clark D H, Crawford D F and Green A J 1975 *Australian Journal of Physics Astrophysical Supplement* **37** 1
[2] Gaensler B M 1998 *ApJ* **493** 781–792
[3] Stupar M and Parker Q A 2011 *MNRAS* **414** 2282–2296
[4] Case G L and Bhattacharya D 1998 *ApJ* **504** 761–772
[5] Kaspi V M, Manchester R N, Johnston S, Lyne A G and D'Amico N 1996 *AJ* **111** 2028
[6] Zavlin V E, Pavlov G G and Shibanov Y A 1996 *A&A* **315** 141–152
[7] Mewe R, Gronenschild E H B M and van den Oord G H J 1985 *A&AS* **62** 197–254
[8] Baskill D S, Wheatley P J and Osborne J P 2005 *MNRAS* **357** 626–644
[9] Reis R C, Wheatley P J, Gänsicke B T and Osborne J P 2013 *MNRAS* **430** 1994–2001
[10] Buccheri R *et al.* 1983 *A&A* **128** 245–251
[11] Brazier K T S 1994 *MNRAS* **268** 709
[12] Parker Q A *et al.* 2005 *MNRAS* **362** 689–710
[13] Güver T and Özel F 2009 *MNRAS* **400** 2050–2053
[14] Witham A R *et al.* 2006 *MNRAS* **369** 581–597
[15] Cardelli J A, Clayton G C and Mathis J S 1989 *ApJ* **345** 245–256
[16] La Palombara N, Mignani R P, Hatziminaoglou E, Schirmer M, Bignami G F and Caraveo P 2006 *A&A* **458** 245–257
[17] Trümper J E and Hasinger G 2008 *The Universe in X-Rays*

Journal of Physics: Conference Series **769** (2016) 012004 doi:10.1088/1742-6596/769/1/012004

Radio-interferometric and near-IR observations of PSR J1357−6429

Aida Kirichenko[1], Dmitry Zyuzin[1], Yury Shibanov[1,2], Peter Shternin[1], Andrey Danilenko[1], Sergey Zharikov[3], Maxim Voronkov[4], Dmitry Barsukov[1,2] and Ronald Mennickent[5]

[1]Ioffe Institute, 26 Politekhnicheskaya st., St. Petersburg 194021, Russia

[2]Peter the Great St. Petersburg Polytechnic University, 29 Politekhnicheskaya st., St. Petersburg 195251, Russia

[3]Observatorio Astronómico Nacional SPM, Instituto de Astronomía, Universidad Nacional Autónoma de Mexico, Ensenada, BC, Mexico

[4]CSIRO Astronomy and Space Science, Australia Telescope National Facility, PO Box 76, Epping, NSW 1710, Australia

[5]Department of Astronomy, Universidad de Concepcion, Casilla 160-C, Concepcion, Chile

E-mail: aida.astro@mail.ioffe.ru

Abstract. PSR J1357−6429 is a young radio pulsar detected in X-rays and γ-rays. We performed radio-interferometric and high spatial resolution near-infrared observations of the pulsar. Based on the obtained radio data, we estimated the most accurate pulsar position, RA = 13:57:02.525(14) and Dec = 64:29:29.89(15). Using the new and archival radio data, we obtained the 90 per cent pulsar proper motion upper limit $\mu < 106$ mas yr^{-1}, which corresponds to the pulsar transverse velocity $v_{tr} \lesssim 1300$ km s^{-1}. The near-infrared imaging revealed a point source which we propose as a pulsar near-infrared counterpart candidate. It is confidently detected in the J and K_s bands, with $J = 23.51 \pm 0.24$ and $K_s = 21.82 \pm 0.25$. The respective dereddened near-infrared fluxes are compatible with the long-wavelength extrapolation of the pulsar X-ray spectrum. If the candidate is the true counterpart, by this property PSR J1357−6429 would be similar to the nearby middle-age pulsar PSR B0656+14. In this case, both pulsars demonstrate an unusually high near-infrared efficiency relative to the X-ray efficiency as compared to other pulsars detected in both ranges.

1. Introduction

PSR J1357−6429 is a young (characteristic age $\tau = 7.3$ kyr) and energetic (spin-down luminosity $\dot{E} = 3.1 \times 10^{36}$ ergs s^{-1}) 166 ms pulsar [1]. At a distance of 2.5 kpc, estimated from its dispersion measure [1], it is one of the nearest young pulsars known. The pulsar is visible in X-rays, where it powers a tail-like pulsar wind nebula. A fainter pulsar plerion extended up to a few tens of arcminutes is detected in the X-ray, TeV and radio domains [2–6].

The optical source detected at the pulsar X-ray positions with the ESO Very Large Telescope (VLT) in 2009 was proposed as a pulsar optical counterpart candidate [7]. However, the candidate demonstrated a 5σ offset from the pulsar radio-interferometric position measured with the Australia Telescope Compact Array (ATCA) in 2000 [1]. The offset implied that the pulsar should have a very high transverse velocity of ≈ 2000 km s^{-1} assuming the distance of

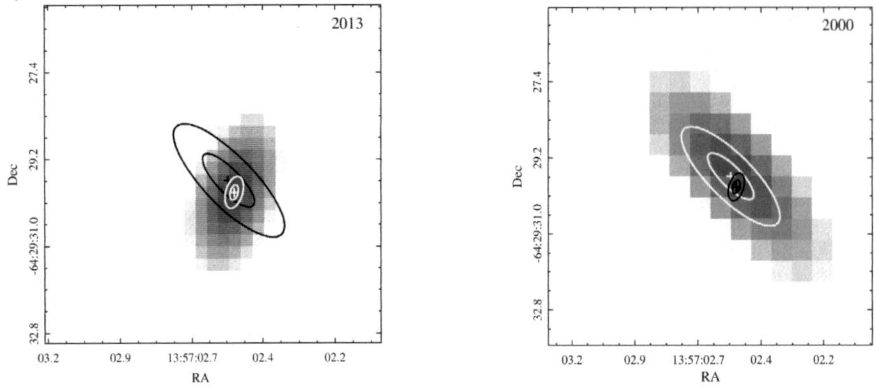

Figure 1. $8'' \times 8''$ ATCA image fragments with PSR J1357−6429 in the centre obtained in the 2013 (*left*) and 2000 (*right*) epochs with pixel scales of $0\farcs25$ and $0\farcs5$, respectively. For a given epoch, the derived pulsar position and the 40 and 90 per cent position uncertainties are shown by the white cross and ellipses, while the black cross and ellipses are related to the other epoch. The images were used for absolute astrometry.

2.5 kpc [7]. A similar estimate was obtained based on the difference between the pulsar radio and X-ray positions [8].

In order to verify the pulsar nature of the optical source and check if its velocity is indeed that high, we repeated the radio-interferometric observations with the ATCA and obtained near-infrared (IR) observations of the pulsar field using high spatial resolution techniques available at the VLT. We also performed an independent analysis of the archival radio-interferometric data obtained with the ATCA in 2000 [1].

For details of all observations, data reduction and analysis, see Refs. [9] and [10].

2. Radio-interferometric observations

We observed PSR J1357−6429 with the ATCA in 2013. Standard data reduction including Radio Frequency Interference (RFI) flagging, calibration and imaging was performed on the data with the `MIRIAD` package [11]. The data were obtained in the 1.078−3.126 GHz range. To account for a possible bias between the frequencies in the astrometrical analysis, we divided the total 2.048 GHz bandwidth into four 512 MHz sub-bands, which were analysed separately. The total bandwidth for the data obtained from the ATCA 2000 observations is 128 MHz. Given the narrower band as compared to the 2013 data, no splitting was applied to the 2000 data.

We obtained the resulting mean coordinates for the pulsar position in the ATCA 2013 data, RA = 13:57:02.525(14) and Dec = −64:29:29.89(15) (J2000). The derived position appears to be different from the previously reported values RA = 13:57:02.43(2) and Dec = −64:29:30.2(1) based on the ATCA 2000 observations [1] implying that the pulsar has moved between 2000 and 2013 epochs. To check the pulsar shift significance, we remeasured the pulsar position on the reprocessed ATCA 2000 data. The derived coordinates are RA = 13:57:02.546(76) and Dec = −64:29:29.64(55). The position uncertainties appear to be considerably larger than the reported ones (see Ref. [1]). As our uncertainties are consistent with the estimate based on the synthesised beam size and the pulsar signal-to-noise ratio (S/N), we conclude that the published pulsar position errors [1] were severely underestimated (see Ref. [9] for details).

In Fig. 1, we show the pulsar full-band images obtained for two epochs. The pulsar positions for the two observational epochs with the respective error ellipses at 40 and 90 per cent confidence levels are overlaid on the images. No significant pulsar shift is visible and only an upper limit of $\approx 1\farcs3$ (90 per cent confidence) can be established.

To reliably determine the proper motion, we employed a more robust technique of relative

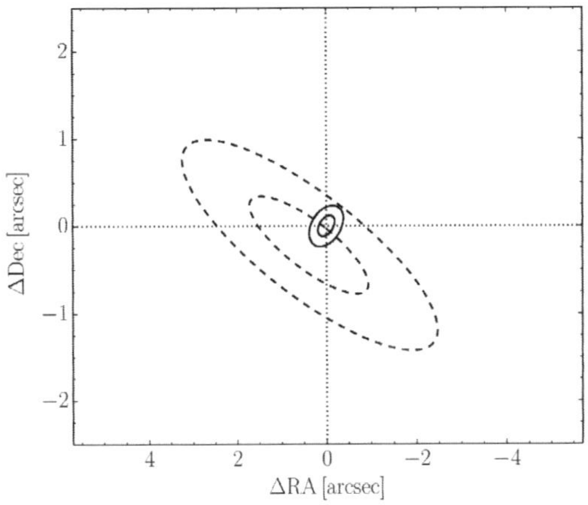

Figure 2. 40 and 90 per cent pulsar position uncertainty ellipses after relative astrometry. The 2013 and 2000 positions are shown by solid and dashed ellipses, respectively. The positions are shown relative to the mean pulsar position at the 2013 epoch. The offsets differ from those presented in Fig. 1 for the absolute astrometry (see text for details).

astrometry. We used the four sub-band images and one full-band image from the 2013 and 2000 data sets, respectively. The five images were aligned together using a custom-built routine. For referencing, we used nine bright point-like sources detected in all images with the S/N $\gtrsim 30$.

In Fig. 2, the mean of the pulsar positions in the aligned 2013 sub-bands is compared with the 2000 position. We find that the arrangement of the pulsar error ellipses on epochs 2000 and 2013 slightly differs from that provided by the absolute astrometry (cf., Fig. 1), while the pulsar shift is insignificant and its upper limit is virtually the same as obtained by the absolute astrometry (see Ref. [9] for details). Accounting for the time-base of 12.76 yr between the two ATCA observations, the 90 per cent upper limit on the pulsar proper motion is $\mu < 106$ mas yr^{-1}. It corresponds to the pulsar transverse velocity $v_{tr} \lesssim 1300$ km s^{-1} for the distance of 2.5 kpc.

3. Near-IR observations: a counterpart candidate

The first near-IR observations of the pulsar field were performed in the J, H and K_s bands with the Nasmyth Adaptive Optics System (NAOS) and Coude Near-Infrared Camera (CONICA) (in short NaCo) at the VLT/UT4 unit in the period of 2012–2013.

High spatial resolution provided by the adaptive optics system allowed us to find a pulsar near-IR counterpart candidate labelled by C' in Fig. 3. It overlaps with the pulsar position based on the ATCA 2013 observations (solid-line ellipse in the K_s-band image) and with the 1σ error ellipses of the pulsar X-ray positions obtained from the *Chandra* 2005 and *Chandra* 2009 observations (dashed-line ellipses in the K_s-band image). The source coordinates are RA=13:57:02.52 and Dec=−64:29:30.15. Its 1σ uncertainty for both coordinates is $0\farcs22$ in the K_s band. The object is detected with $J = 23.51\pm0.24$ and $K_s = 21.82\pm0.25$. There is also a hint of the source in the H band, but the full width at half maximum is worse by a factor of 1.5 in this band, meaning that a confident detection cannot be made. The candidate is unresolved from bright stars labelled A and A' in the previous seeing-limited VLT optical images [7].

One of the ways to support the pulsar nature of the candidate is to compare its colours with those of ordinary stars. However, such analysis appeared to be uninformative in our case, since $J - K_s$ colour of the object C' does not appear peculiar and is compatible with those of the

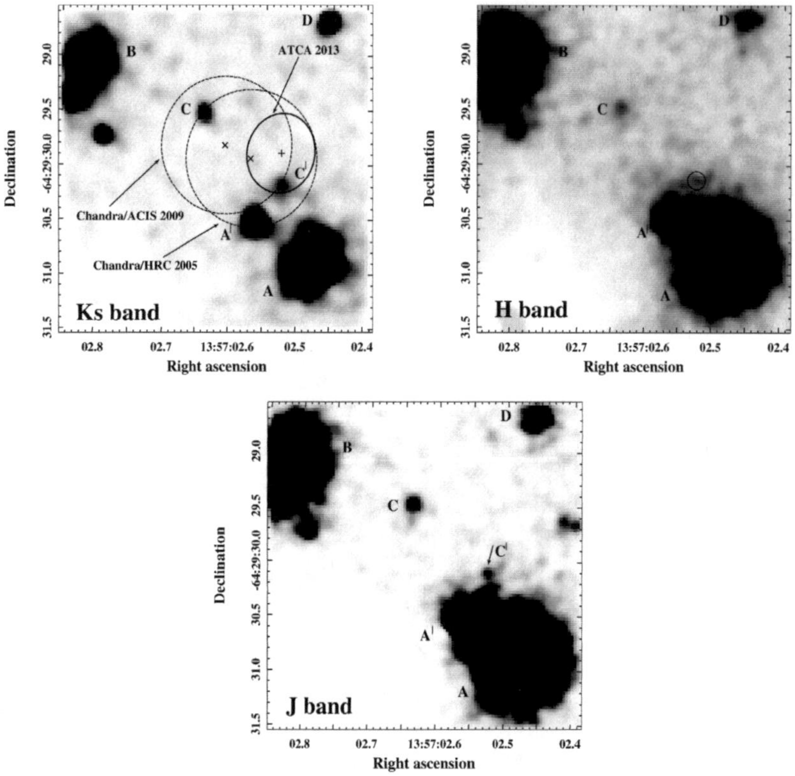

Figure 3. Image fragments of PSR J1357−6429 field obtained with the VLT/NaCo in K_s (*top-left*), H (*top-right*) and J (*bottom*) bands. The ×-points with dashed-line circles and the + with the solid-line ellipse in the *top-left* panel show the pulsar positions and their 1σ uncertainties from the two *Chandra* X-ray and new ATCA radio observations, respectively. A, A', B, C and D label objects as in [7]. C' labels the pulsar new counterpart candidate.

the M−L-dwarf sequence. The same property is found for other pulsars such as Crab, Vela and PSR B0540−69 (see Ref. [10] for details).

Another possibility to establish the C' nature is to compare its IR fluxes with the pulsar X-ray spectral data. As seen from Fig. 4, the near-IR fluxes of the counterpart candidate are consistent within 1σ uncertainties with the long-wavelength extrapolation of the pulsar X-ray spectrum. The same situation is observed for the middle-aged pulsar B0656+14 [12, 13].

We also performed the analysis of spindown power conversion efficiencies to the X-ray and K_s-band emission for the pulsars detected in both ranges. The efficiencies are defined as $\eta_X \equiv L_X/\dot{E}$ and $\eta_{K_s} \equiv L_{K_s}/\dot{E}$, where L_X and L_{K_s} are the X-ray (0.5−8 keV) nonthermal and K_s luminosities, respectively. The plot in Fig. 5 demonstrates a strong correlation between Log η_{K_s} and Log η_X. The only pulsar, which shows an excess of the near-IR efficiency over the fit, is PSR B0656+14. It was discussed that the excess could point to the presence of a passive post-supernova fall-back disk around the pulsar [13]. However, it could also represent an intrinsic property of the pulsar magnetospheric emission. If the object C' is confirmed as the PSR J1357−6429 near-IR counterpart, it will be another example of a pulsar with a peculiar near-IR efficiency. This is important for understanding the origin of the near-IR and optical emission of pulsars.

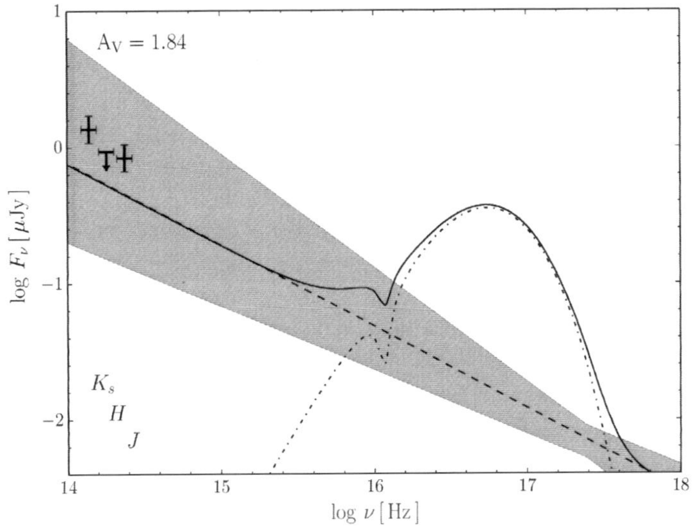

Figure 4. Tentative multi-wavelength unabsorbed spectrum of PSR J1357−6429. The X-ray part is described by the magnetized NS hydrogen atmosphere and power law models and is taken from [7]. The solid, dash-dotted and dashed lines are the total X-ray spectrum and contributions of the atmospheric and power law components, respectively. The grey region shows uncertainties of the power law component. The X-ray spectrum is extrapolated to the near-IR. The JK_s fluxes and the H-band upper limit for the object C' dereddened with the plausible extinction value A_V [7] presented in the plot are shown by bold error bars.

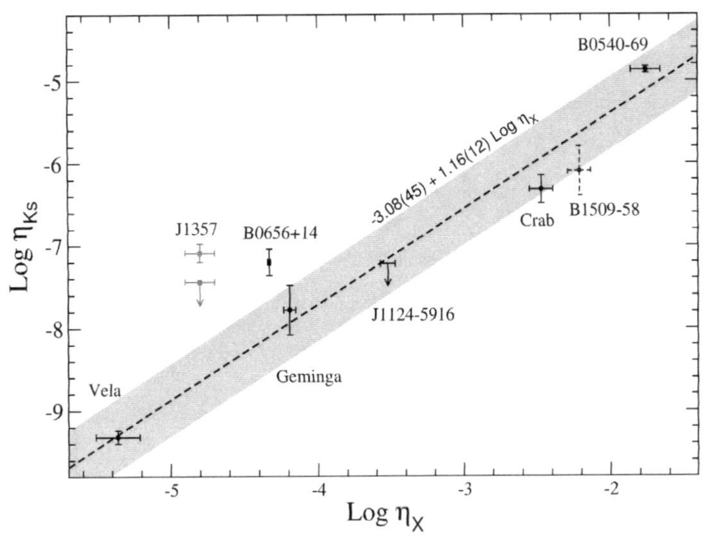

Figure 5. K_s vs. X-ray efficiencies for pulsars observed in both ranges. The dashed line and grey region show the best fit and its 1σ uncertainty to the data for four firmly detected near-IR pulsars (marked by black circles) excluding B0656+14. PSR J1357−6429 is represented by the near-IR counterpart candidate C' and by the detection limit of our observations.

4. Conclusions

We analysed the new radio and near-IR observations of PSR J1357−6429. Based on the new and archival radio-interferometric data, we did not detect any proper motion of the pulsar and estimated the upper limit on its transverse velocity $v_{tr} \lesssim 1300$ km s^{-1} for the distance of 2.5 kpc. The new radio-interferometric position implies that the optical counterpart candidate reported before [7] can now be disclaimed. However, we propose another candidate for the pulsar J1357−6429 counterpart detected in the near-IR. The candidate position overlaps with the new radio-interferometric position of the pulsar. Optical observations of PSR J1357−6429 field using high spatial resolution imaging instruments, like the *Hubble Space Telescope*, would be useful to verify the pulsar nature of the object C'. Further proper motion measurements in the radio and near-IR would be also valuable.

Acknowledgments

The work was supported by the Russian Science Foundation, grant 14-12-00316. The Australia Telescope Compact Array is part of the Australia Telescope National Facility which is funded by the Commonwealth of Australia for operation as a National Facility managed by CSIRO.

References

[1] Camilo F, Manchester R N, Lyne A G, Gaensler B M, Possenti A, D'Amico N, Stairs I H, Faulkner A J, Kramer M, Lorimer D R, McLaughlin M A and Hobbs G 2004 *ApJL* **611** L25–L28
[2] Chang C, Pavlov G G, Kargaltsev O and Shibanov Y A 2012 *ApJ* **744** 81
[3] Esposito P, Tiengo A, de Luca A and Mattana F 2007 *A&A* **467** L45–L48
[4] Lemoine-Goumard M, Zavlin V E, Grondin M H, Shannon R, Smith D A, Burgay M, Camilo F, Cohen-Tanugi J, Freire P C C, Grove J E, Guillemot L, Johnston S, Keith M, Kramer M, Manchester R N, Michelson P F, Parent D, Possenti A, Ray P S, Renaud M, Thorsett S E, Weltevrede P and Wolff M T 2011 *A&A* **533** A102
[5] Zavlin V E 2007 *ApJ* **665** L143–L146
[6] HESS Collaboration 2011 *A&A* **533** A103
[7] Danilenko A, Kirichenko A, Mennickent R E, Pavlov G, Shibanov Y, Zharikov S and Zyuzin D 2012 *A&A* **540** A28
[8] Mignani R P, Shearer A, de Luca A, Moran P, Collins S and Marelli M 2011 *A&A* **533** A101
[9] Kirichenko A, Shibanov Y, Shternin P, Johnston S, Voronkov M A, Danilenko A, Barsukov D, Lai D and Zyuzin D 2015 *MNRAS* **452** 3273–3280
[10] Zyuzin D, Zharikov S, Shibanov Y, Danilenko A, Mennickent R E and Kirichenko A 2016 *MNRAS* **455** 1746–1750
[11] Sault R J, Teuben P J and Wright M C H 1995 *Astronomical Data Analysis Software and Systems IV* (*Astronomical Society of the Pacific Conference Series* vol 77) ed Shaw R A, Payne H E and Hayes J J E p 433
[12] Shibanov Y A, Zharikov S V, Komarova V N, Kawai N, Urata Y, Koptsevich A B, Sokolov V V, Shibata S and Shibazaki N 2006 *A&A* **448** 313–326
[13] Durant M, Kargaltsev O and Pavlov G G 2011 *ApJ* **743** 38

First intermediate flare from SGR 1935+2154

A V Kozlova[1], G L Israel[2], D S Svinkin[1] and D D Frederiks[1]

[1] Ioffe Institute, 26 Polytekhnicheskaya, St. Petersburg, 194021, Russia
[2] INAF - Osservatorio Astronomico di Roma, via Frascati 33, I- 00040 Monteporzio Catone, Roma, Italy

E-mail: ann_kozlova@mail.ioffe.ru

Abstract. The first intermediate flare from newly discovered SGR 1935+2154 was detected and localized by four Interplanetary network (IPN) spacecraft on 2015 April 12. Among the observing instruments, only Konus-*Wind* gamma-ray burst spectrometer (KW) was able to measure high-resolution light curves and multi-channel energy spectra of the flare. We report on the results of temporal and spectral analyses of the KW data, the flare energetics, a search for Quasi-Periodic Oscillations in the light curve, and, finally, discuss the source distance estimate based on the distribution of double blackbody spectral fit parameters.

1. Introduction

Observations of Soft Gamma Repeaters (SGRs) have already more than 30 years of history. They were discovered through the detection of reccurent short (~ 0.1 s) bursts of hard X-rays/soft Gamma-rays by the Konus instrument aboard the Venera 11-14 spacecraft. The first bursts were initially classified as a subtype of gamma-ray bursts (GRB), one with a short duration and a soft spectrum [1].

Up to now, we know that such numerous short bursts, with a total energy release $E_{\rm tot}$ of $\sim 10^{38}$–10^{40} erg, are the most common manifestation of SGR bursting activity, but there are two other, more rare types of bursts emitted by SGRs: giant flares and intermediate bursts. The extraordinary giant flares are the most intense Galactic events. They are characterized by short, hard initial pulse having a huge energy release of $\sim 10^{44}$–10^{46} erg followed by a long-duration decaying tail modulated with the neutron star rotation period. So far, only three giant flares have been observed from three out of the fifteen confirmed SGRs [2] on 1979 March 5, 1998 August 27 and 2004 December 27. The high-fluence intermediate bursts are intermediate in terms of duration, peak luminosity and energy between the short bursts and the giant flares. Such events are characterized by durations of few seconds–few tens of seconds; the brightest of them have $E_{\rm tot}$ up to $\sim 5 \times 10^{42}$ erg [3], less than that of the giant flare, though orders of magnitude larger than that of the most common short SGR bursts. Only a few dozen intermediate bursts are known to date and properties of such outstanding events are of special interest.

The soft gamma-ray repeater SGR 1935+2154 was discovered on 2014 July 5 through a series of three short bursts [4], detected by the *Swift*/BAT. Several days later, its persistent pulsating X-ray counterpart was discovered by the *Chandra* X-ray Observatory [5] with a pulse period of ~ 3.2 s. The location of the SGR as determined by *Swift*/XRT [4] sits very close to the geometric centre of the Galactic supernova remnant (SNR) G57.2+0.8 [6] with known distance of about 9.1 kpc [7]. In 2015 February *Fermi*/GBM and *Swift*/BAT observed a weak burst activity from the source —the first one since the discovery. More recently, the first intermediate flare from SGR 1935+2154 was detected and localized by four Interplanetary network (IPN) spacecraft [8]: *INTEGRAL*, *Wind*, *Mars-Odysey*, and *MESSENGER*. Among the observing instruments, only Konus-*Wind* gamma-ray burst spectrometer (KW) [9] was able to measure high-resolution light curves and multi-channel energy spectra of the flare.

The rest of this paper is organized as follows. In Section 2, we present results of the KW timing analysis and detailed spectroscopy. We report on the search for Quasi-Periodic Oscillations (QPOs) in the light curve in Section 3. Finally, in Section 4, we conclude with a few remarks, compare our results to the previous ones derived for the short and intermediate bursts from different SGRs and discuss the source distance estimate based on the distribution of double blackbody spectral fit parameters.

2. Data analysis

The SGR burst triggered detector S2 of the Konus-*Wind* γ-ray spectrometer at $T_0 = 41064.683$ s UT (11:24:24.683) on 2015 April 12. The burst time histories were recorded in three energy bands: G1 (20–80 keV), G2 (80–300 keV), and G3 (300–1200 keV) with a time resolution of 2 ms from $T_0 - 0.512$ s to $T_0 + 0.512$ s and of 16 ms afterwards.

As observed by Konus-*Wind*, the burst light curve shows a single pulse which starts at $\sim T_0 - 0.062$ s with a sharp (<10 ms) rise and decays to background level at around $T_0 + 1.680$ s. The total burst duration $T_{100} = 1.742$ s was determined at the 5σ level in the G1+G2 energy band (20–300 keV). The corresponding values of T_{90} and T_{50} defined as the times during which 90 % and 50 % of the total burst counts have been accumulated are 1.412 ± 0.016 s and 0.654 ± 0.016 s, respectively. The start of the T_{90} (T_{50}) interval specified by the time at which 5 % (25 %) of the total counts have been reached. The time history of the burst is shown in Figure 1.

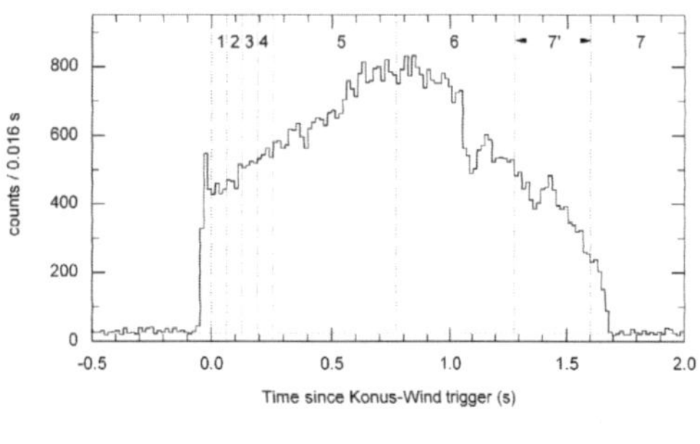

Figure 1. Time history of the SGR burst recorded by Konus-*Wind* in the 20–1200 keV energy range with 16 ms resolution. The count rates are dead-time corrected; the dashed line indicates the background level. The vertical dotted lines denote the intervals over which the time-resolved spectra 1–7 were accumulated. 7' represent a 0.32 s-long part of spectrum 7, which contains ~ 90% of counts in this spectrum.

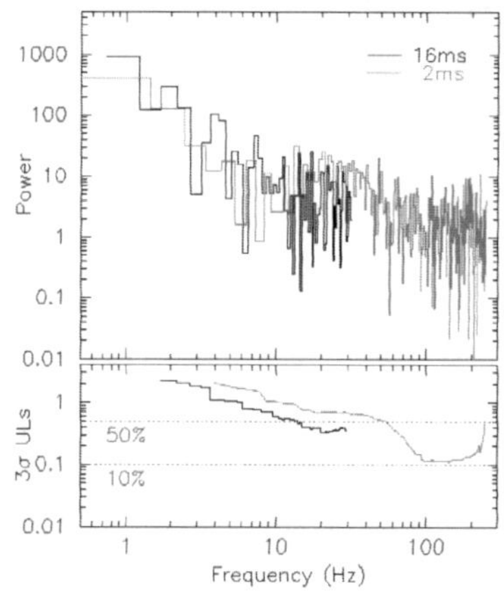

Figure 2. Power spectrum produced from the KW light curve in the G1+G2 energy range (upper panel). Curves representing the upper limits to the non-detection of pulsations are in the lower panel. The two horizontal dashed lines represent the 10 % and 50 % upper limits to the pulsed fraction. Lines of different colours correspond to the different time resolutions.

Table 1. 2BB model spectral fit results (in the 20–250 keV band). R_{km} is the radius of the emitting region in km and D_{10} is the distance to the source in units of 10 kpc. Errors are quoted at the 68 % confidence level.

Spectra	Accumulation interval	kT_1 (keV)	$norm_1$ R^2_{km}/D^2_{10}	kT_2 (keV)	$norm_2$ R^2_{km}/D^2_{10}	χ^2/dof
1	0.0 –0.064	$7.7^{+0.8}_{-1.1}$	299^{+123}_{-70}	$14.9^{+5.9}_{-2.6}$	$8.2^{+14.5}_{-7.1}$	4.8/15
2	0.064–0.128	$6.3^{+0.9}_{-1.1}$	621^{+486}_{-202}	$13.1^{+1.9}_{-1.4}$	$25.1^{+29.3}_{-17.9}$	8.3/15
3	0.128–0.192	$4.5^{+1.0}_{-1.0}$	1830^{+3730}_{-986}	$10.7^{+0.9}_{-0.6}$	$88.9^{+36.8}_{-32.6}$	15.0/15
4	0.192–0.256	$8.4^{+1.1}_{-2.6}$	246^{+134}_{-56}	$14.1^{+14.1}_{-2.8}$	$12.7^{+82.1}_{-12.5}$	8.4/15
5	0.256–0.768	$5.4^{+0.9}_{-0.8}$	769^{+571}_{-260}	$11.6^{+0.5}_{-0.3}$	$93.1^{+20.3}_{-22.9}$	29.3/26
6	0.768–1.280	$7.0^{+0.4}_{-0.6}$	449^{+108}_{-72}	$13.6^{+0.9}_{-0.6}$	$32.9^{+14.0}_{-11.5}$	36.4/28
7	1.280–9.472[a]	$7.7^{+0.6}_{-1.0}$	165^{+23}_{-60}	$14.3^{+5.1}_{-3.4}$	$3.7^{+7.8}_{-2.0}$	23.2/30
1–7	0.0 –9.472[a]	$6.4^{+0.4}_{-0.4}$	455^{+73}_{-55}	$12.4^{+0.4}_{-0.4}$	$43.5^{+8.8}_{-8.1}$	37.0/30

[a] The model normalizations for spectra 7 and 1-7 are calculated using intervals 1.280 – 1.680 s and 0.0 – 1.680 s, respectively.

During the burst the instrument measured seven multichannel energy spectra covering a wide energy range from 20 keV to 14 MeV; for the spectra accumulation intervals see Table 1. The raw count rate spectra were rebinned in order to have at least 10 counts per energy bin, and fitted using XSPEC, version 12.8 [10]. We use only the 20 to 250 keV fitting interval since no emission was detected at a higher energies. The spectral analysis was performed by applying two spectral models, which have been shown to be the best-fits to the broadband spectra of SGR bursts (e.g. [11, 12, 13, 14]). The first one is a sum of two blackbody functions with the normalization proportional to the surface area (`bbodyrad+bbodyrad` in XSPEC; abbreviated as 2BB model throughout this paper). The second model is a power law with an exponential cutoff (CPL), parametrized as E_p: $f(E) \propto E^{\alpha} \exp(-(2+\alpha)E/E_p)$, where E_p is the peak energy in the νF_ν spectrum. We also tried to fit the spectra to a single blackbody function and found that this model may be rejected on statistical grounds ($\chi^2 = 288/32$ dof for time-integrated spectrum).

A summary of the Konus-*Wind* spectral fits is presented in Tables 1 and 2. The time-integrated spectrum is best fitted with the CPL model ($\chi^2 = 32.4/31$ dof) with an index α of ~ 0.2 and $E_p \sim 36$ keV. The 2BB model applied to this spectrum also yields a reasonably good fit ($\chi^2 = 37/30$ dof). The temperatures of the cool and hot BB components are $kT_1 \simeq 6.4$ and $kT_2 \simeq 12.4$ keV, and the corresponding radii of the emitting areas (calculated at the distance of 9.1 kpc) are $R_1 \simeq 19.4$ km and $R_2 \simeq 6.0$ km, respectively. The low-kT BB contribution to the total flux in the 20–200 keV range is about 27%. We note that since no burst emission was detected after $T_0+1.680$ s, the 2BB model normalizations for spectrum 7 and the time-integrated spectrum are calculated using intervals 1.280–1.680 s and 0.0–1.680 s, respectively. The time-resolved spectral analysis using the 2BB model shows significant variation of the parameters during the burst, while the variations of the CPL model parameters do not differ much from those found for the time-integrated spectrum.

From the CPL spectral fits, we estimate the total energy fluence of the burst S to be $(2.50\pm0.03)\times10^{-5}$ erg cm^{-2} and the peak energy flux F_{max} to be $(2.15\pm0.13)\times10^{-5}$ erg cm^{-2} s^{-1} in a 16 ms time interval starting at $T_0 + 0.800$ s. Both values are measured in the 20–200 keV

Table 2. CPL model spectral fit results (in the 20–250 keV band). Errors are quoted at the 68 % confidence level.

Spectra	Accumulation interval	α	E_p (keV)	χ^2/dof
1	0.0 –0.064	$0.25^{+0.46}_{-0.42}$	$34.3^{+1.6}_{-1.9}$	6.9/16
2	0.064–0.128	$-0.38^{+0.38}_{-0.35}$	$31.2^{+2.2}_{-2.7}$	8.1/16
3	0.128–0.192	$-0.13^{+0.39}_{-0.37}$	$30.9^{+2.0}_{-2.4}$	16.4/16
4	0.192–0.256	$0.70^{+0.43}_{-0.40}$	$37.4^{+1.3}_{-1.4}$	8.8/16
5	0.256–0.768	$0.37^{+0.12}_{-0.12}$	$37.3^{+0.5}_{-0.5}$	26.7/27
6	0.768–1.280	$0.05^{+0.12}_{-0.12}$	$37.5^{+0.6}_{-0.6}$	31.0/29
7	1.280–9.472	$0.75^{+0.46}_{-0.43}$	$32.9^{+1.1}_{-1.2}$	25.8/31
1–7	0.0 –9.472	$0.20^{+0.08}_{-0.08}$	$35.7^{+0.3}_{-0.3}$	32.4/31

energy range and quoted errors are at the 90 % confidence level. Assuming isotropic emission and the distance to the source of 9.1 kpc, the total energy release in the burst E_{tot} is $\sim 2.5 \times 10^{41}$ erg and the peak luminosity, L_{max}, $\sim 2.1 \times 10^{41}$ erg s^{-1}.

3. Search for QPOs

Quasi-periodic oscillations discovered in soft-gamma repeaters [15, 16, 17] are expected to help us to study the properties of matter in neutron stars. Through detection of the frequencies of neutron star oscillations, it might be possible to deduce neutron star masses and radii, the equation of state, and so on (e.g., [18, 19]).

We searched for pulsations in the data total energy range by using the Fast Fourier Transform method without finding any statistically significant signal. Following the recipe described in [20] we derived the upper limits on the pulsed fraction at a 3σ confidence level of $> 50\%$ for frequencies in the 5–60 Hz range and 10–30% between 60 and 250 Hz. The results are presented in Figure 2.

4. Discussion

We have carried out a detailed study of the spectral and temporal properties of the first intermediate burst from the recently discovered SGR 1935+2154 observed by Konus-*Wind*. The rather long duration of the burst along with the large measured energy fluence put it in the class of "intermediate" SGR bursts. We found that the burst spectra in the 20–250 keV energy range are fitted well by both CPL and two-component 2BB models. Previously, these two models were successfully used to describe the SGR bursting emission over broad energy ranges. The temperatures and radii we obtained for the 2BB model are similar to those of other SGR sources ($kT_1 \sim$ 3–7 keV, $R_1 \sim$ 15–35 km and $kT_2 \sim$ 10–20 keV, $R_2 \sim$ 2–7 km [14, 21, 22, 23]). Specifically, the derived 2BB model parameters are in agreement with the results reported previously for intermediate bursts ($kT_1 \sim 4.8$ keV, $R_1 \sim 30$ km and $kT_2 \sim 9.0$ keV, $R_2 \sim 5.7$ km [12, 24]).

The E_p values obtained from the CPL model fits, $E_p \sim$ 30–37 keV, are typical of SGR bursts, but there seems to be a difference across SGR sources regarding the power-law indices α. Our result on α is comparable to those obtained for SGR 0501+4516 [23] and SGR 1900+14 [11], and is significantly different than the one found for SGR J1550-5418 [14]. Moreover, the measured spectral parameters are consistent with results obtained using *Fermi*/GBM observations for much weaker and shorter bursts detected during the SGR 1935+2154 activity in 2015 February.

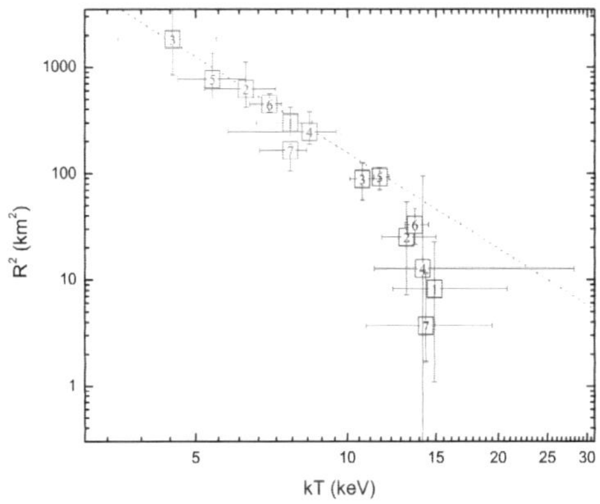

Figure 3. Square of radii of the emitting areas as a function of their temperatures. The soft BB component is shown in red and the hard BB —in blue. Source distance of 10 kpc is assumed. We also plot the $R^2 = kT^{-3}$ relation to guide the eye.

Using the 2BB fits described in Section 2, we calculated the soft and hard BB luminosities for each of the seven burst spectra. Almost all derived luminosities are over 10^{40} erg s^{-1}, and hence there is just some hint for the saturation effect of the low-kT BB luminosity as it was previously noted for SGR 1900+14 [24]. In order to further investigate this trend, we studied the R^2 versus kT distribution (see Figure 3). The sharp edge in the distribution of the data described by the $R^2 = kT^{-3}$ relation indicates the presence of the saturation. So we can use the magnetic Eddington luminosity formula derived in [25]

$$L_{\rm Edd,B} \approx 2 \times 10^{40} \left(\frac{B}{B_{\rm QED}}\right)^{4/3} \left(\frac{R}{R_{\rm NS}}\right)^{2/3}$$

where $B_{\rm QED} \simeq 4.4 \times 10^{13}$ G is magnetic field critical value and $R_{\rm NS}$ is a neutron star radius for which we assume a typical value of 10 km. It can be rewritten in terms of the distance d and peak flux $F_{\rm max}$ into

$$\left(\frac{d}{\rm kpc}\right) \simeq 0.4 \times \left(\frac{F_{\rm max}}{10^{-5}\,{\rm erg\,cm^{-2}\,s^{-1}}}\right)^{-1/2} \left(\frac{kT_{\rm break}}{\rm keV}\right)^{5/4} \left(\frac{B_{\rm surf}}{10^{14}\,{\rm G}}\right)^{1/4} \left(\frac{R_{\rm NS}}{10\,{\rm km}}\right)^{5/8}$$

Now we can derive an approximate value for the distance by using the saturated flux recorded for the source and the magnetic field strength inferred by timing analysis. For the magnetic field we used a value of $\sim 2.2 \times 10^{14}$ G [26]. $kT_{\rm break}$ is the energy at which the data in the R^2 versus kT distribution start departing from the relation kT^{-3}. With the value of $kT_{\rm break}$ for the studied burst lying in the 12–15 keV range we derive the distance estimate of 7.4–9.8 kpc. However, given that we have seen only one bright intermediate burst from this source, we cannot be sure that the luminosity in the April 12 burst is close to the maximum observable from SGR 1935+2154 and a brighter burst would situate the source closer to us. So, in this work, we estimate the SGR 1935+2154 distance to be $<$ 10.0 kpc, in agreement with that of the Galactic supernova remnant G57.2+0.8.

References

[1] Mazets E and Golenetskii S 1981 *Ap&SS* **75** 47–81
[2] Olausen S A and Kaspi V M 2014 *ApJS* **212** 6
[3] Frederiks D D, Golenetskii S V, Palshin V D, Aptekar R L, Ilyinskii V N, Oleinik F P, Mazets E P and Cline T L 2007 *Astronomy Letters* **33** 1–18
[4] Cummmings J, Barthelmy S, Chester M and Page K 2014 *ATel #6294*
[5] Israel G, Rea N, Zelati F, Esposito P, Burgay M, Mereghetti S, Possenti A and Tiengo A 2014 *ATel #6370*
[6] Gaensler B 2014 *GCN Circular 16533*
[7] Pavlović M, Urošević D, Vukotić B, Arbutina B and Göker Ü 2013 *ApJS* **204** 4
[8] Golenetskii S *et al.* 2015 *GCN Circular 17699*
[9] Aptekar R *et al.* 1995 *SSRv* **71** 265–72
[10] Arnaud K 1996 *ADASS* **101** 17
[11] Feroci M, Caliandro G, Massaro E, Mereghetti S and Woods P 2004 *ApJ* **612** 408–413
[12] Olive J F, Hurley K, Sakamoto T, Atteia J L, Crew G, Ricker G, Pizzichini G, Barraud C and Kawai N 2004 *ApJ* **616** 1148–58
[13] Lin L *et al.* 2012 *ApJ* **756** 54
[14] van der Horst A *et al.* 2012 *ApJ* **749** 112
[15] Israel G L, Belloni T, Stella L, Rephaeli Y, Gruber D E, Casella P, Dall'Osso S, Rea N, Persic M and Rothschild R E 2005 *ApJ* **628** L53–L56
[16] Strohmayer T E and Watts A L 2005 *AAS* **37** p 1497
[17] Huppenkothen D *et al.* 2014 *ApJ* **787** 128
[18] Doneva D D, Gaertig E, Kokkotas K D and Krüger C 2013 *Phys. Rev. D* **88** 044052
[19] Andersson N and Kokkotas K D 1996 *Phys. Rev. Lett.* **77** 4134–37
[20] Israel G L and Stella L 1996 *ApJ* **468** p 369
[21] Nakagawa Y *et al.* 2007 *OUP* **59** 653–78
[22] Esposito P *et al.* 2008 *MNRAS* **390** L34–L38
[23] Lin L *et al.* 2011 *ApJ* **739** 87
[24] Israel G *et al.* 2008 *ApJ* **685** 1114–28
[25] Paczynski B 1992 *Acta Astronomica* **42**(3) 145–53
[26] Israel G L *et al.* 2016 *MNRAS* **457** 3448–56

Radiation pressure influence on the motion of hazardous asteroids

A Martyusheva[1]

[1]The Central Astronomical Observatory of the Russian Academy of Sciences, Pulkovskoe chaussee 65, Saint-Petersburg, 196140, Russia

E-mail: alex.mart13@gmail.com

Abstract. The deviations under radiation pressure influence were calculated for three asteroids 308635 (2005 YU55), 367943 Duende (2012 DA14) and 357439 (2004 BL86) that had a close encounter with the Earth. Calculations were made for different values of the assumed albedo of the asteroids. Deviations in orbital motion of the asteroids before the next close encounter with the Earth are also presented.

1. Introduction

In recent decades, investigation and identification of potentially hazardous Near Earth Asteroids are one of the most important problems in astronomy. For hazard level evaluation firstly it is necessary to determine the orbit of an object. During the orbit modeling it is important to consider all possible factors influencing on its variation, including radiation pressure. Although radiation effects in motion of asteroids, as known, are small, at relatively large time intervals they can cause significant changes in parameters of orbital motion. By changing albedo of the asteroid's reflecting surface, in the future one can change its orbit.

The deviations in orbital motion of hazardous asteroids 308635 (2005 YU55), 367943 Duende (2012 DA14) and 357439 (2004 BL86) under radiation pressure influence were calculated. Estimation of radiation pressure effect was produced by means of specially developed program based on numerical integration of motion equations by Everhart method [1].

The five-body system the Sun — Jupiter — an asteroid (a material point; only gravity affects on it) — an asteroid (a spherical body having a finite radius, a certain density and optical coefficient; radiation pressure and gravity affect on it) — the Earth was selected to solve this problem. The system splits into four independent models of three bodies of different configuration and allows revealing the Earth influence on the asteroid motion, which is important in connection with the asteroid and comet hazard. The comparison of these models allows one to calculate the deviation of asteroid positions relative to each other, caused by radiation pressure influence. The work [1] is devoted to a detailed description of the model.

Each of asteroids, selected for investigation, had a close encounter with the Earth that is of particular interest. Three values were calculated for each asteroid: the displacement along the heliocentric radius-vector Δr, the displacement along the orbit Δl and the total displacement Δd. As a comparison calculations were made in case of increased reflectivity of asteroids using the bright coating of surface and also in case of absolutely white body.

2. Results

The orbital and physical parameters of the asteroid 367943 Duende (2012 DA14), used for taking into account the solar radiation pressure, are following:
$e = 0.089428$ — eccentricity [2],
$a = 0.910316$ AU — semi-major axis [2],
$H_V = 21.0$ — absolute magnitude [2],
$D = 40$ m — diameter [3],
$\rho = 2330$ kg/m^3 — silicon density under normal conditions (L-class asteroid),
Hence $\delta = 0.28$ — albedo in the band V, derived from the asteroid's absolute magnitude and its diameter by using the formula [4]

$$\lg D = 3.122 - 0.5\lg\delta - 0.2H_V \quad (1)$$

In general, the optical model of asteroid surface is based on the assumption of predominance of diffuse nature of reemission, and the continuity condition for solar energy flux is: $\alpha + \rho + \delta = 1$, where α is the absorption coefficient, $\rho + \delta$ is the reflection coefficient composed of the mirror reflection ρ and the diffuse reflection δ. The optical coefficient for the diffuse reemission is: $k = \alpha + \rho + 13\delta/9$ [5]. For a natural body it was accepted that there is no mirror reflection ($\rho = 0$), and the diffuse reflectance δ is the geometric albedo of an asteroid. Thus, for extended spherical diffusely reflecting body the optical coefficient is: $\alpha = 1 - \delta$, $k = \alpha + 13\delta/9 = 1 + 4\delta/9$.
Therefore $k = 1.12$ — optical coefficient.

In the next 20 years the maximum deviations of 2012 DA14, caused by radiation pressure for different values of the assumed albedo are summarised in Table 1:

Table 1. The maximum deviations of 2012 DA14 in the next 20 years.

Deviation	$\delta = 0.28$	$\delta = 0.80$ (bright body)	$\delta = 1.0$ (absolutely white body)
Δr	21.8 ± 0.1 km	26.5 ± 0.1 km	28 ± 0.1 km
Δl	270.4 ± 0.1 km	328.4 ± 0.1 km	347.7 ± 0.1 km
Δd	270.6 ± 0.1 km	328.6 ± 0.1 km	348 ± 0.1 km

The orbital and physical parameters of the asteroid 308635 (2005 YU55), used for taking into account the solar radiation pressure, are following:
$e = 0.430609$ — eccentricity [6],
$a = 1.157390$ AU — semi-major axis [6],
$H_V = 21.9$ — absolute magnitude [6],
$D = 400$ m — diameter [7],
$m = 10^{11}$ kg — mass [8],
$\rho = 3.0$ g/cm^3 — density,
Hence $\delta = 0.02$ — albedo, derived from (1),
$k = 1.00(8)$ — optical coefficient.

In the next 20 years the maximum deviations of 2005 YU55, caused by radiation pressure for different values of the assumed albedo are summarised in Table 2:

Table 2. The maximum deviations of 2005 YU55 in the next 20 years.

Deviation	$\delta = 0.02$	$\delta = 0.80$ (bright body)	$\delta = 1.0$ (absolutely white body)
Δr	16.6 ± 0.1 km	22.4 ± 0.1 km	23.7 ± 0.1 km
Δl	47.2 ± 0.1 km	63.6 ± 0.1 km	67.3 ± 0.1 km
Δd	49.3 ± 0.1 km	66.4 ± 0.1 km	70.3 ± 0.1 km

The orbital and physical parameters of the asteroid 357439 (2004 BL86), used for taking into account the solar radiation pressure, are following:
$e = 0.403073$ — eccentricity [9],
$a = 1.502202$ AU — semi-major axis [9],
$H_V = 19.3$ — absolute magnitude [9],
$D = 325$ m — diameter [10],
$\rho = 1.6$ g/cm^3 — density [11],
Hence $\delta = 0.32$ — albedo, derived from (1),
$k = 1.14$ — optical coefficient.

In the next 20 years the maximum deviations of 2004 BL86, caused by radiation pressure for different values of the assumed albedo are summarised in Table 3:

Table 3. The maximum deviations of 2004 BL86 in the next 20 years.

Deviation	$\delta = 0.32$	$\delta = 0.80$ (bright body)	$\delta = 1.0$ (absolutely white body)
Δr	32.2 ± 0.1 km	38.4 ± 0.1 km	40.6 ± 0.1 km
Δl	109.4 ± 0.1 km	130.5 ± 0.1 km	138.2 ± 0.1 km
Δd	109.8 ± 0.1 km	130.9 ±0.1 km	138.6 ± 0.1 km

3. Conclusion

These results allowed one to estimate the possible effect on asteroids by changing their albedo. Before the next encounter of asteroids 308635 (2005 YU55), 367943 Duende (2012 DA14) and 357439 (2004 BL86) with the Earth (using the bright coating of surface) possible total deviation under radiation pressure influence will be approximately 199.2 ± 0.1 km (2075 year), 509 ± 0.1 km (2046 year), 79 ± 0.1 km (2027 year), respectively. Thus, changes in orbital motions of the asteroids under the influence of such non-gravitational effect as radiation pressure can be significant enough to affect the risk assessment of objects' collision during close encounters with the Earth.

References

[1] Martyusheva A. A., Petrov N. A., Polyakhova E. N. Vestnik of SPbU, Vol.2 (60), Issue 1, pp. 135-147, 2015.
[2] http://ssd.jpl.nasa.gov/sbdb.cgi?sstr=2012%20DA14;old=0;orb=0;cov=0;log=0;cad=0#phys_par

[3] http://www.jpl.nasa.gov/news/news.php?release=2013-063
[4] Vinogradova, T. A., Zheleznov N. B., Kuznetsov V. B., Chernetenko Yu. A., Shor V. A. Tr. Inst. Prikl. Astron. Ross. Akad. Nauk, Issue 9, 2003.
[5] Polyakhova E. N., Shmyrov A. S. Vestnik of SPbU, Vol. 2, Issue 8, pp. 87–104, 1994.
[6] http://ssd.jpl.nasa.gov/sbdb.cgi?sstr=2005%20YU55;old=0;orb=0;cov=0;log=0;cad=0#elem
[7] http://echo.jpl.nasa.gov/asteroids/2005YU55/2005YU55_planning.html
[8] http://minorplanetcenter.net/blog/close-but-no-cigar-the-flyby-of-asteroid-2005-yu55/
[9] http://ssd.jpl.nasa.gov/sbdb.cgi?sstr=2004%20BL86;old=0;orb=0;cov=0;log=0;cad=0#phys_par
[10] http://solarsystem.nasa.gov/news/display.cfm?News_ID=48702
[11] http://www.johnstonsarchive.net/astro/astmoons/am-357439.html

Tectonics of the terrestrial litosphere in spherical harmonics

A V Mokhnatkin[1,2], S D Petrov[2], V L Gorshkov[1,2], N V Shcherbakova[1,2], S S Smirnov[1,2] and D A Trofimov[2]

[1] Main Astronomical Observatory, Russian Academy of Sciences, St.Petersburg, Russia
[2] St.Petersburg State University, St.Petersburg, Russia

E-mail: artspace3@mail.ru, s.d.petrov@spbu.ru, vigor@gao.spb.ru

Abstract. According to the classical tectonics theory, lithosphere is divided into more than twenty plates that are considered as absolutely rigid. However, modern astronomical and geodetic observations reveal disagreements with the standard theory. First of all, there are already more than two hundreds plates and newer ones are still being discovered. Then, it was found out that lithospheric plates can not be regarded as absolutely rigid, since they are evidently subject to deformations. This work is aimed at expansion of velocities of the global geodetic network stations into vector spherical harmonic series on the assumption of division of the velocity field into toroidal and poloidal components. Spherical harmonics up to fourth degree are obtained for corresponding components and a preliminary attempt of their geophysical interpretation is submitted.

1. Introduction

Current plate motion models are used for construction of the terrestrial reference system and are based on fitting of the plate rotation parameters to velocities of individual stations [1]. There are several ways to find high precision site coordinates and, accordingly, their velocities, such as Global Navigation Satellite Systems (GNSS), Very Long Baseline Interferometry (VLBI), Satellite or Lunar Laser ranging. In order to facilitate the lithospheric motion models, plates are considered as absolutely rigid bodies, moving along the spherical Earth surface, though at the same time plates have internal deformations, which are indicated by significant values of residual velocities after the motion model reduction. The problem of the reference system selection also remains for stations velocities as the lithosphere is considered generally to be fixed for the global models. Despite the accepted conception of plate boundaries as infinitely thin areas, which follows from the representation of lithospheric plates as absolutely rigid objects, observations indicate extended boundaries, which are subject deformations. These plate motion issues come from assumptions made for modern purely kinematic models. Current empirical models can be improved if internal deformations would be added to the station velocities interpretations. However, the question of the general lithospheric motion laws is still open, but nowadays it is possible to use the spherical harmonics to search for features in motion of lithospheric blocks.

2. GNSS station velocity data

Horizontal velocities with respect to the International Terrestrial Reference Frame (ITRF2008) of more than two thousand stations computed by Jet Propulsion Laboratory (USA)[1] were used as input data. This data set includes positions and velocities of most of the world GNSS stations with open access to their observations. The orbits of GPS satellites and their clock offsets were calculated in the IGS08 system by use of roughly 80 global GPS receivers, GIPSY software package were used for postprocessing. The station velocities were computed after breaks were rejected from series of position observations according to the chi-squared statistic. The velocities for 2490 stations were used, four of them excluded from further consideration as they were situated on a glacier and have a very high annual velocity value in horizontal projection. The rest stations velocities are represented in Fig. 1.

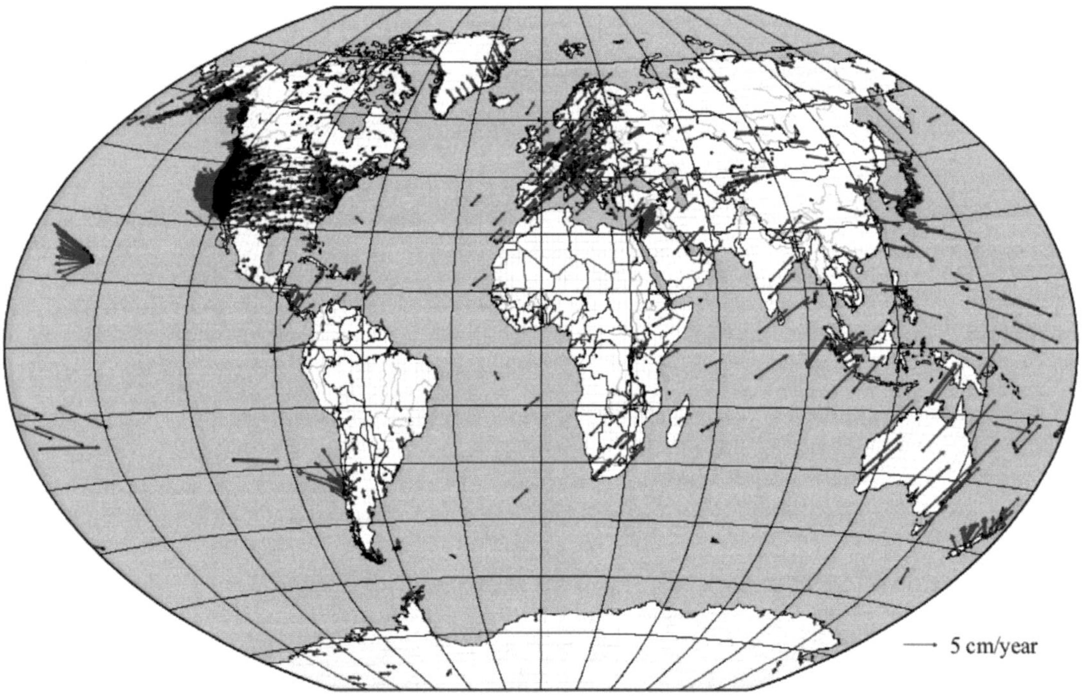

Figure 1. Horizontal velocities of 2486 GNSS stations by JPL

3. Vector spherical harmonics

The vector decomposition into series of spherical harmonics can be done based on the Helmholtz theorem for irrotational, poloidal and toroidal components of the velocity vector. The Helmholtz representation was used by previous authors [2] for horizontal velocities in the hotspot reference frame. We accepted that the zonal and meridional velocity field of stations is solenoidal and considered it on the sphere as the sum of toroidal and poloidal components. All this components are expanded into series of scalar spherical harmonics using expressions similar to the case in [3]. The proposed method imposes restrictions on the vector field, it does not need to be solenoidal – spreading zones, for example, obviously are areas of divergence of the vector velocity field. Nevertheless, proposed method is formal and has demonstrative physical interpretation.

[1] http://sideshow.jpl.nasa.gov/post/series.html

4. Results

The results are shown in Fig. 2-5. Meridional and zonal velocity components for spherical functions up to degree four are shown in Fig. 2. Some notable features of lithospheric tectonics can be seen there, such as a relatively high value of the Australian plate velocity. The velocity field for spherical harmonics of degree one is shown in Fig. 3. It can be seen that the distribution of the zonal velocity has a pronounced dipole structure. Velocity fields for spherical functions of degree two and three are respectively shown in Fig. 4 and Fig. 5, which exhibit that the velocity distribution has a more meridional character than zonal.

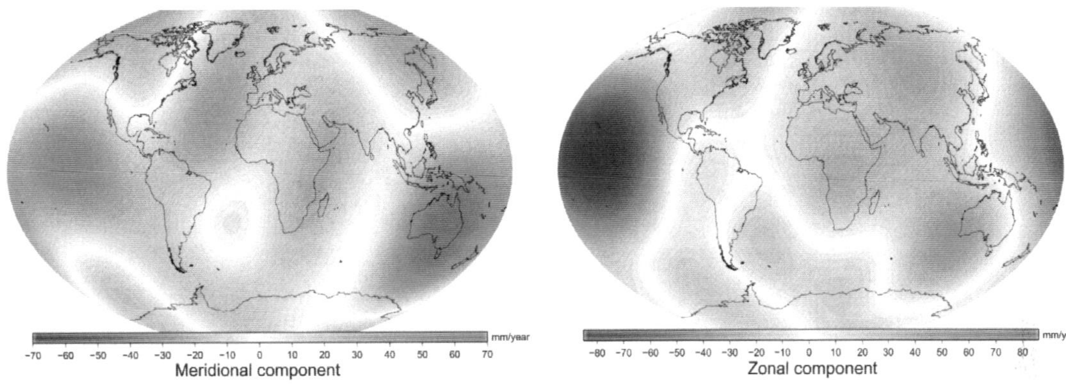

Figure 2. General meridional and zonal velocity components

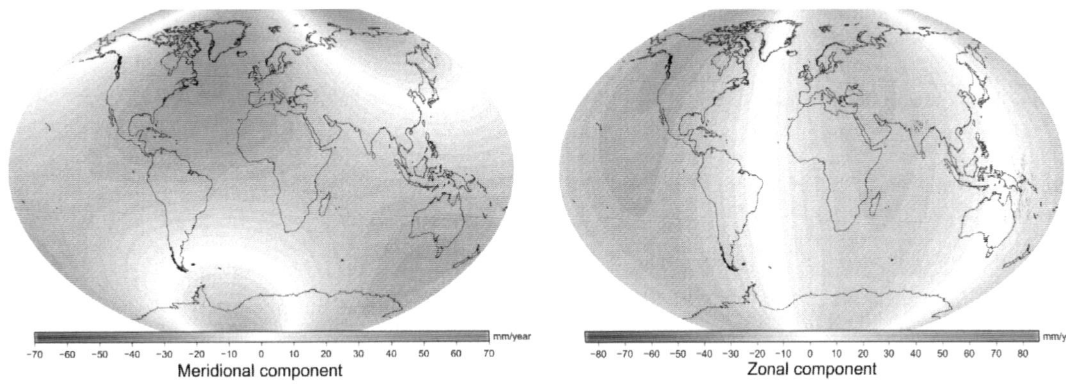

Figure 3. Meridional and zonal velocity components of degree one

5. Conclusions

An expansion of GNSS station velocities into series of spherical harmonics was performed. The results for poloidal and toroidal velocity components are generally consistent with the known features of the lithospheric tectonics. They yielded a visual representation of the large-scale lithospheric tectonics. The aim of further research is tectonophysical interpretation of the results, the elicitation of possible linkages of large-scale tectonic motions and the geoid variations. This research has bee supported by the SPSU Grant nr. 6.37.343.2015.

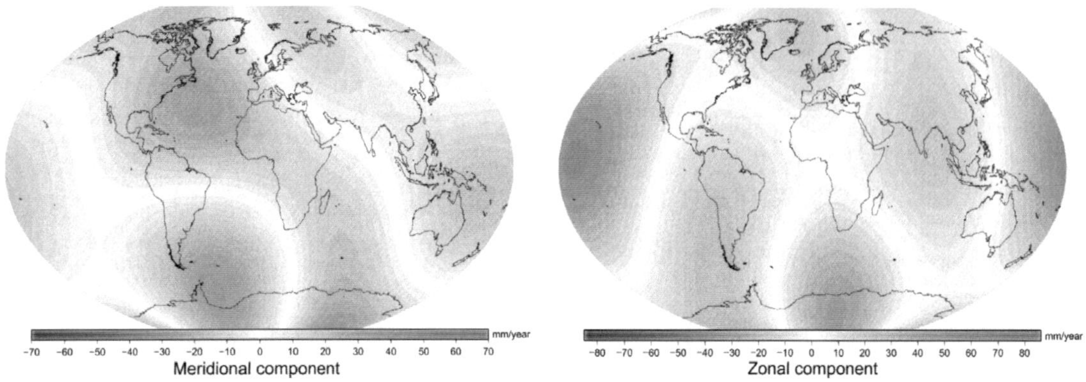

Figure 4. Meridional and zonal velocity components of degree two

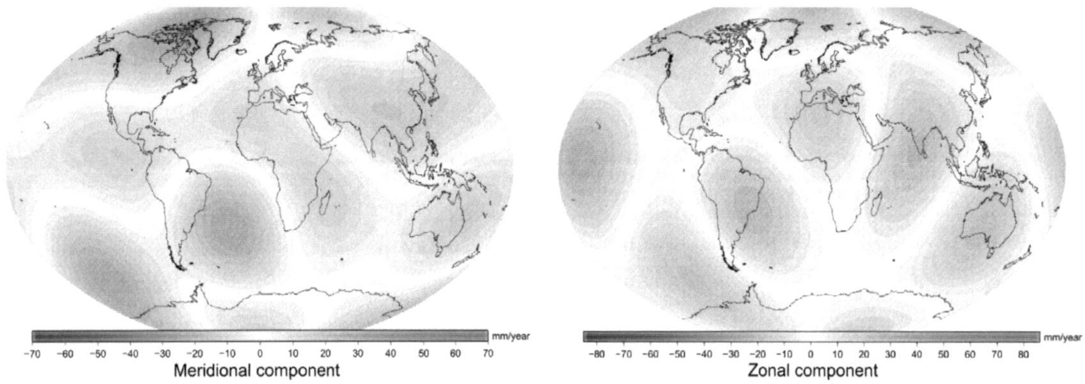

Figure 5. Meridional and zonal velocity components of degree three

References

[1] Altamimi Z, Métivier L and Collilieux X ITRF2008 plate motion model *J. Geophys. Res.* 2012 Vol. **117** Issue B7

[2] Bercovici D and Wessel P A continuous kinematic model of plate-tectonic motions *Geophys. J. Int.* 1994 **119** 595-610

[3] Vityazev V V and Tsvetkov A S UCAC4: Stellar kinematics with vector spherical functions *Astron. Nachr.* 2010 No. **8** 760-764

Journal of Physics: Conference Series **769** (2016) 012008 doi:10.1088/1742-6596/769/1/012008

Magnetic structures propagating in non-equilibrium relativistic plasma of pulsar wind nebulae

A E Petrov and A M Bykov

Ioffe Institute, 26 Politekhnicheskaya st., St. Petersburg 194021, Russia

E-mail: alexey.e.petrov@gmail.com

Abstract. The kinetic model of highly non-equilibrium relativistic electron-positron plasma is used to study dynamical magnetic structures in pulsar wind nebulae (PWNe). The evolution equation which describes a propagation of a long-wavelength magnetosonic type perturbation of small but finite amplitude is derived. The wavelength is assumed to be longer than the scattering length of the background positrons and electrons. The rates of scattering of electrons and positrons by the stochastic magnetic field fluctuations are distinguished but the difference is supposed to be small compared with the gyrofrequencies of particles. The phase velocity, the dissipation rate and the dispersion length of the magnetic pulse are studied as the functions of plasma parameters and the scattering rates of electrons and positrons. The model being confronted to observations can help to determine the pulsar wind composition, particle distribution and non-thermal pressure in PWNe.

1. Introduction

Investigations of physical mechanisms responsible for conversion of energy of a magnetized rotating neutron star into a pulsar wind (PW) and for particle acceleration at the termination shock (TS) of a PW are of great interest for the high energy astrophysics [1, 2]. The studies of these phenomena require modeling of highly non-equilibrium processes in relativistic magnetized plasma of pulsar wind nebulae (PWNe). An important issue here is to determine the values of the plasma parameters, like its pressure and particle energy distribution.

One of the most intriguing problems here is the composition of the relativistic PW. The PW plasma is thought to be predominately consisted of pairs, but the ions also may present. The question of the ion component presence in the PWs is important for understanding of physics of these objects. Because the particle energy in the upstream flow is determined by the Lorentz-factor of the flow, even small by the number density fraction of ions can affect the processes, responsible for the energy conversion in the wind. If the ratio of the ion concentration to the pair concentration exceeds the ratio of the electron mass to the ion mass, then the ion component contains the predominant fraction of the PW energy [3]. The presence of ions can affect the particle acceleration at the TS and, as a consequence, the observed spectra of emission [3].

An adequate interpretation of the observed dynamic structures in PWNe allows to put some constraints on relativistic plasma composition, particle distribution function and pressure [4–6]. The observed structures in PWN also may shed light on the nature of the giant gamma-ray flares in the Crab nebula [7]. Owing to the predominately synchrotron nature of the nebular emission, the observed structures may be related to the propagation of magnetic field perturbations.

The kinetic consideration of the magnetic field perturbation propagation in PWN plasma allows to derive an evolution equation, which describes the local dynamics of the structures. The obtained parameters of this equation describing the features and the evolution of the structures should depend on the plasma parameters and the parameters of particle scattering by the stochastic magnetic field fluctuations. In section 2 we provide a sketch of the procedure of the evolution equation derivation. We present the coefficients of this equation — the velocity of propagation, the dissipation rate, the dispersion coefficient. In sections 3-4 we consider the dependencies of these coefficients on the plasma parameters and the scattering rates. We discuss possible constraints on the plasma parameters, which could be imposed via the comparison of the observations with the model results using the obtained relations. In section 5 the presented results are summarized.

2. Local dynamics of magnetic structures

As like as in [8], we consider a propagation of a local magnetic field perturbation. Due to locality, the one-dimensional analysis is performed: a plane wave propagating along the Ox axis across the mean quasi-stationary magnetic field $\mathbf{B}_0 \parallel Oz$ is studied. The system of kinetic equations for the distribution functions f_p and f_e of plasma components — positrons and electrons — is treated (hereafter index p marks functions related to positrons, e — to electrons). The regime of the strong scattering of particles by the stochastic magnetic field fluctuations is considered, which is expressed via the non-zero collision operators. These are taken in the relaxation time approximation with different scattering frequencies ν_1 and ν_2 in the form $-\nu_1\left(f_p - f_p^{iso}\right) + \nu_2\left(f_e - f_e^{iso}\right)$ in the equation for the positron component and in the same form but with the opposite sign in the equation for the electron component. Here f_α^{iso}, $\alpha = (p, e)$ denotes the isotropic part of the distribution function of the component α.

The difference between the typical frequencies of scattering is introduced in order to take into account possible influence of the admixture of an ion component in the PWN plasma on the processes studied here. For instance, in [3] the authors presented the model of acceleration of pairs by the resonant cyclotron absorption of high harmonics of the ion-cyclotron waves (ICWs) emitted by ions. These waves propagate in the pair-ion plasma across the magnetic field and have an electrostatic component [3]. They are elliptically polarized and left-handed, and the direction of their electric field vector rotation matches the direction of the positrons' gyration [3]. Due to that, the positrons absorb the ICWs more effectively than the electrons [3]. A similar mechanism can affect the scattering of pairs by the stochastic electromagnetic field fluctuations in PWN plasma. This may cause the difference between the typical frequencies of scattering for positrons and electrons, which are thought to be equal in pure pair plasma. Thus, in order to build a simplified model of studied processes in the pair-ion plasma, we introduce different scattering frequencies. This is the only manifestation of the ion component influence in the discussed model: ions are not included explicitly here.

We consider the processes, whose typical frequencies ω are small in comparison with $\nu_{1,2} \leq \Omega = eB_0/m_e c\gamma$ (here e is the positron charge, m_e is the mass of an electron, c is the light velocity, γ — the Lorentz-factor of the particle). We suppose hereafter that $\nu_{1,2}$ are proportional to the gyrofrequency Ω:

$$\omega \ll \nu_{1,2} = a_{1,2}\Omega \leq \Omega \tag{1}$$

The solution of the system of kinetic equations is sought in the form of

$$f_\alpha = f_0 + \delta f_{\alpha 0} + \delta f_{\alpha x} \sin\theta\cos\phi + \delta f_{\alpha y} \sin\theta\sin\phi + \delta f_{\alpha z}\cos\theta \tag{2}$$

where $\alpha = p, e$. Function f_0 is a non-perturbed (isotropic in the momentum space) distribution function, taken to be equal for positrons and electrons, $\delta f_{\alpha 0}$ is the perturbation of the isotropic part of the distribution function, and the other terms give the anisotropic perturbation. Here θ and ϕ are the polar and the azimuthal angles in the momentum space.

In the considered case the transverse propagation corresponds to the wave mode with the dispersion equation $c^2k^2/\omega^2 = \epsilon_{yy} - \epsilon_{xy}\epsilon_{yx}/\epsilon_{xx}$, where ϵ_{ij} are the components of the dielectric permittivity tensor, k is the wave number. The electric field of this wave mode is polarized in the Oxy plane.

In the framework of the kinetic approach one has to calculate the current response on the magnetic field perturbation to get the evolution equation. This current response $\mathbf{j}$ should be substituted into the Maxwell equation for **rot B**. In the geometry under consideration the y-component of this equation is of interest, thus, the j_y component has to be derived. The derivation procedure is similar to the presented in [8] one.

The long-wavelength limit $ck/\nu_{1,2} \ll 1$ is considered, thus, one can expand the components of the dielectric permittivity and conductivity tensors in the power series in ck/Ω and retain only a few terms of the lowest orders. The dispersion and dissipative effects are of great interest for the problem studied here, thus, the weak dissipation and dispersion are included via retaining the terms of the second and third orders of ck/Ω in the conductivity tensor components σ_{ij}. The amplitudes of the perturbations are supposed to be small enough to allow keeping only the lowest order nonlinear terms in the derived current response.

At the first stage of the derivation the system of kinetic equations is linearized, and their Fourier transforms (FTs) are taken. After the substitution of (2) into the derived equations, one can obtain system of eight algebraic equations for $f_{\alpha 0,\omega k}$, $f_{\alpha x,\omega k}$, $f_{\alpha y,\omega k}$, $f_{\alpha z,\omega k}$, where $\alpha = (p, e)$, via multiplying of the equations on 1, $\sin\theta\cos\phi$, $\sin\theta\sin\phi$ and $\cos\theta$ and integration of the result over angles. The solution of this system provides the FT of the distribution function response on the electromagnetic field perturbation. This allows to evaluate the components of the linear current response FT and therefore the components of the σ_{ij} and ϵ_{ij} tensors.

We suppose here that $\delta\nu^2 = (\nu_1 - \nu_2)^2/\Omega^2 \ll 1$. While deriving the solution of the dispersion equation and the current response we expand them into the power series in $\delta\nu^2$ and keep only the lowest-order terms of zeroth and first orders. Also, all calculations are performed in the ultrarelativistic limit. The solution of the general dispersion equation is sought in the form:

$$\omega = Vk - i\tilde{\chi}k^2 - \mu k^3 \tag{3}$$

Using (1) and the condition $ck \ll \nu_{1,2}$, one can perform the third-order expansions of σ_{ij} mentioned above. After substitution of the results in the dispersion equation one can substitute there the equation for ω (3). After disregarding of all the terms of the fourth and elder orders of ck/Ω and equating of the coefficients at each power of ck/Ω to zero, the equations for the coefficients of (3) can be obtained. The derived equations for the phase velocity V, the dissipation coefficient $\tilde{\chi}$, and the dispersion coefficient μ are:

$$\frac{V^2}{c^2} = u^2 = \frac{1}{3} + U_0^2 = \frac{1}{3} + \frac{2}{3}\frac{1+2\beta}{(1+2\beta)^2 - 2\beta\delta\nu^2} \tag{4}$$

$$\tilde{\chi} = X_0\frac{c^2}{2u^2}U_0^2\left(U_0^2 - \delta\nu^2 u^2\frac{c_A^2}{c^2}\left(2 + \frac{c_A^2}{c^2}\right)\right) \tag{5}$$

$$\begin{aligned}\mu = A_0\frac{c^3}{2u}U_0^4\left[1 - 4a^2\left(1 - \frac{1}{3u^2}\right) + \delta\nu^2\left(6 - 2\frac{c_A^2}{c^2} - \frac{c_A^4}{c^4}\right) + 8a^2\delta\nu^2\left(\frac{c_A^2}{c^2} - \frac{2}{3}\right)\right] + \\ + 2a^2\delta\nu^2 A_0 c^3 u U_0^2\left[\frac{c_A^4}{c^4} - \frac{U_0^4}{u^4}\left(3 - 2\frac{c_A^2}{c^2}\right)\right] + Vd_\chi^2\end{aligned} \tag{6}$$

where d_χ^2, U_0^2, X_0 and A_0 are defined as follows:

$$d_\chi^2 = \delta\nu^2 X_0^2 \frac{c_A^2}{3u^2}\left[\left(2u^2-\frac{1}{3}\right)\left(\frac{1}{3u^2}-2\frac{c_A^2}{c^2}\right)-u^2\left(1-\frac{c_A^2}{c^2}\right)^2\right]-$$
$$-\frac{\chi X_0}{18u^4}\left[1+6u^2-27u^4-\delta\nu^2\left(2+15u^2-81u^4\right)+\delta\nu^2\left(1-\frac{c_A^2}{c^2}\right)^2\left(27u^4-3u^2\right)\right]- \quad (7)$$
$$-\delta\nu^2\frac{\chi X_0}{9u^4}\left(1-\frac{c_A^2}{c^2}\right)\left(1+9u^2-54u^4\right)-\frac{\chi^2}{2c^2u^2}$$

$$U_0^2 = \frac{2}{3}\frac{1+2\beta}{(1+2\beta)^2-2\beta\delta\nu^2}; \qquad X_0 = 10a\frac{\omega_p^2}{\omega_B^3}\left\langle\gamma^2\right\rangle U_0^2; \qquad A_0 = 6\frac{\omega_p^2}{\omega_B^4}\left\langle\gamma^3\right\rangle U_0^2 \quad (8)$$

Here $a = (\nu_1+\nu_2)/2\Omega$, $\omega_p = \left(4\pi e^2 n_0/m_e\right)^{1/2}$ is the plasma frequency, $n_0 = \int f_0 d^3p$ is the equilibrium number density of each kind of pairs, where integration is performed over the momentum space, and $\omega_B = eB_0/mc$. The angle brackets denote the averaging over the distribution function: $\langle(...)\rangle = \int (...) f_0 d^3p/n_0$. The typical velocity c_A — the ultrarelativistic analogue of the Alfven velocity — is defined as $c^2/c_A^2 = 1+2\beta$, where $\beta = 8\pi P/B_0^2$ and P is the particle pressure. We also introduce the dispersion length $a_D = |\mu|/V$.

Using the obtained equations for conductivities, one can derive the FT of the linear part of the current response. To express it only via the magnetic field of the perturbation b one may use the FTs of the Maxwell equations.

At the second stage the nonlinear correction to the current response is evaluated. The procedure is analogous to the discussed in [8] one. Again the system of partial differential equations is reduced to the algebraic system via the Green functions evaluations and the expansions into series over small ω/Ω, $\omega/\nu_{1,2}$ parameters, such that only the terms of the first non-vanishing order in the non-linear current response are kept. Finally, after the substitution of the sum of the linear and non-linear current responses into the y-component of the Maxwell equation for **rot B**, one can obtain the evolution equation for the magnetic field perturbation. We present it in the dimensionless form:

$$\partial_\tau h + \partial_\xi h + \frac{\mu}{|\mu|}\partial_\xi^3 h + \lambda h \partial_\xi h = \chi \partial_\xi^2 h \quad (9)$$

Here $h = b/B_0$, while λ and $\chi = \tilde{\chi}/Va_D$ — dimensionless coefficients; $\tau = Vt/a_D$, $\xi = x/a_D$ — dimensionless time and coordinate. Coefficient λ is a cumbersome functional of f_0 depending on B_0 and on the frequencies $\nu_{1,2}$.

3. The phase velocity of structures and the pressure of plasma

The dependence of the phase velocity (4) on the ratio of the particle pressure to the magnetic field pressure $\beta = 8\pi P/B_0^2$ is of interest for the physics of PWN plasma. In Figure 1 we show the values of the phase velocity V at various β. The results are presented for $\delta\nu = 0$ and $\delta\nu = 0.4$. As one can see from (4), the dependence of the phase velocity on $\delta\nu$ is monotonic for $0 \leq \delta\nu \leq 1$ and for small $\delta\nu^2 \ll 1$ is not very strong, which is demonstrated in Figure 1.

The velocity of propagation of a small amplitude perturbation described by (9) is mainly defined by the phase velocity. From (4) and Figure 1 one can see that the value of V significantly decreases with the growth of β. For $\beta \to 0$ it tends to the limit $V = c$, while at large β it approaches to $c/\sqrt{3}$. This sensitivity of the phase velocity to the particle pressure can allow to impose the constraints on the PWN plasma pressure via the comparison of the results of the modeling with the observed velocities of propagation of dynamic structures.

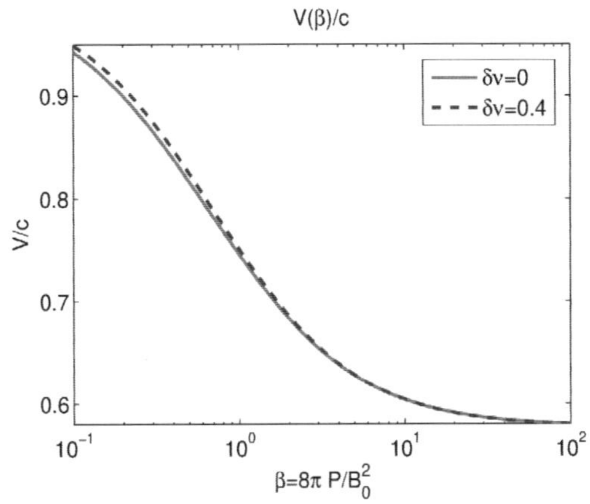

Figure 1. The phase velocity V for various values of $\beta = 8\pi P/B_0^2$ at $\delta\nu = 0$ and $\delta\nu = 0.4$.

4. The dissipation coefficient and the ion component of plasma

The dependence of the dissipation coefficient (5) on $\delta\nu = (\nu_1 - \nu_2)/\Omega$ also may help to get some constraints on the plasma parameters. This is illustrated in Figure 2, were the values of the dimensionless dissipation coefficient χ at various $\nu_{1,2}$ are presented. Figures show the values of χ at various $\nu_1 = \nu_2 + \delta\nu\Omega$: $\delta\nu$ is varied, while ν_2 and Ω are fixed.

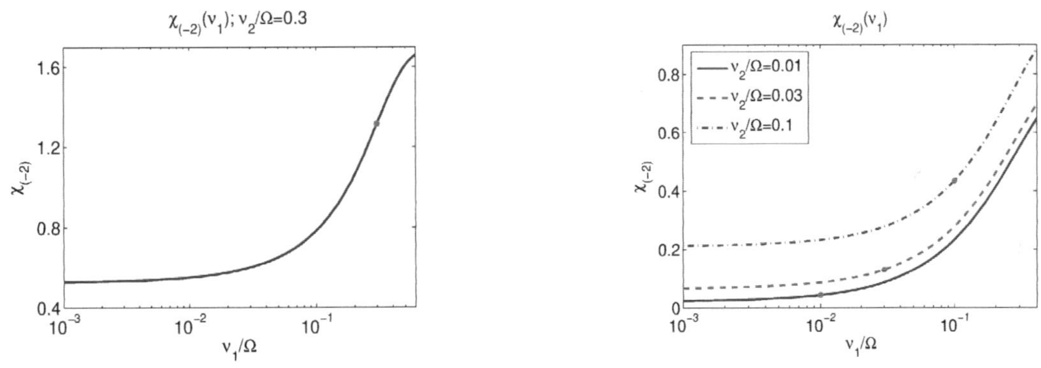

Figure 2. The value $\chi_{(-2)} = \chi/10^{-2}$, where χ is the dimensionless dissipation coefficient, at various ν_1 for $\beta = 8\pi P/B_0^2 = 10$, $B_0 = 200\mu$G. The energy distribution is the broken power-law with the indices $s_1 = 1.35$ for $100 \leq \gamma \leq \gamma_b = 7.7 \times 10^5$, $s_2 = 2.2$ for $\gamma_b < \gamma \leq \gamma_{max} = 10^9$. The value of ν_2 is fixed, the red points at each curve mark the values of χ for $\nu_1 = \nu_2$.

One can see that the increasing of the absolute value of $\delta\nu$ provides remarkable variation of χ. Indeed, in the case of $\nu_2 = 0.3\Omega$ (Figure 2a) the increasing of ν_1 by 0.3Ω gives growth of χ by a fourth (in comparison with the value for $\nu_1 = \nu_2 = 0.3\Omega$), while for the shift with $\delta\nu = -0.2$ one finds a 1.7 times reduction of χ. For the case of $\nu_2 = 0.1\Omega$ (Figure 2b, dot-dashed grey line) increasing of ν_1 by 0.3Ω leads to twofold increase of χ, and the tenfold reduction of ν_1 (down to $\nu_1 = 0.01\Omega$) results in nearly twofold decrease of χ. For the cases of smaller values of ν_2 (see Figure 2b) increasing of ν_1/Ω by $\delta\nu = 0.2 - 0.3$ provides even more significant variations of χ: it becomes $3 - 4$ times larger in the case of $\nu_2 = 0.03\Omega$ (dashed green line) and about 10 times larger in the case of $\nu_2 = 0.01\Omega$ (solid blue line).

The present consideration is limited by the approximation $\delta\nu^2 \ll 1$, but even in the framework of this treatment one can find significant variations of χ caused by the deflection of $\delta\nu$ from zero. Thus, if the influence of the ion component on the scattering processes shortly described in Section 2 can cause sufficient variations of the frequency of scattering of positrons, χ can be significantly affected. The value of χ determines the character of evolution of the magnetic field perturbation, described by (9): the evolution of its profile depends on the relation of the dispersion and dissipative terms [9]. Also, the dissipation coefficient, obviously, defines the characteristic distance of damping of the perturbation. The synthetic synchrotron images of the models of nebulae with the considered magnetic structures giving the spatial distributions of the synchrotron emission intensity of nebular particles should reflect such manifestations. On the other side, the observations reveal dimming and deceleration of the wisps [4]. Due to these reasons, modeling of magnetic structures, propagating in PWN plasma, and comparison of the synthetic synchrotron images with the observations may help to get observational constraints on the composition of plasma. This may give an answer to the question of presence of ions in the PWs. The crucial issue here is how large the shift of the frequency of scattering due to the influence of the ions can be. It requires modeling of the processes of scattering.

5. Conclusions

In this paper the kinetic model of local dynamics of magnetic field perturbation in highly non-equilibrium relativistic PWN plasma is briefly considered. The analysis of the obtained equations for quantities defining the features and the evolution of magnetic structures is presented. The dependencies of these parameters on the pressure of PWN plasma and the parameters of scattering of pairs by the stochastic magnetic field fluctuations are treated. These dependencies can allow to get some observational constraints on the PWN plasma parameters. In particular, the dependence of the phase velocity of the perturbations on the particle pressure of plasma can provide constraints on the plasma pressure. Also, on the base of the mechanism of acceleration of pairs in the pulsar wind presented in [3], we suppose that the presence of the ion component in the pulsar wind can affect the parameters of scattering of pairs. If the typical frequencies of scattering are affected strongly enough, the substantial change of the dissipation coefficient value can occur. Thus, the constraints on the particle pressure and the composition of the PWN plasma could be imposed via the comparison of the simulated spatial distributions of the nebular synchrotron emission intensity with the observations. In order to implement this for the problem of plasma composition, the modeling of the processes of scattering by the stochastic magnetic field fluctuations is required. Thus, the kinetic modeling of the dynamical structures, like wisps, can help to answer some questions about the PWN plasma parameters important for the hot issues of mechanisms responsible for the energy conversion in the relativistic PWs.

Acknowledgments

A. E. Petrov acknowledges support from RSF grant 16-12-10225 for making possible the calculations, whose results are demonstrated in Figure 2.

References

[1] Arons J 2012 *Space Sci. Rev.* **173** 341-67
[2] Bühler R and Blandford R 2014 *Rep. Prog. Phys.* **77** 6
[3] Amato E and Arons J 2006 *ApJ.* **653** 325-38
[4] Hester J J 2008 *Annu. rev. astron. astrophys.* **46** 127-55
[5] Porth O, Komissarov S S and Keppens R 2013 *MNRAS* **431** L48-52
[6] Olmi B, Del Zanna L, Amato E and Bucciantini N 2015 *MNRAS* **449** 3149-59
[7] Bykov A M, Pavlov G G, Artemyev A V and Uvarov Yu A 2012 *MNRAS* **421** L67-71
[8] Petrov A E and Bykov A M 2014 *J. Phys.: Conf. Ser.* **572** 012005
[9] Petrov A E and Bykov A M 2015 *J. Phys.: Conf. Ser.* **661** 012010

Journal of Physics: Conference Series **769** (2016) 012009 doi:10.1088/1742-6596/769/1/012009

Astrometric and photometric investigation of asteroids (3100) Zimmerman, (3504) Kholshevnikov, (4592) Alkissia, (4619) Polyakhova

S N Petrova[1], **A V Devyatkin**[1], **D L Gorshanov**[1]

[1] The Central Astronomical Observatory of the Russian Academy of Sciences, Pulkovskoe chaussee 65, Saint-Petersburg, 196140, Russia

E-mail: stalkered@yandex.ru

Abstract. Observations of the chosen understudied asteroids of the main belt, named after astronomers of Pulkovo observatory and St.Petersburg State University (3100) Zimmerman, (3504) Kholshevnikov, (4592) Alkissia, (4619) Polyakhova were made. The data was processed and light curves of the asteroids were obtained. Axial rotation periods of the observed objects were discovered for the first time. Accuracy of the orbital elements of the asteroids was estimated on the basis of obtained observations and MPC data.

1. Introduction

Among 75 asteroids named after astronomers of Pulkovo observatory and St.Petersburg State University the axial rotation periods are discovered only for 35 of them and no characteristics except the absolute magnitude are known for 22 of them. In this investigation astrometric and photometric observations were made for the chosen understudied asteroids (3100) Zimmerman, (3504) Kholshevnikov, (4592) Alkissia, (4619) Polyakhova. During the period from October 2013 to March 2014 two telescopes of Pulkovo observatory (ZA-320M and MTM-500M) and two telescopes of astronomical observatories of Irkutsk and Kuban State Universities were used for the observations. All the asteroids in question are the main belt asteroids and they all were discovered by Nikolay Stepanovich Chernykh in the Crimean Astrophysical Observatory (CrAO).

The aim of the investigation was to obtain the light curves of the chosen insufficiently explored asteroids, to measure their axial rotation periods by means of frequency analysis [1] as well as to try to improve the orbital elements of the objects in question.

2. Astrometric study

EPOS software systems [2] were used to process the astrometric data. As the chosen objects have a long observational history the addition of our new observations that were made in a relatively short period of time hasn't made a significant improvement of the orbital elements [3][4][5][6]. Although these results prove that the orbits of the chosen objects are stable and the asteroids are not hazardous.

Table 1. Orbital elements improvement for (3100) Zimmerman for the epoch J2457400.5.

Elements	Input	Improved	Correction	RMS
M, (°)	210.465350	210.465390	0.000040	0.0000204
ω, (°)	94.593270	94.593285	0.000015	0.0000761
Ω, (°)	27.077110	27.077048	-0.000062	0.0000729
i, (°)	2.820810	2.820797	-0.000013	0.0000032
e	0.08738650	0.08738592	-0.00000058	0.000000039
a, (AU)	2.259161200	2.259161217	0.000000017	0.0000000055
q, (AU)	2.061741010	2.061742343		

Table 2. Orbital elements improvement for (3504) Kholshevnikov for the epoch J2457400.5.

Elements	Input	Improved	Correction	RMS
M, (°)	166.335030	166.335002	-0.000028	0.0000170
ω, (°)	165.008000	165.008065	0.000065	0.0001447
Ω, (°)	118.853770	118.853696	-0.000074	0.0001439
i, (°)	2.047070	2.047085	0.000015	0.0000034
e	0.17944990	0.17945044	0.00000054	0.000000055
a, (AU)	3.100686099	3.100686103	0.000000003	0.0000000146
q, (AU)	2.544268289	2.544266614		

Table 3. Orbital elements improvement for (4592) Alkissia for the epoch J2457400.5.

Elements	Input	Improved	Correction	RMS
M, (°)	132.610090	132.610046	-0.000044	0.0000107
ω, (°)	305.821510	305.821322	-0.000188	0.0005312
Ω, (°)	73.981360	73.981605	0.000245	0.0005302
i, (°)	0.430540	0.430532	-0.000008	0.0000035
e	0.16121240	0.16121178	-0.00000062	0.000000046

a, (AU)	3.197454600	3.197454557	-0.00000062	0.000000046
q, (AU)	2.681985270	2.681987228		

Table 4. Orbital elements improvement for (4619) Polyakhova for the epoch J2457400.5.

Elements	Input	Improved	Correction	RMS
M, (°)	162.423140	162.423317	0.000177	0.0000729
ω, (°)	195.155510	195.155124	-0.000386	0.0001480
Ω, (°)	266.835790	266.836000	0.000210	0.0001283
i, (°)	1.818180	1.818183	0.000003	0.0000040
e	0.03082380	0.03082322	-0.00000058	0.000000051
a, (AU)	2.688744900	2.688744835	-0.000000065	0.0000000094
q, (AU)	2.605867565	2.605869048		

Where: M — the mean anomaly; ω — the argument of perihelion; Ω — the ascending node; i — the inclination; e — the eccentricity; a — the semi-major axis; q — the perihelion distance; RMS — the root mean square error of the defined correction.

3. Photometric study

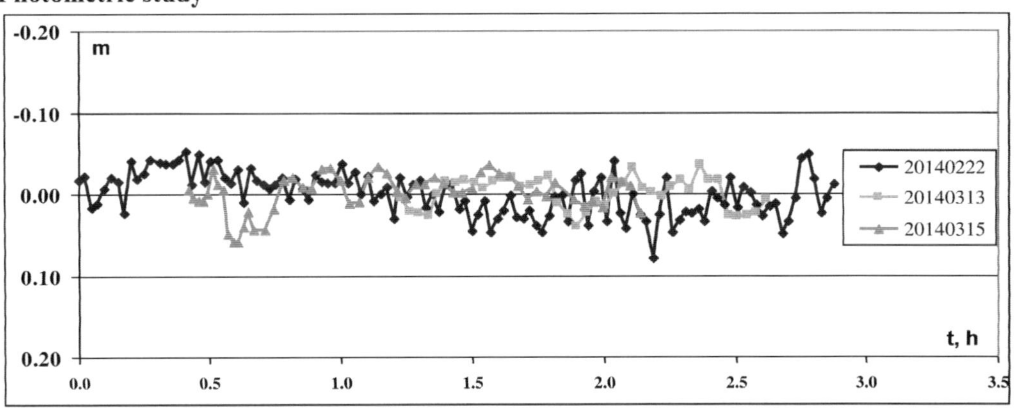

Figure 1. The light curve of the asteroid (3100) Zimmerman.

Asteroid (3100) Zimmerman (1977 EQ1 = 1975 TJ2 = 1975 WV1 = 1982 UJ6) was discovered on March 13, 1977. This asteroid was named in honor of Nikolai Vladimirovich Zimmerman (1890-1942) who had worked in Pulkovo observatory since 1915 and until the end of his life. In 1938-1942 N.V. Zimmerman had been the head of the observatory's astrometric department and since 1937 he also had been a professor of the Leningrad State University.

By this stage of the investigation the axial rotation period hasn't been obtained due to the flat shape of the object's light curve. That could mean that the asteroid could have a near-spherical shape. The observations of (3100) Zimmerman will be continued.

Journal of Physics: Conference Series **769** (2016) 012009	doi:10.1088/1742-6596/769/1/012009

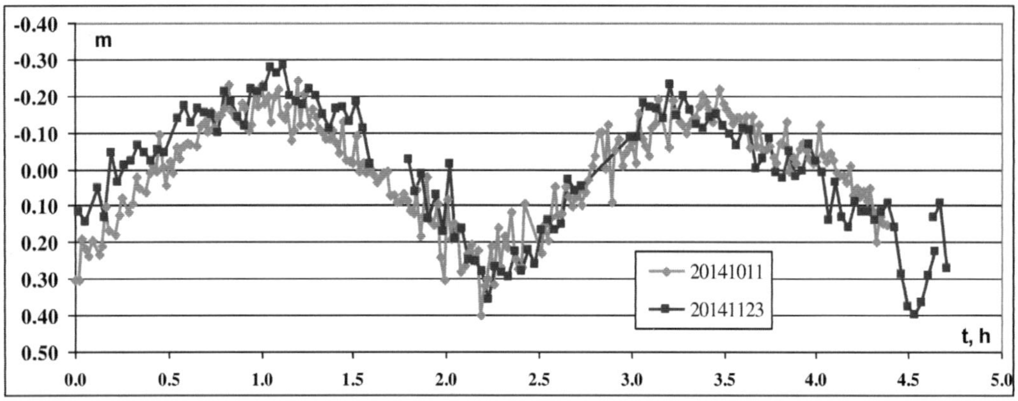

Figure 2. The light curve of the asteroid (3504) Kholshevnikov.

Asteroid (3504) Kholshevnikov (1981 RV3 = 1959 TN = 1975 RH2 = 1980 KT1 = 1980 LD1) was discovered on September 3, 1981. This asteroid was named in named in honor of Konstantin Vladislavovich Kholshevnikov (b. January 19, 1939) who is the head of the celestial mechanics department of St.Petersburg State University. K.V. Kholshevnikov is also a distinguished scientist, an academician of Russian Academy of Natural Sciences and an author of more than 150 science papers including 9 monographs and textbooks.

The period of axial rotation for the asteroid (3504) Kholshevnikov was established firstly in 2012 during our previous study and it is equal to 4.7201 ± 0.0001 hours [7]. During this study the period has been confirmed.

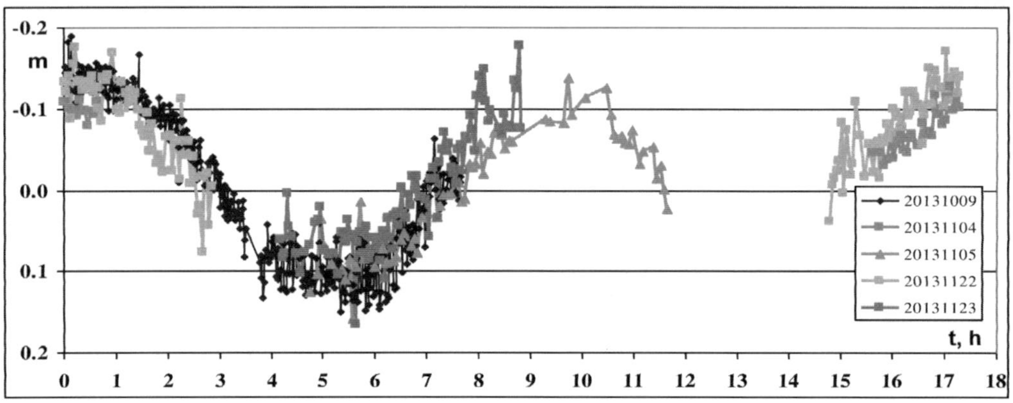

Figure 3. The light curve of the asteroid (4592) Alkissia.

Asteroid (4592) Alkissia (1979 SQ11 = 1951 WE1 = 1978 NT7) was discovered on September 24, 1979. This asteroid was named in honor of Aleksey Alekseevich Kiselev (1922-2013) who had worked in the Laboratory of astrometry and stellar astronomy of Pulkovo observatory since 1956 and until the end of his life. The Apparent Motion Parameters (AMP) method was developed by him.

In this study the axial period of the asteroid (4592) Alkissia was established for the first time and it equals 17.330±0.001 hours.

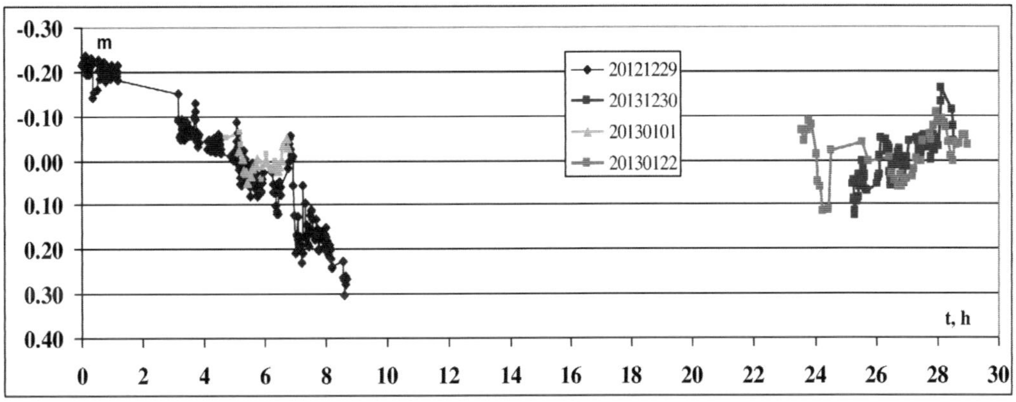

Figure 4. The light curve of the asteroid (4619) Polyakhova.

Asteroid (4619) Polyakhova (1977 RB7 = 1966 CW = 1988 DH) was discovered on September 11, 1977. This asteroid was named in honor of Elena Nikolayevna Polyakhova (b. December 16, 1935). She is the assistant professor of the celestial mechanics department of St.Petersburg State University who has worked in the University since 1964. E.N. Polyakhova also is an author of more than 200 science papers including several monographs and textbooks.

The period of axial rotation of the asteroid was not possible to determine with sufficient accuracy at this stage of the investigation; according to preliminary data it can be estimated as 29.91 ± 0.01 hours. The observations of (4619) Polyakhova will be continued.

4. Conclusion

In this study the light curves of all four chosen asteroids are obtained. The orbital period of (4592) Alkissia was discovered for the first time (17.330±0.001 hours) and the orbital period of (3504) Kholshevnikov that was discovered in our previous study was confirmed (4.7201±0.0001). According to available data the period for (4619) Polyakhova can be presumably estimated as 29.99±0.01. An attempt to improve the orbital elements of the chosen asteroids was made.

References

[1] Scargle J D 1982 Studies in Astronomical Time Series Analysis. II. Statistical Aspects of Spectral Analysis of Unevenly Spaced Data *Ap.J.* **263** p 835-53

[2] L'vov V N et al. 2012 The use of the Epos software package for research of the Solar System objects *Solar System Research* **46** issue 2 p 177-9

[3] http://minorplanetcenter.net/db_search/show_object?object_id=3504&page=2

[4] http://minorplanetcenter.net/db_search/show_object?object_id=4619&page=2

[5] http://minorplanetcenter.net/db_search/show_object?object_id=3100&page=2

[6] http://minorplanetcenter.net/db_search/show_object?object_id=4592&page=2

[7] Petrova S N, Devyatkin A V, Gorshanov D L et al 2013 Astrometric and photometric investigation of asteroids (857) Glasenappia, (2323) Zverev, (3504) Kholshevnikov *Izv. GAO v Pulkove* **220** p 65-70

[8] Devyatkin A V, Gorshanov D L, Kouprianov V V et al 2010 Apex I and Apex II Software Packages for the Reduction of Astronomical CCD Observations *Solar System Research* **44** № 1 p 68–80

Magnetic field amplification and high-energy emission properties of compact galactic clusters

V I Romansky[1], P E Gladilin[1], A M Bykov[1] and S M Osipov[1]

[1]Ioffe Physical-Technical Institute of the Russian Academy of Sciences, Saint-Petersburg, Russia,

E-mail: romanskyvadim@gmail.com, peter.gladilin@gmail.com

Abstract. The observed appearance of the high-energy gamma-ray emission from galactic star clusters is determined by the geometry of the cluster and the distribution of the energy sources inside the cluster. A large number of massive stars with strong winds and supernova explosions lead to the formation of multidirectional plasma flows, shocks and strong magnetohydrodynamic turbulence. Magnetic turbulence is the main factor which determine the maximal energies of the accelerated particles and the power of the source. Here we discuss an effect of the efficient magnetic field amplification at the supernova shocks due to the cosmic-ray driven plasma instabilities on the gamma-ray and neutrino fluxes from massive star clusters. We show here that the photon flux attenuation by the $e^{\pm}$ pair production is important for this kind of sources in the photon energy range $10^{15} - 10^{16}$ eV. The ratio of the gamma-ray to the neutrino fluxes is a declining function of the distance to the source at 5-15 kpc and the photon flux can be several times lower than the corresponding neutrino flux from the hadron interactions.

1. Introduction

High-energy emission processes are associated with the most powerful events in the Galaxy. Gamma-ray observations with modern space- and ground-based telescopes have revealed that supernova remnants (SNRs) are the most likely sources of cosmic rays (CRs) at least up to 100 TeV. [1, 2, 3, 4, 5]. The problem of the sources of galactic CRs above 100 TeV is still under debate. Among the others possible sources OB-associations and compact clusters of young massive stars have a potential of CR acceleration well above the PeV energy regime [6, 7, 8]. Moreover the *Fermi* gamma-ray observatory has reported a discovery of an extended gamma-ray source which is associated with the very rich Cygnus OB2 association [9]. Supernovae in compact clusters of young massive stars were argued [8, 10] to be efficient accelerators of CRs above 10 PeV and are the candidate sources to produce a few PeV neutrino events detected by the *Ice Cube* observatory [11].

Efficient particle acceleration in young massive clusters can take place at the SNR and stellar winds shocks, in binary systems at colliding shocks and in collective processes such as multiple shocks interactions. Generation of the magnetohydrodynamic turbulence leads to the fast growth of the magnetic fields in the vicinity of the SNR shocks and stellar winds, that in turn lead to the increasing efficiency of particle acceleration. Non-thermal emission in shell-type structures near molecular clouds may signal about hadronic interactions of the accelerated particles with the interstellar matter (ISM). Modern telescopes such as H.E.S.S., Fermi LAT and Chandra have high sensitivity and resolution that is enough to observe the gamma-ray emission from compact star clusters.

Recently Chandra has detected X-rays emission from supercompact star clusters Arches and Quintuplet which are only 25-30 pc from the galactic center [12]. Fermi LAT and H.E.S.S. have detected hard X-rays and γ-emission from the star cluster Westerlund 1, which is $3.5 - 4$ kpc from the Earth [13, 14].

The progress in numerical simulations of the stellar evolution and the rise of the efficiency of observations provided much better understanding of the physics of star-forming regions. Careful modeling of the emission and acceleration processes requires realistic model that should account for the non-linear feedback of the CRs particles to the acceleration process. In order to consistently describe our non-linear model of high-energy sources in star-forming regions this paper is divided into two parts. The first part is devoted to the description of the semi-analytical model of the magnetic field amplification via CR-ray driven instabilities. This model can be efficiently applied to the numerical simulations of the SNR's shocks and multiple shocks systems in compact star clusters. In the second part we discuss the specific properties of the high-energy gamma-rays and neutrino emission spectra from colliding shock flows systems (CSF) in galactic clusters.

2. The model of magnetic field amplification

Recently E.Amato, P.Blasi and D.Caprioli have developed the non-linear semi-analytical model of the particle acceleration at the SNR shocks, which is useful to calculate feedback effects of the CR particles [15, 16]. We have extended this model to the case of CSF systems that could often occur in star clusters and OB-associations [17, 18].

The model allows to get self-consistently the spectral energy distribution of the accelerated protons $f(x)$ and velocity profile at the shock $u(x)$ for an arbitrary environment parameters (Mach number, the size of the system and density) and diffusion properties of the medium.

Assume that we know $f(x)$ and $u(x)$ from the model above. Let us consider the CRs current, calculated from the proton distribution function, f(x):

$$J_{cr}(x) = -e \int_{p_{min}}^{p_{max}} D(x,p) \frac{\partial f}{\partial x} p^2 dp \tag{1}$$

where $D(x,p)$ - is the diffusion coefficient, p_{min} and p_{max} - are the minimum and the maximum momentum of the particle. The growth rates for the three CR-driven instabilities are derived from the dispersion relation for modes propagating along the mean magnetic field in the magnetized background plasma.

This dispersion relation which includes short-wavelength Bell instability was derived in Bykov et.al. [19]:

$$\omega^2 \mp \omega i k k_c \frac{\alpha_t}{4\pi\rho} \left[\frac{1}{2} \frac{eA(x,k)}{J_{cr}(x)} + \frac{3}{2} \right] - k^2 v_a^2 \pm$$
$$\pm k k_c v_a^2 \left(1 + \frac{k_t}{B_{ls}(x,k^*)} \right) \left[\frac{eA(x,k)}{J_{cr}(x)} - 1 \right] = 0, \tag{2}$$

where ρ is the background plasma density, $v_a = B_{ls}(x,k)/\sqrt{4\pi\rho(x)}$ is the Alfven speed, $k_c = 4\pi|J_{cr}(x)|/(cB_{ls}(x,k))$ is Bell's critical wavenumber at position x in the precursor of the shock, $B_{ls}(x,k)$ is the mean magnetic field, which is defined as the sum of the long-wavelength harmonics and the ambient homogeneous field; $\pm$ signs correspond to the two circularly polarized modes. The value of k^* is determined by the effective resonant condition $k^* c p_{min}(x) = eB_{ls}(x,k^*)$; $A(x,k)$, k_t and α_t are defined in [19] and [20].

The relation 2 was obtained from the averaged and linearized momentum equation for the magnetized background plasma (see eq. 11-14 in [19]). Considering only Bell instability we put $k_t = 0$ and $\alpha_t = 0$ in equation 2 and obtain the simplified dispersion relation in the form

$$\omega^2 - k^2 v_a^2 \pm k k_c v_a^2 \left[\frac{eA(x,k)}{J_{cr}(x)} - 1\right] = 0, \tag{3}$$

The modes growth rates $\Gamma(x,k)$ could be found through the roots of the equation above:

$$\Gamma(x,k) = 2 \cdot Im[\omega(x,k)] \tag{4}$$

where the "2" accounts for the fact that the energy in turbulence is proportional to the square of the amplitude of B.

To calculate the spectral energy density of the magnetic turbulence $W(x,k)$ one should take into account the energy balance equation. Let us consider the energy balance without the flux of the energy through k-space toward larger k and spectral energy of the dissipation rates. Therefore, the energy balance equation is [20]:

$$u(x)\frac{\partial W}{\partial x} + \alpha_g W(x,k)\frac{du(x)}{dx} = \Gamma(x,k)W(x,k) \tag{5}$$

where $\alpha_g = 3/2$.

The equation 5 is solved for magnetic turbulence $W(x,k)$ starting from the following initial condition: $W(x,k) = \frac{B_0^2}{4\pi}\frac{k^{-5/3}}{\int_{k_{min}}^{k_{max}} k^{-5/3}dk}$. Once we have $W(x,k)$ we are able to calculate the effective, self-generated, magnetic field:

$$B_{eff} = \sqrt{4\pi \int_{k_{min}}^{k_{max}} W(x,k')dk'} \tag{6}$$

The presented model of magnetic field calcultion allows to simulate self-consistently non-linear cases of the diffusive shock acceleration (DSA) mechanism. The efficiency of the particle acceleration increases with taking into account feedback effects. Given the large system size $L \approx 3-5$ pc, the high compression ratio of the shock $\sigma = u_1/u_2 \approx 8-10$ and the high velocity of the SNR shock $U_{snr} \approx 5000-7000$ km/s one can get the values of the magnetic field up to $700-800\,\mu G$ through the discussed procedure. Numerical simulations have shown that the maximal energies of the particles accelerated by the SNR's shock with such magnetic fields can reach $Z \cdot (10^{15}-10^{16})$ eV, where Z is the charge of the particle [21, 22]. For the case of the acceleration by converging flows in the CSF system the maximum can be shifted to $Z \cdot 10^{17}$ eV for a several hundreds years [23].

3. The properties of gamma-ray and neutrino spectra

Magnetic fields determine maximal energies of the accelerated particles and emission properties of the sources as well. In order to get realistic gamma-ray and neutrino emission spectra from compact galactic clusters we have added the discussed magnetic field amplification procedure to the model of the acceleration and emission of CSF systems.

Consider a SNR shell colliding with the stellar wind of a nearby massive star. For a certain period the system will be an efficient accelerator: it will produce high-energy particles bouncing between converging flows. The important features of the particle acceleration in CSFs are: (i) the production of a piece-wise power-law particle distribution with a very hard spectrum of confined particles (with $\Gamma < 2$) at the high-energy end just before a break, and (ii) an increase in the maximum energy of the accelerated particles up to $10^{16}-10^{17}$ eV. CSF system also increases the acceleration efficiency, compared to that obtained with DSA at an isolated SNR shock of the same speed.

To model the spectra of accelerated particles in CSFs, a non-linear, time-dependent model was constructed in Bykov et al. [17]. From this model with the procedure of magnetic field amplification described in the Section 2 one may calculate particle spectra and escape fluxes [17, 24].

To provide realistic gamma-ray emissivity of the galactic and extragalactic source one should account for the pair-production effect. Energetic photons interact with extragalactic background light (EBL) photons and cosmic microwave background (CMB) radiation field. This interaction leads to the significant suppression of the gamma-ray emission at the highest energy range at the distances ($\geq$ 15 kpc) from the source. Cosmic gamma-gamma opacity strongly depends on the distance to the source of the gamma-ray photon. The optical depth of a gamma-ray photon at observed energy E_γ, emitted by a source at redshift z due to this process is given by [25]:

$$\tau_{\gamma\gamma}(E_\gamma, z) = \int_0^z dz' \frac{dl}{dz'} \int_{-1}^1 d\mu \frac{1-\mu}{2} \cdot$$
$$\cdot \int_{\varepsilon_{th}}^{\infty} d\varepsilon \, n_\varepsilon(\varepsilon, z')(1+z')^3 \sigma_{\gamma\gamma}(\beta', z') \quad (7)$$

where $n_\varepsilon(\varepsilon, z)$ is the number density of background photons with energy ε at redshift z, $\varepsilon_{th} = 2(m_e c^2)^2/E_\gamma(1-\mu)(1+z)$ is the pair-production threshold energy, $\sigma_{\gamma\gamma}(\beta', z')$ – is the gamma-gamma cross-section, $\beta' = (1-\varepsilon_{th}/\varepsilon)^{1/2}$, ε – is the energy of background photon, $\mu = cos\theta$, θ – is the angle between directions of two photons. Gamma-gamma-opacity $\tau_{\gamma\gamma}$ integrated over μ for a wide energy range for the distances 1, 5, 10 and 15 kpc from the Earth is shown on the top of the figure 1. Extragalactic optical-infrared background radiation density n_ε was adopted from [26].

The critical influence of CMB photons occurs at PeV energy range of gamma-rays. Spectrum starts to be totally suppressed when $\tau_{\gamma\gamma} >> 1$, that is the case of extragalactic sources with $d_s \gg 15$ kpc. The EBL from young stars in the cluster would affect the spectrum of the long-distance source only in the 100 GeV – 10 TeV energy regime. This influence is negligible for galactic sources because of the small $\tau_{\gamma\gamma} < 10^{-3}$.

We have calculated neutrino and gamma-ray emission spectra produced by p-p cosmic rays interactions that could come from the one active starforming region with the CSF system. The velocity of the SNR's shock is assumed to be 6000 km/s, the velocity of the stellar wind is 4000 km/s, magnetic field in the vicinity of the SNR's shock is amplified through the model in Section 2 up to 800 μG and the calculated maximal energy of accelerated protons is $2 \cdot 10^{16}$ eV.

For certain galactic sources there would exist a specific influence of gamma-gamma opacity on the resulting spectral shape. In a specific case, when the source is located at $5 \leq d_s \leq 15$ kpc from the Earth and has a spectrum extending to the highest energies $10^{16} - 10^{17}$ eV, one could observe a recess in the gamma spectrum at the PeV energy band. Indeed, $\tau_{\gamma\gamma}$ has a maximum at $\approx 3 \cdot 10^{15}$ eV and then smoothly falls down. If the cut-off of the spectrum is above $\approx 10^{16}$ eV after the "gamma-suppression band" there will exist the second hump in the gamma-ray spectrum (see figure 1, bottom panel). Depending on the spectrum shape the second "hump" can be less or even greater than the first one. Note, that for an extragalactic sources this effect will not appear because of the total suppression of the gamma-ray emission for $E \geq 10^{14}$ eV and very sharp cut-off of the spectrum. Such behavior is also impossible for the closest sources with $d_s \leq 3$ kpc because $\tau_{\gamma\gamma}$ is too small and doesn't affect the PeV gamma-ray spectrum.

It is interesting to note that in this specific case in a little energy band near 10^{15} eV the neutrino flux will exceed corresponding gamma-ray emission up to an order of magnitude.

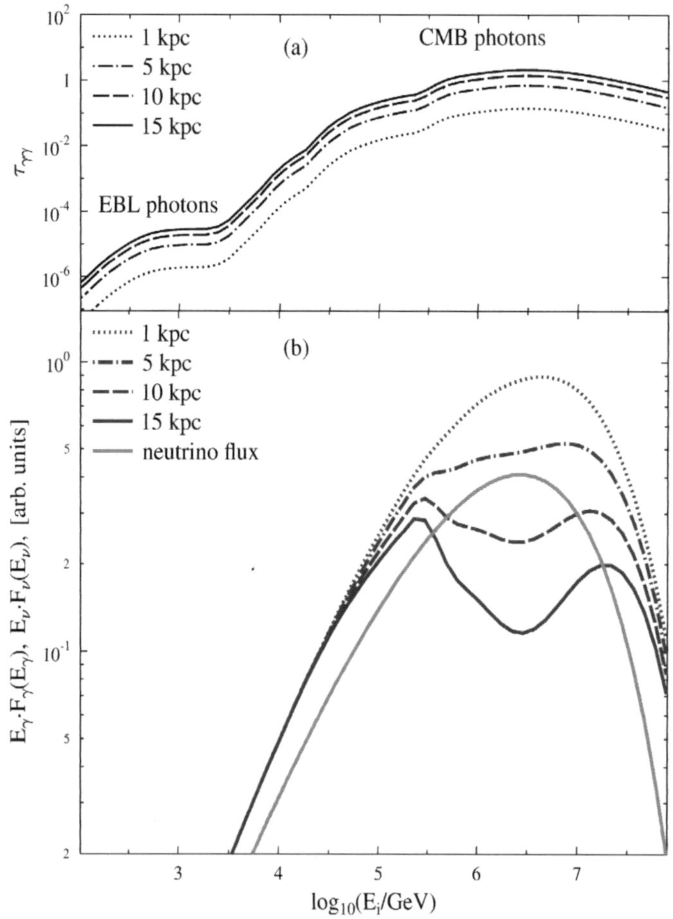

Figure 1. (a): the gamma-gamma opacity for a wide energy range integrated over μ for distances 1, 5, 10 and 15 kpc from the Earth; **(b):** the gamma-ray (blue) and neutrino (magenta) high-energy spectra from CSF systems at different distances from the Earth according to the top frame. For a long-distance sources the neutrino flux exceeds corresponding gamma-ray flux at $10^{15} - 10^{16}$ eV.

This feature is extremely important for the physical interpretation of observations of the high-energy emission from galactic clusters and powerful sources which can accelerate particles up to $10^{16} - 10^{17}$ eV.

It should be noted that typical SNRs can hardly produce such a specific spectral shape because of the steep cut-off that usually takes place at $E_c \leq 10^{14}$. Only the source with the hard spectrum at the highest energy range like SNRs of IIb-IIn types and CSF systems could provide specified spectral shape [17, 22]. Future high-energy observations of galactic sources will be able to check this behavior of the high-energy gamma-ray spectrum. Compact massive clusters Arches and Quintuplet situated at approximately 8.5 kpc from the Earth seem to be the most suitable sources to test this model.

4. Conclusion

We have presented the semi-analytical model of the efficient magnetic field amplification in the vicinity of the SNR's shock. Magnetic fields having amplified to 700-800 μG allow protons and heavy nuclei to be accelerated up to $10^{16} - 10^{17}$ eV in compact star clusters. Gamma-ray emission spectra from high-energy sources at 5-15 kpc from the Earth will be strongly affected by gamma-gamma interactions with CMB photons. We have shown that the gamma-ray flux at the energy range of $10^{15} - 10^{16}$ eV demonstrates the non-monotonous behavior and can be several times lower than the corresponding neutrino flux. Future measurements of this difference between gamma-ray and neutrino emission spectra could be an additional tool for the estimation of the distance to the source.

Acknowledgments

V.I. Romansky, P.E. Gladilin and S.M. Osipov are supported by RFBR grant 16-32-00578-mol-a for the young scientists which was used for numerical simulations via non-linear model. A.M. Bykov acknowledges support from RSF grant 16-12-10225 which was used to develop the magnetic field amplification procedure.

References

[1] Helder E A, Vink J, Bykov A M, Ohira Y, Raymond J C and Terrier R 2012 *Space Sci. Rev.* **173** 369–431 (*Preprint* 1206.1593)
[2] Bell A R, Schure K M, Reville B and Giacinti G 2013 *MNRAS* **431** 415–429 (*Preprint* 1301.7264)
[3] Blasi P 2013 *Nuclear Physics B Proceedings Supplements* **239** 140–147 (*Preprint* 1211.4799)
[4] Amato E 2014 *International Journal of Modern Physics D* **23** 1430013 (*Preprint* 1406.7714)
[5] Blandford R, Simeon P and Yuan Y 2014 *Nuclear Physics B Proceedings Supplements* **256** 9–22 (*Preprint* 1409.2589)
[6] Bykov A M and Toptygin I N 2001 *Astronomy Letters* **27** 625–633
[7] Ferrand G and Marcowith A 2010 *A&A* **510** A101 (*Preprint* 0911.4457)
[8] Bykov A M 2014 *Astron. Astroph. Reviews* **22** 77 (*Preprint* 1511.04608)
[9] Ackermann M, Ajello M and Allafort A 2011 *Science* **334** 1103
[10] Bykov A M, Ellison D C, Gladilin P E and Osipov S M 2015 *MNRAS* **453** 113–121 (*Preprint* 1507.04018)
[11] Aartsen M G e a 2014 *Phys. Rev. Letters* **113** 101101 (*Preprint* 1405.5303)
[12] Wang Q D, Dong H and Lang C 2006 *MNRAS* **371** 38–54 (*Preprint* astro-ph/0606282)
[13] Abramowski A, Acero F, Aharonian F and Akhperjanian 2012 *A&A* **537** A114 (*Preprint* 1111.2043)
[14] Muno M P, Law C, Clark J S, Dougherty S M, de Grijs R, Portegies Zwart S and Yusef-Zadeh F 2006 *ApJ* **650** 203–211 (*Preprint* astro-ph/0606492)
[15] Amato E and Blasi P 2005 *MNRAS* **364** L76–L80 (*Preprint* arXiv:astro-ph/0509673)
[16] Caprioli D, Amato E and Blasi P 2010 *Astroparticle Physics* **33** 307–311 (*Preprint* 0912.2714)
[17] Bykov A M, Gladilin P E and Osipov S M 2013 *MNRAS* **429** 2755–2762 (*Preprint* 1212.1556)
[18] Bykov A M, Gladilin P E and Osipov S M 2011 *Memories of Italian Astronomical Society* **82** 800 (*Preprint* 1111.2587)
[19] Bykov A M, Osipov S M and Ellison D C 2011 *MNRAS* **410** 39–52 (*Preprint* 1010.0408)
[20] Bykov A M, Ellison D C, Osipov S M and Vladimirov A E 2014 *ApJ* **789** 137 (*Preprint* 1406.0084)
[21] Malkov M A and O'C Drury L 2001 *Reports on Progress in Physics* **64** 429–481
[22] Ptuskin V, Zirakashvili V and Seo E S 2010 *ApJ* **718** 31–36 (*Preprint* 1006.0034)
[23] Gladilin P E, Bykov A M and Osipov S M 2014 *J. Phys.: Conf. Series* **572** 012003
[24] Gladilin P E, Bykov A M, Osipov S M and Romanskiy V I 2015 *J. Phys.: Conf. Series* **661** 012004
[25] Dwek E and Krennrich F 2013 *Astroparticle Physics* **43** 112–133 (*Preprint* 1209.4661)
[26] Franceschini A, Rodighiero G and Vaccari M 2008 *A&A* **487** 837–852 (*Preprint* 0805.1841)

Possibility of the BF Ori rotation slowdown due to the orbital synchronization

S G Shulman

St.-Petersburg State University, 7-9 Universitetskaya Nab., St.-Petersburg, 199034, Russia
Pulkovo Astronomical Observatory of RAS, 65/1 Pulkovskoye chaussee, St.-Petersburg, 196140, Russia

E-mail: sgshulman@gmail.com

Abstract. BF Ori is an UX Ori type star with a very slow projected rotational velocity. We analyzed a possibility of the slowdown due to the synchronization in close binary systems. We hypothesized that BF Ori has a companion and estimated its properties as it might be massive and close enough to make this slowdown possible. The conclusion that the rotational velocity of BF Ori is apparently not a result of the tidal interaction with a low mass companion is made.

1. Introduction

UX Ori type stars are mainly Herbig Ae/Be stars observed with a small angle between a line of view and a circumstellar disk plane [1, 2]. From the estimates of a disk opening angle [3] we can conclude that the angle between the line of view and the star's rotation axis is not less then 55°. Hence an equatorial rotation velocity might be close to the observed projected one.

Usually UX Ori type stars have projected rotational velocities about 100–200 km/s [4]. The only known exclusion is BF Ori with the projected rotational velocity of 39 ± 9 km/s [5].

We studied a theory that the rotation of BF Ori was slowed as a result of a tidal interaction with a hypothesized low mass companion (a hypothesis about low mass companion near BF Ori was already represented in the paper Grinin et. al. 2010 [6]).

2. Pre-main sequence model grids

For the detailed analysis of the tidal interaction effects we used pre-main sequence model grids of Siess et. al. 2000 [7]. These models include values of stellar radii, luminosities, moments of inertia, and convective envelopes masses for various ages which where used in the computations.

Early stages of the tracks from these pre-MS model grids were omitted to conform this canonical models with models computed in terms of an accretion paradigm. So, we started computations not from the begging of Siess tracks but from the birthline like it was recommended by Palla and Stahler [8].

3. The tides

An orbital motion of the component produces a periodical perturbation potential. This potential influences on the orbit and the rotation of the star. As a result secular changes of the orbit exist. Usually this perturbation potential does not lead to sensible changes itself. But there are various inner processes in the star which make the influence of the component much stronger [9].

We studied main processes which make the tidal interaction stronger. These processes were suggested by Zahn [9, 10], Press [11] and Tassoul [12, 13]. They are based on a viscosity friction in turbulent convective envelops, an exciting of low frequency modes of the star's oscillation, a fully developed turbulence driven by a tidal shear in stars with radiative envelopes and tidally driven meridional currents. Magnetic torques are not discussed because BF Ori has a very low magnetic field [5].

BF Ori is a young intermediate mass object ($M = 2.5M_{\odot}$, age $\simeq 3Myr$ [5]). As we can see in Siess models, BF Ori had a convective envelope for the first 2.5 Myr and was radiative after it. The analysis of tidal interactions lead us to conclusion that Zahn's mechanism with a viscosity friction in the convective envelope should be used in the beginning of the evolution. The less effective mechanism connected with a tidal shear is suitable for the last half a million years.

In our case an exiting of low frequency modes of the star's oscillation is ineffective. Tassouls' hydrodynamical mechanism has some restrictions on the use [13] which do not allow to apply it for BF Ori.

In Zahn's theory [9, 10] the angular rotation velocity of the star and the semi-major axis of its orbit vary with time according to

$$\begin{aligned} -\frac{d}{dt}(I\Omega) &= \frac{6}{t_F} q^2 M R^2 \left(\frac{R}{a}\right)^6 \lambda_2(\Omega - \omega), \\ \frac{da}{dt} &= -a\frac{12}{t_F} q(1+q) \left(\frac{R}{a}\right)^8 \lambda_2 \left(1 - \frac{\Omega}{\omega}\right), \end{aligned} \tag{1}$$

where

$$t_F = \left[\frac{MR^2}{L}\right]^{\frac{1}{3}}, \tag{2}$$

$$\lambda_2 \approx 0.019\alpha^{\frac{4}{3}} \sqrt{\frac{3160}{3160 + \eta^2}}, \qquad \eta = 2(\Omega - \omega)t_F. \tag{3}$$

Here I is a moment of inertia, R is a radios of the star, L is a luminosity of the star, ω is an angular velocity of the orbital motion, q is a mass ration, t_F is a friction time (the smaller this time, the more efficient the dissipation), $\alpha \approx 2$ is a mixing length.

When the star does not have any convective envelope another physical mechanism is taken into attention. Press et. al. 1975 [11] stated that local tidal driven shares are turbulent unstable. Hence the viscosity in these turbulent shares is important for the tidal effects. We obtain the following expression for the secular variation of the angular rotation velocity

$$-\frac{d}{dt}(I\Omega) = \frac{224}{75} \frac{K_\mu}{R_T(1-e^2)^{9/2}} q^2 M R^2 \left(\frac{R}{a}\right)^9 (\Omega - \omega)^2. \tag{4}$$

Here $R_T \approx 20$ is the effective Reynolds number , $K_\mu \approx 0.025$ is a dimensionless parameter, and e is an orbital eccentricity.

4. Results

We computed rotational velocities corresponding to the age of BF Ori (3.2 Myr) using various initial velocities, semi-major axes and mass ratios. Figures 1–3 show the time variations of the rotational speed and its dependencies on the model parameters.

We assume that the angular momentum of the star changes only as a result of a tidal interaction. This assumption is reasonable because other ways of angular momentum loss from young stars are possible only when stars have strong magnetic fields.

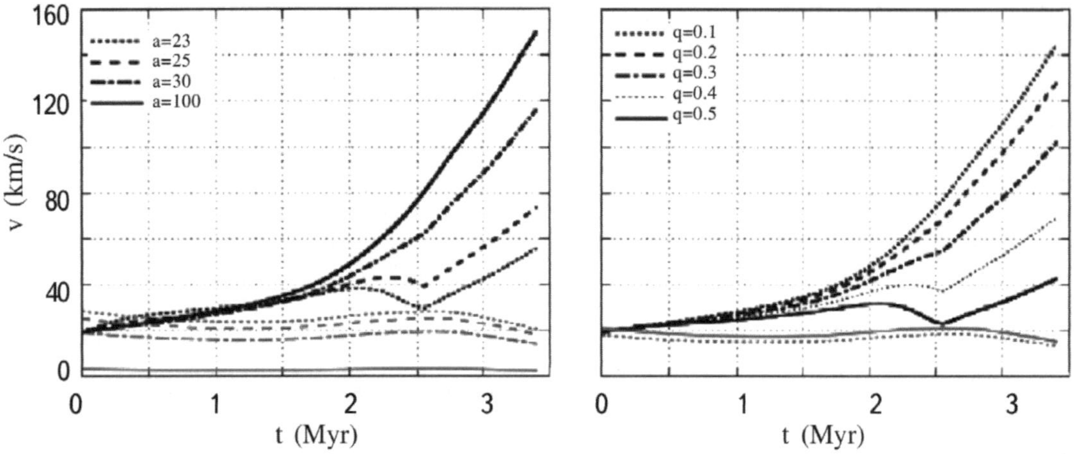

Figure 1. Temporal variations of the rotational speed. On the left panel the results for the mass ratio $q = 0.25$ and different semi-major axes are shown. On the right panel the data for various mass ratios and the semi-major axis equal to 30 Solar radii are demonstrated. On the both panels black curves correspond to the star rotation speed and gray lines show the speed of a fully synchronized star.

The moment of inertia decreases during the pre-MS evolution. This leads to an acceleration of the rotation. Under some conditions Zahn's dissipation mechanism can counteract the acceleration. After the reaching the age of 2.5 Myr the star becomes almost fully radiative. Zahn's mechanism does not slowdown the star any more. The rotation becomes much faster.

Figures 2 and 3 show that the minimal achievable velocity weakly depends on the initial velocity and heavily depends on the mass ratio. The observed projected rotation velocity $\approx$ 40km/s may be reached when the mass ratio is $0.2 - 0.5$ (when the mass ratio is $0.2 - 0.3$ there must be a minimum possible angle between the line of sight and the star's rotation axis). With these mass ratios the companion will be of $0.5 - 1.3 M_\odot$. The star of such a mass and the age of 2–3 Myr is usually a T Tau star with strong emission lines of metals. In the spectrum of BF Ori we do not see these lines. So it is hardly believable that BF Ori has a companion with mass ratio $q = 0.2 - 0.5$ which slowded it.

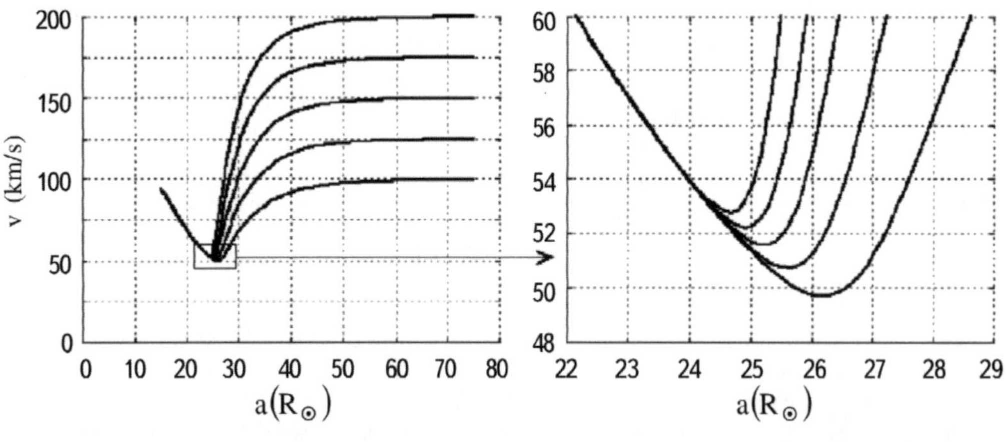

Figure 2. The semi-major axis dependence of rotational velocities in the cases of different initial velocity values. The mass ratio $q = 0.3$.

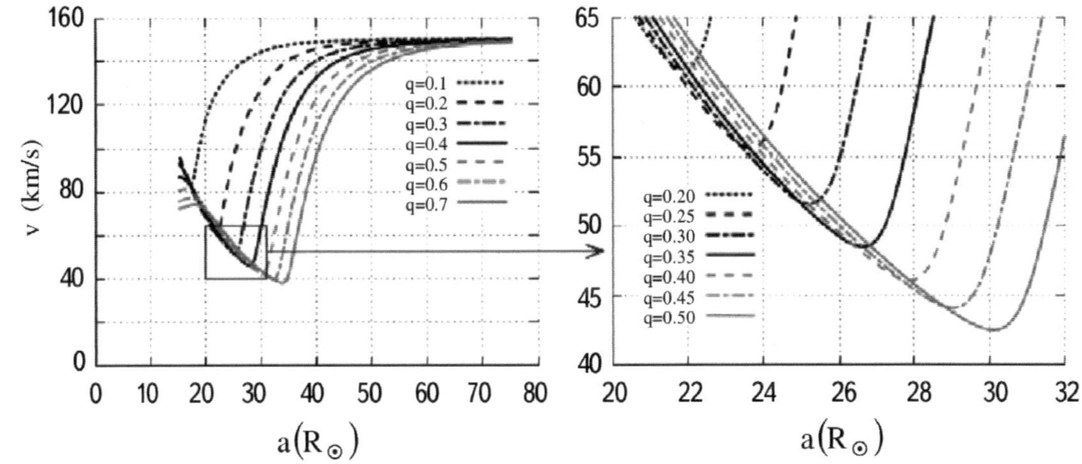

Figure 3. The semi-major axis dependence of rotational velocities for systems with different mass ratios.

5. An orbital momentum

We studied one more hypothesis explaining the slow rotation of BF Ori. When a star is forming from a protostars cloud the angular momentum shares between the star, a disk, a companions etc. It is easy to see that a small body on an orbit around the star may have an orbital momentum compared to the angular momentum of the star.

The semi-major axis of a companion with such properties near a typical UX Ori applies

$$\frac{a}{R} = 0.2\left(\frac{I}{MR^2}\right)^2 \frac{R}{R_\odot}\frac{M_\odot}{M}\frac{1}{q^2}. \tag{5}$$

for BF Ori with $R = 3R_\odot$ and $M = 2.5M_\odot$ it is equal to

$$\frac{a}{R} = 2.5 \cdot 10^{-3}\frac{1}{q^2}. \tag{6}$$

Hence, a low-mass companion with $q = 0.01$ and $a = 25R$ has the angular moment equal to the star's one.

6. Conclusion

We calculated the tidal cause slowdown of the rotational velocity for various mass ratios and semi-major axes. We found that the stars rotation on the pre-MS stage of evolution may be significantly slowed. It is shown that the observed rotational velocity of BF Ori ≈ 40 km/s may be reached when the mass ratio is larger then 0.2. We think it is hardly possible because a star of the mass greater then $0.5M_\odot$ at the age of 3 Myr is a T Tau star. The T Tauri star near BF Ori could be detected by its spectral features. A low-mass T Tauri star may be undetected yet, but it must be extremely particular condition for a significant slowdown.

We also estimated a mass and semi-major axis of a low-mass hypothetical companion with an orbital momentum equal to the typical UX Ori angular momentum. We find that a companion with the mass ratio about 0.01 and the semi-major axis of 25 star radii has the orbital momentum compared to the angular momentum of the star. This may explain the rotational velocity of BF Ori in the assumption that protostars momentum was shared between the star and the companion.

A detailed analysis is presented in the paper [14].

We expect an interferometric observation of BF Ori to clear up the disks inclination and solve the question about a duality of the star.

Acknowledgments
This work was partially supported by the Russian Foundation for Basic Research (RFFI), grant 15-02-09191.

References
[1] Grinin V P et al 1991 *Astrophys. Space Sci.* **186** 283–298
[2] Natta A, Grinin V P, Mannings V and Ungerechts H 1997 *Astrophys. J.* **491** 885–890
[3] Natta A, Grinin V and Mannings V 2000 *Protostars & Planets IV* (Tuscon, AZ: Ubiversity of Arizona Press) p 559
[4] Grinin V P and Kozlova O V 2000 *Astrophysics* **43** 239-244
[5] Alecian E et al 2013 *Mon. Not. Roy. Astron. Soc.* **56** 1001–1026
[6] Grinin V P, Rostopchina A N, Barsunova O Yu and Demidova T V 2010 *Astrophysics* **53** 367-372
[7] Siess L, Dufour E and Forestini M 2000 *Astron. Astrophys.* **358** 593–599
[8] Palla F and Stahler S W 1993 *Astrophys. J.* **418** 414–425
[9] Zahn J-P 1977 *Astron. Astrophys.* **57** 383–394
[10] Zahn J-P 2008 *EAS Publ. Ser.* **29** 67–90
[11] Press W H, Wiita P J and Smarr L L 1975 *Astrophys. J.* **202** L135–L137
[12] Tassoul J-L 1987 *Astrophys. J.* **322** 856–861
[13] Tassoul M and Tassoul J-L 1997 *Astrophys. J.* **481** 363–368
[14] Shulman S G 2016 *Astrophysics* **59** 20–30

WASP-17b - possibly the first moving away from parent star exoplanet. International campaign on its observations

E N Sokov[1,2]

[1] Central Astronomical Observatory at Pulkovo of Russian Academy of Sciences, Pulkovskoje shosse, 65, St Petersburg, 196140, Russia

[2] Special Astrophysical Observatory, Russian Academy of Sciences, Nizhnij Arkhyz, Russia, 369167

E-mail: jenias06@gmail.com

Abstract. We present detected moving away from the parent star WASP-17b exoplanet and organized international campaign on its observations. The difference between the predicted moment of a mid of transit and the observed (*O-C*) reached ~55 minutes. In the period from 2011 to 2015 yy. there was a non-linear increasing of the (*O-C*) data, which may point on an increasing of the orbital period of WASP-17b. Difference between duration of the transit obtained at 2011 year and 2015 year is ~15 minutes. This fact also point on moving away from the star WASP-17b exoplanet.

1. Introduction

The transiting WASP-17b exoplanet was discovered by [1] based on WASP-South project (Pollacco et al. 2006). The parent star WASP-17 (α =15h 59m 51s; $\delta = -28°\ 03'\ 42''$) is a V=11.6 mag F6 class star in Scorpius. WASP-17b has 1.932 R_{Jup} [3] and 0.486 M_{Jup} (Anderson et al., 2010). The period of rotation WASP-17b predicted by [1] is 3.735438 days.

WASP-17b is a first discovered exoplanet that has retrograde orbit (Anderson et al., 2010). This assumption was made based on high-precision radial velocities measurements of WASP-17 obtained with the CORALIE and HARPS spectrographs and Rossiter-McLaughlin effect (Fig. 1). Subsequently, the Rossiter-McLaughlin effect of WASP-17b have observed in [4] and confirmed the retrograde orbit of the planet with calculated angle of inclination of the orbit $\beta = 148.5°$.

2. Observations

At the end of 2011 year we had drew attention on WASP-17b as TTV (Transit timing Variations) candidate. (*O-C*) diagram, presented on Exoplanet Transit Database (http://var2.astro.cz/ETD) (Fig. 2) indicated strong deviations from the theory. And we included it in the program of a search exoplanets having timing variations.

In a project, created in Pulkovo Observatory (Saint-Petersburg, Russia), for a search new extrasolar planets based on Transit timing variations (TTV) were involved 3 professional and private observatory, located in the Southern Hemisphere: Observatorio Astronómico de Córdoba - Universidad Nacional de Córdoba (Estación Astrofísica de Bosque Alegre 1.54-mtelescope, Argentina); Complejo Astronómico El Leoncito Observatory (CASLEO, Argentine) (2.15-mtelescope); Phil Evans private observatory (Rarotonga island) (Meade LX-200 telescope). Transit times of carried out observations with calculated O-C data (with the period P = 3.735433 days) are given Table 1.

After obtained four new observations of WASP-17b transits (Table 1) we detected non-linear increasing of *O-C* in time (Fig. 3). Thus, on the summer of 2015 y. *O-C* reached nearly 1 hour.

Table 1. Transit times

Date	HJD mid (2400000+)	Epoch	Observatory	O-C, days
2012-07-08	56116.87412	417	Phil Evans Observatory	0.0176
2013-05-29	56441.86391	504	Phil Evans Observatory	0.0247
2015-06-04	57177.75602	701	Estación Astrofísica Bosque Alegre Observ.	0.0365
2015-06-19	57192.69937	705	Estación Astrofísica Bosque Alegre Observ.	0.0381

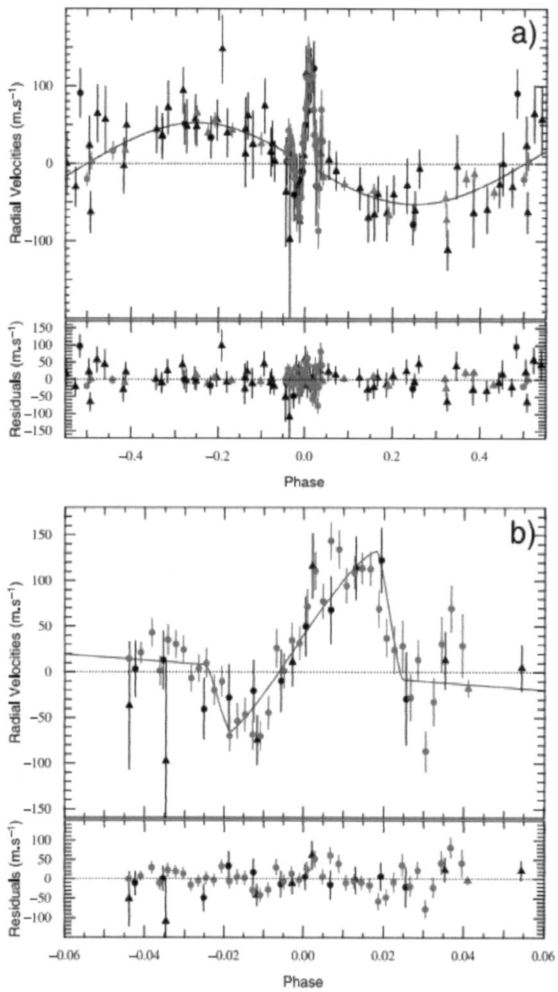

Figure 1. Fit results for WASP-17b. On a) and b) dark circles represent the RM effect taken with CORALIE, while dark triangles picture the remaining CORALIE measurements; grey dots show the HARPS RM data, red triangles are the remained HARPS points. [4].

The increasing of the duration of WASP-17b transits in time (D = 262.8 minutes from [1] (Fig. 4) and D = 278.3±1.5 minutes from observations carried out on Estación Astrofísica Bosque Alegre Observatory on 2015-06-19 (Fig. 5)) also points on moving away from the parent star WASP-17b.

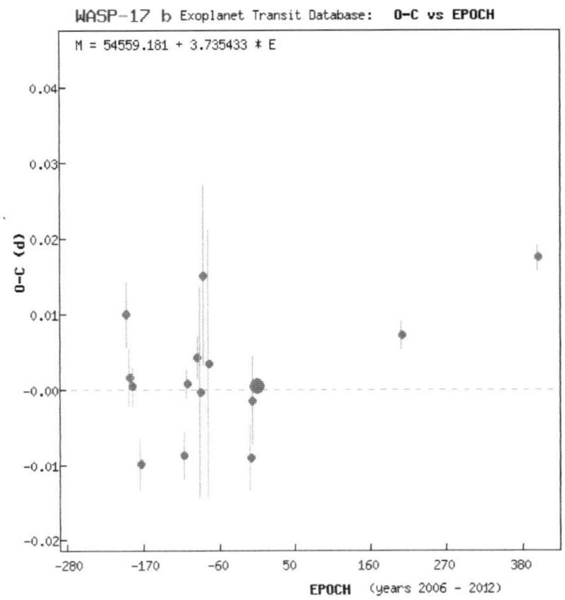

Figure 2. *O-C* diagram of WASP-17b transit times on the end of 2012 year.

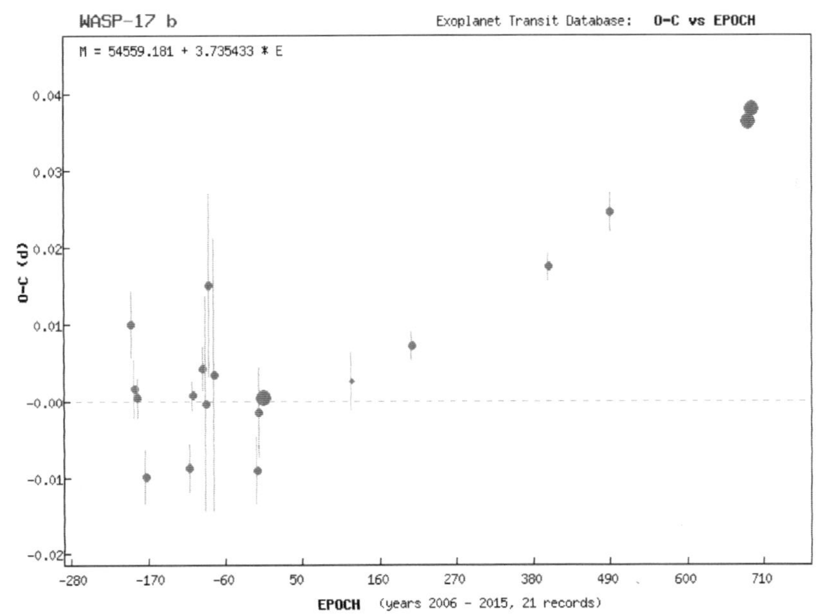

Figure 3. *O-C* diagram of WASP-17b transit times on the mid of 2015 year.

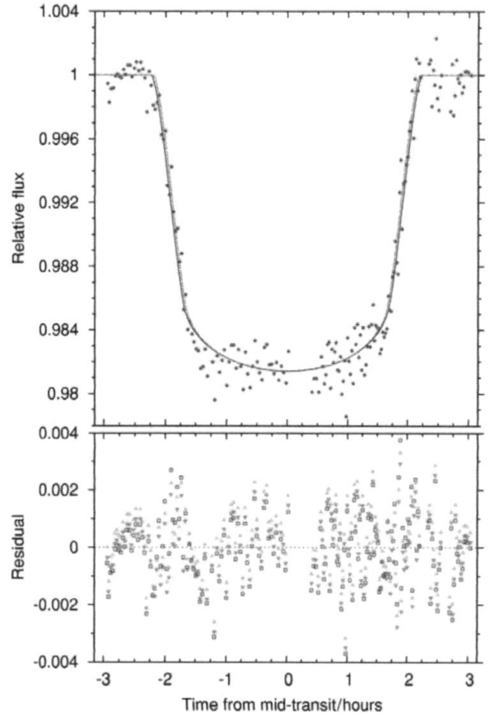

Figure 4. A light-curve of WASP-17b from [1]. Calculated duration of the transit D = 262.8 minutes.

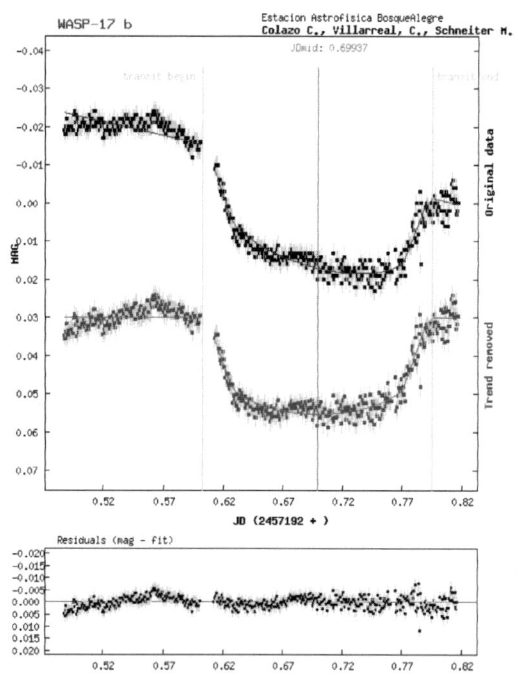

Figure 5. A light-curve of WASP-17b observations carried out on Estación Astrofísica Bosque Alegre Observatory on 2015-06-19. Calculated duration of the transit D = 278.3 minutes.

3. Conclusion

Based on all photometrical observations uploaded in Exoplanet Transit Database (ETD) including last 4 obtained in the international observable campaign of WASP-17b, we found a significant nonlinear increasing of *O-C* data. This might indicate perturbations caused by 3-th body in the system. This is can be planet or faint star located on distant orbit.

We plan to carry out new high-precision photometrical observations of WASP-17b transits and radial velocities observations.

4. Acknowledgments

The creation of international observable campaign of WASP-17b and the analysis of photometrical data were supported by the Russian Science Foundation (project 14-50-00043, area of focus Exoplanets).

References

[1] Anderson D. R. et al. 2010 *ApJ* **709** 159

[2] Anderson D. R. et al. 2011 *MNRAS* **416** 2108

[3] Southworth J. et al. 2012 *MNRAS* **426** 1338

[4] Triaud A. H. M. J. et al. 2010 *A&A* **524** A25

[5] http://var2.astro.cz/ETD

Conditions for the naked singularity formation in generalized Vaidya spacetime

V D Vertogradov

Department of Theoretical Physics, Herzen University, St. Petersburg 191186, Russia

E-mail: vitalii.vertogradov@yandex.ru

Abstract. The gravitational collapse of generalised Vaidya spacetime is considered. It is known that the endstate of gravitational collapse, as to whether a black hole or a naked singularity is formed, depends on the mass function $M(v, r)$. Here we give conditions for the mass function which corresponds to the equation of the state $P = \alpha\rho$ where $\alpha \in \left(0, \frac{1}{3}\right]$ and according to these conditions we obtain either a black hole or a naked singularity at the endstate of gravitational collapse. Also we give conditions for the mass function when the singularity is gravitationally strong. Also we provide the metric which is the analogue of Vaidya metric in case of rotation.

1. Introduction

In recent articles of A.A. Grib, Yu.V. Pavlov and V.D. Vertogradov [1, 2] properties of geodesics for particles with negative energy or Penrose geodesics in Kerr metric have been studied. It has been shown that all such geodesics appear in the ergosphere from a region inside the gravitational radius. However Kerr black holes are eternal ones. In real case one needs to consider the question about the gravitational collapse to learn the nature of Penrose geodesics. One of the opportunities for existence of such geodesics is the naked singularity formation during the gravitational collapse.

In the beginning we decided to consider the simplest case of gravitational collapse - the case of spherically symmetric collapse, in particular - the case of gravitational collapse of generalized Vaidya spacetime which will be treated in this article.

It is worth mentioning that Papapetrou [3] showed that Vaidya metric (also known as the radiating Schwarzschild spacetime) breaks the cosmic censorship principle. So Vaidya metric is the earliest example of cosmic censorship violation.

Joshi[4] showed that result of gravitational collapse, as to whether a black hole or a naked singularity is formed, depends on the initial data. In the resent paper [5] it has been shown that the endstate of gravitational collapse of generalized Vaidya spacetime depends on the mass function $M(v, r)$. Also in Ref. [6] the attempt has been made to show that the result of gravitational collapse of Vaidya-DeSitter spacetime is the naked singularity. However in this papers the question about the apparent horizon formation has not been considered. But in the case of Vaidya-DeSitter spacetime the time of the apparent horizon formation is less than the time of the singularity formation, so the result of such collapse is the black hole.

We consider the gravitational collapse of thin radiating shells. The first shell collapses down at the central singularity at $r = v = 0$ where $M(0, 0) = 0$. During the collapse of other shells the mass function is growing and when the last shell collapses down then the mass function becomes well-known Schwarzschild mass. Also we are only interested in shell-focusing singularities. Shell-crossing singularities are gravitationally weak and we are not interested in them. If there is a family of non-spacelike future-directed geodesics which originate at the central singularity in past and the time of singularity formation is less than the time of the apparent horizon formation then the result of such gravitational collapse is the naked singularity. If there is no such a family of geodesics or the time of the apparent horizon formation is less than the time of singularity formation then the result is the black hole. It is worth mentioning that here we consider only locally naked singularities. It means that the apparent horizon is formed earlier than the geodesic crosses the last collapsing thin shell. If it is not so then the singularity is global naked one. But here we won't consider global naked singularities.

In this paper we give conditions for the mass function and corresponding to these conditions we obtain either a naked singularity or a black hole as a result of gravitational collapse. We consider the matter which satisfies the equation of the state $P = \alpha\rho$ where $\alpha \in (0, \frac{1}{3}]$. Also we give conditions for mass function and corresponding to these conditions we obtain the gravitationally strong singularity. If we follow Tipler definition which was given in the paper[7]: a singularity is termed to be gravitationally strong or simply strong if it destroys by stretching or crushing any object which falls into it. If it does not destroy any object this way then the singularity is termed to be gravitationally weak.

In Sec. 2 we give basic information about generalized Vaidya spacetime. In Sec. 3 we present conditions for the mass function when we obtain either a naked singularity or a black hole. In Sec. 4 we give conditions for the mass function when the singularity is gravitationally strong. In sec.4 we provide the metric which is the analogue of Vaidya metric in the case of rotation.

The system units $G = c = 1$ will be used in this paper. Dash and dot denote partial derivatives $\frac{d}{dr}$, $\frac{d}{dv}$ respectively. Values $g_{\alpha\beta}$, $\Gamma^{\alpha}_{\beta\gamma}$, $R_{\alpha\beta}$ are metric, Christoffel and Ricci tensors components respectively. Greek letters are equal to $0, 1, 2, 3$.

2. Generalized Vaidya Spacetime

Generalized Vaidya spacetime corresponds to the combination of two matter fields I and II types and in general case is given by:[8]

$$ds^2 = -e^{2\psi(v,r)}\left(1 - \frac{2M(v,r)}{r}\right)dv^2 + 2\varepsilon e^{\psi(v,r)}dvdr + r^2 d\omega^2\,,$$
$$d\omega^2 = d\theta^2 + \sin^2(\theta)d\varphi^2\,, \tag{1}$$

here $M(v,r)$ - the mass function depending on coordinates r and v which corresponds to advanced/retarded time, $\varepsilon = \pm 1$ - ingoing/outgoing radiating thin shells respectively.

So we are interested in gravitational collapse then we put $\varepsilon = +1$. Also with suitable choice of coordinates we can put $\psi(v,r) = 0$. So (1) now has the form:

$$ds^2 = -\left(1 - \frac{2M(v,r)}{r}\right)dv^2 + 2\varepsilon dvdr + r^2 d\omega^2\,. \tag{2}$$

Now let us write down non vanishing covariant and contravariant metric components:

$$g_{00} = -\left(1 - \frac{2M(v,r)}{r}\right), g_{01} = 1\,, g_{22} = r^2\,, g_{33} = r^2\sin^2(\theta)\,. \tag{3}$$

$$g^{01} = 1\,, g^{11} = \left(1 - \frac{2M(v,r)}{r}\right), g^{22} = \frac{1}{r^2}\,, g^{33} = \frac{1}{r^2\sin^2(\theta)}\,. \tag{4}$$

Now we can write down non-vanishing Christoffel components:

$$\begin{aligned}
\Gamma^0_{00} &= \frac{M - M'r}{r^2}\,, \\
\Gamma^0_{22} &= -r\,, \\
\Gamma^0_{33} &= -r\sin^2(\theta)\,, \\
\Gamma^1_{00} &= \frac{\left(1 - \frac{2M}{r}\right)(M - M'r) + \dot{M}r}{r^2}\,, \\
\Gamma^1_{10} &= -\gamma^0_{00}\,, \Gamma^1_{22} = 2M - r\,, \\
\Gamma^1_{33} &= \sin^2(\theta)\,(2M - r)\,, \\
\Gamma^2_{12} &= \gamma^3_{13} = \frac{1}{r}\,, \\
\Gamma^2_{33} &= -\sin(\theta)\cos(\theta)\,, \\
\Gamma^3_{23} &= ctg(\theta)\,.
\end{aligned} \tag{5}$$

Now let us write down non-vanishing Ricci components:

$$\begin{aligned}
R_{01} &= \frac{M''}{r}\,, \\
R_{00} &= \frac{(2M - r)M'' + 2\dot{M}}{r^2}\,, \\
R_{22} &= 2M'\,, \\
R_{33} &= \sin^2(\theta)2M'\,.
\end{aligned} \tag{6}$$

Non-vanishing components of Einstein tensor are:

$$\begin{aligned}
G^0_0 = G^1_1 &= -\frac{2M'}{r^2}\,, \\
G^1_0 &= \frac{2\dot{M}}{r^2}\,, \\
G^2_2 = G^3_3 &= -\frac{M''}{r}\,.
\end{aligned} \tag{7}$$

We can write down the energy momentum tensor in the following form[5, 8]:

$$T_{\mu\nu} = T^{(n)}_{\mu\nu} + T^{(m)}_{\mu\nu}\,, \tag{8}$$

where the first term corresponds to the matter field I type and the other one corresponds to the matter field II type[11].

Now let us write down the expression of the energy momentum tensor:

$$\begin{aligned}
T^{(n)}_{\mu\nu} &= \mu L_\mu L_\nu\,, \\
T^{(m)}_{\mu\nu} &= (\rho + P)(L_\mu N_\nu + L_\nu N_\mu) + P g_{\mu\nu}\,, \\
\mu &= \frac{2\dot{M}}{r^2}\,, \\
\rho &= \frac{2M'}{r^2}\,,
\end{aligned}$$

$$\begin{aligned} P &= -\frac{M''}{r}\,, \\ L_\mu &= \delta^0_\mu\,, \\ N_\mu &= \frac{1}{2}\left(1-\frac{2M}{r}\right)\delta^0_\mu - \varepsilon\delta^1_\mu\,, \\ L_\mu L^\mu &= N_\mu N^\mu = 0\,, \\ L_\mu N^\mu &= -1\,. \end{aligned} \tag{9}$$

Here P - pressure, ρ - density and L, N - two null vectors.

This model must be physically reasonable so the energy momentum tensor should satisfies weak, strong and dominant energy conditions. It means that ρ must be positive and for any non-spacelike vector v^α:

$$T_{\alpha\beta}v^\alpha v^\beta > 0\,, \tag{10}$$

and the vector $T_{\alpha\beta}v^\alpha$ must be timelike.

Strong and weak energy conditions demand:

$$\mu \geq 0\,, \rho \geq 0\,, P \geq 0\,. \tag{11}$$

The dominant energy condition imposes following conditions on the energy momentum tensor:

$$\mu \geq 0\,, \rho \geq P \geq 0\,. \tag{12}$$

Also let us introduce following equations and notation. The equation of radial null geodesic has form:

$$\frac{dv}{dr} = \frac{2r}{r-2M}\,. \tag{13}$$

The existence of a family of non-spacelike future-directed geodesics which originate at the central singularity in past is defined by sign $\frac{dv}{dr}$. If the sign is "'+"' then such family exists if the sign is "' - "' then such family does not exist. Also let us introduce the following notation X_0 as:

$$x_0 = \lim_{v\to 0, r\to 0} \frac{v}{r}\,. \tag{14}$$

3. The Endstate of Gravitational Collapse

In the beginning let us consider the simplest case when the equation of the state is:

$$P = \frac{\rho}{3}\,. \tag{15}$$

Then if we use (9) then we obtain the mass function:

$$\begin{aligned} 3M''(v,r)r + 2M'(v,r) &= 0\,, \\ M(v,r) &= C(v) + D(v)r^{\frac{1}{3}}\,. \end{aligned} \tag{16}$$

Pressure and density are given by:

$$\begin{aligned} \rho &= \frac{2}{3}\frac{D(v)}{r^{\frac{8}{3}}}\,, \\ P &= \frac{2}{9}\frac{D(v)}{r^{\frac{8}{3}}}\,. \end{aligned} \tag{17}$$

Strong, weak and dominant energy conditions demand:

$$\begin{aligned} D(v) &\geq 0\,, \\ \dot{C}(v) + \dot{D}(v)r^{\frac{1}{3}} &\geq 0\,, \\ C(v) &\geq 0\,. \end{aligned} \tag{18}$$

Also the condition $M(0,0) = 0$ demands:

$$C(0) = 0\,. \tag{19}$$

The equation of the apparent horizon is given by:

$$r = 2C(v) + 2D(v)r^{\frac{1}{3}}\,. \tag{20}$$

Now we can see that when $v = 0$ is the time of the singularity formation then the equation of the apparent horizon is given by:

$$r = 2D(0)r^{\frac{1}{3}}\,. \tag{21}$$

If $D(0) > 0$ then $r > 0$ and we have that the time of the apparent horizon formation is less than the time of singularity formation and in this case we have a black hole as a result of gravitational collapse. Hence naked singularity is formed only then when $D(0) = 0$.

Now let us consider the question about existence of a family of non-spacelike future-directed geodesics which originate at central singularity $r = v = 0$ in the past. Let us substitute the mass function (16) into (13):

$$\frac{dv}{dr} = \frac{2r}{r - 2(C(v) + D(v)r^{\frac{1}{3}})}\,. \tag{22}$$

Now if we consider the limit at $r \to 0, v \to 0$ in (22) we obtain the following conditions for mass function and according to these conditions we have naked singularity as a result of gravitational collapse:

$$\begin{aligned} D(0) &= 0\,, \\ \lim_{v\to 0, r\to 0} \frac{C(v)}{r} &= a \geq 0\,, \\ \lim_{v\to 0, r\to 0} \frac{D(v)}{r^{\frac{2}{3}}} &= b \geq 0\,, \\ 2a + 2b &< 1\,, \end{aligned} \tag{23}$$

where a, b - arbitrary constants.

So when conditions (23) are satisfied then we have naked singularity as a result of collapse.

Now let us consider more general case when the equation of the state is given by:

$$P = \alpha\rho\,, \alpha \in \left(0, \frac{1}{3}\right)\,. \tag{24}$$

Here we use (9) and obtain the mass function:

$$M(v,r) = C(v) + D(v)r^{1-2\alpha}\,. \tag{25}$$

Now if we use (25) we obtain expression for the pressure and the density:

$$P = 2\alpha\,(1 - 2\alpha)\,\frac{D(v)}{r^{2\alpha+2}}\,,$$
$$\rho = 2\,(1 - 2\alpha)\,\frac{D(v)}{r^{2\alpha+2}}\,,$$
$$\mu = 2\frac{\dot{C}(v) + \dot{D}(v)r^{1-2\alpha}}{r^2}\,. \tag{26}$$

We can easily see that in general case strong, weak and dominant energy conditions impose the same requirements as in case when the equation of state is $P = \frac{1}{3}\rho$. Also it is not difficult to show that conditions for the mass function are the same lake in previous case.

So we have given conditions for the mass function when the equation of the state is $P = \alpha\rho\,, \alpha \in (0, \frac{1}{3}]$ and according to these conditions we obtain the naked singularity as a result of gravitational collapse. Now we give conditions for mass function when the singularity is gravitationally strong.

4. The Strength of the Singularity

In this section we use the definition which was given in the paper[9]. The singularity is gravitationally strong if:

$$\lim_{\tau\to 0} \tau^2\psi > 0\,,$$
$$\psi = R_{\alpha\beta}K^\alpha K^\beta\,,$$
$$K^\alpha = \frac{dx^\alpha}{d\tau}\,, \tag{27}$$

here τ - affine parameter, K^α is tangent vector to geodesic at the singularity. In the paper[5] it has been shown that in our case the equation (27) can be written in the following form:

$$\lim_{\tau\to 0} \tau^2\psi = \frac{1}{4}X_0^2\left(\chi - \frac{4\alpha\mu^2}{2\alpha - 1}X_0^{2\alpha-1}\right)\,, \tag{28}$$

where χ is an arbitrary constant and Expression for X_0 was given in 14.

Now let us substitute (25) into (28) we obtain:

$$\frac{1}{4}X_0^2\left(\chi - \frac{4\alpha\left(\frac{2(C(v)+D(v)r^{1-2\alpha}}{r^2}\right)^2}{2\alpha - 1}X_0^{2\alpha-1}\right)\,. \tag{29}$$

Now it is not difficult to obtain conditions for the mass function (25) when the singularity is strong:

$$\chi \geq 0\,,$$
$$\lim_{v\to 0} \dot{D}(v) = +\infty\,,$$
$$\lim_{v\to 0} \dot{C}(v) = +\infty\,,$$
$$\lim_{v\to 0, r\to 0} \frac{\dot{C}(v) + \dot{D}(v)r^{1-2\alpha}}{r^2} = +\infty\,. \tag{30}$$

Now let us give an explicit example when the singularity is strong. When either $C(v)$ or $D(v)$ in the form $\lambda^2 v^\gamma$, where λ is real constant, $\gamma \in (0, 1)$.

5. The Analogue of Vaidya Metric

In this section we provide the metric which is the analogue of Vaidya metric in the case of rotation. For this purpose we consider Kerr metric which is given by:

$$ds^2 = -\left(1 - \frac{2Mr}{\rho^2}\right) dt^2 - \frac{4amr\sin^2(\theta)}{\rho^2} dtd\varphi + \frac{\rho^2}{\delta} dr^2 + \rho^2 d\theta^2 + (r^2 + a^2 + 2a^2 Mr\sin^2(\theta))\sin^2(\theta) d\varphi^2 , \quad (31)$$

here M-the mass of the black hole, a-its angular momentum, $\rho^2 = r^2 + a^2\cos^2(\theta)$ and $\delta = r^2 - 2Mr + a^2$.

Now let us go to a new coordinate v by using following formula:

$$v = t + \int \sqrt{\frac{r^3}{(r - 2M)\delta}} dr . \quad (32)$$

Then we obtain the metric which is given by:

$$ds^2 = -\left(1 - \frac{2M(v,r)r}{\rho^2}\right) dv^2 - 2\left(1 - \frac{2M(v,r)r}{\rho^2}\right)\sqrt{\frac{r^3}{(r - 2M(v,r))\delta}} dvdr - \frac{4aM(v,r)r\sin^2(\theta)}{\rho^2} d\varphi \left(dv + \sqrt{\frac{r^3}{(r - 2M(r,v))\delta}} dr\right) + \frac{a^2\cos^2\theta}{\delta} dr^2 + \rho^2 d\theta^2 + (r^2 + a^2 + 2a^2 M(v,r) r\sin^2(\theta))\sin^2(\theta) d\varphi^2 . \quad (33)$$

Now it is not hard to see that when $a \to 0$ then the metric becomes Vaidya metric:

$$ds^2 = -\left(1 - \frac{2M(v,r)}{r}\right) dv^2 - 2dvdr + r^2 d\Omega^2 . \quad (34)$$

We can use this metric to consider the gravitational collapse in the case of rotation but we obtain very difficult equations.

6. Conclusion

In this paper we have considered the endstate of gravitational collapse of generalised Vaidya spacetime in terms of either a black hole or a naked singularity. Also we have given conditions for the mass function when the naked singularity is formed and also conditions when the singularity is strong.

The naked singularity formation at the end state of gravitational collapse of spherically symmetric object is one of the opportunities for explanation of Penrose geodesics nature. If the naked singularity was formed in case of gravitational collapse with rotation then we would can explain emergence in the ergosphere from a region inside gravitational radius not only Penrose geodesics but so-called white hole geodesics which was classified in the paper[10].

Also we provided the new metric which is the analogue of Vaidya metric in the case of rotation.

Acknowledgments

The author says thanks to professor D.Ph.-M.Sc. A.A Grib for scientific discussion and this work was supported by RFBR grant 15-02-06818-a and "Dynasty Foundation".

References

[1] Grib A A Pavlov Yu V and Vertogradov V D 2014 *Mod. Phys. Lett. A* **29** 14501
[2] Vertogradov V D 2015 *Grav. and Cosm.* **21** 171
[3] Papapetrou A 1985 in *A random walk in relativity and cosmology* Wiley Eastern New Delhi
[4] Joshi S P 2007 *Gravitational Collapse and Spacetime Singularities* Cambridge University Press 273
[5] Maombi M D Goswami R and Maharaj S D 2014 arXiv:1407.4309 10
[6] Wagh S M and Maharaj S D 2008 arXiv:gr-qc/9903083 9
[7] Nolan C B 1999 *Phys. Rev. D* **60** 024014
[8] Wang A and Wu Yu 1999 arXiv:gr-qc/9803038 5
[9] Tipler F J 1977 *Phys. Lett. A* **64** 8
[10] Grib A A and Pavlov Yu V 2015 *Grav. and Cosm.* **21** 13
[11] Hawking S W and Ellis G F R 1973 *The Large Scale Structure of Space-Time* Cambridge University Press 432

Simulation of the X-Ray Beam Absorption by the ABS-Plastic Filled with Different Metallic Additives

I A Miloichikova, S G Stuchebrov, D A Verigin, A A Krasnykh, I B Danilova
Tomsk Polytechnic University, Tomsk, Russia

E-mail: miloichikova@gmail.com

Abstract. This article is a part of the work on developing new materials for manufacturing filaments for fused deposition modeling (FDM). The computations of depth dose distributions for gamma-radiation in ABS plastic filled with lead and zinc additives of various concentration were performed via Monte Carlo technique and are represented in graphic form.

1. Introduction

Methods based on the use of gamma-active isotopes are an integral part of modern medicine [1-4]. There are always problems associated with planning procedures of radiation treatment of a patient when working with ionizing radiation. They control the required dose on the area of interest and minimize the dose load to critical organs. Dosimetric planning of treatment is mainly realized with methods based on computer simulations of the process [5]. However, these methods do not accurately take into account the characteristics of the emitter, which may depend on many factors. Availability of technology, allowing one to create phantoms for simulation of the interaction of gamma radiation with different parts of the patient's body, would develop experimental methods for planning such medical procedures. Not only does such technology require a method of manufacturing a tissue-equivalent material but also a method for quick creation of the items from them. Modern rapid prototyping technology fulfills all these requirements [6]. The solution would be to obtain filaments designed to work with 3D-printers based on the technology of fusing layering (Fused Filament Fabrication, FFF). The main parameters for the design are specified density and, accordingly, a certain nature of the interaction with gamma radiation.

It is necessary to theoretically estimate the interaction of modified ABS-plastic with gamma radiation based on the radiation source cobalt-60 to conduct research in this area.

2. Materials and methods

2.1. Test materials

ABS-plastic doped with various additive concentrations of zinc or lead was used as test materials in this study. The density of the modified materials was calculated using the formula:

$$\rho = \frac{\rho_{ABS} \cdot \rho_{add}}{k \cdot \rho_{ABS} + (100 - k) \cdot \rho_{add}}, \quad (1)$$

where ρ_{ABS} – density of ABS plastic; ρ_{add} - density of the material that is used as an additive; k - mass concentration of additive in the test material.

Figure 1 shows the dependence of the investigated material density on corresponding mass concentration of zinc and lead.

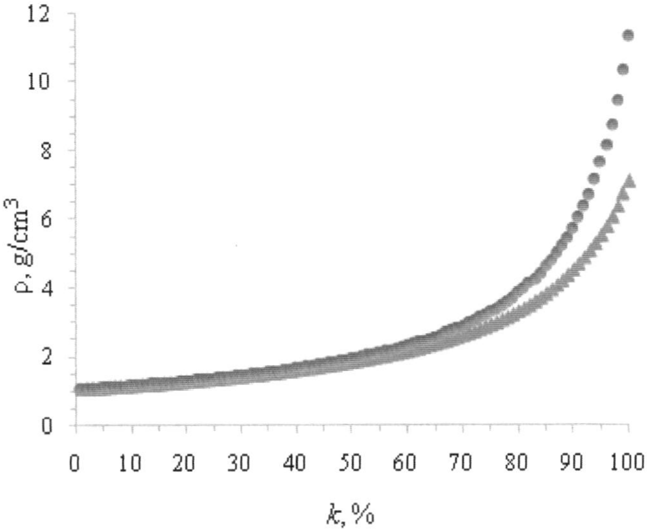

Figure 1. The dependence of the investigated material density on corresponding mass concentration of ● – zinc and ▲ – lead.

Specific materials densities which corresponds to the density of certain tissues and organs of the human body (1.06, 1.1, 1.5, 2.0, 2.5 g/cm^3) were chosen for calculations [7].

2.2. Software

"Computer laboratory (PCLab)" version 9.6 was used to simulate the interaction of gamma radiation with a modified ABS plastic in the mode "Phantom". This mode gives three-dimensional spatial distribution of the absorbed energy in the detector by calculation using Monte Carlo method [8].

2.3. The radiation source and the scheme of the experiment

Beam therapy system "Theratron 780" was selected as the source of gamma radiation. This system is based on the radionuclide cobalt-60. The software "Computer Laboratory (PCLab)" has a package for simulation of the source. The following parameters were used for the simulation: the spatial source of cobalt-60 of cylindrical shape with diameter 15 mm and height 20 mm. The source was placed in the steel box: the thickness of the side wall was 3 mm, of the upper end was 10 mm, and of the lower end was 1 mm. The box with the source was in the lead shield. Radiation from the source was emitted through a tungsten collimator. Figure 2 shows a scheme of the source with a constant collimator of therapy device Theratron 780.

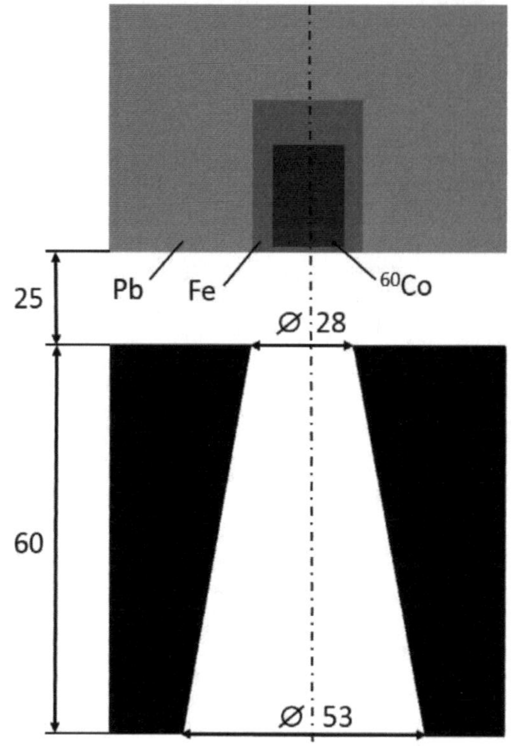

Figure 2. Scheme the source with a constant collimator of therapy device Theratron 780.

The emission spectrum of the source has 5 lines of gamma radiation of cobalt-60: 0.3469, 0.8263, 1.3324, 1.1732, 2.1588 MeV, and the line of electrons of internal conversion: 1.16488, 1.32414 MeV [8].

3. Results and analysis of the experimental results

Depth dose distributions of photon beam radiation in ABS plastic with additives of zinc or lead were obtained in this study. Figure 3 shows the calculated distribution of photons in a transverse plane of the test materials with densities of 1.1 and 2.5 g/cm3 at a depth of 2 mm. The calculated parameters were normalized by the maximum value in the detector.

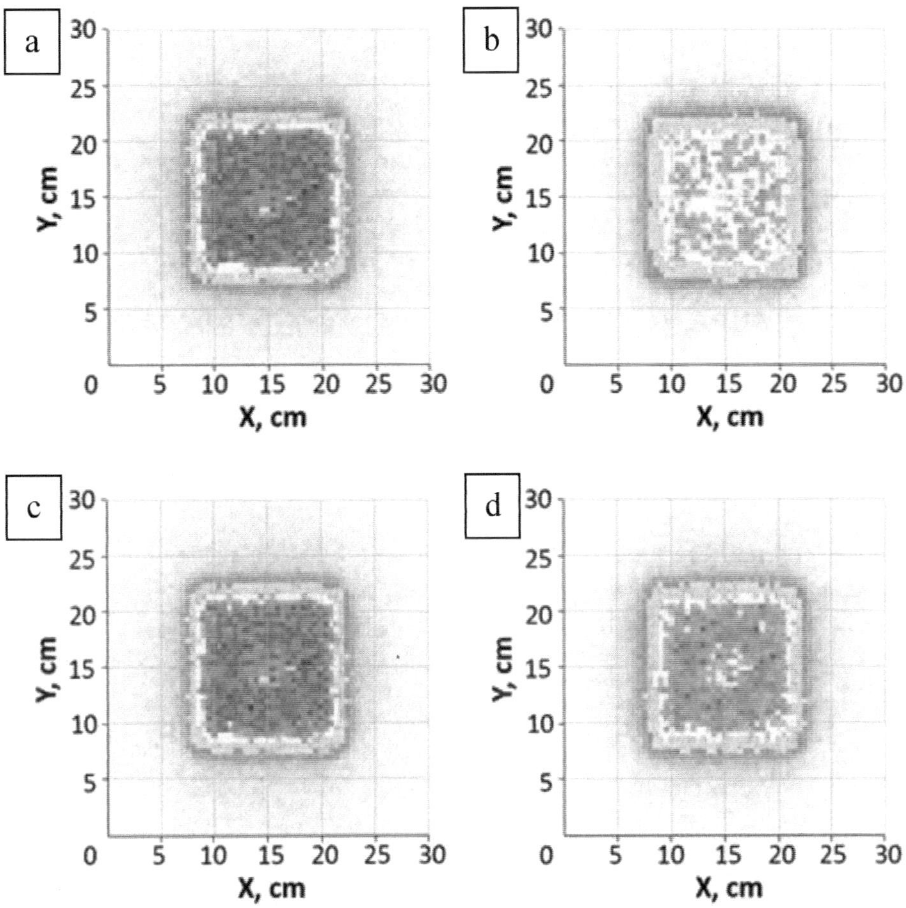

Figure 3. Photons distribution in the transverse plane:
a – ABS plastic with lead additive, density 1.1 g/cm^3;
b – ABS plastic with zinc additive, density 1.1 g/cm^3;
c – ABS plastic with lead additive, density 2.5 g/cm^3;
d – ABS plastic zinc additive, density 2.5 g/cm^3.

Figure 3 shows that the effect of various additives on the interaction of photon radiation with modified plastic increased with the increase of materials density. Dose distributions for zinc and lead additives were similar at low concentrations of additives. However, the dose distributions were significantly different for the cases of a large amount of additives.

Then depth dose distributions for the test materials with zinc and lead additives were calculated. Figure 4 shows the resulting dependences.

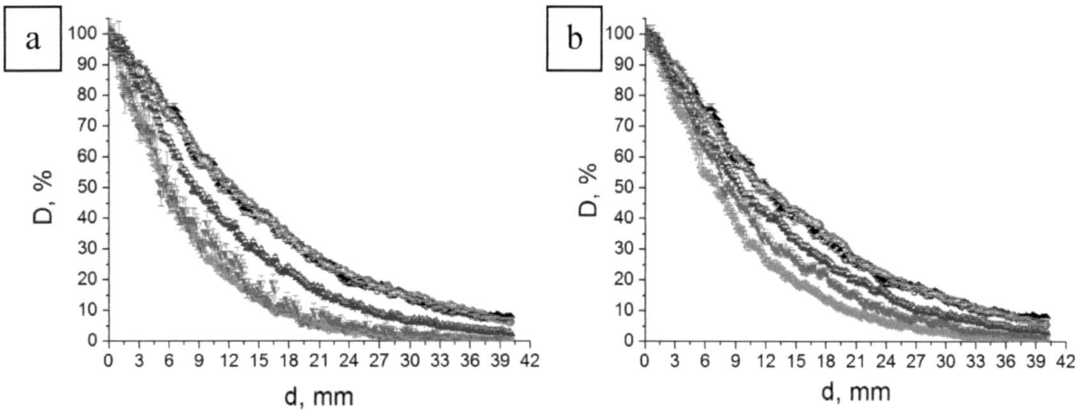

Figure 4. The absorbed dose depth distributions in test materials a – lead and b – zinc of different densities: ◆–1.06 g/cm^3, ◑ –1.1 g/cm^3; ▲ – 1.5 g/cm^3; ▼ –2.0 g/cm^3; ◁ – 2.5 g/cm^3.

Figure 5 shows the difference between the interaction of photons with the test materials of the same density, but with different metal additives.

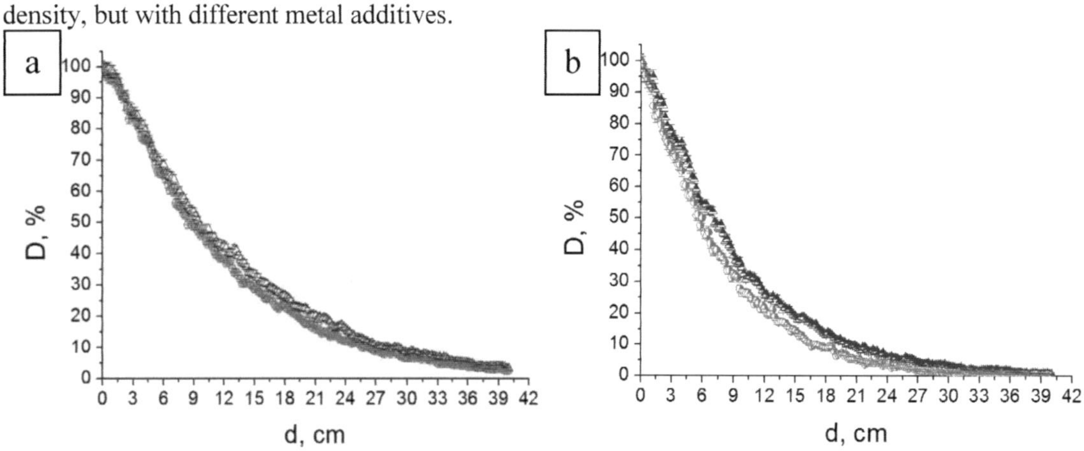

Figure 5. The absorbed dose depth distribution for ABS plastic with ◑ – lead and ▲ – zinc additives: a – density 1.5 g/cm3, b – density 2.5 g/cm3.

4. Summary

Absorbed dose depth distribution for the interaction of gamma radiation with the test materials of ABS plastic having different metal additives with different concentrations were calculated during the investigation. The calculations have shown the differences in the nature of the radiation interaction with the test materials of the same density, but with the different additives. This is especially true in the case of high densities. This occurs because of the complex combination of different factors that influence on the gamma rays interaction with the materials. On the one hand, the metal with a lower density has higher concentrations. However, this metal has a smaller total cross section of interaction with gamma radiation. It means that it is necessary to focus not on the mass density, but on the Hounsfield scale when selecting materials for the creation of phantoms simulating the gamma radiation interaction with different parts of the patient's body.

Since Hounsfield scale should be mainly determined experimentally, the next stage of this work will be: the 3D printing of modified ABS plastic with various additives of lead and zinc; the measurement of the actual density of the samples; obtaining experimental dependences of the Hounsfield scale on the concentration of metallic additives.

Acknowledgements
This work was partly supported by Competitiveness Enhancement Program of Tomsk Polytechnic University.

References

[1] Podgorsak E B 2005 Radiation oncology physics: a handbook for teachers and students (Vienna: IAEA) 657 p
[2] Musabaeva L I, Lisin V A and Velikaia V V 2014 Radiation-induced reactions in breast cancer patients after breast-conserving surgery with intraoperative and external beam radiotherapy *Radiatsionnaia biologiia, radioecologiia* **54(5)** pp 474-478 [in Russian]
[3] Maloth K N, Velpula N, Ugrappa S and Kodangal S 2014 Radioisotopes: an overview *Int J Case Rep Images* **5(9)** pp 604–609
[4] Loboda A.N. and Markevich V.E. 2014 The gamma-glutamyl transpeptidase contents in serum and urine in newborn as an indicator of disturbance kidney function due to asphyxia *Modern technologies in Medicine* **6(4)** pp 102-106 [in Russian]
[5] International Atomic Energy Agency 2005 Commissioning of radiotherapy treatment planning systems: testing for typical external beam treatment techniques (IAEA-TECDOC-1459) (Vienna: IAEA) 67 p
[6] Christ S, Schnabel M, Vorndran E, Groll J and Gbureck U 2015 Fiber reinforcement during 3D printing. *Materials Letters* **139** pp 165–168.
[7] Information on http://www.scrollseek.com/training/densitiesofdifferentbodymatter.html
[8] Bespalov V I 2015 Computer laboratory (Version 9.5) (TPU: Tomsk) 115 p [in Russian]

Journal of Physics: Conference Series **769** (2016) 012015 doi:10.1088/1742-6596/769/1/012015

Electron detachment process in collisions of negative hydrogen ions with hydrogen molecules

O V Aleksandrovich, A K Belyaev, A S Tiukanov

Department of Theoretical Physics and Astronomy, Herzen University, St. Petersburg 191186, Russia

atiukanov@herzen.spb.ru

Abstract. Investigation of the nonadiabatic nuclear dynamics in collinear and perpendicular collisions of negative hydrogen ions with hydrogen molecules is performed by the branching classical trajectories method. Quantitative assessment of the detachment probability H^- in collisions with H_2 is obtained for energy range 0.5 – 30 eV and the decay mechanisms are analyzed.

1. Introduction

Collisions of negative hydrogen ions with hydrogen molecules are of fundamental interest as a prototype of negative-ion – molecule scattering with the presence of an autoionization channel. In addition, they are of practical importance for the formation and decay of negative ions in plasma, including various astrophysical applications [1].

Theoretical studies of the collision complex H_3^- have more than a 50-year history. In spite of the fact that H_3^- is the simplest polyatomic negative ion, a study of its electronic structure and nuclear dynamics by strict quantum-mechanical methods encounters severe difficulties. This is mainly because of instability of the H_3^- complex with respect to an electron emission. As a result, the ground and electronically excited states of H_3^- are quasi-stationary in a broad region of the configuration space. Conventional *ab initio* calculations of quasi-stationary states are not possible, and alternative theoretical approaches should be used. Such approaches are well developed and successfully applied to quasi-stationary states of diatomic molecules [2], however, their generalization to polyatomic molecules is complicated.

A treatment of the nuclear dynamics of $H^- + H_2$ collisions by strict quantum-mechanical methods also meets severe difficulties. The main reason is in accounting non-adiabatic transitions both between discrete states and between discrete-continuous states. Such transitions are usually neglected or taken into account by means of restricted ways [3-5].

In the present study, nonadiabatic transitions of both types are taken into account. The nonadiabatic nuclear dynamics is studied by means of the branching classical trajectory method which takes into account nonadiabatic transitions between discrete states. Nonadiabatic transitions between discrete and continuous states are taken into account via quasi-stationary widths as a part of the local complex Potential Energy Surfaces (PESs) calculated by means of the Generalized Diatomics-In-Molecule (GDIM) method [6] for the ground and excited states of H_3^- [7]. It should be pointed out that electronic structure calculations are performed «on-the-fly», which does not require performing any analytic fit of the PESs. This approach allows us to perform a nonadiabatic treatment for an arbitrary nuclear configuration, which ultimately makes it possible to calculate the cross section and the rate constant of the process, not only for a distinguished geometries, as it was performed in the past [4,5].

2. Potential energy surfaces

The potential energy surface (PES) of the ground electronic state of H_3^- has been calculated by various methods [6-13] (see also references therein), including standard *ab initio* methods [8-10,13], although standard *ab initio* calculations are reliable only for stationary states. The PESs calculated by this way can be used for nuclear dynamics at low-energy collisions, when nonadiabatic effects are negligible. For electronically excited states (which are predominantly quasi-stationary) and for the ground state in the regions of quasi-stationarity, PESs cannot be calculated correctly by standard *ab initio* methods. An attempt of the *ab initio* calculation of the first excited state was performed in [10], where the result was a PES of the neutral quasimolecule H_3 with an additional free electron. In case of quasi-stationarity for the ground and excited states, nonadiabatic effects for discrete-continuous transitions can be accounted within the complex potential energy approximation by an imaginary part of a complex PES, the so-called quasi-stationary width.

The H_3^- PESs (as well as the nonadiabatic couplings [3]) were calculated in the full three-dimensional configuration space including the quasi-stationary regions by the conventional Diatomics-In-Molecule (DIM) method [11, 12] and also by the Generalized Diatomics-In-Molecule (GDIM) method [4,6,7]. It should be noted that the GDIM, in contrast to the standard DIM method allows one to calculate not only the real part of the local (and, in general, nonlocal) complex potentials, but their imaginary parts responsible for the decay of quasi-stationary states. In this regard, in order to account for the nonadiabatic effects of both types we have used the local complex PESs of the ground and the first excited states, the PESs calculated by GDIM [7]. In the regions of stationarity of the ground state, the GDIM PES is in a good agreement with the *ab initio* PES [8,9,13], although the former has slightly lower accuracy that does not significantly affect on the nuclear dynamics.

3. Nuclear dynamics

As mentioned in the Introduction, the branching classical trajectory method [18] is used for studying of the nonadiabatic nuclear dynamics. The method takes into account various nonadiabatic effects. Nowadays, it is generally accepted that the classical trajectory method is an efficient tool to analyze of elementary processes and calculate their quantitative characteristics. In particular, there are many papers, including those devoted to collisions $H^- + H_2$, which demonstrate the high efficiency of the method of classical trajectories, see, e.g., [14-17] and references therein, as a tool to obtain a good quantitative agreement with both accurate quantum-mechanical results and experimental data.

The idea of the branching classical trajectory method implemented in this paper is the following. The classical trajectory starts on the PES of the ground state in the asymptotic region with an asymptotic weight (population). When a local minimum of a splitting between the ground and first excited states PESs is found along a trajectory, the trajectory splits into two trajectories: a child and a parent ones. A population of the child and the parent trajectories are determined by the nonadiabatic transition probability calculated within the Landau-Zener model. Such branching can occur many times, that allows one to take into account nonadiabatic transitions between discrete states. To calculate a nonadiabatic transition probability within the Landau-Zener model, we used a novel formula obtained in [18]

$$P_{if}^{LZ} = exp\left(-\frac{\pi}{2\hbar}\sqrt{\frac{Z_{if}^3}{\ddot{Z}_{if}}}\right), \qquad (1)$$

where Z_{if} is a minimum of an adiabatic PES splitting along a trajectory, and $\ddot{Z}_{if}$ is the second derivative with respect to time.

In the case when a trajectory passes a quasi-stationarity region, the system can make a nonadiabatic transition from a discrete state to continuum states (an electron detachment). The nonadiabatic transition probability P_d reads [4]

$$P_d(E_c) = 1 - exp\left(-\frac{1}{\hbar}\int_0^\infty \Gamma[Q(t)]dt\right), \qquad (2)$$

where Γ is the quasi-stationary width of the PES along the individual trajectory, Q is the vector of generalized coordinates. The total probability value is then obtained by averaging over all trajectories with regard to their populations.

The generalized coordinates and the equations of motion, given in the paper [19], are used for the classical trajectory calculations. To visualize the trajectories in the case of collinear and perpendicular configurations the Jacobi coordinates were used. The equations of motion are integrated numerically by the fourth-order extrapolation Adams method with fixed step. To start integration, the fourth-order Runge-Kutta method is used. The accuracy of the numerical solution was estimated by the total energy.

4. Results

The examples of the computed classical trajectories for collinear and perpendicular collisions along PESs for the ground and the first-exited states are shown in Figs. 1 and 2. In the same figures, the quasi-stationary regions of the ground and the first-exited states are shown by the colored contour plots as well. The Jacobi internal coordinates are used for these figures: *r* is an internuclear distance in the hydrogen molecule, and *R* is a distance between the nucleus of the hydrogen ion and the center of mass of the target diatomic molecule. It is seen that almost each trajectory along the ground state PES passes the regions of quasi-stationarity where the quasi-stationary width is nonzero. In this case, the electron detachment (discrete-continuum transition) probability is nonzero.

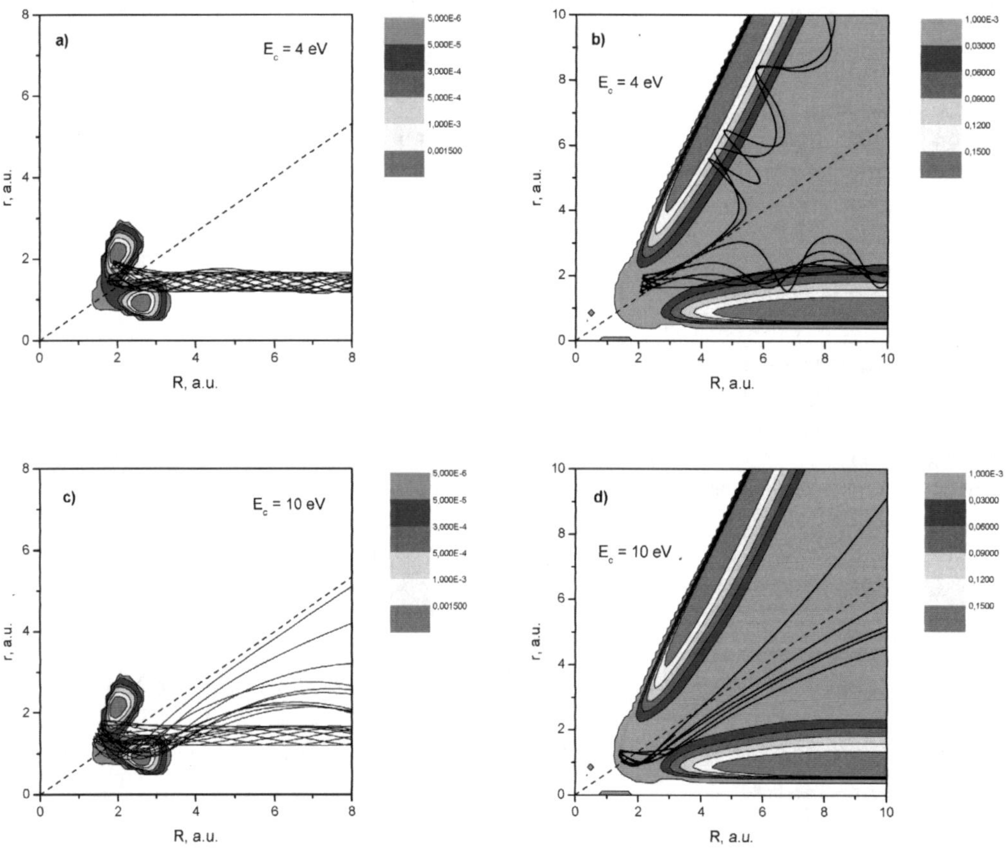

Figure 1. Classical trajectories in Jacobi coordinates along PESs of the ground (a, c) and the first excited states (b, d) for collinear collisions at the energies of 4 eV and 10 eV. The quasi-stationary widths (in atomic units) for the ground and the excited states are shown by the colored contour plots.

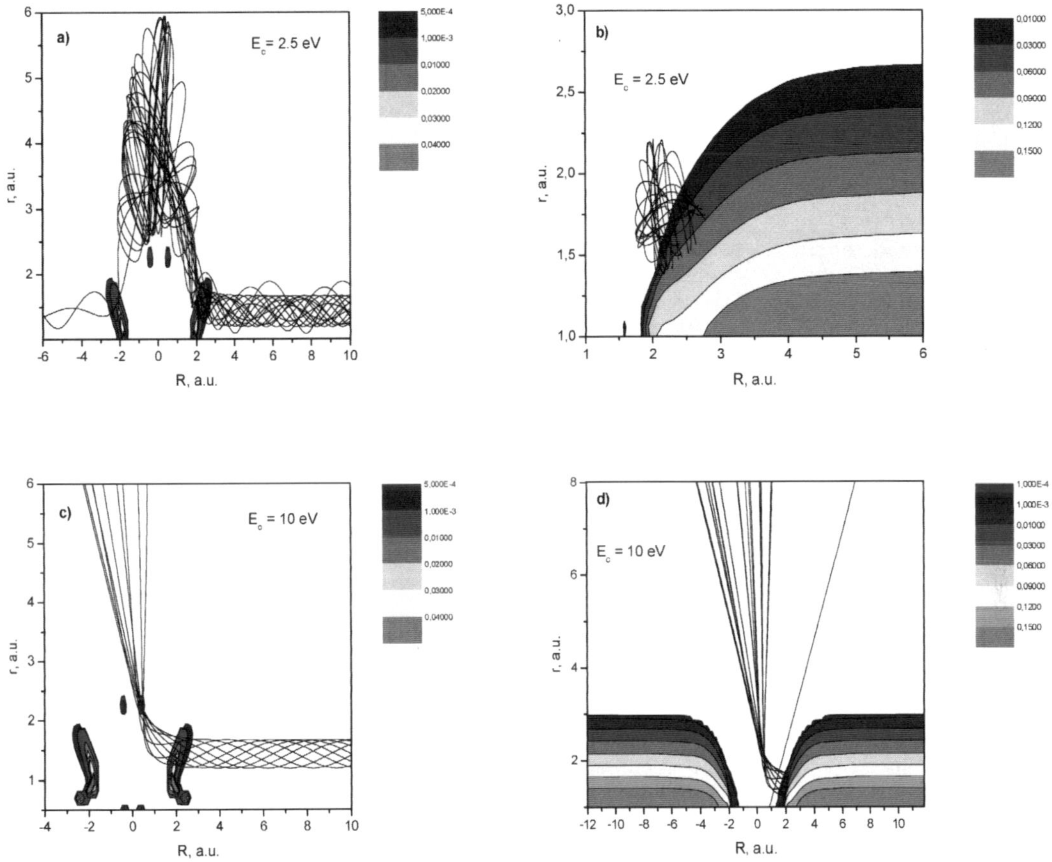

Figure 2. Classical trajectories in Jacobi coordinates along PESs of the ground (a, c) and the first excited states (b, d) for perpendicular collisions at the energies of 2.5 eV and 10 eV. The quasi-stationary widths (in atomic units) for the ground and the excited states are shown by the colored contour plots.

Figure 3 represents the calculated probabilities of a negative-hydrogen-atom decay in collinear collisions with a hydrogen molecule. The solid line 1 shows the result due to the direct mechanism: the detachment processes is based on quasi-stationarity of the ground state only. It is seen that the electron detachment process due to the direct mechanism has the energy threshold about 1 eV, the energy at which the trajectories get reaching the region of quasi-stationarity of the ground state. With increasing of the collision energy, the trajectories pass the regions with higher values of the quasi-stationary width. This increases the electron detachment probability. The probability has a maximum is around 2.5 eV. The further increasing of collision energy reduces time for a trajectory being in the quasi-stationary region, finally, this reduces the detachment probability. Note the peaks of the probability, in particular, the peak in the vicinity $E_c = 1$ eV. That peaks are the result of classical trajectories resonances due to the nuclei motion between two valleys of the ground state PES. These resonances involve the quasi-stationary regions in these valleys, and consequently increase the residence time of the anion in the quasi-stationary region.

Figure 3 also shows the electron detachment probability due to the alternative mechanism in addition to the direct mechanism (the dashed line). The alternative mechanism includes nonadiabatic transitions between the ground and the exited discrete states and the quasi-stationarity of the excited states. The quasi-stationary width of the exited state is much larger than that for the ground state, therefore the detachment probability of the excited state is higher than the detachment probability for the ground state. However, at low collision energies E_c, the nonadiabatic transitions probability from the ground state to the excited state is relatively small, and the contribution of the alternative mechanism is negligible (see. Fig. 3). With increasing the collision energy, the nonadiabatic transition probability between the ground and first excited states increases, and the contribution of the alternative mechanism becomes more significant.

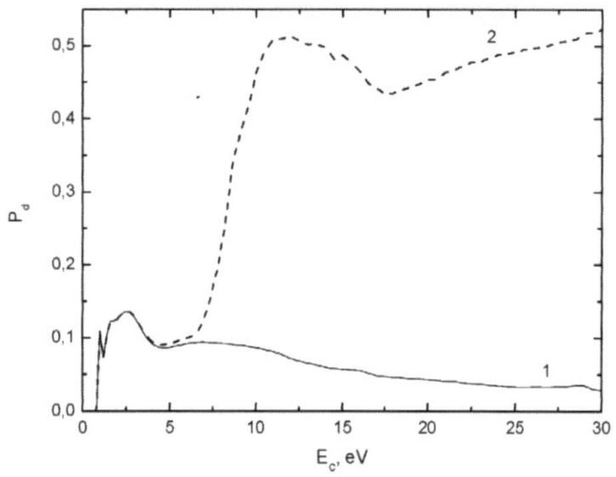

Figure 3. The negative hydrogen ion detachment probability in collinear collisions with hydrogen molecules as a function of the collision energy: 1 – calculation based only the quasi-stationarity of the ground state only; 2 – calculation taking into account the quasi-stationarity of both the ground and the first excited states of the anion, as well as non-adiabatic transitions between these discrete states.

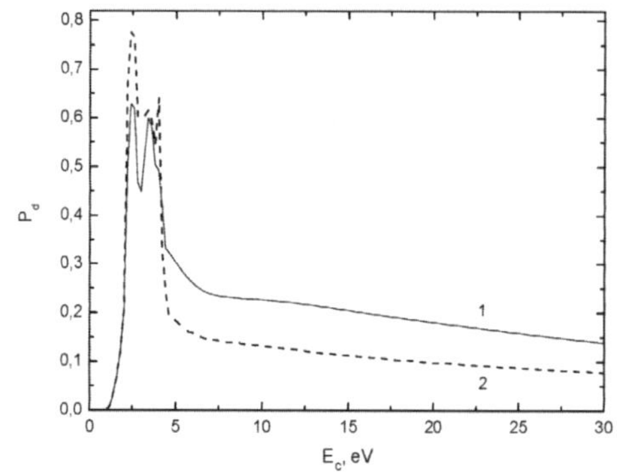

Figure 4. The same as in Fig. 3, but for perpendicular collisions.

At collision energies below 4 eV, the trajectories in the excited state are trapped in the potential well, where the quasi-stationary width significantly lower than in the asymptotic region. At the energy of 4 eV, dissociative channels get open for the excited state having a large value of quasi-stationary widths, resulting a significant increase for the electron detachment probability (see. Fig. 3). At energies greater than 7 eV, the alternative mechanism becomes dominant and determines the electron detachment process.

Figure 4 shows the detachment probability of the negative hydrogen ions at perpendicular collisions with hydrogen molecules. As in the case with collinear collisions, the threshold value of the detachment probability for the perpendicular geometry is about 1 eV. In the energy range from 2 to 4 eV the detachment probability reaches a maximum and has a resonance structure. At energies of 4 eV and higher, the probability begins to decrease. An interesting feature is the fact that the inclusion of an excited state in this case leads to reduce the probability, as compared with the detachment probability due to the quasi-stationary ground state (switched off nonadiabatic interaction between the ground and excited states). This can be explained by the fact that after the opening of the dissociative channels for the excited state, trajectories pass mainly through the region in which the width of the quasi-stationary has negligible values, that leads to small values of the electron detachment probability.

Both Figs. 3 and 4 shows that the electron detachment probabilities can reach substantial values, up to the value of 0.8, even at low collision energies. This clearly indicates that the electron-detachment channel should be taken into account in considerations of all processes in collisions of negative hydrogen ions with hydrogen molecules.

5. Conclusion

The investigation of the nonadiabatic nuclear dynamics in collinear and perpendicular collisions of a negative hydrogen ion with a hydrogen molecule is performed by the branching classical trajectories method. The quantitative assessment of the probability for decay of H^- in collisions with H_2 is obtained and the decay mechanisms is analyzed. It is proved that for low-energy collisions (from 1 to 5 eV) the direct mechanism associated with the quasi-stationarity of the H_3^- ground state dominates, whereas at high energies (higher than 5 eV), the alternative mechanism related to nondiabatic transitions between the ground and the excited discrete electronic states and with the quasi-stationarity of the excited state provides more significant contribution.

Acknowledgments
The work was supported by the Russian Foundation for Basic Research (grant № 14-03-91337).

References

[1] Ayouz M, Lopes R, Raoult M et al. 2011 *Phys. Rev. A.* **83** 052712
[2] Domcke W 1991 *Phys. Rep.* **208** 97-188
[3] Belyaev A K, Tiukanov A S 1999 *Chem. Phys. Lett.* **302** 65
[4] Belyaev A K, Tiukanov A S, Domcke W 2009 *Phys. Scripta.* **80** 048124
[5] Belyaev A K, Tiukanov A S, Toropkin A I 2012 *Rus. J. Phys. Chem.* B **6** 229
[6] Belyaev A K, Tiukanov A S, Domcke W 2002 *Phys. Rev. A.* **65** 012508
[7] Belyaev A K, Tiukanov A S, Domcke W 2006 *Chem. Phys.* **325** 378
[8] Stärck J, Meyer W 1993 *Chem. Phys.* **176** 83
[9] Ayouz M, Dulieu O, Guerout R et al. 2010 *J. Chem. Phys.* **132** 194309
[10] Kabbaj O K, Volatron F, Malrieu J P 1988 *Chem. Phys. Lett.* **147** 353
[11] Belyaev A K, Colbert D T, Groenenboom G C, Miller W H 1993 *Chem. Phys. Lett.* **209** 309
[12] Belyaev A K, Tiukanov A S 1997 *Chem. Phys.* **220** 43
[13] Panda A N, Sathyamurthy N 2004 *J. Chem. Phys.* **121** 9343
[14] Bowman J M, Kuppermann A 1971 *Chem. Phys. Lett.* **12** 1
[15] Persico M, Granucci G 2014 *Theor. Chem. Acc.* **133** 1526
[16] Wang D, Jaquet R 2013 *J. Phys. Chem. A.* **117**, 7492
[17] Gol'denberg M Ya, Kuznetsov I. M. 1982 *Rus. Usp. Khimii.* **LI** 1075
[18] Belyaev A K, Lebedev O V 2011 *Phys. Rev. A.* **84** 014701
[19] Karplus M, Porter R N, Sharma R D 1965 *J. Chem. Phys.* **43** 3259

FTIR study of secondary structure of bovine serum albumin and ovalbumin

K V Abrosimova, O V Shulenina, S V Paston

Department of Molecular Biophysics and Polymer Physics,
St. Petersburg State University, Petergof, Ulyanovskaya st. 3, St.Petersburg, Russia

E-mail: pilgr111m@gmail.com

Abstract. Proteins structure is the critical factor for their functioning. Fourier transform infrared spectroscopy provides a possibility to obtain information about secondary structure of proteins in different states and also in a whole biological samples. Infrared spectra of egg white from the untreated and hard-boiled hen's egg, and also of chicken ovalbumin and bovine serum albumin in lyophilic form and in aqueous solution were studied. Lyophilization of investigated globular proteins is accompanied by the decrease of α-helix structures and the increase in amount of intermolecular β-sheets. Analysis of infrared spectrum of egg white allowed to make an estimation of OVA secondary structure and to observe α-to-β structural transformation as a result of the heat denaturation.

1. Introduction

Proteins structure determines their functioning. Variation of environment conditions, intermolecular interactions or even a slight chemical modification can lead to the changes in the secondary and higher level structures of proteins [1]. For example, the irreversible transition of α-helix regions in the normal prion protein into β-sheets occurs as a result of its contact with the infectious isoform of prion protein and leads to the formation of amyloid plaques [2].

Bovine serum albumin (BSA) and ovalbumin (OVA) are widely studied as model globular proteins [3, 4]. Furthermore the interest to serum albumin structure is caused by its property to bind ions and large molecules and consequently to realize transport and regulate the blood composition [5]. To investigate the protein secondary structure Fourier transform infrared (FTIR) spectroscopy is actively used [3, 5, 6]. Among the advantages of this method are very small amount of a sample (~1mg), short time of spectrum record, possibility to measure the IR spectrum of crystalline, amorphous or liquid substance and so the opportunity of minimal sample preparation procedure [3]. Moreover a single IR spectrum of composite object contains information about all of the infrared active molecules present in the object. Indeed, some IR spectral features caused by the structural transitions in proteins can be observed in whole biological samples, such as nails, hair, blood and tissues [7–11], and there are hopeful attempts to use them as biomarkers of a broad range of diseases [8–11]. One of the obstacles in this way is the difficulty in interpretation of spectrum of complex multicomponent sample and the determination of correspondence between changes in vibrational bands and alterations in biochemical properties of constituents.

In the present study the IR spectra of the egg white from the untreated and hard-boiled hen's egg, as well as chicken OVA in lyophilic form and in aqueous solution were obtained and analyzed to get information about the influence of the sample content on the protein secondary structure. The impact of lyophilization on the protein conformation was checked also on BSA.

2. Experimental section

2.1. Materials

OVA (chicken egg albumin) and BSA were purchased from Sigma. Proteins solutions (30 mg/ml) were prepared with bidistilled water. The egg whites were taken from the untreated and boiled for 10 min. food hen's egg.

2.2. FTIR spectra measurement

Spectra were obtained using a FTIR spectrometer Nicolet 8700 by ATR method [3]. The spectral data within the range of 4000 to 500 cm^{-1} were recorded, and 512 scans were averaged for each spectrum with a spectral resolution of 2 cm^{-1}. The spectrum of the background was recorded first and it was subtracted from the spectra of samples automatically. The spectrum of water was registered for further subtraction from the spectra of liquid samples (proteins solutions and intact egg white). The liquid samples were covered with a cap to prevent evaporation. The software supplied with the spectrometer and OriginPro were used for data processing.

3. Results and discussions

The resulting absorption spectra (see for example Figure 1) were smoothed by FFT filtering and then the area between 1700 and 1600 cm^{-1} (Amid I) was isolated for further analysis. This band is for the most part due to the C=O stretch vibrations of the peptide linkages and highly sensitive to the protein secondary structural components. Quantitative estimation of protein secondary structure is based on the assumption that protein can be considered as a linear sum of the definite fundamental secondary structural elements: α-helix, β-sheet, β-turn and random (or unordered) [3, 6]. After the baseline correction the decomposition of Amid I band to components of Gaussian contours was done (Figure 2). The points of minimum of the second derivative, additionally smoothed by FFT filtering, determined centres of Gaussians. The area of each Gaussian contour representing a contribution of definite type of the secondary structure [6] (Figures 3, 4).

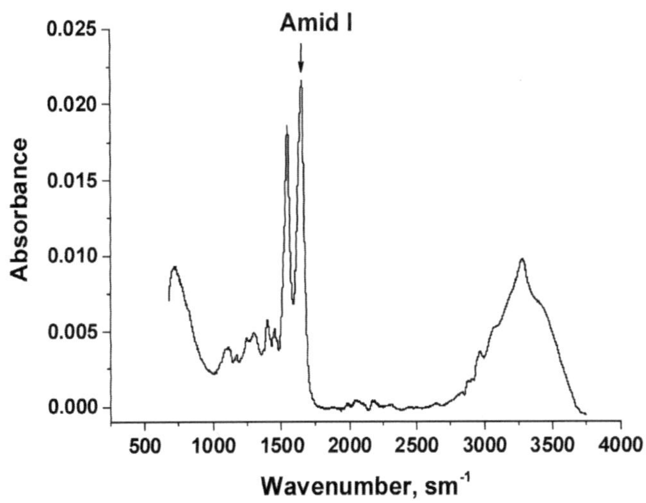

Figure 1. IR spectrum of BSA solution after water subtraction.

Journal of Physics: Conference Series **769** (2016) 012016 doi:10.1088/1742-6596/769/1/012016

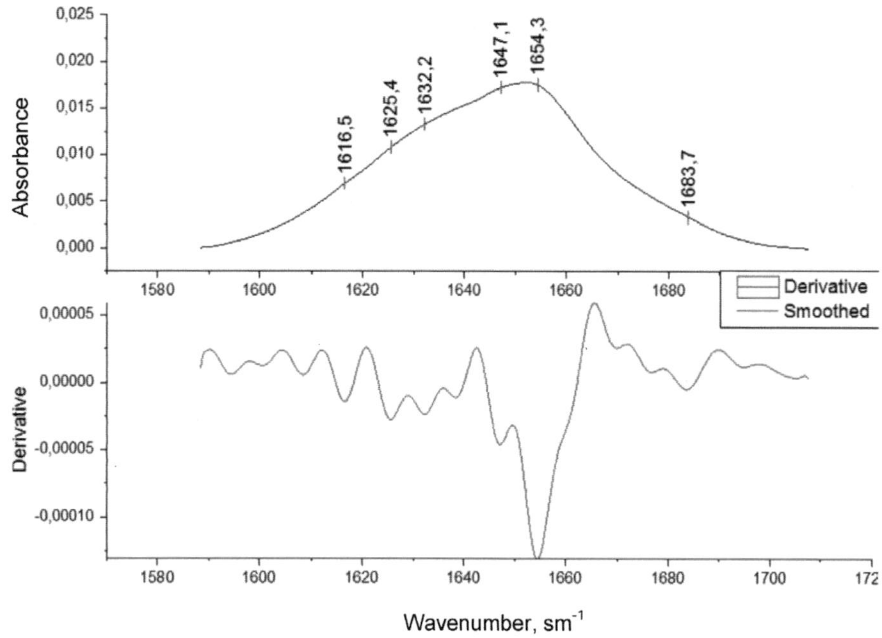

Figure 2. Amid I band and its second derivative.

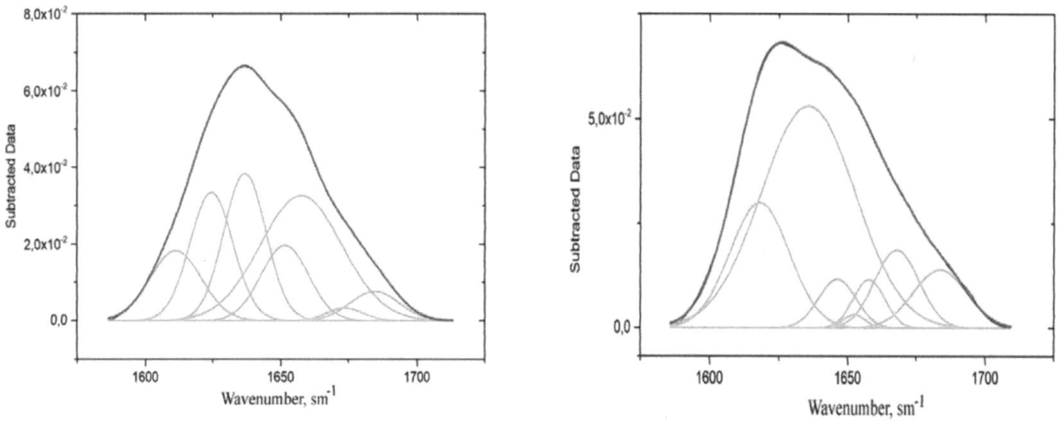

a

Gaussian center, cm-1	Area, %	Secondary structure
1611.2	11.7	indefinite
1624.5	17.9	β-sheet
1636.8	20.0	β-sheet
1651.5	11.5	random
1657.5	32.9	α-helix
1673.7	1.4	β-turn
1684.5	4.5	β-turn

b

Gaussian center, cm-1	Area, %	Secondary structure
1618.0	19.3	indefinite
1636.6	54.1	β-sheet
1646.2	4.8	random
1652.2	0.8	random
1657.8	4.0	α-helix
1667.9	9.0	β-turn
1683.6	8.1	β-turn

Figure 3. Decomposition of Amid I of egg white from the untreated (a) and hard-boiled (b) hen's egg.

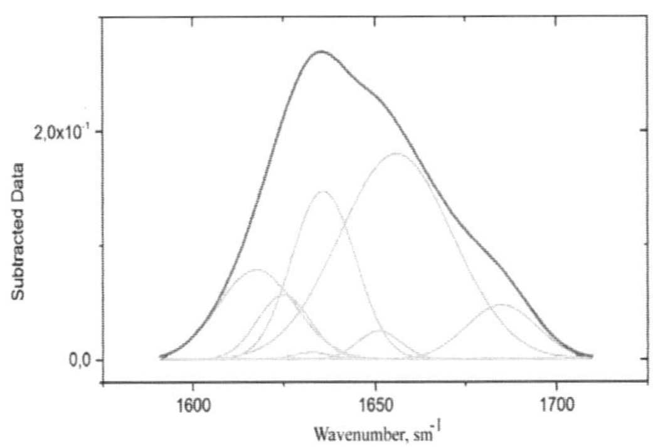

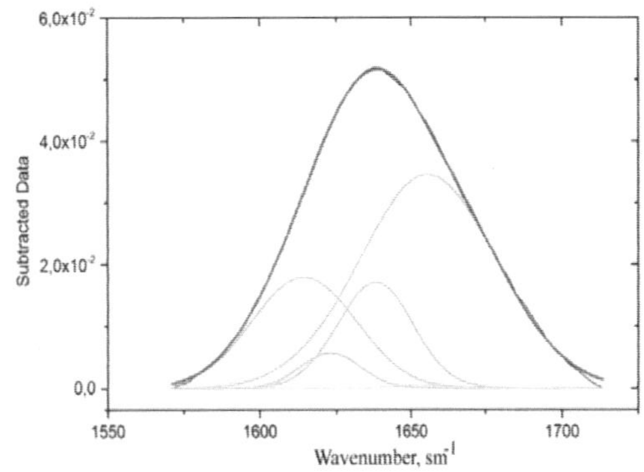

a

Gaussian center, cm^{-1}	Area, %	Secondary structure
1617.5	13.9	indefinite
1625.0	6.9	β-sheet
1632.7	0.6	β-sheet
1635.8	21.3	β-sheet
1651.0	2.5	random
1655.8	47.1	α-helix
1684.8	7.7	β-turn

b

Gaussian center, cm^{-1}	Area, %	Secondary structure
1614.4	22.3	indefinite
1623.0	4.1	β-sheet
1638.0	15.7	β-sheet
1650.3	0.2	random
1655.6	57.5	α-helix

Figure 4. Decomposition of Amid I of OVA in lyophilic form (a) and in aqueous solution (b).

Table 1. Percentage of secondary structures of egg white and OVA.

	Untreated egg white[a]	Lyophilized OVA[a]	OVA in aqueous solution[a]	OVA in aqueous solution [12][b]	OVA in aqueous solution [13][c]	Egg white from hard-boiled egg[a]	Denatured OVA in aqueous solution [12][b]	Denatured OVA in aqueous solution [13][c]
α-helix	33%	47%	58%	31.2%	41%	4%	27.8%	16%
β-sheet	38%	29%	20%	22.6%	34%	54%	37.0%	51%
β-turn	6%	8%	-	n.d.[d]	12%	17%	n.d.	20%
random	12%	3%	0%	n.d.	13%	6%	n.d.	13%

[a] Present work, absolute error 1%.
[b] From circular dichroism data
[c] From FT Raman spectroscopy data
[d] not definite

Table 2. Percentage of secondary structures of BSA (absolute error 1%).

	Lyophilized BSA[a]	BSA in aqueous solution[a]	BSA in aqueous solution [14][b]	BSA in crystal [15][c]
α-helix	48%	53%	54%	68%
β-sheet	25%	14%	18%	17%
β-turn	8%	4%	n.d.[d]	n.d.
random	4%	16%	n.d.	15%

[a] Present work, absolute error 1%.
[b] From circular dichroism data
[c] From X-ray diffraction data
[d] not definite

Let's consider the results of decomposition of Amid I in spectra of egg white and OVA (Figure 3). The presence of indefinite vibrations can be explained by the contribution of amino acids side chains motions [16]. In the Table 1 the data obtained in the present work compared with results of other researches. It should be noted that in the egg white the amount of OVA is approximately 60% [17] and so we see a significant difference between the percentage of secondary structures of egg white and OVA. Nevertheless the tendency of α-to-β structural transformation, observed at the heat denaturation in OVA solutions [12, 13], revealed also in the whole egg white.

Influence of aqueous surrounding on the proteins structure disclosed in different ways. Water stabilizes globular form of proteins by the hydrophobic and hydrophilic interactions [1]. Protein can exist in a molecular solution if it has a sufficient number of dissociated groups on its surface [18]. Water evaporation from the solution (lyophilization) leads to protein aggregation with the increase in amount of β-sheets (due to the intermolecular hydrogen bonds) and α-helix decrease, just so it follows from the experiment data for both OVA (Table 1) and BSA (Table 2). At the same time water molecules compete for the hydrogen bonds with amino and carboxyl groups of peptide bonds and thus destabilize the helical structure of protein. Crystallization of proteins allows to remove solvent and to preserve the initial protein structure [1]. That is why the most helical form of BSA is observed in crystal [15].

4. Conclusions

Lyophilization of water-soluble globular proteins BSA and OVA is accompanied by the decrease of α-helix structures and the increase in amount of intermolecular β-sheets. It is an indicator of protein aggregation.

Analysis of FTIR spectrum of egg white allowes to make an estimation of OVA secondary structure and to observe α-to-β structural transformation after the heat denaturation.

5. Acknowledgements

Authors are grateful to Dr. A. Polyanichko for usefull discussion.

This work was supported by SPbSU (11.37.290.2015).

Scientific research were performed at the Center for Optical and Laser Materials Research of Research park of St.Petersburg State University.

References

[1] Lehninger A L 1982 *Principles of Biochemistry* vol 1 (Worth Publisher)
[2] Dobson Ch M 2001 *Philos Trans R Soc Lond B* **356** 133
[3] Chalmers J, Griffiths P 2002 *Handbook of vibrational spectroscopy* (Chichester: Wiley)
[4] Yohannes G, Wiedmer S K, Elomaa M, Jussila M, Aseyev V, Riekkola M-L 2010 *Analytica Chimica Acta* **675** 191
[5] Murayama K, Tomida M 2004 *Biochemistry* **43** 11526
[6] Kong J, Yu Sh 2007 *Acta Biochimica et Biophysica Sinica* **39** 549
[7] Kreplak L, Doucet J, Dumas P, Briki F 2004 *Biophysical Journal* **87** 640
[8] Gniadecka M, Nielsen O F, Christensen D H, Wulf H Ch 1998 *Journal of Investigative Dermatology* **110** 393
[9] Jackson M, Mantsch H H 1997 *Journal of Molecular Structure* **408/409** 105
[10] Movasaghi Z, Rehman Sh, Ihtesham ur Rehman 2008 *Applied Spectroscopy Reviews* **43** 134
[11] Mostaço-Guidolin L B, Bachmann L 2011 *Applied Spectroscopy Reviews* **46** 388
[12] Hu H Y, Du H N 2000 *J Protein Chem.* **19** 177
[13] Ngarize S, Herman H, Adams A, Howell N 2004 J Agric Food Chem. **52** 6470.
[14] *Albumin from Bovine Serum. Product information.* Sigma-Aldrich
[15] Peters Jr T 1995 *All About Albumin. Biochemistry, Genetics, and Medical Applications* (Elsevier)
[16] Barth A 2007 Biochimica et Biophysica Acta **1767** 1073

[17] Huntington J A, Stein P E 2001 *Journal of Chromatography B* **756** 189
[18] Zhang F, Skoda M W A, Jacobs R M J, et al. 2008 *Phys Rev Lett.* **101** 148101

Journal of Physics: Conference Series **769** (2016) 012017 doi:10.1088/1742-6596/769/1/012017

Conformations of polyelectrolyte macromolecules with different charge density in solutions of different ionic strengths

O A Dommes[1], O V Okatova[1] and G M Pavlov[1,2]

[1] Institute of Macromolecular Compounds Russian Academy of Sciences, Bolshoy pr. 31, St.Petersburg 199004, Russia
[2] Saint Petersburg State University, Universitetskaya nab. 7/9 St. Petersburg199034, Russia

oljaspb552@gmail.com

Abstract. Studies of charged polymer chains are interesting in both fundamental and applied aspects. Especially, polyelectrolytes attract huge attention of researchers due to their ability to form interpolymer complexes with synthetic and biopolymers. The study was carried out on the fractions of hydrophilic copolymers of N-methyl-N-vinyl acetamide and N-methyl-N-vinyl amine hydrochloride of different degrees of polymerization and of different charge density using methods of molecular hydrodynamics. Hydrodynamic and conformational characteristics as well as molar masses of isolated molecules were estimated. In addition, the intrinsic viscosity of fractions was studied at the extreme ionic strengths – in distilled water (~10^{-6}M) and in 6M NaCl. Scaling relations for intrinsic viscosity, sedimentation and translational diffusion coefficients with molar mass were obtained. Conformational behavior of macromolecules with different linear charge density was compared.

1. Introduction

Polyelectrolytes are polymers which are able to dissociate in polar solvents. As a result, a charged polymer chains and mobile counterions appear [1, 2]. A recently new field of science is now developing – biomedical macromolecular systems. Many of such polymeric systems carry electric charges. The systematic studies of the relationship of chemical structure and biological activity of water-soluble neutral and charged polymers are carried out in the Institute of Macromolecular Compounds RAS in the laboratory of hydrophilic polymers (head corr. RAS, Prof. E.F. Panarin) [3]. A series of synthetic polymers exhibiting various types of biological activities have been synthesized. One of the important physical-chemical tasks in the investigation of these polymers is to determine the molar mass and to establish the Kuhn-Mark-Houwink relations (scaling relations) between the molar masses and the hydrodynamic characteristics.

In current research fractions of the hydrophilic copolymers of N-methyl-N-vinyl acetamide and N-methyl-N-vinyl amine hydrochloride (MVAA–MVAH) with average composition 95.6: 4.4% and 92:8 mol. % were studied using the methods of molecular hydrodynamic. The chemical structure of its recurrent units is shown below:

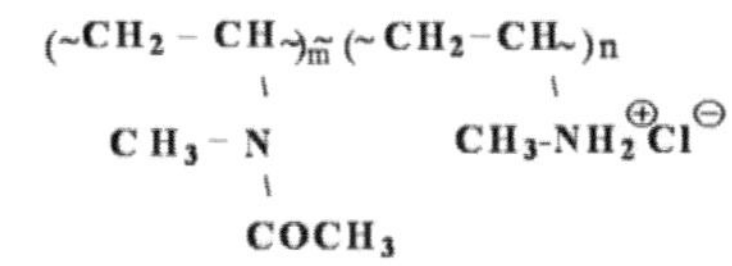

The homopolymer PMVAA and copolymers (MVAA–MVAH) with average composition 56: 44% were investigated earlier [4,5].

2. Experimental part

The hydrodynamic characteristics of charged macromolecules unambiguously correlate with the molecular characteristics only under the conditions when the polyelectrolyte effects are practically suppressed. This information was gained from the results of studying a homologous series of copolymers by methods of molecular hydrodynamics, namely, by studying viscous flow, translational diffusion, and velocity sedimentation of dilute solutions of the copolymers in 0.2 M aqueous NaCl.

The solution viscosity was measured in an Ostwald capillary viscometer under the conditions of negligible corrections for the kinetic energy and gradient dependence.

The translational diffusion was studied with a Tsvetkov polarizing diffusometer [6] in a metal cell with a Teflon centerpiece [7]. The optical path length was 3 cm. The diffusion coefficients D were determined from the slope of the linear dependences of the diffusion boundary dispersion σ^2 on time: $D = (1/2)\partial\sigma^2/\partial t$.

The velocity sedimentation was studied with a Beckman XLI analytical ultracentrifuge in a two sector cell with the optical path length of 12 mm at the rotor speed 40 000 rpm. The sedimentation interference scans were processed with Sedfit program [8] using the model of continuous distribution c(s) in which the solution of the general Lamm equation for velocity sedimentation of an ensemble of specifically noninteracting globular molecules, ensuring the best coincidence of the calculated and experimental profiles of the sedimentation boundary, is sought for. The Sedfit program allows obtaining the sample distribution with respect to sedimentation coefficients. The concentration dependence of the sedimentation coefficients was examined. The sedimentation coefficients at infinite dilution s_0 and the concentration coefficients k_s characterizing the isolated macromolecules were determined from the linear approximations $s^{-1} = s_0^{-1}(1 + k_s c + \ldots)$ [9].

The set of the experimental hydrodynamic quantities ($[\eta]$, D_0, s_0, k_s) can be transformed into the matrix of hydrodynamic invariants and molar masses, which are calculated directly from the experimental quantities without using any model concepts [6, 10]. In so doing, the only assumption is that the translational friction coefficients and chain lengths in the transport methods used are assumed to be equivalent. The hydrodynamic invariant A_0 was calculated by the relationship $A_0 = (R[D]^2[s][\eta])^{1/3}$, and the molecular weight – by the Svedberg relationship [11] $M_{sD} = R[s]/[D]$, where $[s] = s_0\eta_0/(1 - \upsilon\rho_0)$ is the intrinsic sedimentation coefficient, $[D] = D_0\eta_0/T$ is the intrinsic diffusion coefficient, and R is the gas constant. The intrinsic sedimentation and diffusion coefficients, like the intrinsic viscosity, are the quantities characterizing isolated macromolecules. Note that, for fractions of each polymer, the parameters A_0 vary insignificantly. This fact allows us to perform further interpretation of the hydrodynamic quantities. The hydrodynamic characteristics (intrinsic viscosity $[\eta]$, diffusion coefficient D_0, sedimentation coefficient at infinite dilution s_0, concentration sedimentation coefficient k_s), molar masses (M_{sD}) and hydrodynamic invariants (A_0) for investigated copolymers MVAA–MVAH in 0.2M NaCl at 25 °C are presented in tables 1 and 2. Specific partial volume value: υ=0.828 cm^3/g.

Table 1. The hydrodynamic characteristics, molar masses and hydrodynamic invariants for copolymers MVAA–MVAH 95.6: 4.4 mol % in 0.2M NaCl at 25 °C.

	$[\eta]$, cm^3/g ±1.4%[a]	$D_0\cdot10^7$, cm^2/s ±1.5%	$s_0\cdot10^{13}$, s ±5.3%	k_s, cm^3/g ±19.1%	$M_{sD}\cdot10^{-3}$, g/mol ±6.8%	$A_0\cdot10^{10}$, erg/(mol·K)
1	22	6.4	0.92	-	21	3.26
2	37	4.81	1.19	-	36	3.49
3	57	3.06	1.66	113	82	3.33
4	90	2.20	2.18	191	150	3.41
5	141	1.54	2.53	168	250	3.28
6	176	1.39	2.88	247	300	3.44
7	194	1.15	3.30	249	430	3.28
8	252	0.98	3.95	346	600	3.41

[a] Average mean square deviations are given.

Table 2. The hydrodynamic characteristics, molar masses and hydrodynamic invariants for copolymers MVAA–MVAH 92:8 mol % in 0.2M NaCl at 25 °C.

	$[\eta]$, cm^3/g ±2.2%[a]	$D_0\cdot10^7$, cm^2/s ±1.7%	$s_0\cdot10^{13}$, s ±4.0%	k_s, cm^3/g ±13.3%	$M_{sD}\cdot10^{-3}$, g/mol[b] ±5.7%	$A_0\cdot10^{10}$, erg/(mol·K)
1	34	3.94	1.19	-	45	2.97
2	47	3.26	1.45	72	65	3.14
3	67	2.73	1.74	111	97	3.30
4	77	2.20	2.07	157	141	3.19
5	114	1.80	2.42	170	190	3.28
6	140	1.43	2.78	192	290	3.21
7	219	0.99	3.53	322	540	3.10
8	261	0.94	3.69	355	600	3.32

[a] Average mean square deviations are given.

The intrinsic viscosity is a basic value of molecular physics of polymers, which is most frequently used to characterize linear polymers, especially synthetic. The classical Huggins ($\eta_{sp}/c=[\eta]+k'[\eta]^2c+\ldots$) and Kraemer ($\ln\eta_r/c=[\eta]+k''[\eta]^2c+\ldots$) plots are common method for the evaluation of intrinsic viscosity value ($[\eta]$), where c is the polymer concentration in the solution, $\eta_{sp}=\eta_r-1=(\eta/\eta_0)-1$, η and η_0 are the dynamic viscosities of the solution and solvent, respectively. The Huggins and Kraemer relationships are mathematically equivalent when $c\rightarrow0$. Therefore, it is reasonable to consider Huggins and Kraemer plots together, like one twin plot. Such plots were used to obtain $[\eta]$ values on 0.2M NaCl solutions, where polyelectrolyte effects are practically suppressed (tables 1,2). However, these plots become nonlinear in case of charged polymers and do not allow to determine the $[\eta]$ value. But, these plots become the test for the presence of charges in linear chains. At the same time, a more "simple dependencies", namely: (η_{sp} vs. c) and ($\ln\eta_r$ vs. c) have similar shape both for the neutral and charged polymers. The initial slope of "simple plots" at $c\rightarrow0$ is regarded as estimate of the $[\eta]$ value, characterizing the individual macromolecules of any kind [12,13]: $[\eta]=\lim_{(c\rightarrow0)}(\eta_{sp}/c)=\lim_{(c\rightarrow0)}(\ln\eta_r/c)$. Such plots for copolymers with 4.4 mol.% of charged monomers are presented below in figire 1. In both plots the initial slope increases with the increase of molar mass.

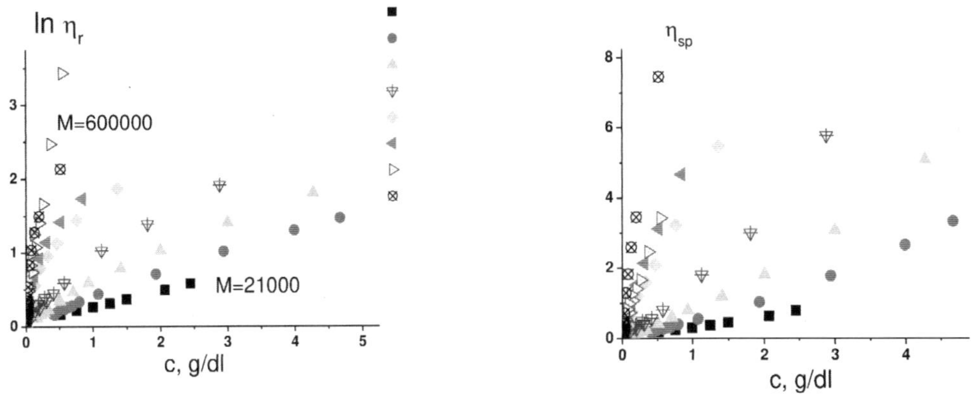

Figure 1. Plots for copolymers MVAA–MVAH 4 mol % in distilled water (ionic strength ≈10^{-6}M)

The copolymers MVAA–MVAH of different charge density and homopolymer MVAA in different solutions were compared. The dependencies lg [η]M_L vs lg M/M_L (where M_L is the mass of the unit of length of the macromolecule) are presented below in Figure 2 (a, b).

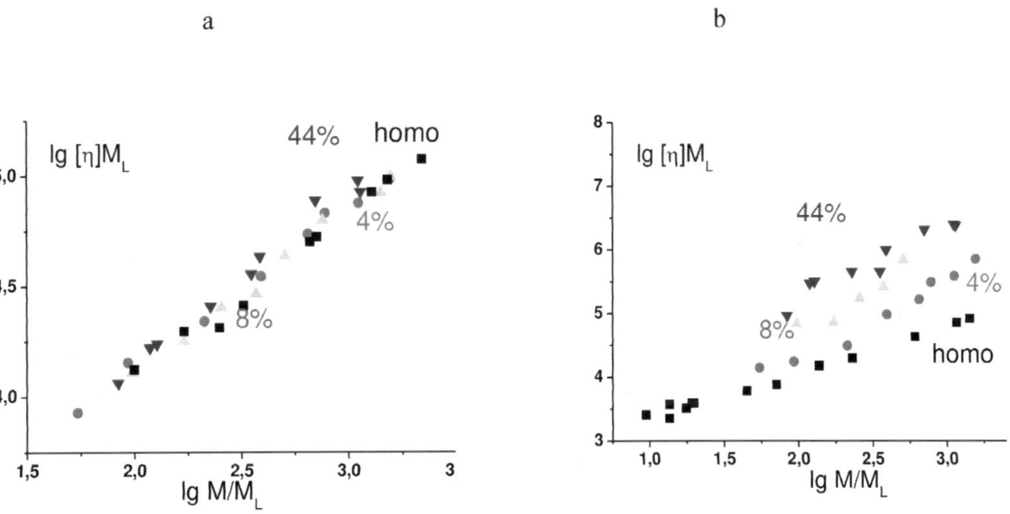

Figure 2. Comparison of copolymers MVAA–MVAH of different charge density and homopolymer MVAA in different solutions: a – 0.2M NaCl, b – distilled water (~10^{-6}M).

The conformational behavior of the macromolecules of different charged density in sodium chloride is practically the same. In the pure water solution (≈10^{-6} M) there is sufficient difference. Firstly, the [η]·M_L, i.e. the hydrodynamic volume per unit of length, increases with the increase of charge density. Secondly the slope increases, i.e. the rigidity of macromolecules increases.

3. Conclusions

Copolymers MVAA–MVAH were investigated and their hydrodynamic characteristics, molar masses and hydrodynamic invariants were obtained. The recent approach was applied to quantitative estimation of intrinsic viscosity at very low ionic strength (~10^{-6}M). It was demonstrated that this approach means to adequate evaluation of such crucial characteristic of linear macromolecules as the intrinsic viscosity. Copolymers of different charge density and different ionic strength of solution were compared. The charged macromolecules of MVAA-MVAH occupied much larger volume in pure water solutions (~10^{-6} M) than in NaCl (0.2 M) solutions where polyelectrolyte effects are practically suppressed.

Acknowledgments
The polymers were synthesized in the laboratory of hydrophilic polymers of the Institute of Macromolecular Compounds. Authors are very thankful to corr. RAS, Professor E. F. Panarin and to I. I. Gavrilova for our collaboration.

References
[1] Dautzenberg H, Jaeger W, Kötz J, Philipp B, Seidel C and Stscherbina D. 1994 Polyelectrolytes: Formation, Characterization, and Application. Hanser-Gardner Publications: Munich, p. 343
[2] Dobrynin A V and Rubinstein M. 2005 Theory of polyelectrolytes in solutions and at surfaces *Prog. Polym. Sci.* **30** No. 11 pp. 1049-1118
[3] Panarin E F, Lavrov N A, Solovskiy M V, Shalnova L I. 2014 Polymers - carriers of biologically active substances St. Petersburg: Professija ISBN: 978-5-91884-058-0
[4] Pavlov G M, Okatova O V, Mikhailova A V, Ulyanova N N, Gavrilova I I, Panarin E F. 2010 Conformational parameters of poly(N-methyl-N-vinylacetamide) molecules through the hydrodynamic characteristics studies. *Macromolecular Bioscience* **10**. p. 790-797
[5] Okatova O V, Gavrilova I I, Ulyanova N N, Panarin E F, Pavlov G M. 2012 Hydrodynamic, molecular, and conformational characteristics of macromolecules of a random copolymer of N-methyl-N-vinylacetamide and N-methyl-N-vinylamine hydrochloride. *Russian Journal of Applied Chemistry* **85**. No. 8. pp. 1239–1246
[6] Tsvetkov V N 1986 Rigid-Chain Polymeric Molecules, Leningrad: Nauka
[7] Lavrenko P N and Okatova O V 1977, *Vysokomol. Soedin.* Ser. A vol. **19** no. 11, pp. 2640–2644
[8] Schuck P 2000 *Biophys. J.*, **78**, 1606–1619
[9] Tsvetkov V N, Eskin V E, Frenkel S Ya 1964 Structure of macromolecules in solutions, Moscow: Nauka
[10] Pavlov G M, Panarin E F, Korneeva E V et al 1990 *Vysokomol. Soedin.* Ser. A vol. **32** no. 6 pp. 1190–1997
[11] Svedberg T and Pedersen K O 1940 The Ultracentrifuge, New York: Oxford Univ. Press
[12] Pavlov G M, Gubarev A S, Zaitseva I I and Sibileva M A 2006 *Journal of Applied Chemistry* **79**, p.1423
[13] Pavlov G M, Okatova O V, Gubarev A S, Gavrilova I I, Panarin E F 2014 *Macromolecules* **47** (8), p. 2748

Morphological analysis of red blood cells by polychromatic interference microscopy of thin films

A A Dyachenko[1], L I Malinova[2], V P Ryabukho[1,3]

[1]Saratov State University, 83 Astrakhanskaya ul., Saratov, 410012, Russia
[2]Saratov State Medical University, 112 Bolshaya Kazachia ul., Saratov, 410012, Russia
[3]Institute of Precision Mechanics and Control of Russian Academy of Science, 24 Rabochaya ul., Saratov, 410028, Russia

E-mail: Anton-1047@yandex.ru

Abstract. Red blood cells (RBC) distribution width (RDW) is a promising hematological parameter with broadapplications in clinical practice; in various studies RDWhas been shown to be associated with increased risk of heart failure (HF) in general population. It predicts mortality and other major adverse events in HF patients. In this report new method of RDWmeasurement is presented. It's based on interference color analysis of red blood cells in blood smear and further measurement of its optical thickness. Descriptive statistics of the of the RBC optical thickness distribution in a blood smear were used for RDW estimation in every studied sample. Proposed method is considered to be avoiding type II errors and minimizing the variability of measured RDW.

1. Introduction

Red blood cell (RBC) distributional width (RDW) is a numerical measure of the variability in size of circulating erythrocytes. It is a hematological parameter of an increasing interest due to its prognostic value in various cardiovascular disorders [1,2].

Routine clinical methods of RDW measurement are performed in different hematology analyzers. RDW is a parameter, which lacksinternational standardization: analyzers from different manufacturers may produce different RDW values in the same blood samples. On one hand, different measuring technologies and algorithmic approaches used by hematology analyzers can lead to differences in RDW. On the other hand upper mentioned differences may result from unpredictable flow induced blood cell deformations [2].

This paper represents the tentative results of an alternative RDW measurement method development. We consider it to be avoiding type II errors and minimizing the variability of measured parameter.

Proposed method is based on coherent effects in conventional optical microscopy of microobjects and allows to measure object'soptical thickness.The method is based on white light interferometry, as a non-contact measurement technique and on interference color analysis of microscopic images of studied objects. The quantitative relationship between interference colors of image and object optical thickness is used in various fields of optical measurements [3]. Presented method of RDW measurement uses optical microscope and color digital camera to capture interference microscopic images of the studied object, which thickness can be determined by the color analysis of captured image of the object in every pixel of captured image.

2. Red cell Distribution Width (RDW)

The clinicians have recently witnessed massively growing interest in the RDW, especially after progressive increase of data demonstrated prognostic significance of RDW in heart failure patients. Usually RDW is defined as the standard deviation of the RBC volume distribution curve, but practically is reported as a percentage relative to mean RBC volume.

In our study RDW was calculated as following:

$$RDW = \frac{\sigma}{M} \cdot 100\% , \tag{1}$$

where σ is standard deviation of measured thicknesses of RBC and M is mean value of RBC thickness in studied blood samples.

3. Method of measurement of RBC optical thickness

The object of the study was RBC monolayer prepared on highly reflective surface. This RBC monolayer can be considered as a thin transparent film. The incident on the film light is reflected by both upper and lower boundaries of the film. If optical thickness of film is less than half of length of longitudinal coherence of incident light interference of the light is observed. If light source of polychromatic radiation is used interference colors and hues are appeared in image of the film [4]. These colors and hues mainly depend on optical thickness of illuminated film. Therefore interference color analysis can be used for measurement of optical thickness of thin film.

We use interference color analysis for estimation of RBC monolayer optical thickness with further going RDW calculation.To be more specific the main principle of developed method is comparison of image interference color of research object with color of numeric simulated interference pattern. Method of optical thickness measurement of thin film consists of three main steps:

- simulation of interference pattern;
- registration and processing interference image of research objects;
- computer analysis of registered images and estimation of object's optical thickness.

Physical process of formation and recording of interference pattern in case of polychromatic illumination of thin film can be considered with taking in to account spectral properties of light source and spectral sensitivity of recording photo elements of digital camera. We used the spectral representation for oscillations of interfering light waves to describe influence of spectral properties of light and matrix photodetector on the forming of interference patternof thin film. This representation allows using light spectrum of the source in an explicit form. Using this representation the interfering fields can be defined as

$$E_1(t) = \int_0^{\infty} r_1(\omega) \cdot g(\omega) \cdot \exp(i\phi_1) \cdot \exp(i\omega t) d\omega , \tag{2}$$

$$E_2(t) = \int_0^{\infty} r_2(\omega) \cdot g(\omega) \cdot \exp(i\phi_2) \cdot \exp\left(i\omega \left(t - \frac{\Delta}{c} \right) \right) d\omega , \tag{3}$$

where $g(\omega)$ is amplitude frequency spectrum of incident on the thin film light, $r_1(\omega)$ and $r_2(\omega)$ are amplitude reflection coefficients of the thin film's boundaries, ϕ_1 and ϕ_2 are phase shifts of the reflected light waves, Δ is optical path difference of the interfering light waves reflected by boundaries of the film.

The interference equation in spectral representation can be obtained using equation (2) and (3) in the following form

$$I(\Delta) \sim \left\langle \left|E_1 + E_2\right|^2 \right\rangle = \int_0^\infty r_1^2(\omega)\cdot g^2(\omega)\cdot d\omega + \int_0^\infty r_2^2(\omega)\cdot g^2(\omega)\cdot d\omega + \\ + 2\cdot \mathrm{Re}\left(\exp(i(\phi_1 - \phi_2)) \int_0^\infty r_1(\omega)\cdot r_2(\omega)\cdot g^2(\omega)\cdot \exp\left(i\omega \frac{\Delta}{c} \right) \right)\cdot d\omega \quad (4)$$

where angular brackets $\langle \cdots \rangle$ define time averaging operation due to temporary inertia of the photodetector. Equation (4) allows computing and simulating color interference pattern in polychromatic light.

To consider process of color calculating of interference images we used the RGB (Red, Green, Blue) color model. This choice is explained by using this color model in system of image recording by digital matrix photodetectors. In this model color image consists of three monochromatic patterns that are called chrominance channels [5]. Each of the pattern is registered in monochromic light - red, green and blue. The color in any pixel of resultant image is formed by combination of intensity level in each chrominance channel.

To create RGB image we should calculate three matrix arrays. Each of them describes interference pattern that is formed after transmission of interfering fields through on of the color light filters in matrix photodetector. So the spectral transmission of color light filter should be introduced into equation (4) and it will look as following

$$I(\Delta) = \int_0^\infty r_1^2(\omega)\cdot f_{ch}^2(\omega)\cdot g^2(\omega)\cdot d\omega + \int_0^\infty r_2^2(\omega)\cdot f_{ch}^2(\omega)\cdot g^2(\omega)\cdot d\omega + \\ + 2\cdot \mathrm{Re}\left(\exp(i(\phi_1 - \phi_2)) \int_0^\infty r_1(\omega)\cdot r_2(\omega)\cdot f_{ch}^2(\omega)\cdot g^2(\omega)\cdot \exp\left(i\omega \frac{\Delta}{c} \right) \right)\cdot d\omega \quad (5)$$

where $f_{ch}(\omega)$ is amplitude transmission spectrum of corresponding chrominance channel ch.

Assuming that $r_1(\omega) = r_2(\omega) = Const = 1/2$ and ignoring multiple light reflections in the film, equation (5) can be reduced to the form that is used in computer simulation of interference pattern. This equation also can be changed to the form in terms of wavelengths λ:

$$I_{ch}(\Delta) \sim \int_0^\infty f_{ch}^2(\lambda)\cdot g^2(\lambda)\cdot \left(1 + \cos\left(\frac{2\pi}{\lambda}\Delta + \pi \right) \right) \frac{d\lambda}{\lambda^2}. \quad (6)$$

In our work simulation was performed in range of optical thickness of 0 – 2000 nm. Light spectrum of the light source can be calculated using Planck radiation law [6] for temperature of the radiation 2500 K. The values of wavelengths are over the range of 0 – 1000 nm. For spectral amplitude transmission of light filters $f_{ch}(\omega)$ we use Gaussian function

$$f_{ch}(\lambda) = \exp\left(\frac{(\lambda_{0ch} - \lambda)^2}{\Delta\lambda_{ch}^{\ 2}} \right), \quad (7)$$

where λ_{0ch} is central wavelength of transmission spectrum of chrominance channel; $\Delta\lambda_{ch}$ is width of spectrum contour of the channel.

Parameters of transmission spectrum of chrominance channel are presented in table 1.

Table 1. Parameters of transmission spectrum of chrominance channel

Channel	λ_{0ch}, nm	$\Delta\lambda_{ch}$, nm
Red	610	39
Green	540	56
Blue	460	58

The resultant interference signal and color interference pattern for thin linear optical wedge are shown on figure 1. This interference image demonstrates color dependence vs optical thickness in range of 0 – 2000 nm.

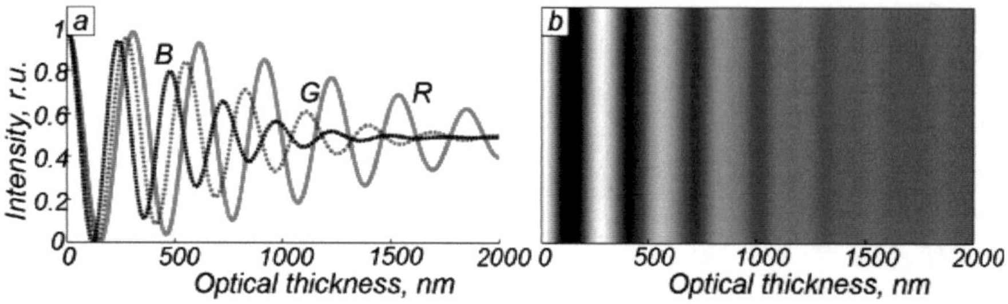

Figure 1. Results of numerical simulation: simulated interference signal for each chrominance channel (a), color interference image of thin linear optical wedge (b).

Speaking about studied object (RBC monolayer prepared on reflective substrate), it is necessary to take into account properties of this reflective substrate. In case of preparing blood smear on glass substrate light reflection from lower boundary object-glass is negligibly small. So interference colors of object images are practically undetectable. The usage of substrate with highly reflective surface, for instance polished silicon plate as in our work, instead of a glass platecould significantly increase light reflection from lower object boundaryand enable to obtain interference images of the RBC with high color contrast.

Interference RBC images were obtained by means of the optical microscope Axio Imager 2 (Carl Ceiss, Germany) with microobjectiveEpiplan-Neofluar 100x/0.75 HD DIC and captured by digital camera Nikon D80.Examples of these imagespresent in figure 2.Captured interference image were stored as a digital image andoptical thickness of RBC was measured.

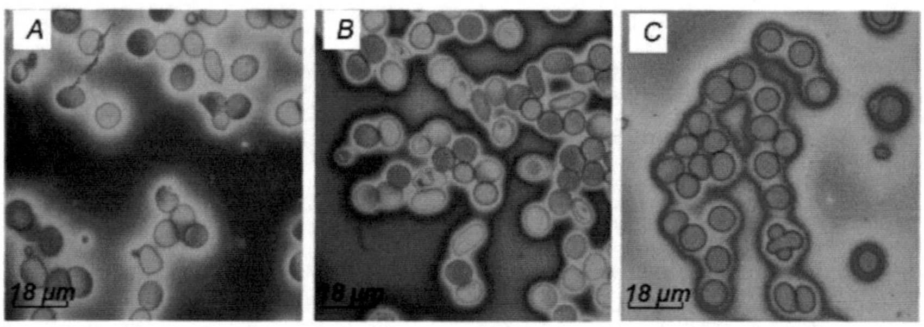

Figure 2. Interference images of formed blood elements (primary red blood cells) obtained by optical microscopy (Samples A, B and C respectively).

Main idea of the procedure for determining of the optical thickness is comparison of color in chosen pixel of experimental image with computer simulated interference scale. The pixel of image is characterized by three values of light intensity (in RGB chrominance channel).To measure optical thickness in this pixel it is necessary to compare color in the pixel with simulated color scale. Determination on this scale the point with similar color values and using known relationship between color and optical thickness one can be measured RBC optical thickness of in each pixel of experimental image.

4. Results

All blood samples in our work are obtained from patients with chronic heart failure (Stage IIa, 3NYHA functional class) during acute hyperglycemia test at base point (0, 10 and 60 min points).

Interference images of studied samples obtained by optical microscope are shown on figure 2. Image *A* captures from a blood smear prepared at base point (before 40% glucose solution administration), images *B* and *C* – from blood smear prepared at 10 and 60 minutes after 40% glucose solution administration respectively. Measurements of RBC optical thickness were performed in the central zone of the blood smear. Histograms of optical thickness distribution in each samples present in figure 3. Red cell Distribution Width was calculated, results are presented in figure 3.

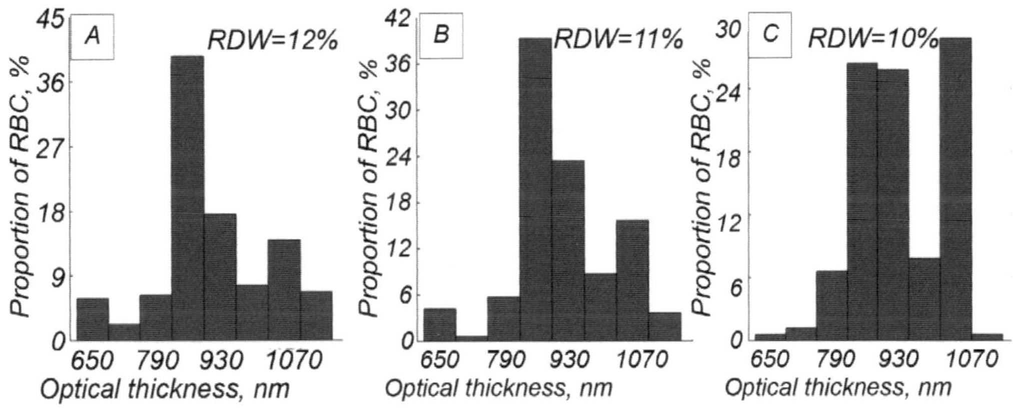

Figure 3. Optical thickness distribution in researched blood samples.

To validate obtained results we compare received data with those obtained by reference method – blood test using automatic hematology analyzers ABS Micros and Datacell-16. Optical thickness of RBC reveal strong positive correlation with MCV (mean cell volume), that is a measure of the average volume of a red blood cell and RDW ($R = 0.84$, 0.79 $p < 0.05$). Variability of studied method was 0.2 compare to 0.5 and 0.4 (automatic hematology analyzers ABS Micros and Datacell-16).

5. Conclusion

Received data elucidate interference microscopy as a promising method for RDW assessment in routine clinical practice. Among obvious merits of proposed method the simplicity of experimental set-up and procedure of computer image processing should be mentioned. Thickness profile measurement of thin blood smears by white light interferometry allows receiving more stable results. This method is potentially high sensitive promising tool to assess morphological and functional RBC properties as a prognostic biomarkers in HF patients.

References

[1] H Uyarel, M Ergelen, G Cicek, etc. 2011 Red cell distribution width as a novel prognostic marker in patients undergoing primary angioplasty for acute myocardial infarction *Coronary Artery Disease* **vol 22** is 3. pp 138-144

[2] LMalinova, L Akhmadullina 2013 Red blood cell distribution width and peripheral blood cells parameters in patients with heart failure and a reduced ejection fraction vs heart failure with "preserved" ejection fraction *European Journal of Heart Failure* **vol 12** is 1. p S269

[3] K Kitagawa 2013 Thin-film thickness profile measurement by three-wavelength interference color analysis *Applied Optics* **vol 52** is 10. pp 1998-2007

[4] M Born, E Wolf 1970 *Principles of optics* (Oxford: Pergamon Press)

[5] V V Lychagov, V P Ryabukho, A L Kalyanov, I V Smirnov 2012 Polychromatic low-coherence interferometry of stratified structures with digital interferogram recording and processing *Journal of Optics*, **vol 14** is 1. p 015702

[6] D V Sivukhin 2002 *Obshchiy Kurs Fiziki. T.4. Optika.*(Moscow: Fizmatlit)

Extended kinetic model of real-time polymerase chain reaction process

A A Fedorov[1], D G Sochivko[2], D A Varlamov[3] and V E Kurochkin[1]

[1] Institute for Analytical Instrumentation, Russian Academy of Sciences, St. Petersburg, Russia
[2] Syntol JSC, Moscow, Russia
[3] All-Russia Research Institute of Agricultural Biotechnology, Russian Academy of Sciences, Moscow, Russia

E-mail: f_aa@mail.ru

Abstract. Real-time polymerase chain reaction (real-time PCR) is the main molecular genetic method used for qualitative and quantitative analysis of specific nucleic acid sequences in many areas of biomedical research. Theoretical study of PCR models allows to estimate the influence of various reaction components and parameters, and to determine the unknown parameter values by approximating the experimental real-time PCR curves. An extended kinetic model of real-time PCR is presented. The model takes into account the enzyme activity based on Michaelis-Menten kinetics, the hybridization of complementary DNA fragments, the presence of a fluorescent probe used for detection of the reaction products, and the temperature dependence of primers and probe hybridization.

1. Introduction

Currently, real-time polymerase chain reaction (real-time PCR) is the main molecular genetic method used for qualitative and quantitative analysis of specific nucleic acid sequences in many areas of biomedical research. The feature that distinguishes real-time PCR among other analytical methods is extremely high sensitivity and specificity, permitting detection and analysis of samples containing as few as single molecules of target sequence. Despite the fact that the overall molecular mechanism of the reaction is well known, new articles suggesting quantitative models of various complexity are published regularly [1-3]. The on-going research is stimulated by the fact that real-time PCR is a multistage reaction involving about dozen key molecules; therefore the analysis of its kinetics is a non-trivial task which requires a certain degree of simplification.

The main approach to model the real-time PCR process is based on the application of chemical kinetic equations. To do this, one has to create a chemical framework of PCR in the desired level of detailing, write down the reaction equations, derive from them the system of differential equations and then solve it, usually by numerical methods. This model study allows to estimate the influence of various reaction components and parameters, and to determine the unknown parameter values by approximating the experimental real-time PCR curves.

The review of simple kinetic models describing the major reaction components is given in [4]. The model suggested there includes the reversible primer hybridization, the hybridization of the complementary DNA strands, and Michaelis-Menten kinetics of DNA fragment synthesis. However, most of the models don't take into account two aspects of the actual real-time PCR. The first aspect is the presence of the reporter probe that is used to detect reaction product in many real-time PCR realizations. The second one is the fact that hybridization is a temperature-dependent process. Indeed, the PCR primers and probes are designed to work at the specific temperature which is determined by their melting curve [5]. In the present paper, we introduce the model of real-time PCR which includes the reporter probe and hybridization temperature dependence.

2. The model description

We have developed an extended kinetic model of real-time PCR. The system takes into account enzyme activity based on Michaelis-Menten kinetics, the hybridization of complementary DNA fragments, the presence of the fluorescent probe used for detection of the reaction products, and temperature dependence of primers and probe hybridization. The model reactions are shown in Fig. 1.

A

$$F + \mathrm{Pr} \underset{k_{02}}{\overset{k_{01}}{\Leftrightarrow}} S$$

$$F + Z \underset{k_{12}}{\overset{k_{11}}{\Leftrightarrow}} FZ$$

$$FZ + \mathrm{Pr} \underset{k_{02}}{\overset{k_{01}}{\Leftrightarrow}} SZ$$

$$S + Z \underset{k_{12}}{\overset{k_{11}}{\Leftrightarrow}} SZ$$

B

$$S \xrightarrow{k_m} P_1$$

$$SZ \xrightarrow{k_m} P_{1Z} + M$$

C

$$F + F \xrightarrow{k_2} P_2$$

$$F + S \xrightarrow{k_2} P_2 + \mathrm{Pr}$$

$$F + FZ \xrightarrow{k_2} P_2 + Z$$

$$F + SZ \xrightarrow{k_2} P_2 + \mathrm{Pr} + Z$$

$$S + S \xrightarrow{k_2} P_2 + 2\mathrm{Pr}$$

$$S + FZ \xrightarrow{k_2} P_2 + Z$$

$$S + SZ \xrightarrow{k_2} P_2 + \mathrm{Pr} + Z$$

$$FZ + FZ \xrightarrow{k_2} P_2 + 2Z$$

$$FZ + SZ \xrightarrow{k_2} P_2 + 2Z + \mathrm{Pr}$$

$$SZ + SZ \xrightarrow{k_2} P_2 + 2Z + 2\mathrm{Pr}$$

Figure 1. The model reactions. (A) reversible hybridization processes, (B) fragment synthesis by Taq polymerase, (C) irreversible complementary chains hybridization.

The model includes the following reactants:

F – amplified DNA fragment,
Pr – primer,
Z – TaqMan fluorescent reporter probe,
S – complex of fragment F and primer Pr (non-fluorogenic substrate for polymerase),
FZ – complex of fragment F and probe Z,
SZ – complex of fragment F, primer Pr and probe Z (fluorogenic substrate for polymerase),
P_1 – product of polymerase activity (double-strand fragment produced by synthesis),
P_{1Z} – fluorescent product of polymerase activity,
P_2 – duplex (double-strand fragment produced by hybridization of complementary chains),
M – fluorescent reporter molecule (produced during synthesis of P_1 from SZ by polymerase).

Differential equation system corresponding to the reactions above was derived:

$$\begin{aligned}
\frac{\partial[F]}{\partial t} &= k_{02}[S]+k_{12}[FZ]-k_{01}[F][\mathrm{Pr}]-k_{11}[F][Z]-k_2[F]([F]+[S]+[FZ]+[SZ])\\
\frac{\partial[\mathrm{Pr}]}{\partial t} &= k_{02}([S]+[SZ])-k_{01}[\mathrm{Pr}]([F]+[FZ])+k_2[S]([F]+[S]+[FZ]+2[SZ])+k_2[SZ]([F]+[FZ]+[SZ])\\
\frac{\partial[Z]}{\partial t} &= k_{12}([FZ]+[SZ])-k_{11}[Z]([F]+[S])+k_2[SZ]([F]+[S]+[FZ]+[SZ])+k_2[FZ]([F]+[S]+[FZ])\\
\frac{\partial[S]}{\partial t} &= k_{01}[F][\mathrm{Pr}]-k_{11}[Z][S]-k_2[S]([F]+[S]+[FZ]+[SZ])-k_m[S]\\
\frac{\partial[FZ]}{\partial t} &= k_{11}[F][Z]-k_{01}[FZ][\mathrm{Pr}]-k_2[FZ]([F]+[S]+[FZ]+[SZ])\\
\frac{\partial[SZ]}{\partial t} &= k_{01}[FZ][\mathrm{Pr}]+k_{11}[S][Z]-[SZ](k_{02}+k_{12})-k_2[SZ]([F]+[S]+[FZ]+[SZ])-k_m[SZ]\\
\frac{\partial[P_2]}{\partial t} &= k_2([F](0.5[F]+[S]+[FZ]+[SZ])+[S](0.5[S]+[FZ]+[SZ])+[FZ](0.5[FZ]+[SZ]))\\
\frac{\partial[P_1]}{\partial t} &= k_m[S]\\
\frac{\partial[P_{1Z}]}{\partial t} &= k_m[SZ]\\
\frac{\partial[M]}{\partial t} &= k_m[SZ]
\end{aligned} \tag{1}$$

The temperature dependence of primer and probe hybridization was introduced through the dissociation constant k_{02} and k_{12}. It is known that the association rate of oligonucleotides is weakly dependent on temperature, in contrast to the dissociation rate. To estimate the dissociation constant at the certain temperature, the melting curve of the primer or the probe was used.

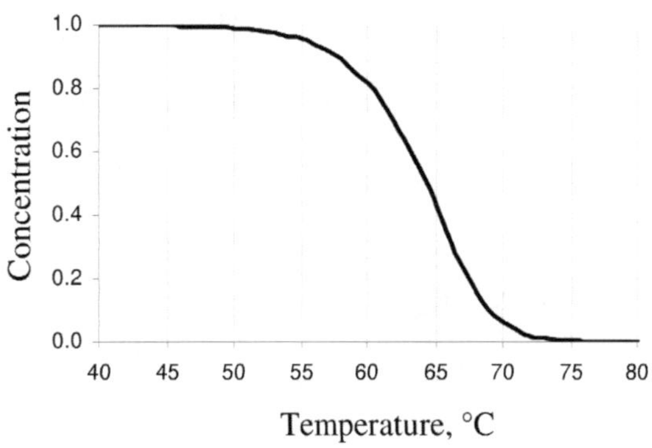

Figure 2. The melting curve of the model PCR primer. Normalized DNA hybrid concentration is plotted vs. temperature.

The melting curve represents the steady-state fraction of DNA hybrid assessed in the given temperature range. The melting curve can be obtained using fluorescent probes of different kinds. For example, two complementary oligonucleotide fragments can be mixed with an intercalating fluorescent dye and heated in a PCR machine. At lower temperatures, the fluorescent signal will be maximal as all the chains will be in the hybridized state. At higher temperatures, fluorescence will reach minimum as all the double-stranded complexes will dissociate. The normalized curve obtained is presented in Fig. 2. This curve can be well approximated by generalized logistic function with parameters a_1, a_2, T_M (a_1 – slope parameter, a_2 – asymmetry parameter, T_M – midpoint parameter) which allows calculating the hybridized fragment fraction $M(T)$ at any temperature T (Eq. 1). Next, the dissociation rate constant can be calculated from equilibrium condition reached during melting curve analysis (a_3 is equal to the concentration of melted oligonucleotides) assuming that association constant k_1 does not depend on temperature (Eq. 2):

$$M(T) = \frac{1}{\left(1 + e^{-a_1(T-T_M)}\right)^{a_2}}, \tag{2}$$

$$k_{-1} = k_1 a_3 \frac{\left(1 - M(T)\right)^2}{M(T)}. \tag{3}$$

Enzymatic synthesis of products by Taq polymerase was described by Michaelis-Menten kinetics of k_m rate constant (V_{max} – maximum rate, K_M – Michaelis constant, $[S]$ – substrate S or SZ concentration):

$$k_m = \frac{V_{\max}}{K_M + [S]}. \tag{4}$$

The resulting system of kinetic equations was solved and analyzed by numerical simulation using the C++ language. The selection of parameters of the system was carried out on the basis of literature data and own experimental results: initial DNA concentration – 10^{-11}M, primer and probe concentration $2\cdot10^{-7}$M (unless otherwise specified), k_{01} =10^6M^{-1}s^{-1}, k_2 =$3\cdot10^6$M^{-1}s^{-1}, k_{11} =10^6M^{-1}s^{-1}, a_1=–0.31, a_2=3.4, a_3=$2\cdot10^{-7}$M, $V_{\max}$ = $1\cdot10^{-8}$M·s^{-1}, K_M = $2\cdot10^{-9}$ M.

3. Results

In contrast to the most existing models of real-time PCR counting DNA fragments synthesized during reaction, product registration in our model is carried out indirectly by the number of fluorescent reporter molecules M generated in the reaction. This allows us to assess the influence of probe kinetics on the PCR curve shape and amplitude. We analyzed the PCR curves for product P_1 and reporter M at several probe Z concentrations (Fig. 3A). The concentration of the reporter at the PCR plateau is lower than the concentration of the product when they are added to the reaction in the same amount. When we look at the curves normalized to the plateau level we can see that the reporter curves saturate slightly earlier than the product curve (Fig. 3B).

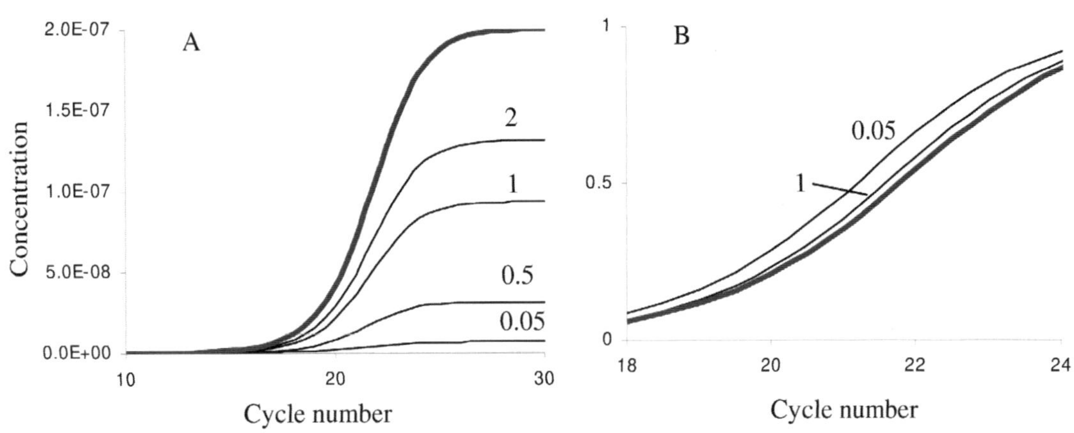

Figure 3. The PCR curves for product (thick line) and reporter (thin lines). (A) absolute concentrations, (B) normalized curves. The initial probe concentration relative to the primer concentration is denoted.

We analyzed the effect of temperature on the product synthesis. The primer melting curve is shown in Fig. 2, the corresponding approximated T_M from Eq. 1 was 69°C. Reaction temperature T was varied, and the resulting PCR curves were depicted in Fig. 4. Reaction kinetics started to change as the temperature reached 70°C when the melting curve shows just 10% of the primer is hybridized. At higher temperatures the reaction declined dramatically.

Next, we studied how the reaction curve amplitude is affected by the relative melting temperatures of the primers and the probe. Reaction temperature T was 60°C. Primer melting temperature T_M was fixed at 69°C, whereas T_M of the probe was varied from 60 to 80°C. The reporter concentration curves are presented at Fig. 5. Surprisingly, the curve amplitude differed just slightly for all melting temperatures equal or higher than primer T_M. This is in contrast to the general belief that the probe melting temperature should be about 10 degrees higher than the one for the primer. Normalized curves were identical for all the temperatures studied (data not shown).

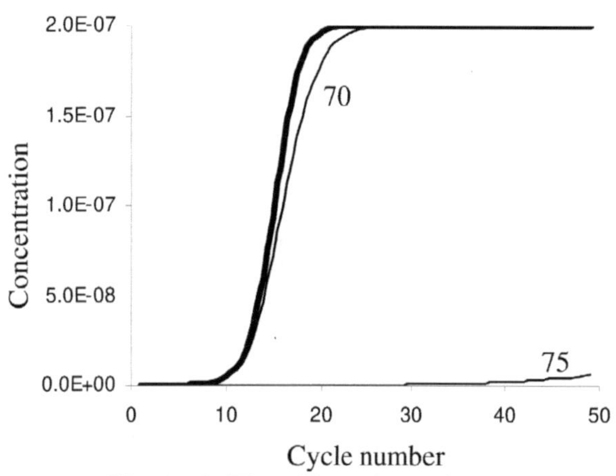

Figure 4. The effect of reaction temperature T on the product synthesis. Thick line – T = 65°C and lower. Thin lines – T is denoted in the panel.

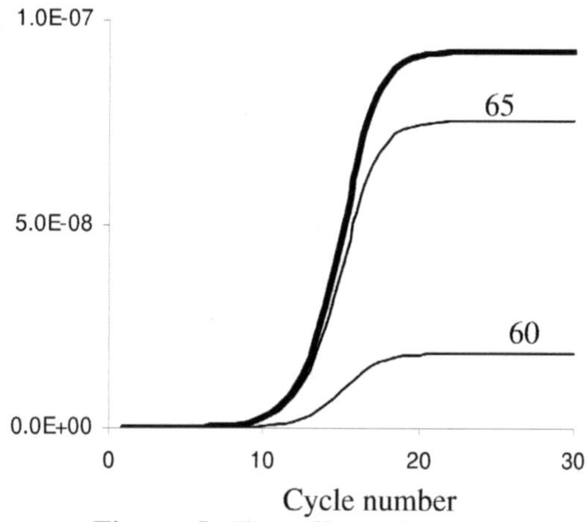

Figure 5. The effect of probe melting temperature T_M on the PCR curve amplitude. Thick line – T_M is 70°C and higher. Thin lines – T_M is denoted in the panel.

4. Conclusion

The extended kinetic model of real-time PCR presented here was designed to study how reaction results depend on various reagents and reaction conditions. The model parameters have clear meaning and allow experimental or theoretical evaluation. The results of modeling demonstrate realistic behavior characteristic for the actual real-time PCR curves.

Acknowledgements

The work was supported by the Ministry of education and science of the Russian Federation, agreement 14.579.21.0012 (ID # RFMEFI57914X0012).

References

[1] Mehra S and Hu W S 2005 *Biotechnol Bioeng* **91** 848-860
[2] Cobbs G 2012 *BMC bioinformatics* **13** 203
[3] Marimuthu K, Jing C and Chakrabarti R 2014 *Biophys J* **107** 1731-1743
[4] Sochivko D G, Fedorov A A, Varlamov D A, Kurochkin V E and Petrov R V 2016 *Dokl Biochem Biophys.* **466** (in press)
[5] SantaLucia Jr J 2007 *PCR Primer Design* (New York: Humana Press) chapter 1 pp 3-33

Computer modelling of the microstructure of the trabecular bone fragments for the study of stress-strain state

Y N Lastovkina[1] **, T V Kolmakova**[1,2]
[1] National Research Tomsk State University, Tomsk, Russia
[2] Institute of Strength Physics and Materials Science (ISPMS) SB RAS , Tomsk, Russia

caterinka4@gmail.com, kolmakova@ftf.tsu.ru

Abstract: The paper presents the computer models of the structure of the trabecular bone tissue. The approach to the construction of computer models of fragments of cancellous bone tissues was offered. The model fragment of the trabecular bone tissue was built based on the data of structure of the real bone fragments, taking into account the orientation of the trabeculae of bones, their length and thickness.

1. Introduction

The problems of mechanics arising during production of the bone implants are resolved on the basis of the investigation of the mechanics of bone tissue and comprise the development of the approaches and modelling techniques of their complex hierarchy structure, the estimation of the effective mechanical characteristics, the development of the basic requirements to the alternate materials of the implants regarding mechanics.

The bone is the bio composite material of the complex structural arrangement of components that determines the anisotropy of it mechanical properties [1-4]. "On each structural level a biological tissue may be considered as a composite material", "The heterogeneity of the mechanical properties of bone tissue over the zones of tibia cross-section is caused by both the differences in structure and the heterogeneity of the biochemical composition."[1]. "Mechanical properties of bones are largely determined by their microstructure." [2]. "At the micron scale and above, bone tissue is a hierarchical composite." "Reflecting the anisotropy of its microstructure, the elastic and strength properties of human cortical bone are anisotropic."[3]. "This microstructural directionality gives trabecular bone anisotropy of mechanical properties." "Thus, trabecular bone is classified from an engineering materials perspective as a composite, anisotropic, open porous cellular solid." [4].

Any skeletal bone (macro level) is composed of the compact (cortical) and spongy (trabecular) bone tissue. The nanolevel of the bone tissue is the fibers consisting of a great number of collagenous fibrillas and the associated hydroxyapatite crystals. The microstructure of the cortical bone tissue is a bunch of the structural elements, i.e. the osteons formed by concentrically interleaved bone layers with the different location focus of the collagen- mineral fibers [3]. The microstructure of the spongy bone is formed by the directed location of the trabeculae [3]. The direction of location of the collagen-mineral fibers of the osteons and trabeculae of the bone tissue varies depending on the anatomical location of the bone fragments while being determined by the effected load [1, 4, 5].

The purpose of the present research is the computer modelling of the spongy bone tissue fragments in view of their structural features determined by the anatomical location.

Journal of Physics: Conference Series **769** (2016) 012020 doi:10.1088/1742-6596/769/1/012020

2. Computer modeling of the microstructure of the trabecular bone tissue

The model fragments of trabecular bone tissue are built on the basis of data of ellipses of structure of the real fragments of bone (Fig. 1), taken from literature [6, 7]. H_1 and H_2 are semiaxis of the ellipse of structure corresponding to the halves of lengths of main and secondary trabeculae; θ_H is the angle showing the direction of the ellipse of structure, i.e. the direction of location of the main trabeculae of the spongy bone tissue.

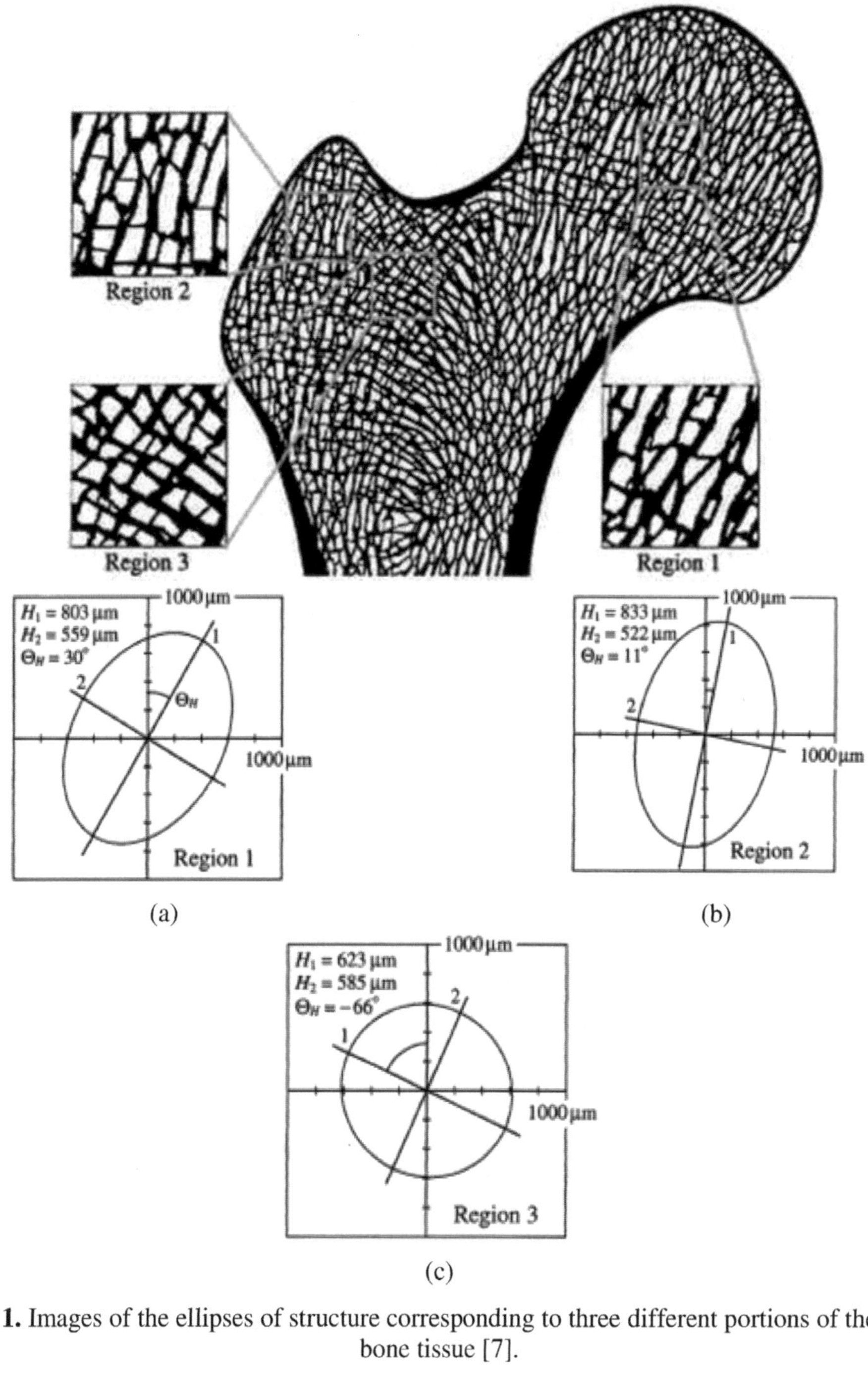

Figure 1. Images of the ellipses of structure corresponding to three different portions of the spongy bone tissue [7].

The trabecular node including the trabeculae, orthogonally located relative to each other, is considered as the structural unit of the geometric model (Fig. 2).

The trabecular node is built based on the parameters H_1, H_2, $H_3 = H_2$ and the values of the average thickness of the trabeculae [8].

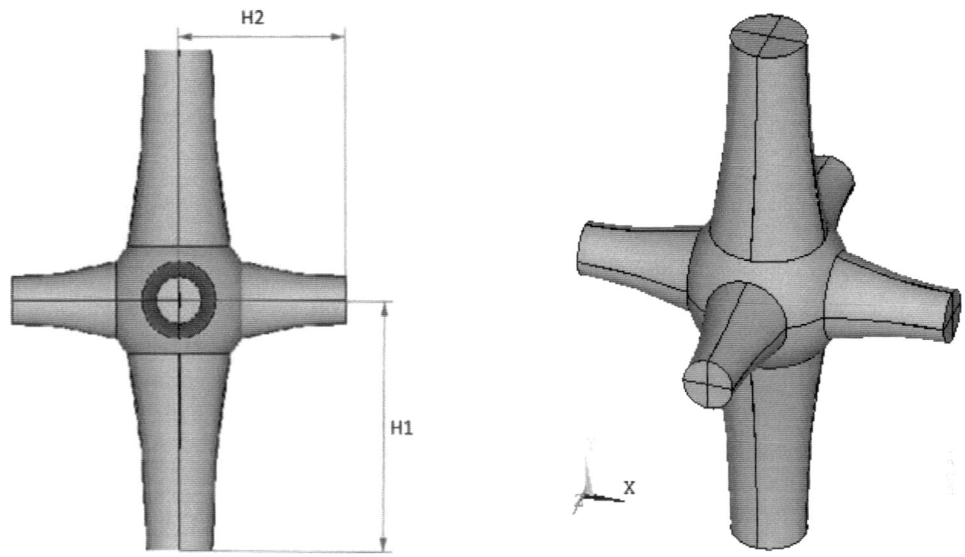

Figure 2. Trabecular node.

The geometric model is built with use the trabecular nodes copying in three directions, given the angle θ_H, in compliance with the ellipses of structure of natural bone fragments (Fig. 1).

Figure 3 shows the geometric models of the spongy bone tissue fragments with the different anisotropy of the structure corresponding to the ellipses of structure presented in Figure 1.

The developed approach allows building the computer models of trabecular bone fragments, in which the length and thickness of the main and secondary trabeculae may vary. These facts, and the consideration of the bone trabeculae direction by means of turn through θ_H, distinguishes the presented models from the model [8] and allows to more closely replicate the microstructure of natural fragment.

In future it is planned to conduct the calculations of the stress-strained state of fragments of the spongy bone tissue. Similar calculations are needed to develop recommendations for the selection and establishment of mechanically compatible implants intended to replace the fragments of spongy bone in orthopedics.

3. Conclusions

The approach to the construction of computer models of fragments of cancellous bone tissues was offered. The model fragment of the trabecular bone tissue was built based on the data of structure of the real bone fragments, taking into account the orientation of the trabeculae of bones, their length and thickness.

The computer models of the fragments of structure of the spongy bone tissue are required to perform calculations of the stress-strain state, which analysis will allow development of the recommendations for creation of the mechanically compatible implants, ensuring the retention of the stress-strain state existing in the bone macrovolume during implantation.

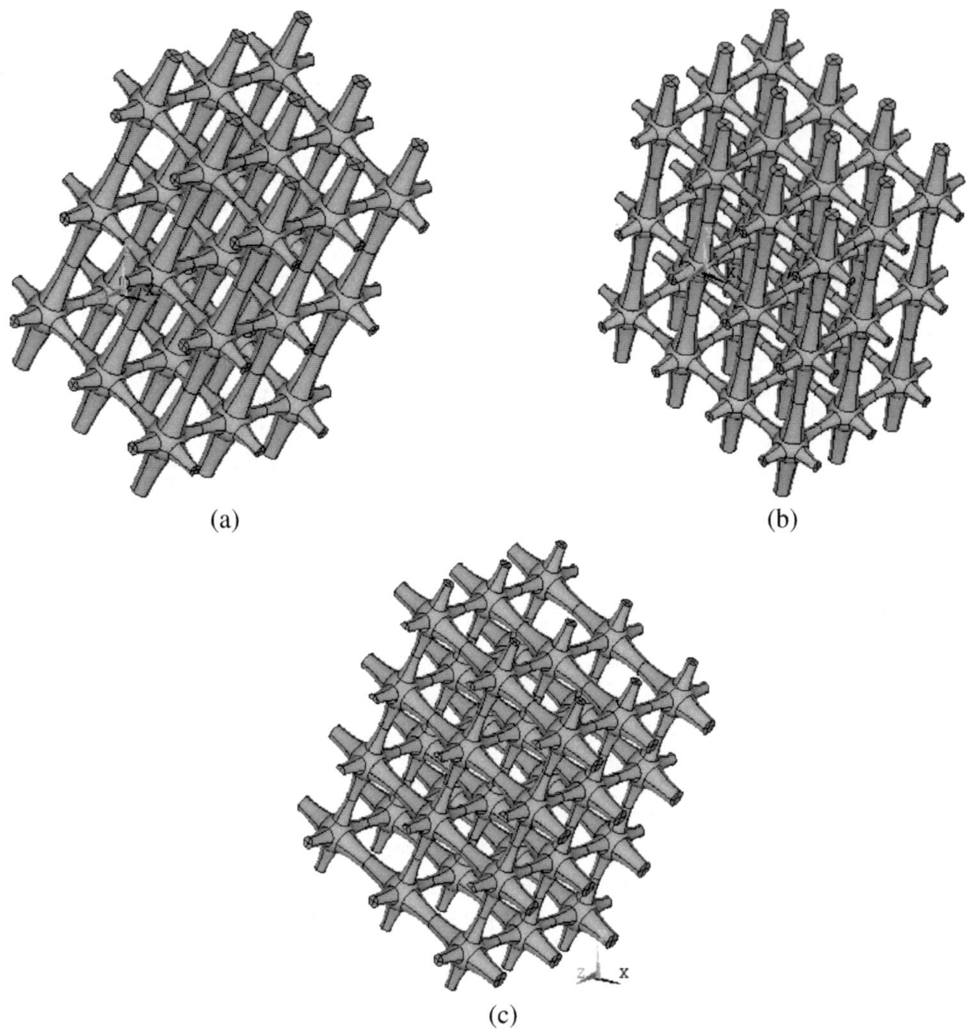

Figure 3. The geometric models of the spongy bone tissue fragments.

Acknowledgments
This research has been performed within RFMEFI60714X0069 Project and financial support of the Ministry of Education and Science of the Russian Federation.

References

[1] Knets I 1999 *Acta Bioeng. Biomech.* **1** 55
[2] Sevostianov I, Kachanov M 2000 *J. Biomech.* **33** 881
[3] Keaveny T M, Morgan E F and Yeh O C 2009 Biomedical Engineering and Design Handbook vol 1, ed Myer Kutz (New York: McGraw-Hill) chapter 9 pp 221–243.
[4] Keaveny T M, Morgan E F, Niebur G L and Yeh O C 2001 Annu. Rev. Biomed. Eng. **3** 307
[5] Reilly D T, Burstein A H 1974 *J. Bone Joint Surg.* **56a** 1001
[6] Cowin S C 1986 *J. Mech. Materials.* **5** 251
[7] Tsubota K, Adachi T, Tomita Y 2002 *J. Biomech.* **35** 1541
[8] Dagan M, Gefen A 2004 *Med. Biol. Eng. Comput.* **42** 549

The interaction of DNA with piperazine derivatives of benzoimidazophthalazine

D N Osinnikova, E B Moroshkina and V I Travkina

St. Petersburg State University, 1 Ul'ianovskaia str., Peterghof, St. Petersburg 198504, Russia

E-mail: osinnikovadasha@yandex.ru

Abstract. The interaction of DNA molecule with new synthetic compounds, piperazine derivatives of benzo[4,5]imidazo[1,2-a]phthalazine, was investigated by the spectral, hydrodynamic and optical methods. The thermodynamic parameters of the interaction and the stoichiometry of the complexes were determined by spectrophotometric titration. A mode of binding and structure of the complexes were determined by analyzing the changes in the intrinsic viscosity and the optical anisotropy of the macromolecule upon complexation. It is shown that an increase in the intrinsic viscosity and the optical anisotropy with a small content of the ligand in the complex is caused by the increase in the thermodynamic rigidity of macromolecules upon the formation of the complex. Increase of the contour length of the macromolecule in the complex does not occur, that indicates nonintercalative mode of binding the piperazine derivatives of benzoimidazophthalazine with DNA.

1. Introduction

The mechanism of biological activity of various heterocyclic compounds: antibiotics, alkaloids, synthetic compounds, involves the formation of the complex with the DNA molecule [1]. This fact determines the relevance of studies of the interaction of new synthetic compounds with planar polycyclic chromophore with DNA *in vitro*.

There are several modes of binding of these compounds with DNA as monomers or as aggregates on the surface of the double helix [2, 3]. The most biologically significant binding is a monomeric ligand binding with DNA. In this case the chromophore intercalates into double helix, or it locates in the minor groove of the DNA double helix. Previously, it was shown on the interaction of DNA with derivatives of actinocine [4], xanthone [3], phenazine [5] that the localization of molecule of the ligand upon binding is determined by the ionic conditions [3] and its structural features [3-5].

Spectral and thermodynamic methods of research do not allow to distinguish between both types of monomeric ligand binding. It is necessary to study the changes in the macromolecular structure of DNA upon binding with the ligand to determine the structure of the complex. The local untwisting of the double helix and its elongation are typical changes upon the intercalation [6, 7]. The viscometry is commonly used method to determine the increase of the contour length of the double helix. The increase in reduced viscosity of DNA solution: $(\eta_r\text{-}1)/C_{DNA}$, where η_r – relative viscosity, C_{DNA} – DNA concentration, upon formation of the complex is considered as a sign of intercalation [8, 9]. However, the change of the reduced viscosity of the solution of the macromolecule in the presence of low molecular weight ligand does not allow to make a conclusion about the presence of certain changes in macromolecular parameters. This requires the determination of the value of the intrinsic viscosity $[\eta] = lim\ (\eta_r\text{-}1)/C_{DNA}$, when $C_{DNA} \rightarrow 0$, which is directly related to the size of the macromolecule.

In the case of high molecular DNA changes in intrinsic viscosity can occur not only as a result of changes in the contour length, but also due to changes in thermodynamic rigidity of the macromolecule [10]. Therefore, to detect changes in the contour length of the macromolecule more research is needed.

In this work the binding of new synthetic compounds, derivatives benzoimidazophthalazine (figure 1), with DNA was studied by the spectral, hydrodynamic and optical methods. Thermodynamic parameters of binding of compounds with DNA and the stoichiometry of the investigated complexes were determined by spectrophotometric titration (SPT). Changes of the macromolecular structure of DNA in the formation of DNA-ligand complex were determined by using a parallel measurement of the intrinsic viscosity and the optical anisotropy of the complexes [11]. The use of these two methods allows us to make unambiguous conclusions about the presence or absence of extension of the DNA double helix upon binding.

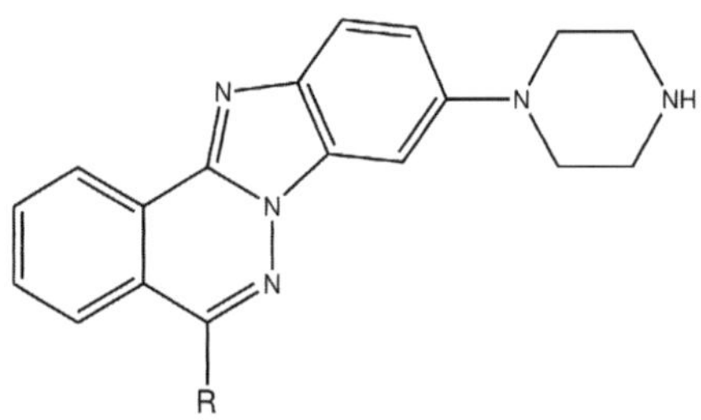

Compound 1 - R = $-CH_3$; Compound 2 - R= 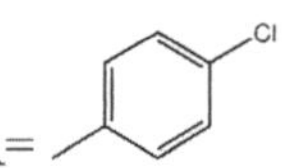

Figure 1. The structure of investigated compounds

2. Materials and methods

Calf thymus DNA "Sigma" with molecular weight M=22•10^6 Daltons, defined by the value of the intrinsic viscosity $[\eta]$ in water solution of 0.15 M NaCl [12]. The DNA concentration was determined spectrophotometrically by the absorption of DNA hydrolyzate at λ= 290 and 270 nm [13]. Benzoimidazophthalazine derivatives were synthesized at the Research Institute of Hygiene, Occupational Pathology and Human Ecology FMBA of Russia. Synthesis and properties of these compounds have been described previously [14]. Compounds were dissolved in ethanol. Complexes were prepared by adding an alcoholic solution of the compound in water-salt solution of DNA. The concentration of ethanol in the final solution did not exceed 1%. The ionic strength of the solution μ = 0.001. The stoichiometry of the DNA-ligand complex (r) was determined from the binding curve resulting from the spectrophotometric titration. The absorption spectra of the solutions were recorded with a spectrophotometer Shimadzu UV-1800.

The mode of binding of compounds with DNA was determined by analysing of changes in intrinsic viscosity and optical anisotropy of the macromolecule during the formation of the complex by the method described previously [11]. According to the model of free-jointed chain, the intrinsic viscosity of the macromolecule is related to the contour length L and the length of the Kuhn segment A by the Flory formula [10]:

$$[\eta] = F\frac{(LA)^{3/2}}{M}\gamma^3, \tag{1}$$

where F - Flory constant for a given polymer-solvent system, M – molecular weight of the macromolecule, γ– the coefficient of linear swelling. Then

$$\frac{L_r}{L_0} = \left(\frac{[\eta]_r}{[\eta]_0}\right)^{2/3} \cdot \frac{A_0}{A_r}, \tag{2}$$

here and below the codes "r" and "0" denote characteristics of the complex and free DNA, respectively.

To determine the change in contour length of the macromolecule during formation of the complex it is necessary to independently determine the change in the length of the Kuhn segment A. To this end, we determined the value of the Peterlin ratio [15] for the free DNA and its complexes with compounds 1 and 2 using dynamic birefringence. It is proportional to the optical anisotropy of the statistical segment of the macromolecule (α_1-α_2):

$$\frac{\Delta n}{g(\eta - \eta_0)} \sim (\alpha_1 - \alpha_2) = (a_{\|} - a_{\perp})\frac{A}{d}, \tag{3}$$

where Δn – the value of birefringence, g – the gradient of flow velocity, η и η_0 – viscosity of the solution and solvent, ($a_{\|}$-$a_{\perp}$) – optical anisotropy of the monomer unit, d – thickness of the monomer unit.

Intrinsic viscosity was determined by using graphical extrapolation of the reduced viscosity to zero concentration. To preserve the stoichiometry of the complex in the process of the concentration measurement, the values of the ionic strength of the solution and equilibrium concentration of free ligand (C_{free}) is maintained equal to the corresponding values in the initial solution.

The value of birefringence was determined using experimental optical system [16]. Solution viscosity was determined using magnetic rotational viscometer [17].

3. Results and discussion

3.1. Spectrophotometric titration

The changes in spectral properties of ligand in the presence of DNA in solution are the main features of the interaction of compound with DNA. Both compounds have long-wavelength absorption band in the 280-450 nm region. In the presence of DNA decrease in intensity and a bathochromic shift of this absorption band were observed. The bathochromic shift increases with raising concentration of DNA in solution (figure 2).

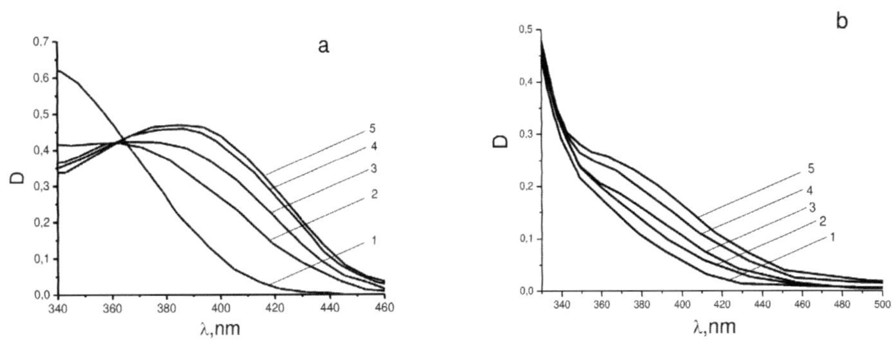

Figure 2. The absorption spectra of compounds 1 (a) and 2 (b) when C_{lig} = 2.5•10^{-5}M in the presence of DNA at different ratios of molar concentrations of ligand and DNA(C_{lig}/ C_{DNA}) in 0.001M NaCl: a) C_{DNA} = 0 (1), C_{lig}/ C_{DNA} = 1.1 (2), 0.6 (3), 0.3 (4), 0.1 (5); б) C_{DNA} = 0 (1), C_{lig}/ C_{DNA} = 5.0 (2), 0.7 (3), 0.3 (4), 0.2 (5)

Based on the SPT data the binding curves (dependencies of r on C_{free}) were constructed (figure 3).

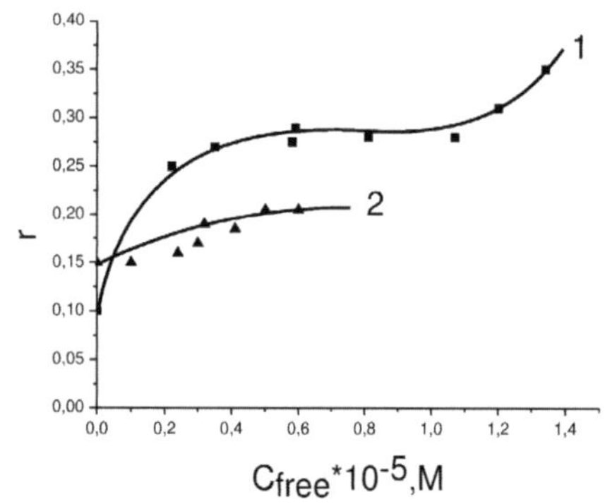

Figure 3. The binding curves: compound 1 (1) and compound 2 (2) in 0.001 M NaCl

The second, weaker type of binding occurs only with the large excess of free ligand in solution. By using the model with excluded places of binding [18] the binding constants (k) and the number of places of binding per pair of DNA bases (n) of first, the energetically stronger binding, were determined: $k_1 = (19\pm2)\bullet10^5$ M^{-1}, $n_1 = 0.32\pm0.05$, $k_2 = (5.5\pm0.5)\bullet10^5$ M^{-1}, $n_2 = 0.20\pm0.5$ for compounds 1 and 2, respectively.

3.2. Viscometry and dynamic birefringence

To determine the mode of binding, intrinsic viscosity and an optical anisotropy of DNA-ligand complexes were measured with different concentrations of the ligand in the complex under existence of only a primary binding mode. Figure 4 shows the dependence of the reduced viscosity of the solution from DNA concentration for the free DNA and for DNA-ligand complexes. These dependencies were used to determine the corresponding intrinsic viscosity. Their linearity indicates isoionic dilution and preservation of the stoichiometry of the complexes upon dilution of initial solutions. Obtained in this way values of intrinsic viscosity of free DNA and its complexes with ligands are shown in table 1.

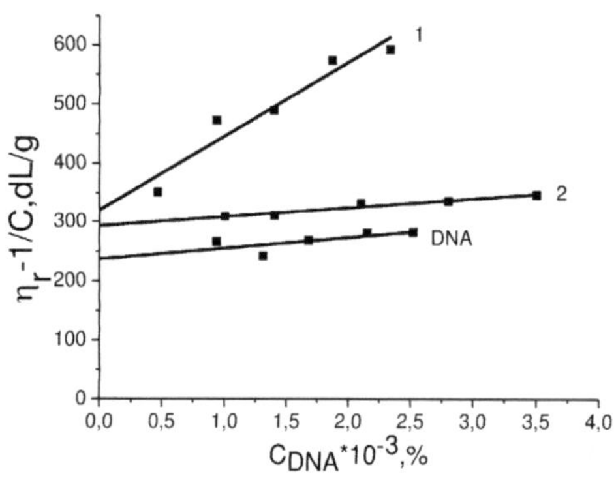

Figure 4. Dependence of η_r-1/C on C_{DNA} (1) and its complexes with compound 1 at $r = 0.1$ (2), $r = 0.2$ (3) and the compound 2 at $r = 0.1$ (4) in 0.001 M NaCl

The value of birefringence, Δn, was determined for the same solutions and Peterlin ratio (equation 3) was calculated. The results are shown in table 1.

Table 1. Hydrodynamic and optical parameters of DNA and its complexes with compounds 1 and 2

	r	$[\eta]$ (m^3/kg)	$\Delta n/g(\eta - \eta_0) \cdot 10^7$ (m·s^2/kg)
DNA	-	24.0±0.5	26±1
DNA-compound 1	0.1	32.0±0.5	33±1
	0.2	32.5±0.5	34±1
DNA-compound 2	0.1	29.0±0.5	31±1

The value of the Peterlin ratio does not depend on the contour length of the macromolecule. Its increase in the complex formation may be caused by the increase of the thermodynamic rigidity and/or the increase of the average optical anisotropy of the monomeric unit by the addition of the ligand molecule. In the case of intercalation binding, the average optical anisotropy of the monomeric unit can be calculated by the equation:

$$\left(a_{\parallel} - a_{\perp}\right)_r = \frac{\left(a_{\parallel} - a_{\perp}\right)_0 + r \cdot \left(a_{\parallel} - a_{\perp}\right)_l}{1 + r}, \tag{4}$$

where $\left(a_{\parallel} - a_{\perp}\right)_l$ – optical anisotropy of the ligand in the axes of its main polarizabilities.

In the case of an alternative method of binding, when the ligand molecule is located in the minor groove of the DNA double helix, the average optical anisotropy of the monomeric unit will be expressed by the equation:

$$\left(a_{\parallel} - a_{\perp}\right)_r = \left(a_{\parallel} - a_{\perp}\right)_0 + r\left(a_{\parallel} - a_{\perp}\right)_l \cdot \frac{3\cos^2\beta - 1}{2}, \tag{5}$$

The orientation angle of the plane of the ligand chromophore relative to the axis of the macromolecule segment β in the case of intercalation is equal to zero, since the plane of intercalated chromophore in the double helix is perpendicular to the axis of the DNA double helix. In the case of binding in the minor groove this angle is 45° [19]. The optical anisotropy of the monomeric unit of free DNA and the ligand molecule in its own axes can be approximately calculated by using the scheme of tensor additivity of polarizabilities of valence bonds.

When $r = 0.1$ in both cases a change in average optical anisotropy of a monomeric unit $\frac{\left(a_{\parallel} - a_{\perp}\right)_r}{\left(a_{\parallel} - a_{\perp}\right)_0}$ is slightly: for intercalation model the ratio is 1.04, and for the model of groove binding - 1.035. The relative change in length of the Kuhn segment A_r/A_0 upon complex formation can be calculated by equation:

$$\frac{A_r}{A_0} = \left(\frac{\Delta n}{g(\eta - \eta_0)}\right)_r \Big/ \left(\frac{\Delta n}{g(\eta - \eta_0)}\right)_0 \bullet \frac{\left(a_{\parallel} - a_{\perp}\right)_0}{\left(a_{\parallel} - a_{\perp}\right)_r}, \tag{6}$$

The results of calculation of this value for the intercalation and groove binding models are presented in Table 2.

Using the obtained values of A_r/A_0 to determine the value of change of the macromolecule contour length (L_r/L_0) during the formation of the complex from the experimental values of the intrinsic viscosity (equation 1) shows that this value stays practically unchanged (table 2), which indicates the nonintercalative type of binding of investigated compounds with DNA. We assume that these compounds bind to the minor groove of DNA.

Journal of Physics: Conference Series **769** (2016) 012021 doi:10.1088/1742-6596/769/1/012021

Table 2. Changes in the macromolecular parameters of the DNA in the formation of complexes, calculated from experiment and theory for different models

	r	$(A_r/A_0)_{exp}$		$(L_r/L_0)_{exp}$		$(L_r/L_0)_{theor}$	
		Intercalation	Groove binding	Intercalation	Groove binding	Intercalation	Groove binding
DNA-comp. 1	0.1	1.22±0.05	1.22±0.05	0.99±0.05	0.99±0.05	1.1	1
	0.2	1.22±0.05	1.22±0.05	1.08±0.05	0.99±0.05	1.2	1
DNA-comp. 2	0.1	1.15±0.05	1.15±0.05	0.99±0.05	0.99±0.05	1.1	1

4. Conclusions

The studies show that the benzoimidazophthalazine derivatives containing residue of the piperazine in 9th position interact with the DNA molecule by forming the equilibrium complexes in a water-salt solution at $\mu = 0.001$. In the formation of the complex, an increase of the intrinsic viscosity and the optical anisotropy of the macromolecule are observed. These changes are caused by an increase in the thermodynamic rigidity of macromolecules in the formation of the complex. Increasing the contour length of the macromolecule is not happening. Therefore, these compounds do not intercalate into the double helix of DNA upon binding. Probably, the ligand molecules are located in one of the grooves of DNA double helix, presumably in a small one.

Thus, an increase in the characteristic viscosity, and especially in the reduced viscosity, during the formation of complexes of DNA with a variety low molecular weight compounds is not unambiguous evidence of the intercalation binding of these compounds with a macromolecule.

References

[1] Ferguson L R, Denny W A 2007 *Mutation Research* **623** 14–23
[2] Moroshkina E B 2007 *Vestnik SPbGU Seria 4(Rus)* **2** 17-23
[3] Moroshkina E B, Bogdanov A A and Kolonistova M O 2009 *Zhurnal Strukturnoi Khimii Vypusk 4(Rus)* **50** 1027-1033
[4] Moroshkina E B, Kvachadze E A, Ovchinnikov D V and Glibin E N 2005 *Russ. J. of Physical Chemistry Suppl. 1* **79** 54-59
[5] Moroshkina E B and Saf'yannikova M G 1999 *Biophysics* **44** 425-429
[6] Lerman L S 1961 *J. Mol. Biol.* **3** 18-24
[7] Waring M J 1970 *J. Mol. Biol.* **54** 247-258
[8] Bera R, Sahoo B K, Glosh K S and Dasgupta S 2008 *Int. J. Biol. Macromol.* **42** 14-21
[9] Shelton A H, Rodger A and McMillin D R 2007 *Biochemistry* **46** 9143-9154
[10] Fox T and Flory P 1950 *J. Polymer Sci.* **6** 74
[11] Veselkov A N, Moroshkina E B, Soboleva O I and Frisman E V 1984 *J. Mol. Biol. (Rus)* **18** 481-487
[12] Eigner J and Doty P 1965 *J. Mol. Biol.* **12** 549-580
[13] Spirin A S 1958 *Biokhimiya (Rus)* **23** 656-662
[14] Kuznetsov V A, Shubin K M, Schipalkin A A and Petrov M L 2006 *Tetrahedron* **62** 14-17
[15] Peterlin A 1950 *J.Polymer Sci.* **12** 45-52
[16] Frisman E V and Cvetkov V N 1952 *ZhETF (Rus)* **23** 690-702
[17] Frisman E V, Shchagina L V and Vorob'ev V I 1965 *Kolloidn. J.* **27** 130
[18] McGhee J D and von Hippel P H 1974 *J. Mol. Biology* **86** 469-482
[19] Tuite E and Norden B 1994 *J. Amer. Chem. Soc.* **116** 7548-7556

Device for Investigation of Mechanical Tension of Isolated Smooth Muscle Vessels and Airway Segments of Animals

A Aleinik[1], N Karpovich[1], N Turgunova[1], A Nosarev[1, 2]
[1]National Research Polytechnic University Institute of Physics and Technology Tomsk, Russia, Lenin Avenue, 30, 634050
[2]Siberian State Medical University, Department of Biophysics Tomsk, Russia, Moscowski Trakt, 2, 634050

E-mail: ousa@tpu.ru

Abstract. For the purpose of testing and the search for new drug compounds, designed to heal many human diseases, it is necessary to investigate the deformation of experimental tissue samples under influence of these drugs. For this task a precision force sensor for measuring the mechanical tension, produced by isolated ring segments of blood vessels and airways was created. The hardware and software systems for the study of changes in contractile responses of the airway smooth muscles and blood vessels of experimental animals was developed.

1. Introduction

Blood vessels play an important role in the regulation of blood pressure and blood flow distribution. Cell-to-cell communication provides coordination of processes in complex organisms. It is important to understand the mechanisms of regulatory processes of vessels and their interaction in vessel networks in health and disease. However, vessel responses *in vivo* depend on a complicated interaction between the different cell types of the vessel wall, factors released from nerve endings, metabolites, hormones and mechanical factors like shear stress from the surrounding tissue [1]. The complexity of the system limits the interpretation of experimental results in mechanistic terms. Therefore, *in vitro* methods have been introduced, where many of these factors can be controlled. The endothelium provides vasodilator effects of circulating hormones. It metabolizes various vasoactive substances, coverts angiotensin I to angiotensin II and secretes the potent vasodilators prostacyclin and EDRF (NO) and the vasoconstrictor peptide endothelin-1. The balance between these mediators determines the responses of the cardiovascular system in diseases such as hypertension, atherosclerosis and myocardial infarction 2]. Moreover, it may tonically suppress vessel constriction even without direct activation [3, 4]. The impaired endothelial function is a major cause of many cardiovascular disorders [5].

Investigation of the vessel segment reactions to the application of external stimuli such as vasoactive substances, pressure, etc., requires different methods. Several methods for the study of vessel segments have been developed. One method employs ring vessel segments threaded on two hooks or wires and the other cannulated vessel segments fixed on two pipettes. In *in vitro experiments*, vessels are studied in physiological saline solution. Ca_2^+ is an important ion, because it plays a central role in the contraction [6]. H^+ is another important ion, because small changes in pH induce considerable changes in diameter of vessel.

The fundamental mechanisms for regulation of contractile activity of vascular smooth muscle and the airways are currently actively investigated to find new drugs to correct pathological states. Smooth muscle hyperactivity is a key mechanism of cardiovascular diseases and pathological conditions associated with the development of an inflammatory response under the influence of radiation exposure to the body [7-9]. Research of contractile activity of smooth muscle in biomedical experiments is not conceivable without the use of high-precision force sensors available, that can measure mechanical tension of small diameter bronchi and blood vessels. So, the actual task is to develop a hardware and software system for medical and biological experiments in this field.

The most common method of measuring contractile force is by the use of a force transducer. Such transducers provide real time, continuous readings and can be combined with software to create plots and data analyses. It is very easy to calibrate such devices. Advancements in technology development now allow to measure even microscale forces accurately.

Capacitive sensors can directly measure various quantities - motion, chemical composition, electric fields and many other variables, which can be converted into a movement or dielectric constant. The principle of operation of capacitive sensors for measuring mechanical quantities is based on the measurement of displacements. Movement of the one electrode with respect to the two fixed ones creates a variable capacitor.

Theoretically, reducing the value of clearance can increase the sensitivity of the sensor to high values. However, it is necessary to consider the electrical and mechanical conditions, which limit the maximum sensitivity. One of the major advantages of the capacitive sensor is that to move one of the plates, a very small value of the force is required. Another advantage is stability and weak influence of the ambient temperature and pressure, as well as high reliability and linear response.

2. Materials and methods

For investigation of mechanical tension of isolated smooth muscle vessels and airway segments of animals the capacitive sensor is used. A microcontroller based device using ATMega-16 microcontroller for the capacitance force sensor has been designed and developed. The block diagram of the device is shown in Figure 1.

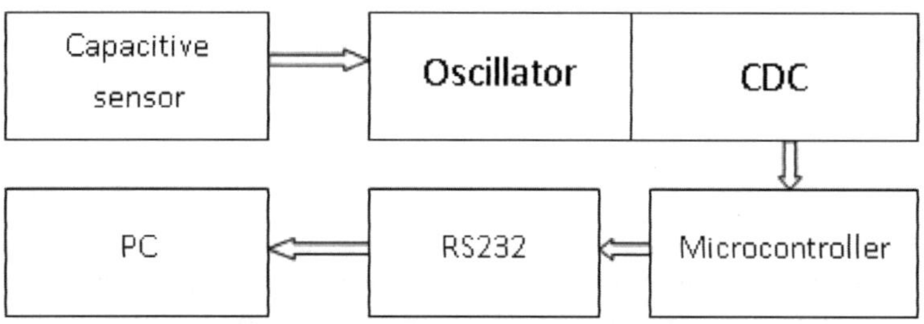

Figure 1. Block diagram of the device

To accurately measure the capacitance changes AD7745 chip is used, which is a capacity-digital converter with high resolution. The measured capacitance is connected directly to the input of the chip. The structure includes a 24-bit ADC with 0.01% linearity. Measurement accuracy is $\pm$ 4 fF.

Calibration of the device was made using weights of a known mass. The capacitance value for each sample was measured. The results of calibration are shown in Figure 2.

Journal of Physics: Conference Series **769** (2016) 012022 doi:10.1088/1742-6596/769/1/012022

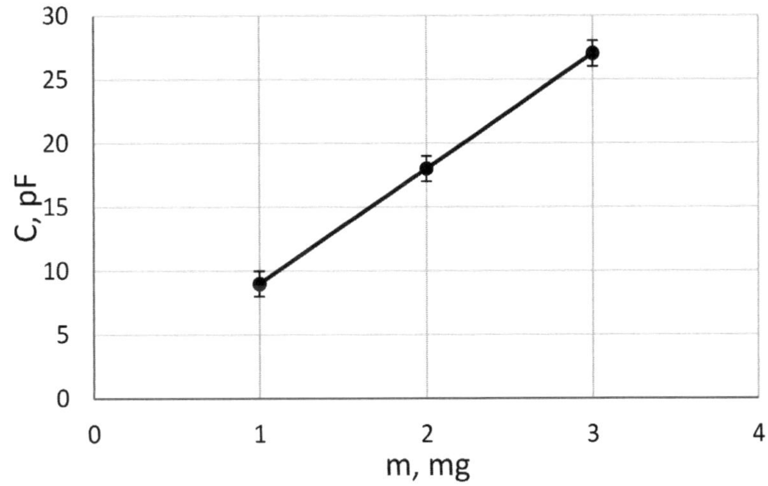

Figure 2. Calibration curve of the device

Form of the curve confirms the efficiency of the device. Next, the device was tested on the real segments vessels. Material for the study was provided by the Department of Biophysics and Functional Diagnostics Siberian State Medical University, Russia, Tomsk. We used mature outbred Wistar rats weighing 180-240 g (n = 13) and guinea pigs weighing 380-420 gr (n = 13). Trachea and bronchi were purified from the surrounding tissue in the bath with Krebs solution at room temperature, after which the ring segments were prepared - width 1.5-2 mm; diameter - 1-1.5 mm.

If necessary, the epithelium was mechanically removed, by rotation of the wooden spatula in the lumen segment for 1 min before performing the experiment. Before the study the intact segments were tested in Krebs potassium solution (40 mM). Contractile response values were evaluated as a percentage of the amplitude of the reference contraction. Increasing the KCl concentration leads to smooth muscle cell membrane depolarization, opening of voltage-dependent calcium channels, and a subsequent contraction of bronchial smooth muscle. Injection of L-cysteine reduces the mechanical tension of segments Figure3.

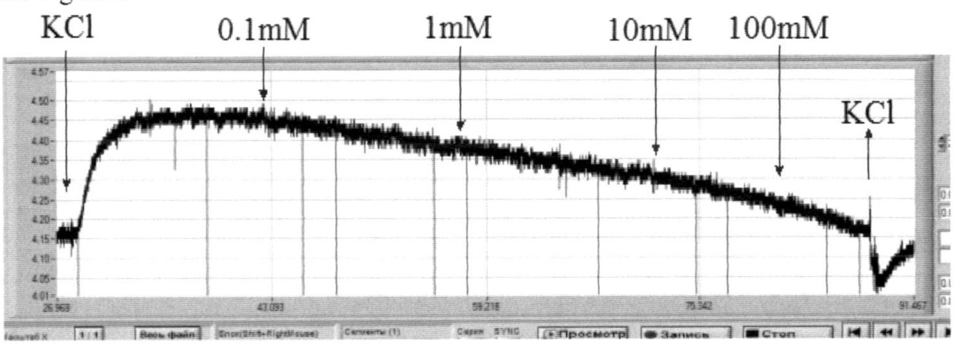

Figure 3. Effect of L cysteine on bronchus segments tension of guinea pig. The arrows show the injection of L cysteine in concentrations 100 µm, 1000 µM, 10 mM and 100 m

There is a difference in contractility of the guinea pigs intact bronchial segments and with impaired epithelium Figure 4. Injection of L-cysteine has a relaxing effect on the whole concentration range. Similar experiments were performed using rat trachea segments Figure 5.

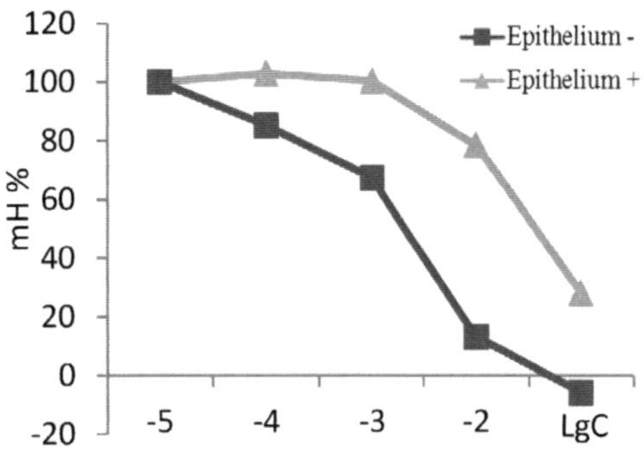

Figure 4. The dependence of the mechanical tension of isolated bronchial segments on (LgC) concentration of L-cysteine .

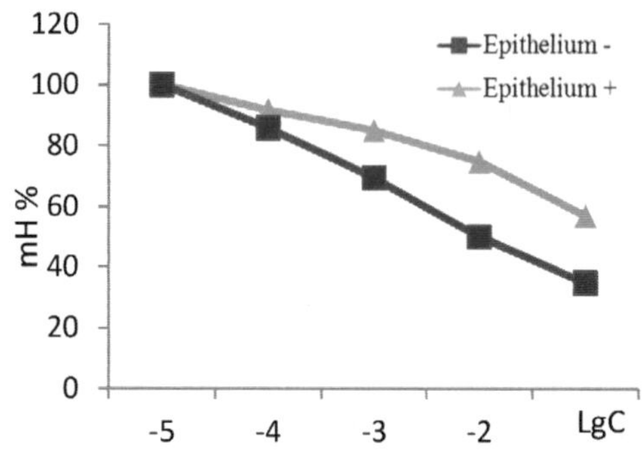

Figure 5. The dependence of the mechanical tension of isolated trachea segments on (LgC) concentration of L-cysteine

A comparison of the relaxing segment responses with intact epithelium and nude one showed that the epithelium contributes to the regulation of modulating tension of the bronchus smooth muscle wall.

3. Results and discussion

The developed device allows to measure the tension of the airways and blood vessels segments with good accuracy. The operation principle is based on the use of capacitive sensor. The contractile properties of airway smooth muscle determine the responsiveness in isometric conditions. The responses to exogenous stimulus are dependent on the intrinsic tone of the airway preparation. When this intrinsic tone is low, contraction is observed, in the case of high tone, relaxation appears. Thus, hydrogen sulphide donor L-cysteine, in concentrations 0,1-100mM exerts a relaxing action on isolated segments of rats and guinea pigs airways. Bronchial epithelium of guinea pigs decreases the effect of hydrogen sulfide donor on the relaxation, while the tracheal epithelium of rats enhances dilation. According to the literature, the relaxation of smooth muscles of the airways depends on opening of different types of potassium channels. These results show that hydrogen sulfide is very important in the development of pathological disorders of the airways and contribute to finding effective ways of pharmacological correction of these conditions.

4. Conclusions

The device is developed for measuring the mechanical tension, produced by isolated ring segments of blood vessels and airways. The device allows to measure the value of forces to within 0.1 mN. The pathophysiology of animal airway was measured using the device. The functional properties of airway smooth muscle are very important in understanding the properties of the airways. There are important differences between species. Different influence of epithelium in rats and pigs can be linked to the density of potassium channels in the membrane of the epithelium and smooth muscle cells. Nevertheless, studies *in vitro* offer the opportunity to test the side effects of novel pharmacological compounds. The developed model allows to study the effects of drugs, inflammatory mediators, and epithelial cells on the responsiveness of airway smooth muscle and blood vessels *in vitro*.

Acknowledgements

This work was supported by Program on National Research Tomsk Polytechnic University Competitiveness Enhancement.

References

[1] Schubert R 2005 Isolated Vessels, *Practical Methods in Cardiovascular Research.* pp 198-211

[2] Vane J 1994 The Croonian Lecture. The endothelium: maestro of the blood circulation, *Philos Trans R Soc Lond B Biol Sci.* **343** pp 225– 246

[3] Dora K, Doyle M, Duling B 1997 Elevation of intracellular calcium in smooth muscle causes endothelial cell generation of NO in arterioles, *Proc Natl Acad Sci U S A.* **94** pp 6529– 6534

[4] Gaynullina D, Lubomirov L, Sofronova S, Kalenchuk V, Gloe T, Pfitzer G, Tarasova O, Schubert R 2013 Functional remodeling of arterial endothelium during early postnatal development in rats, *Cardiovasc Res.* **99** pp 612– 621

[5] Vallance P, Chan N 2001 Endothelial function and nitric oxide: clinical relevance, *Heart.* **85** pp 342– 350

[6] Bukoski R, Shearin S, Jackson W, Pamarthi M 2001 Inhibition of Ca2+-induced relaxation by oxidized tungsten wires and paratungstate *J Pharmacol Exp Ther.* **299** pp 343–350

[7] Mulvany M, Halpern W 1977 Contractile properties of small arterial resistance vessels in spontaneously hypertensive and normotensive rats *Circulation Research* **41(1)** pp. 19-26

[8] Richard L, Twort H, Cameron I, Jeremy P 1989 A comparison of contractile function in large and small pulmonary arterial vessels of the rat Quarterly *Journal of Experimental Physiology* **74** pp 947-950

[9] Gainullina D, Tarasova O, Kiryukhina O, Shestopalov V, Panchin Y 2014 Endothelial function is impaired in conduit arteries of pannexin1 knockout mice Biology Direct. 9 pp 2-5 Extending the Capacitive Input Range of the AD7745/AD7746 Capacitance-to-Digital Converter 2012 *Application Circuit Note* **CN-0129** - p. 4

Journal of Physics: Conference Series **769** (2016) 012023 doi:10.1088/1742-6596/769/1/012023

Interaction between cationic agents and small interfering RNA and DNA molecules

I N Unksov[1], A V Slita[2], A V Petrova[2, 3], I Pereviazko[1], V M Bakulev[1], V I Rolich[1], A B Bondarenko[1, 2], N A Kasyanenko[1]

[1] St.Petersburg State University, Petrodvorets, Uluanovskaya str.,1 St. Petersburg, 198504, Russian Federation

[2] Research Insitute of Influenza, Prof. Popova str., 15/17, St. Petersburg, 197022, Russian Federation

[3] Peter the Great St.Petersburg Polytechnic University, Polytechnicheskaya str., 29, St. Petersburg, 195251, Russian Federation

E-mail: ivonijikt@gmail.com

Abstract.
Azobenzene containing surfactant AzoTAB was used for investigation of binding in cationic-agent + nucleic acid in NaCl salt aqueous solutions. Two nucleic acids, macromolecular DNA and small interfering RNA, were examined upon the interaction with the surfactant. For DNA the interaction was studied using spectral methods and the methods of viscometry and flow birefringence measurement. For siRNA the possibility of surfactant-based delivery was checked *in vitro*.

1. Introduction

The interaction between cationic agents and nucleic acids DNA and RNA has been studied extensively for two main reasons. Firstly, the condensation that occurs for the solutions of DNA molecules in presence of cations has appealed to scientific interest since the 1970s [1, 2]. Being a transition from the rigid and charged molecule to the compact form, this process presents a change in conformational parameters of DNA and was studied for a variety of cationic agents: multivalent ions [3, 4], polymers and surfactants [5, 6], lipids. Secondly, the compact nanoparticles, formed upon the interaction of nucleic acids with some agents, are employed in biomedicine as carriers for the delivery of nucleic acids into cells. Such nanoobjects have been studied for DNA and also for RNA, since the delivery of small interfering RNAs (siRNAs) is regarded as a powerful therapeutic tool and has been extensively studied since the demonstration of RNA interference (RNAi) in 1998[7]. A delivery system for nucleic acid is required to neutralize negative charges of phosphates for facilitation of the cellular uptake, protection from nucleases and from elimination via clearance is necessary for prolongation of therapeutic activity, especially for short and unstable RNAs. For siRNA delivery cationic lipids are widely used [8, 9], but some of them demonstrate high toxicity [9].

Surfactants, whose amphiphilic molecules lower surface tension and form micelles in solution, also have been used for the creation of nanocarriers [10], and have been employed for enhancing modification of delivery agents [11]. The interaction of surfactants with nucleic acids in solution has been scrupulously studied for DNA-surfactant systems [12, 13, 14, 15]. Addition of nucleic acids to the solution induces micellization of surfactant at concentrations lower than critical micelle concentration of surfactant, and micellar aggregates neutralize charges of phosphates, inducing intramolecular contacts. For DNA of high molecular weight it was shown that the addition of cationic surfactant leads to conformational changes of DNA in solution, and compact nanoparticles, which are promising for the delivery, were observed [12]. These conformational transitions can be controlled using photosensitive surfactants, but their toxicity should be taken into consideration [16].

The goal of our study is to obtain novel data on binding of azobenzene containing cationic surfactant AzoTAB to DNA and siRNA *in vitro*. For siRNA the conditions of the formation of nanostructures in solution are determined; we examined the efficiency of cell delivery for these systems.

2. Materials and methods

We used Sigma Aldrich calf thymus (CT) DNA and Dharmacon GE siRNA. For delivery study negative control siRNA-AF546 (with AlexaFlour-546 fluorescent label) was used. Molar concentrations of nucleic acids were calculated from average molecular weight per phosphate of DNA 330 g/mol and of siRNA 321 g/mol. Nucleic acids were dissolved in Milli-Q Type I ultrapure water or in Sigma ultrapure water. Surfactant samples (figure 1) were provided to us by Prof. S. Santer (University of Potsdam, Germany). We used C4-Azo-C6-TMAB (trimethylammonium bromide) for experiments with CT-DNA, and C4-Azo-C6-TEAB (triethylammonium bromide) for experiments with siRNA (transfection studies). Azobenzene containing surfactants undergo trans- to cis-isomerisation under UV irradiation (353 nm). Reverse cis-trans transition is induced by visible light (453 nm) or can be realized in darkness. Delivery of siRNA + surfactant systems was tested on MDCK cell line.

For spectral studies we used SF-56 spectrophotometer (OKB-Spectr, Russia). A low-gradient viscometer of Zimm-Crothers type [17] was used for viscosity measurements, and atomic force microscopy (AFM) studies were carried out using NanoScope IVa (Veeco) in tapping mode in air. Samples were prepared by placing solutions mixed with $MgCl_2$ on freshly cleaved mica surface for 2 min followed by washing the samples with 0.5 – 1 ml of distilled water and drying them in vacuum.

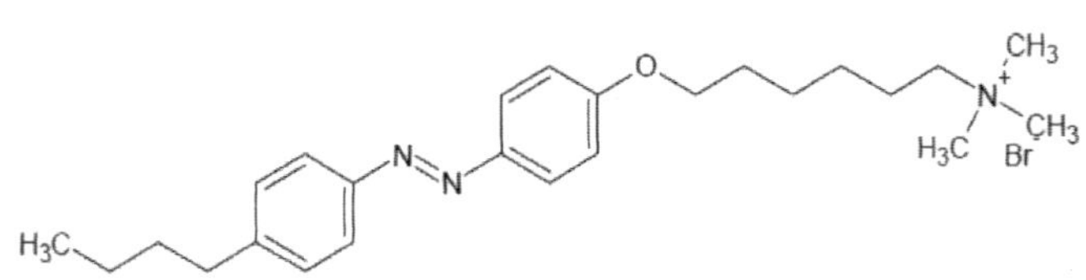

Figure 1. Structure of C4-Azo-C6-TMAB (AzoTMAB). AzoTEAB has ethyl groups instead of methyls on the polar 'head' of the surfactant.

3. Results and discussion

3.1. Spectral studies

UV-Spectra of the surfactant (figure 2a and 2b) show the small shift in absorption maximum (importantly, nucleic acids have no absorption on these wavelengths) upon the addition of nucleic acid, which induces the formation of micelles for the concentrations lower than critical micelle concentration (CMC), and the shift in the same direction (to higher wavelength for cis- isomers and to lower wavelength for trans-isomers) occurs for the concentrations of surfactants higher than CMC in absence of nucleic acids. The addition of La^{3+} ions (figure 2b) to the solution interferes in binding of surfactant to nucleic acid, and the shift is insufficient in this case (figure 2b line c). The addition of La^{3+} to the surfactant solution without DNA induces no shift in surfactant absorption maximum (control, figure 2b line d).

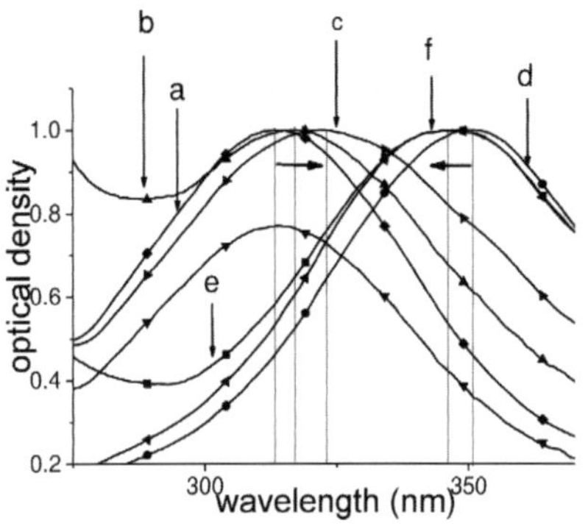

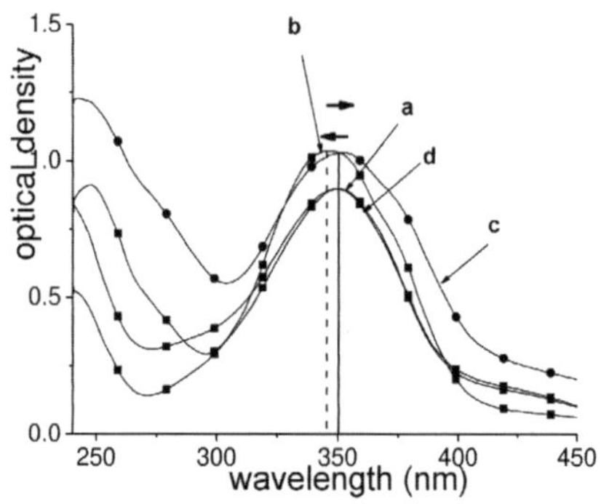

Figure 2a. Normalized spectra: a – cis- surfactant AzoTEAB in 5 mM NaCl solution, C < CMC; b – the same upon addition of siRNA, C(siRNA) = 16 μM; c – cis-surfactant, C > CMC; d – trans-surfactant in 5 mM NaCl solution, C < CMC; e – the same upon addition of siRNA, f – trans-surfactant, C > CMC.

Figure 2b. Normalized spectra: a – trans- surfactant AzoTMAB in 5 mM NaCl solution, C < CMC; b – the same upon addition of CT-DNA, C(DNA) = 60 μM; c – DNA + trans- surfactant upon the addition of $LaCl_3$; d – trans-surfactant upon the addition of LaCl3 (control). a and d are normalized to 0.9 to distinguish them from a and b.

Furthermore, the shift in surfactant absorption maximum is also observed upon addition of denatured DNA (figure 3a) to the surfactant solution, indicating that inner sites of the molecule, which are not exposed in the native DNA structure, play no role in inducing the surfactants binding, and the destruction of DNA secondary structure seems not to prevent it. This adds the data we obtained from spectra of circular dichroism (CD) of DNA (figure 3b), which show no sufficient changes upon the binding for the studied values of $z = C(\text{surfactant}) / C(\text{DNA})$, and also indicate external binding of the surfactant (to the phosphates of nucleic acid).

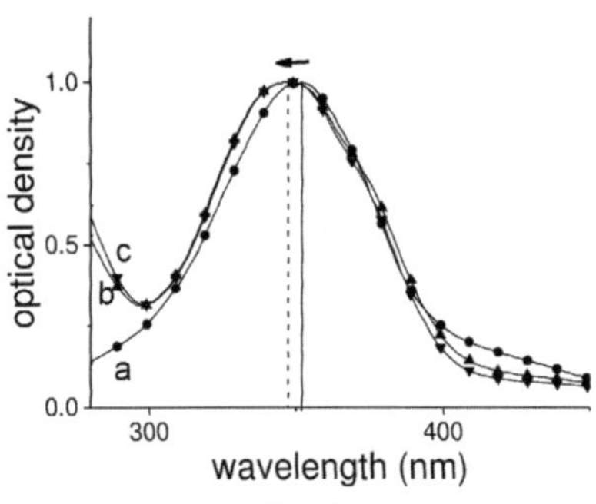

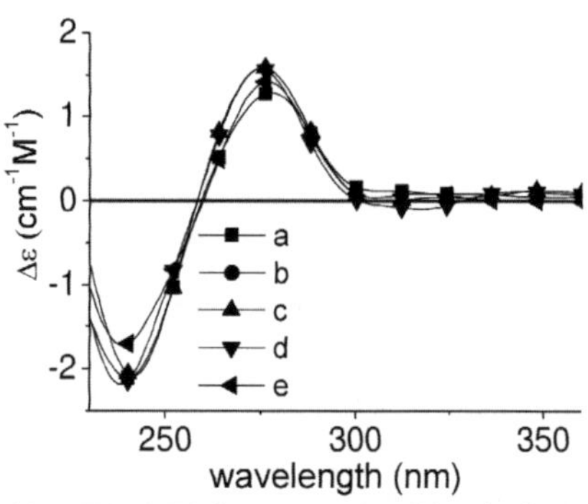

Figure 3a. Normalised spectra: a – trans- surfactant (AzoTMAB) in 5 mM NaCl solution, C < CMC; b – trans- surfactant + native DNA, C(DNA) = 125 μM; c – trans- surfactant + denatured DNA.

Figure 3b. CD spectra for DNA + surfactant (AzoTMAB) in 5 mM NaCl, C(DNA) = 242 μM, $z = C(\text{surf.}) / C(\text{DNA})$: a – DNA without surfactant ($z = 0$), b – 0.1 DNA + cis- surf.; c – 0.3 DNA + trans-surf.; d – 0.3 DNA + cis- surf.; e – 0.4 DNA + cis- surf.

3.2. DNA-surfactant interaction

We estimated the change in reduced viscosity $(\eta_r - 1)/C$ (DNA), where η_r – relative viscosity (solution viscosity / solvent viscosity), of the DNA + surfactant solutions for increasing concentration of the surfactant (ratio z = C(surf.) / C(DNA is increasing, C(DNA) remains constant). The measurement of flow birefringence provides us with the anisotropy of optical polarizability of DNA coil in the solution $(\gamma_1 - \gamma_2)$, where 1 and 2 are two molecular axis. From the values of reduced viscosity and $(\gamma_1 - \gamma_2)$ of the molecule, we calculated optical anisotropy of the DNA molecule statistical segment $(\alpha_1 - \alpha_2)$ (detailed information in [12]). We observed the decrease in relative $(\alpha_1 - \alpha_2)$ with the growing C(surfactant) in the DNA solution in low salt concentration 5 mM NaCl (figure 4a). Importantly, it was followed by the decreasing viscosity (figure 4b). The results allowed us to demonstrate and estimate the conformational changes of DNA upon the interaction with the surfactant (trans- and cis- isomers). We also tested the DNA + surfactant solutions in high salt concentration 1 M NaCl. These systems do not demonstrate similar decrease in viscosity and optical anisotropy. However, these novel results for the complex two-component systems require further investigation both for lower and higher salt concentrations.

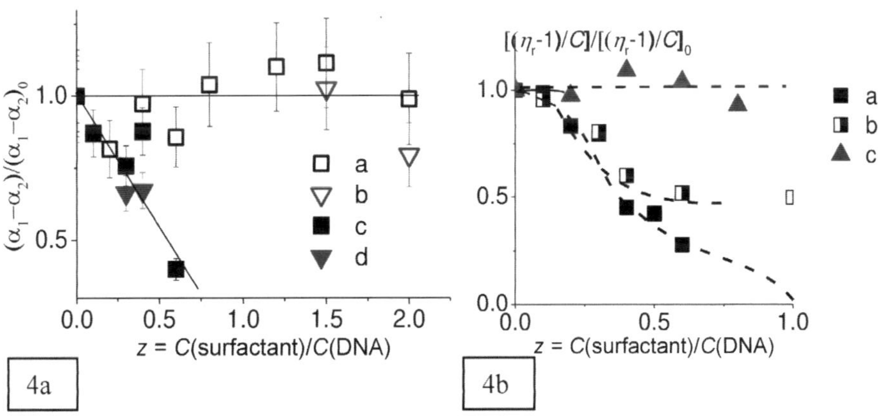

Figure 4a. Changes in relative segmental optical anisotropy ($(\alpha_1 - \alpha_2)_0$ is for DNA solution) with the increase of *z* value: a – in 1 M NaCl for trans- surfactant; b – in 1 M for cis- surfactant; c – in 5 mM for trans-; d –in 5 mM for cis-.

Figure 4b Changes in relative reduced viscosity: a – in 5 mM for trans- surfactant; b – in 5 mM for cis-; c – in 1 M for trans-.

3.3. Delivery studies

For the transfection test we prepared siRNA + surfactant (AzoTEAB) solutions. Starting this experiment, we tested solutions in 0.15 M NaCl and in 5 mM NaCl. However, for higher salt concentration 0.15 M NaCl we observed no delivery in studied conditions. This indicates that due to neutralizing of nucleic acid's phosphates by Na^+ ions higher salt concentration (higher ionic strength) is less favorable for the binding and, consequently, for the formation of nanocarriers.These systems were containing two concentrations of siRNA 25 μM and 10 μM, and the concentration of the surfactant (AzoTEAB) in its trans- form was varied (*z* = *C*(surf.) / *C*(DNA) = 0.8, 3.25, 5, 12).

After incubating the similar volume (V = 0.175 ml) of samples with cells in plates (for samples with 25 μM of siRNA we had 1500 ng of siRNA in each well, and for samples with 10 μM of siRNA we had 600 ng in each well), they were washed with PBS, and the delivery was observed (figure 5, 6). Since the final quantities of siRNA per each well in fig. 5a and 5b were different, the intensity of fluorescence cannot be compared, but the result shows that for both concentrations of nucleic acid

Journal of Physics: Conference Series **769** (2016) 012023 doi:10.1088/1742-6596/769/1/012023

complexes are formed, and the delivery occurs. An increasing quantity of delivered siRNA is demonstrated for the growing surfactant concentration (growing z), together with the increase in toxicity. Importantly, we observed no delivery for z = C(surf.) / C(DNA) = 0.8, that confirms the absence of formed nanocarriers for the smaller ratios.

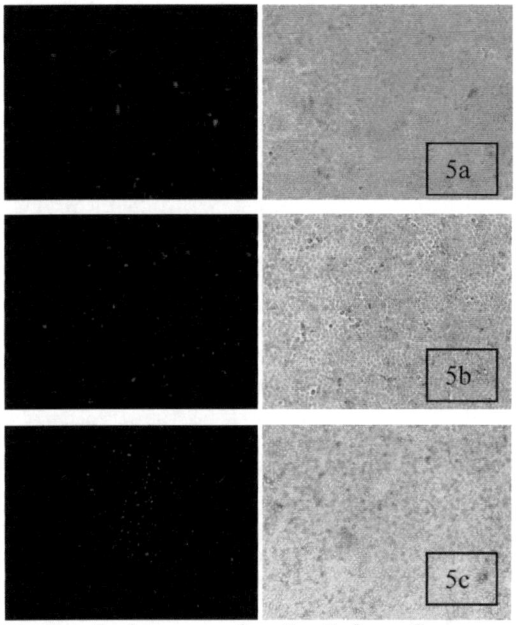

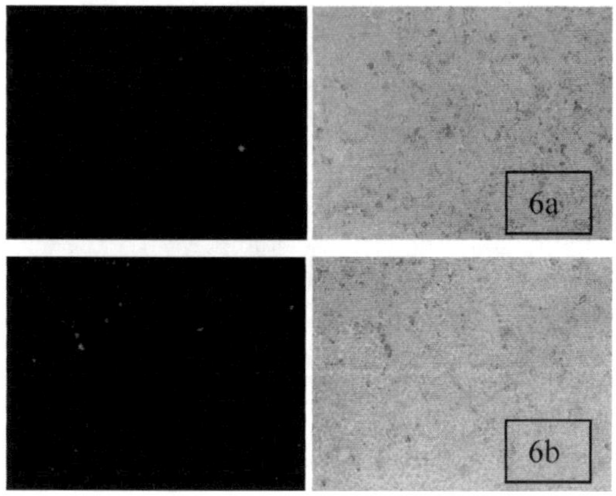

Figure 5. Images of cells (left – fluorescence, right – visible light) after treatment with fluorescently labeled siRNA (C(siRNA) = 25 μM) + surfactant: a – z = C(surf.) / C(NA) = 3; b – z = 5; c – z = 12

Figure 6. The same for C(siRNA) = 10 μM: a – z = C(surf.) / C(NA) = 5; b – z = 6.5.

The formation of nanostructures (figure 7) is also indicated by AFM images obtained for siRNA + surfactant in 5 mM NaCl, C(siRNA) = 1 mM with the addition of $MgCl_2$ (C($MgCl_2$) = 0.5 mM) on mica surface. However, possible interaction between cationic surfactant and mica (see figure 7b) does not allow regarding the observed structures as fully equivalent to those formed in solution.

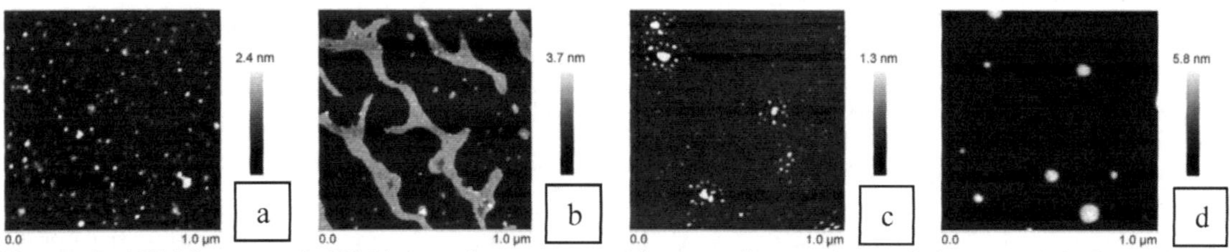

Figure 7. AFM images of siRNA + surfactant: a – without surfactant; b – z=0.6; c – z=1.6; d – z=8

4. Conclusion

We examined binding of cationic agent (azobenzene containing surfactant AzoTAB) to the molecules of nucleic acids siRNA and DNA using UV-Vis spectroscopy. For DNA macromolecules of high molecular weight we also carried out a combined viscometry/flow birefringence study to examine DNA conformational changes upon binding with trans- and cis- isomers of the surfactant. For siRNA + surfactant systems we aimed to determine the conditions of gene vectors' creation by AFM observation and by testing the delivery on MDCK cell line.

References

[1] Gosule L C and Shellman J A 1976 *Nature* **259** 333–35
[2] Wilson R W and Bloomfield V A 1979 *Biochemistry* **18** 2192–96.
[3] Ma C and Bloomfield V A 1994 *Biophys J.* **67** 1678
[4] Kasyanenko N A, Mukhin D A and Perevyazko I Yu 2010 *Polymer Science Ser. C* **52(1)** 122
[5] Kasyanenko N A, Lysyakova L A, Dribinskii B A, Zolotova Yu I, Nazarova O V, Panarin E F 2012 *Polymer Science Ser. C* 54(1) 57
[6] Zinchenko A, Pyshkina O., Lezov A Sergeyev V and Yoshikawa K 2008 In: *DNA interactions with polymers and surfactants* (Hoboken, NJ: John Wiley & Sons, Inc.)
[7] Fire A, Xu S, Montgomery M K, Kostas S A, Driver S E and Mello C C 1998 *Nature* **391(6669)** 806–11
[8] Rothdiener M, Muller D, Castro P G, Scholz A, Schwemmlein M, Fey G, et al. 2010 *J. Control. Release* **144** 251-58.
[9] Khatri N, Baradia D, Vhora I, Rathi M and Misra A 2014 *AAPS PharmSciTech* **15 (6)** 1630–43
[10] Rodrigues L R 2015 *Journal of Colloid and Interface Science* **449** 304-16
[11] De Backer L, Braeckmans K, Stuart M C, Demeester J, De Smedt S C and Raemdonck 2015 *J. Control. Release*. 2015 **206** 177-86
[12] Kasyanenko N, Lysyakova L, Ramazanov R, Nesterenko A, Yaroshevich I, Titov E, Alexeev G, Lezov A and Unksov I 2015 *Biopolymers* **103** 109–22
[13] Zakrevsky Y, Kopyshev A, Lomadze N, Morozova E, Lysyakova L; Kasyanenko N and Santer S 2011 *Phys Rev E* **84** 021909
[14] Diguet A, Mani N K, Geoffroy M, Sollogoub M and Baigl D 2010 *Chem. Eur. J.* **16** 11890–96
[15] Le Ny A L M and Lee C T 2009 *Biophys Chem* **142** 76–83
[16] Zakrevsky Y, Roxlau J, Brezesinski G, Lomadze N and Santer S 2014 *J. Chem. Phys.* **140** 044906
[17] Frisman E V, Schagina L V, Vorobiev V I 1965 *Biorheology* **2** 189–94

Journal of Physics: Conference Series **769** (2016) 012024 doi:10.1088/1742-6596/769/1/012024

Modelling of hydrogen thermal desorption spectrum in nonlinear dynamical boundary-value problem

E K Kostikova and Yu V Zaika

Institute of Applied Mathematical Research, Karelian Research Centre, Russian Academy of Science, 11 Pushkinskaya street, Petrozavodsk, 185910, Russia

E-mail: zaika@krc.karelia.ru

Abstract. One of the technological challenges for hydrogen materials science (including the ITER project) is the currently active search for structural materials with various potential applications that will have predetermined limits of hydrogen permeability. One of the experimental methods is thermal desorption spectrometry (TDS). A hydrogen-saturated sample is degassed under vacuum and monotone heating. The desorption flux is measured by mass spectrometer to determine the character of interactions of hydrogen isotopes with the solid. We are interested in such transfer parameters as the coefficients of diffusion, dissolution, desorption. The paper presents a distributed boundary-value problem of thermal desorption and a numerical method for TDS spectrum simulation, where only integration of a nonlinear system of low order (compared with, e.g., the method of lines) ordinary differential equations (ODE) is required. This work is supported by the Russian Foundation for Basic Research (project 15-01-00744).

1. Introduction

Interaction of hydrogen and its isotopes with solids has many applications [1–7]. It is sufficient to mention power engineering, protection of metals from hydrogen corrosion, design of chemical reactors, rocket and missile engineering. Hydrides help to retain substantial amounts of hydrogen: environmental-friendly energy source. Hence high expectations are attached to hydrogen batteries and motors that avoid using high pressures or low temperatures. Some special topics are considered in [8, 9]. Enthusiasts speak not only of hydrogen energy but also of hydrogen economy [5]. Transfer parameters depend on the process characteristics of producing the material batch, and one needs effective algorithms for processing measured curve instead of focusing on 'tabular data'. We focus on the thermal desorption method and take into account only the limiting factors and the information capacities of the TDS experiment explicated below.

2. Mathematical model of TDS-experiment

We consider hydrogen transfer through a test metal sample (plate of thickness ℓ). For brevity we speak of a metal plate, although it may be a multialloy. Suppose heating is relatively slow and uniform so that the diffusion flow can be assumed to be proportional to the concentration gradient. The material is sufficiently homogeneous to neglect interaction of H with traps (micro-defects of the crystal structure). Let us assume a standard model for diffusion:

$$c_t(t,x) = D(T)c_{xx}(t,x), \quad (t,x) \in Q_{t_*}, \qquad (1)$$

where t is time, $Q_{t_*} = (0, t_*) \times (0, \ell)$; $c(t, x)$ is concentration of dissolved atomic hydrogen. Diffusion coefficient D is an Arrhenius function of temperature $T(t)$: $D = D_0 \exp\{-E_D/[RT(t)]\}$, D_0 and E_D are the preexponential factor and activation energy.

More detailed transfer models are known. For example, different diffusion channels can be considered: transcrystalline, the grain boundary diffusion, along the defects, with interchange between them. However, a significant increase of number of unknown a priori coefficients leads to the complex problem of parameter estimation. TDS-experiment provides limited amount of information. Transfer by different diffusion channels is impossible to distinguish; therefore we use integral parameters (effective coefficients). The main difficulty of computations is due to dynamic boundary conditions (not the diffusion equation). Suppose the plate contacts with gas H_2 and external surface is potential barrier (see [2, pp. 177–206]). Taking into account (de)sorption processes, we obtain the following boundary conditions:

$$c(0, x) = \bar{c}(x), \quad x \in [0, \ell], \ t \in [0, t_*], \tag{2}$$

$$c_0(t) = g(T)q_0(t), \quad c_\ell(t) = g(T)q_\ell(t), \tag{3}$$

$$\dot{q}_0(t) = \mu s(T)p_0(t) - b(T)q_0^2(t) + Dc_x(t, 0), \tag{4}$$

$$\dot{q}_\ell(t) = \mu s(T)p_\ell(t) - b(T)q_\ell^2(t) - Dc_x(t, \ell), \tag{5}$$

$b(T) = b_0 \exp\{-E_b/[RT]\}$, $s(T) = s_0 \exp\{-E_s/[RT]\}$. Here $c_0(t) \equiv c(t, 0)$, $c_\ell(t) \equiv c(t, \ell)$ are boundary bulk concentrations of diffusing atomic hydrogen; $q_0(t)$, $q_\ell(t)$ are concentrations on the surfaces ($x = 0, \ell$); $g(T)$ is the coefficient of local equilibrium between concentrations on the surface and near-surface bulk; μ is the kinetic coefficient; $s(T)$ is the parameter indicating that only a small part of incident hydrogen find its way on the surface (for brevity, we call s the adhesion coefficient and keep in mind that this coefficient describes the result of concurrent physisorption, dissociation, and chemisorption of molecular gas on the surface); $p_0(t)$, $p_\ell(t)$ are pressures of molecular gas. Although hydrogen can be both in atomic and molecular forms, we use atoms as the unit of amount: $[c] = 1/\text{cm}^3$, $[q] = 1/\text{cm}^2$, $[Dc_x] = [J] = 1/\text{cm}^2\text{s}$ $(J = bq^2)$. According to the kinetic gas theory, μp is the number of H_2 particles colliding with unit area per unit time; then μsp is the flux density of atoms precipitating on the surface. This quantity is a general parameter, applicable without distinguishing the processes. The quick dissolution model (3) is obtained from the more universal formulas

$$k^-(T)q_0(t) - k^+(T)c_0(t) = -D(T)c_x(t, 0), \quad k^-(T)q_\ell(t) - k^+(T)c_\ell(t) = D(T)c_x(t, \ell).$$

Coefficients k^-, k^+ describe the intensity of the dissolution processes in the bulk and transfer to surface. If these processes are significantly faster than diffusion (for considered temperature range) then it is assumed that $Dc_x \approx 0$. Hence we obtain (3), where $g = k^-/k^+$. If the surface is isotropic (in terms of $E_{k^-} \approx E_{k^+}$) then the coefficient $g(T)$ has weak temperature dependence. We can formally use the Arrhenius equation $g(T) = g_0 \exp\{-E_g/[RT\}]$, $E_g = E_{k^-} - E_{k^+}$. Here activation energy E_g is not necessarily positive. In addition, we can take into account degree of the surface covering and near-surface bulk saturation:

$$k^-(T)\big[1 - c_{0,\ell}(t)c_{\max}^{-1}\big]q_{0,\ell}(t) - k^+(T)\big[1 - q_{0,\ell}(t)q_{\max}^{-1}\big]c_{0,\ell}(t) = \mp D(T)c_x\big|_{0,\ell}.$$

The value $\theta(t) = q(t)/q_{\max}$ means the degree of the surface covering. Flux density of atom H adsorption (dissociative hydrogen chemisorption on the surface) can be simulated by formula $\mu s(T)p(t)[1 - \theta(t)]^2$ in the balance equations (4), (5). But we have $\theta \ll 1$ for low level of hydrogen. It's in good agreement with quadratic desorption, linear diffusion equation, $D \neq D(c)$. In the context of current TDS-experiment it is convenient to take $[p] =$ Torr and so $\mu(T) = (2\pi mkT)^{-1/2} \approx 2.484 \cdot 10^{22}/\sqrt{T}$. Here we have $[\mu] = 1_{H_2}/(\mathrm{Torr\,cm^2 s})$, $[T] = \mathrm{K}$, k is the Boltzmann constant, m is mass of a hydrogen molecule. Variation of the kinetic constant with temperature ($\mu \propto 1/\sqrt{T}$) is usually neglected against exponential relation for $s(T)$.

Saturation means that pressure $\bar{p}$ of H_2 on both sides of the plate is constant at $T = \mathrm{const}$. Then (after a time) saturation concentration ($\bar{c}$) of dissolved atomic diffusivity hydrogen is established. Assuming zero derivatives in (3)—(5), we obtain $\bar{c} = \Gamma\sqrt{\bar{p}}$, $\Gamma \equiv g\sqrt{\mu s/b}$ ($\bar{c} = g\bar{q}$). Consequently the model is according to the Sieverts law adequacy range ($\bar{c} \propto \sqrt{\bar{p}}$).

Let us clarify the TDS-experiment. Plate of the material is placed into a chamber. Hydrogen gas is pumped into the chamber under relatively high pressure. The sample is heated to hasten sorption until the sample absorbs enough hydrogen (equilibrium gas saturation). Then heating current is shut off and the material is cooled. In such a case, the physico-chemical processes slow down sharply and significant amount of hydrogen remains inside the sample. The sample is monotonically and rather slowly reheated in vacuum. The heating rate is low and constant. Molecular hydrogen pressure in the chamber is measured by a mass-spectrometer. The desorption flux from the surface is defined by pressure. The desorption flux density is denoted by $J(t) = b(t)q^2(t)$ (usually $J(t) \propto p(t)$). From now on the contracted notation will be used for simplification: $b(t) \equiv b(T(t))$, $D(t) \equiv D(T(t))$, $s(t) \equiv s(T(t))$. Parameter g of rapid dissolution (local surface-bulk equilibrium) is taken to be constant ($E_{k^-} = E_{k^+}$) near the TDS-peak. Symmetry conditions are fulfilled for the TDS method:

$$p(t) = p_{0,\ell}(t),\ q(t) = q_{0,\ell}(t),\ c_0(t) = c_\ell(t),\ D(t)c_x(t,0) = -D(t)c_x(t,\ell),\ \bar{c}(x) = \bar{c} = \mathrm{const}.$$

The experiment end time t_* is defined by the degassing condition: $p \approx 0,\ c(t_*, x) \approx 0$.

3. Integrodifferential equation

Suppose the TDS-degassing model: $r(t) \equiv \mu s(t)p(t)$, $J(t) \equiv b(t)q^2(t)$,

$$c_t = D(t)c_{xx}\,,\ c(0,x) = \bar{c},\ c_{0,\ell}(t) = gq(t),\ \dot{q}(t) = r(t) - b(t)q^2(t) + D(t)c_x(t,0).$$

Vacuum system is assumed to be powerful enough. Hence resorption is neglected ($r(t) = 0$) for degassing. Values of $J(t)$ are input data for the inverse problem of parametric identification.

In the paper we only present numerical simulation of TDS-spectrum $J = J(T)$. It is needed as the iterative component of the identification algorithm. Heating is usually linear ($T(t) = T_0 + vt$). Heating rate v is less than $1\,\mathrm{K/s}$. Heating is stopped (although degassing is continued) if upper temperature limit $T(t) = T_{\max}$ is reached.

Let us formulate the mathematical problem. Difference schemes for the boundary-value problem (with different types of defects) solution was presented in [10]. But a curve of TDS-spectrum is determined by surface concentration ($J = bq^2$). Hence, we try to avoid the iterative solution of the boundary-value problem for the current approximations of the model parameters D_0, E_D, b_0, E_b, s_0, E_s, g. For this aim we perform the problem transformation. Finally, the modelling requires only integration of a nonlinear system of low order ordinary differential equations (ODE).

We replace time by $t' = \int_0^t D ds$ (keep previous notation t):

$$c_t(t,x) = c_{xx}(t,x),\ c(0,x) = \bar{c},\ c_{0,\ell} = gq(t), \tag{6}$$

$$c_x|_0 = -c_x|_\ell = \dot{q}(t) + [J(t) - r(t)]D^{-1}(t). \tag{7}$$

Let $q(t)$ be functional parameter, formula (7) is additional relation to the linear problem (6). Let us change variables so that boundary conditions in (6) become homogeneous:

$$\hat{c} = c(t,x) - gq(t),\ \hat{c}_t(t,x) = \hat{c}_{xx}(t,x) + f(t),\ f(t) = -g\dot{q}(t),\ \hat{c}(0,x) = \hat{\varphi}(x) = 0,\ \hat{c}|_{0,\ell} = 0.$$

Write the solution using function of instantaneous source (Green's function):

$$\hat{c}(t,x) = \int_0^t \int_0^\ell G_1(x,\xi,t-\tau) f(\tau)\, d\xi\, d\tau,\ G_1(x,\xi,t) = \frac{2}{\ell}\sum_{n=1}^{\infty} \exp\Big\{-\frac{n^2\pi^2}{\ell^2}t\Big\} \sin\frac{n\pi x}{\ell} \sin\frac{n\pi\xi}{\ell}.$$

Dynamical boundary conditions contain $\hat{c}_x(t,0)$:

$$\hat{c}_x|_0 = -\frac{4g}{\ell}\int_0^t \dot{q}(\tau)\sum{}' \exp\Big\{\frac{n^2\pi^2}{\ell^2}(\tau - t)\Big\}\, d\tau, \quad \sum{}' \equiv \sum\nolimits_{n=1,3,5...}.$$

The series is divergent for $\tau = t$ so that term-by-term integration is meant. For original time t we obtain $c_x(t,0) = \hat{c}_x(t,0) = c_x(t,\ell) = \hat{c}_x(t,\ell)$,

$$\dot{q}(t) = r(t) - b(t)q^2(t) - \frac{4gD}{\ell}\sum{}'\int_0^t \dot{q}(\tau)\exp\Big\{-\frac{n^2\pi^2}{\ell^2}\int_\tau^t D(s)\, ds\Big\}\, d\tau. \qquad (8)$$

The equation and the original boundary-value problem are equivalent. This means that the solution $q(t)$ uniquely determines the solution $c(t,x)$. An analogy to functional-differential equations of neutral type is due to impossibility to eliminate derivative $\dot{q}$ in the right side of the equation due to divergent series. We are interested in the time interval $[t_1,t_2] \subset (0,t_*)$ near activity peak of thermal desorption (measurements for t near to $0, t_*$ are uninformative). Resorption is neglected ($r(t) = 0$) for powerful vacuum system, thus the equation is simplified.

Dimensionless parameters are more convenient for numerical modelling. We use replaces: $t' = \int_0^t D(s)ds/\ell^2$, $x' = x/\ell$, $v = q/\bar{q}$ ($\bar{c} = g\bar{q}$). But notation of t is not changed. We obtain:

$$\dot{v}(t) = -\tilde{b}(t)v^2(t) - 4g\ell\sum{}'\int_0^t \dot{v}(\tau)\exp\{-n^2\pi^2[t-\tau]\}d\tau, \quad \dot{v} \equiv \frac{dv}{dt},\ \tilde{b}(t) \equiv \frac{b(t)\ell^2\bar{q}}{D(t)}. \qquad (9)$$

Here $v(0) = 1$ is the initial condition. We only use first k terms of the series. Denote

$$z_i(t) = \int_0^t \dot{v}(\tau)\exp\{-(2i-1)^2\pi^2[t-\tau]\}\, d\tau.$$

Differentiate $z_i(t)$ with respect to t and use $\dot{v}(t)$ from formula (9) ($\max n = 2k-1$):

$$\dot{z}_i = -\tilde{b}v^2 - \big[(2i-1)^2\pi^2 + 4g\ell\big]z_i - 4g\ell(z_1 + ... + z_{i-1} + z_{i+1} + ... + z_k).$$

The result is the following system of ordinary differential equations:

$$\frac{d}{dt}\begin{pmatrix} v \\ z_1 \\ ... \\ z_k \end{pmatrix} = -\tilde{b}(t)v^2(t)\begin{pmatrix} 1 \\ 1 \\ ... \\ 1 \end{pmatrix} - 4g\ell\begin{pmatrix} 1 & 1 & ... & 1 \\ \alpha_1 & 1 & ... & 1 \\ ... & ... & ... & ... \\ 1 & 1 & ... & \alpha_k \end{pmatrix}\begin{pmatrix} z_1 \\ z_2 \\ ... \\ z_k \end{pmatrix},$$

$\alpha_i \equiv 1 + (2i-1)^2\pi^2/(4g\ell)$, $v(0) = 1$, $z_i(0) = 0$ (matrix size is $(k+1)\times k$). The system is rapidly integrated using classic Runge-Kutta fourth-order method. We make calculations using Scilab 5.5. The system order is not large (compared with, e. g., the method of lines) due to measurement error is usually not less than 10 % and the exponent function contains n^2.

Initial saturation by hydrogen and coefficient $\tilde{b}(t)$ of quadratic nonlinearity are focused. Initial saturation is conducted at high temperature $T = \overline{T} = \text{const}$ and pressure $\bar{p} = \text{const}$ (to stimulate sorption). After establishing of equilibrium concentration we obtain

$$\mu(\overline{T})s(\overline{T})\bar{p} = b(\overline{T})\bar{q},\ \ \bar{c} = g\bar{q}\ (g = \text{const})\ \Rightarrow \bar{c} = gb_0^{-1/2}\sqrt{\mu s\bar{p}}\,|_{\overline{T}}\exp\{E_b[2R\overline{T}]^{-1}\},$$

$$\tilde{b}(t) = \bar{q}\,b(t)\ell^2D^{-1} = \bar{c}\,b(t)\ell^2[gD(t)]^{-1} = \tilde{b}_0\exp\{-E_{\tilde{b}}[RT(t)]^{-1}\},$$

$$E_{\tilde{b}} \equiv E_b - E_D,\ \tilde{b}_0 \equiv \ell^2D_0^{-1}\sqrt{b_0\mu s\bar{p}}\,|_{\overline{T}}\exp\{E_b[2R\overline{T}]^{-1}\}.$$

Formally let $\tilde{b}(t) \equiv \tilde{b}(T(t))$ be Arrhenius parameter, here $E_b - E_D$ is activation energy. Usually $E_b > E_D$ and surface processes are faster speeded up on heating. ($\dot{b} = \ldots E_b \dot{T}$, $\dot{D} = \ldots E_D \dot{T}$), thus, $E_{\tilde{b}} > 0$. The pre-exponential factor $\tilde{b}_0$ depends on all parameters D_0, b_0, E_b, s_0, E_s (except E_D). Hence the coefficient $\tilde{b}$ changed monotonically and limited $0 < \tilde{b}^- \leq \tilde{b}(t) \leq \tilde{b}^+$.

The first approximation represents qualitative situation (factor n^2 in exponent for the second term of the series is an order of magnitude greater the similar factor in the first term):

$$\dot{v}(t) = -\tilde{b}(t)v^2(t) - \varkappa \int_0^t \dot{v}(\tau) \exp\{-\pi^2[t-\tau]\} d\tau, \ \varkappa \equiv 4g\ell,$$

$$\begin{cases} \dot{v}(t) = -\tilde{b}(t)v^2(t) - \varkappa z(t), \ z \equiv z_1, \\ \dot{z}(t) = -\tilde{b}(t)v^2(t) - (\varkappa + \pi^2)z(t), \quad v(0) = 1, \ z_1(0) = 0. \end{cases} \tag{10}$$

The qualitative behaviour of the motion on phase plane (v, z) is presented in Fig. 1. Dash-and-dot lines show parabolas $\dot{v}, \dot{z} = 0$: $z = -\tilde{b}(t)\varkappa^{-1}v^2$, $-\tilde{b}(t)[\varkappa + \pi^2]^{-1}v^2$.

Field of phase rate of the system (10) is nonstationary. But its change is monotonous and limited due to $\tilde{b}(t) \in [\tilde{b}^-, \tilde{b}^+]$. Parabolas are moved left (for $E_b > E_D$) or right (for $E_b < E_D$). The second approximation of equation (9) is considered in the same way:

$$\begin{cases} \dot{v}(t) = -\tilde{b}(t)v^2 - \varkappa z_1(t) - \varkappa z_2(t), \\ \dot{z}_1(t) = -\tilde{b}(t)v^2 - [\varkappa + \pi^2] z_1 - \varkappa z_2, \\ \dot{z}_2(t) = -\tilde{b}(t)v^2 - \varkappa z_1 - [\varkappa + 9\pi^2] z_2. \end{cases} \tag{11}$$

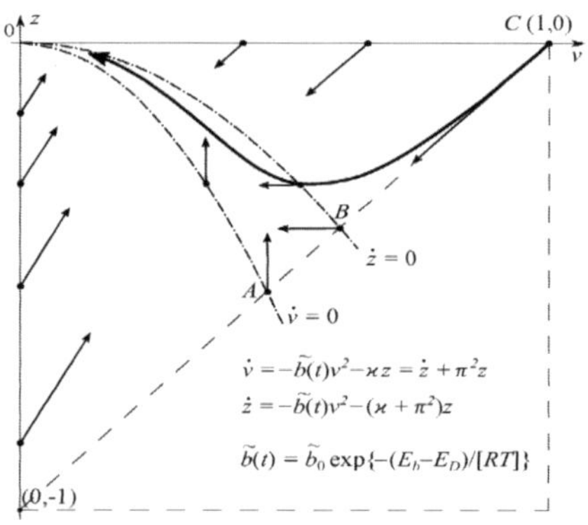

Figure 1. Motion on phase plane

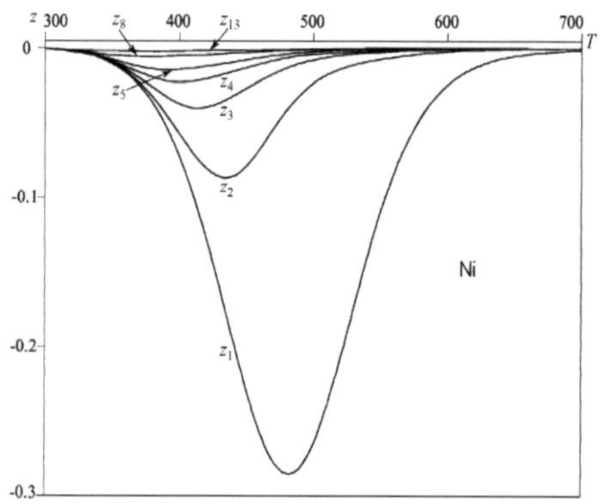

Figure 2. Decrease of $|z_i(t)|$

4. Computer simulation results

Data for nickel [4] are used: $D = 7.5 \times 10^{-3} \exp\{-40\,000/RT\}$ [cm^2/s]; $g = 100$ [1/s]; $b = 1.53 \times 10^{-14} \exp\{-43\,200/RT\}$ [cm^2/s]; $s = 1.8 \times 10^{-2} \exp\{-61\,400/RT\}$; $\bar{p} = 37.4$ [Torr]; $\ell = 0.02$ [cm]; $\overline{T} = 770$ [K], $T(0) = 295$ [K]; $\dot{T} = 0.5$ [K/s]. Variation of parameters provides an opportunity to estimate the degree of sensitiveness ('derivatives') of TDS-spectrum to a selected limiting factors (Fig. 3–Fig. 6). We solved the system of 10–15 equations. In Fig. 6 we notice rise of an extra peak structure due to interplay between diffusion and desorption activation energies. Warning, this effect is usually interpreted by traps with different binding energies.

Conclusion

The paper presents a distributed boundary-value problem of thermal desorption with nonlinear dynamical boundary conditions for modelling the degassing TDS-spectrum of a hydrogen-presaturated structural material. The variation of parameters allows estimating the TDS-spectrum's sensitivity to changes in the material service conditions and its composition.

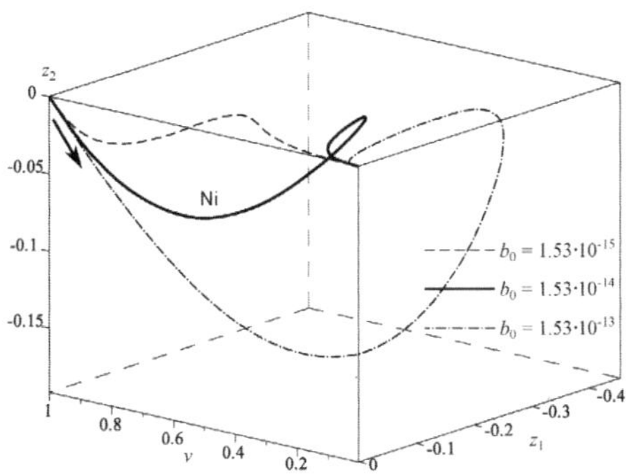

Figure 3. System (11), effect of b_0

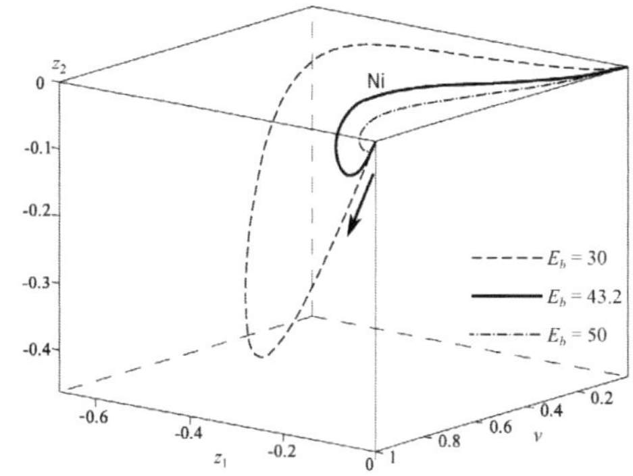

Figure 4. System (11), effect of E_b

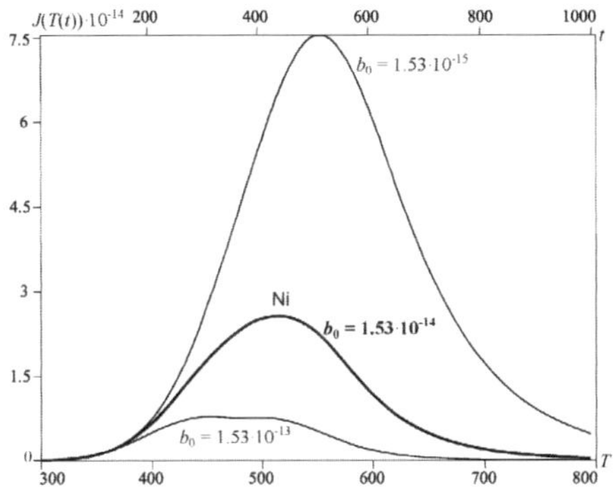

Figure 5. TDS-spectrum, effect of b_0

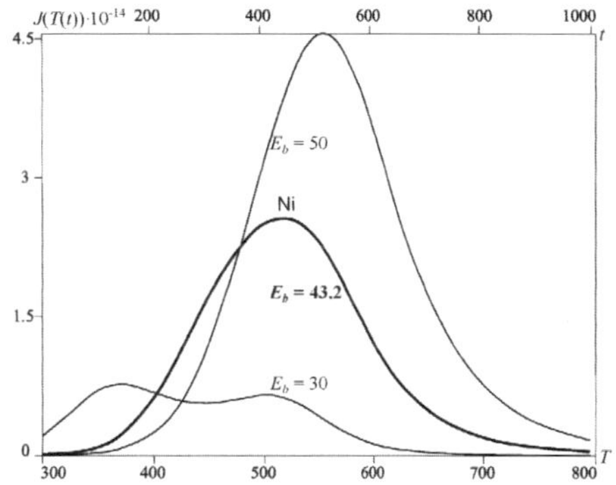

Figure 6. TDS-spectrum, effect of E_b

The algorithm only requires integration of a nonlinear system of low order ordinary differential equations (compared, e.g., with the method of lines). The results of computer simulations using experimental data for nickel are presented. Researchers and experts in hydrogen interaction with solids are thus enabled to directly estimate transfer parameters using current experimental data, and no special software development is required for solving of boundary-value problem.

The main physical sense of the presented results is the following. Unexpectedly at first sight, flux decreases as the desorption coefficient increases. Taking into account initial saturation we conclude that the increase of b prevents saturation to a greater extent than it enhances degassing. We notice rise of an extra peak structure due to interplay between diffusion and desorption activation energies (but this effect is usually interpreted by traps with different binding energies).

References

[1] Alefeld G and Völkl J Eds. 1978 *Hydrogen in metals* (Berlin: Springer–Verlag)
[2] Zakharov A P Ed. 1987 *Interactions of hydrogen with metals* (Moscow: Nauka)
[3] Pisarev A A *et al* 2008 *Hydrogen permeability through metals* (Moscow: Nauka)
[4] Yukhimchuk A A Ed. 2009 *Hydrogen isotopes. Basic and applied research* (Sarov: RFNC VNIIEF)
[5] Ball M and Wietschel M Eds. 2009 *The hydrogen economy* (Cambridge Univ. Press)
[6] Hirscher M Ed. 2010 *Handbook of hydrogen storage: new materials for future energy storage* (Wiley–VCH)
[7] Lototskyy M V Yartys V A, Pollet B G and Bowman R C Jr. *et al* 2014 Metal hydride hydrogen compressors: a review *Int. J Hydrogen Energy* **39** pp 5818–5851
[8] Evard E A, Gabis I E and Yartys V A 2010 Kinetics of hydrogen evolution from MgH2: experimental studies, mechanism and modelling *Int. J. Hydrogen Energy* **35** pp 9060–9069
[9] Zaika Yu V and Rodchenkova N I 2013 Hydrogen-solid boundary-value problems with dynamical conditions on surface *Math. Modelling* (New York: Nova Sci. Publishers) pp 269–302
[10] Zaika Yu V, Kostikova E K 2014 Computer simulation of hydrogen thermodesorption *Advances in Materials Sci. and Applications* **3** 3 pp 120–129

Inverse problem of laser correlation spectroscopy for analysis of polydisperse solutions of nanoparticles

E Nepomnyashchaya[1], E Velichko[1,2], E Aksenov[1]

[1]Saint Petersburg State Polytechnic University, Saint Petersburg, Russia
[2]Saint Petersburg National Research University of Information Technologies, Mechanics and Optics, St. Petersburg, Russia

E-mail: elina.nep@gmail.com

Abstract. A new algorithm for the solution of the inverse problem of laser correlation spectroscopy is suggested. The algorithm allows one to analyse sizes of nanoparticles in polydisperse solutions. Experimental results demonstrating the efficiency of our approach are presented.

1. Introduction

Laser correlation spectroscopy (LCS) is one of few methods that allow a highly precise determination of nanoparticle sizes in solutions, which is important for the chemical and food industry and for medicine [1]. Other advantages of laser correlation spectrometers are their small sizes, a high speed, and low requirements to the conditions of measurements [2].

However, the commercially available laser correlation spectrometers are designed to analyze particle sizes only is one-component solutions and do not allow one to analyze solutions containing particles of several sizes. Determination of particle sizes in polydisperse solutions is a challenging task for the researchers. The complexity of identifying and accurately determining several particle sizes in polydisperse solutions is caused by the necessity to solve the inverse incorrect problem of LCS, and therefore the development of new algorithms for analyzing polydisperse solutions is highly important.

The goal of our study was to develop an algorithm for solving the inverse problem of laser correlation spectroscopy aimed at determining particle sizes in polydisperse solutions.

2. Experimental

A typical scheme of a laser correlation spectrometer used in our investigations is presented in figure 1 [3–5].

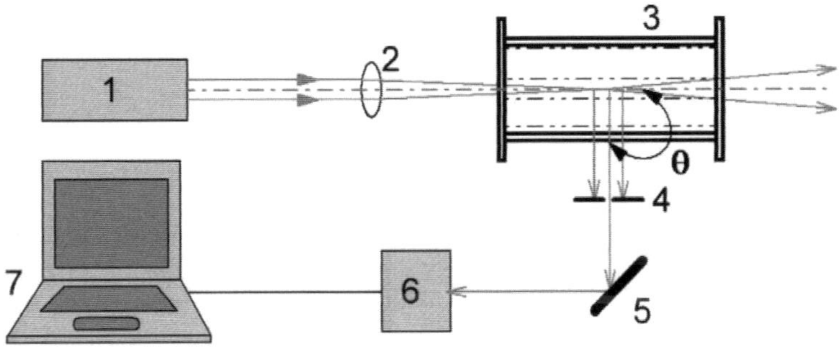

Figure 1. Typical scheme of laser correlation spectrometer. 1 — laser radiation source; 2 — focusing lens; 3 — cell with the solution; 4 — diaphragm; 5 — mirror; 6 — photoelectron multiplier; 7 — computer

The light from the laser scatters from particles of the solution of interest (3). The scattered light which contains information on the velocity of particles is registered by a photomultiplier (6). The signal from the photomultiplier is transmitted to a computer (7) and is analyzed by using special software.
Then the correlation analysis of the signal is carried out. The autocorrelation function can be approximated by [6]

$$|g^{(1)}(\tau)| = \int_0^{\infty} F(\Gamma)e^{-\Gamma\tau}\,d\Gamma, \tag{1}$$

where Γ is the diffusion broadening, and F(Γ) is the contribution of the radiation component scattered by particles of one size to the total intensity. The diffusion broadening is related to the diffusion coefficient D by

$$\Gamma = Dq^2, \tag{2}$$

where $q = (4\pi n/\lambda)\sin(\theta/2)$ is the scattering vector, n is the refractive index of the medium, λ is the wavelength, and θ is the angle of scattering.
By using the Stokes-Einstein formula

$$D = k_b T/6\pi\eta R \tag{3}$$

one can calculate the radius of the particles under study. Here, η is the viscosity of the medium, k_b is the Boltzmann constant, T is the temperature, and R is the particle radius.
Therefore, the solution of Eq. (1) and estimation of Γ are the main tasks in the mathematical processing of a scattered signal. Equation (1) is referred to as the Fredholm integral equation. Because of the presence of random noise in the experimental data, its solution belongs to the class of incorrect problems. This means that even a small error in the initial data will lead to the accumulation of errors in the course of solution, and the result will be incorrect.

3. The algorithm

There are a variety of methods for solving the Fredholm integral equation. The operation of modern spectrometers is typically based on the method of cumulants. This method allows one to estimate particle sizes with a high accuracy only for monodisperse solutions. There are also solutions relying on the Tikhonov regularization method which allow a more detailed analysis, and are more suitable for the determination of particle sizes in polydisperse solutions. So the Tikhonov regularization method was used as a basis for the algorithm design for the solution of the inverse problem of LCS.
In order to solve Eq. (1), it is presented in the form of a set of N points and, hence, we obtain a system of N equations which is written in the matrix form as

$$A\boldsymbol{f} = \boldsymbol{g}. \tag{4}$$

Since in practice only approximate values of the right-hand sides of system of equations (4) are known $\boldsymbol{g} \approx \tilde{\boldsymbol{g}}$, it is necessary to solve the problem of the form

$$||A\boldsymbol{f} - \tilde{\boldsymbol{g}}||^2 + \alpha\Omega(\boldsymbol{f}) \to min. \tag{5}$$

where $\alpha > 0$ is the smoothing parameter, and $\Omega(\boldsymbol{f})$ is the stabilizing functional which is selected for each task separately. The minimization of this type stabilizes the solution of the system, thereby improving its conditionality and increasing the accuracy of solution. The stabilizing functional was chosen as $\Omega(\boldsymbol{f}) = ||\boldsymbol{f}||^2$.

Below we present the implementation of the algorithm for solving the inverse problem of laser correlation spectroscopy based on the Tikhonov regularization method.

The algorithm is as follows:

1. set the initial (sufficiently high) value of α equal to 1% of the maximum diagonal element of matrix A;
2. solve system of equations (5) by the modified Gauss method and find the solution $\boldsymbol{f}$;
3. calculate the convergence $||A\boldsymbol{f} - \tilde{\boldsymbol{g}}||^2$ (reduce α by 90% after the first iteration and return to step 1);
4. compare $||A\boldsymbol{f} - \tilde{\boldsymbol{g}}||_i^{\,2} < 0{,}1 \cdot ||A\boldsymbol{f} - \tilde{\boldsymbol{g}}||_{i-1}^{\,2}$, if the answer is "yes", reduce α by 90% and return to step 1, if the answer is "no", go to step 5;
5. check for the presence of negative components of solution $\boldsymbol{f}$. If $f_i < 0$ are present, take $f_{min} = 0$ and return to step 1. The corresponding component is excluded from further calculations;
6. if $f_i < 0$ are absent, check the number of points at the function peak $n > N$ (set prior to the solution), if the answer is "yes", take $f_{min} = 0$ and return to step 1, if the answer is "no", the calculation is completed.

The last step is a modification suggested in our study. The aim of this modification is to provide determination of not only greatly differing but also slightly differing particle sizes.

Other important features of our algorithm which improve the accuracy of determining particle sizes are

- not only the points with the minimal values but also the points with the weakest influence on the convergence are removed from the signal (in contrast to the conventional method described in [7]);
- the procedure for setting and correcting the regularization parameter α is described.

4. Discussion

The algorithm we suggested was tested on model signals from the particles with specified sizes of 48 nm (40% of particles in the solution), 46 nm (20% of particles in the solution) and 43.5 nm (30% of particles in the solution). A random noise from 1% to 10% was added to the model signals. Figure 2 a shows the result of calculations based on the algorithm suggested.

Journal of Physics: Conference Series **769** (2016) 012025 doi:10.1088/1742-6596/769/1/012025

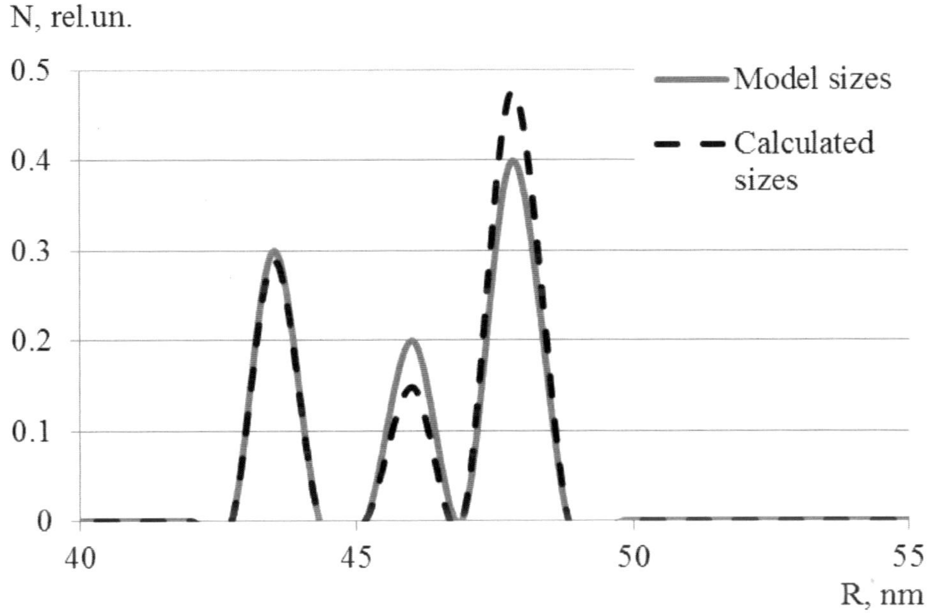

Figure 2. Model sizes and calculated distributions for particle sizes 48 nm (40%), 46 nm (20%) and 43.5 nm (30%) for a noise level of 5%

The solution of the inverse problem based on our algorithm resulted in a highly precise estimation of particle sizes in polydisperse solutions containing up to three components (the error was 6% at a confidence level of 95%). In addition, as one can see from Figure 2, the method we propose has a sufficiently high spatial resolution, owing to which the particle with the sizes differing by only 2.5 nm can be differentiated. The commercially available programs for LKS signal analysis do not allow one to distinguish between the particles with such close sizes.

The algorithm was also verified in experimental studies. Water solutions of latex microspheres, albumin and metal nanoparticles were investigated. The results of analysis were compared with the results of mathematical processing by "DynaLS". "DynaLS" is a special program for a Photocor laser correlation spectrometer but it can also be used for a large variety of commercial laser correlation spectrometers. The mathematical tools of this program are also based on regularization methods.

Figure 3 shows the results of analysis of a polydisperse mixture of latex microspheres with three sizes calculated by our program and by "DynaLS". We used latex microspheres with specified sizes of 30 nm, 45 nm, and 75 nm.

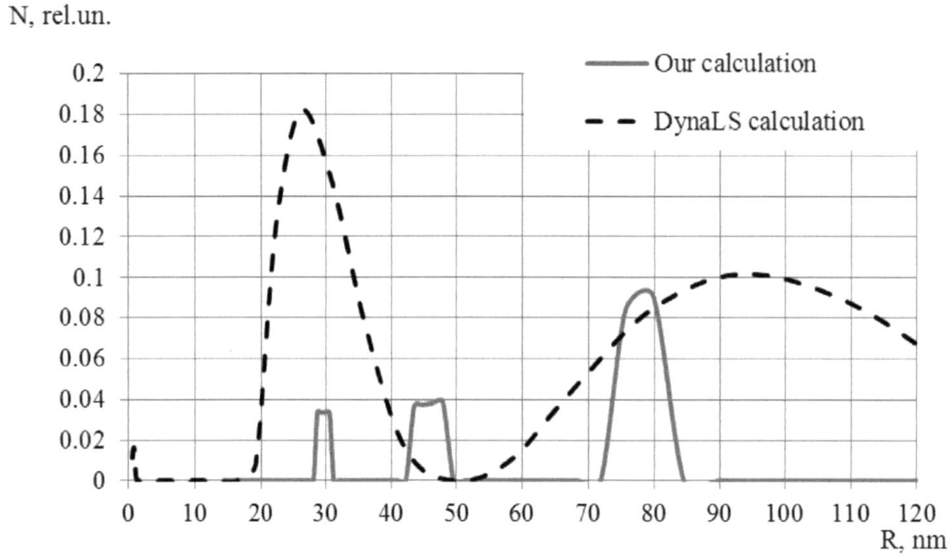

Figure 3. Experimental determination of latex microsphere sizes in a polydisperse mixture

It is evident that the calculations based on our algorithm are characterized by a high accuracy, whereas the calculations employing the commercial program give incorrect results.

5. Conclusion

The algorithm for the solution of the inverse problem of laser correlation spectroscopy aimed at analysis of polydisperse solutions is suggested. The results obtained in our study have demonstrated a high efficiency of our algorithm for analysis of polydisperse solutions.

References

[1] Linegar K L 2008 Applications of dynamic light scattering in chemical engineering: polymers, proteins, and liquid crystals *USA: Fac. of the Graduate School of the U. of Maryland*

[2] Kato H, Nakamura A, Takahashi K and Kinugasa Sh 2012 Accurate size and size-distribution determination of polystyrene latex nanoparticles in aqueous medium using dynamic light scattering and asymmetrical flow field flow fractionation with multi-angle light scattering *J. Nanomaterials.* **2** 15–30

[3] Nepomnyashchaya E, Velichko E, Aksenov E and Bogomaz T 2014 Optoelectronic method for analysis of biomolecular interaction dynamics *J. of Phys.: Conf. Ser.* **541** 1–5

[4] Broillet S, Szlag D, Bouwens A, Maurizi L, Hofmann H, Lasser T and Leutenegger M 2014 Visible light optical coherence correlation spectroscopy *J. O.S.A.* **22** 18 21944–57

[5] Andreev A N and Lazarenko A G 2014 Measurement of Particle Dimensions in Colloidal Solutions Using the Correlation Spectroscopy Technique *J. Telecom. and Radio Eng.* **73** 18 1671–78

[6] Shuai M, Jin Sh, Thomas J C, Xinjun Z, Wei L and Xianming S 2012 Improved inversion procedure for particle size distribution determination by photon correlation spectroscopy *J. Appl. Opt.* **51** 25 6220–6

[7] Golovitski A P 2001 Inverse problems of experimental physics *SPb SPBGTU*

Phase diagram analysis of noise-induced transition in an autocatalytic reaction

Tuan Minh Pham, Yu. P. Virchenko

Belgorod National Research University, 14 Studencheskaya, 308015 Belgorod, Russia

E-mail: physicsidea@gmail.com

Abstract. We present the analysis of the genetic model of autocatalytic chemical reactions proposed by Arnold et al. This model, when subjected to multiplicative white noise modelling environmental fluctuations, can undergo a sudden change from a unimodal state to a bimodal one, while no such transition occurs if the noise is absent. Here, this noise-induced transition is studied analytically by investigating the so-called critical surface in the three-dimensional parameter space.

1. Introduction

Noise-induced phenomena have become a subject of widespread interest in recent decades, because of their observations in various nonlinear systems in physics, chemistry and biology [1,2]. In particular, autocatalytic reaction occurring in a random environment

$$A + X + Y \underset{k_1}{\overset{k_2}{\leftrightarrows}} 2Y + A^* , \qquad B + X + Y \underset{k_3}{\overset{k_4}{\leftrightarrows}} 2X + B^* ,$$

is one of the most studied models in chemistry. The reaction dynamics is governed by the following stochastic differential equation (SDE) for the concentration $x(t)$,

$$dx(t) = [\alpha - x(t) + \lambda x(t)(1 - x(t))]\, dt + \sigma x(t)(1 - x(t)) dw(t) , \quad x(t) \in [0, 1] \tag{1}$$

where $\alpha \in [0, 1]$, $\lambda \in \mathbb{R}$, $\sigma > 0$. This SDE is interpreted in the Stratonovirch sense according to the well-known results of Van Kampen [3], Wong-Zakai [4] and Blankenship [5]. Such chemical interpretation was proposed by Arnold, Horsthemke and Lefever in [6].

It was shown in [1] that this equation also describes the time dependence of gene frequency in a haploid population where there are only two possible alleles with corresponding frequencies x and $1 - x$. For this reason, the model was named "genetic model".

The solutions of Eq.(1) form the Markovian diffusion process $x(t) \in [0, 1]$ at any initial condition $x(0) \in [0, 1]$. In thermodynamic equilibrium states, this process has the stationary probability density

$$p(x) = \frac{C}{x(1-x)} \left(\frac{x}{1-x} \right)^{\beta} \exp\left\{ \frac{2}{\sigma^2} \left(\frac{\alpha - 1}{1 - x} - \frac{\alpha}{x} \right) \right\} , \tag{2}$$

$$C = \frac{e^{2/\sigma^2}}{2K_{-\beta}\big(4\sigma^{-2}\sqrt{\alpha(1-\alpha)}\big)}\left[\frac{1-\alpha}{\alpha}\right]^{\beta/2}, \quad \beta = \frac{2(2\alpha+\lambda-1)}{\sigma^2},$$

here $K_{-\beta}(.)$ is the modified Bessel function of the second kind with index $(-\beta)$ defined as

$$K_{-\beta}(z) = \frac{1}{2}\int_{-\infty}^{\infty} e^{-z\operatorname{ch} u+\beta u}du\,, \quad \operatorname{Re} z > 0\,.$$

2. The analysis of critical surface

In our system coupled to a fluctuating environment, a phase transition can occur whenever the density function $p(x)$ qualitatively changes. In fact, this function may have different mode numbers, depending on the value of parameters λ, σ^2, α. Therefore, the three-dimensional parameter space is divided into domains in such a way that there is a certain mode number in each of them. This partition, on the analogy of thermodynamics, is called the phase diagram, while the surface Σ dividing these domains is called the critical surface. Our paper is devoted to the surface investigation.

Since any changes in the mode number of $p(x)$ corresponds to changes in the solution number of the equation $dp(x)/dx = 0$ which can be written as

$$S(x) \equiv \alpha - x + \lambda x(1-x) - \frac{\sigma^2}{2}x(1-x)(1-2x) = 0\,, \qquad x \in (0,1)\,, \tag{3}$$

we find the degeneracy condition of solutions. This is equivalent to the existence of such a multiple root x_0 that $S(x_0) = S'(x_0) = 0$. Then apply the Euclid algorithm for polynomials $S(x)$ and $S'(x)$. As a result, we obtain the equation of the critical surface:

$$P(\lambda,\sigma^2,\varepsilon) \equiv \lambda^4+\lambda^2\Big(1-5\sigma^2-\sigma^4/2\Big)-\lambda\varepsilon(9\sigma^4+18\sigma^2-4\lambda^2)-4\sigma^2\Big(1-\sigma^2/4\Big)^3-27\sigma^4\varepsilon^2 = 0\,, \tag{4}$$

where $\varepsilon = \alpha - 1/2$. At this time, the root x_0 is given by

$$x_0 = \frac{1}{2} - \frac{2\lambda(1+2\sigma^2)+18\varepsilon\sigma^2}{\Delta}\,, \quad \Delta = 4\lambda^2+3\sigma^4-12\sigma^2\,.$$

In order that x_0 be the bifurcation point of $p(x)$, it is necessary and sufficient that $0 < x_0 < 1$ which is equivalent to the following inequality

$$G_+(\lambda,\sigma^2,\varepsilon)G_-(\lambda,\sigma^2,\varepsilon) \geq 0\,, \tag{5}$$

by introducing the hyperbolas

$$G_\pm(\lambda,\sigma^2,\varepsilon) \equiv \Delta \pm 4\big[\lambda(2\sigma^2+1)+9\varepsilon\sigma^2\big] = 0\,. \tag{6}$$

The inequality (5) defines the allowable domain for the location of the points of the critical surface, whose boundaries are the hyperbolas $G_\pm(\lambda,\sigma^2,\varepsilon) = 0$. It is crucial as the surface (4) is not connected.

Let Σ_ε be the intersection of the surface defined by Eq.(4) with the plane $\varepsilon =$ const and (λ_*,σ_*^2) be the coordinates of the contacting points between the curve Σ_ε and the ellipse $\Delta = 0$. Then

$$\lambda_* = -\frac{9\varepsilon\sigma_*^2}{1+2\sigma_*^2}\,. \tag{7}$$

If exclude λ_* from this relation by using $\Delta = 0$, the single-valued function $\sigma_*^2(\varepsilon)$ is derived from the equation

$$4(\sigma_*^2 - 1)^3 = 27\sigma_*^2(1 - 4\varepsilon^2)\,, \quad \sigma_*^2 > 1 \tag{8}$$

due to the convexity of the left-hand side function.

Now, we represent the polynomial $P(\lambda, \sigma^2, \varepsilon)$ as Taylor's power series of $(\lambda - \lambda_*)$ and $(\sigma^2 - \sigma_*^2)$, which, after introducing the variable substitution

$$\lambda = \lambda_* + \rho\cos\varphi\,, \quad \sigma^2 = \sigma_*^2 + \rho\sin\varphi\,, \tag{9}$$

has the form

$$P(\lambda, \sigma^2, \varepsilon) = -3(\sigma_*^2 - 1)^2(\rho\sin\varphi)^2\,Q_2(z) + \frac{1}{6}(\rho\sin\varphi)^3\,Q_3(z) + \frac{1}{16}(\rho\sin\varphi)^4\,Q_4(z)\,,$$

where $z = \mathrm{ctg}\varphi$, $z_* = \mathrm{ctg}\varphi_* = \lambda_*/3\sigma_*^2$, $Q_2(z) = (z - z_*)^2$, $Q_4(z) = (4z^2 - 1)^2$ and

$$Q_3(z) = 8z_*(7\sigma_*^2 - 1)z^3 - 6(\sigma_*^2 + 5)z^2 + 18z_*(1 + \sigma_*^2)z + \frac{3}{2}(\sigma_*^2 - 3)\,.$$

The Eq.(4) now becomes

$$\frac{1}{16}\rho^2 Q_4(z)\sin^2\varphi + \frac{1}{6}\rho Q_3(z)\sin\varphi - 3(\sigma_*^2 - 1)^2 Q_2(z) = 0\,, \tag{10}$$

and its solutions read

$$\rho_\pm(\varphi) = \frac{4}{3Q_4(z)\sin\varphi}\Big(- Q_3(z) \pm \sqrt{Q_3^2(z) + 27(\sigma_*^2 - 1)^2 Q_4(z) Q_2(z)}\,\Big)\,. \tag{11}$$

These two functions describe the curve Σ_ε at such φ that $\rho_\pm(\varphi) \geq 0$. This means that Σ_ε is defined by $\rho_+(\varphi)$ for $\varphi \in [0, \pi]$ and by $\rho_-(\varphi)$ for $\varphi \in [-\pi, 0]$. From Eq.(11), it is not difficult to show that

$$\rho_+(\varphi) = \frac{\sqrt{5}|Q_3(\pm 1/2)|}{12(z^2 - 1/4)^2}\big(1 + o(1)\big) \quad \text{when} \quad z \to \pm\frac{1}{2}\,.$$

Lemma 1. Σ_ε is a biconnected curve consisting of two components Σ_+ and Σ_- where $\Sigma_+ \equiv \rho_+(\varphi)$ at $\varphi \in [\psi, \pi - \psi]$, $\psi = \mathrm{arcctg}\,(1/2)$, while Σ_- is the glue of simple curves

$$\rho_+(\varphi), \quad \varphi \in [0, \psi)\,; \quad \rho_-(\varphi), \quad \varphi \in [-\pi, 0]; \quad \rho_+(\varphi), \quad \varphi \in (\pi - \psi, \pi].$$

□ This statement follows from the fact that connected curve component must be defined as a continuous function and both $\rho_\pm(\varphi)$ tend to the same finite limit at $\varphi \to 0, \pi$. ■

Theorem 1. Σ_+ is composed of two branches stitched at the contacting point (λ_*, σ_*^2).
□ In the neighborhood of the point (λ_*, σ_*^2), the asymptotic formula of Σ_+, in terms of local Cartesian coordinates u and v, has the form

$$u = \mathrm{const}|v|^{2/3} \quad \text{when} \quad v \to 0\,. \tag{12}$$

In fact, from Eq.(11), we find

$$\rho_+(\varphi) = \varkappa(z - z_*)^2 + O\left((z - z_*)^3\right) \quad \text{when} \quad z \to z_*\,, \quad \varkappa = \frac{81\sigma_*^4(1 + z_*^2)^{1/2}}{4(\sigma_*^2 - 1)}\,.$$

Hence $\rho_+(\varphi)$ is presented by $u^2 + v^2 = \varkappa^2 \arctan^4(v/u)$ where $\varphi - \varphi_* = \arctan(v/u)$ and consequently it follows Eq.(12). ■

In order to establish which component of Σ_ε corresponds to the critical curve, we find the following expressions of $G_\pm(\lambda, \sigma^2, \varepsilon) = 0$ in the polar coordinates (ρ, φ)

$$\rho^{(\pm)}\big[4(1 \pm \sin 2\varphi) - \sin^2\varphi\big] + 2\big[2\cos\varphi(2(\lambda_* \pm \sigma_*^2) \pm 1) + \sin\varphi(3\sigma_*^2 \pm 4\lambda_* - 6 \pm 18\varepsilon)\big] = 0\,.$$

Then such angles $\varphi^{(\pm)}$ that $\rho^{(\pm)}(\varphi^{(\pm)}) = 0$ are given by

$$z^{(\pm)} = \frac{6z_* \mp 3(\sigma_*^2 - 2)}{2(2\sigma_*^2 + 1) \pm 12\sigma_*^2 z_*} \tag{13}$$

where $z^{(\pm)} = \operatorname{ctg}\varphi^{(\pm)}$. Note that the denominator is greater than zero due to $\sigma_*^2 > 1$.

Theorem 2. Σ_+ is the intersection of the critical surface Σ and the plane $\varepsilon = \text{const}$.
□ It can be seen from Eq.(4) and Eq.(6) that (0,0) and (λ_*, σ_*^2) are two common points of Σ_ε and the hyperbolas $G_\pm = 0$. Furthermore, the former is the intersection of Σ_- and $G_\pm = 0$ (as $\rho_-^2(\varphi_0) = \lambda_*^2 + \sigma_*^4$ where $\operatorname{ctg}\varphi_0 = \lambda_*/\sigma_*^2$), while the latter is the intersection of Σ_+ and $G_\pm = 0$ $(\rho_+(\varphi_*) = 0)$.

We now prove that the component Σ_- is contained inside the domain $G_+G_- < 0$. In fact, the derivative $(d\sigma^2/d\lambda)$ of the implicit function $\sigma^2(\lambda)$ defined by Eq.(4) vanishes at $(0,0)$ while those derivatives of functions $\sigma^2(\lambda)$ described by $G_\pm = 0$ are

$$\left(\frac{d\sigma_\pm^2}{d\lambda}\right)_{(0,0)} = [3(1 \mp 3\varepsilon)]^{-1} \neq 0\,.$$

This means that for the points of Σ_-, being close enough to $(0,0)$, $G_+G_- < 0$ since $G_\pm(\lambda, 0, \varepsilon) = \pm 4\lambda + O(\lambda^2)$ when $\lambda \to 0$. Moreover, because (0,0) is the unique common point of Σ_- and $G_\pm = 0$, if the inequality $G_+G_- < 0$ is satisfied in the neighborhood of (0,0), it definitely holds for all others points (λ, σ^2) of Σ_-. Consequently, Σ_- is confined in this domain.

In contrast, the half-plane $(\lambda, \sigma^2 > 0)$ is divided by the component Σ_+ into two separate parts. Whenever one of the hyperbolas $G_\pm = 0$ crosses this component, it must go from one part to another part when φ continuously changes, passing through the contacting point ($\rho^{(+)}(\varphi^{(+)}) = 0$ and $\rho^{(-)}(\varphi^{(-)}) = 0$ respectively). However, this is impossible because of the existence of the cusp point $(0, \varphi_*)$. Since tangent of Σ_+ at this point is a ray radiating from it, for the proof of this fact it is sufficient that $\varphi^{(-)} > \varphi_* > \varphi^{(+)}$ i.e. $z^{(+)} < z_* < z^{(-)}$ which follows from Eq. (13). Therefore, only Σ_+ fulfills the condition (5) and represents the critical curve. ■

As an illustration for the obtained result, we demonstrate here the critical curve in the case of $\alpha = 1/2$, which is defined as

$$\lambda^2 = \frac{1}{2}\left[\frac{\sigma^4}{2} + 5\sigma^2 - 1 - (2\sigma^2 + 1)^{3/2}\right].$$

The inverse function $\sigma^2(\lambda)$ is depicted in Pic. 1.

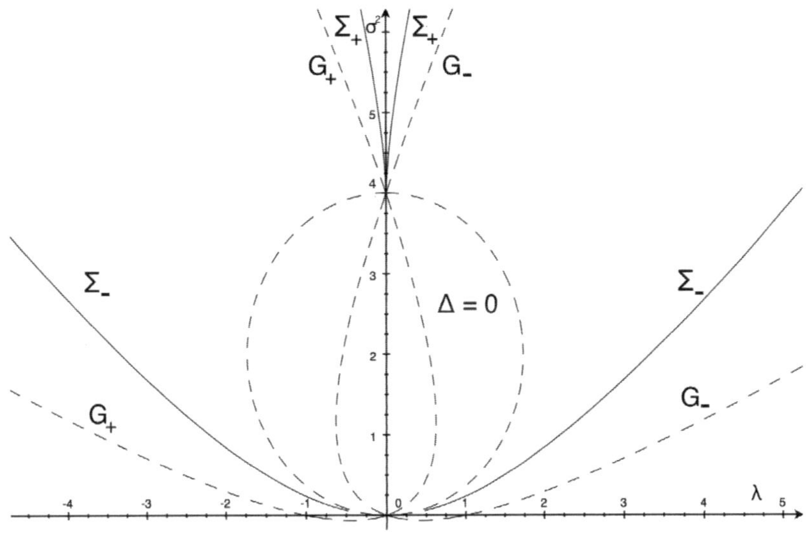

Pic. 1. Phase diagram in the symmetrical case $\alpha = 1/2$. The components Σ_+ and Σ_- are presented by the solid lines, while the hyperbolas $G_\pm = 0$ and the ellipse $\Delta = 0$ are drawn by dotted lines.

3. Conclusion

In contrast to the previous work of Horsthhemke et al., the complete analysis of the critical surface Σ of the model is obtained. This surface divides the parameter space into two domains where qualitatively different stationary dynamical regimes of the system are observed. From the physical point of view, for sufficiently slow variations of the parameters, the switch between these regimes represents a transition between two "phases": the unimodal one and the bimodal one. The dynamical regime in the bimodal phase consists of temporal intervals consequentially replacing each other and having random lengths. In these intervals reactant relative concentrations fluctuate nearby the two maxima of the density function $p(x)$.

4. References

[1] Horsthemke W and Lefever R 2006 *Noise-Induced Transitions: Theory and Applications in Physics, Chemistry, and Biology* (Berlin: Springer)
[2] Landa P S and McClintock P V E 2000 *Physics Reports* **323** 1
[3] Van Kampen N G I 1981 *J. Stat. Phys.* **24** 175
[4] Wong E and Zakai M 1965 *Ann. Math. Stat.* **36** 1560
[5] Blankenship G and Papanicolaou G C 1978 *SIAM J.Appl.Math.* **34** 437
[6] Arnold L, Horsthemke W and Lefever R 1978 *Zs. Phys. B* **29** 367

Band structure of one-dimensional photonic crystal containing two negative index materials

K V Pravdin[1] and I Yu Popov[1]

[1] ITMO University, 49 Kronverkskiy Ave., St. Petersburg, 197101, Russia

E-mail: construeman@gmail.com

Abstract. A layered system composed of two types of metamaterials is examined. We assume the metamaterials are isotropic, homogeneous, dispersive and non-absorptive medias. We confine our attention by the case of identical permittivity and permeability of the metamaterial with the dispersion described by a single Lorentz contribution. For equal layers' widths and several sets of metamaterials parameters we obtain the photonic crystal band structures of the system. We consider the situation, when permittivities (and, correspondingly, permeabilities) of the metamaterials coincide by absolute values and are opposite in sign. In this situation we observe absence of reflection phenomenon, which holds true with changing the metamaterials parameters.

1. Introduction

Materials, which have a negative refractive index for certain frequencies, are known as negative index materials (NIMs) or metamaterials [1]. New covered surfaces and cloaking materials can be created by using NIMs [2]. The superlens with the resolving power, exceeding the diffraction limit [3], also needs the NIMs. Among systems with NIMs (NIM systems), layered NIM systems are widely known [3-10]. The layered systems are considered also as one-dimensional photonic crystals (1DPCs) [11, 12]. Numerous investigations of 1DPC were carried out in recent years [13, 14]. But most of these ones consider systems without the frequency dispersion. The goal of our work is to describe the photonic band gap (PBG) structure for the periodic layered NIM system (photonic crystal) filled with two types of metamaterials. We assume the metamaterials are the isotropic, homogeneous, dispersive and non-absorptive medias, the permittivity and permeability stand equal and are described by using a single Lorentz contribution [4, 15]. Earlier, we have studied the similar NIM system composed of alternated metamaterial and vacuum layers [16]. We obtained absence of reflection in the case, which we called NIM situation, when the permittivity and permeability of the metamaterial are equal to the opposite their vacuum value, i.e. -1 instead of $+1$. Now we consider the system with two types of metamaterials, and we need to widen the NIM situation concept. The NIM situation is a case, when permittivities (and permeabilities) of the two types of the metamaterials coincide by absolute values but are opposite in sign. We are interesting in the electric field for this case and it's dependence on metamaterials parameters.

2. Periodic NIM system

We study the periodic (infinite) NIM system composed of parallel layers (figure 1). Unit vectors $\mathbf{e}_1$, $\mathbf{e}_2$ of the Cartesian basis $\{\mathbf{e}_1,\mathbf{e}_2,\mathbf{e}_3\}$ set a plane of layer's surfaces. Unit vector $\mathbf{e}_3$ sets the x_3 axis. We assume the translation invariance along the plane of layer's surfaces. There are two types of layers. The first one is Δ_1 in width and filled with a metamaterial (M_1). The second one is Δ_2 in width and filled with another metamaterial (M_2). Metamaterials of neighbor layers have different optical properties, which are described below. The layers alternate each other. Value $\Delta_1 + \Delta_2$ is a period of the system. Then, the system is a 1DPC.

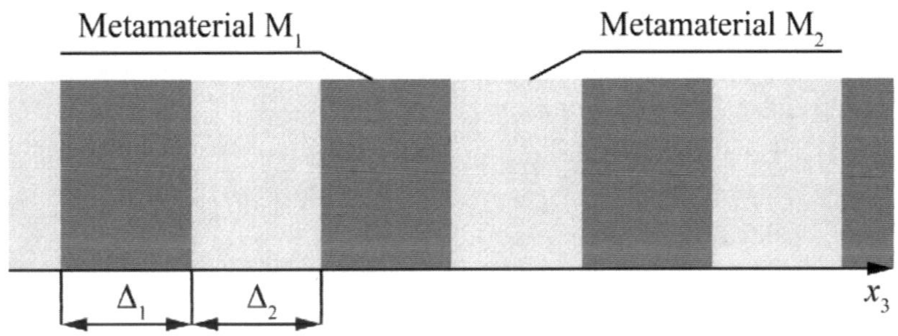

Figure 1. The periodic NIM system composed of parallel alternated layers filled with the metamaterials M_1 and M_2. There is the translation invariance along the plane of layer's surfaces.

We consider the Maxwell's equations in a differential form

$$\frac{\mathrm{d}\mathbf{D}}{\mathrm{d}t}(\mathbf{x},t)=\nabla\times\mathbf{H}(\mathbf{x},t),\quad \frac{\mathrm{d}\mathbf{B}}{\mathrm{d}t}(\mathbf{x},t)=-\nabla\times\mathbf{E}(\mathbf{x},t),\quad \nabla\cdot\mathbf{D}(\mathbf{x},t)=0,\quad \nabla\cdot\mathbf{B}(\mathbf{x},t)=0,$$

where $\mathbf{x}=\{x_1,x_2,x_3\}$ is a vector located in the Cartesian basis $\{\mathbf{e}_1,\mathbf{e}_2,\mathbf{e}_3\}$, ∇ is the Hamilton operator, $\times$ is a cross product symbol as well as a numerical product symbol, $\cdot$ is an inner product symbol as well as a matrix product symbol. Also, we consider the auxiliary field equations

$$\mathbf{D}(\mathbf{x},t)=\varepsilon_0\mathbf{E}(\mathbf{x},t)+\mathbf{P}(\mathbf{x},t),\quad \mathbf{B}(\mathbf{x},t)=\mu_0\left[\mathbf{H}(\mathbf{x},t)+\mathbf{M}(\mathbf{x},t)\right],$$

$$\mathbf{P}(\mathbf{x},t)=\varepsilon_0\int_{t_0}^{t}\boldsymbol{\chi}_e(\mathbf{x},t-s)\cdot\mathbf{E}(\mathbf{x},s)\mathrm{d}s,\quad \mathbf{M}(\mathbf{x},t)=\int_{t_0}^{t}\boldsymbol{\chi}_m(\mathbf{x},t-s)\cdot\mathbf{H}(\mathbf{x},s)\mathrm{d}s,$$

where ε_0 and μ_0 are the electric and magnetic constants, and $\varepsilon_0\mu_0=1/c^2$, where c is the speed of light in vacuum. $\boldsymbol{\chi}_e(\mathbf{x},t)$ and $\boldsymbol{\chi}_m(\mathbf{x},t)$ are the electric and magnetic susceptibility tensors. We use the causality condition $\boldsymbol{\chi}_e(\mathbf{x},t)=\boldsymbol{\chi}_m(\mathbf{x},t)=0$ for $t<t_0$ (we assume $t_0=0$) and the passivity condition [4].

The considered system composed of layers divided by planes. The boundary conditions for each one are presented in the general form as follows:

$$(\mathbf{E}_1-\mathbf{E}_2)\times\mathbf{e}_3=\mathbf{0},\quad (\mathbf{H}_1-\mathbf{H}_2)\times\mathbf{e}_3=\mathbf{0},\quad (\mathbf{D}_1-\mathbf{D}_2)\cdot\mathbf{e}_3=0,\quad (\mathbf{B}_1-\mathbf{B}_2)\cdot\mathbf{e}_3=0. \tag{1}$$

$\mathbf{E}_j=\mathbf{E}_j(\tilde{\mathbf{x}},t)$, $\mathbf{H}_j=\mathbf{H}_j(\tilde{\mathbf{x}},t)$, $\mathbf{D}_j=\mathbf{D}_j(\tilde{\mathbf{x}},t)$, and $\mathbf{B}_j=\mathbf{B}_j(\tilde{\mathbf{x}},t)$ stand for the limits with $\mathbf{x}\to\tilde{\mathbf{x}}$, where $\mathbf{x}=\mathbf{x}^\perp+x_3\mathbf{e}_3$, $\tilde{\mathbf{x}}=\mathbf{x}^\perp+\tilde{x}_3\mathbf{e}_3$, $\mathbf{x}^\perp=\{x_1,x_2,0\}$ and $x_3\to\tilde{x}_3$, j is the index distinguishing limits calculated on different layer's surface sides ($j=1,2$). We use the Fourier transform with t time

$$\hat{f}(\omega)=\int_{-\infty}^{+\infty} exp[-\mathrm{i}\omega t]f(t)\mathrm{d}t,\quad f(t)=\frac{1}{2\pi}\int_{-\infty}^{+\infty}\exp[\mathrm{i}\omega t]\hat{f}(\omega)\mathrm{d}\omega, \tag{2}$$

and the Fourier transform with x_1 and x_2 coordinates

$$f_{\mathbf{k}^\perp}(x_3)=\int_{-\infty}^{+\infty}\int_{-\infty}^{+\infty}\exp[\mathrm{i}(k_1x_1+k_2x_2)]f(\mathbf{x})\mathrm{d}x_1\mathrm{d}x_2=\int_{R^2}\exp[\mathrm{i}\mathbf{k}^\perp\cdot\mathbf{x}^\perp]f(\mathbf{x})\mathrm{d}\mathbf{x}^\perp, \tag{3}$$

$$f(\mathbf{x})=\frac{1}{(2\pi)^2}\int_{-\infty}^{+\infty}\int_{-\infty}^{+\infty}\exp[-\mathrm{i}(k_1x_1+k_2x_2)]f_{\mathbf{k}^\perp}(x_3)\mathrm{d}k_1\mathrm{d}k_2=\frac{1}{(2\pi)^2}\int_{\mathrm{R}^2}\exp[-\mathrm{i}\mathbf{k}^\perp\cdot\mathbf{x}^\perp]f_{\mathbf{k}^\perp}(x_3)\mathrm{d}\mathbf{k}^\perp,$$

where k_1 and k_2 are the coordinates of the two-dimensional wave vector $\mathbf{k}^\perp=\{k_1,k_2,0\}$ with the length $\kappa=|\mathbf{k}^\perp|=\sqrt{k_1^2+k_2^2}$ and the unit vector $\mathbf{e}_{\mathbf{k}^\perp}$, i.e. $\mathbf{k}^\perp=\kappa\mathbf{e}_{\mathbf{k}^\perp}$, $\mathbf{k}=\{k_1,k_2,k_3\}$ is the full wave vector.

We consider the isotropic homogeneous metamaterials in layers. Therefore, the electric and magnetic permeabilities do not depend on coordinates of the $\mathbf{x}$ vector, i.e. $\boldsymbol{\varepsilon}(\mathbf{x},\omega)=\varepsilon(\omega)\mathbf{U}$, $\boldsymbol{\mu}(\mathbf{x},\omega)=\mu(\omega)\mathbf{U}$, where $\mathbf{U}$ is the 3×3 unit matrix. Also, we assume that the metamaterials are the dispersive non-absorptive medias. Then, the susceptibilities consist of a sum of Lorentz contributions [14]. We deal with a single dispersive Lorentz contribution [4]. We assume that the permittivity and permeability of each metamaterial stand equal and

$$\varepsilon(\omega)=\mu(\omega)=1-\frac{\Omega^2}{\omega^2-\omega_0^2}, \tag{4}$$

where Ω and ω_0 are the plasma and resonance frequencies, respectively. Note that each metamaterial behaves like a NIM if the frequency (ω) is inside the interval $(\omega_0,\omega_{\mathrm{ZERO}})$ (NIM interval), where $\omega_{\mathrm{ZERO}}=\sqrt{\omega_0^2+\Omega^2}$, $\varepsilon(\omega_{\mathrm{ZERO}})=\mu(\omega_{\mathrm{ZERO}})=0$, and $\varepsilon(\omega_0+0)=\mu(\omega_0+0)=-\infty$. We denote ω_{NIM} as the NIM frequency, which leads to the NIM situation, i.e. $\varepsilon_1(\omega_{\mathrm{NIM}})=-\varepsilon_2(\omega_{\mathrm{NIM}})$ and $\mu_1(\omega_{\mathrm{NIM}})=-\mu_2(\omega_{\mathrm{NIM}})$. It is given by the following formula:

$$\omega_{\mathrm{NIM}}=\frac{1}{\sqrt{2}}\left\{\omega_{0,1}^2+\omega_{0,2}^2+\frac{\Omega_1^2+\Omega_2^2}{2}+\left[\left(\omega_{0,1}^2-\omega_{0,2}^2+\frac{\Omega_1^2-\Omega_2^2}{2}\right)^2+\Omega_1^2\Omega_2^2\right]^{1/2}\right\}^{1/2},$$

where parameters with the $_1$ and $_2$ subscripts correspond to the M_1 and M_2 metamaterials, respectively.

The Fourier transform (2) of the Maxwell's equations leads to the following Helmholtz equation for the electric field function $\mathbf{E}(\mathbf{x},t)$:

$$\nabla\times\nabla\times\hat{\mathbf{E}}(\mathbf{x},\omega)=(\omega/c)^2\varepsilon(\omega)\mu(\omega)\hat{\mathbf{E}}(\mathbf{x},\omega). \tag{5}$$

By using the Fourier transforms (2) and (3) for the Helmholtz equation (5), the boundary conditions (1) and the Floquet-Bloch theorem [17, 18] we obtain the following dispersion equation, which is similar to one obtained earlier for the NIM system composed of alternated metamaterial and vacuum layers [16]:

$$\left(\mathrm{e}^{-\mathrm{i}\theta(\Delta_1+\Delta_2)}\right)^2-\left[\frac{\overset{+}{\sigma}_{1,2}\overset{+}{\sigma}_{2,1}}{4}\left(1+\mathrm{e}^{\mathrm{i}2k_3^{(1)}\Delta_1}\mathrm{e}^{\mathrm{i}2k_3^{(2)}\Delta_2}\right)+\right.$$

$$\left.+\frac{\overset{-}{\sigma}_{1,2}\overset{-}{\sigma}_{2,1}}{4}\left(\mathrm{e}^{\mathrm{i}2k_3^{(1)}\Delta_1}+\mathrm{e}^{\mathrm{i}2k_3^{(2)}\Delta_2}\right)\right]\mathrm{e}^{-\mathrm{i}k_3^{(1)}\Delta_1}\mathrm{e}^{-\mathrm{i}k_3^{(2)}\Delta_2}\mathrm{e}^{-\mathrm{i}\theta(\Delta_1+\Delta_2)}+1=0, \tag{6}$$

where θ is a yet undefined wave vector, called the Bloch wave vector (the analogous vector in quantum mechanics is named quasimomentum), $\sigma_{j,p}^{\pm}=\left(\varepsilon_j k_3^{(p)}\pm\varepsilon_p k_3^{(j)}\right)/\varepsilon_j k_3^{(p)}$ with indices $j=1$ and $p=2$, or $j=2$ and $p=1$. Here ε_j and $k_3^{(j)}$ are calculated for the M_j metamaterial. The dispersion equation (6) is the same for TE and TM modes. It follows from the equality of the permittivity and permeability (4). Therefore, we have the identical PBG structure for TE and TM modes.

3. Band structure

To study the PBG structure of the considered 1DPC we use a numerical approach. We consider equality (6) and seek for such values of ω and κ that equality (6) holds true with any value of θ (it is sufficient to consider values of θ belonging to interval $(0, 2\pi/\Delta)$). There are two cases for each metamaterial: if $(k_3)^2 > 0$, where $(k_3)^2 = |\mathbf{k}|^2 - |\mathbf{k}^{\perp}|^2 = (\omega/c)^2 \varepsilon(\omega)\mu(\omega) - \kappa^2$, then there is a radiative regime; else if $(k_3)^2 < 0$, then there is an evanescent regime. We fix the constants $\Delta_1 = \Delta_2 = 10$ ηm. The following four cases for the $\omega_{0,1}$, Ω_1, $\omega_{0,2}$, and Ω_2 parameters are considered:

(a) $\omega_{0,1} = 30$ THz, $\omega_{0,2} = 60$ THz, $\Omega_1 = 90$ THz, $\Omega_2 = 90$ THz (see (a) in figure 2);

(b) $\omega_{0,1} = 30$ THz, $\omega_{0,2} = 30$ THz, $\Omega_1 = 90$ THz, $\Omega_2 = 180$ THz (see (b) in figure 2);

(c) $\omega_{0,1} = 30$ THz, $\omega_{0,2} = 60$ THz, $\Omega_1 = 90$ THz, $\Omega_2 = 180$ THz (see (c) in figure 2);

(d) $\omega_{0,1} = 30$ THz, $\omega_{0,2} = 60$ THz, $\Omega_1 = 180$ THz, $\Omega_2 = 90$ THz (see (d) in figure 2).

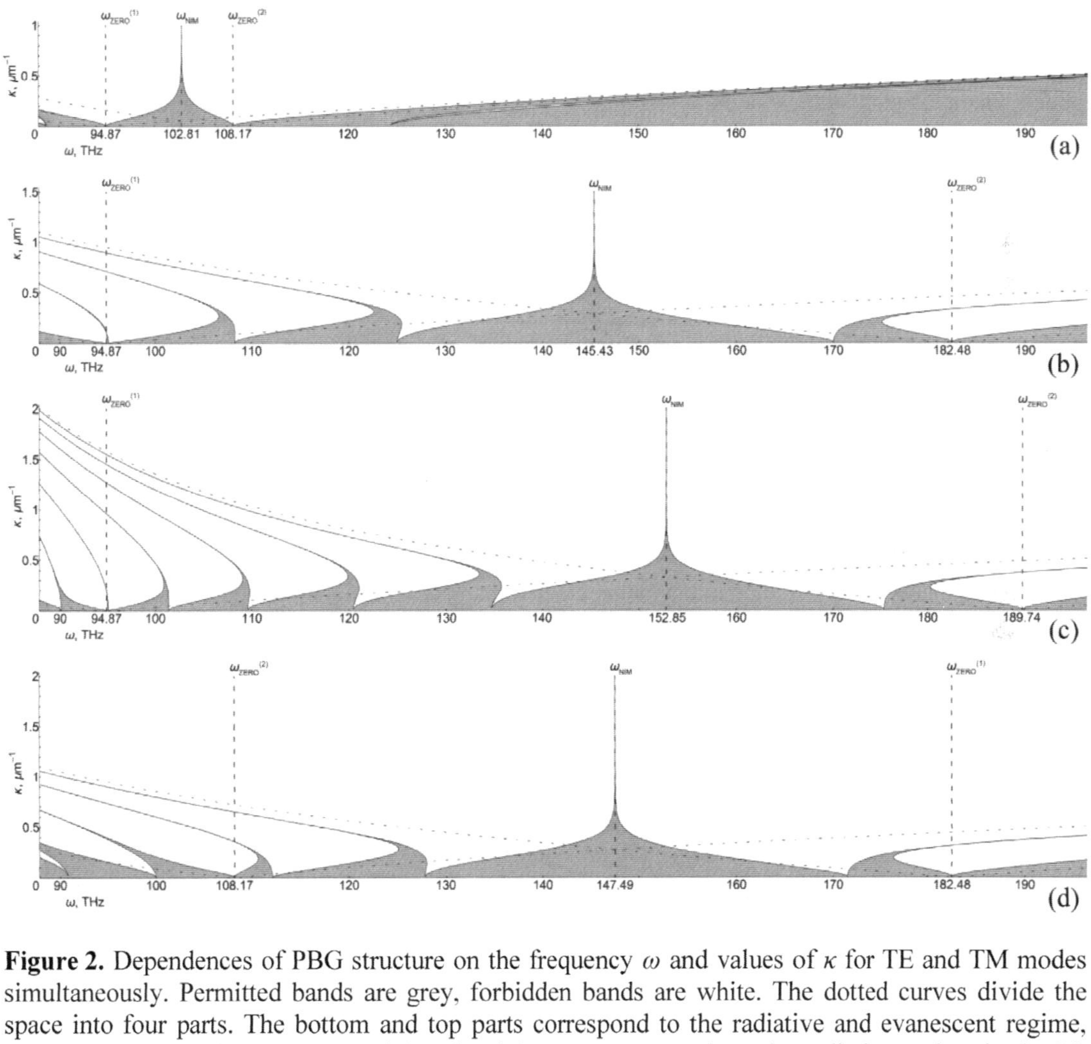

Figure 2. Dependences of PBG structure on the frequency ω and values of κ for TE and TM modes simultaneously. Permitted bands are grey, forbidden bands are white. The dotted curves divide the space into four parts. The bottom and top parts correspond to the radiative and evanescent regime, respectively, in the both metamaterials. The right part corresponds to the radiative regime in the M_1 metamaterial and the evanescent regime in the M_2 metamaterial for the (a), (b), and (c) cases, and vice versa for the (d) case. The left part corresponds to the evanescent regime in the M_1 metamaterial and the radiative regime in the M_2 metamaterial for the (a), (b), and (c) cases, and vice versa for the (d) case. The vertical straight dotted lines show the $\omega_{ZERO}^{(1)}$ and $\omega_{ZERO}^{(2)}$ frequencies in M_1 and M_2 metamaterials, respectively, and the ω_{NIM} frequency.

In all cases, we observe permitted bands near the NIM frequency for any value of κ, which become narrower and converge to the ω_{NIM} frequency. The NIM frequency is located between $\omega_{\mathrm{ZERO}}^{(1)}$ and $\omega_{\mathrm{ZERO}}^{(2)}$ frequencies. Then, one of two metamaterials behave like a NIM and the second one behave like a PIM (positive index material). These permitted bands keep on existing with changing the metamaterials parameters ω_0, Ω, and depend only on ω_{NIM} frequency. The same effect was observed for the NIM situation in the similar NIM system composed of alternated metamaterial and vacuum layers [16].

4. Electric field

Now we are interesting in solutions of the Helmholtz equation in the NIM situation. We obtained them earlier for the general case of the frequency, when we examined a periodic NIM system composed of two alternated layers filled with metamaterial and vacuum, respectively [16]. In the NIM situation for TE and TM modes they are expressed for the radiative regime as follows:

$$E_1(x_3,\omega_{\mathrm{NIM}},\kappa) = -4C(\omega_{\mathrm{NIM}},\kappa)\left(e^{\mathrm{i}\rho\Delta_1} - e^{\mathrm{i}\rho\Delta_2}e^{-\mathrm{i}\theta(\Delta_1+\Delta_2)}\right)e^{-\mathrm{i}\rho\Delta_1}e^{\mathrm{i}\rho x_3}, \tag{7}$$

$$E_2(x_3,\omega_{\mathrm{NIM}},\kappa) = -4C(\omega_{\mathrm{NIM}},\kappa)\left(e^{\mathrm{i}\rho\Delta_1} - e^{\mathrm{i}\rho\Delta_2}e^{-\mathrm{i}\theta(\Delta_1+\Delta_2)}\right)e^{\mathrm{i}\rho\Delta_1}e^{-\mathrm{i}\rho x_3}, \tag{8}$$

and for the evanescent regime as follows:

$$E_1(x_3,\omega_{\mathrm{NIM}}) = 4C(\omega_{\mathrm{NIM}},\kappa)\left(e^{\rho\Delta_1} - e^{\rho\Delta_2}e^{-\mathrm{i}\theta(\Delta_1+\Delta_2)}\right)e^{-\rho\Delta_1}e^{\rho x_3}, \tag{9}$$

$$E_2(x_3,\omega_{\mathrm{NIM}}) = 4C(\omega_{\mathrm{NIM}},\kappa)\left(e^{\rho\Delta_1} - e^{\rho\Delta_2}e^{-\mathrm{i}\theta(\Delta_1+\Delta_2)}\right)e^{\rho\Delta_1}e^{-\rho x_3}, \tag{10}$$

where $C(\omega,\kappa)$ is an unknown function, which can be obtained from initial conditions, and $\rho = \rho(\omega_{\mathrm{NIM}})$ is a value of the following function:

$$\rho(\omega) = \sqrt{|k_3|^2} = \sqrt{\left||\mathbf{k}|^2 - |\mathbf{k}^{\perp}|^2\right|} = \left|\omega^2\left(1 - \frac{\Omega^2}{\omega^2 - \omega_0^2}\right)^2 - \kappa^2\right|^{1/2}.$$

The solutions (7) and (8) are correct only if ${\omega_{\mathrm{ZERO}}}^{(1)} > {\omega_{\mathrm{ZERO}}}^{(2)}$ ((d) case), i.e. if M_1 and M_2 metamaterials behave like a NIM and PIM, respectively. Else, if ${\omega_{\mathrm{ZERO}}}^{(1)} < {\omega_{\mathrm{ZERO}}}^{(2)}$ ((a), (b), and (c) cases), i.e. if M_1 and M_2 metamaterials behave like PIM and NIM, respectively, and, hence, solutions (7) and (8) should be changed by adding the "–" sign to the ρ value. Solutions (9) and (10) are correct for the both conditions.

Note that solution (7), as well as solutions (8)-(10), has only one term responsible for propagation and no terms responsible for reflection. This effect is observed only when the dispersion equation (6) holds true, i.e. only for permitted bands. As we observed in the third section, the frequency ω_{NIM} belongs to a permitted band for any value of κ. Therefore, although the system is layered, there is no reflection at the NIM frequency. Indeed, the refractive indices in neighbor layers coincide with absolute values and are opposite in sign. Then, as we observe, the system is totally translucent at the NIM frequency, and the waves differ only by a sign of the wave phase in the radiative regime (7)-(8), and by the increasing or decreasing of its values in the evanescent regime (9)-(10).

5. Conclusion

In this paper we solved the problem of describing the photonic band gap structure of the periodic layered NIM system (1DPC) with equal layer widths, near the NIM frequency. We assumed that the permittivity and permeability of the metamaterial stand equal and are defined with the single Lorentz contribution. By the numerical approach, we observed that there is a permitted band with the "center" at the NIM frequency, i.e. at the NIM frequency the permitted band is not restricted by "angle of the wave direction". The effect holds true with changing the metamaterials parameters. Also, we obtained expressions for the scalar field function in the permitted bands. In the NIM situation they contain only one term responsible for propagation and no terms responsible for reflection, i.e. there is no reflection at the NIM frequency for any "wave direction". This property can be useful for metadevice construction (see, e.g., [19]).

The photonic band gap structure is similar to obtained earlier in [16], where we examined the periodic system, composed of alternated metamaterial and vacuum layers. In [16] the NIM frequency, for which we observe a total translucence of the system, is defined by parameters of only the metamaterial (plasma and resonance frequencies). In this paper the NIM frequency is defined by parameters of the both metamaterials. Thus, from a practical point of view there are more ways to produce a NIM system with a certain NIM frequency. Also, the system containing only metamaterial layers seems to be more practical than the similar one containing furthermore vacuum layers.

6. Acknowledgements
This work was partially financially supported by the Government of the Russian Federation (grant 074-U01), by the Ministry of Education and Science of the Russian Federation (GOSZADANIE 2014/190, Projects No 14.Z50.31.0031 and No. 1.754.2014/K), by grant MK-5001.2015.1 of the President of the Russian Federation, by ITMO University from centralized resources (grant 415829).

References
[1] Burgos S, Waele R, Polman A and Atwater H 2010 *Nat. Mater.* **9** 407
[2] Dubinov A E and Mytareva L A 2010 *Phys. Usp.* **53** 475
[3] Iyer A and Eleftheriades G 2009 *IEEE. Trans. Antennas Propag.* **57** 1720
[4] Gralak B and Tip A 2010 *J. Math. Phys.* **51** 052902
[5] Collin R 2010 *PIER* **B19** 233
[6] Pravdin K and Popov I 2013 *Nanosystems: Phys. Chem. Math.* **4** 570
[7] Pravdin K and Popov I 2015 *OPA* **48** 31
[8] Lai K, Tsang L and Huang C 2005 *Micro. Opt. Tech. Lett.* **44** 86
[9] Maksimovic M, Hammer M and Jaksic Z 2008 *Proceedings of SPIE* **6896** 689605
[10] Potemkin A, Poddubny A, Belov P and Kivshar Yu 2012 *Phys. Rev. A.* **86** 023848
[11] Pendry J *Phys. Rev. Lett.* 2000 **85** 3966
[12] Wu L, He S, Shen L 2003 *Phys. Rev. B* **67** 235103
[13] Pandey G, Thapa K, Srivastava S, Ojha S 2008 *PIER M* **2** 15
[14] Feng X, Li H 2013 *Eur. Phys. J. D* **67** 40157
[15] Tip A 2004 *Phys. Rev. E* **69** 016610
[16] Pravdin K and Popov I 2014 *Nanosystems: Phys. Chem. Math.* **5** 626
[17] Floquet G 1883 *Ann. Ecole Norm. Sup.* **12** 47
[18] Bloch F 1929 *Z. Phys.* **52** 555
[19] Kivshar Y 2015 *Nanosystems: Phys. Chem. Math.* **6** 346

Modelling of zirconium alloy hydrogenation

Yury V Zaika and Natalia I Rodchenkova

Institute of Applied Mathematical Research, Karelian Research Centre,
Russian Academy of Sciences, Pushkinskaya Street 11, Petrozavodsk, Russia

E-mail: zaika@krc.karelia.ru, nirodchenkova@krc.karelia.ru

Abstract. Zirconium alloys are the construction materials for critical elements in active zones of nuclear power reactors. During the operation of reactors such materials are subject to hydrogenation. Hydrogenation results in a decrease of alloy plasticity and cracking resistance. The formation of brittle hydrides at crack tips can result in severe embrittlement. One of the most important requirements for the reactor's active zone materials is low hydrogen absorptivity. The mathematical model of hydride layer formation and growth is developed. The problem is to determine the dynamics of the free boundary of phase interface and the distributions of hydrogen concentration in hydride and in solution. Iterative computational algorithm for solving the nonlinear boundary-value problem with the Stefan condition based on implicit difference schemes is developed.

1. Introduction

The interest in the interaction of hydrogen and its isotopes with metals, alloys and intermetallic compounds is multifarious [1–8]. It is sufficient to mention problems in power production, protection of metals from hydrogen corrosion, chemical reactor design, rocket production. Enthusiasts have been speaking about hydrogen economy along with hydrogen energy [4]. Some mathematical models of dehydrogenation and hydrogen permeability connected with this article are presented in [9–11]. One of the most important requirements for the reactor's active zone materials (made of zirconium alloys) is low hydrogen absorptivity since hydrogen embrittlement may cause zirconium cladding damage. Depending on the hydrogen content and operation temperature, hydrogen may be present in zirconium alloys as a solid solution or as hydrides. Hydrides have the greatest embrittlement effect on alloys as they can form and enlarge cracks.

We briefly describe the experiment (see details in [12]). A plate is placed into a chamber. The plate is polished on one side. The other side is nearly impermeable to hydrogen due to the presence of impurities, oxides and such. We disregard butt ends. The temperature of the sample and input pressure ≈ 2 atm. are constant during the experiment (sample cooling is carried out).

The alloy absorbs hydrogen quite well. A thin volume layer at which the hydrogen distribution can be considered uniform at a relatively high input pressure $p = \text{const}$ (≈ 2 atm.) is selected. Diffusion resistance starts to set only at a certain initial depth ℓ_0. When the concentration of dissolved hydrogen reaches a certain value, this layer forms the initial hydride crust. Further transport of hydrogen proceeds through the growing hydride layer. The problem is to model the dynamics of the moving boundary of phase transition and to estimate the concentration distribution in hydride and in solution.

2. Mathematical model

The notations are the following. L is the plate thickness; ℓ_0 is the thickness of the layer that absorbs hydrogen relatively easily and there is no diffusion resistance inside this layer yet (future initial hydride crust); $u(t)$ is the concentration of atomic hydrogen (H) in ℓ_0-layer ($1_{\rm H}/{\rm cm}^3$); Q is the concentration ($1_{\rm H}/{\rm cm}^3$), when $u(t)$ becomes equal to Q the lattice is rearranged and the solution is converted to hydride; $c(t,x)$ is H concentration in $(L-\ell_0)$-layer; μ is the kinetic constant. During the experiment the temperature and input pressure are constant ($T = {\rm const}$, $p = {\rm const}$). The notations are schematically represented in Fig. 1. The α-phase is the H solution in Zr–1Nb alloy and δ-phase is the hydride wherein atomic H is also dissolved and diffuses.

According to the gas kinetic theory the particle flux density J_p incoming to the surface (in this case the flux of H_2 molecules) is connected with the pressure p by the Hertz–Knudsen formula: $J_p = p/\sqrt{2\pi mkT}$ (k is the Boltzmann constant, m is the mass of H_2 molecule). In the context of the experiment it is convenient to select the following units $[x,\,\ell,\,L] = {\rm cm}$, $[p] = {\rm Torr}$. Then we numerically obtain the dependence $J_p = \mu p$, $\mu(T) \approx 2.474\cdot 10^{22}/\sqrt{T}$ ($[\mu] = 1_{\rm H_2}/({\rm Torr\,cm^2\,s})$, $[T] = {\rm K}$). Since the atomic hydrogen diffuses, for consistency we will calculate the flux in H atoms: $J_p = 2\mu p$. Only a small part of H will however be absorbed into the alloy: $J_{\rm abs} = 2\mu s p$ ($s \ll 1$). Multiplier s is the fraction of incoming H that gets to the near-to-surface volume. We combine the elementary stages of physisorption, chemosorption, dissociation and dissolution itself into one stage: s is the effective coefficient of absorption.

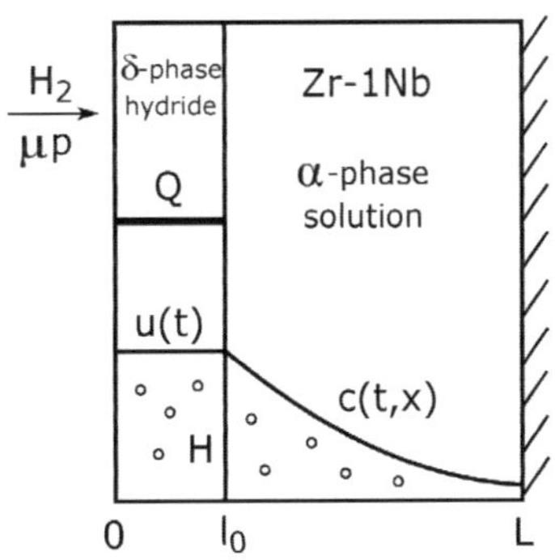

Figure 1. Stage I.

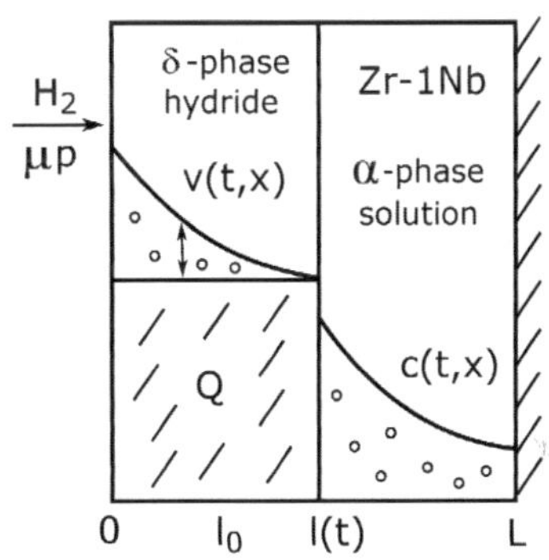

Figure 2. Stage II.

Stage I: preliminary stage — H dissolution in Zr–1Nb

For the diffusion $(L-\ell_0)$-layer we formulate the standard boundary-value problem:

$$\frac{\partial c}{\partial t} = D\frac{\partial^2 c}{\partial x^2},\ x\in(\ell_0,L),\ t>0;\quad c(0,x)=0,\ x\in[\ell_0,L];\quad c(t,\ell_0)=u(t),\ \frac{\partial c}{\partial x}\Big|_L = 0. \tag{1}$$

The boundary condition $c(t,\ell_0) = u(t)$ indicates that the H distribution with respect to sample thickness is continuous, while $\partial_x c|_L = 0$ indicates that the plate side $x = L$ is impermeable. Hereinafter, we assume that coefficients depend on the temperature in an Arrhenius way: $D = D_0\exp\{-E_D/[RT]\}$, $s = s_0\exp\{-E_s/[RT]\}, \ldots$ The temperature $T = {\rm const}$ is maintained throughout one experiment.

From the fluxes balance we derive the ordinary differential equation (ODE) for the concentration $u(t)$:

$$\dot u(t)\ell_0 = 2\mu s p - b u^2 + D\frac{\partial c}{\partial x}\Big|_{\ell_0},\quad u(0)=0. \tag{2}$$

The meaning of this equation is the following. The amount of H atoms that has been absorbed through cm^2 during 1s due to the hydrogen pressure is $2\mu sp$, but there are the opposing desorption flux bu^2 (b is the effective coefficient of recombination) and the diffusion outflow. The imbalance of these flux densities determines the accumulation of hydrogen atoms within the ℓ_0-layer ($\dot{u}\ell_0$). This equation should be considered together with (1) since $u(t)$ determines the boundary concentration in (1). When p is not too large (without hydride formation), in the equilibrium situation (when all derivatives are zero) we have

$$2\mu sp - b\bar{u}^2 = 0 \Rightarrow \bar{u} = \Gamma\sqrt{p}, \quad \Gamma \equiv \sqrt{2\mu s/b}.$$

Therefore, under static conditions the dynamics of (2) is consistent with the Sieverts law $\bar{u} \propto \sqrt{p}$, where Γ is the solubility coefficient. Let us emphasize that we speak about the dissolved diffusion-movable atomic hydrogen. In the "saturation–degassing" experiment the total absorption of hydrogen is taken into account, including reversible capture, hydride phases ... — the coefficient Γ may have another meaning and numerical value. Technically it is easy to add the reversible capture of H atoms in the $(L - \ell_0)$-layer by different types of "traps" (defects of the material), but in this problem we consider the reversible capture to be a minor factor.

Stage II: hydride formation and phase boundary movement

At some time instant the concentration $u(t)$ becomes equal to the threshold concentration $Q = Q(T)$, at which the lattice is rearranged and hydride is formed (δ-phase). We consider that this transition process is relatively fast. Thus, $u(t) = Q \Rightarrow \ell_0$-crust turns into a hydride. We set a new time zero ($t = 0$) and model the growth of the hydride phase. By this time we have $c(0, x) = \varphi(x)$, $\varphi(\ell_0) = Q$ (the distribution of dissolved hydrogen is taken from the previous stage). Now the hydrogen begins to diffuse through the hydride ℓ_0-layer. Denote by $v(t, x)$ the H concentration in the δ-layer. We emphasize that $v(t, x)$ is the concentration of diffusing hydrogen, the total concentration is equal to $Q + v(t, x)$. Notations are schematically represented in Fig. 2. We put down the diffusion equations:

$$\frac{\partial v}{\partial t} = D_* \frac{\partial^2 v}{\partial x^2}, \; x \in (0, \ell(t)), \quad v(0, x) = 0, \; x \in [0, \ell_0], \tag{3}$$

$$\frac{\partial c}{\partial t} = D \frac{\partial^2 c}{\partial x^2}, \; x \in (\ell(t), L), \quad c(0, x) = \varphi(x), \; x \in [\ell_0, L]. \tag{4}$$

Here, D_* is the H diffusion coefficient in the new material (in the hydride, in δ-phase). We write the "input-output" boundary conditions similarly to stage I:

$$2\mu s_* p - b_* v^2(t, 0) = -D_* \frac{\partial v}{\partial x}\Big|_{x=0}, \quad \frac{\partial c}{\partial x}\Big|_{L} = 0. \tag{5}$$

In the hydride, the concentration gradient appears almost immediately since it is much harder to diffuse through the δ-phase. There is no "cumulative" layer (the analogue of the ℓ_0-layer in stage I). The free phase boundary $x = \ell(t)$ is now movable and two conditions at the interface $x = \ell(t)$ are needed to "sew together" the diffusion equations.

We begin with the Stefan type condition describing the dynamics of the phase boundary:

$$\left[v(t, \ell(t)) + Q - c(t, \ell(t))\right]\dot{\ell}(t) = -D_* \frac{\partial v}{\partial x}\Big|_{\ell(t)} + D \frac{\partial c}{\partial x}\Big|_{\ell(t)}.$$

A jump of concentrations occurs: $v(t, \ell) + Q > Q > c(t, \ell)$, $\ell = \ell(t)$, $t > 0$. Still, the concentration $v(t, \ell)$ is negligible (the alloy "does not resist" the penetration of hydrogen). We assume that

$v(t,\ell(t))=0$. The incoming flux from the δ-phase almost completely goes to hydride formation (movement of the boundary $x=\ell(t)$) and to the solution (α-phase).

Thus, we accept the following conditions on the moving free boundary:

$$v(t,\ell(t))=0, \quad [Q-c(t,\ell(t))]\dot{\ell}(t)=-D_*\frac{\partial v}{\partial x}\Big|_{\ell}+D\frac{\partial c}{\partial x}\Big|_{\ell}. \tag{6}$$

At the beginning $[\ldots]=0$ $\big(c(0,\ell_0)=\varphi(\ell_0)=Q\big)$, and then $(t>0)$ a jump of concentrations occurs $\big([\ldots]>0\big)$ and $\dot{\ell}(t)>0$. The front moves with the concentration Q (total) towards $x=L$. Passing to the limit $(t\to+\infty)$ we have $\ell(t)\to L$, but phase boundary movement slows down appreciably. The concentration distribution $c(t,x)$ tends to uniformity and slowly increases up to the limit value Q, when one can practically neglect the Zr–1Nb layer.

3. Computational algorithm

Stage I. The main computational challenges are to solve the nonlinear boundary-value problem with moving phase boundary (Stage II). Therefore, we only briefly describe here the general scheme for Stage I. The authors approximate the diffusion equation (1) (using four-point pattern) and ODE (2) by implicit difference schemes of second–order accuracy. The algorithm is of iterative nature due to the model nonlinearity. Moving on to the next time layer, we use explicit schemes for the diffusion equation to calculate the initial approximation for the first three grid nodes with respect to the spatial variable. Knowing the current approximations of the concentration in the boundary nodes $(x=\ell_0,L)$, we used the Thomas algorithm for solving the system of algebraic equations with tridiagonal matrix. After correcting the concentration values in internal nodes, the concentration in the boundary node $x=\ell_0$ is recalculated as a positive root of quadratic equation (from the difference approximation of the boundary condition (2)). The iterations proceed until the distribution $c(t,x)$, $x\in[\ell_0,L]$ becomes stable. Then we move on to the next time layer.

Stage II. Transformation of the boundary-value problem

The authors chose the following variant: the moving segments $[0,\ell(t)]$ and $[\ell(t),L]$ are transformed into [0,1]. Then uniform grids are chosen. We introduce a change of variables with a new time zero $(t_0=0)$: $(t,x)\leftrightarrow(t,y)$, $x=\ell(t)y$, $\hat{v}(t,y)=v(t,x(t,y))$; $(t,x)\leftrightarrow(t,z)$, $x=\ell(t)+z[L-\ell(t)]$, $\hat{c}(t,z)=c(t,x(t,z))$.

The boundary-value problem (3)–(6) for zirconium alloy hydrogenation after the change of variables is

$$\frac{\partial\hat{v}}{\partial t}=\frac{D_*}{\ell^2(t)}\cdot\frac{\partial^2\hat{v}}{\partial y^2}+\frac{\dot{\ell}(t)y}{\ell(t)}\cdot\frac{\partial\hat{v}}{\partial y},\ y\in(0,1),\quad \hat{v}(0,y)=0,\ y\in[0,1], \tag{7}$$

$$\hat{v}(t,1)=0,\quad 2\mu s_* p-b_*\hat{v}^2(t,0)=-\frac{D_*}{\ell(t)}\cdot\frac{\partial\hat{v}}{\partial y}\Big|_{y=0}; \tag{8}$$

$$\frac{\partial\hat{c}}{\partial t}=\frac{D}{[L-\ell(t)]^2}\cdot\frac{\partial^2\hat{c}}{\partial z^2}+\frac{\dot{\ell}(t)(1-z)}{L-\ell(t)}\cdot\frac{\partial\hat{c}}{\partial z},\ \hat{c}(0,z)=\psi(z),\ \frac{\partial\hat{c}}{\partial z}\Big|_{z=1}=0; \tag{9}$$

$$[Q-\hat{c}(t,0)]\dot{\ell}(t)=-\frac{D_*}{\ell(t)}\cdot\frac{\partial\hat{v}}{\partial y}\Big|_{y=1}+\frac{D}{L-\ell(t)}\cdot\frac{\partial\hat{c}}{\partial z}\Big|_{z=0}. \tag{10}$$

The initial distribution $\psi(z)$ is taken from the previous stage. Formally, one can forget about the physical meaning of $\ell(t)$ as the phase bound and consider it as a functional parameter. Model coefficients in the boundary-value problem are nonlinear functionals with respect to solutions $\hat{v}$, $\hat{c}$. Indeed, if we consider the functions $\hat{v}(t,y)$, $\hat{c}(t,z)$ as formally known we get a nonlinear functional differential equation for $\ell(t)$: $\dot\ell(t) = G(t, \ell(\cdot))$ from the Stefan condition (10). All model coefficients are unambiguously determined using the solution $\ell(t)$. Therefore, the computational algorithm will be based on the implicit scheme and will have an iterative character. Iterations will be related to more precise definition of $\dot\ell(t)$ at every time layer. We use a standard four-point pattern to approximate the diffusion equations (7), (9) and then solve the resultant linear equation systems applying the Thomas algorithm. We now describe how to "sew together" the diffusion equations using the Stefan condition (10). We have to restrict ourselves to the case of the initial stage only ($\dot\ell(0) = \ldots$, we proceed similarly on the next time layers) to avoid cumbersome technical manipulations.

We first assume that $\dot\ell_0 \equiv \dot\ell(0) = 0$ (later on we use the value from the previous layer). The parameter $\dot\ell_0$ is subject to iterative improvement, the value of ℓ_0 is known.

We fix the time step Δt and suppose that $\dot\ell = 0$, $\ell = \ell_0$ in the boundary-value problem (7)–(8). At the initial stage we assume $\hat v = 0$ at $y = y_{1,2}$ in the standard three-point pattern. We have a quadratic equation in (8) for $\hat v(t_1, 0)$ and choose the positive root. Find the distribution (approximation) $\hat v(t_1, y)$, $y \in [0, 1]$ using the Thomas algorithm. Go back to the improvement of $\hat v(t_1, 0)$ in virtue of (8) (the values of $\hat v$ at $y = y_{1,2}$ used to calculate $\partial_y \hat v|_{y=0}$ have already changed). Proceed with this process until the distribution $\hat v(t_1, y)$ becomes stable. The time step is small enough, so a few iterations will suffice.

Move on to the boundary-value problem (9) returning to the previous time layer ($t = 0$). The missing boundary condition is determined from the Stefan equation (10): $\dot\ell_0 = 0$, $\hat v = 0$ $\Rightarrow \partial_z \hat c|_{z=0} = 0$. We proceed with the iterative calculations until the distribution $\hat c(t_1, z)$, $z \in [0, 1]$ becomes stable. Note that the input concentration will then go down $\hat c(t_1, 0) < Q$.

As a result of these calculations we get $[\ldots] > 0$ and a positive right-hand side into the Stefan equation (10). We find a new approximation of $\dot\ell_0 > 0$ and return to the beginning of the described subalgorithm. After some iterations we obtain a steady-state value of $\dot\ell_0$ and determine $\ell_1 = \ell_0 + \dot\ell_0 \Delta t$. We go on to the layer $t_1 = \Delta t$ and begin with the following equality $\dot\ell_1 = \dot\ell_0$. Next, we specify the value of $\dot\ell_1$ and the corresponding distributions $\hat v(t_2, y)$, $\hat c(t_2, z)$ according to the scheme drafted above.

Numerical algorithm testing. The iterative algorithm is developed on the basis of implicit difference schemes and the Thomas algorithm. Computational experiments confirming the algorithm efficiency have been conducted. The input data, for example, are $L = 6 \cdot 10^{-2}\,\mathrm{cm}$, $\ell_0 = 1.3 \cdot 10^{-3}\,\mathrm{cm}$, $p = 1520\,\mathrm{Torr}$, $T = 593\,\mathrm{K}$, $Q = 6.4 \cdot 10^{20}\,\mathrm{cm}^{-3}$, $D_0 = 2.2 \cdot 10^{-3}\,\mathrm{cm^2\,s^{-1}}$, $E_D = 35 \cdot 10^3\,\mathrm{J\,mol^{-1}}$, $D_{*0} = 1.5 \cdot 10^{-3}\,\mathrm{cm^2\,s^{-1}}$, $E_{D_*} = 59 \cdot 10^3\,\mathrm{J\,mol^{-1}}$, $b = 5 \cdot 10^{-23}\,\mathrm{cm^4\,s^{-1}}$, $b_* = 3 \cdot 10^{-24}\,\mathrm{cm^4\,s^{-1}}$, $s = 7 \cdot 10^{-6}$, $s_* = 6 \cdot 10^{-7}$.

The possibility to analyze the distribution dynamics of dissolved hydrogen concentrations across the hydride and the alloy layers is shown in Fig. 3.

Conclusions

Zirconium alloys are the construction materials for critical elements in active zones of nuclear power reactors. During the operation of reactors such materials are subject to hydrogenation. Hydrogenation results in a decrease of alloy plasticity and cracking resistance. The formation of brittle hydrides at crack tips can result in severe embrittlement. One of the most important requirements for the reactor's active zone materials is low hydrogen absorptivity.

Since dissolved atomic hydrogen represents an extremely mobile interstitial phase, it is practically impossible in experimental studies to determine time-dependent hydrogen distribution in the material (even within the experiment time scale) based on the available external measurements.

The article presents a model taking into account nonlinear dynamics of adsorption-desorption processes and the movement of free phase boundary (hydride–metal). In the asymptotic (under chemical equilibrium conditions) the model corresponds to the experimental Sieverts law of

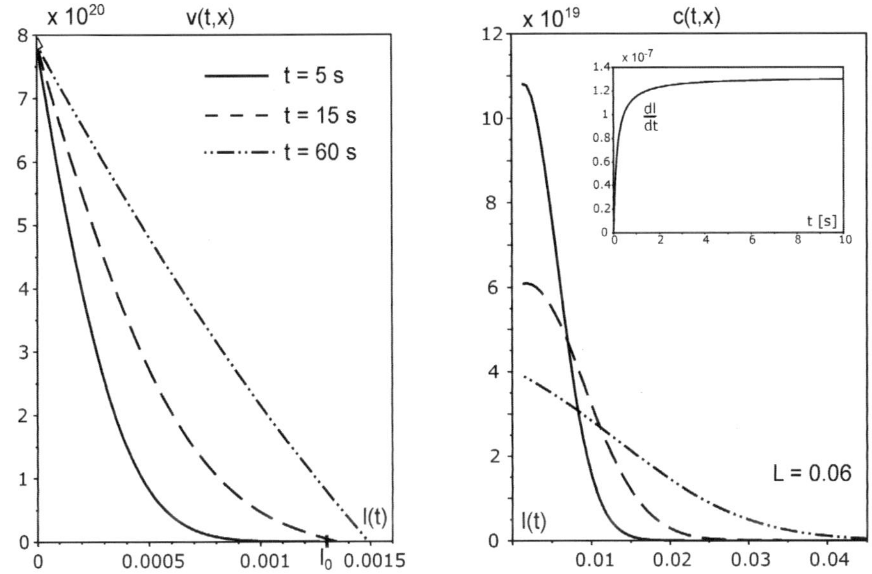

Figure 3. Distributions of hydrogen concentration across the hydride and the alloy layers.

solubility. The algorithm that allows to numerically model the dynamics of phase boundary and distribution of dissolved hydrogen concentrations across the hydride and the alloy layers is briefly presented. Particularly important is the model-based forecast on the time horizon much exceeding experimental capabilities.

Acknowledgments
This work was supported by the Russian Foundation for Basic Research, Project 15-01-00744.

References

[1] Alefeld G and Völkl J (Eds) 1978 *Hydrogen in Metals* (Berlin; New York: Springer) vol I Basic Properties p 427, vol II Application-oriented Properties p 387
[2] Pisarev A A, Tsvetkov I V, Marenkov E D and Yarko S S 2008 *Hydrogen Pearmeability through Metals* (Moscow: MEPhI) p 144
[3] Varin R A, Czujko T and Wronski Z S 2009 *Nanomaterials for solid state hydrogen storage* (Springer) p 338
[4] Ball M and Wietschel M (Eds) 2009 *The hydrogen economy* (Cambridge University Press) p 646
[5] Hirscher M (Ed) 2010 *Handbook of Hydrogen Storage* (Wiley–VCH) p 353
[6] Evard E A, Gabis I E and Yartys V A 2010 Kinetics of hydrogen evolution from MgH2: experimental studies, mechanism and modelling *Int. J. Hydrogen Energy* **35** 9060–69
[7] Lototskyy M V, Yartys V A, Pollet B G and Bowman R C Jr. 2014 Metal hydride hydrogen compressors: a review *Int. J. Hydrogen Energy* **39** 5818–51
[8] Indeitsev D A and Semenov B N 2008 About a model of structure-phase transformations under hydrogen influence *Acta Mechanica* **195** 295–304
[9] Zaika Yu V and Rodchenkova N I 2009 Boundary-value problem with moving bounds and dynamic boundary conditions: diffusion peak of TDS-spectrum of dehydriding *Appl. Math. Modelling* **33** 3776–91
[10] Zaika Yu V and Bormatova E P 2011 Parametric identification of a hydrogen permeability model by delay times and conjugate equations *Int. J. Hydrogen Energy* **36** 1295–1305
[11] Zaika Yu V and Rodchenkova N I 2013 Hydrogen-solid boundary-value problems with dynamical conditions on surface *Math. Modelling* (New York: Nova Sci. Publishers) 269–302
[12] Pushilina N S, Kudiiarov V N, Laptev R S, Lider A M and Teresov A D 2015 Microstructure changes in Zr–1Nb alloy after pulsed electron beam surface modification and hydrogenation *Surf. Coat. Technol.* **284** 63–68

Direct measurement of elastic modulus of InP nanowires with Scanning Probe Microscopy in PeakForce QNM mode

P Geydt[1*], M Dunaevskiy[2, 3], P Alekseev[2], J-P Kakko[4], T Haggrén[4], E Lähderanta[1] and H Lipsanen[4]

[1] Laboratory of Physics, Lappeenranta University of Technology, 53851 Lappeenranta, Finland
[2] Ioffe Physical Technical Institute, RAS, 194021 Saint-Petersburg, Russia
[3] National Research University of Information Technologies, Mechanics and Optics (ITMO), Saint-Petersburg 197101, Russia
[4] Department of Micro- and Nanosciences, Micronova, Aalto University, P.O. Box 13500, FI-00076 Aalto, Finland

*Corresponding author's email: Pavel.Geydt@gmail.com

Abstract. In this manuscript, we present the study of elastic properties of InP nanowires with help of scanning probe microscope in advanced PeakForce Tapping® regime. The measuring method was developed in order to investigate the Young's modulus of these cone-shaped structures with significant accuracy. The difference in InP elasticity for wurtzite phase and zinc-blende phase was revealed. It was shown that elastic modulus of InP nanowires significantly increases from 60 GPa to more than 100 GPa when diameter of a nanowire is reduced below 50 nm. The core-shell model for InP nanowire was used for the explanation of this effect.

1. Introduction

Due to the worldwide increase of electric power demands and modern requirements for clean energy, much attention is paid to renewable Solar technologies. For example, the photovoltaic cell covered with semiconductor nanowires (NWs) is an interesting architecture proposed for next-generation Solar energy panels [1]. Its diverse designs and characteristics are extensively studied with significant attention to features of their electrical conductivity [2], surface states and charge trapping effects [3], type and uniform properties of shell passivation material [4] and necessary level of doping [5]. Additionally, it is vital to consider the NW's mechanical strength [6], since nanowires need to be covered by viscous transparent protective layer and withstand operational stresses under temperature expansion.

While electric properties of the NW array can be partially tested with a massive pad (i.e. a microplate electrode), it is still difficult to access the NWs individually. Seemingly, only awareness of all mentioned parameters of individual NW can help to design and develop NW-based devices. Although, measuring their elastic properties in contact appears even more laborious and accessible only by devices enabling nanoscale resolution. Among all available mechanical parameters, specifically the Young's modulus is the most important criterion, because it characterizes the preservation of technological quality of the NWs considering their reversible deformations.

Multiple publications describing tools and methods to study the elastic characteristics of one-dimensional systems have been recently published [7 - 10]. Moreover, few works were focused on semiconductor nanowires, e.g. ZnO [11] and GaAs [12], but surprisingly no reports revealed the elastic properties of InP NWs. In fact, InP is considered more advantageous, e.g. for Solar cells, due to the reduced density of surface states [13]. Capability to produce photocurrent for InP NW-based cell has recently been extended above efficiency of 13% [14] and further beyond theoretical expectation of efficiency limit [15]. It should also be noted, that crystal structure of InP NWs can be either wurtzite (WZ) or zinc-blende (ZB), which was confirmed by electron diffraction experiments. Wurtzite InP is an exotic material, whose elastic properties have not been experimentally studied yet. We suggested semicontact Scanning Probe Microscopy (SPM) as a proper tool for this investigation.

Considering the previously collected experience on difficulties with positioning, calibration and interpreting of the data, we would like to report our experimental results for the measurements of elasticity of inclined cone-shaped InP NWs, obtained with PeakForce Tapping® mode SPM. In this manuscript, our observations over two series of InP NWs with wurtzite crystal structure and zinc-blende structure will be demonstrated with suggestions concerning the usage of PeakForce measuring method.

2. Sample preparation and Experiment

The arrays of inclined nanowires were fabricated by vapour-liquid-solid (VLS) mechanism in metalorganic vapour phase epitaxy (MOVPE) chamber. Golden droplets with diameter ~20 nm were sputtered on hot silicon substartes, followed by controlled growth of conical InP NWs. The growth direction, diameters and lengths of the nanowires were controlled by duration of vapour source flow and temperature regime. Thickness of obtained InP NWs appeared gradually declining from approximately 160 nm for diameter of a lower base to approximately 20 nm for diameter of an upper part, i.e. near the Au caps. Lengths of InP NWs were approximately 3 μm while distance between adjacent structures exceeded value of 3 μm to secure them from unintentional encounter. NWs were not intentionally doped in order to study the properties of genuine InP material.

Figure 1. (a) Optical image of an array of InP NWs at magnification 200x with A-shaped SPM cantilever and reflection spot of scattered red laser beam. (b) HRTEM image of wurtzite InP NW near the Au cap, and inset (c) presenting the corresponding electron diffraction pattern for WZ InP phase. (d) 3D SPM image of inclined InP NW and a broken NW lying nearby on the substrate. (e) SEM image of inclined InP NW, where the conical shape and a lower “thick base” are distinguishable.

Crystal structure and sizes of InP NWs were investigated by scanning electron microscope (SEM), transmission electron microscope (TEM) and scanning probe microscope (SPM) (See Figure 1). It should be noted, that NWs had specific "thick bases" at the growth contact point on the wafer (seen on Fig. 1 (d)), so their lever's length (leg of bending) for further correct calculation of elasticity should be reduced by approximately 0.2 micron (See the lower part of the NWs on Fig. 1 (d) and Fig. 2). All SPM measurements were performed in standard room conditions without intentional sample's pretreatment.

Experimental setup

At first, it may be emphasized, that NWs can be observed and targeted even with help of standard optical microscope with magnification 100x. It does not seem possible to detect the NW's structure and geometry in details, but location of single freestanding nanoobject can be noticed as one or two dark pixels revealed by the sensitive matrix of the optical microscope OMV Nikon coupled with BRUKER Multimode 8 microscope used in our work. Due to the distinctive distance between the NWs, it was possible to distinguish separate individual NWs having length ~3 μm with only tens of nanometers in cross-section (See black spots on Fig. 1 (a)). Reason here is associated with distortion of light by such "surface defect" (i.e. NW) mounted on the smooth flat wafer's surface having roughness ~1 nm for 10x10 μm^2. Later these NWs were successfully targeted with SPM probe, considering the known geometry of the probe tip in such a way that tip's apex was engaging the sample's surface successfully already in few microns to the certain desired NW.

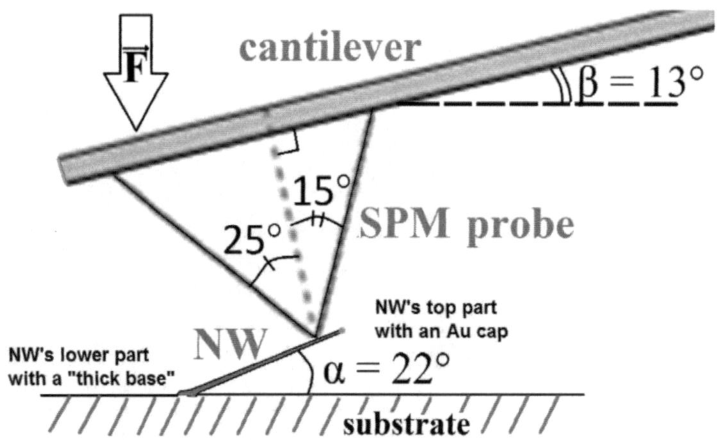

Figure 2. Experimental scheme reflects the realistic scale for SPM probe above inclined InP NW mounted on the wafer and corresponding angles of their contact interaction. The specified angles α ~ 22° and β = 13° are being reduced upon increasing of the bending force F in our experiment.

Schematics of the measuring setup is shown on Figure 2. The incline angle of an SPM cantilever mounted in a probe holder is 12°. Addition to bending angle ≈1° is specified by the manufacturer (Bruker Inc.) for the SPM probes used in this study. NW-substrate angle α was ~22°.

Methodic of the measurements

Elastic properties of inclined NWs were studied previously by the "multiple points method" (also called as method of "Force-Load curves") [12]. The main advantages of this method are that it can be operated on most of the SPM systems and it provides data considerably fast. However, this method requires precise knowledge of the lever distance from the NW's base, while it seems difficult to determine the coordinates of "fixing point", where the NW's lower base begins to be detached from the wafer. Moreover, small amount of data points in one experimental cycle is a significant drawback from the viewpoint of repeatability of these measurements.

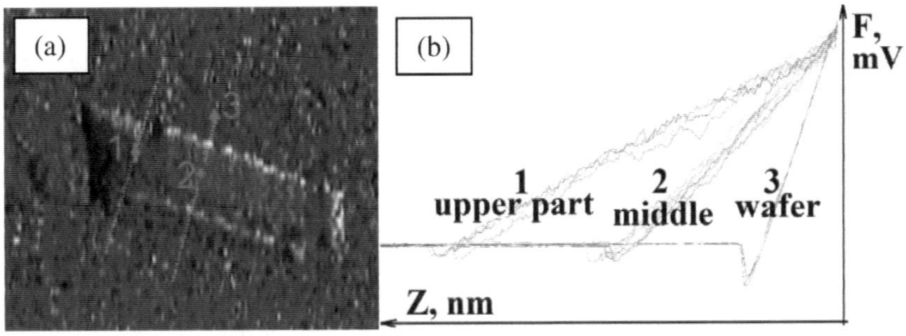

Figure 3. Multiple points method of studies allowed obtaining the data values of elasticity. (a) SPM image of inclined InP NW; (b) *F(z)* force-load curves measured at the substrate (3) and at the upper (1) and middle (2) parts of the NW, corresponding to certain NW's lever lengths.

The measurements of force-load curves *F(z)* can be performed only in a limited amount of positions (See Figure 3), due to the sample drift and lateral bending of the nanostructure, causing moderate accuracy. Moreover, the uncertainty of location of lever base's starting point causes additional error in the calculated elastic modulus, leading to decrease of preciseness of each measurement. The force-load curves measurements were used only for comparison of data with the elasticity measured with PeakForce mode. Still, force-load curves provided strong correlation of results at least by the order of values with Young's modulus of the InP NWs from PeakForce mode.

Utilization of the PeakForce QNM mode
From emergence of the PeakForce Tapping® technique in 2009 it became possible to make quantitative force estimations after processing the data of SPM experiments, even for fragile samples, e.g. standing NWs. Later, with the help of newer PeakForce Quantitative Nanomechanics (QNM) mode [16] it became possible to obtain values of elastic modulus in the range between 100 Pa - 20 GPa simultaneously with mapping of a material's surface Topography. Method provides eight data channels simultaneously to diversely describe the samples (See Figure 4). Unfortunately, even QNM cannot be applied to straightforward study of the Young's modulus of inclined NWs, because "Deformation" data used in calculation of DMT elastic modulus channel (based on Derjaguin-Muller-Toporov model of elastic contact) would represent the bending of a NW instead of the material's Deformation. However, there is a way to calculate the proper elastic modulus, if a spring constant of the SPM cantilever would be accounted with correct lever length and precisely registered force of probe-sample interaction.

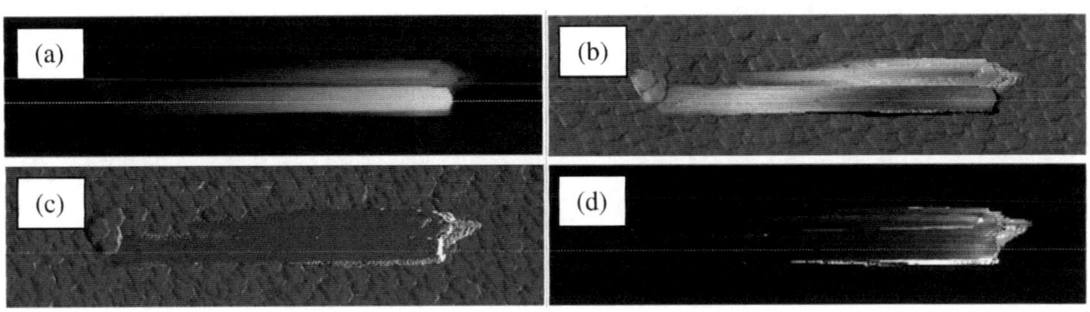

Figure 4. SPM images of inclined InP NW, with lower base on the left. Comparison of data from PeakForce mode channels used to quantify the elasticity: (a) Topography, (b) PeakForce, (c) PeakForce Error and (d) Deformation. Image sizes are 3x0.75 μm. PeakForce Setpoint was normalized to 0.2 nN, while distinct color gradient on (b) is further explained on Figure 5.

At the start, we calibrated the cantilever's spring constant k_{Probe} and the SPM photodetector's deflection sensitivity for relatively soft cantilevers. SPM probe was chosen experimentally from the probes with spring constants k in the range 0.06 - 2.80 N/m. For example, the "SCM-PIT" cantilever's k was evaluated as 0.438 N/m by the calibration protocol. Probe tip was coated by manufacturer with electrically conductive metal layer, so the tip's apex was wide enough (tip radius > 20 nm) to establish a stable contact and decrease the sliding effect away from the NW's sloping surface. Then, one of the NWs which was oriented towards the SPM cantilever was chosen from the OMV and Topography imaging, and imaging area was additionally rotated by few necessary degrees in such a way that "fast scanning direction" of the SPM met the NW's long axis, i.e. the profile of the plane of bending down movement. It is important that NW's base should be situated on the left of the image when cantilever is on the right (as seen on Fig. 1), so that inclination of the NW ($\alpha = 22°$) should partly compensate the inclination of the SPM cantilever ($\beta = 13°$) (See scheme on Fig. 2). Thus, data channels for tip movement direction from left to right become sources of valuable data for further calculations. It is required to establish the negligible sample drift achievable by constant temperature and measurements started always from the same side of the scan image, e.g. up or down. Otherwise, it seems beneficial to fix the "slow scanning direction". Lateral resolution of the scans was 1024x100 for scan size 3x0.375 micron, so that at least one of the horizontal profile lines would have a width ~4 nm which is considerably smaller than diameter of the NWs. This was considered important for eliminating of the parasitic sliding effect when SPM tip does not properly track the upper face of the inclined NW. Here we need also to highlight that fabricated NWs appeared to have a conical shape instead of cylindrical, which was considered by adding $\gamma \approx 1.5°$ of cone angle to numerical model of the NW's shape and bending, additionally correcting the Young's modulus measurement.

3. Results

Series of recorded profiles for necessary data channels (i.e. Topography, Deformation and Peak Force) were used for calculation of inverted spring constant $1/k(L)$ along the NW's length L, when Deformation channel was divided by the corresponding measured Peak Force channel values (PF) (See horizontal profile lines on Fig. 4). PeakForce mode indicates a deviation from the ideal presetted force parameter (i.e. Setpoint), which is recorded separately as Peak Force Error channel. Topography channel (correspondent to height of the structure under applied bending force) provided us information about the height profile line of our interest (See Fig. 4(a)) and about the location of the NW lever's beginning.

To additionally increase the accuracy of our measurements, we summarized data for following bending forces applied to the NWs: 100 pN, 200 pN, 500 pN, 1 nN, 2 nN, 5 nN, 10 nN, 20 nN, 50 nN, 100 nN and 200 nN. These three orders of magnitude in range of the utilized Setpoint force of probe-sample interaction led to few distinctive observations. Firstly, bending of the NW (controlled by Topography and Deformation) is linearly dependent from the PF in the areas, where spring constants of a NW and a cantilever were comparable. Secondly, increasing of the force did not affect the shape of the Peak Force channel profile line $PF(L)$ on the area where it was acquired, while curves had same features further described in Section 3.2. Finally, increasing of the force to extensive Setpoint values above 20 nN did not lead to non-linear deformations inside the NW. Instead of reaching the shear strain value and irreversibly deform the structure, the SPM probe slided off from it, and NW further became bended in lateral plane. This resulted in observation that the displayed form of the NW was observed thinner at the NW's top part on certain extent depending from the force. We did not account the profile area where bending was unstable and indicated lateral movement. Therefore, even for the relatively high value of the force F = 50 nN we were able to obtain a shortened profile of the NW without the segment responsible for the upper part of the NW. Forces above 100 nN were used simultaneously with increasing the PeakForce tapping amplitude (from 150 nm to 300 nm) only to brake the NWs. A broken NW lying on a smooth wafer (Seen on Fig. 1(d)) provided us with information about the NW's conical shape $D(L)$ and the cone angle $\gamma \approx 1.5°$. In such a procedure, NWs did not fly off far from the growth location, but were securely attached to the wafer by the van der Waals forces and surface adhesion.

In order to improve preciseness of our calculations, diameters of the NWs were primarily controlled by averaging the SEM data for separate NWs in an array, then compared and proofed on each certain broken NW (by the Topography profile on smooth wafer surface), and finally verified for each NW by modeling auto fit of calculated parameter of middle diameter D_{mid}. Thereby, diameter still could be slightly enlarged, which correlates with RMS roughness of the wafer surface ~5 nm for 5x5 μm^2 area (See Fig. 1 (d)) and insignificant own roughness of the NWs. Although, summarized estimation of this error for values of middle diameter D_{mid} is approximately 1 nm.

Processing of the acquired PeakForce data and curves

It should be specified, that synchronization time moment (Sync parameter) needs to be manually corrected to perform correct PeakForce QNM measurements of a highly bendable structure. It is explained by the fact that cantilever requires additional time to reach the lowest (bended) position. Thus, "*Sync distance New*" parameter must be manually designated to be close or equal to "*Sync distance QNM*" parameter, which cannot be achieved by *Autoconfig* feature of NanoScope SPM software. The value would be adjusted from ~77.5 msec to ~84 msec, correspondingly with the Setpoint force, when it is being increased from 0.5 nN to 50 nN. Otherwise, all parameters of a PeakForce curve would be estimated for incorrect moment of time. Unfortunately, this time-shift can only be found experimentally for each individual NW and force. It cannot be adjusted properly beforehand, since SPM probe's first engagement with the sample is done usually on the spacious area of the hardly deformable wafer. We have further intention to develop such calibration procedure of Sync parameter in the SPM's software.

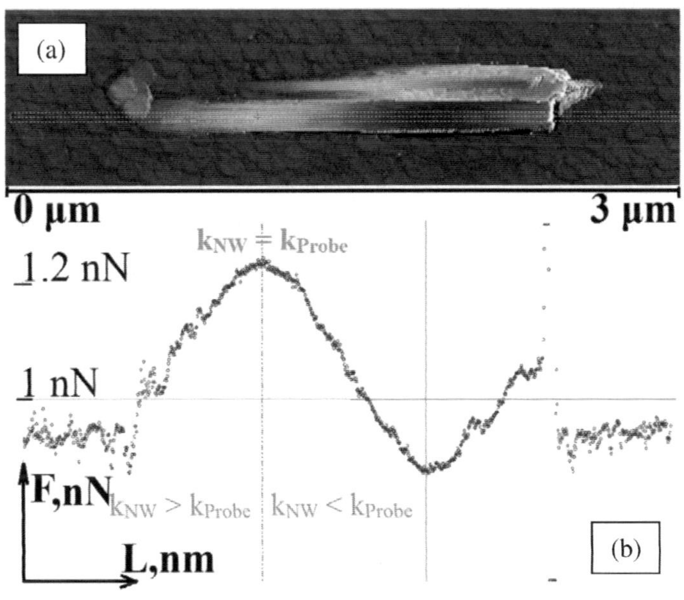

Figure 5. (a) Peak Force data channel and (b) corresponding *PF(L)* values for the bended InP NW. PeakForce Setpoint was normalized to 1 nN. The curve represents 1024 data points of elasticity over one line. It should not be confused with the PeakForce curve itself associated with only one data point.

As can be seen from the curve on Figure 5 read from left to right, negligible noise level is noticed for wafer and this deviation is further recorded into the Peak Force Error channel. Then SPM probe starts to tap on the stiff base of the NW, which can not be deformed by the soft probe. The microscope's feedback controller system preliminary increases the force applied in every measurement point of the elevated base. Deformation here is solely related with bending of the cantilever ($k_{NW\_base} >> k_{Probe}$). Then a bending lever is approached, which causes a gradual increase of the measured *PF* value, but the nanostructure is still stiffer than cantilever ($k_{NW} > k_{Probe}$). Increase of the presented curve is the result of the lever rule, and at distinct moment this curve have maxima. Here the spring constants are equal and Deformation registered by SPM photodetector is divided equally between bending deflection of the SPM cantilever and the NW's bending (**k_{NW} = k_{Probe}**). It should be noted, that according to the QNM calibration requirements specified by the device manufacturer, the PeakForce QNM mode is capable to measure the Young's modulus when stiffness of the probe is equal or similar to stiffness of the measured material. Actual informative bending of the NW occurs mainly on the right from the maxima, i.e. in the region where the NW is softer then the cantilever ($k_{NW} < k_{Probe}$). Then curve decreases further until the minima, where sliding becomes predominant and measurement is not stable anymore. The more right part of the curve represents bending of the SPM cantilever's base back side and one spike due to the short encounter with side plane of the cantilever. Thereafter, one can conclude that only limited part of the *PF(L)* curve, i.e. slope closer to the area where spring constants are equal, should be used to calculate the k_{NW} and the resulting elastic modulus. Still, it was already enough to obtain more than 300 data points from the profiles for each applied force. Then, the following formula (1) valid for calculation of stiffness of the conical balk (with round section) was used to calculate the Young's modulus *E* of the NWs:

$$\frac{Deformation}{PF} = \frac{1}{E}\frac{4}{3\pi}\frac{16}{D_{mid}^4}\frac{x^3}{(1+a(2x-L))(1-aL)^3} \qquad (1)$$

Here D_{mid} is the average diameter of the nanowire, L is the length of the nanowire, $a = \gamma/D_{mid}$ is a conical factor associated with a cone angle γ, and x is a coordinate of the SPM probe's contact with the nanowire.

Summarizing the elastic strength of the InP nanowires

Finally, we can compare the calculated elastic modulus data for 10 different NWs in Table 1, where first five values are shown for ZB, latest five are given for WZ. Corresponding error values are based primarily on inaccuracy of the measured diameters of the NWs. In our model, elasticity was estimated for the point of middle diameter D_{mid}, i.e. approximately near the middle of a NW. Its location was found from the *D(L)* profile for already broken NW, and it was also situated on the linear slope of *PF(L)* curve (See Fig. 5(b) where $k_{NW} < k_{Probe}$), and further proofed by model fitting. Since diameter *D(L)* is linearly dependent from the length *L*, so *PF(L)* ≡ *PF(D)*, and *E* is a property of a NW's material.

Table 1. Estimated values of Young's moduli for 10 measured ZB and WZ InP NWs.

	Middle diameter, nm	**Young's modulus, GPa**	Estimated Error, GPa
ZB series	60	60	25
	60	54	20
	54	70	30
	50	150	60
	50	80	30
WZ series	50	105	40
	42	170	70
	40	230	90
	40	340	130
	36	250	100

Young's modulus for InP ZB bulk material is known to be 61 GPa. This was approved experimentally for ZB InP NW with middle diameter >80 nm. However, according to our results, elasticity of the InP material increases with decreasing of NW diameter. Partly this can be explained in terms of the core-shell model, previously used to explain similar trend for GaAs [11] and ZnO [12] nanowires. In fact, shell atoms are dominating in the properties with decreasing of the diameter of the NWs less then ~50 nm, so stiffness of the nanostructure increases, if surface material is rigid.

Thus, it seems that shell material of InP NWs is stiffer than its core. This effect can be partially associated with properties and amount of oxidized surface material in the overall volume of the NW, since NWs were kept in air for two months before the experiments. However, the drastic difference in elasticity was observed specifically between WZ and ZB phases of InP NWs, and we were able to convince that WZ phase of InP gives strongly rigid contribution to the elastic properties of InP material.

4. Conclusions

We made measurements of elasticity of InP NWs with relatively new experimental method and developed a protocol for data treatment, which gave highly reproducible data. This can be qualified as an attempt to make reproducible accurate and fast tests with SPM in PeakForce QNM mode, which can be further used to organize a SPM-based station for assessment of elasticity of versatile bendable 1D structures. Elastic modulus of InP NWs was studied for the first time. Moreover, Young's modulus of wurtzite InP has been obtained for property indexing issues for the first time. Results show that wurtzite InP NWs are at least two times stiffer than zinc-blende InP NWs. Young's modulus increases when material's size decreases, which was observed for the NWs with diameters less than 50 nm. These considerations are proposed to improve design and characterization of NW-based structures.

References

[1] Parkinson P, Lee Yu-H, Fu L, Breuer S, Tan H and Jagadish Ch 2013 *Nano Lett.* **13** 4
[2] Duan X, Huang Yu, Cui Y, Wang J and Lieber C M 2001 *Nature* **409** 6816
[3] Mikkelsen A and Lundgren E 2013 *Surf. Sci.* **607**
[4] Vugt L K, Veen S J, Bakkers E P A M, Roest A L and Vanmaekelbergh D 2005 *J. Am. Chem. Soc.* **127** 35
[5] Zhang W, Lehmann S, Mergenthaler K, Wallentin J, Borgström M T, Pistol M-E and Yartsev A 2015 *Nano Lett.* **15** 11
[6] LaPierre R R, Chia A C E, Gibson S J, Haapamaki C M, Boulanger J, Yee R, Kuyanov P, Zhang J, Tajik N, Jewell N and Rahman K M A 2013 *Phys. Stat. Sol. RRL* **7** 10
[7] Song J, Wang X, Riedo E and Wang Z L 2005 *Nano Lett.* **5** 10
[8] Stan G, Ciobanu C V, Thayer T P, Wang G T, Creighton J R, Purushotham K P, Bendersky L A and Cook R F 2009 *Nanotechnology* **20** 035706
[9] Gordon M J, Baron Th, Dhalluin F, Gentile P and Ferret P 2009 *Nano Lett.* **9** 525
[10] Kushima A, Huang J Yu and Li J 2012 ACS Nano 6 11
[11] Chen C Q, Shi Y, Zhang Y S, Zhu J and Yan Y J 2006 *Phys. Rev. Lett.* **96** 075505
[12] Alekseev P A, Dunaevskii M S, Stovpyaga A V, Lepsa M and Titkov A N 2012 *Semiconductors* **46** 641
[13] Mohan P, Motohisa J and Fukui T 2005 *Nanotechnology* **16** 2903
[14] Wallentin J, Anttu N, Asoli D, Huffman M, Åberg I, Magnusson M H, Siefer G, Fuss-Kailuweit P, Dimroth F, Witzigmann B, Xu H Q, Samuelson L, Deppert K and Borgström M T 2013 *Science* **339** 6123
[15] Krogstrup P, Jørgensen H I, Heiss M, Demichel O, Holm J V, Aagesen M, Nygard J and Morral A F 2013 *Nature Photonics* **7** 4
[16] Sahin O and Erina N 2008 *Nanotechnology* **19** 445717

Journal of Physics: Conference Series **769** (2016) 012030 doi:10.1088/1742-6596/769/1/012030

Structural and morphological features of ultrathin epitaxial InSb films in AlAs matrix

D A Kolotovkina[1,2], **A K Gutakovskii**[1,2]

[1]A. V. Rzanov Institute of Semiconductor Physics, SB RAS, Novosibirsk, Russia

[2]Novosibirsk National Research State University, Novosibirsk, Russia

E-mail: koldarya@yandex.ru

Abstract. This work presents results of the investigation of structural and morphological features of epitaxial InSb layers in the AlAs matrix. Our research group used transmission electron microscopy (TEM). The specimens were grown by molecular beam epitaxy and prepared in the cross section (110) and plan view foils (100). We found a formation of the embedded epitaxial layer of solid solution $In_xAl_{1-x}Sb_yAs_{1-y}$ in the AlAs matrix during precipitation of In and Sb on the AlAs surface. The embedded layer had continuous area (wetting layer) and islands. The study revealed two types of islands in the epitaxial layer the first having coherent interfacing with the matrix lattice and the second a relaxed island. We estimated concentration of In, Sb in the solid solution by the indirect method. We used the method of geometric phase to analyze the distribution of misfit dislocation cores on the interface. Every misfit dislocation was formed by two close 60^0-dislocations with the Burgers vectors like $\frac{a}{2}$<110>. The sum Burgers vector of the dislocation pair was in the plane of the interface.

1. Introduction

Today the system AlAs-InSb-AlAs intrigues scientists because of the interesting features of its energy spectrum. Quantum dots (QD) of indium antimonide formed in the aluminum arsenide matrix are of the greatest interest, a structure of QD energy spectrum suppresses photon and phonon scattering of electrons. Theoretically it can provide a long life time (up to a few milliseconds) of exciton spin polarization (excitons localized in QD) [1]. In this case theory allows to predict an indirect QD energy-band structure, namely, the energy-band structure of the first type with the main electron state belonging to X-minimum of conduction band. It leads to the increased life time of the exciton up to few milliseconds that gives a chance to study the spin relaxation of exciton.

The theoretical calculations [1] point to energy spectrum of the first type of the system InSb/AlAs even if mixing of the materials of QD and matrix takes place during the process of growth. Thus, the InSb/AlAs system allows to form QD having different sizes and composition, so it gives an opportunity to study parameters of the exciton spin relaxation with various values of exchange interaction between an electron and the hole.

Structural data of the system are necessary for a correct interpretation of future experimental results. Unfortunately, at present any data about formation and structure of the InSb QD in the wide-band-gap AlAs matrix are absent in literature. A goal of the work is to study structural features and morphology of epitaxial indium antimonide layers in the wide-band-gap aluminum arsenide matrix by transmission electron microscopy (TEM).

2. Objects of investigation

The object of the investigation was GaAs-AlAs-InSb-AlAs-GaAs (001) heterosystem, grown by molecular beam epitaxy (MBE) on "RIBER compact 21T" (Figure 1). For electron microscopy (EM) investigations we chose six samples, which had different MBE modes (atom layer epitaxy or ALE and simultaneous deposition or standard MBE) and conditions (an order of the alternation, a number of the deposition cycles of In-Sb and deposition time of the flows of indium and antimony for each cycle). An effective thickness of the epitaxy layer depended on the growth conditions and was 1.3-2.0 ML. The main growth parameters of epitaxy are in Table 1.

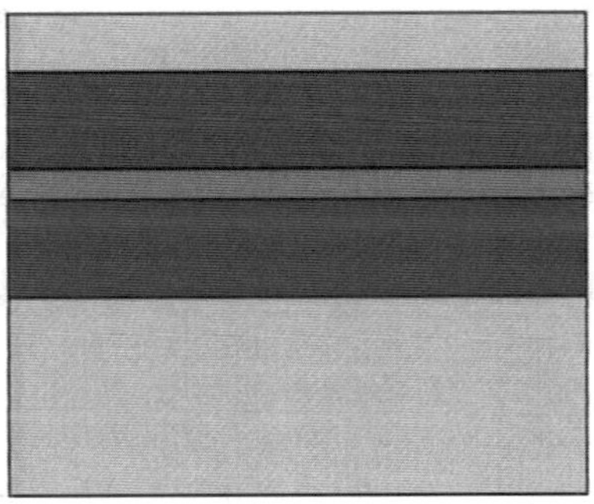

Figure.1. The schematic image of the part of the InSb/AlAs system.

All samples were prepared in the cross sections (CS) (110) and plane view (PV) (001) foils by means of ion etching for CS and chemical etching using standard solution $3H_2SO_4 : 1H_2O_2 : 1H_2O$ for PV.

Table 1

The deposition parameters of molecular beams on the AlAs (001) surface

Specimen	**938**	**939**	**940**	**941**	**1025**	**1027**
Mode of epitaxy	ALE	ALE	ALE	ALE	ALE	MBE
The expected thickness of the InSb layer, ML	1.5	2.5	2.0	1.3	1.5	1.5

3. Experimental results and their discussion

We carried out the investigations by transmission electron microscopy (TEM) in the scanning mode (STEM), in the two-beam diffraction contrast mode and in the multi-beam interference contrast mode (HREM) using JEOL-4000EX and TITAN 80-300 microscopes.

For the analysis of the structural features we used EM images filtered from "noise", associated with non-elastic scattering. We obtained two types of the cross sectional images. The first (Figure 2) is typical only for sample № 940. The second type is characteristic for the other samples (Figure 3). The analysis showed a formation of continuous (wetting) layer with islands for all specimens. The wetting layer is coherent interfaced with the crystal lattice of the matrix, its average thickness was 3-4 nm having statistical deviation about 0.5 nm. The experimental thickness of the epitaxial layer in the AlAs matrix differed from the expected thickness, so we supposed formation of the solid solution $In_xAl_{1-x}Sb_yAs_{1-y}$ in the matrix instead of the binary compound. We measured the interplanar distance on the obtained HREM images (Figure 2) using digital diffraction patterns. The interplanar distances in continuous areas of the embedded epitaxial layer differed from the corresponding values of the matrix less than 3%, indicating the concentration of indium and antimony in the solid solution is low. To estimate chemical composition of $In_xAl_{1-x}Sb_yAs_{1-y}$ we used Vegard's law. The possible values of «x» and «y» in the solid solution do not exceed 0.2, if «x» is equal «y».

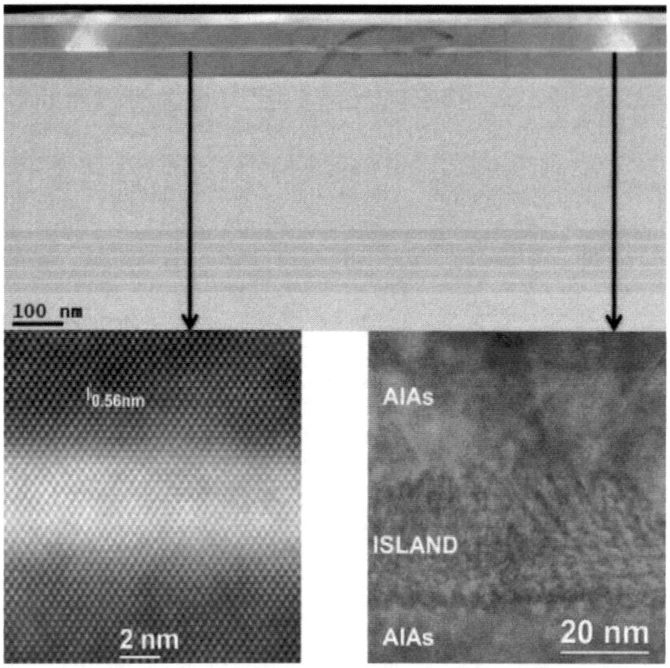

Figure 2. The cross sectional dark field EM images of sample № 940 (the top), where large islands in the AlAs matrix appear. The arrows point to STEM (the left) and HREM (the right) images of the according cross sectional areas. Titan 80-300, 300 keV and JEOL-4000EX, 400 keV.

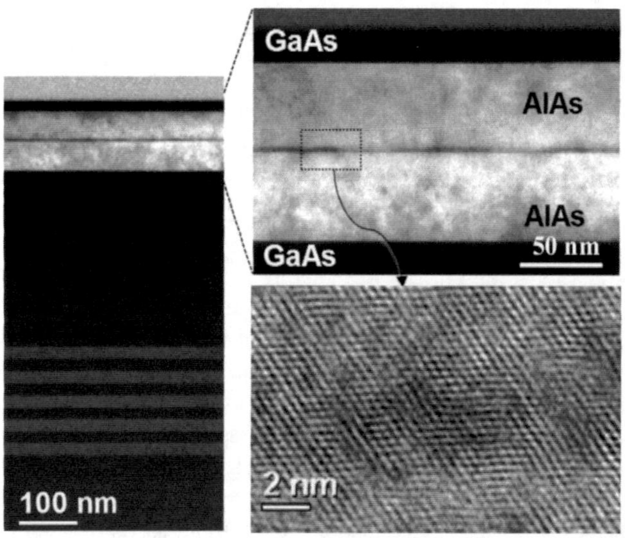

Figure 3. The characteristic cross sectional bright field EM images of samples № 938, № 939, № 941, № 1025, and № 1027. The dotted rectangle on the enlarged image of the epitaxial layer corresponds to an island. The arrow points to the typical HREM image of the islands. JEOL-4000EX, 400 keV.

We got TEM images of the PV foils to study an island film which had the definite MBE conditions (Figure 4, Figure 5). These images revealed small and large islands in the epitaxial layer. The small islands have the contrast, which is characteristic for coherent inclusions, meaning their crystal lattice is interfaced with the lattice of the matrix coherently (Figure 4a, Figure 5a, Figure 5e, Figure 5f, Figure 5i, and Figure 5l). The large islands are relaxed, so parallel moiré pattern originated in the TEM images of these islands (Figure 4b, Figure 5b, Figure 5c, Figure 5g, and Figure 5j). Structural defects in a volume of the islands result in the break and local non-parallelism of the moiré pattern's fringes.

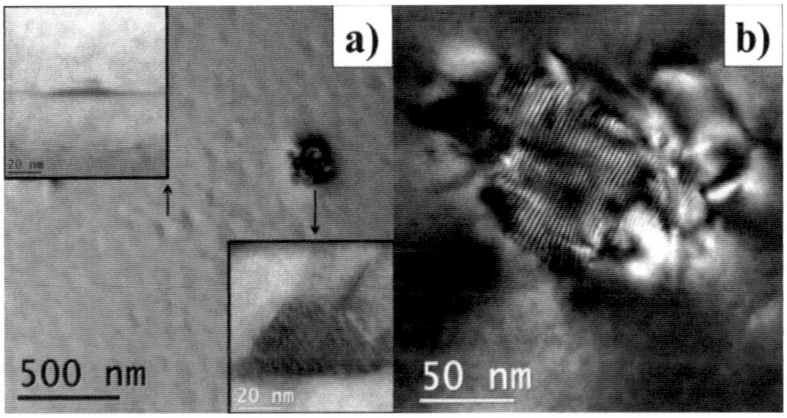

Figure 4. The bright (a) and dark (b) field EM images of the island film (sample № 940): the arrows correspond to typical large and small islands. The characteristic cross sectional images of the islands are at insertions. JEOL-4000EX, 400 keV.

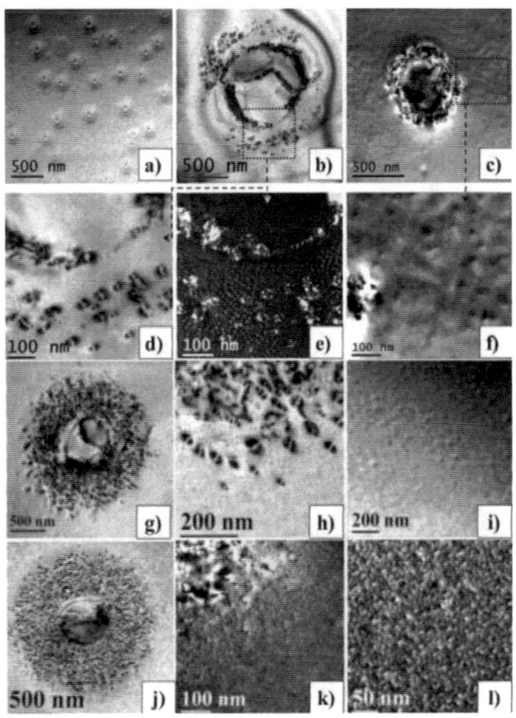

Figure 5. The dark field EM images of the island films of samples № 938 (a), № 939 (c, f), № 941 (b, d, e), № 1025 (g, h, i), and № 1027 (j, k, l). Figure 5d and Figure 5e, obtained under the precise Bragg condition and the "weak beam" condition correspondingly, present the enlarged image of the rectangle in Figure 5b. Figure 5f is the enlarged image of the rectangle in Figure 5c. PV (001), JEOL-4000EX, 400 keV.

For specimen № 938 the contrast on the island TEM images (Figure 5a) indicates the existence of a disk-shaped "pedestal" under each island. The contrast increases along the outer perimeter, meaning the lateral surface of the island has a step form. The islands have coherent interfacing with the crystal lattice of the matrix and uniform distribution. In this case, we did not detect the large islands shown in Figure 2, Figure 4, and Figure 5 during the TEM study. In other specimens the small coherent islands were localized near the large islands.

We estimated the average lattice constant of the relaxed islands ($\mathbf{a}_d$) using the moiré pattern period and digital diffraction patterns: $\mathbf{a}_d \approx 5.96$ Å (specimen № 939), $\mathbf{a}_d \approx 6.15$ Å (specimen № 940), $\mathbf{a}_d \approx 5.87$ Å (specimen № 941), $\mathbf{a}_d \approx 5.90$ Å (specimen № 1025), and $\mathbf{a}_d \approx 6.00$ Å (specimen № 1027). An evaluation of the chemical composition of the formed solid solution $In_xAl_{1-x}Sb_yAs_{1-y}$ was done assuming that the hetero-epitaxial stress was relaxed and Vegard's law was kept. If «x» is equal «y», then values of «x» and «y» in $In_xAl_{1-x}Sb_yAs_{1-y}$ do not exceed 0.35 (specimen № 939), 0.59 (specimen № 940), 0.25 (specimen №941), 0.27 (specimen № 1025), and 0.40 (specimen № 1027).

We used the geometric phase analysis to visualize the distribution of misfit dislocation cores on the interface (Figure 6). The Burgers vectors were determined by classic method with the Burgers contour (Figure 7).

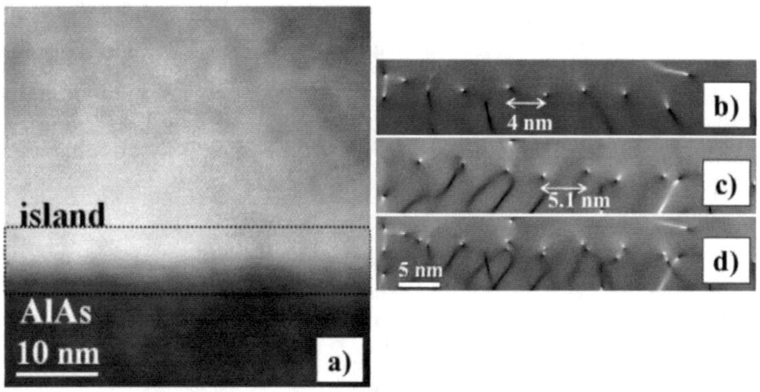

Figure 6. The dark field TEM image of interface of large island and matrix (specimen № 940) (a) and the distribution map of interplanar distances for (1-11) (b) and (-111) (c). Figure 6d is superposition of Figure 6b and Figure 6c. We used the geometric phase analysis in the dotted rectangle in Figure 6a to obtain Figure 6b and Figure 6c.

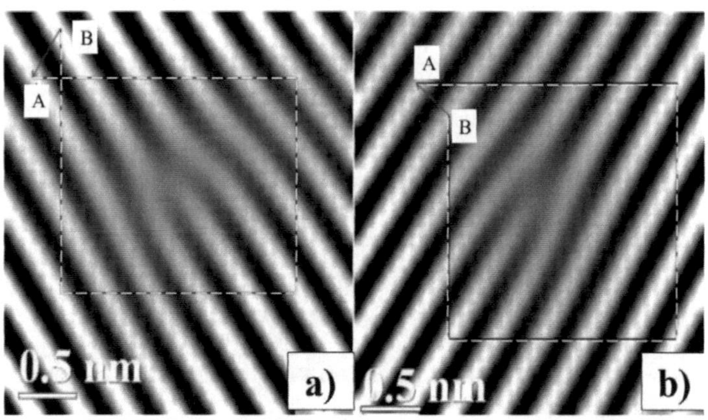

Figure 7 .The digital images, which were filtered near misfit dislocation core using reflexes (1-11) (a) and (-111) (b). The Burgers contour of the misfit dislocation of specimen № 940 (the dotted line) and its Burgers vector (the solid line).The letters "A" and "B" point to start and end of the Burgers contour.

The method revealed, that a base type of dislocations in the InSb/AlAs system was 60^0-dislocation with the Burgers vectors like $\frac{a}{2}$<110>, but the dislocation cores were splintered. Every misfit dislocation was formed by two close 60^0-dislocations with the Burgers vectors like $\frac{a}{2}$<110>. The sum Burgers vector of the dislocation pair was situated in the plane of the interface. Thus the dislocation pair was similar to edge misfit dislocation.

4. Conclusion

We revealed origination of the embedded $In_xAl_{1-x}Sb_yAs_{1-y}$ epitaxial layer in the AlAs matrix (instead of a pure binary compound InSb) during the precipitation of In and Sb on the AlAs surface (and the following deposition of aluminum arsenide layer). Its characteristics depend on the MBE conditions, but the continuous epitaxial layer with islands was grown in all cases. To use ALE is not a critical requirement in this case, because the growth mechanism was uniform for specimen № 1025 (ALE) and № 1027 (standard MBE). The main morphological and structural features of the formed epitaxial film are:

1. There is the wetting layer with possible concentration of In («x») и Sb («y») up to «x»=«y»≈0.2. The continuous layer has coherent interfacing with the crystal lattice of matrix.
2. In sample № 938 we found small islands with uniform distribution on the wetting layer's surface. They have coherent interfacing with the AlAs lattice too. Their average size and density are 100 nm and 10^8 cm^{-2} accordingly.
3. In the other samples the small coherent islands were localized around large islands.
4. The epitaxial film contains incoherent (relaxed) large islands with possible concentration of In («x») и Sb («y») up to «x»=«y»≈0.35 (specimen № 939), «x»=«y»≈0.59 (specimen № 940), «x»=«y»≈0.25 (specimen № 941), «x»=«y»≈ 0.27 (specimen № 1025), and «x»=«y»≈0.40 (specimen № 1027). The parallel moiré pattern appears in the TEM images of these islands. The defects of crystal structure originated in specimen № 940.

Acknowledgments

TEM studies were performed using the equipment of CCU "Nanostructures". The part of the work carried out with the support of the Ministry of Education and Science of the Russian Federation (project ID RFMEFI62114X0004) and RSCF (project No 14-22-00143)

References

[1] Shamirzaev T 2011, Type-I semiconductor heterostructures with an indirect-gap conduction band, *J. Semiconductors* vol. **45** No. 1 pp. 96 – 102

Synthesis of composite structures on the basis of $GaBO_3$ and $FeBO_3$ trigonal crystals

S M Mustafaeva[1], M P Matvienko[1], Yu A Mogilenec[1], E M Maksimova[1], I A Nauhatsky[1], M B Strugatsky[1], K A Seleznyova[1], S V Yagupov[1]

[1] V.I. Vernadsky Crimean Federal University, 4 Vernadsky Avenue, Simferopol, Republic of Crimea, 295007 Russia

E-mail: mogilenecya@cfuv.ru

Abstract. Film of iron borate $FeBO_3$ is of a great interest as a model object for various scientific studies and is a promising material for practical applications in different branches of experimental science and engineering. Thin magnetic films of $FeBO_3$ on $GaBO_3$ single crystal substrate were prepared by a liquid phase epitaxy route. In order to observe a mechanism of film formation we have carried out optical microscopy studies. X-ray diffraction analysis has confirmed the existence of the $FeBO_3$ layer on the $GaBO_3$ substrate.

1. Introduction

Iron borate, $FeBO_3$, is one of the most interesting dielectric magnetic materials. This is due to a unique set of its properties: magnetic, magneto-acoustic, optical, magneto-optical, resonance, etc, see e.g. [1-6]. From the viewpoint of magnetic properties $FeBO_3$ is an easy-plane transparent antiferromagnet with a weak ferromagnetism, the Néel temperature being 348 K. It has a rhombohedral calcite structure with D^6_{3d} space group [7]. Due to the features of the magneto-crystalline structure, characteristics of a thin near-surface layer of iron borates radically differ from those of the volume [8, 9]. We assume that thin iron borate magnetic films are appropriate objects to study the effects of surface magnetism by magneto-optics techniques. Thus the aim of the present work is to develop synthesis technique and to prepare $FeBO_3$ films on a single crystal optically transparent diamagnetic substrate. To achieve this goal it is necessary to:

- choose and prepare a high-quality single crystal to be used as a substrate for synthesis a thin layer of iron borate;
- choose the synthesis technique, determine charge compositions and temperature modes for obtaining iron borate films.

Iron borates can be grown by two routes: (i) from the gas phase and (ii) from the solution in the melt. The first technique allows to synthesize bulk crystals [10], and the second one to obtain thin hexagonal plates [11, 12]. For the purposes of the present work substrates of a high structural perfection are required. The solution in the melt technique has proved to be most appropriate to synthesize required single crystals of a high quality [12, 13]. Thus it should be used for the preparation of the substrate. As a most promising technique for synthesizing thin iron borate films a liquid phase epitaxy (LPE) has been selected.

In the present work we develop synthesis technique, obtain composite structures on the basis of iron borate crystals, carry out optical microscopy and X-ray diffraction (XRD) studies of the synthesized samples.

2. Experimental

Gallium borate $GaBO_3$ single crystals have been used as a substrate to synthesize thin iron borate film. The $GaBO_3$ solution in the melt crystallization was carried out in the Ga_2O_3 - B_2O_3 - PbO - PbF_2 system.

Iron borate $FeBO_3$ film has been synthesized by LPE in the Fe_2O_3 - B_2O_3 - PbO - PbF_2 system.

The charge compositions and temperature modes for both crystallizations have been chosen based on the results of differential thermal analysis and probe method [12].

Control of temperature mode of crystallization was carried out using laboratory-developed microprocessor system. Experiments can be performed in a temperature range from room temperature to 1150°C, maintaining the accuracy of the temperature ±0.2°C.

In order to study steps and to determine a mechanism of film formation we have carried out studies using optical microscope.

The unit cell parameter *c* [7] of the film have been determined by means of XRD analysis with a DRON-3 diffractometer using monochromatic copper radiation (K $\lambda_{\alpha 1}$ = 1.54051 Å).

3. Results and discussions

One of the most important conditions for a successful synthesis of a magnetic film is a selection and preparation of an appropriate substrate. Gallium and iron borates, $GaBO_3$ and $FeBO_3$, both have calcite structure with space group D^6_{3d} [14], therefore, their unit cells are identical. According to the Royer-Freidel, a fusion of two crystals in a regular orientation is possible if the corresponding parameters of unit cells will not differ by more than 15%. Table 1 shows the comparative characteristics of $GaBO_3$ and $FeBO_3$ structures.

Table 1. Comparative characteristics of $GaBO_3$ and $FeBO_3$ structures

Crystal	Space group	Crystal system	Unit cell parameters in the hexagonal system [15]	The difference in the parameters
$FeBO_3$	D^6_{3d}	trigonal	a = 4.62 Å c = 14.49 Å	Δa = 1.29 % Δc = 2.13 %
$GaBO_3$			a = 4.56 Å c = 14.18 Å	

As one can see from Table 1, the unit cell parameters of gallium and iron borates differ from each other by no more than 2%. It makes $GaBO_3$ a promising candidate for use as a substrate for the magnetic film $FeBO_3$.

The synthesis of the $GaBO_3$ substrate starts with a charge preparation; fusing the charge in a crucible at 900°C in the muffle furnace and setting of the crucible with the solution melt in the growing furnace. Crystallization has been carried out according to predetermined temperature mode. It consists of the following steps:

- heating of the furnace;
- homogenization of the solution melt at 950°C during 24 hours (At the beginning of this step a seed holder was dipped and rotated in the solution melt. During the homogenization it served as a mixer.);
- sharp, with a rate of 280°C per hour, lowering the temperature down to 810°C in order to avoid the emergence of "parasitic" phases, such as Ga_3BO_6;

- slow, with a rate of 0.3°C per hour, lowering the temperature down to 760 °C for nucleation and $GaBO_3$ crystal growth (At the end of this step the seed holder with the synthesized crystals was extracted from the solution melt, see Figure 1a. These crystals on the holder will be used as a substrate for the $FeBO_3$ film.);
- cooling of the furnace.

The sizes of the crystals are in the range of 1 to 3 mm in the cross section and 0.01 to 0.06 mm thick. The synthesized $GaBO_3$ samples are suitable for use as a substrate for the $FeBO_3$ film.

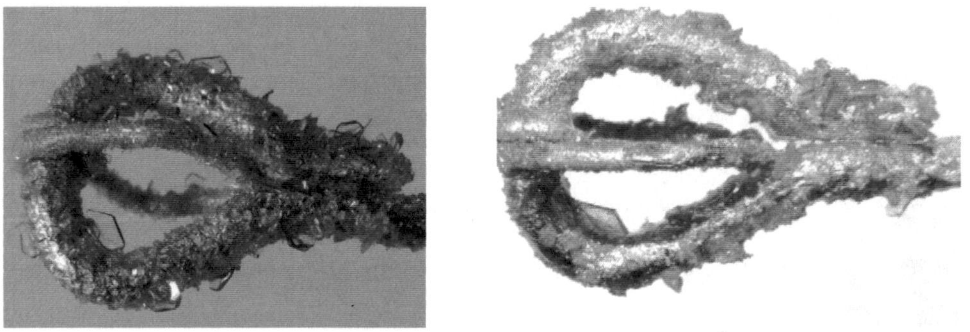

a b

Figure 1. The seed holder with almost transparent $GaBO_3$ single crystals (a), and green online film of $FeBO_3$ on the $GaBO_3$ substrate (b).

For the LPE synthesis of the $FeBO_3$ film, first, we have to prepare a solution melt. For this purpose we have used the same procedure as for $GaBO_3$ synthesis, vide supra. After setting the crucible with solution melt for $FeBO_3$ in the growing furnace the general crystallization procedure was as follows:

- heating of the furnace;
- homogenization of the melt at 950°C during 24 hours;
- sharp, with a rate of 260°C per hour, lowering the temperature down to 810-835°C in order to avoid the emergence of "parasitic" phases, such as Fe_3BO_6 and α-Fe_2O_3 (During this step a seed holder with $GaBO_3$ substrate (was prepared earlier and shown in Figure 1a) was dipped in the solution melt at 835°C.);
- keeping the temperature 820°C during 0.5-3 hours;
- slow, with a rate of 3°C per hour, lowering of the temperature down to 800°C (At the end of this step the seed holder has been extracted from the crucible, see Figure 1b. As one can see, $GaBO_3$ crystals became green.);
- cooling of the furnace.

The second one important condition for the successful epitaxial synthesis is to choose appropriate temperature modes. Crystallization regimes suitable for obtaining epitaxial film of $FeBO_3$ were determined empirically. We have carried out a series of crystallizations with and without rotation of the seed holder, testing different temperatures modes and time of keeping the seed holder at 820°C. Figure 2 shows optical microscope pictures of the obtained samples. As one can see from Figure 2a, microcrystals of iron borate (0.006 - 0.012 mm in cross section) have the same orientation on the surface of a gallium borate. This indicates that the $FeBO_3$ film grows copying the $GaBO_3$ substrate structure. Analysis of Figure 2 allows concluding that the mechanism of the film formation is island growth [16]. Our preliminary XRD studies of the synthesized samples have allowed determining a value of the hexagonal parameter c=14.48±0.01 Å of the film. This value is in a good accordance with those previously determined for the $FeBO_3$ crystal, see Table 1, that confirms the existence of the $FeBO_3$ layer on the $GaBO_3$ substrate. The detailed results on our XRD studies will be published elsewhere. The obtained epitaxial film of iron borate, see Figure 1b and 2d, has a thickness of 0.003 to 0.006 mm.

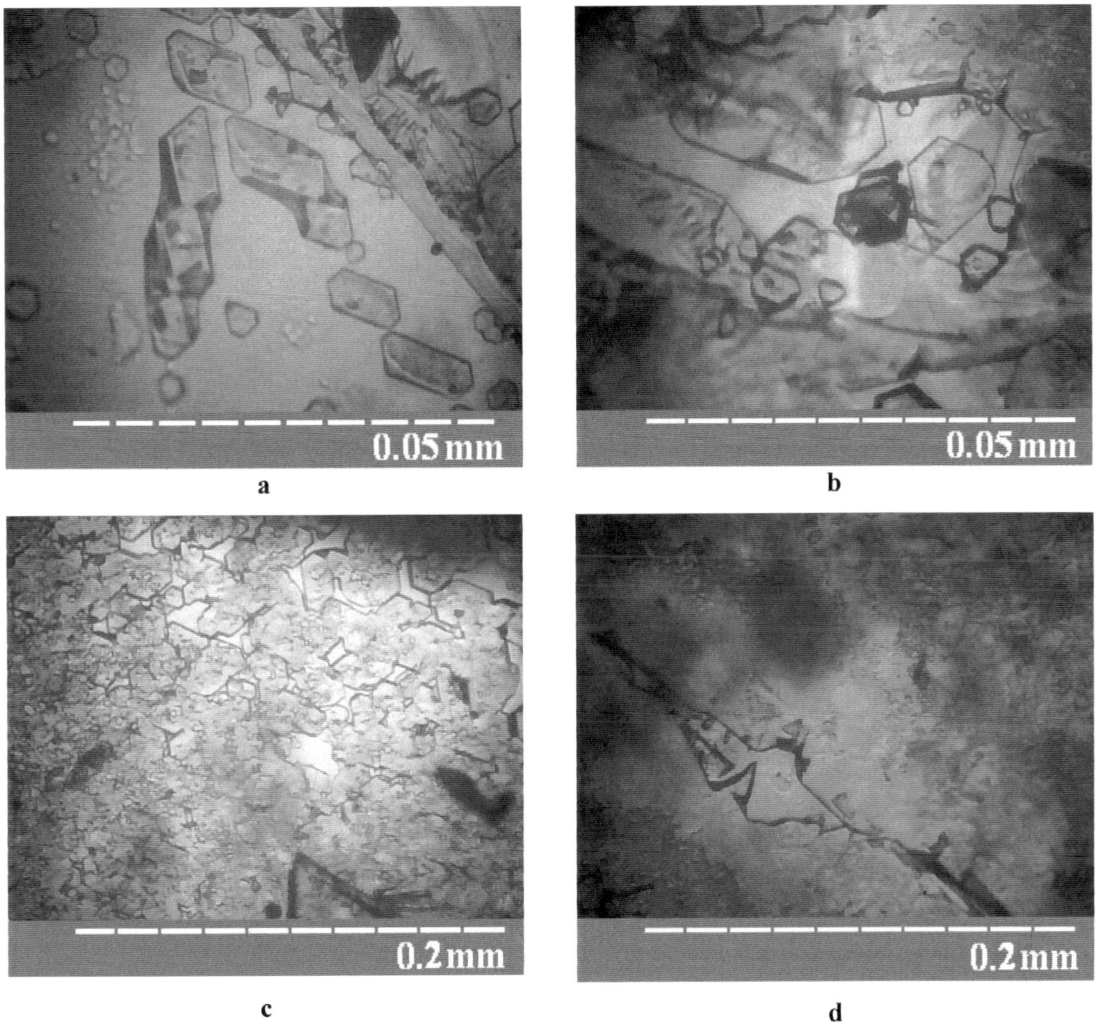

Figure 2. Optical microscope pictures showing different stages of a formation of the $FeBO_3$ film: a coalescence of islets (a), a formation and filling of channels (b), a process of accretion of individual sections of the film $FeBO_3$ (c), a final stage of the formation of the film of iron borate on the $GaBO_3$ substrate (d).

4. Conclusions

As a result, we have developed synthesis technique and obtained the film of $FeBO_3$ on the diamagnetic substrate using LPE technique. We have succeeded in the selection and preparation of the single crystal for the substrate. It is shown that gallium borate $GaBO_3$ is an appropriate material for this purpose. In the framework of the present work a series of the experiments has been carried out in order to determine crystallization regimes suitable for the film $FeBO_3$ growth.

The optical studies of the prepared composite structures have allowed us to study in details different stages of the film formation and to conclude that the growth mechanism is island growth. The XRD analysis of the obtained samples confirms the existence of the iron borate film on the $GaBO_3$ substrate.

Acknowledgement
This work was partially supported by Russian Foundation for Basic Research and Republic of Crimea in the framework of scientific project Grant No.15-42-01008 "p_юг_a".

References
[1] Eibshutz M, Lines M E 1973 *Phys. Rev.* B **7** 4907
[2] Afanasiev D, Razdolski I, Skibinsky K M, Bolotin D, Yagupov S V, Strugatsky M B, Kirilyuk A, Rasing Th, Kimel A V 2014 *Phys. Rev. Let.* **112** 147403
[3] Strugatsky M B, Skibinsky K M, Tarakanov V V, Khizhnyi V I 2002 *JMMM* **241** 330
[4] Malakhovskii A V, Edelman I S, 1976 *Phys. Status Solidi* B **74** K145
[5] Jantz W, Sandercock J R, Wettling W 1976 *J. Phys.*C **9** 2229
[6] Velikov L V, Prokhorov A S, Rudawevskii E G, Seleznev V N 1974 *JETP* **66** 1847
[7] Diehl R, Jantz W, Nolang B I and Wettling W 1984 *Current Topics in Materials Science* ed E Kaldis (New-York: Elsevier) vol **11** p 241
[8] Zubov V E, Krinchik G S, Seleznyov V N, Strugatsky M B 1990 *JMMM* **86** 105
[9] Maksimova E M, Nauhatsky I A, Strugatsky M B, Zubov V E 2010 *JMMM* **322** 477
[10] Pankratov A K, Strugatsky M B, Yagupov S V 2007 *Sci. Notes of Taurida National University Ser. Phys.* **20(59)** 64
[11] Strugatsky M B, Yagupov S V 2007 *Sci. Notes of Taurida National University Ser. Phys.* **19(58)** 76
[12] Yagupov S, Strugatsky M, Seleznyova K, Maksimova E, Nauhatsky I, Yagupov V, Milyukova E, Kliava J 2015 *Phys. A* **121** 179
[13] Seleznyova K, Strugatsky M, Yagupov S, Postivei N, Artemenko A, Kliava J 2014 *Phys. Status Solidi* B **251** 1393
[14] Bernal L, Struck C W and White T G 1963 *Acta Cryst.* **16** 849
[15] Yagupov S, Maksimova E, Nayhatsky I, Yagupov V, Milyukova E, Seleznyova K, Strugatsky M 2014 *Abs. of Int. Conf. on Oxide Materials for Electronic Engineering* (Lviv: Lviv Polytechnic National University) p 207
[16] Maissel L I and Glang R 1970 *Hand Book of Thin Film Technology* (New York: McGraw Hill Book co)

Theoretical investigation of thin-film multiferroic heterostructures based on a width-modulated slot transmission line

A A Nikitin[1,2], A A Nikitin[1,2,3], A B Ustinov[1,2], E Lähderanta[2], A Stashkevich[3,4]

[1]Department of Physical Electronics and Technology, St. Petersburg Electrotechnical University, St. Petersburg, 197376 Russia
[2]Department of Mathematics and Physics, Lappeenranta University of Technology, Lappeenranta, 53850 Finland
[3]International laboratory "MultiferrLab", ITMO University, St. Petersburg, 197101 Russia
[4]LSPM (CNRS-UPR 3407), Universite Paris 13, Sorbonne Paris Cite, 93430 Villetaneuse, France

E-mail: alexeynikitin1@gmail.com

Abstract. The microwave properties of multiferroic magnonic crystals based on a width-modulated slot-line were investigated theoretically for the first time. Dispersion characteristics of spin-electromagnetic waves propagating in the heterostructures were calculated. It was shown that band gaps in the spin-electromagnetic wave spectrum are controlled with both electric and magnetic fields.

I. Introduction

One of the most common ways to implement a magnonic crystal (MC) is a periodic modulation of a waveguiding structure. Nowadays an increased interest to study such structures for microwave applications is evident. From one hand they are a convenient model for the investigation of linear and nonlinear wave effects [1]. From other hand new functional microwave devices can be created with MCs. It can be generators [2], phase shifters [3], and other [4].

Development of frequency-agile materials for microwave devices has led to appearance of artificial multiferroics [5]. The multiferroics are usually fabricated by a combination of ferrite and ferroelectric materials so as to obtain structures in the form of multilayers, pillars, spheres, wires and other. A distinctive feature of the multiferroics is dual tunability of their physical properties. The tunability exists basically due to two effects. The first one is magnetoelectric effect based on mechanical interaction between ferrite and ferroelectric crystal lattices [6]. The second one is effect of electrodynamic coupling of spin waves (SWs) and electromagnetic waves (EMWs) in the ferrite-ferroelectric structures [7]. These coupled waves are called hybrid spin-electromagnetic waves (SEWs).

The purpose of the present work is the theoretical investigation of thin-film multiferroic heterostructures based on a width-modulated slot transmission line. This periodic heterostructure can be considered as a dual tunable multiferroic magnonic crystal.

II. Topology of the magnonic crystal

The investigated periodic structure is shown in figure 1. It consists of several layers: a dielectric substrate (layer $j = -2$), a ferroelectric film (layer $j = -1$), a ferromagnetic film (layer $j = 1$) and a dielectric substrate (layer $j = 2$). Thickness and dielectric permittivity of a layer with number j will be designated as d_j and ε_j , respectively. The slot transmission line has the form of a narrow slot between two infinitely thin perfectly conducting metal electrodes. Bias voltage can be applied to these electrodes in order to change the polarization of the ferroelectric film.

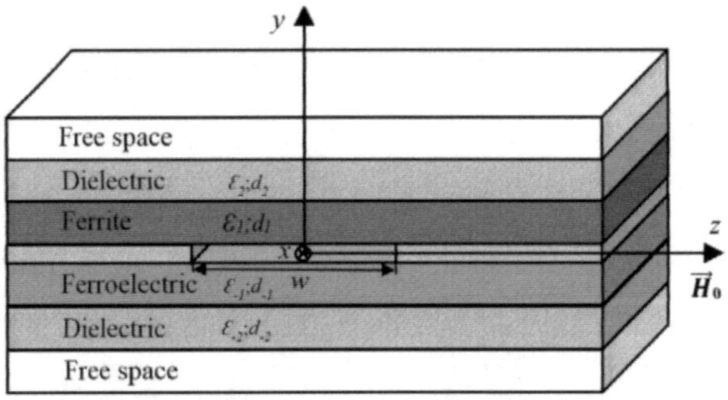

(a)

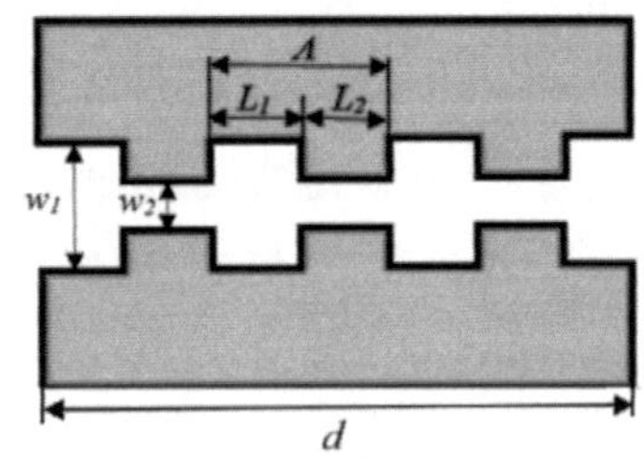

(b)

Figure 1. A cross section of the thin-film multiferroic structures based on a slot-line (a) and a topology of electrodes of the slot-line (b).

As is shown in Fig. 1b a slot width varied with a period Λ in order to implement microwave MC. It consists of sections of a length L_1 and a width w_1, and sections of a length L_2 and a width w_2. We assume that the SEW propagates along the x-axis in the tangentially magnetized structure.

III. Theoretical model of the thin-film multiferroic magnonic crystals

Development of the theoretical model of the structure was carried out in several steps. As the first step we calculated dispersion characteristics of SEWs in the ferrite-ferroelectric structure based on the slot transmission lines with different width. Dispersion relation for SEWs in the periodic structures formed with multiferroic slot-line sections with different width of the slot were obtained by coupled-mode approach. Following this method, a periodic variation of the slot width is considered as a disturbance. It leads to a coupling between the unperturbed waves. Therefore the SEW dispersion in the magnonic crystal is described by the following expression [8]:

$$\cos K\Lambda = \cos k_1 L_1 \cos k_2 L_2 - \frac{k_1^2+k_2^2}{2k_1 k_2} \sin k_1 L_1 \sin k_2 L_2 \quad (1)$$

where K is Bloch wave vector, k_1 and k_2 are wave numbers of SEWs for wide and narrow sections of the slot transmission line. These wave numbers were derived with a method of approximate boundary conditions developed in the work [9]. This work shows that such a method allows to calculate dispersion characteristics of SEWs in planar multiferroic heterostructures with high accuracy.

As the second step, we calculated frequency response with the transfer-matrix method described in details in the work [10]. The advantage of this method is a possibility of modelling S-parameters for the periodic finite-length structure taking into account the losses during the wave propagation. In this method, the travelling waves in a narrow and wide sections of the slot line and reflecting waves at the transition between the sections are described by T-matrices. Propagation of SEWs in the narrow section of the slot-line is described by the matrix:

$$T_1 = \begin{bmatrix} \exp[(-ik_1 + \alpha_1)d_1] & 0 \\ 0 & 1/\exp[(-ik_1 + \alpha_1)d_1] \end{bmatrix}. \quad (2)$$

Reflection of wave from the boundary between narrow and wide slot-line sections is described by the matrix:

$$T_2 = \begin{bmatrix} 1/(1-G) & G/(1-G) \\ G/(1-G) & 1/(1-G) \end{bmatrix}. \quad (3)$$

Propagation of SEWs in the wide section of the slot-line is described by the matrix:

$$T_3 = \begin{bmatrix} \exp[(-ik_2 + \alpha_2)d_2] & 0 \\ 0 & 1/\exp[(-ik_2 + \alpha_2)d_2] \end{bmatrix}, \quad (4)$$

and reflection of wave from the boundary between wide and narrow sections of the slot-line is described by the matrix:

$$T_4 = \begin{bmatrix} 1/(1+G) & -G/(1+G) \\ -G/(1+G) & 1/(1+G) \end{bmatrix}. \quad (5)$$

Here G is a reflection coefficient from the boundary between the sections, $\alpha_1(\omega)$ and $\alpha_2(\omega)$ are damping decrements of SEWs in wide and narrow sections, respectively. To determine the damping decrements we use the following approximate expression [11]:

$$\alpha(\omega) = |\frac{\partial k(\omega)}{\partial H}| \cdot \Delta H + |\frac{\partial k(\omega)}{\partial \varepsilon}| \cdot \varepsilon_{-1} \cdot tan\delta, \quad (6)$$

where ΔH is the half-width of the ferromagnetic resonance curve, ε_{-1} is the permittivity of the ferroelectric layer, and $tan\delta$ is the dielectric loss tangent of the ferroelectric

Propagation of SEWs at the period length Λ can be described by serial multiplication of the four T-matrix:

$$T_\Lambda = T_1 \cdot T_2 \cdot T_3 \cdot T_4. \quad (7)$$

Therefore, propagation of SEWs at the overall MC with the number of periods N is described by the following expression:

$$T_\Sigma = T_\Lambda^N. \quad (8)$$

Finally, the transmission coefficient of the whole MC can be obtained from the following equation:

$$S_{21} = 20 \cdot \log(1/T_{\Sigma_{11}}). \quad (9)$$

IV. Numerical analysis

Dispersion characteristics of SEWs and transmission characteristics of the thin-film multiferroic periodic structure were investigated with the theoretical model discussed in the previous Section. We assume that the structure consists of an epitaxial yttrium iron garnet (YIG) film, a gadolinium gallium garnet (GGG) substrate, the slot-line electrodes, a polycrystalline barium strontium titanate (BST) film, and a sapphire substrate. The calculations were carried out for the typical parameters of these materials.

For numerical simulation we assumed that the dielectric layer with number "-2" has parameters of the sapphire substrate with thickness d_{-2} = 500 μm and permittivity ε_{-2} = 10. The layer "-1" has parameters of the polycrystalline film of a BST with d_{-1} = 2 μm and ε_{-1} = 1500. Parameters of a YIG film with number "1" and a GGG substrate with number "2" are d_1 = 13.6 μm, ε_1 = 14, and d_2 = 500 μm, ε_2 = 12, respectively.

Figure 2 shows dispersion characteristics for thin-film multiferroic structures based on a slot-line without modulation. These dependences were calculated for slot-lines having different width of the slot *w* of 50 and 75 μm, a magnetization $4\pi M_0$ = 1750 Oe, and a magnetic field H= 1350 Oe.

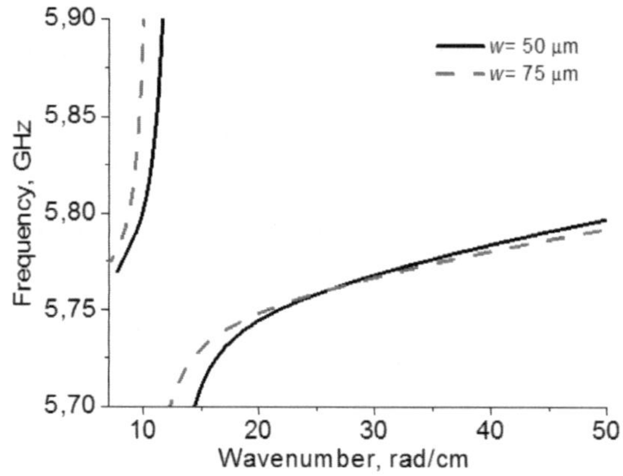

Figure 2. Dispersion characteristics of SEWs for slot-lines having different width of the slot *w* as indicated.

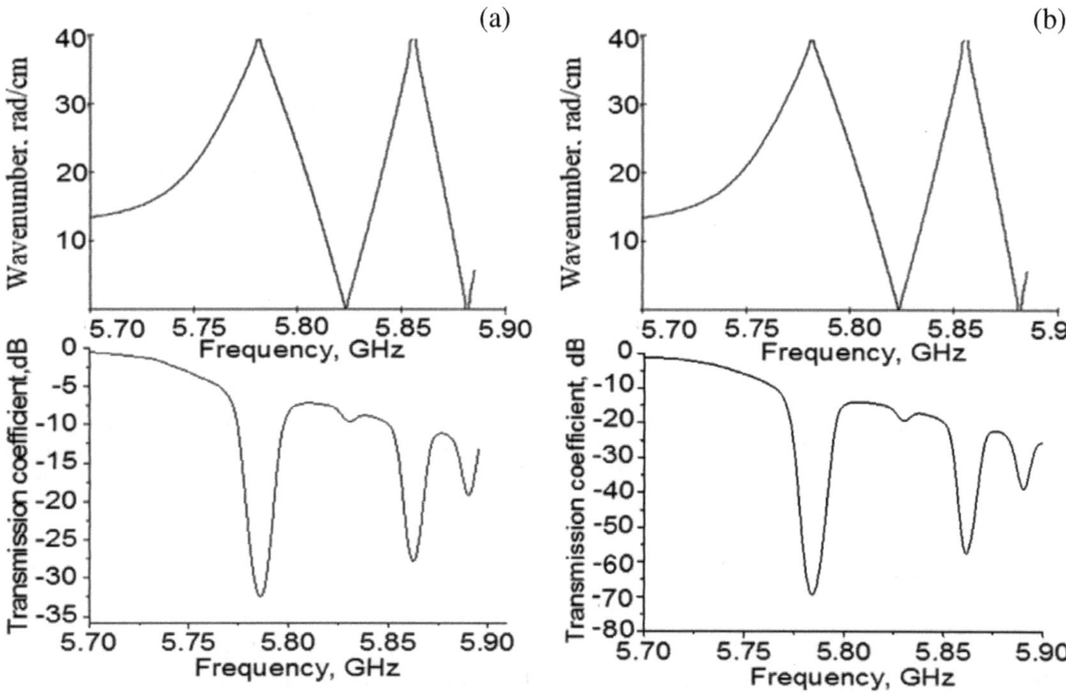

Figure 3. SEW dispersion characteristics (top graphs) and the amplitude-frequency characteristics (bottom graphs) for different number of the period N = 10 (*a*) and 20 (*b*).

Dispersion characteristics of SEWs and amplitude-frequency characteristics (AFCs) of the multiferroic magnonic crystal for the slot width of w_1=50 μm, w_2=75 μm are shown in figure 3. The numerical simulation was carried out for the fixed value of period Λ = 500 μm but for different number of the periods, namely N = 10 (Fig. 3(*a*)) and N = 20 (Fig. 3(*b*)). The results show that an increase in the number of the period (or in another words in the length of the structure) results in the increasing of transmission losses at the frequencies of the band gaps. For example, if the structure consists of 20 periods, the first band gap provides losses of -69 dB. Thus, one can obtain different AFCs with the necessary attenuation of SEWs by changing the length of the multiferroic MC.

Electric tuning of frequency responses are shown in figure 4. An electric field applied to the slot-line electrodes produces a reduction of the ferroelectric film permittivity ε_{-1}. It provides an increase in group velocity of the electromagnetic wave in the slot-line. Calculation parameters are the same like in previous case except that permittivity ε_{-1} was changed from 1500 (solid line) to 1000 (dashed line).

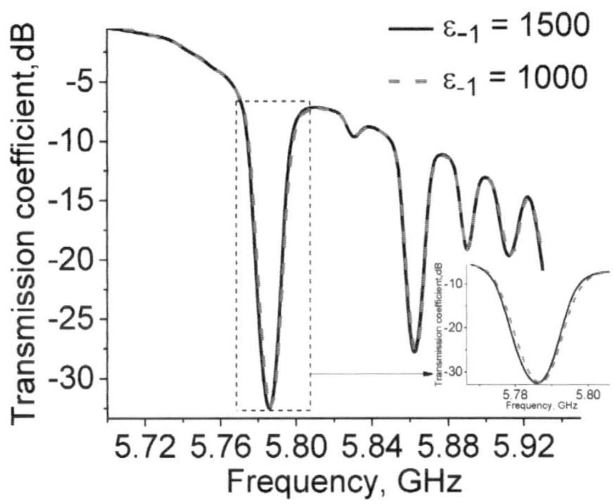

Figure 4. Amplitude-frequency characteristics for different ferroelectric permittivity.

As is seen from figure 4, a decrease in ε_{-1} leads to the shift of the band gap toward the higher frequencies. A width of band gaps reduces with frequency increasing. This behaviour is determined by the efficiency of hybridization electromagnetic and spin waves. The difference between transmission spectra shown in Fig. 4 reaches values of 880 kHz with the ferroelectric film permittivity varying from 1500 to 1000. Such a change in permittivity can be obtained by applying control voltage about 100 V to the slot-line electrodes. In practice, we can detect this electrical tuning range by the vector network analyzer. It allows to measure the transmission spectra with an accuracy of 1 Hz.

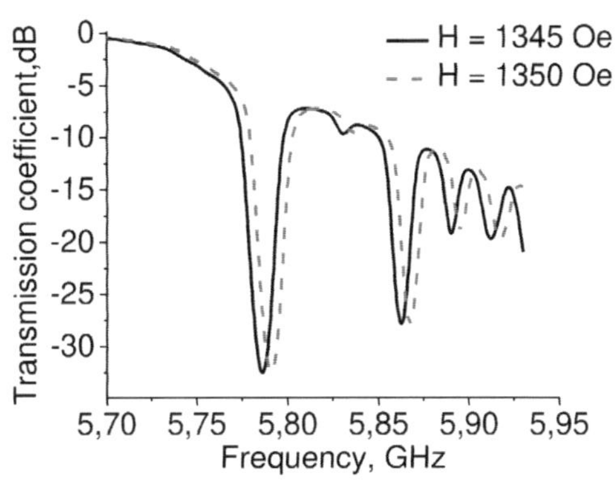

Figure 5. Amplitude-frequency characteristics for different bias magnetic fields.

Similar characteristics were simulated for different values of the external magnetic field *H* of 1345 Oe (solid line) and 1350 Oe (dashed line). They are shown in figure 5. An increase of the external magnetic field *H* leads to the shift of the SEWs spectrum toward higher frequencies. Therefore, the area of the effective hybridization of EMW and SW demonstrates an up-frequency shift also.

It was shown that a significant advantage of the multiferroic MCs is a possibility to control dispersion and transmission characteristics by both electric and magnetic fields. The advantage of electrical tuning is high speed of operation. It means that this method allows to shift band gaps for a few MHz (Fig.4). If it is necessary to control the transmission spectra in relatively wide ranges of frequencies magnetic tunability can be used, as it was shown in Fig. 5.

V. Conclusion

Transmission coefficients of thin-film ferrite-ferroelectric structure with the width-modulated slot-line were analyzed. It was shown that there are band gaps in transmission coefficient for such structures. Location of these gaps can be tuned by the external electric and magnetic fields. It can be concluded that investigated structures could find application for development of new tunable microwave devices.

Acknowledgments

The work at SPbETU was supported in part by the Russian Foundation for Basic Research (Grant #15-32-20357 mol_a_ved) and by Ministry of Education and Science of Russian Federation (Project "Goszadanie"). The work at ITMO University was supported by the Government of the Russian Federation (Grant 074-U01). The work at LUT was supported by Academy of Finland.

References

[1] Ustinov A B, Grigorieva N Y, Kalinikos B A 2008 *JETP Lett.* **88(1)** 31–35
[2] Grishin S V, Beginin E N, Morozova M A, Sharaevskii Yu P, Nikitov S A 2014 *J. Appl. Phys.* **115** 053908
[3] Zhu Y, Chi K H, Tsai C S 2014 *Appl. Phys. Lett.* **105** 022411
[4] Serga A A, Chumak A V, Hillebrands B 2010 *J. Phys. D: Appl. Phys.* **43** 264002
[5] Vaz C A F and Staub U 2013 *Journal of Materials Chemistry C* **1(41)** 6731-6742
[6] Shastry S, Srinivasan G et al 2004 *Physical Review B* **70(6)** 064416
[7] Demidov V E, Kalinikos B A and Edenhofer E 2002 *J. Appl. Phys.* **91** 10007
[8] Yariv A and Yeh P 1984 *Optical waves in crystals* (New York: Wiley)
[9] Nikitin A A, Ustinov A B et al 2015 *Journal of applied physics* **118** 183901
[10] Seshadri S R 1986 *J. Appl. Phys.* **60** 1758
[11] Nikitin A A, Ustinov A B, Semenov A A and Kalinikos B A 2012 *Technical Physics* **57(7)** 994-997.

Investigation of the dielectric properties and defectoscopy of nanocomposites based on silica and polymers reinforced with carbon nanotubes

C S Osokin[1], M K Eseev[1], A A Goshev[1], P Horodek[2,3], S N Kapustin[1], A GKobets[2,4], A S Volkov[1]

[1]Northern Arctic Federal University named after M.V. Lomonosov, Severnaya Dvina Emb. 17, 163002, Arkhangelsk, Russia
[2]Joint Institute for Nuclear Research, Dubna, Moscow region, Russian Federation;
[3]Institute of Nuclear Physics Polish Academy of Sciences, Krakow, Poland
[4]Institute of Electrophysics and Radiation Technologies NAS of Ukraine, Kharkov, Ukraine

E-mail: m.eseev@narfu.ru

Abstract. This work presents the results of experimental studies of the properties of nanocomposites based on silica and polypropylene reinforced with carbon nanotubes by dielectric relaxation and positron annihilation spectroscopy. On the basis of these results the technique of diagnosis and control of the investigated materials are proposed. This work was supported by the project of the Ministry of Education of Russian Federation №3635 "Investigation of the nanocomposites properties at controlled modification of the structure by reinforcement with carbon nanotubes".

1. Introduction

Active research of composite materials which include nanostructures such as carbon nanotubes (CNTs) as filler is being conducted widely around the world. Due to the large surface area of CNT and significant differences in the physical properties between nanoobjects and macrostate material, the properties of nanocomposites are not additive characteristics of each phase and can radically differ from those of each of its components [1]. The data on the study of electrical and other properties of nanocomposites with CNTs is provided in the review [2, 3]. There is a particularly urgent task to study strongly nonlinear dependence of the composite properties on the concentration of the filler, which is necessary for the selection of the optimum functional properties of materials (strength, thermal, dielectric and other characteristics). Such dependence was observed in works [4,5], where a sharp increase of electromagnetic waves absorption and dispersion occurred at the concentration of CNTs of about 0.05%. The data on the influence of impurities on dielectric properties of CNTs of different composites at microwave frequencies associated with the search for promising materials for "stealth"-technologies is presented in the literature [6]. These papers [7,8] investigate microwave characteristics at a controlled modification of carbon nanotubes and nanocomposite materials based on them. We were also interested how the positrons will annihilate in the materials reinforced with CNTs.
For this purpose, the composite consisted of sintered silicon glue (matrix) was prepared. It was reinforced with multi-walled CNTs at concentrations of 0% and 1% by weight, with an aspect ratio of $10\text{-}10^2$ by the method of chemical vapor deposition.

We also studied electrical properties of composite based on polypropylene and epoxy matrix considering orientation effects of CNTs. Nanotubes were preliminarily separated in 1:1 acetone-alcohol solution with CNTs concentration of ~0.05% by weight exposed to ultrasound.

2. Experimental studies of silica reinforced with CNTs by method of positron annihilation spectroscopy

Positron annihilation spectroscopy (PAS) is a method usually recommended as a successive for detection of open volumes in materials. In the aim of verification of PAS potential application in further studies of our nanocomposites, the measurements of Doppler Broadening of Annihilation Line (DB) were performed. Variable-energy positron beam available at the LEPTA facility [9] at Joint Institute for Nuclear Research in Dubna was used in experiment. This is a small linac which uses frozen Ne as the moderator of positrons emitted from ^{22}Na isotope. Then it allows them to accelerate to demanded energies within the range between 50 eV and 30 keV and to implant them into the sample in the form of a beam. Controllability of positron energies makes it possible to implant particles at a precise depth. DB measurements consisting of registration of energetic spectrum of gamma quanta emitted in annihilation processes were done using HpGe detector with energy resolution of 1.2 keV at 511 keV. Analysis of obtained spectra results in extracting of so called S parameter. It is defined as the ratio of area under the central part of 511 keV line to the total area below this line. S parameter represents a quantity related to the amount and size of free volume. More information about the method is presented e.g. here [10,11].

In Figure1 the S parameter depending on the energy for reference sample without CNTs (circles) and sample containing 1% of CNTs (squares) are presented. It should be noticed that differences between these dependencies are well visible. The values of S parameter for specimen containing CNTs are much lower. Similar tendency was observed by Chen et. al. [12]. In this case the decreasing S parameter was explained as the presence of additional carbon nanofiber particles which have no free volume.

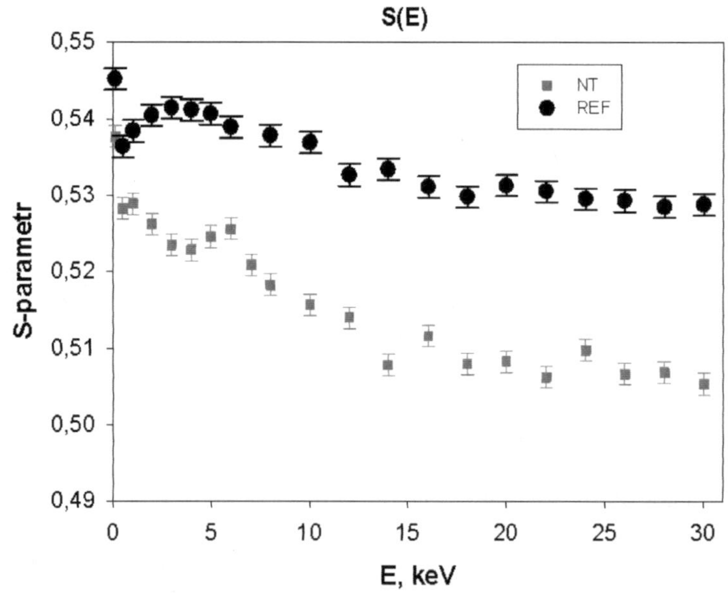

Fig. 1.The dependence of S-parameter of annihilation on positron energy for ceramic samples. Squares (NT) stand for CNTs concentration of 1%, and circles (REF) stand for thesample without CNTs

3. The frequency dependence of the polypropylene nanocomposite electrical properties

Electrical characteristics of the samples were studied by the dielectric relaxation spectroscopy method [2,3], according to which the sample is placed between the plates of the capacitor. The sample was exposed to an alternating electric field with a frequency varying in the range of 0.01 Hz - 1 MHz. The

measured values were permittivity ε and conductivity σ. The study of electrical properties of composites was accomplished using broadband dielectric spectrometer Novocontrol concept 80.
Composites, which were parallel layers of polypropylene with CNTs as filler, were placed in capacitor parallel and perpendicular to the plates. Data on the electrical properties were also compared with layered polypropylene without CNTs and polypropylene without preferred direction.
Figures 2 and 3 show samples conductivity and permittivity dependence on the frequency of the field. It can be seen that addition of CNTs enhances orientation properties of the sample.

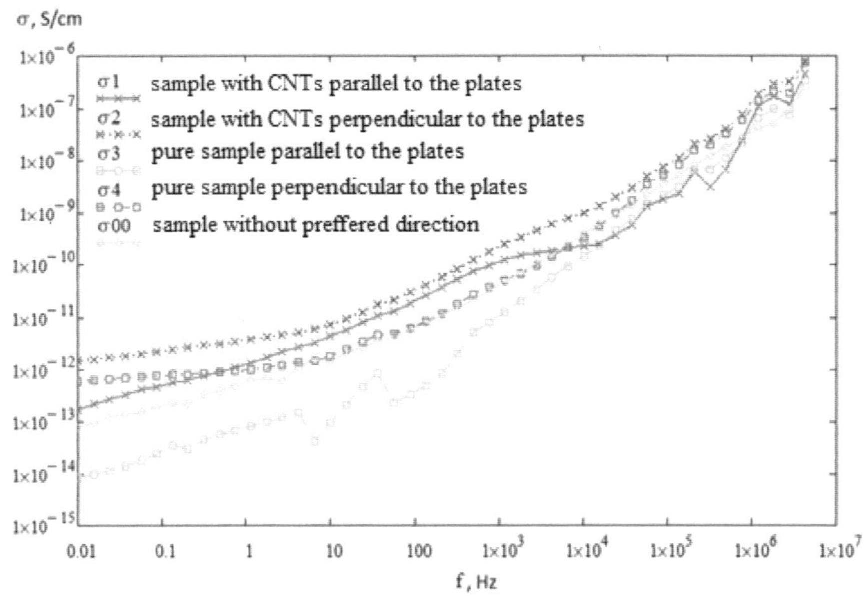

Fig. 2. Conductivity of samples: $\sigma 1, \sigma 2$ – polypropylene with CNTs parallel and perpendicular to plates; $\sigma 3, \sigma 4$ – pure polypropylene parallel and perpendicular to plates; $\sigma 00$ – polypropylene without preferred direction

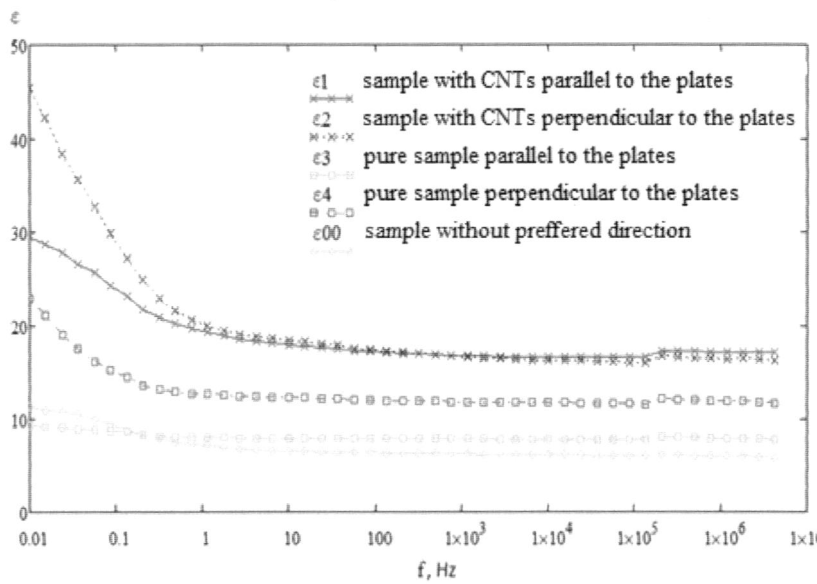

Fig. 3. Permittivity of samples: $\varepsilon 1, \varepsilon 2$ – polypropylene with CNTs parallel and perpendicular to plates; $\varepsilon 3, \varepsilon 4$ – pure polypropylene parallel and perpendicular to plates; $\varepsilon 00$ – polypropylene without preferred direction

The results obtained may find application in relatively simple studying of composites for the presence of preferred direction respectively to the electrical properties.

4. The frequency dependence of the epoxy nanocomposite electrical properties

The samples for these series of experiments were epoxy reinforced with carbon nanotubes at different concentrations – 0%, 0.05%, 0.2% and 0.5% by weight. A part of samples was placed in the capacitor with a constant field and voltage of 30 kV during the solidification of epoxy. The other part solidified freely without electric field. All samples then were studied by the dielectric relaxation spectroscopy method mentioned above.

Figure 4a shows the frequency dependence of the conductivity of samples which solidified freely. The lines with rectangles stand for samples placed in the installation in the position in which they solidified and the lines with triangles stand for samples turned at angle of 90 degrees to its solidification position. The black, red and blue lines stand for concentrations of CNTs – 0%, 0.05% and 0.5% respectively.

It can be seen that the conductivity greatly increases for the samples, which were turned, especially in the low frequency area. It can be explained by the inhomogeneous distribution of CNTs in samples obtained during the solidification in gravitational field. The CNTs, deposited on the bottom of the sample, open the conductivity channel. The graphs also show clear dependence on CNTs concentration.

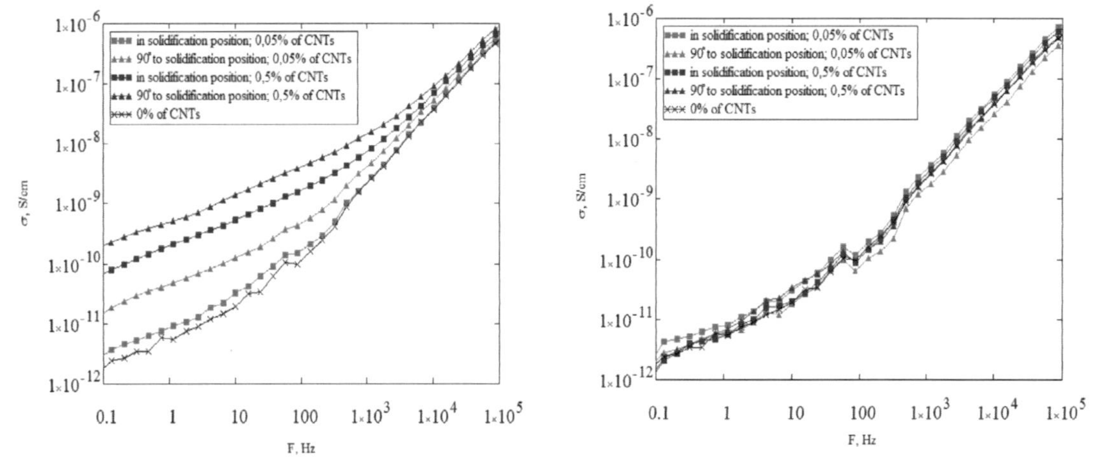

a) b)

Fig. 4.The conductivity of a) freely solidified samples; b) samples solidified in electric field; Squares stand for samples in the position in which they solidified, rectangles stand for samples turned at the angle of 90 degrees to its solidification position; black – 0% of CNTs, red – 0.05% of CNTs, blue – 0.5% of CNTs.

Figure 4b shows the frequency dependence of the conductivity of samples which solidified in constant electric field. The lines designations are the same as the previous ones.

Here we cannot see any clear rotational or even concentration dependence of samples conductivity. Consequently the conclusion can be made that the distribution of nanotubes in the samples which solidified in electric field is more homogeneous.

5. Conclusion

- The results obtained show orientation dependence of electrical properties of the polypropylene samples with CNTs layers.
- On middle and low frequencies conductivity and permittivity of samples with CNTs considerably exceed respective characteristics of pure propylene samples.
- The electric field results in more homogeneous distribution of CNTs in epoxy sample during the solidification process.
- The results obtained show the possibility to detect CNTs in different samples using positron annihilation spectroscopy methods.

References

[1] Eletskii A V 1997 Phys. Usp. **40**(9) 899-924
[2] Eletskii A V, Knizhnik A A, Potapkin BV, Kenny J M2015 Phys.Usp.**58**(3) 209
[3] Rakov E G 2013 Rus. Chem. Rev. **82**(1) 27
[4] Eseev M K, Goshev A A, Horodek P, KapustinS N., KobetsA G, Osokin C S Methods of diagnostics of nanocomposites based on silica-reinforced carbon nanotubes 2015 Book of Abstract «12th International Conference Advanced Carbon NanoStructures ACNS'2015» (Saint-Petersburg, Russia),p. 226.
[5] GoshevA Aet al 2015 J. Phys.: Conf. Series **643**012126
[6] Muradyan VE, Sokolov E A, Babenko S D, Moravsky A P 2010 Tech.Phys. **55**(2) 242
[7] Usanov D A, Skripal' A V, Romanov A V 2011 Tech.Phys. **56**(1) 102
[8] Usanov D A, Skripal' A V, Romanov A V 2014 Tech.Phys.**59**(6) 873
[9] Sidorin A A et al2013 Materials Science Forum**733** 291
[10] Puska M J, Nieminen R M 1994 Rev. Mod. Phys. **66** 841
[11] Krause-Rehberg R., Leipner S. H. 1998 Positron annihilation in semiconductors (Springer, Berlin)
[12] Chen H Met al 2011 AIP Conf. Proc.**1336** 444

Melt-spun Fe-Co-P-B metglas: structure, crystallization kinetics, magnetic properties

D A Prahova[1], A M Grishin[1, 2], V S Ignahin[1], L A Lugovskaya[1] and R N Osaulenko[1]

[1] Petrozavodsk State University, Petrozavodsk 185000, Karelian Republic, Russia
[2] INMATECH Intelligent Materials Technology, SE-127 51 Skärholmen, Sweden

E-mail: darja_pro@mail.ru

Abstract. Crystallization kinetics and magnetic properties were studied in series of rapidly solidified metallic glasses $Fe_{80-\xi}Co_{\xi}P_{14}B_6$ with ξ = 25, 32, 35 and 40 at. %. Differential scanning calorimetry (DSC) was employed to determine crystallization temperature and to demonstrate the incubation phenomenon at isothermal annealing. Fitting DSC traces recorded in non-isothermal and isothermal regimes, we obtained the same activation energy Q. It describes atomic transfer across crystal/amorphous interface in Kolmogorov-Johnson-Mehl-Avrami (KJMA) model of the transition of metallic glasses into crystalline state. As-quenched metglas ribbons have amorphous structure and superior soft magnetic properties: coercive field 4 A/m, ac magnetic permeability as high as 110 000 at 60 Hz.

1. Introduction

Unique combination of functional properties as high saturation magnetization, permeability and magnetostriction, low coercive field and high mechanical strength and corrosion resistance makes metallic glasses to be indispensable material for magnetic screens and high frequency transformer cores, magnetic field, tensile and angular position sensors. Although a lot of new filmy and bulk metallic glasses were discovered recently, magnetosoft Fe-, Ni, and Co-based amorphous alloys remain to be the reference materials to engineer preassigned nanocrystalline structure and to explore its relation to magnetic properties. Early it was shown that the replacement of Ni by Co in the well-known $Fe_{40}Ni_{40}P_{14}B_6$ composition raises significantly a thermal stability of metglas [1, 2]. In this paper we investigate how Fe-to-Co concentration ratio defines properties of as-quenched and crystallized $Fe_{80-\xi}Co_{\xi}P_{14}B_6$ alloys.

2. Experimental

Preparation technique of 20-30 μm thick metallic Fe-Co-P-B glass ribbons by a single roller melt spinning process has been described in details elsewhere [1]. They were annealed in protective Ar atmosphere with different heating rates α = 2.5, 5, 10, 20 and 40 K/min using *Hitachi* STA7300 simultaneous thermal analyzer. Crystalline structure of as-spun and annealed ribbons was examined by X-ray diffraction in CuK_{α} and MoK_{α} radiations. Magnetic hysteresis loops were recorded with *M-H* tracer in magnetic field up to 100 A/m in the frequency range from 20 to 600 Hz.

3. Results and discussion

3.1 Thermal Stability. Differential scanning calorimetric study was performed to determine the temperature range of practical use of fabricated metallic glasses. Annealing process was carried out in two regimes: steady (isochronal) heating with a constant rate α = 2.5, 5, 10, 20 and 40 K/min and isothermally – fast heating with a rate of 40 K/min to a predetermined temperature T and expose specimen at reached temperature for prolonged time.

Sharp exothermic peaks appear at DSC traces for all the samples (see figure 1). As shown in figure 2, at isochronal annealing, peak's temperature T_{max} is higher for specimens with a higher Fe content.

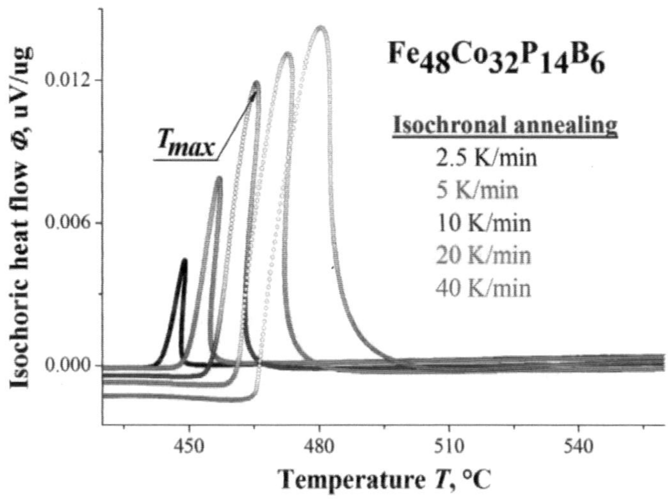

Figure 1. DSC traces for as-spun $Fe_{48}Co_{32}P_{14}B_6$ ribbons annealed at different heating rates α.

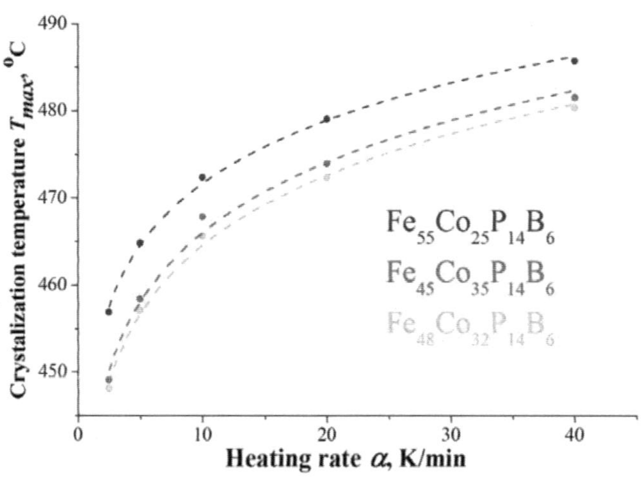

Figure 2. Dependence of the position of crystallization peak T_{max} on heating rate α for three different metallic glasses.

The activation energy Q of metglas crystallization process was determined in Kolmogorov-Johnson-Mehl-Avrami (KJMA [3]) model (figure 3). Q was found to be 46430 K, 46650 K, and 51850 K for $Fe_{45}Co_{35}P_{14}B_6$, $Fe_{48}Co_{32}P_{14}B_6$, and $Fe_{55}Co_{25}P_{14}B_6$ alloy, respectively.

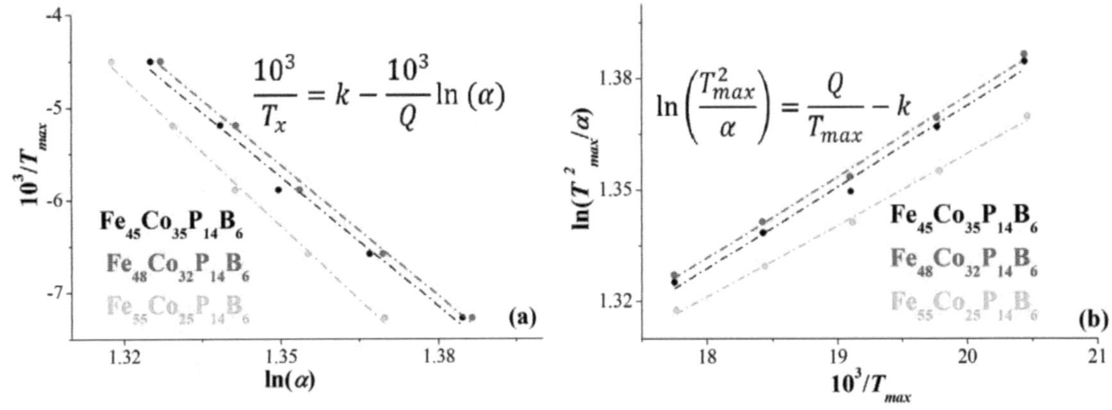

Figure 3. Kissinger [4] plots of crystallization peak temperature T_{max} at different heating rates for three metallic Fe-Co-P-B glasses.

At isothermal annealing DSC traces in figure 4 reveal the incubatory effect. Ribbon with the nominal composition $Fe_{45}Co_{35}P_{14}B_6$ was fast heated up with a rate of 40 K/min to the preassigned temperature and kept there for a long time. At isochronal annealing with α = 40 K/min, crystallization peak appears at T_{max} = 754.7 K. Isothermal annealing performed at lower temperatures T: 700.1, 707.7, 714.5, 719.2, 724.6, 729.5, and 733.9 K leads to the retardation of the appearance of the exothermic peak.

The incubatory effect is a time delay t_{inc} between the appearance of exothermic peak during isochronal and isothermal annealing processes. As shown in inset to figure 4, experimental temperature dependence of incubation time $t_{inc}(T)$ strictly follows exponential curve calculated within KJMA model. Fitting parameters Q appear to be the same both in isochronal and isothermal heating regimes of crystallization for all three samples under study.

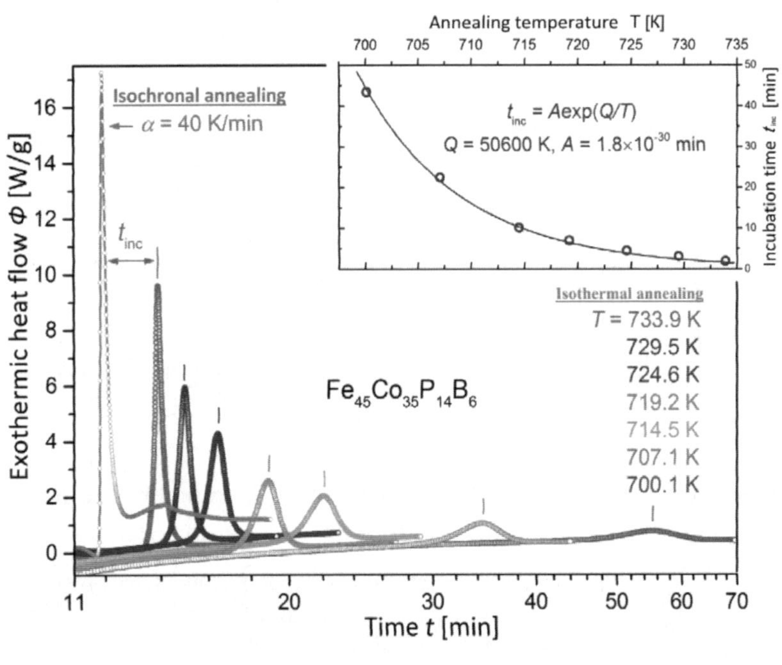

Figure 4. DSC traces of isothermal annealing of as-spun $Fe_{45}Co_{35}P_{14}B_6$ amorphous alloy.

3.2 The structure. Figure 5 (a) shows X-ray diffraction pattern of as-spun $Fe_{55}Co_{25}P_{14}B_6$ ribbon recorded in MoK_α radiation. Scattering factors as well as ionic radii of Fe and Co atoms are very close, thus XRD patterns of ribbons with different Fe-to-Co concentration ratio look similar to each other.

As-quenched ribbons are X-ray amorphous. Very broad peaks around 2Θ = 20°, 34°, 52° and one shoulder at 41° might be relied upon, respectively, to (110), (211), (220), and (222) diffraction reflections of α-FeCo *bcc* phase. To carry out phase analysis we employed C. Finbak - B.E. Warren method of atomic pair distribution function $D(r)$ (PDF). From XRD pattern in figure 5 (a) we calculated experimental $D(r)$ and compared this PDF with theoretical $D(r)$ functions we built for α-FeCo and (Fe, Co)$_3$(P, B) crystalline structures. $D(r)$ for (Fe, Co)$_3$(P, B) gives much better correspondence to the experiment. Therefore, (Fe, Co)$_3$(P, B) we consider as nuclei of an extant parent phase in amorphous matrix.

After annealing, a lot of new Bragg reflections appear in CuK_α XRD spectrum in figure 5 (b). This pattern is a superposition of *bcc* α-FeCo (ICSD-56273) and *bct* (body-centered tetragonal) (Fe, Co)$_3$(P, B) phases. Percent composition of crystallized phases was determined using Rietveld refinement method [5]. To describe (Fe, Co)$_3$(P, B) structure we assumed isomorphic Co substitution of Ni atoms in $Fe_{40}Ni_{40}P_{14}B_6$ (ICSD-614312) compound. It was proven that during the isochronal annealing up to 873 K percentage of (Fe, Co)$_3$(P, B) phase rapidly increases for Fe-rich metallic glasses at the expense of α-FeCo phase.

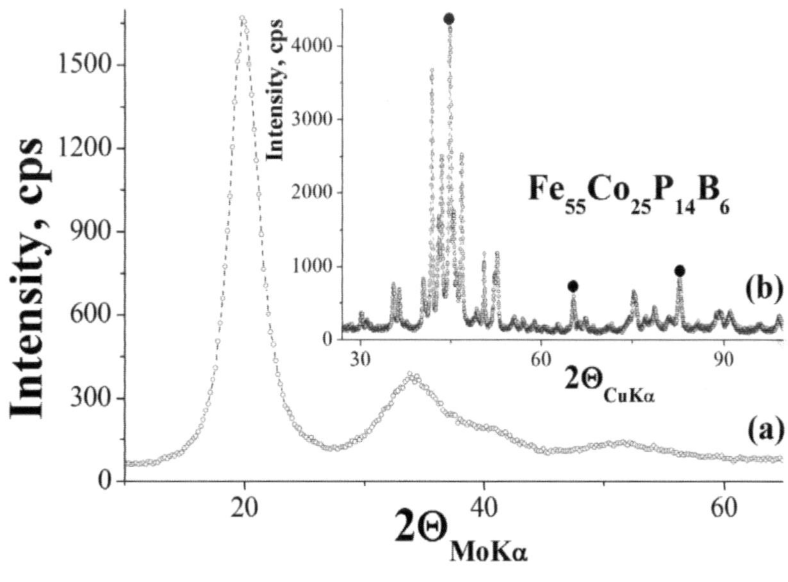

Figure 5. X-ray diffraction patterns of $Fe_{55}Co_{25}P_{14}B_6$ ribbon in MoK_α and CuK_α radiations. **(a)** – as-spun specimen, **(b)** – ribbon isochronally annealed at α = 40 K/min. XRD spectrum shows a mixture of α-FeCo (marked with circular symbols) and (Fe, Co)$_3$(P,B) phases.

3.3 Magnetic properties. Figure 6 presents an example of hysteresis *M–H* loops recorded at 20 Hz for as-quenched $Fe_{45}Co_{35}P_{14}B_6$ and $Fe_{55}Co_{25}P_{14}B_6$ glassy ribbons. The coercive force H_c was found to be 8-10 A/m. Hysteresis losses grow with a frequency of ac-signal. It is clearly shown in Figure 7 from the frequency dependence of coercive force. Differential permeability calculated as a derivative dM/dH of magnetization curve $M(H)$ at $H = H_c$ appears to be higher than 110 000 at 60 Hz.

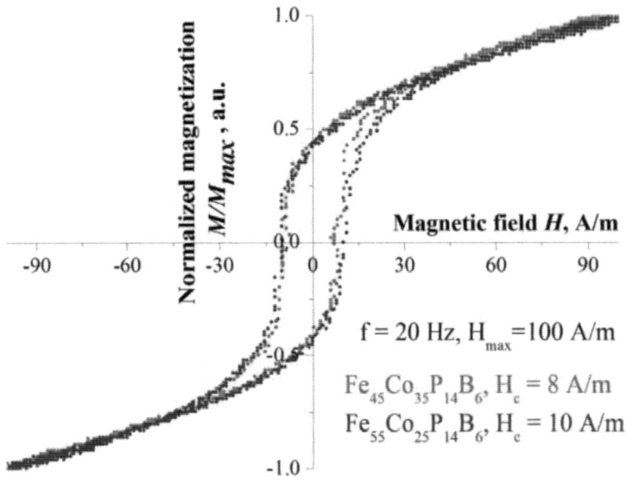

Figure 6. Hysteresis *M–H* loops of the as-quenched $Fe_{45}Co_{35}P_{14}B_6$ and $Fe_{55}Co_{25}P_{14}B_6$ melt-spun glassy ribbons recorded at 20 Hz in magnetic field parallel to ribbons' surface.

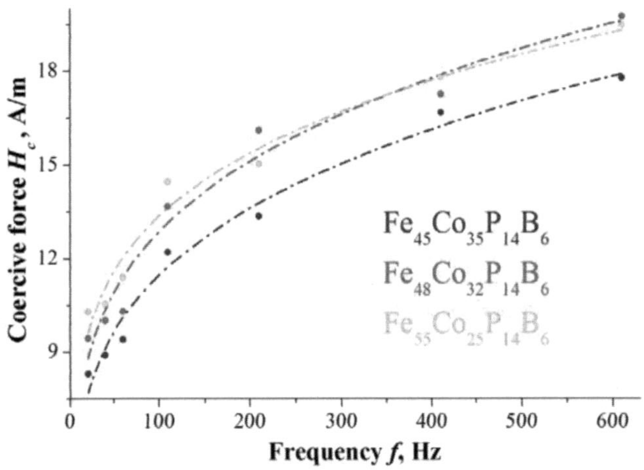

Figure 7. Frequency dependence of coercive force in as-quenched amorphous Fe-Co-P-B metallic glasses.

4. Conclusions

We studied crystallization kinetics during isothermal and non-isothermal heating of rapidly quenched melt-spun Fe-Co-P-B metallic glasses. Kolmogorov-Johnson-Mehl-Avrami model gives adequate description both of these processes. Crystallization temperature is higher in Fe-rich metallic glasses indicating their higher thermal stability. Activation energy Q appeared to have the same value to nicely fit crystallization at fast isochronal annealing and incubation effect at long isothermal heating.

Acknowledgements

This work was supported by the RTL-Service JSC, Petrozavodsk State University and the Ministry for Education and Science of the Russian Federation (contract 14.574.21.0059, unique identifier RFMFIBBB14X0164).

References

[1] Hollmark M, Tkatch V I, Grishin A M and Khartsev S I 2001 Processing and Properties of Soft Magnetic $Fe_{40}Co_{40}P_{14}B_6$ Amorphous Alloy *IEE Transactions on Magnetics* **37** 2278 – 2280

[2] Tkatch V I, Grishin A M and Khartsev S I 2002 Delayed nucleation in $Fe_{40}Co_{40}P_{14}B_6$ metallic glass *Materials Science and Engineering* **A337**, 187 – 193

[3] A.N. Kolmogorov, Izv. Akad. Nauk USSR, Ser. Matem. **1**, 355 (1937) (in Russian); W.A. Johnson, K.F. Mehl, Trans. Am. Inst. Mining. Met. Eng. **135**, 416 (1939); M. Avrami, J. Chem. Phys. **7**, 1103 (1939); **8**, 212 (1940); **9**, 177 (1941)

[4] Kissinger H E 1956 J. Res. Nat. Bur. Stand. **57**, 217

[5] Rietveld H M 1969 A profile refinement method for nuclear and magnetic structures *Journal of Applied Crystallography* **2** (2): 65–71

Controlling of immiscible liquids fluid in a capillary reactor – from continuous to segmented flow

L B Matyushkin[1], R C Mbwahnche[2], O A Ryzhov[1]

[1] Postgraduate, St. Petersburg State Electrotechnical University "LETI", Pr. Popova 5, St.Petersburg, 197376 Russia

[2] Student, St. Petersburg State Electrotechnical University "LETI", Pr. Popova 5, St.Petersburg, 197376 Russia

E-mail: o.a.ryzhov@gmail.com

Abstract. Experimental variation of the droplets form in the flow of immiscible fluids in a capillary with the inner diameter of 0,9 mm is presented. The possibility of the transition from continuous to segmented flow by varying the velocity of precursors and carrier liquids is demonstrated. It is shown that the rate of variation of one precursor in respect to other makes it possible to vary the ratio of mixing components. Simulation results for Ψ-shaped mixer in COMSOL Multiphysics software are also presented.

1. Introduction

Traditional synthesis of monodisperse nanoparticles in chemical flask has a number of shortcomings that limit the scaling of such systems for industrial applications. The solution to this problem is to turn from batch technology to flow synthesis. In comparison with traditional chemical synthesis in a flask, the flow reactor type allows increasing synthesis rate of nanoparticles. The quality improvement also occurs [1]. Apart from its obvious superiority in large scale production, flow reactor has a number of additional advantages: a) more effective heat control and mass transfer, b) the ability to control the cooling rate, b) a better synthesis reproducibility, d) lower reagents consumption.

The capillary is a central element in such reactors, where precursors are dispersed and mixed with other reagents. Fluid transfer in the channel is usually described by the Reynolds number Re, which allows determining the transition from laminar to turbulent flow:

$$\mathrm{Re} = \frac{\rho v D}{\eta}, \tag{1}$$

where ρ is density, v is the characteristic flow velocity, D is the hydraulic diameter (for a circular tube it is equal to geometric diameter), η – dynamic viscosity of environment. Due to small diameter, microreactor channel normally corresponds to Reynolds criterion which is less than the critical value and eliminates the uncontrolled emergence of turbulent flow so that mixing occurs mainly by molecular diffusion or by means of special taps.

Mixing is achieved by the creation of strong vortex fields that are generated inside the microchannel. In paper [2] different tools and mixing mechanisms in microfluidics were presented. Of course, each design has its advantages and disadvantages, and the most appropriate mixing tool must be selected according to specific application. The authors of [3] believe that there is no single approach when choosing mixer that meet all requirements and are suitable for all applications. However, from the productivity point of view and production cost, different technologies and mixing concepts can be described. Microfluidic mixers can be classified as *active*: that utilize external energy for mixing, or *passive*: where mixing occurs due to the microfluidic device geometry or the functional structures of a certain configuration.

The mixing degree improves the quality of synthesized particles [4].There are various types of passive mixers (Figure 1), which are quite simple to implement.

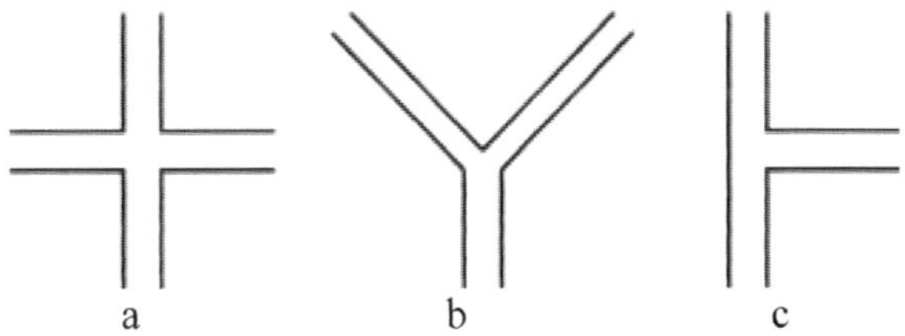

Figure 1 – Types of flow reactor mixers: a –focused flowchannel, b –Y-shapechannel, c –T-shapedchannel

All nanoparticles flow synthesis reactors can be divided into several groups with different types of stream creation, each of which will affect the flow shape of the reactants inside the capillary (Figure 2).

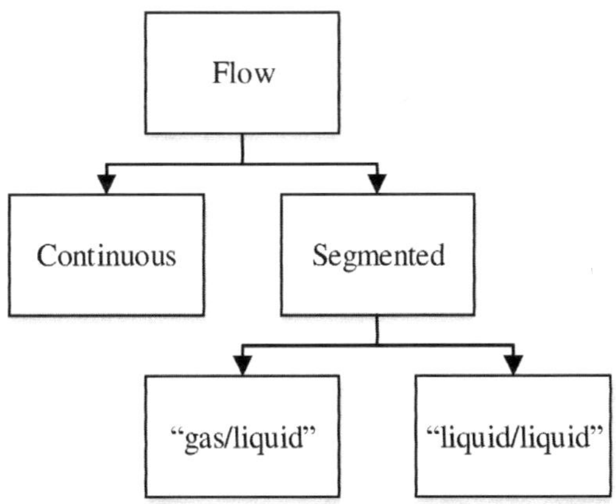

Figure 2 – The diagram illustrating continuous and segmented modes of flow in channels with a circular cross section

In the first reactors continuous flow was utilized but it had several disadvantages: large polydispersity of the obtained nanoparticles in consequence of precursors laminar flow as well as gradual pollution of the channel due to contact with the channel wall. The example of CdS nanoparticles synthesis [5] was demonstrated with a complete channel blockage thirty minutes from the start of the synthesis. The same paper presents another technique for reagents flow forming in a capillary using inert gas to separate the flow of liquid precursor to bullet-separated bubbles of gas. In such reactors the problem of nanoparticles polydispersity was solved but contact with the walls leading to pollution of the channel was still inevitable. In paper [6] segmented flow was formed by using a carrier fluid that does not mix with the precursor solution, but also wets the walls of the channel. In this way the problem of polydispersity of obtained nanoparticles was solved as well as the problem associated with channel pollution. The segmented flow type of reactor has been used to synthesis a wide range of materials: nanoparticles of iron oxides [7], Ag [8], $CaCO_3$ [9], $BaSO_4$ [10], zeolites [11], TiO_2 [12], γ-AlOOH, β-FeOOH [13], and more complex structure types like core/shell, such as SiO_2/Au[14] and Au/Ag/Au [15].

The present paper describes the implementation of nanoparticle synthesis in the liquid/liquid segmented flow [16]. For the binary compounds nanoparticles synthesis, one must use the upgraded Y-shape mixer (Figure 1, b). Two side capillary inlets meant for two precursors and the middle inlet for the liquid carrier altogether forms a Ψ-shaped mixer (Figure 3, a).

2. Experiment

One of the stages in the flow reactor development for nanoparticle synthesis is the simulation of flow for certain construction of reactor. We use for such purpose COMSOL Multiphysics software with Microfluidics Module. There are two main groups of methods of flow simulations: interface tracking methods (ITM) and the dispersion methods. For large-scale problems with a large number of bubbles, drops and solid particles, a less computational resources-demanding model is required. The so-called dispersive methods which track the volume proportion attributable to each phase allows one to use less computing resources. With the help of the dispersion method, one can model for example, a circulating boiling film which is often used in food, pharmaceutical and chemical processing. The interface tracking methods is designed to simulate flow of two different immiscible fluids separated by well-defined surface. These methods are usually applied to simulate the formation of bubbles or drops; mixers, turbulators; separated oil streams, water or gas. Methods to track the interface are accurate and give a clear picture of the flow field (velocity, pressure and force of surface tension). However, they require large computational resources, making them not always preferable. These methods for interface tracking in general are better suited for microfluidic applications where monitoring a small number of drops or bubbles are essential.

Methods of interface tracking includes: level set method; phase field method; two-phase flow with moving mesh. All of these methods allow accurate tracking of the interface between two immiscible liquids. They take into account the difference in density and viscosity of the two fluids, and the influence of surface tension and gravity. Using the phase-field maps, we obtained volume fraction in the mixing of the two precursors (side capillaries) and liquid carrier (middle capillary). Figure 3 presents the simulation results using the same pressure in all three capillaries. As it can be seen, mixed precursors replace the carrier fluid both in forward and reversed direction. In the reversed flow direction, substitution occurs in a weaker manner.

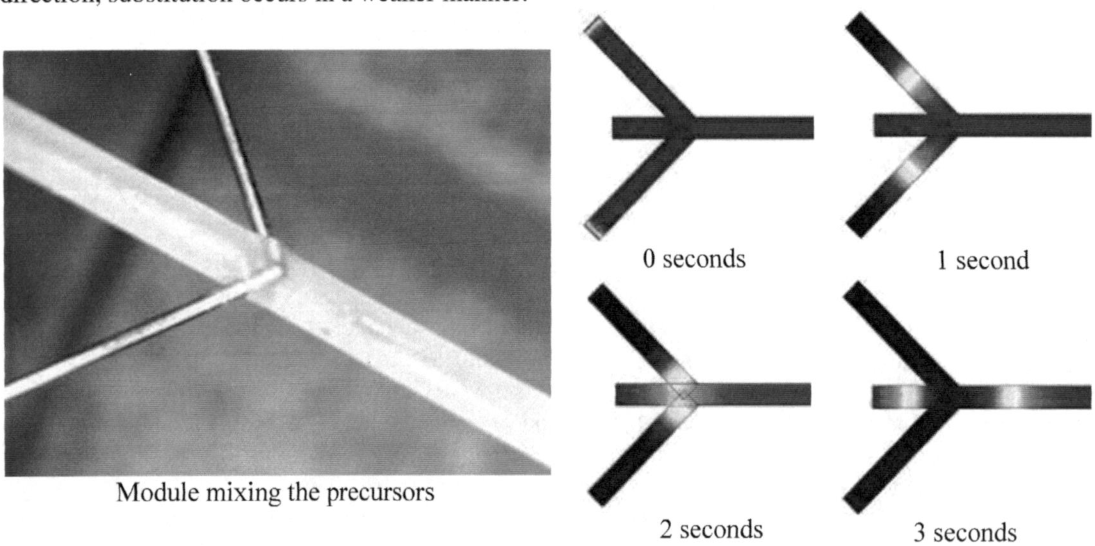

a b

Figure 3 –Module mixing the precursors(a) and simulation result (b) of volume fraction using the same pressure in all three capillaries inlets

In this research, we carried out experiments to find the optimal flow rates of reagents necessary for synthesis. The capillary was a flexible polytetrafluoroethylene tube having an inner diameter of 0,9 mm. Fluids rates were varied with the help of syringe pumps. For experiments water-based inks of different colours were used in the role of precursor and hexane as a liquid carrier. Figure 4 shows an image of the channel specified in the listed fluids.

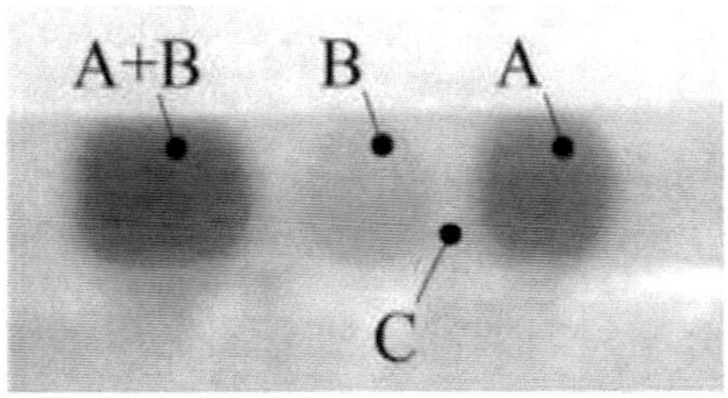

Figure 4 – Flow elements in the PTFE capillary: A and B – model precursors, C– a liquid media

Figure 5 presents results of an experiment where the precursors and carrier fluid are supplied at the same rate. The smaller the feed rate of the precursors, the smaller the droplet size of the precursor mixture. But at the rate of 1.2 μl/sec mixing occurs only partially as it was evidenced by the colour of drops (yellowish tint). Based on these results, the best rate to obtain segmented flow is in the range of 10 to 20 μl/sec.

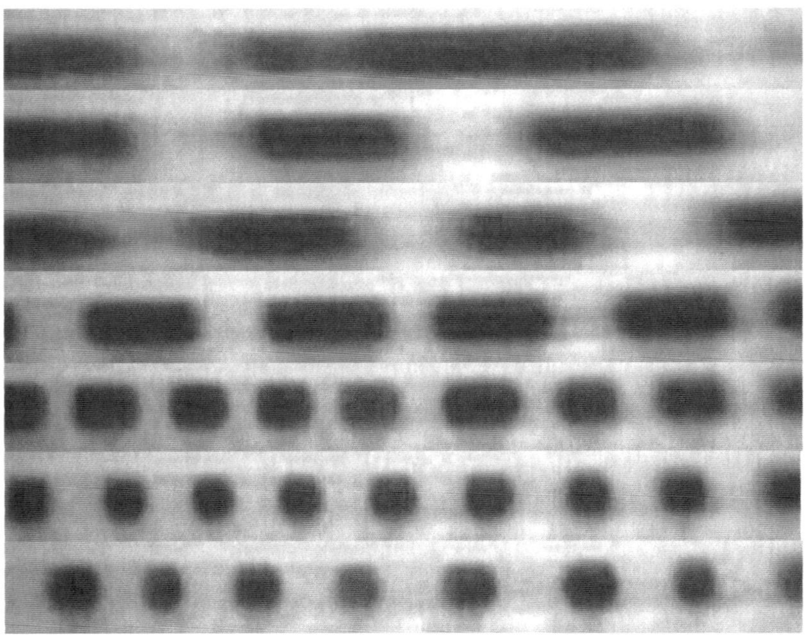

Figure 5 –Shape evolution of the segmented flow during simultaneously changing of precursors feedrate A, B and liquid media C. Numbers marked flow rate in μl/sec

Figure 6 presents the results of an experiment where the rate of one precursorwas changed in relation to another. The images show that there is a complete mixing (no color alternation of droplets), but because of the relative change in the feed rate of the precursor A in relation to the B there is observed changing of color tone and the size of the output drops.

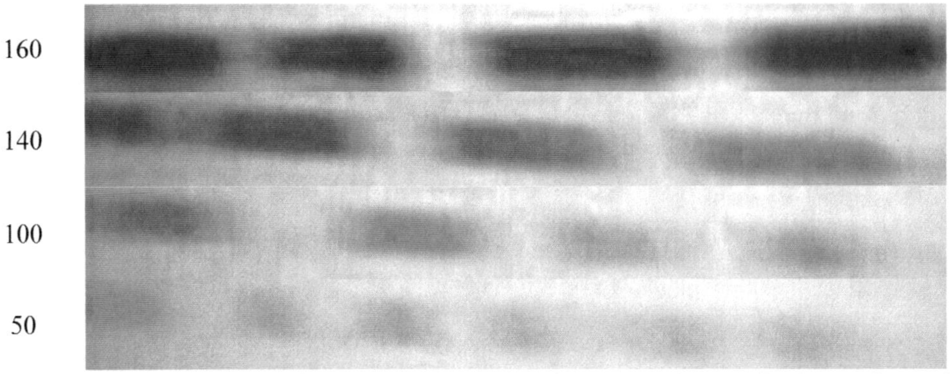

Figure 6 – Evolution of the shape of the segmented stream due to reducing the feed rate of one of the precursors (U_A) at constant speed of the other precursors and liquid media ($U_C = U_B = 160$ μl/sec). The numbers indicate the flow rate of U_A in μl/sec

Figure 7 presents results of the change in flow rate of the fluid carrier relative to the constant feed rate of precursors, which leads to a change in droplet size of the mixed precursor solution due to the change in the fraction of carrier fluid in the capillary.

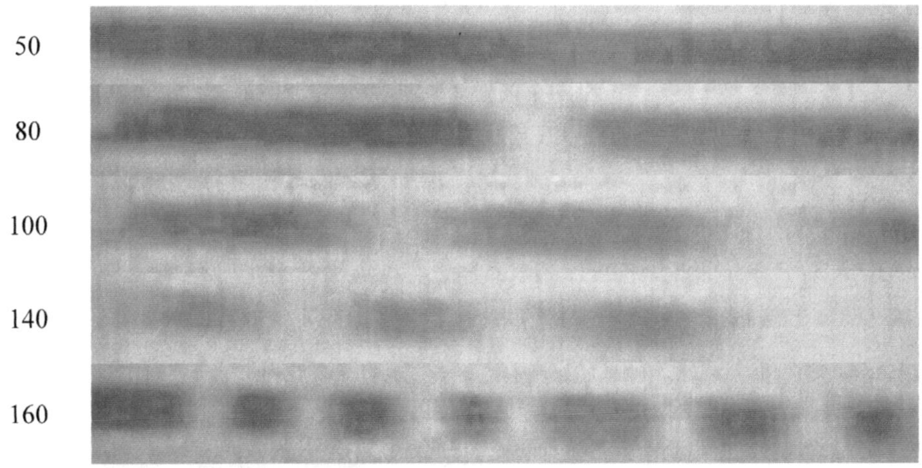

Figure 7 –The experimental results of change in flow rate of the carrier liquid at a constant flow rate of precursors ($U_A = U_B = 160$ μl/sec). The numbers indicated the flow rate of the carrier liquid in μl/sec

After analysing the above results, it can be assumed that the optimum velocity at which there is complete mixing of the precursors and the separation of the stream into droplets with carrier fluid are rates from 10 to 50 μl/sec.

3. Results

As a result, modeling and experimental research has shown that varying the flow rate of the carrier and precursor fluids output of Ψ-shaped mixer gives additional variation of flux: 1) a transition from a continuous flow to segmented type, 2) mixing the components in different ratios, while maintaining a segmented or continuous flow form. The obtained results are used to select flow rates of the synthesis reactor nanoparticles.

Acknowledgements
The authors acknowledge financial support from the Russian Research Foundation (project 14-15-00324).

References

[1] Nightingale A.M., de Mello J.C. 2013 Adv. Mater, V. 25, pp. 1813–1821
[2] Suh Y.K., Kang S.A. 2010 Micromachines№1. pp. 82–111
[3] Hardt S., Drese K.S., Hessel V., Schönfeld F. 2005 MicrofluidNanofluid №1, pp. 108–118
[4] Zhao C. X. et al. 2011 Chemical Engineering Science T. 66. №. 7, pp. 1463-1479.
[5] Shestopalov I., Tice J. D., Ismagilov R. F 2004 Lab on a Chip. T. 4., №. 4, pp. 316-321.
[6] Nightingale A. M. et al. 2011 Lab on a Chip. T. 11., №. 7, pp. 1221-1227
[7] Doh I., Erdem E. Y., Pisano A. P. 2012 Applied Physics Letters T. 100, №. 7, pp. 074106
[8] Nightingale A. M. et al. 2011 Lab on a Chip T. 11, №. 7, pp. 1221-1227
[9] Sotowa K. I. et al. 2007 Chemical engineering & technology T. 30, №. 3, pp. 383-388
[10] Jeevarathinam D. et al. 2011 Chemical Engineering Journal T. 173, №. 2, pp. 607-611
[11] Pan Y. et al. 2009 Chem. Commun №. 46, pp. 7233-7235
[12] Krishnadasan S. et al. 2007 Lab on a Chip T. 7, №. 11, pp. 1434-1441
[13] Hoang P. H., Park H. S., Kim D. P. 2011 Journal of the American Chemical Society T. 133, №. 37, pp. 14765-14770
[14] Duraiswamy S., Khan S. A. 2010 Nano letters. T. 10, №. 9, pp. 3757-3763
[15] Knauer A. et al. 2011 Chemical Engineering Journal T. 166, №. 3, pp. 1164-1169
[16] Aleksandrova O.A., et al 2015 Proceedings of the 2015 IEEE North West Russia Section. Young Researchers in Electrical and Electronic Engineering Conference. February, 2-4, p. 13

Increasing the quantum efficiency of GaAs solar cells by embedding InAs quantum dots

R A Salii[1], S A Mintairov[1], A M Nadtochiy[1,2], A S Payusov[1,2], P N Brunkov[1,2], M. Z. Shvarts[1] and N A Kalyuzhnyy[1]

[1] Ioffe Institute, Russia, Polytechnicheskaya str. 26, 194021 St.-Petersburg

[2] St Petersburg Academic University, St Petersburg, Russia, Hlopina str. 8/3, 194021 St.-Petersburg

E-mail: rino.art@gmail.com

Abstract. Development of Metalorganic Vapor Phase Epitaxy (MOVPE) technology of InAs quantum dots (QDs) in GaAs for photovoltaic applications is presented. The growth peculiarities in InAs-GaAs lattice-mismatched system were considered. The photoluminescence (PL) intensity dependences on different growth parameters were obtained. The multimodal distribution of QDs by sizes was found using AFM and PL methods. GaAs solar cell nanoheterostructures with imbedded QD arrays were designed and obtained. Ones have been demonstrated a significant increase of quantum efficiency and photogenerated current of QD solar cells due to photo effect in InAs QD array (0.59 mA/cm^2 for AM1.5D and 0.82 mA/cm^2 for AM0).

1. Introduction

Technology to form self-organized InAs quantum dots (QDs) in GaAs was researched intensively within the framework of the Molecular Beam Epitaxy (MBE) technique, which ensures ample feasibilities for controlling the properties of the QD arrays [1]. For light-emitting devices (like lasers, micro-resonators etc.) grown using the MBE, QDs have already demonstrated the improvement of a number of parameters [2, 3]. Recently the great interest for using InAs QDs into the photovoltaic devices have been appeared. The InAs QDs imbedded into GaAs solar cell (SC) allow realizing two important photovoltaic concepts. First of all it is intermediate-band (IB) solar cell [4] which promised to create the equivalent of a triple-junction solar cell in a single-junction solar cell. Secondly, InAs QDs allow obtaining photo-current balance into the high-effective multijunction (MJ) solar cells based on lattice-matched $Ga_{0.51}In_{0.49}P/Ga_{0.99}In_{0.01}As/Ge$ heterostructure [5] due to extending $Ga_{0.99}In_{0.01}As$ subcell spectral sensitivity and thus increasing its photogenerated current. But the growth of SC heterostructures has been widely studied as applied to the Metalorganic Vapor Phase Epitaxy (MOVPE) technique while the technique for the formation of multilayer QD arrays was investigated in detail regarding the MBE. Consequently, the first IB solar cell prototype [6] was made using InAs QDs grown by MB in GaAs SC [7]. The highest efficiency for QD SC achieved by MBE was 18.32% in 2009 [8]. However in all cases the increase in photocurrent obtained due QDs, as compared to a control cell without QDs, is very small and has been accompanied by a reduction in the voltage with an overall reduction in efficiency.

The first successes with QD SC grown by MOVPE was a were demonstrated by [9-11], where QD SCs with an efficiency of 18.7% (AM1.5G, 1 sun) was achieved due to five layers of QDs. The strain-compensation GaP layer technology allow demonstrating in [12] GaAs solar cell, made by MOVPE with InAs QDs, with a higher efficiency (14.3%) than a control cell manufactured without QDs. However, in spite of demonstrated high sub-bandgap photon absorption by the QDs, the spectral dependence of quantum efficiency in GaAs absorption spectral range was not perfect.
This paper focuses on the research and development of the formation process of InAs QDs in GaAs matrix using MOVPE technique. The main purpose of this investigation was to obtain high quality GaAs solar cell structure with good spectral response and, at the same time, with a strong sub-bandgap photon absorption by the QDs.

2. Experimental procedure

All experimental and device structures have been obtained using the MOVPE technique on an R&D installation with a horizontal reactor at low pressure (100mbar). Metal-organic compounds were used as sources of Group III elements (trimethylgallium (TMGa), trimethylaluminum (TMAl) and trimethylindium (TMIn)). Arsine (AsH_3) was used as a source of the Group V element (As). Experiments were carried out on (100) GaAs substrates misoriented to the (111) direction by 6°.
The main task in the development of epitaxial technology for the GaAs–InAs system is determining the dependence of the InAs growth rate (GR) on the molar flow of In source under various conditions. Solving this problem in such system is not so trivial because of the lattice mismatch between GaAs and InAs materials which makes it impossible to obtain defect free layers on GaAs substrates and subsequently determine the growth rates both by *ex-situ* and optical *in-situ* methods [13], which are effective for lattice matched systems. Thus, to find the dependence of the InAs GR on In molar flow we studied the photoluminescence (PL) spectra of InGaAs/GaAs quantum wells (QWs). The QW structures were $Al_{0.3}Ga_{0.7}As/GaAs/In_xGa_{1-x}As/GaAs/Al_{0.3}Ga_{0.7}As$. The growth temperature of QW was varied in the range from 500 to 700°C, In composition "x" was varied from 5% to 30%.
In order to investigate the growth peculiarities of InAs QDs in GaAs matrix the structures for PL and atomic force microscopy (AFM) measurements were grown. There were two types of structures: $Al_{0.3}Ga_{0.7}As$/GaAs/GaAs cap/InAsQDs/GaAs/$Al_{0.3}Ga_{0.7}As$ (a-heterostructure) and uncovered InAsQDs/GaAs/$Al_{0.3}Ga_{0.7}As$ (b-heterostructure) for PL and AFM measurements respectively. Structures in series of experiments with a-heterostructure were differed by quantity of InAs material, and by GaAs cap layer growth parameters.
To obtain the PL spectra of the a-heterostructures, a 532nm Nd:YAG laser of 350mW power was used as a source of radiation. A study of the surface topography of b-heterostructure was carried out using the AFM method on a Dimension 3100 (Veeco) microscope at room temperature.
As a result of the optimization of the PL intensity, the growth conditions of QD arrays have been established. These QD arrays were embedded into the active area of the GaAs SC structure (QD SC). The QD SC structures were based on a GaAs p-i-n diode having 10-15 InAs QD layers in the i-region with a 3 µm-thick base, 0.5 µm-thick emitter and $Al_{0.8}Ga_{0.2}As$ and $Al_{0.3}Ga_{0.7}As$ layers for the window and the back surface field (BSF) layers respectively.
Based on the heterostructures grown, photodiodes with QDs with a configuration of the top metal contact without a grid shadowing the surface have been manufactured. A control photodiode without QDs has also been manufactured. This construction of a photodiode is suitable for spectral response measurements, as there is no shadowing of a surface by contact grid.
The multifunctional installation at the Ioffe Institute has been used for measuring the External Quantum Efficiency (EQE) and the reflectance of the solar cells.

3. Results and discussion

Based on the measured PL spectra from described above experimental QW heterostructures, the linear dependences of the In composition in the $In_xGa_{1-x}As$ solid solution on the TMIn /(TMGa+ TMIn) ratio in the gas phase were obtained in a wide range of growth temperatures from 500 to 700°C (Fig. 1). It was found that In atoms are more effectively incorporated at lower temperatures.

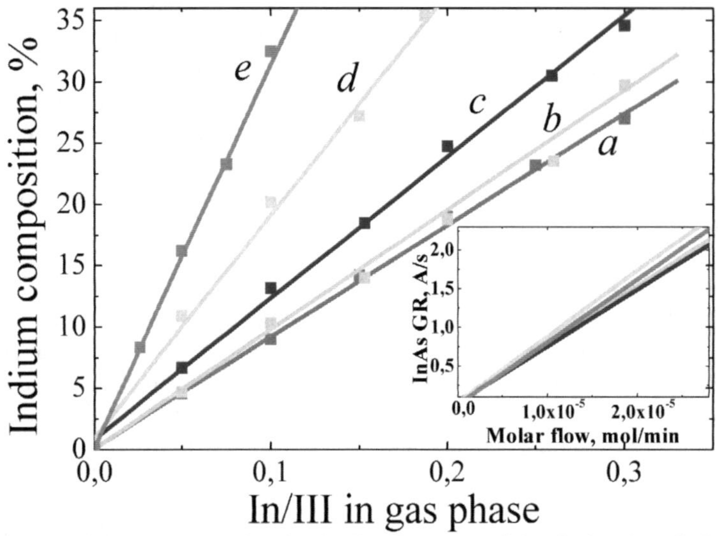

Figure 1. In concentration in the $In_xGa_{1-x}As$ solid solution in relation to the amount of indium in the gas phase at temperatures of (a) 700, (b) 600, (c) 550, (d) 525 and (e) 500 °C. On inset: the dependences of the InAs growth rate on the TMIn molar flow.

Using the obtained data about In composition in the $In_xGa_{1-x}As$ solid solution, the InAs growth rates were calculated from the relation:

$$V_{InAs} = \frac{V_{GaAs}x}{1-x} \quad (1)$$

where V_{GaAs} is the GaAs growth rate.

As a result the dependences of the InAs growth rate on the In molar flow were determined for each growth temperature (Fig. 1 on inset). These curves demonstrate that there is no strong influence exerted by the growth temperature on the InAs deposition rate in the range from 500 to 700°C.

These dependences were used for growing heterostructures with InAs QDs. In our work [14] the b-heterostructures grown on GaAs substrate with (100) orientation and different misorientation (2°,6°, 10°) were investigated and it was shown that 6° allow perfect GaAs underlayer surface planarity. The AFM image for samples grown on GaAs surface with 6° misoriatation is presented in fig. 2 a. Statistic data (fig. 2 b) shows the multimodal distribution of QDs by size.

Moreover, PL spectra from b-heterostructures (fig.3) show two states of QDs (marked as QD1 and QD2). Both states are clearly visible on each curve, while QD1 becomes more pronounced with decreasing measurement temperature. The presence of two peaks at different wavelengths may be indicative of the QDs of different sizes. Thus, this proves the effect of the multimodal distribution of quantum dots by size. The optical properties of a bimodal QD array are considered in detail in [15]. The evolution of PL spectra can also be compared with those produced in the articles [16, 17].

In our work [14] the a-heterostructures grown on different condition were studied by PL and growth rate and growth temperature of InAs QDs for 6° misoriented GaAs substrates were optimized. Here we also investigate the influence of GaAs cap layer growth parameters on PL intensity of QDs.

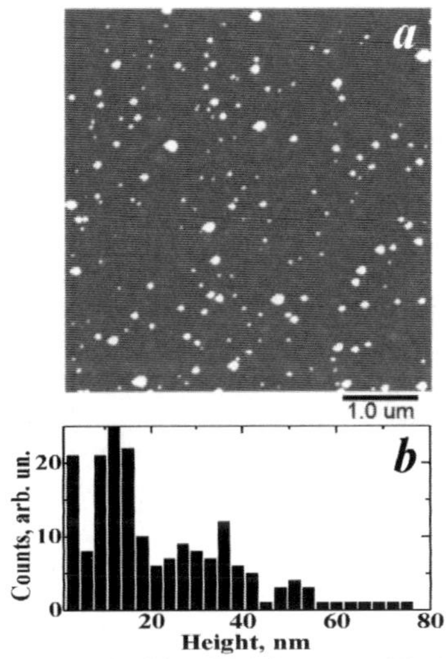

Figure 2. (a) AFM image of InAs QDs with a thickness of 1.7 ML deposited on a GaAs (100) surface with misorientation of 6° (b) statistics of QDs distribution by size.

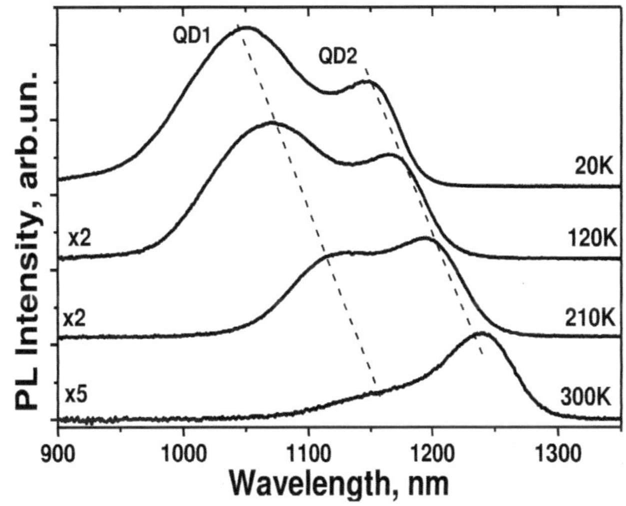

Figure 3. PL spectrum evolution with temperature at excitation level of 10 W/cm^2. GaAs cap layer thickness is 7nm.

For QD b-heterostructures grown on 6° misoriented substrates at 520°C the conditions under which the maximum PL intensity is observed depend of the on the amount of deposited InAs material under different growth conditions were found (Fig. 4). It represents a typical case of a Stranski Krastanov growth mode. Quantum dots formation occurs at a certain critical thickness of the deposited material [18]. This critical thickness depends on the difference between lattice parameter of deposited material and matrix. InAs has a quite narrow technological window for forming QDs on GaAs due to the large lattice constant difference. At the optimum critical thickness of the QDs formation the maximum of PL intensity will be observed. For QDs, deposited at sufficiently low growth rate (V_{InAs} = 0,0625 ML/s) and covered with 7 nm GaAs cap layer at growth rate of 4 A/s maximum position of PL intensity corresponds to ~ 1.7 monolayer of InAs (Fig. 2, b-curve).

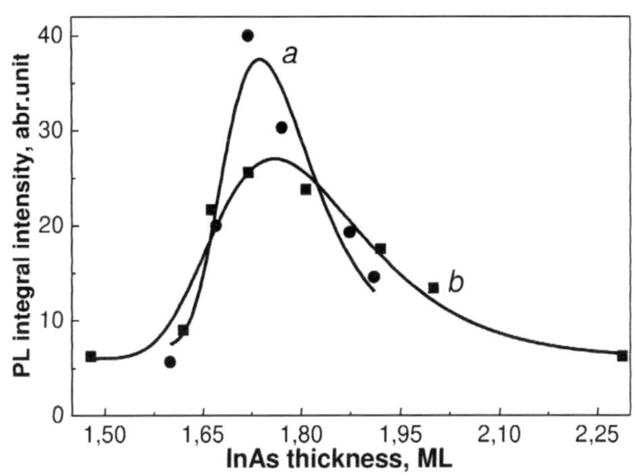

Figure. 4 – Dependences of the integrated PL intensity on InAs thickness grown at 0.0625 ML/s with different GaAs cap layer growth rate: (a) – 2 Å/s, (b) – 4 Å/s.

However, we can see that the GaAs cap layer growth rate is very critical factor affecting the maximum PL intensity of QDs. After reducing the growth rate of the GaAs cap layer from 4 Å/s to 2 Å/s the PL intensity significantly increased (Fig.2 a-curve). PL intensity increased by ~60% by reducing caplayer growth rate by half, that indicate the importance of the cap material quality for optical properties of QDs.

The optimal QD growth conditions determined were used for forming multilayer QD arrays (10 and 15 QD layers) and for embedding them into the GaAs solar cell heterostructure (Fig.5). No strain balance layers have been used. So, to avoid the strain, wide spacers of 35 nm have been grown on the GaAs caplayers. Based on the heterostructures grown, control photodiodes according to description presented in "experimental procedure" part have been manufactured.

Spectral dependencies of the external quantum efficiency (EQE) and the reflection have been obtained using such control diodes. As a result, a spectral dependence of the internal quantum efficiency (IQE) for both samples with QD arrays and for reference solar cell without QDs has been determined (Fig.6). In the region below λ= 880nm, the spectral dependency of the IQE for a QD SC with 10 QD rows (10xQDs) is at the same level as for the reference SC. This indicates the high quality of the GaAs p-n junction embedding 10xQDs layers into the active region even without the application of strain-balance technology. For the QD SC with 15xQDs a slight drop in the IQE in the spectral range below λ= 880nm is observed.

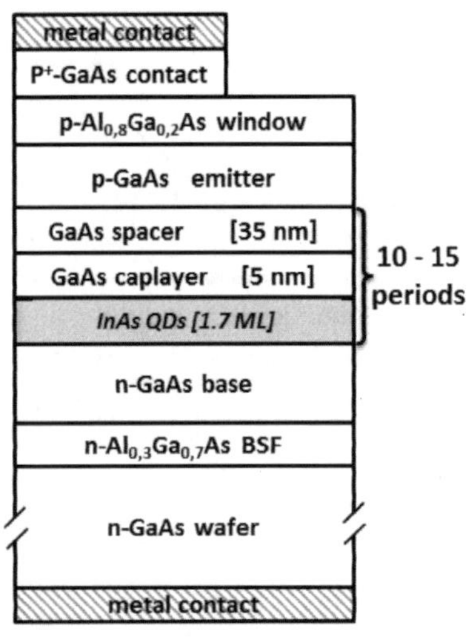

Figure 5. Schematic of solar cell structure with array of QDs.

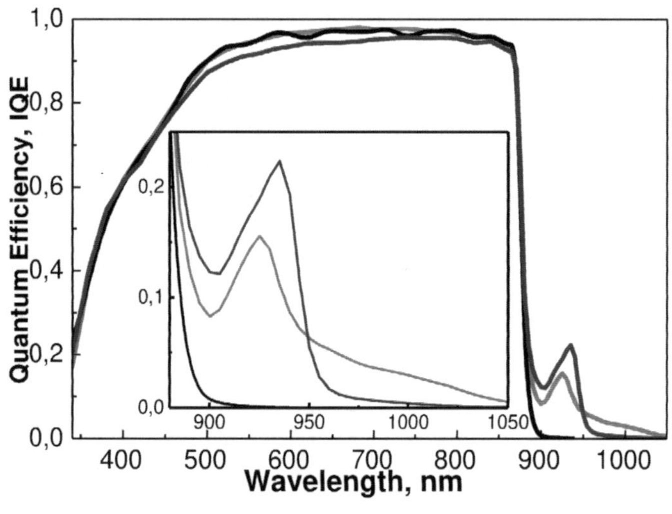

Figure 6. Spectral dependences of Internal Quantum Efficiency (IQE) for QD solar cells with 10-layer (red line) and 15-layer (blue line) QD arrays and for a reference solar cell without QD (black line).

The photogenerated current values obtained by integrating the spectral dependence with the IQE in the range of 880nm–1200nm for the heterostructure of a QD SC with 10xQDsc is 0.58 mA/cm^2 for the AM1.5D spectrum and about 0.74mA/cm^2 for that of the AM0. This corresponds to about 0.06mA/cm^2 and 0.07mA/cm^2 per a QD layer for AM1.D and AM0 spectra, respectively, that is the highest value so far achieved in InAs QDs embedded in pure GaAs. For the 15xQD SC, photogenerated current (λ= 880-1200nm) is 0.59 mA/cm^2 for the AM1.5D spectrum and 0.82 mA/cm^2 for that of the AM0.

4. Conclusions

The peculiarities of the epitaxial growth in the InAs–GaAs lattice-mismatched system were considered and MOPVE technology of InAs QDs growth on GaAs misoriented substrates was developed. The critical influence of the GaAs cap layer growth parameters on InAs QD PL intensity was demonstrated. Using developed technology SC structures with 10 and 15 layers of QDs were grown and control photodetectors were created. These QD SCs show an excellent sub-bandgap photon absorption in QDs (0.59 mA/cm^2 for the AM1.5D spectrum and 0.82 mA/cm^2 for that of the AM0) without noticeable degradation of quantum efficiency within GaAs spectral range in comparison with reference SC without QDs.
We believe that developed MOPVE technology of InAs QDs growth is attractive for improving photocurrent balance in the lattice matched multi-junction SCs and is also useful for IB SCs.

Acknowledgements

The authors wish to thank Professor Antonio Luque from Polytechnic University of Madrid for valuable discussions of the results and N. Kh. Timoshina for the spectral dependence measurements.
This work has been supported by the contract No. 14.B25.31.0020 from the Russian Ministry of Education and Science (resolution No. 220).

References

[1] Ledentsov N N, Ustinov V M, Shchukin V A, Kop'ev P S, Alferov Zh I and Bimberg D 1998 *Semiconductors* **32** 343
[2] Mikhrin S S et al. 2005 *Semicond. Sci. Technol.* **20** 340
[3] Kryzhanovskaya N V, Mukhin I S, Moiseev E I, Shostak I I, Bogdanov A A, Nadtochiy A M *et al* 2014 *Opt. Express* **22** 25782
[4] Luque A and Marti A 1997 *Phys. Rev. Lett.* **78** 5014
[5] Green M A, Emery K, Hishikawa Y, Warta W and Dunlop E D 2014 *Progr. Photovolt.: Res. Appl.* **22** 701
[6] Luque A, Martí A, Stanley C, López N, Cuadra L, Zhou D *et al* 2004 *Journal of Applied Physics* **96** 903
[7] Ramiro I, Martí A, Antolín E, Luque A 2014 *IEEE Journal of Photovoltaics* **4** 736
[8] Blokhin S A, Sakharov A V, Nadtochy A M, Pauysov A S, Maximov M V, Ledentsov N N *et al* 2009 *Semiconductors* **43** 514
[9] Hubbard S M, Cress C D, Bailey C G, Raffaelle R P, Bailey S G, Wilt D M 2010 *Applied Physics Letters* **92** 123512
[10] Guimard D, Morihara R, Bordel D, Tanabe K, Wakayama Y, Nishioka M and Arakawa Y 2010 *Applied Physics Letters* **96** 203507
[11] Tanabe K, Guimard D, Bordel D and Arakawa Y 2012 *Applied Physics Letters* **100** 193905
[12] Bailey C G, Forbes D V, Polly S J, Bittner Z H, Dai Y, Mackos C, Raffaelle R P, Hubbard S M 2012 *IEEE Journal on Photovoltaics* **2** 269
[13] Kalyuzhnyy N A, Mintairov S A, Mintairov M A and Lantratov V M 2009 *Proc. of the 24th European Photovoltaics Solar Energy Conference* (Hamburg: Germany) p 538
[14] Salii R A, Mintairov S A, Brunkov P N, Nadtochiy A M, Payusov A S and Kalyuzhnyy N A 2015 *Semiconductors* **49** 1136
[15] Bansal B 2006 *J. Appl. Phys.* **100** 093107
[16] Fafard S, Wasilewski Z R, Allen C N, Picard D, Spanner M, McCaffrey J P, Piva P G 1999 *Physical Review B* **59** 15368
[17] Kryzhanovskaya N V, Gladyshev A G, Blokhin S A, Maksimov M V, Semenova E S *et al* 2005 *Semiconductors* **39** 1188
[18] Goldstein L, Glas F, Marzin J Y, Charasse M N, Roux G Le 1985 *Appl. Phys. Lett.* **47** 1099

Journal of Physics: Conference Series **769** (2016) 012037 doi:10.1088/1742-6596/769/1/012037

Spectroscopic properties of superparamagnetic Fe_mO_n-SiO_2 nanoparticle colloidal solutions

R S Smerdov[1], T V Bocharova[1], K G Gareev[2]

[1]Saint Petersburg Polytechnic University,
29th Politechnicheskaya Str., Saint-Petersburg, 195251, Russia
[2]Saint Petersburg Electrotechnical University "LETI",
5th Prof. Popova Str., Saint-Petersburg, 197376, Russia

rostofan@gmail.com

Abstract. The investigation of optical and electron spin resonance (ESR) spectroscopic properties are carried out. Six characteristic absorption bands related to distinctive conglomerates of Fe_mO_n formed via sol-gel synthesis process on SiO_2 porous matrix are observed in optical spectra. ESR spectra investigation makes it possible to reveal small characteristic signal associated with magnetostatic interactions between conglomerates of Fe_mO_n nanoparticles.

1. Introduction

The development of novel medical diagnostics methods such as magnetic resonance imaging (MRI), positron emission tomography, X-ray computed tomography has required further research in the field of agents for contrast-enhanced imaging [1]. Composite superparamagnetic FemOn nanoparticles may be considered as negative contrast agents reducing spin-spin relaxation time. The possibility of synthesis of 10 nm Fe_mO_n-SiO_2 composite nanoparticles sustaining magnetic hysteresis properties is of great interest. Furthermore, due to their biocompatibility magnetic nanoparticles may be also used as magnetic drug-targeting agents [2] and as a potential medium for magnetic hyperthermia in high-frequency alternating magnetic field [3]. Optical properties of thin magnetite films are of great importance: such films demonstrate UV absorption and decent transmission in visible region allowing to consider nano-sized magnetite as a promising optoelectronic material [4]. Superparamagnetic nanoparticles FenOm could be used in a wide variety of circumstances due to conjunction of important characteristics with peculiarities of chemical composition and magnetic properties.

2. Experimental methods

Colloidal composites Fe_mO_n-SiO_2 were formed via two stage sol-gel synthesis process. The first stage consisted in tetraethoxysylane (TEOS) sol synthesizing. Aqueous solution of ammonia was used to precipitate the gel. After 72 hours of gel desiccation at room temperature thermal treatment of SiO_2 powders at 300 °C was carried out.

At the second stage emulsification of highly dispersed SiO_2 was conducted by using ultrasonication treatment in aqueous solution of Fe^{2+} and Fe^{3+} chlorides. Then, Fe_mO_n nanoparticles were deposited according to the following reaction:

$$FeSO_4 + 2FeCl_3 + 8NH_3{\bullet}H_2O \rightarrow Fe_3O_4\downarrow + 6NH_4Cl + (NH_4)2SO_4 + 4H_2O$$

The sediment was dried at room temperature after being formed in the solution. In order to analyze chemical composition and magnetic properties of colloidal nanoparticles the sediment was investigated by ESR method.

The investigation presented in this thesis had two distinct objectives. The first objective was to identify parameters of the optical absorption spectra of colloidal solutions of magnetic nanoparticles and to create the structure model of a single composite FemOn-SiO_2 nanoparticle. Specord® 40 Analytik Jena was used to fulfill optical absorption spectra measurements of the samples at the temperature of 300K. ESR spectra were obtained by using ESR-spectrometer JEOL at room temperature. Fe_mO_n powder was irradiated with Co^{60} γ-ray source to dose 10^6 rad.

3. Experimental results

According to P. Levy [5], the shape of the absorption bands is considered to be Gaussian as in equation (1);

$$E(\nu) = K_0 \exp\left[-\frac{4ln2}{U^2}(\nu - \nu_0)^2\right], \qquad (1)$$

where K_0 corresponds to absorbance at the wavelength of an absorption band maximum, U is the half-width of an absorption band (cm^{-1}) and ν_0 is the position of an absorption band maximum (cm^{-1}). The appearance of absorption spectra (figure 1) of colloidal solutions indicates the identity of the chemical composition.

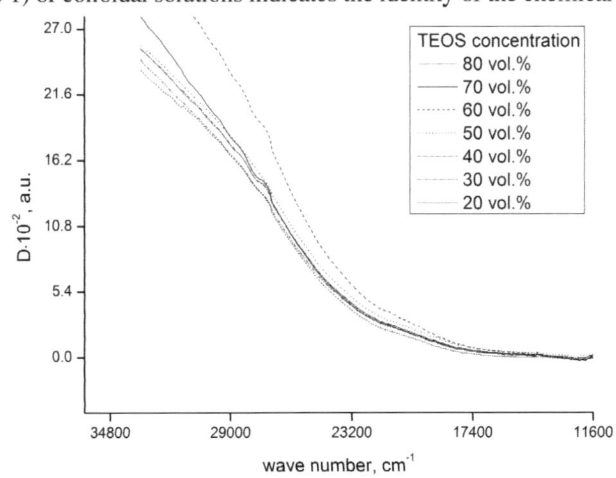

Figure 1. The absorption spectra of colloidal solutions of superparamagnetic nanoparticles Fe_mO_n-SiO_2 in the visible and near UV range; the parameter is the concentration of TEOS during synthesis.

Alternation in the concentration of Fe_mO_n nanoparticles deposited on the SiO_2 gel matrix during the Massart reaction is obvious and the largest concentration of Fe_mO_n nanoparticles is detected while using a solution of TEOS in isopropanol with the concentration of TEOS 60 vol. %.

The result of mathematical simulation of absorption spectra using Origin© 8.1 for colloidal solution with 60 vol. % TEOS concentration is shown on figure 2.

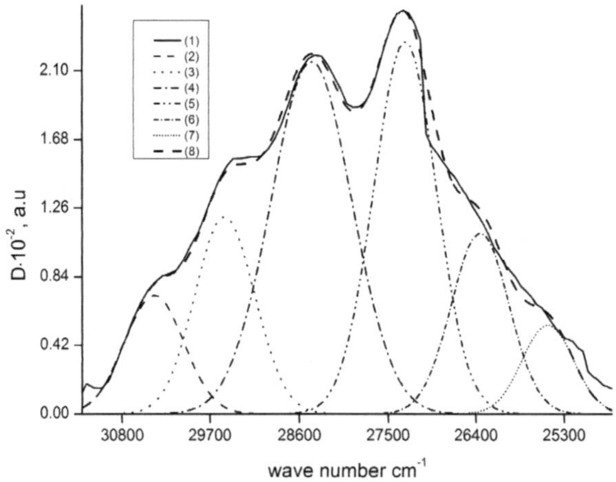

Figure 2. The approximation result of absorption spectrum via Gaussian curves (60 vol. % TEOS concentration): 1 – initial absorption spectrum, 2-7 – absorption bands corresponding to 25130 cm^{-1}, 26220 cm^{-1}, 27300 cm^{-1}, 28480 cm^{-1}, 29530 cm^{-1} and 30480 cm^{-1}, 8 – absorption spectrum approximation.

As one can see in the figure 2 six distinctive absorption bands in UV region corresponding to the following maxima are found: 25130 cm^{-1}, 26220 cm^{-1}, 27300 cm^{-1}, 28480 cm^{-1}, 29530 cm^{-1} and 30480 cm^{-1} with intrinsic half-width remaining constant. The presence of these bands is attributed to Fe_mO_n conglomerates with six characteristic dimensions on the surface of the gel matrix. Absorption maxima observed at a concentration of TEOS 60 vol. % and associated with conglomerates of all sizes are caused by the interaction of clusters (figure 3). Cluster size is estimated using G. Mie Theory [6]. According to equation (2),

$$K_{max}\Gamma = An4\pi R^3/3, \tag{2}$$

where K_0 corresponds to absorbance at the wavelength of an absorption band maximum, Γ – is the half-width of an absorption band (cm^{-1}), A=1,25×102 eV/mm is the optical constant, R – colloidal particles radius, n – concentration of the nanoparticles (cm^{-3}). Calculated radii of clusters turn out to be R = 1,9 nm, 1,8 nm, 2,07 nm, 1,5 nm, 2,57 nm, 3,02 nm with respect to 25130 cm^{-1}; 26220 cm^{-1}; 27300 cm^{-1}; 28480 cm^{-1}; 29530 cm^{-1}; 30480 cm^{-1} absorption bands (at 60 vol.% TEOS concentration).

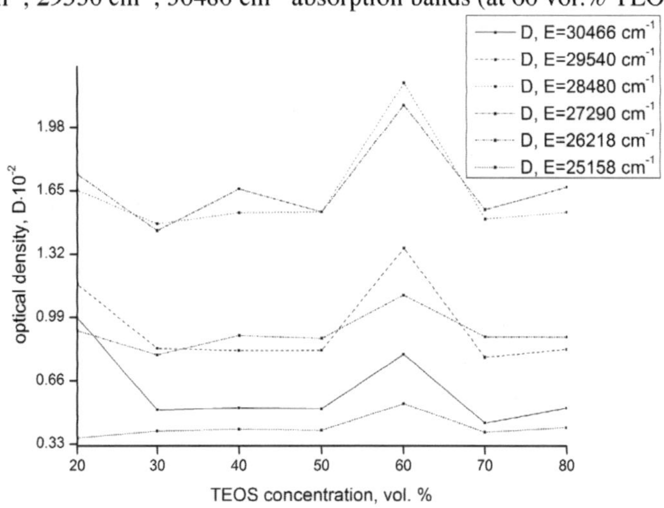

Figure 3. The dependence of the maximum optical density (D_{max}) upon the concentration of TEOS for six characteristic absorption bands.

One can see in figure 4 ESR spectra of initial (signal 1) and γ-irradiated (signal 2) samples of sedimented Fe_mO_n -SiO_2 powders. Initial Fe_mO_n sample reveals slightly asymmetric line in the range of B = 3450 G (g=2.04). Similar signal was obtained in paper [7] dedicated to investigation of temperature and frequency dependences of Fe_mO_n nanoparticle magnetic properties. According to [8] the aforementioned wide ($\Delta B \approx 680$ G) line resulted from relaxation in finite-sized nanoparticles. It is worth mentioning that similar line was observed [9] in ESR spectrum of x Fe_mO_n – (100 - x) SiO_2 amorphous system (x = 0,1...9,5) synthesized via sol-gel technology.

Taking in consideration ESR spectrum of γ-irradiated sample one can see a signal of low intensity located on low energy side of general signal (g=2.04) at the value of g=2.08. Ionizing radiation impact seems to be followed by emergence of radiation defects in silicate matrix. [10]

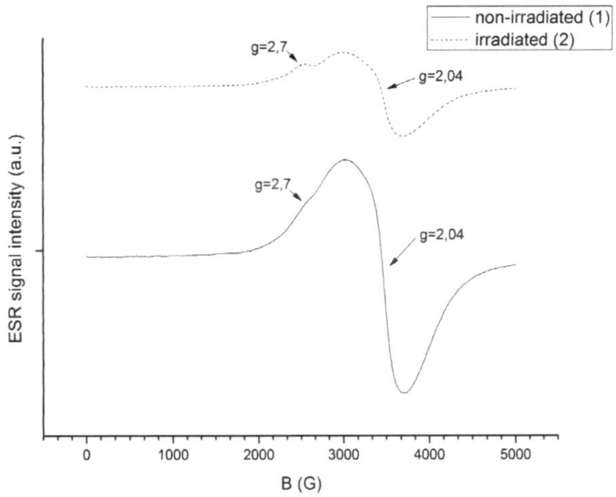

Figure 4. ESR spectra of initial and irradiated Fe_mO_n-SiO_2 powder samples.

4. Conclusions

Composite magnetic nanoparticles Fe_mO_n-SiO_2 are investigated by using optical absorption spectroscopy method with respect to G. Mie theory. The presence of six distinctive absorption bands associated with the formation of Fe_mO_n conglomerates on the surface of the gel matrix is revealed. Approximate estimation of conglomerate dimension correlates with results acquired in [11].

In ESR specta of both non-irradiated and irradiated samples slightly asymmetric line at the value of g=2.04 is found. The existence of this band may be related to radiochemical reactions followed by transformation of Fe^{2+} to Fe^{3+}. The signal at the value of g=2.8 is associated with magnetostatic interactions between conglomerates of Fe_mO_n nanoparticles.

References

[1] Bin Na H, Kim D, Baek JS, Ko MK, Lee JH, et al. 2009 Adv. Mater. **21** 2133–2148
[2] Liao Z, Wang H, Lv R et al. 2011 Langmuir **27** 3100–3105
[3] Kappiyoor R, Liangruksa M, Ganguly R, Puri IK et al. 2010 J Appl. Phys **9** 108
[4] Chang-Neng, et al. 2007 Materials Transactions **48,** *5* 1143–1148
[5] Levy P W 1960 J. American Cerafic Society **8** 389–395
[6] Kreibig U 1976 Appl. Phys. **10** 255–264
[7] Köseoğlu Y, Aktaş B 2004 Phys. Stat. Sol. **1**, *12* 3516
[8] Noginov M M, Noginova N, Amponsah O, Bah R, Rakhimov R, Atsarkin V A 2008 J. of Magn. Magn. Mat. **320** 2282
[9] Tanaka K, Kamiya K, Matsuoka M, Yoko T. 1987 J. of Non-Cryst. Sol. **94**, *3* 365
[10] Bocharova T V, Karapetyan G O, Khaliev V D, Mironov A M. 2006 Opt. Mat. **28** 1296
[11] Wang F, Tang Y, Zhang B, Chen B, Wang Y. J. 2012 Colloid Interface Sci. **386**, *1*, 129

Journal of Physics: Conference Series **769** (2016) 012038 doi:10.1088/1742-6596/769/1/012038

Investigation of light-induced degradation of tandem photoconverters on a-Si:H/μc-Si:H

A S Titov, A S Abramov, D A Andronikov, K V Emtsev, A F Semenov, E E Terukova, S A Yakovlev

Ioffe Institute, Saint-Petersburg, Russia
TF TE Ioffe R&D Center, Saint-Petersburg, Russia

e-mail: titovoz@gmail.com

Abstract. Photo-induced degradation of tandem α-Si: H/μc-Si: H photoconverters under standard light flux densities of 1000 W/m2 (AM1.5G) have been investigated. Spectral and current-voltage characteristics of solar modules produced with standard and modified recipes are presented and analyzed. This research is dedicated to the optimization of crystallization level in intrinsic layer of lower (microcrystalline) cascade.

1. Introduction

Solar cells based on the a-Si:H/mc-Si:H (micromorph) thin-film technology became more widespread in recent years [1]. Record efficiency for one-cascade a-Si:H photoconverters is 10.1%, for double-cascade tandems a-Si:H/mc-Si:H – 12.3% under standard light flux conditions AM1.5G. Thin-film technology allows to produce one of the lowest costs of watt for solar modules on the market [2,3,4]. This is achieved primarily due to the low production cost because of the relatively simple production technology and low raw materials consumption.

Main disadvantage of micromorph technology is photoinduced (light) degradation – a decrease of photoelectrical characteristics under the light exposition [5]. This process is closely connected with the Staebler-Wronski effect – effect of decrease in the photoconductivity of a-Si:H films under continuous light impact [6]. Typical decrease in the efficiency of micromorph cells is 10 to 30% and depends on the number of subelements, thickness and chemical compound of photoactive layers [7,8].

Improvement in the efficiency of micromorph modules can be achieved by two ways – increasing the initial efficiency and reducing the degradation of the modules.

In this article the micromorph modules degradation is researched. Photoconverters in question were produced on TEL Solar (formerly Oerlikon Solar) installation based on a standard recipe (sequence of steps to produce solar cell) and modified one by TF TE Ioffe R&D Center (TFTC). This research is dedicated to the optimization of crystallization level in intrinsic layer of lower (microcrystalline) cascade.

2. Methodic of the experiment

1100x1300mm glass with high optical transmission is used as a top substrate. Different layers are sequentially deposited on the top substrate: front transparent conduction oxide (TCO) layer (ZnO), a-Si:H, mc-Si:H and back TCO contact. A polymeric reflector at the back is used to increase the

Journal of Physics: Conference Series **769** (2016) 012038 doi:10.1088/1742-6596/769/1/012038

operating current. PECVD (Plasma Enhanced Chemical Vapor Deposition) under high pressure method is used to obtain TCO layers - Boron doped ZnO layers with a light scattering degree up to 25-27% and optical transparence (400-1000nm) over 85%. When ZnO layer thickness is 1600-1700 nm, the electrical resistance of the front contact is 16-18 Ohm /□ (Ohm per square), for back layer – 13-15 Ohm /□.

Photocascades are deposited by a PECVD method. It uses a mixture of silane and hydrogen, as well as doping gases trimethylboron (for p-layers) and phosphine (for n-layers). In addition, methane additive is added to the gas mixture to increase the transparency of p-layer of top cascade. CO2 is added to the gas mixture with the reflectors deposition.

Each of the cascades is a p-i-n diode and current flow between them is provided by the recombinative junctions. The pin configuration is used instead of the classical pn-junction due to special material properties of used a-Si:H and mc-Si:H materials. These materials are characterized by a large number of structural defects that serve as traps for electrons and holes, whose presence reduces the diffusion length of charge carriers. Unlike classical crystalline semiconductors a small diffusion length leads to the inability to use any significant thickness of photovoltaic structures and, as a consequence, produce the essential values of the photocurrent. To overcome this fundamental limitation pin-diode circuit is applied, where undoped i-layer is inserted between the doped layers. In this i-layer the absorption of light and separation of charges takes place. Charge transfer is carried out not by the diffusion mechanism, but by the drift in an electric field formed by an energy barrier in the pin-junction. The drift length depends on the electric field, which is determined by the structural parameters: the thickness of the i-layer, the degree of doping of p- and n-layers, the band gap value, the quality of the deposited layers – all parameters of pin-structure deposition process.

Usage of the two cascade scheme can lead to the growth of the cell efficiency by increasing the spectral sensitivity of solar cells, which is achieved with materials with different band gaps as absorbers, as well as with reducing thermalization of photocarriers losses due to the separation of absorption of light in two cascades. In order to ensure separation of photoconverter into individual cell and electrically connect them laser scribing is used - the removal of layers deposited material on a strictly limited area with laser beam. By implementing such a sequence of operations monolithic electrical connection of individual cells is performed in the composition module.

Schematic representation of the structure and electron micrograph of researched cells are presented in Fig. 1 [9].

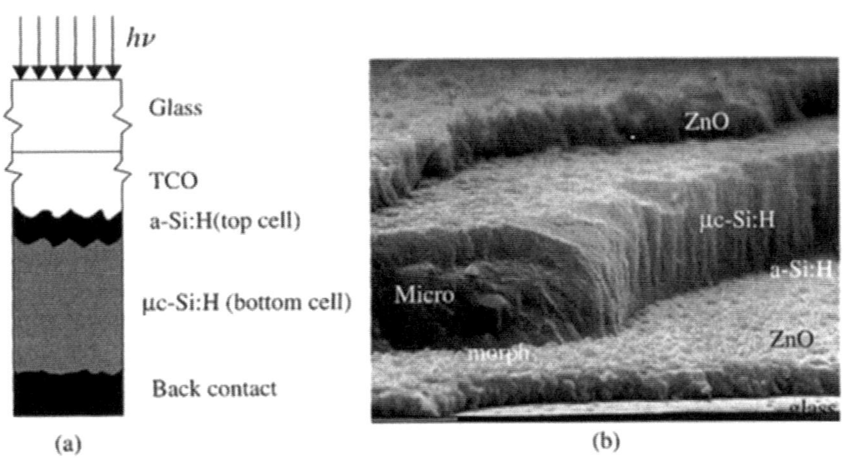

Figure 1. Micromorph solar cell: (a) schematic structure (b) electron micrograph [9].

Optimization of crystallization level (degree) in intrinsic layer of lower (microcrystalline) cascade is performed in this research. This parameter has an optimum value in the range of 55-60% [10, 11], which corresponds to the maximum value of the stabilized power. Increase in crystallinity degree relating to this range leads to an increase of the absorption coefficient in the red and near infrared range and, consequently, to an increase in a short-circuit current, however, negatively affects the open circuit voltage due to increase in number of cracks in the material. Crystal grain boundaries are insufficiently passivated, there is a large number of defects, resulting in increased surface recombination and reduced load voltage. Moreover, there is evidence [12,13] that too high a degree of crystallinity of the microcrystalline silicon layer results in a dark material degradation due to the diffusion of oxygen atoms and water molecules to the grain boundaries, which reduces the open circuit voltage over time. Reduction of the crystallinity degree results in a reduction of the coefficient of absorption in the red and near infrared range and, consequently, reduction of the short circuit current, as well as increase in the level of photo-induced degradation, due to the increase of the volume fraction of the amorphous phase.

The crystallinity level is determined by Raman scattering and calculated as follows: (I520 + I510) / (I520 + I510 + I480), where I520 and I510 is the intensity of Raman lines at frequencies at 520 and 510 cm-1, corresponding to the thermal lattice vibrations of the silicon crystallites, I480 - the intensity of the Raman band at a frequency at 480 cm-1, corresponding to the thermal vibrations of the amorphous matrix. The ratio of intensities of these peaks is taken as the degree of crystallinity of the material. As part of the research optimal characteristics are obtained with a degree crystallinity of i-lower layer at level of 52-53%. In this paper we compare the photovoltaic characteristics of cells with a crystallinity degree of the i-layer in the lower cascade at 66-68% (basic recipe) to the photoelectric characteristics of cells with a crystallinity degree of the i-layer in the lower cascade at 52-53% (modified recipe). In addition to the crystallinity level there is optimization of the deposition rate and area uniformity of deposition. Comparison of these parameters for two recipes is given in Table 1.

Table 1. Parameters of i-layer in the lower cascade for base and modified recipes.

Recipe	Deposition rate, A / s	Area non-uniformity thickness, %	Crystallinity level,%
Base	4,4	18,5	67
Modified	5,6	7,9	53

Thus, we obtained in the recipes with modified i-layers of the lower cascade a lesser crystallinity degree and better thickness uniformity over the area of the glass.

In the research 100 × 100 mm test samples is cut from the full-size modules, in which by the method of laser scribing photoactive part 60 × 66 mm is obtained with an open surface area of 37.95 cm2. It consists of 10 photovoltaic cells of equal size, connected in series. There is a custom solar module 10 × 10 cm on fig. 2.

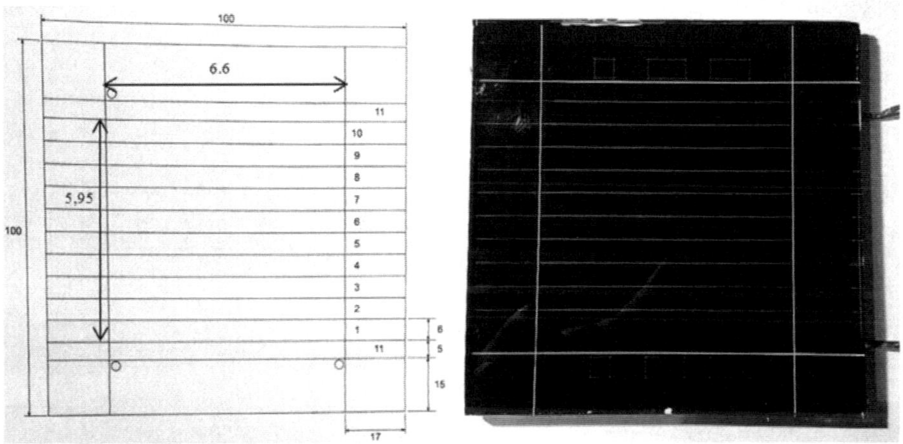

(a) (б)

Figure 2. Mini-module: photo (b) and scheme (a).

Contacts of the test sample are obtained by gluing strips with silver paste. Then the side with the contacts is covered with the laminating film contacts based on the ethylene vinyl acetate, and after that the test cells are subjected to a heat treatment at a temperature 150◦C for 25-30 min to polymerize the film.

Research of the degradation of solar cell is set up using the installation of the light soaking benchmark (LSB), where modules are continuously exposed to the light. In practice, modules degradation tests are carried out with the illumination 600-1000 W/m2 and a temperature no more then 50-60◦C and a total exposure not less than 1000 hours [14]. Since it is very important to maintain a certain temperature to saturate the solar cells, the heat they emit should be scattered evenly. For this purpose an aluminum plate with water cooling is used as the sample holder.

The degraded samples are further investigated with continuous type sun simulator (SUS) and with a system for the study of the external quantum efficiency (EQE). Degradation tests last for about 1000 hours. In the first 100 hours the intervals between measurements are small, since the curve degradation is sufficiently sharp. After that measurements are made with a period of 100 hours or more.

3. Experimental results

The research carried out with the following samples: i305, i738 - two standard recipes, i748, i754, i755 - three different recipes created (modified) in the TFTC.

Average initial parameters obtained with the continuous type solar simulator shown in Table 2.

Table 2. The initial parameters of the elements, where the open circuit voltage - Uoc, short-circuit current - Isc, short-circuit current density - Jsc, efficiency - Eta, fill factor - FF, serial and parallel resistance -Rs and Rp respectively.

Cell ID	Voc [V]	Isc [mA]	Jsc [mA/cm^2]	FF [%]	Eta [%]	Rs [Ω]	Rp [kΩ]
i305	13.33	45.18	11.5	68.77	10.55	35.58	17.54
i738	12.98	42.96	10.93	67.29	9.58	40.54	16.30
i748	13.61	42.30	10.76	71.44	10.46	33.52	10.77
i754	13.72	41.97	10.68	72.24	10.58	32.16	10.08
i755	13.52	40.96	10.42	69.11	10.37	38.04	9.13

The table shows that the initial parameters of the optimized recipes are slightly above ones of base recipes. However, the uniformity of the modified recipes is much better. For example, the standard deviation of the efficiency for the new prescription i754 is 0.21 and for the i305 - 0,81, which is nearly four times higher (see. Table 3). Thus, the total efficiency of the full-scale module (1300 × 1100 mm) is higher.

Table 3. Average parameters of the elements i305 and i754 before degradation.

	Voc [V]	Isc [mA]	Jsc [mA/cm²]	FF [%]	Eta [%]	Rs [Ω]	Rp [kΩ]
i754							
mean	13.72	41.97	10.68	72.24	10.58	32.16	10.08
stddev	0.08	1.03	0.26	1.35	0.21	0.67	5.66
i305							
mean	13.33	45.18	11.50	68.77	10.55	35.58	17.54
stddev	0.17	0.38	0.10	4.87	0.82	6.47	15.02

Then mini-modules of these parties are subjected to degradation tests. The results are shown in Table 4.

Table 4. Parameters of the samples after degradation.

Cell ID	Voc [V]	Isc [mA]	Jsc [mA/cm²]	FF [%]	Eta [%]	Rs [Ω]	Rp [kΩ]
i305	13.30	43.83	11.15	62.73	9.21	45.62	7.98
i738	13.15	41.40	10.54	61.10	8.45	56.54	26.28
i748	13.50	41.83	10.64	64.02	9.18	47.23	10.93
i754	13.60	41.30	10.51	65.85	9.39	46.75	12.40
i755	13.51	41.44	10.55	65.14	9.26	47.37	9.03

After the degradation average values of the optimized recipes appear still with less current and higher voltage, but the fill factor and conversion efficiency have increased significantly in comparison with the standard recipes. Figure 3 shows a comparison of the average current-voltage curves for the recipes i305 - standard and i754 - modified.

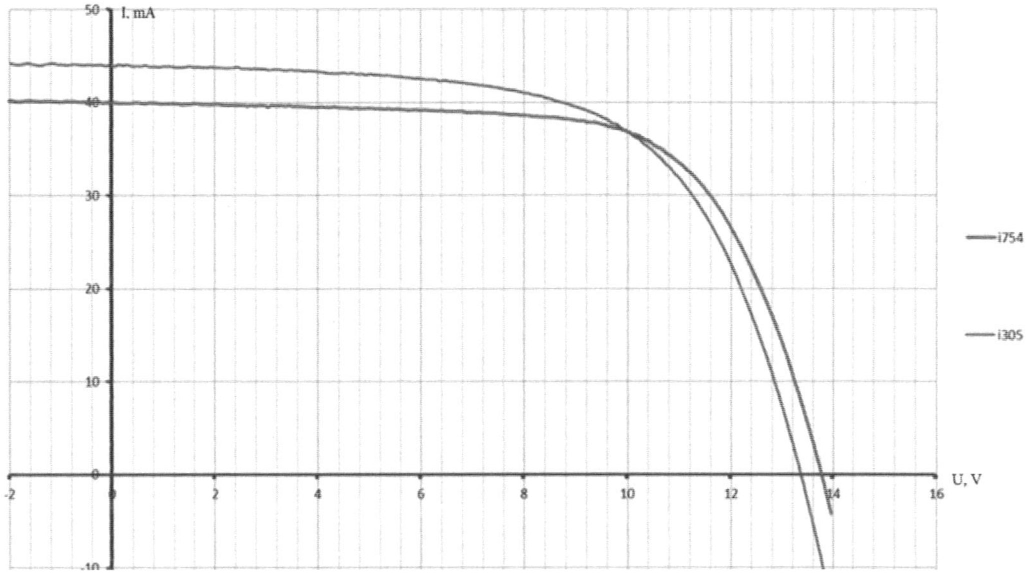

Figure 3. Average current-voltage characteristics of the test samples after degradation.

Studies of external quantum efficiency are shown in Figure 4: the dashed lines are for the i305, solid - for the i754. It can be seen that the lower mc-Si cell substantially does not degrade, compared with a top, a-Si.

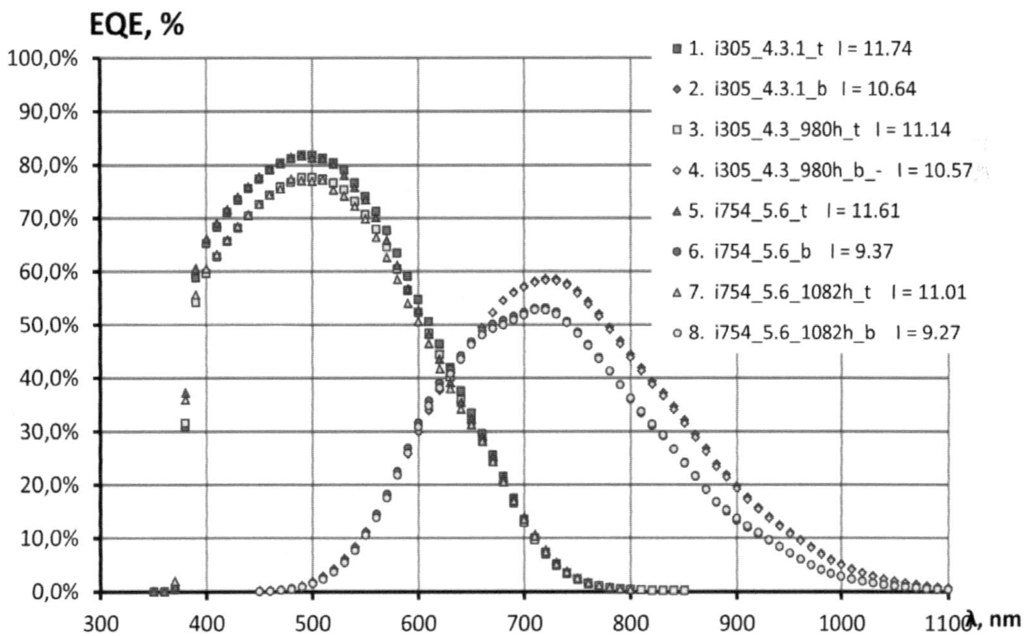

Figure 4. Study of the external quantum efficiency of the samples.

Degradation process itself is described in Table 5 and represented in fig.5.

Table 5. Degradation of the samples.

Glass No.	Δ Pmax, %	Δ Voc, %	Δ Isc, %	Δ FF, %	Δ Rs, %
i305	14.9	1.3	2.6	11.5	38.4
i738	17.0	0.2	3.4	14.0	59.7
i748	11.8	1.3	0.5	10.2	39.1
i754	11.0	0.4	1.5	9.2	48.5
i755	11.4	0.7	0.8	10.0	45.4

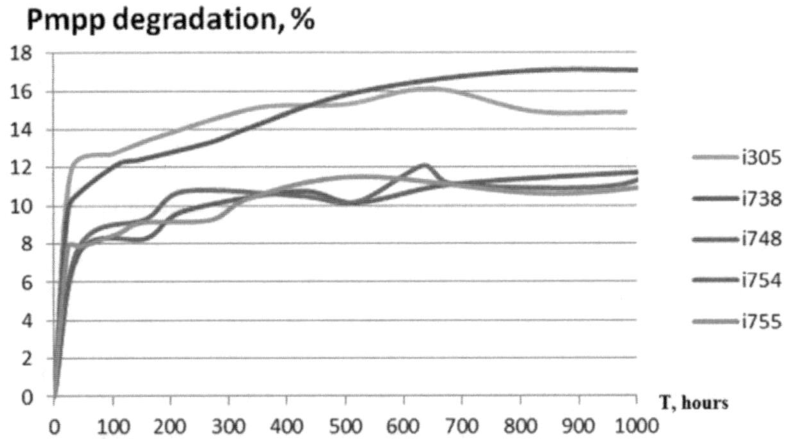

Figure 5. Degradation of the maximum output power of the samples.

From Table 4 and Figure 5 it can be concluded that the modules of the new recipes degrade much less than the old samples - 11-11.8% for the old versus 15-17% for new samples.

4. Conclusion

We investigated the photoinduced degradation of the solar cells based on the tandem structure of a-Si:H/μc-Si:H with a standard illumination – 1000 W/m2 AM1.5. During the tests the spectral and current-voltage characteristics of samples specially produced with the standard and modified in the TFTC recipes measured.

To sum up, while maintaining (or even slightly increasing) the original characteristics of the modules with respect to the basic recipe, the degradation of the modified samples when exposed to sunlight reduced significantly.

References

[1] Photon international (March 2009) p. 170.

[2] M.A. Green, K. Emery, Y. Hishikawa, W. Warta, E.D. Dunlop (2012) *Progr. in Photovolt.: Res. Appl.*, **20**, 606.

[3] R.R. King, A. Boca, W. Hong, X.-Q. Liu, D. Bhusari, D. Larrabee, K.M. Edmondson, D.C. Law, C.M. Fetzer, S. Mesropian, N.H. Karam. (Hamburg, 2009) *Proc. 24th EU PVSEC,* p. 55.

[4] V. Fthenakis, S. Gualtero, R. van derMeulen, H.C. Kim. (2008) *Mater. Res. Soc. Symp. Proc.* v.

1041, p. 25.
[5] C.R. Wronsky. (1997) *Mater. Res. Soc. Symp. Proc.* v. **469**, p. 7.
[6] D.L. Staebler, C.R. Wronski. (1977) *Appl. Phys. Lett.*, **31**, 292.
[7] A. Kolodziej. (2004) *Opto-electron. Rev.*, **12**, 21.
[8] A. Luque, S. Hegedus. (2003) *Handbook of photovoltaic science and engineering*, p. 551.
[9] A. Shah, (2010) *Thin-film silicon solar cells.*
[10] M. Rohde, M. Zelt, O. Gabriel, S. Neubert, S. Kirner, D. Severin, T. Stolley, B. Rau, B. Stannowski, and R. Schlatmann, (2014) *Thin Solid Films,* "Plasma enhanced chemical vapor deposition process optimization for thin film silicon tandem junction solar cells," vol. **558**, pp. 337–343.
[11] S. Hanni, G. Bugnon, G. Parascandolo, J. Escarré, M. Boccard, M. Despeisse, F. Meillaud, and C. Ballif, (2013) *28th European Photovoltaic Solar Energy Conference and Exhibition*, "The role of interfaces in high-efficiency microcrystalline silicon thin-film solar cells".
[12] Y. Mai, S. Klein, R. Carius, H. Stiebig, L. Houben, X. Geng, and F. Finger, (2006) *J. Non. Cryst. Solids* "Improvement of open circuit voltage in microcrystalline silicon solar cells using hot wire buffer layers,", vol. **352**, pp. 1859–1862.
[13] G. Ganguly, G. Yue, B. Yan, J. Yang, and S. Guha, (2006*) IEEE Photovoltaic specialists conference* "Fabrication of Large Area Amorphous Silicon / Nanocrystalline Silicon Double Junction Solar Cells," pp. 1712–1715.
[14] IEC 61646. (2008) *Thin-film Terrestrial Photovoltaic (PV)* Modules- Design, Qualification and Type Approval. 2nd edn.

Processing of ultra-hard coatings based on $AlMgB_{14}$ films

M H Yuzvyuk[1], **V V Putrolaynen**[1] **and A M Grishin**[1,2,3]

[1] Petrozavodsk State University, 185910 Petrozavodsk, Russia
[2] KTH Royal Institute of Technology, SE-164 40 Stockholm-Kista, Sweden
[3] INMATECH Intelligent Materials Technology, SE-127 51 Skärholmen, Sweden

E-mail: qesada21@gmail.com

Abstract First time $AlMgB_{14}$ films were prepared in Ames Lab by pulsed laser deposition technique. In this work, RF magnetron sputtering from a single stoichiometric target was employed to fabricate hard $AlMgB_{14}$ coatings on Si wafer and industrial items. Measurements of nanohardness and elastic Young's modulus were performed to determine reliable strength characteristics of samples. Smooth 3 μm thick $AlMgB_{14}$ films with the RMS surface roughness to be less than 1 nm exhibit hardness of 34 GPa and modulus of elasticity of 230 GPa at 20 mN peak load.

1. Introduction

Development and application of wear-resistant coatings is one of the main directions in the field of protection against wear and tear of the surface layer of cutting tools and saving their geometry at processing. Coatings based on aluminum magnesium boride $AlMgB_{14}$ (BAM films) demonstrate at small contact depths nanohardness up to 51 GPa [1] and have great potential for industrial applications.

For today, the coating was fabricated by pulsed laser deposition (PLD) technique [1, 2], but in terms of industrial application the most promising method is a magnetron sputtering technique. It enables preparation of conformal film coatings onto complex shaped and big size substrates. Several attempts to fabricate the coatings by RF magnetron sputtering were undertaken [3–6]. However, we know only one successful result published to date [7]. This work aims to further develop sputtering technology for ultra-hard $AlMgB_{14}$ coatings fabrication.

2. Materials and methods

BAM films are RF magnetron sputtered from a single 2 inch stoichiometric ceramic $AlMgB_{14}$ target in AJA Orion 5 system. Target was fabricated from elemental aluminum (99.97% pure), magnesium (99.8% pure) and boron (99% pure). The simultaneous synthesis and consolidation of $AlMgB_{14}$ from powder mixtures was performed in a spark plasma sintering apparatus [8].

To obtain films with the required stoichiometric composition and specific properties the parameters of sputtering are modified: target sputtering power (50 – 200 W), substrate temperature (200 – 450°C), target-to-substrate distance (10 – 35 mm), the pressure in vacuum chamber (2 – 5 mTorr of Ar gas), and time of sputtering (30 – 60 min).

Atomic-force microscope (AFM) CMM-2000 and scanning electron microscope (SEM) HITACHI SU1510 are used to control, respectively, films' roughness and stoichiometry by energy dispersive X-ray spectroscopy (EDX). Mechanical tests are carried out using desktop nanohardness tester NHT2-TTX with Berkovich three-sided diamond pyramid tip. Peak load was varied from 1 to 20 mN, loading and unloading time was 30 s, dwelling time was zero. Hardness H and Young's modulus E are obtained as per the Oliver and Pharr method. Effective elastic modulus E_{eff} is instrumentally measured through the indents contact area and unloading stiffness. Young's modulus we calculate $E = E_{eff}(1\text{-}\nu^2)$ assuming BAM film Poisson's ratio ν to be equal to 0.25.

Crystal structure of deposited films is determined by X-ray diffraction (XRD) using CuK_α radiation.

3. Results and discussion

Result of X-ray examination is presented in figure 1. X-ray spectrum shows that BAM film is amorphous, diffraction peaks belong to Si(100) wafer used as a substrate.

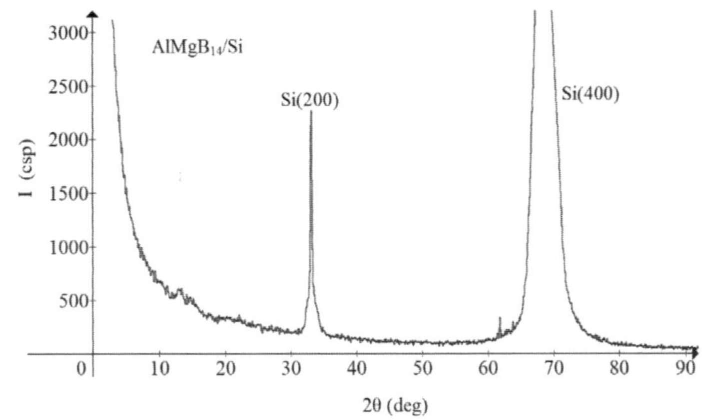

Figure 1. X-ray diffraction spectrum of BAM film deposited on Si(100) wafer.

Films deposited during 40 min without substrate heating at the distance of 35 mm from the target, at 200 W of RF power, and 5 mTorr of Ar gas pressure had low hardness and substantial degree of roughness.

The next series of samples was obtained at target-to-substrate distance equal to 10 – 12 mm and target sputtering power 200 – 250 W. Figure 2 and table 1 present characteristics of EDX spectrum and optical micrograph of island-like surface of sample «RF6» fabricated at these conditions. Sputtered film appeared to be inhomogeneous and besides Al, Mg and B contains also C, N, and O.

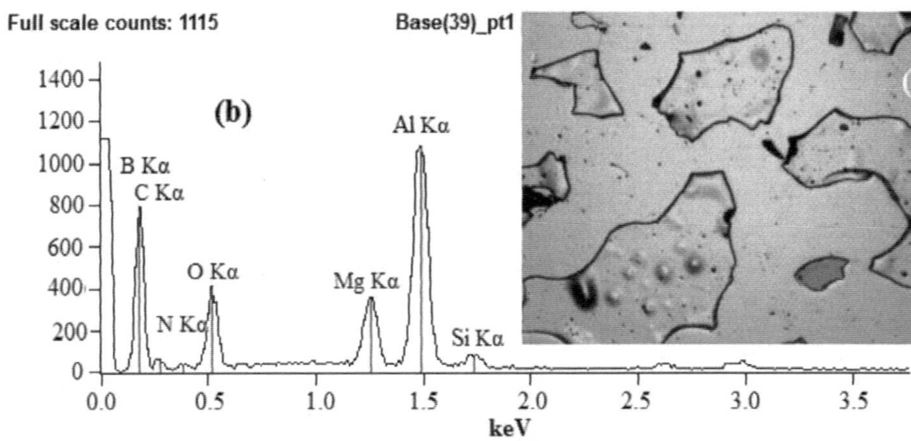

Figure 2. (**a**) - Optical micrograph of the surface of film «RF6» deposited at target sputtering power equal to 250 W and target-to-substrate distance of 12 mm. (**b**) - EDX spectrum of sample «RF6» recorded at acceleration voltage of 10 kV.

Table 1. Ratio of the elements in the sample «RF6».

Element - *Line*	Element Weight %	Weight % Error	Atom %	Atom % Error
B - *K*	46.05	± 0.79	61.53	± 1.07
C - *K*	4.70	± 0.74	5.66	± 0.89
N - *K*	2.12	± 0.92	2.19	± 0.95
O - *K*	13.72	± 0.49	12.39	± 0.44
Mg - *K*	6.76	± 0.23	4.02	± 0.14
Al - *K*	24.79	± 0.36	13.27	± 0.19
Si - *K*	1.86	± 0.12	0.95	± 0.06
Total	100.00		100.00	

Table 2 presents results of nanoindentation test for samples «RF4» and «RF6». Hardness of sample «RF4» at 20 mN peak load in the flat surface outside the «islands» area is 33.7 GPa and Young's modulus is 182 GPa. For the sample «RF6», hardness and Young's modulus slightly vary within the error ranges as, respectively, 39 – 37 GPa and 245 – 270 GPa.

Table 2. Nanoindentation data recorded at 20 mN peak load for samples «RF4» and «RF6» deposited at target-to-substrate distances 35 and 12 mm, correspondingly.

Characteristics	Sample		
	«RF4»	«RF6» within «islands»	«RF6» outside «islands»
Penetration depth h_{max}, μm	0.26	0.24	0.23
Hardness *H*, GPa	33.7 ± 1.6	37 ± 6	39 ± 7
Young's modulus *E*, GPa	182 ± 3	270 ± 40	245 ± 16
Elastic strain index *H/E*	0.19	0.14	0.16

To achieve smooth films' surface we modified processing conditions. Reducing power to 100 – 150 W (target sputtering power density of 5 – 7.5 W/cm^2) sputtered films appeared to be largely smooth. Surface morphology of deposited samples was investigated using AFM at square areas from 1 × 1 μm^2 (figure 3) to 8 × 8 μm^2. Root-mean-square (RMS) surface roughness of 2.6 μm thick film was found to be less than 1 nm.

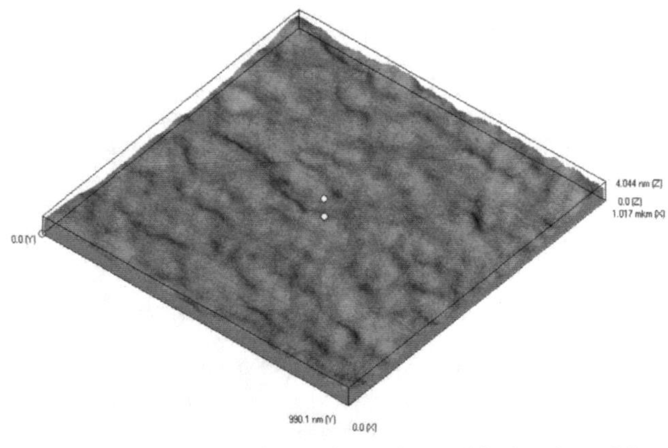

Figure 3. 3D AFM image of sample surface with the size of 1 × 1 μm^2.

The following two stage sputtering procedure enabled accomplishment of ultimate characteristics of $AlMgB_{14}$ films:
- smooth thin template layer was deposited at low target sputtering power of 50 W during 10 – 15 min;
- sputtering of top layer lasted 40 – 50 min at high target sputtering power 100 – 200 W.

Both layers were deposited at the same target-to-substrate distance of 20 mm. Between two processing stages, target sputtering power was changed slowly (1 W/s) increasing to a predetermined level. Films thus obtained had no unevenness. It is worth noting that although a substrate was not specially heated, at short distances it was warmed up by plasma discharge. With a thermocouple, we determined temperature of 7 cm^2 square area of substrate to be 200, 350, and 450° C at, respectively, 100, 150, and 200 W of sputtering RF power.

Figure 4 and table 3 present example of 20 mN load-penetration depth chart and results of nanoindentation tests for «RF8» sample deposited in two stages process at RF power equal to 150 W.

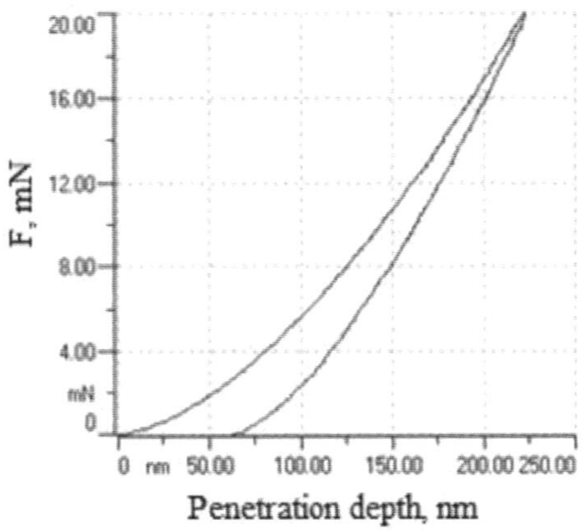

Figure 4. Typical load-penetration depth chart recorded for «RF8» sample at applied peak load $\boldsymbol{F}_{max}$ = 20 mN.

Table 3. Results of nanoindentation tests carried out for sample «RF8» at various loads.

Characteristics	Load F_{max}, mN						
	1	2	3	5	10	15	20
Penetration depth h_{max}, μm	0.036	0.055	0.073	0.095	0.15	0.19	0.23
Hardness H, GPa	61 ± 3	64.0 ± 2.1	59.3 ± 2.7	45.6 ± 1.7	36.4 ± 0.8	35.4 ± 1.1	34.3 ± 1.0
Young's modulus E, GPa	488 ± 17	470 ± 20	402 ± 13	316 ± 6	259 ± 3	245 ± 4	229 ± 3
Elastic strain index H/E	0.13	0.14	0.15	0.14	0.14	0.14	0.15

Data in table 3 are portrayed also in figure 5. They clearly reflect the presence of strong indentation size effect (ISE) found earlier in [1,7]. Both hardness and Young's modulus sharply increase at shallow penetration depths when loads diminish. Hardness and Young's modulus attain the peak values of, respectively, 61 and 488 GPa at 1 mN and reach the saturation of 34.3 and 229 GPa at 20 mN.

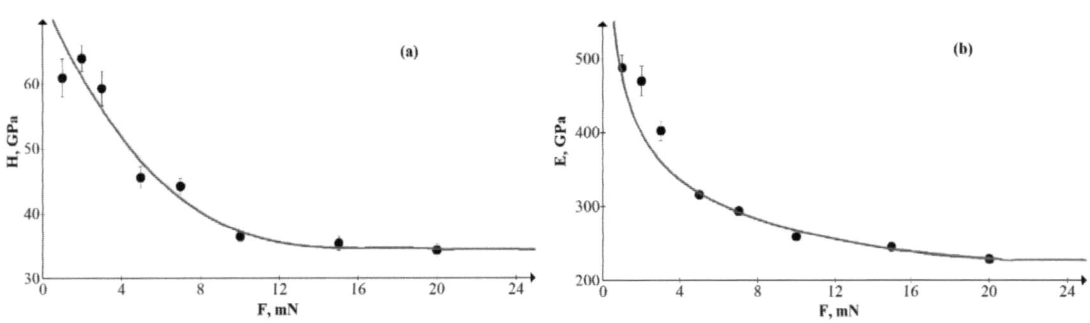

Figure 5. Hardness (**a**) and Young's modulus (**b**) in sample «RF8» at various loads ranged 1 to 20 mN.

H and *E* data for all the indentations made at different loading forces are collected in figure 6. They group themselves along the straight line $H = 0.14\ E$. The ratio of nanohardness H to Young's modulus E is known as the elastic strain index and considered as an indicative measure of the wear resistance of materials. It is commonly assumed that a material with high H/E ratio like 0.1 possesses a better wear resistance than a material with a low ratio $H/E \sim 0.01$. In figure 6 we compared this characteristic of «RF8» film with those in several engineering materials.

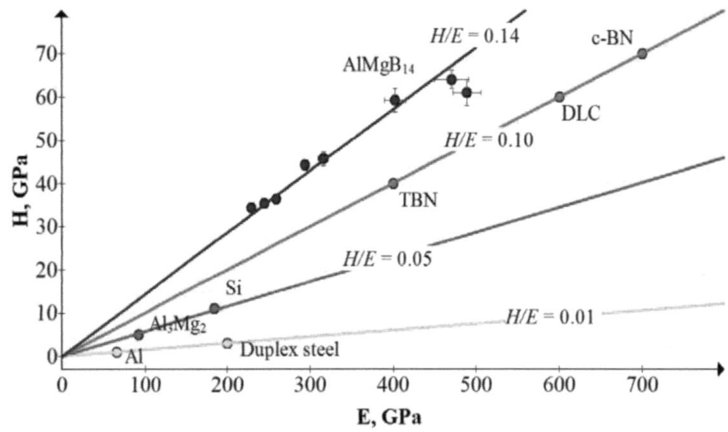

Figure 6. Elastic strain index as high as 0.14 in magnetron sputtered $AlMgB_{14}$/Si(100) film «RF8» compared with H/E ratio in various known materials.

Also, we attempted to coat a surface of some commercial industrial items. Magnetic steel 5 mm in diameter hemispheric ball on the small leg used as a head of special gauge was chosen as an example. Accurate cleaning and polishing of specimen's surface preceded BAM coating. At first, organic contaminants were removed with the ethanol. Then, mechanical polishing of ball's surface was made in two steps:

1 - polishing using paste Pikal® (5 – 10 min);

2 - polishing with clean dry felted fabric (5 – 10 min).

Quantitative AFM analysis of polished surface resulted RMS roughness equal to 20 – 100 nm within required operational area.

Smooth $AlMgB_{14}$ film was deposited to the ball item's surface according to the procedure similar to sample «RF8». The first template layer was deposited at the distance of 18 mm and RF power of 70 W. Subsequently, target sputtering power was increased up to 150W.

Figure 7**a** shows ball's surface is almost completely smooth except few dust particle contaminants. Figure 7**b** confirms that whole ball item's coated surface consists of light elements (B. Al and Mg). Hardness and Young's modulus obtained at 40 mN peak load were found to be 35 – 38 GPa and 250 - 350 GPa, correspondingly.

Figure 7. Secondary electron (**a**) and backscattered electron (**b**) SEM images of $AlMgB_{14}$–coated surface of hemispheric ball item.

4. Conclusion

Hardness of magnetron sputtered $AlMgB_{14}$ films can be enhanced at reduced target-to-substrate distance and increased target sputtering power. Unfortunately, both above mentioned conditions lead to a roughening of coated surfaces. Nevertheless, a wide range of processing parameters exists to find smooth surface – high hardness tradeoff and to achieve ultimate properties of magnetron sputtered BAM coatings required for any special application.

Acknowledgments
Work was carried out in the framework of the Strategic Development Program in Petrozavodsk State University "Complex technology solutions and staffing branches of agricultural, forestry and transport engineering".

References

[1] Tian Y, Bastawros A F, Lo C C H, Constant A P, Russell A M and Cook B A 2003 Superhard self-lubricating $AlMgB_{14}$ films for microelectromechanical devices *Appl. Phys. Lett.* **83** 2781
[2] Cook B A, Tian Y, Harringa J L, Constant A P, Russell A M and Molian P A 2007 Ultra-hard low friction coating based on $AlMgB_{14}$ for reduced wear of MEMS and other tribological components and system *US patent* 7,238,429 B2
[3] Wu Z, Bai Y, Qu W, Wu A, Zhang D, Zhao J and Jiang X 2010 Al–Mg–B thin films prepared by magnetron sputtering *Vacuum* **85** 541
[4] Yan C, Zhou Z F, Chong Y M, Liu C P, Liu Z T, Li K Y, Bello I, Kutsay O, Zapien J A and Zhang W J 2010 Synthesis and characterization of hard ternary AlMgB composite films prepared by sputter deposition *Thin Solid Films* **518** 5372
[5] Wenchao Q U, Aimin W U, Zhanling W U, Yizhen B A I and Xin J 2012 Influence of boron contents on properties of AlMgB films prepared by RF magnetron sputtering *Rare Metals* **31** 164
[6] Liu W, Meng Q-S, Miao Y, Chen F-H and Hu L-F 2012 Preparation and characterization of Al-Mg-B thin films by magnetron sputtering *Adv. Mat. Res.* **465** 112
[7] Grishin A M, Khartsev S I, Böhlmark J and Ahlgren M 2014 Ultra-hard $AlMgB_{14}$ coatings fabricated by RF magnetron sputtering from a stoichiometric target *JETP Letters* **100** 680
[8] Roberts D J, Zhao J and Munir Z A 2009 Mechanism of reactive sintering of $MgAlB_{14}$ by pulse electric current the technology process of hard metals *Int. J. Refract. Met. Hard Mater.* **27** 556

Dark-field spectroscopy of plasmon resonance in metal nanoislands: effect of shape and light polarization

E S Babich[1], S A Scherbak[2], F Heisler[3,4], S D Chervinskii[1,5], A K Samusev[3], A A Lipovskii[1,2]

[1] Department of Physics and Technology of Nanostructures, Peter the Great St. Petersburg Polytechnic University, 29 Polytechnicheskaya, St. Petersburg 195251, Russia

[2] Department of Physics and Technology of Nanoheterostructures, St. Petersburg Academic University, 8/3 Khlopina, St. Petersburg 194021, Russia

[3] The International Research Centre for Nanophotonics and Metamaterials, ITMO University, Kadetskaya line 3b "Red House", St. Petersburg 197101, Russia

[4] Abbe Center of Photonics, Friedrich–Schiller-Universität Jena, Max–Wien-Platz 1, 07743 Jena, Germany

[5] Institute of Photonics, University of Eastern Finland, Yliopistokatu 7, P.O.Box 111, Joensuu 80101, Finland

E-mail: babich.katherina@gmail.com

Abstract. We present the experimental dark-field scattering studies and the simulation of plasmonic properties of isolated silver nanoislands. The nanoislands were fabricated on a soda-lime glass substrate using silver-sodium ion exchange, subsequent thermal poling and annealing of the processed glass substrate in hydrogen. The morphology of the nanoislands was characterized with atomic force microscopy and scanning electron microscopy; the dimensions were 100-180 nm in base and 80-160 nm in height. We measured and modeled dark-field scattering spectra of the silver hemiellipsoidal nanoparticles differing in size and shape. The SPR position varied from 450 nm to 730 nm depending on the particle shape and dimensions. Both experiments and simulation showed a red shift of the SPR for bigger nanoislands of the same shape. Losing the axial symmetry in nanoislands resulted in the resonance splitting, while their elongation led to an increase in the scattering of p-polarized light.

1. Introduction

Unique optical property of metal nanostructures that is the excitation of surface plasmon resonance (SPR) by an incident light-wave [1] is nowadays applied in nonlinear optics [2], photovoltaics [3] and surface-enhanced Raman spectroscopy (SERS) [4]. The last one is of a particular interest due to the promising opportunity of the monomolecular sensing in the case of the coincidence of the SPR frequency with an eigenfrequency of molecules under study [5,6]. Thus, governing the SPR position via control over size and shape of metal nanostructures will allow an effective light manipulation and broadening of a SERS spectral range [7,8]. This paper is dedicated to experimental studies and numerical modeling of the SPR properties of differently shaped isolated silver nanoislands. We grew the nanoislands on an ion-exchanged soda-lime glass by carrying out a thermal poling treatment of silver-enriched glass with the profiled anodic electrode and subsequent annealing of the glass in hydrogen [9].

2. Single silver nanoislands fabrication

The technique to grow an array of isolated silver nanoislands on a glass substrate includes three consequent steps: ion exchange, thermal poling treatment and annealing in a hydrogen atmosphere [9]. According to this technique, we placed a soda-lime glass slide [10] for 20 minutes in $Ag_{0.05}Na_{0.95}NO_3$ melt heated up to 325 °C. This resulted in the enrichment of the subsurface region of the glass with silver ions. The ion-exchanged glass was subjected to thermal poling at 300 °C under applied DC voltage of 500 V. At this stage, the silver ions were buried into the bulk of the glass in the regions where the anodic surface of the glass contacted with used profiled electrode (square net of 300x300 nm^2 deepenings with the periodicity of 5 μm). After the poling for 4 minutes 30 seconds, we treated the sample in a hydrogen atmosphere at the temperature of 250 °C for 30 minutes to reduce Ag^+ ions remained subsurface after the previous step. The out-diffusion of the reduced silver resulted in the growth of silver nanoislands at the positions corresponding to the anodic electrode deepenings in the poling. To avoid a contamination/degradation of the nanoislands in the air, the surface with grown nanoislands was coated with 5 nm thick titania layer using the atomic layer deposition (ALD) technique [11,12].

3. Experiments and modeling

We used an atomic force microscope (AFM) and a scanning electron microscope (SEM) to characterize the shape and the size of the fabricated nanoislands. The scattering spectra characterizing resonant properties of individual nanoislands were measured using a confocal dark-field setup (Figure 1a). In the experiments, each nanoisland was illuminated at oblique incidence by the s- and p-polarized light from a halogen lamp focused at the sample surface with M Plan Apo 10x/0.26 objective lens. The beam waist was approximately 1 μm. Light scattered by the nanoisland was collected with an 100x/0.7 objective lens and then directed to the 1800 gr/mm grating of Horiba LabRAM™ HR UV-VIS-NIR spectrometer.

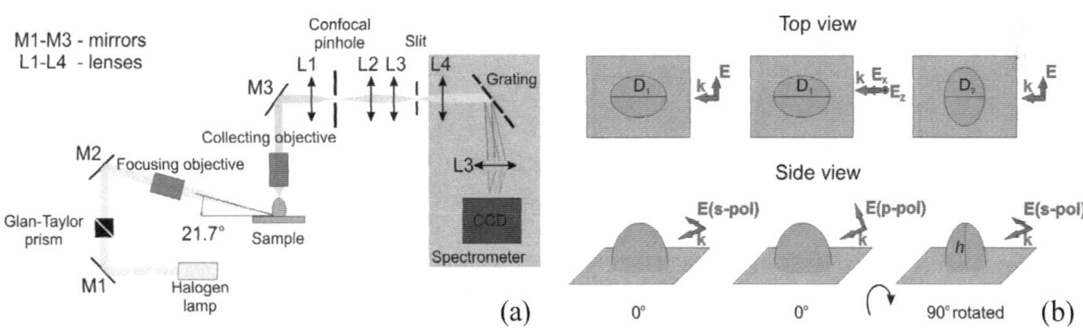

Figure 1. (a) Schematic of the dark-field spectroscopy setup. (b) Three dark-field measurement configurations: the sample in the initial position was lightened by s- (left) and p-polarized (middle) light, and after rotation on 90° it was illuminated by s-polarized light (right), the angle of incidence is 68.3°. **k** and **E** arrows denote the directions of exciting light propagation and polarization, respectively.

To analyze the influence of the incident light-wave polarization and the nanoislands' orientation (relatively to the polarization plane of the wave) on its optical response we used three configurations of the dark-field experiments. As shown in the Figure 1b, firstly the sample was lightened by 1) s- and 2) p-polarized wave, and after 90° rotation around the axis perpendicular to the glass surface, it was also illuminated by 3) s-polarized light.

The results of AFM and SEM measurements and measured scattering spectrum of a typical single silver nanoisland excited by s-polarized light-wave are shown in Figure 2. One can see that the nanoisland is shaped as a prolate hemiellipsoid with a slightly elliptical base.

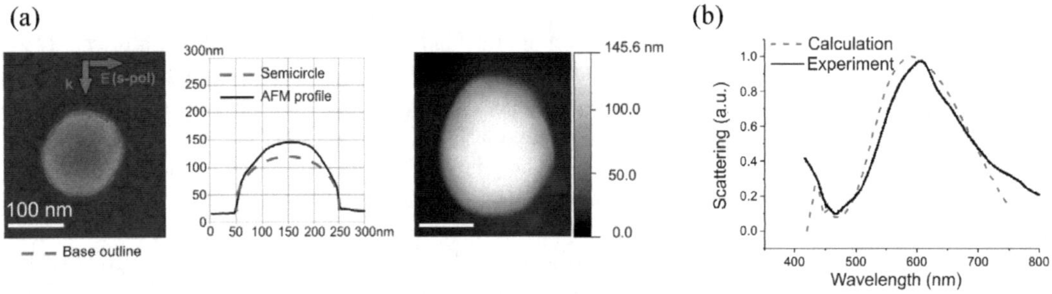

Figure 2. (a) SEM image (left), AFM profile (middle) and height map (right) of a single silver nanoisland. **k** and **E** arrows denote the directions of exciting light propagation and polarization, respectively; (b) experimental and calculated scattering spectra of this nanoisland.

The numerical model reproduced morphological features of each nanoisland; the calculations of the scattering spectra were performed in COMSOL Multiphysics environment (Radio frequency module: electromagnetic waves, frequency domain). In the simulation, the samples were excited by s- and p-polarized plane waves as shown in Figure 1b; the waves' glancing angle 21.7° corresponded to the dark-field measurements configuration (Figure 1a). The scatterer was a hemiellipsoid with silver dielectric permittivity taken from Johnson and Christy data [13], and the substrate had the typical for ion-exchanged glasses dielectric permittivity of 2.5. Figure 2b demonstrates a good agreement between numerical and experimental spectra of the chosen nanoisland.

4. Results and discussion

The SPR position in studied single nanoislands varied from 450 nm to 730 nm depending on the particle shape and dimensions which were from 100 to 180 nm in base diameter and from 80 to 160 nm in height.

In the case of SPR excitation by s-polarized wave we chose axially symmetric nanoislands with height, h, to diameter, D, ratio $h/D\sim0.9$ to estimate the influence of the nanoislands' size on the resonance position. Figure 3 shows that the increase in nanoislands' size resulted in red-shift of the SPR from 600 to 700 nm. At the same time, variations in height measured for a set of nanoislands with base diameter of ~140 nm had no significant influence on the resonant wavelength (Figure 3b). These results are in agreement with the calculated dependences presented in Figure 3. It is worth to note that the account for dielectric coating in the model should move the resonance towards longer wavelengths [14]; this explains the constant shift between the experimental and the calculated SPR positions. The lack of experimental data in Figure 3b is because of the shortage in nanoislands of the same lateral size in the grown array.

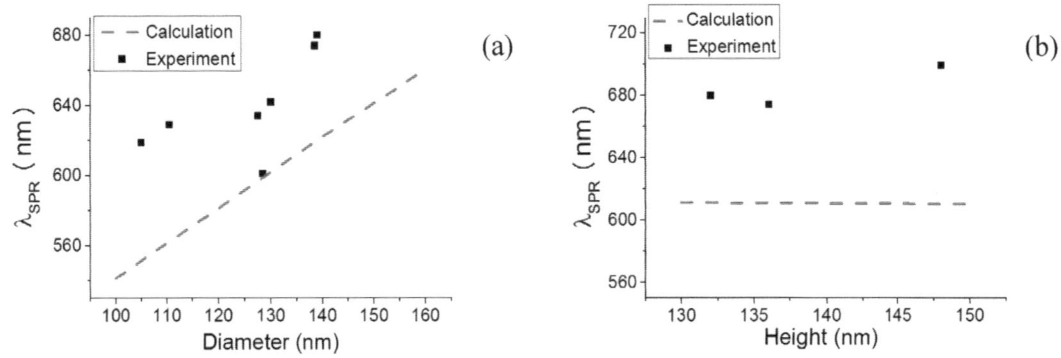

Figure 3. The SPR position versus nanoislands' (a) lateral size (the height to diameter ratio is constant ~ 0.9) and (b) height (the base diameter of the nanoislands is ~140 nm). The dielectric coating is not accounted in the calculations.

In the case of a nanoisland illumination by p-polarized wave, the external electric field has both components: along the substrate surface and normal one. Therefore, the effective dipole excited in the nanoisland should have the both components as well. Since in the geometry of the experiment the dipole excited by normal E-field component substantially does not scatter in the direction of signal collection, mostly the lateral dipole radiation is registered. Thereby, the scattering spectra of the nanoislands illuminated by p-polarized light were expected to be similar to ones in the case of s-polarization. Indeed, the overall shape of scattering spectra and SPR positions measured using s- and p-polarization of the exciting light are similar (Figure 4), and the main difference between these spectra is in intensity.

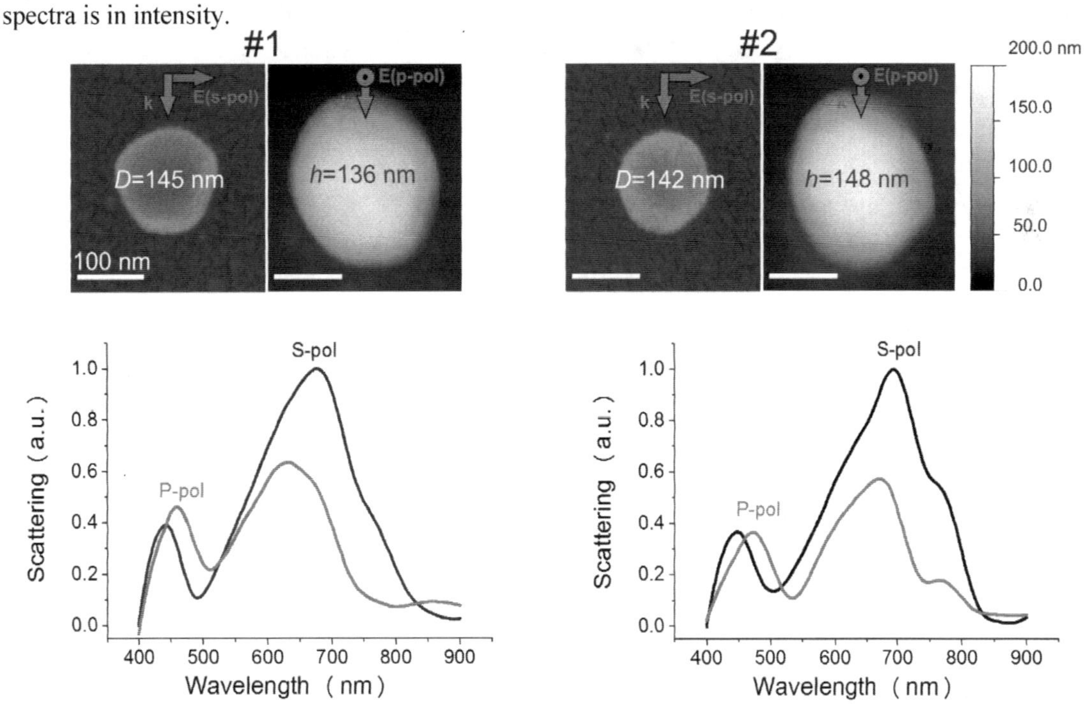

Figure 4. 1st row: SEM (left) and AFM (right) images of single silver nanoislands. **k** and **E** arrows denote the directions of the exciting light propagation and the polarization, respectively. 2nd row: experimental scattering spectra of these nanoislands excited by s- and p-polarized light.

We chose several nanoislands with different eccentricity (Figure 5) to study the influence of the shape in the case of elliptic-base (non-axisymmetric) nanoislands on their spectral characteristics. Even a minor ellipticity of a nanoisland base (nanoisland #1) results in a shift of the SPR position. Pronounced ellipticity affects the shape of scattering spectra, which also depends on the polarization of the incident light-wave (nanoislands #2,3). Figure 5 shows that the spectra corresponding to the excitation by s-polarized wave have a side peak. The 90 degrees rotation of the nanoisland in the lateral plane leads to a redistribution of intensity between this and the main peak. We suppose that two dipoles are excited along minor and major axis of the elliptic cross-section base; that results in appearance of two plasmon oscillation modes with different resonant wavelengths.

Notably, measured resonance positions in scattering spectra of nanoislands lightened by p-polarized wave (Figure 5, 2nd row) almost coincide with ones for 90 degrees rotated nanoislands excited by s-polarized light (Figure 1b). Calculated field distributions (Figure 5, 3rd row) evidence that the effective dipole is excited mainly in the lateral plane of a nanoisland by both polarizations of incident light. This explains previously mentioned coincidence of the spectra measured using s- and p-polarized light-wave. We believe that the appearance of an abnormally large lateral dipole component along the wave propagation direction in the case of p-polarized incident wave is caused by two factors: the wave phase retardation and the high vertical asymmetry of the system substrate-nanoisland. Performed simulation predicts that an elongation of a nanoisland in vertical direction, $h \geq D$, rotates the dipole around the axis lying in the plane of the glass surface. This also predicts an increase in the intensity of the dipolar radiation by elongated nanoislands. We observed this increase in the scattering of p-polarized light by a prolate (strongly elongated) nanoisland (Figure 5, nanoisland #3).

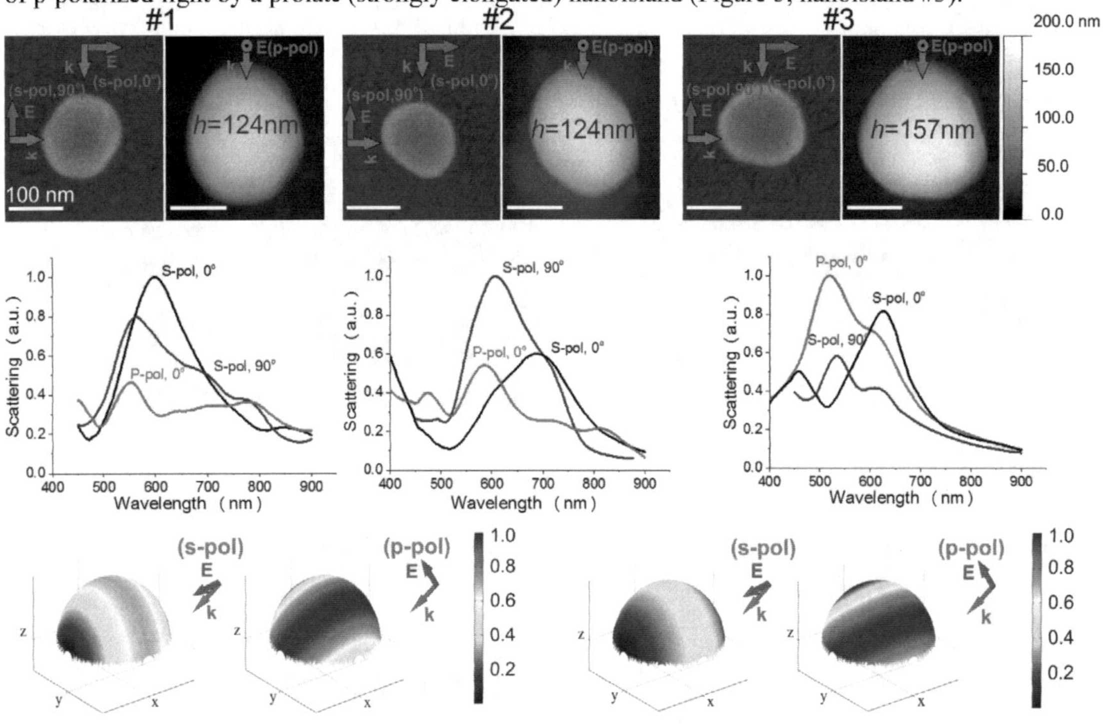

Figure 5. 1st row: SEM (left) and AFM (right) images of three nanoislands with different ratio, l, of major to minor axis of base diameter (#1: l = 1.1, #2: l =1.2, #3: l =1.2) and height, h (#1: h=124 nm, #2: h = 124 nm, #3: h = 157 nm); 2nd row: dark-field scattering spectra of the nanoislands; 3rd row: calculated radiation pattern of two different nanoislands. Modeled nanoislands are axisymmetric with the diameter D = 150 nm and the height h = 120 nm (left) and h = 170 nm (right).

Thus in the standard geometry of scattering measurements, including Raman scattering, regardless to the polarization of the exciting light the main parameter influencing the spectral position of a nanoisland plasmon resonance is its lateral dimension.

5. Conclusions

We applied dark-field scattering spectroscopy to study plasmonic properties of isolated single silver nanoislands. An array of the nanoislands was fabricated using thermal poling of ion-exchanged glass with the profiled electrode followed by the annealing of the glass in hydrogen. The dark-field experiments have demonstrated that an increase in the size of the similarly shaped nanoislands results in linear red shift of the SPR wavelength. At the same time the SPR position did not depend on the height of the nanoislands. In the case of non-axisymmetric nanoislands the appearance of two resonances was registered. These resonances corresponded to the excitation of plasmon oscillations along two axes of the elliptic base of the nanoislands. The increase in the scattering of p-polarized light-wave by strongly prolate nanoislands was predicted in the performed modeling and demonstrated in our experiments.

Acknowledgements

This work was supported by Finnish Academy of Science, the Ministry of Education and Science of the Russian Federation (project #16.1233.2014/K and Program "5-100-2020"). E. S. B and S. D. Ch. are also grateful to the Foundation for Assistance to Small Innovative Enterprises.

References

[1] Mie G 1908 *Ann. Phys.* **25**, 377
[2] Tsutsui Y, Hayakawa T, Kawamura G, Nogami M 2011 *Nanotechnology* **22** (27), 275203
[3] Atwater H A and Polman A 2010 *Nat. Mater.* **9**, 205
[4] Dieringer J A, McFarland A D, Shah N C, Stuart D A, Whitney A V, Yonzon C R, Young M A, Zhang X and Van Duyne R P 2006 *Faraday Discuss.* **132**, 9
[5] Camden J P, Dieringer J A, Wang Y, Masiello D J, Marks L D, Schatz G C and Van Duyne R P 2008 *J. Am. Chem. Soc.* **130** (38), 12616
[6] Etchegoin P G and Le Ru E C 2008 *Phys. Chem. Chem. Phys.* **10**, 6079
[7] Aroca R and Martin F 1985 *J. Raman. Spectrosc.* **16**, 156
[8] Klimov V V 2014 *Nanoplasmonics* (Pan Stanford)
[9] Chervinskii S, Reduto I, Kamenskii A, Mukhin I and Lipovskii A A 2015 *Faraday Discuss.* DOI: 10.1039/C5FD00129C
[10] Menzel-Glaser: Microscope slides
[11] Oates T W H, Losurdo M, Noda S and Hinrichs K. 2013 *J. Phys. D: Appl. Phys.* **46**, 145308
[12] George M 2010 *Chem. Rev.* **110**, 111
[13] Johnson P and Christy R 1972 *Phys. Rev.* B **6** (12), 4370
[14] Scherbak S A, Shustova O V, Zhurikhina V V and Lipovskii A.A. 2015 *Plasmonics* **10**, 519

Journal of Physics: Conference Series **769** (2016) 012041 doi:10.1088/1742-6596/769/1/012041

Nonlinear dynamics of the exciton reflection spectrum

L Yu Belyaev , Yu V Kapitonov , B V Stroganov , Yu P Efimov , Yu K Dolgikh , S A Eliseev , V V Petrov , V V Ovsyankin

St. Petersburg State University, 7/9 Universitetskaya nab., St. Petersburg, 199034 Russia

E-mail: `lenyabelyaev@mail.ru`

Abstract. Nonlinear dynamics of refelction from InGaAs/GaAs quantum well was studied by pump-probe spectroscopy. Brewster geometry allowed to extract the time dependency of radiative width and nonradiative broadening of the excitonic resonance from reflection spectra. Pump pulse action leads to the increase of the nonradiative broadening and decrase of the radiative width. Fast decay component of the additional nonradiative broadening is attributed to the scattering on the pump-induced excitons. Slow component was explained by a scattering either on pump-induced charged carriers, or subsequently formed trions. This assumtion is additionaly proved by the decrase of the radiative width with same decay times caused by the redistribution of oscillator strength between exciton and trion states.

1. Introduction

A3B5 semiconductor quantum wells (QW) are promising objects in the context of research and development of all-optical logic elements [1, 2] due to the high efficiency of the interaction between light and 2D excitons in QW. Molecular-beam epitaxy (MBE) allows one to produce QW demonstrating extremely low inhomogeneous broadening of excitonic resonances [3, 4]. An important question is exciton ensemble behavior in such QW in the case of the intensive excitation. In this work we have studied nonlinear dynamics of the exciton reflection spectrum in pump-probe experiment exploiting picosecond laser pulses. We have observed different relaxation processes, and proposed possible mechanisms explaining these processes by the interaction of excitons with each other and with environment.

2. Experiment and discussion

We have studied MBE-grown structure P561 containing 2 nm $In_{0.02}Ga_{0.98}As$ QW with 60 and 180 nm GaAs barriers. This sample demonstrates lowest inhomogeneous broadening among the whole series of narrow low-indium content samples grown by our group [3]. Such high QW quality makes it possible to precisely determine small pump-induced broadening. However, lower quality samples demonstrate similar qualitative behavior. So in this work we present the most illustrative data for P561 sample. Sample was kept in a closed-loop helium cryostat at the temperature 3.5 K. Pump-probe setup is shown in Fig. 1. Femtosecond Ti:Sapphire laser generates 100 fs pulses with period of 12.5 ns. Laser beam fall on the non-polarizing beam splitter. Splitted probe and pump beams pass through diffraction selectors S1 and S2 shaping the pulses with 1.3 ps duration and full width at half maximum (FWHM) around 0.75 nm. Delay time t between the arrival of pump and probe pulses was scanned from −50 to 610 ps by delay lines. Pump beam falls on the sample normal to the surface and is focussed in 100 μm spot. Probe beam falls on the same sample area at Brewster angle with polarisation in the plane of incidence (P-polarization). Probe beam was focussed in the aperture 1:20 in the 10 μm spot.

The reflection spectrum was detected by a spectrometer equipped with a CCD-detector with spectral resolution around 30 μeV. The spectral position of probe and pump corresponds to the HH-exciton resonance in QW. Pump beam was co-polarized with the probe beam. Measurements were performed in the linear mode on probe intensity at $I_{Probe} = 20\,\mu$W. Dependence of the reflection spectra at t was recorded for different pump intensities I_{Pump} in the $2 \ldots 32$ mW range.

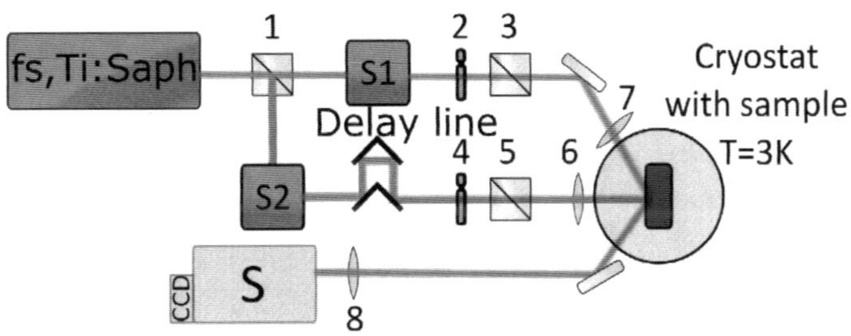

Figure 1. Pump-probe setup. S1,2—Diffraction selector; S—Spectrometer; 1—Beamsplitter; 2,4—$\lambda/2$ plate; 3,5—GlanTaylor prism; 6,7,8—Lens.

Brewster geometry of the probe beam makes possible to supress background reflection from the sample surface. In this case, the reflection spectra $K_R(\omega)$ takes a simple Lorentzian form described by a model of resonance light scattering on the thin layer of two-level systems (oscillators) with Lorentzian frequencies distribution [5]:

$$K_R(\omega) = \frac{\Gamma_R^2}{(\omega - \omega_0)^2 + (\Gamma_R + \Gamma_2 + \Gamma_2^*)^2}, \tag{1}$$

where Γ_R is the radiative linewidth (the rate of the radiative decay of the 2D-excitons polarization), Γ_2 is the homogeneous broadening (the rate of the irreversible phase relaxation of 2D-excitons due to excitons scattering on phonons or some other scattering centers), and Γ_2^* is the inhomogeneous broadening (the rate of the reversible phase relaxation associated with the spread of oscillator frequencies). We also could introduce non-radiative broadening $\Gamma_{NR} = \Gamma_2 + \Gamma_2^*$. Half width at half maximum (HMHW) of the Lorentzian curve corresponds to the rate of the exciton free induction decay $\Gamma_R + \Gamma_2 + \Gamma_2^*$ [6].

Fig. 2,(a) shows reflection spectra for various t at $I_{Pump} = 32$ mW, and approximation by formula (1). At zero delay ($t = 0$) pump pulse induces an additional broadening of the spectrum and a decrease of the reflection coefficient. At positive times ($t > 0$) spectrum relaxes to the initial shape. The signal at the negative delay ($t < 0$) does not depend on the I_{pump}, which proves the absence of relaxation processes with times longer then units of nanoseconds. Reflection spectra fitting by the formula (1) allows to determine the temporal behavior $\Gamma_R(t)$ and $\Gamma_{NR}(t)$:

$$\Gamma_R(t) = \Gamma_{R0} + \Delta\Gamma_R(t), \tag{2}$$

$$\Gamma_{NR}(t) = \Gamma_{NR0} + \Delta\Gamma_{NR}(t), \tag{3}$$

where $\Gamma_{R0} = 33 \pm 2\,\mu$eV and $\Gamma_{NR0} = 29 \pm 2\,\mu$eV are widths of unpertubated resoance and $\Delta\Gamma_R(t)$ and $\Delta\Gamma_{NR}(t)$ – pump induced changes depicted in Fig. 2 (b),(c).

Temporal behavior of $\Delta\Gamma_{NR}(t)$ at $t > 0$ is described by the bi-exponentional decay (Fig. 3,(a)) with fast and slow decay times τ_{NR} presented in Fig. 3,(b). Such behaviour corresponds to the the presence of two types of pump-induced scattering centers with different lifetimes. The possible explanation for the fast τ_{NR} is scattering on the pump-induced excitons. At low pump intensities ($I_{Pump} < 5$ mW) this fast decay time corresponds to the radiative lifetime $T_{R0} = \frac{\hbar}{\Gamma_{R0}} \approx 19.7$ ps (dashed line in Fig. 3,(b)). At higher pump intensities this time increases due to the decrease of Γ_R caused by the pump action (see Fig. 2,(c)).

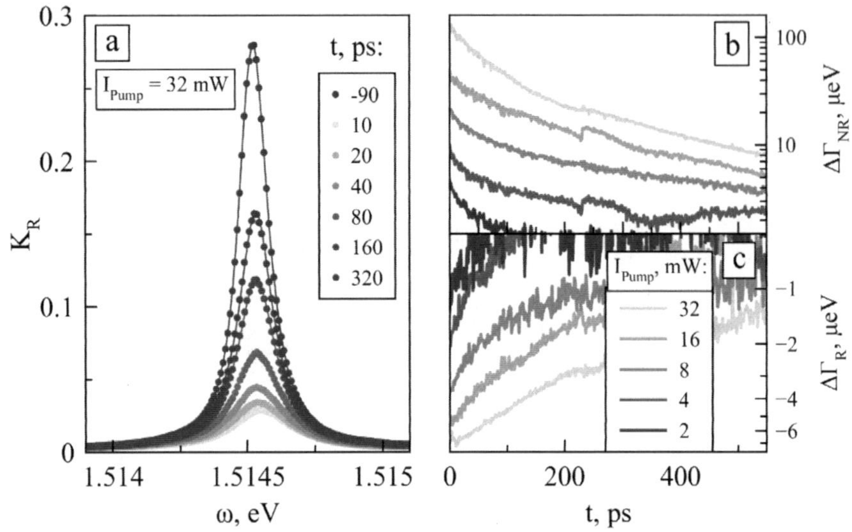

Figure 2. (a) Reflection spectra $K_R(\omega)$ for different pump-probe delay times t at I_{Pump} = 32 mW. Dots – experimental results, solid lines – fit by (1). Decay of the pump induced change of nonradiative broadening (b) and radiative width (c) for different pump intensities I_{Pump} (log scale).

Decrease of Γ_R after the arrival of the pump pulse corresponds to the redistribution of the oscillator strength of the excitonic transition. It could be explained for example by creation of long living nonequilibrium charge carriers leading to the transition of oscillator strength from exciton to trion states. Temporal behavior of $\Delta\Gamma_R(t)$ is described by mono-exponentional decay with time τ_R. The slow τ_{NR} is even slower than τ_R time. Therefore slow decaying component of $\Delta\Gamma_{NR}(t)$ could be partly attributed to the scattering either on nonequilibrium charge carriers, or on trions – bounded states of charge carriers and excitons. But it could also have additional slower decaying components that could not be resolved within t scan range.

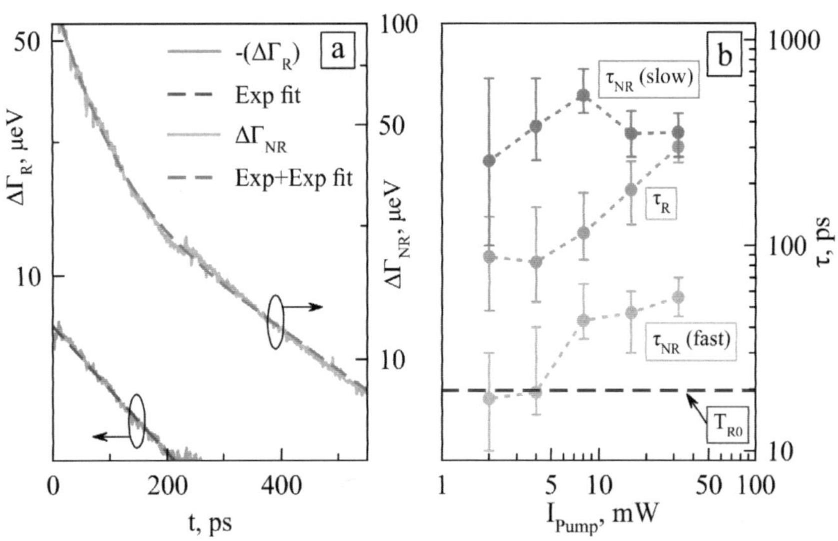

Figure 3. (a) Decay of pump induced changes for I_{Pump} = 32 mW (log scale). Solid lines – experimental data, dashed lines – fit. (b) Characteristic decay times for different I_{Pump} (double log scale).

Similar temporal behavior was observed in $Al_{0.3}Ga_{0.7}As/GaAs$ sample E296 [6], but its analysis was hindered because of the exciton resonance bleaching caused by the accumulation of long-living charge carriers in QW. Recent study of $Al_{0.03}Ga_{0.97}As/GaAs$ quantum wells had shown that in QWs with such low-aluminium containing barriers the parasitic bleaching effect could be suppressed [4].

3. Conclusion

We have studied the nonlinear dynamics of the reflection from InGaAs/GaAs quantum well by pump-probe spectrosocpy. Determined temporal behaviour of the nonradiative broadening of the excitonic resonance could be described by two mechanisms: scattering on the pump-induced excitons and charge carriers (or their excitonic complexes). Formation of nonequilibrium charge carriers is also attributed as a cause of the decrease of the radiative width of resonance.

4. Acknowledgements

The reported study was partially supported by RFBR, research project 15-52-12016 NNIO-A. This work was carried out on the equipment of the SPbU Resource Center "Nanophotonics" (photon.spbu.ru). The authors are thankful to MagicPlot for the convenient tool for analysis and visualization of data.

References

[1] I. Ya Gerlovin , V.V. Ovsyankin, B.V. Stroganov, V.S. Zapasskii. Coherent transients in semiconductor nanostructures as a basis for optical logical operation. 2000 *Nanotecnology* , vol. **11** 383

[2] I. Ya Gerlovin , V.V. Ovsyankin, B.V. Stroganov, V.S. Zapasskii. Nonlinear optical dynamics of semiconductor nanostructures: Feasibility of the photonic quantum gate.2000 *Journal of Luminescence*, **87** 421

[3] S.V. Poltavtsev, Yu.P.Efimov, Yu.K.Dolgikh, S.A.Eliseev, V.V. Petrov, V.V.Ovsyankin. Extremely low inhomogeneous broadening of exciton lines in shallow (In,Ga)As/GaAs quantum wells. 2014 *Solid State Communacitions* **199** 47

[4] I.A. Solovev, V.G. Davydov, Yu.V. Kapitonov, P.Yu. Shapochkin, Yu.P. Efimov, V.A. Lovcjus, S.A. Eliseev, V.V. Petrov and V.V. Ovsyankin, Increasing of AlGaAs/GaAs quantum well robustness to resonant excitation by lowering Al concentration in barriers. 2015 *Journal of Physics: Conference Series* , **643** 012085

[5] M.G.Benedict and E.D.Trifonov. Coherent reflection as superradiation from the boundary of a resonant medium. 1988 *Phys. Rev* **.A38**, 2854

[6] S.V. Poltavtsev, B.V. Stroganov. Experimental investigation of the oscillator strength of the exciton transition in GaAs single quantum wells. 2009 *Phys. Status Solidi* **C 6**, No **2** 483

Journal of Physics: Conference Series **769** (2016) 012042 doi:10.1088/1742-6596/769/1/012042

Absorption of the CO laser sum frequency radiation obtained in a nonlinear crystal $AgGaSe_2$ by molecular gases

O V Budilova, A A Ionin, I O Kinyaevskiy, Yu M Klimachev, A A Kotkov, A Yu Kozlov

P.N. Lebedev Physical Institute of the Russian Academy of Sciences, 119991 Moscow, Russia

Corresponding author: oksana_budilova@mail.ru

Abstract. The broadband laser system based on multi-line Q-switched CO laser was experimentally studied in a nonlinear crystal $AgGaSe_2$. The internal efficiency of sum frequency generation reached 1%. Test experiments on measurement of the absorption of the CO-laser sum frequency radiation by such gaseous substances as nitrous oxide and carbon dioxide were realized. A comparison of the experimental data with the theoretically calculated absorption spectrum of radiation was obtained.

1. Introduction

Because of increasing atmospheric pollution, rapid monitoring of the concentration of gas impurities has become more important, and such monitoring can be done using laser remote sensing of the atmosphere. Maximum information in remote sensing of the atmospheric composition with high spatial and temporal resolution can be obtained only by the optical method using laser, i.e., lidar.

Spectral range of broadband CO laser is very attractive for laser remote gas analysis, because, firstly, it is within the window of atmosphere transparency in the wavelength range from 3 to 5 μm, and secondly, coincides with the absorption bands of a number of pollutant and natural component of the atmosphere. Thirdly, the sensitivity of sensors based on InSb has record values in this spectral range [1]. CO-laser, therefore, is a promising device for various applications in remote laser sensing, spectroscopic gas analysis, medicine etc., where it is able to compete with well-known chemical HF/DF lasers [2-3], the converted frequency CO_2 lasers [4-5] and tunable difference frequency generators [6]. Frequency selective mode of generation of a pulsed CO laser is currently implemented by more than 1000 vibrational-rotational transitions in the fundamental [7] and first-overtone vibrational bands [8]. Opportunities of overtone CO laser (2.5-4.2 μm) for remote gas analysis by differential absorption method were demonstrated in works [9-10]. In addition, it was demonstrated the possibility of using overtone CO laser to determine of the isotopic composition of methane ($^{12}CH_4$, $^{13}CH_4$) in natural air samples with a cavity ring-down technique. A CO overtone sideband laser was utilized to excite a high-finesse cavity, which provides an effective optical absorption path length of 3.6 km [11].

All of the above suggests that the current state of spectroscopic gas analysis requires the search and introduction of new laser IR sources with the possibility of lasing on a large number of narrow spectral lines, in the widest possible frequency range and with a small step of frequency tuning. Emission spectrum of the CO laser can be significantly enriched and expanded from near-to far-IR by frequency conversion in nonlinear crystals. The frequency conversion includes second-harmonic generation (SHG), sum-frequency generation (SFG), and difference frequency generation (DFG).

It was experimentally demonstrated a broadband two-stage frequency conversion of laser radiation by CO laser in a $ZnGeP_2$ crystal with simultaneous generation of several hundred lines in the wavelength range from 2.45 to 6.3 μm [12] and from 2.5 to 8.3 μm [13]. The broadband two-stage frequency conversion of the CO laser radiation was realized in nonlinear $AgGaSe_2$ (AGSe) crystal too [14]. In the experiment ~70 spectral lines of CO laser within range from 4.9 to 6.3 μm were converted into ~70 spectral lines of sum frequencies radiation within range from 2.45 to 2.85 μm. The spectrum of DFG consisted of 70 spectral lines within spectral range from 4.3 to 4.9 μm.

In this paper we study absorption of the sum frequencies radiation in nonlinear crystal AGSe by such gaseous substances as nitrous oxide (N_2O) and carbon dioxide (CO_2).

2. Optical scheme

In our experiment we used the AGSe crystal with length of 12 mm. The same multi-line low-pressure Q-switched CO laser as in [12, 13] was used. The optical scheme of our experiment is presented in figure 1. Low-pressure cryogenic cooling CO laser tube **1** was pumped by DC glow discharge. The peak radiation power was about ~1 kW at pulse duration ~1 μs. Laser cavity was formed by a spherical mirror **2** (radius of curvature 9.0 m) and partially reflective mirror **3** (reflectance ≥90% in the wavelength range from ~5.0 to 6.5 mm). The laser operated in Q-switch mode by a rotating mirror **4**. The aperture diaphragm **5** (diameter 10 mm) installed near a partially reflecting mirror **3**.

To measure parameters of CO laser radiation, part of the laser beam was split off by flat CaF_2 plates **6** onto spherical mirror **7** (radius of curvature 0.25 m). From the spherical mirror, the radiation is directed to measuring average power meter **8** (Ophir-10A), mounted 15 cm in from of spherical mirror **7**. Part of the radiation through the plate **9** of CaF_2 was directed to a photodetector **10** (PEM-L-3) installed in the focal plane of the spherical mirror **7**. Using the power meter we controlled the average power of CO laser radiation.

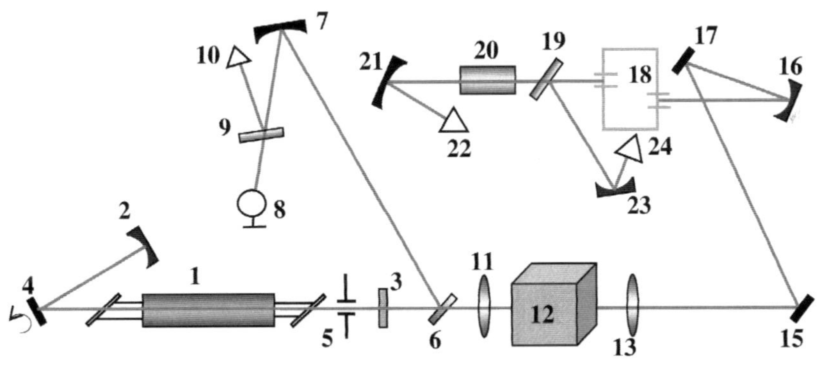

Figure 1. Optical scheme. 1 – CO laser tube; 2 – spherical mirror; 3 – partially reflective mirror; 4 – rotating mirror; 5 – diaphragm; 6,9 – CaF_2 flat plates; 7, 16, 21, 23 – spherical mirrors; 8 – power meters; 10, 22, 24 – photodetectors; 11,13 – lens made of CaF_2; 12 – AGSe crystal; 15, 17 – turning mirrors, 18 – spectrometer IKS-31, 19 – ZnSe flat plate, 20 – gas cell (10 cm)

Multi-line laser radiation was focused by CaF_2 lens **11** with focal length of 60 mm onto AGSe crystal **12**. For optimal focusing of the radiation on the crystal, the crystal holder had 3 translational degrees of freedom and 1 rotational degree of freedom for finding the direction of phase synchronism. To collimate the diverging after the crystal radiation the lens **13** (CaF_2, focal length ~60 mm) was used.

By flat turning mirrors **15** and **17**, and spherical (radius of curvature 0.25 m) **16** mirror, the converted radiation was directed into a spectrometer **18** (IKS-31). Then the radiation passed through the gas cell **20** (length of 10 cm) and using a spherical mirror **21**, it was directed to the photodetector **22** (PDA20H-EC, ThorLabs). In this case, the radiation after leaving the IKS (but before gas cell) was given to the flat plate **19** (ZnSe) on a spherical mirror **23**, from which the radiation was directed to the photodetector **24** (FSG-22-3A1). Shape and amplitude of the CO laser pulse was recorded by using the photodetector **10** on the oscillograph (Tektronix TDS 2014).

The signals from the photodetectors **22** and **24** were registered by oscilloscope Tektronix TDS 5052B. The method to record the SFG spectra was a simultaneous startup of the oscilloscope and the rotation of the diffraction grating (150 grooves/mm) in the IKS-31 from the known initial value of the wavelength. Time dependence, obtained from the photodetectors **22** and **24**, were recalculated to the wavelength dependence.

3. SFG spectra

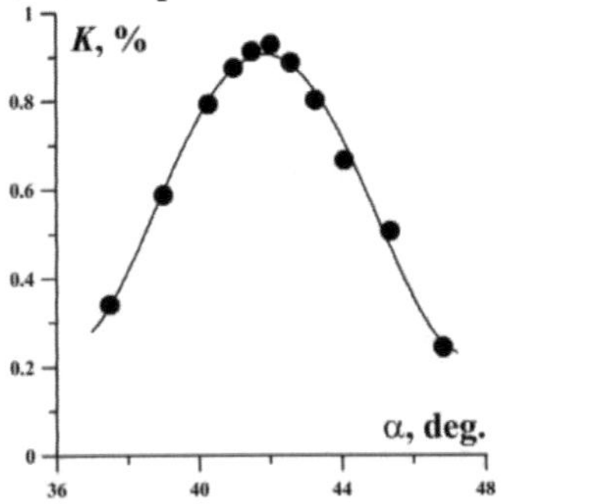

Figure 2. SFG efficiency versus incidence angle of pump radiation on the crystal AGSe.

The internal efficiency of SFG, which considering optical losses on Fresnel reflection from the uncoated faces of the crystal, reaching 1% at the incidence angle of CO laser radiation on crystal α=42° (Figure 2). When we change the incidence angle on the crystal of ± 3°, the efficiency decrease by 2 times, and spectrum of SFG radiation depends on the incidence angle on the crystal and can be changed.

It must be note that frequency conversion efficiency in nonlinear crystal can be increased with increasing input laser intensity. For example, when using a Q-switched and mode-locked regime of the electron-beam sustained discharge CO laser with a short (total pulse duration of about 1 μs) train of nanosecond pulses, the internal efficiency of SHG was increased up to 37% [15].

To measure the absorption of radiation in N_2O, AGSe crystal was mounted so that the incidence angle α of the CO laser radiation on the input facet of the crystal was equal 40.3°. SFG spectra of radiation before gas cell with N_2O (gray) and after gas cell (black) are presented in Figure 3. The average power of CO laser generation was 45.6 mW. The gas pressure in the gas cell was 1 atm at a length of 10 cm, which corresponds to the atmospheric path 1 km at the concentration of N_2O at the level of MAC (maximum allowable concentration).

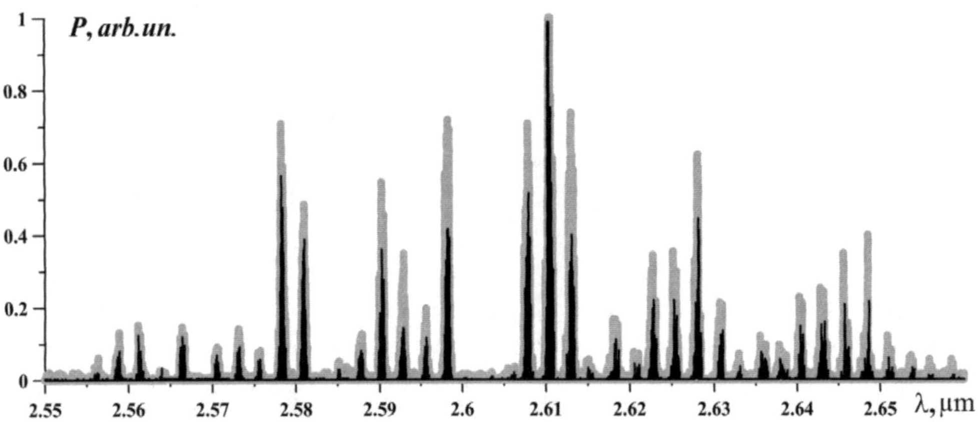

Figure 3. SFG spectra before gas cell (gray) and after cell (black) with 1 atm N_2O

To measure the absorption of radiation in CO_2, crystal was mounted so that the incidence angle α of the CO laser radiation on the input facet of the crystal was equal 42.5°. SFG spectra of radiation before gas cell with CO_2 (gray) and after gas cell (black) are presented in Figure 4. In this experiment, the radiation passed through the cell with gas mixture: 0.3 atm CO_2 and 0.7 atm N_2. The average power of CO laser generation was 47.5 mW.

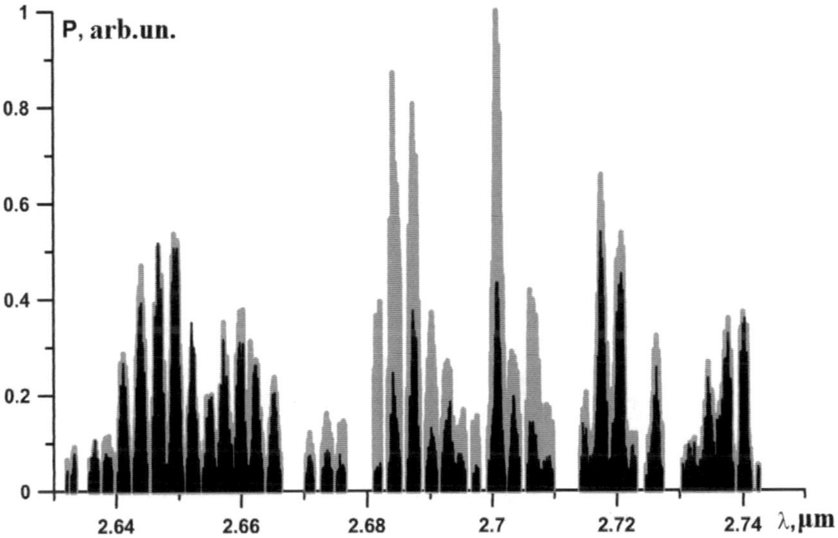

Figure 4. SFG spectra in AGSe: gray – radiation before gas cell, black – after gas cell with 0,3 atm CO_2 and 0,7 atm N_2

Spectra are normalized to the equaled value of the power in arbitrary units at wavelengths that should not be absorbed in the cell. The Figure 4 shows that the radiation up to 2.67 μm and higher 2.71 μm is not practically absorbed, while in the interval from 2.67 to 2.71 μm power of the transmitted radiation decreases significantly. For example, the maximum radiation power at ~ 2.7 μm drops by 2.5 times (from 1 to 0.4).

4. Absorption of SFG

For the calculation of absorption based on experimental data for comparison with absorption, theoretically calculated for our experimental conditions we applied following methodology. Characteristic waveform of a single spike from the spectrum presented in Figure 3, shown in Figure 5.

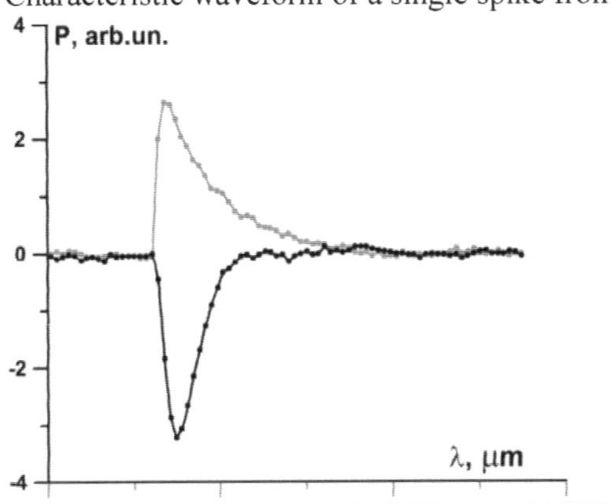

Figure 5. Oscillogram of one spike from the spectrum (for convenience, the radiation that passes through the gas cell (black) is facing down and multiplied by 20).

The gray line corresponds to radiation on the input of the gas cell and black is after passing the gas cell with CO_2 (were multiplied by -20, for visual comparison). Since the shape of the spikes are not the same (different photodetectors are used), it is possible to calculate the absorption we took the integral value of pulses. It should be noted that this pulse shape does not correspond to the real temporal shape of the total pulse of SFG ~1 μs. Then absorption a was calculated by formula:

$$a = \ln (s_2/s_1) - A$$

where s_1 and s_2 are energies of incident and transmitted radiation in each spike, respectively, A is average experimental value of the absorption in the band, which is not absorbed by molecules. Figure 6 shows the values of the absorption for the gas cell with CO_2.

Also it shows the incident of SFG radiation to gas cell (grey line) and SFG radiation after gas cell (black line, radiation multiplied by 5), absorption (striped line) and calculated absorption (black line is facing down) in the wavelength range from 2.699 to 2.704 μm. But it must be note that spectral resolution of the spectrometer was ~ 0.001 μm. It can not allow sufficiently to distinguish complex SFG spectrum and to detect narrow CO_2 absorption lines.

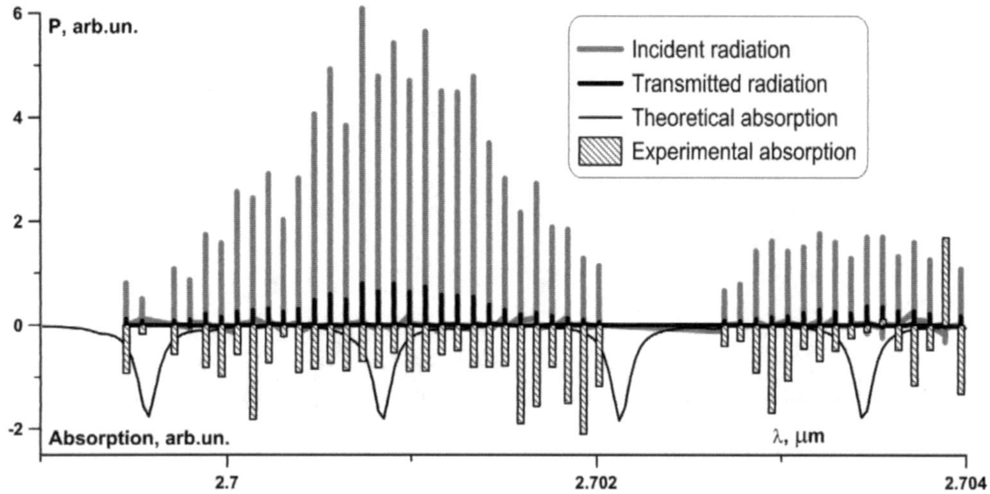

Figure 6. Spectra of incident and transmitted radiation in the range from 2.699 to 2.704 μm, theoretical and experimental absorption

The SFG spectra on the input of the gas cell and after gas cell with CO_2 and N_2O respectively, and also theoretical and experimental graphs of the radiation absorption in the gas cell are presented in Figures 7 and 8.

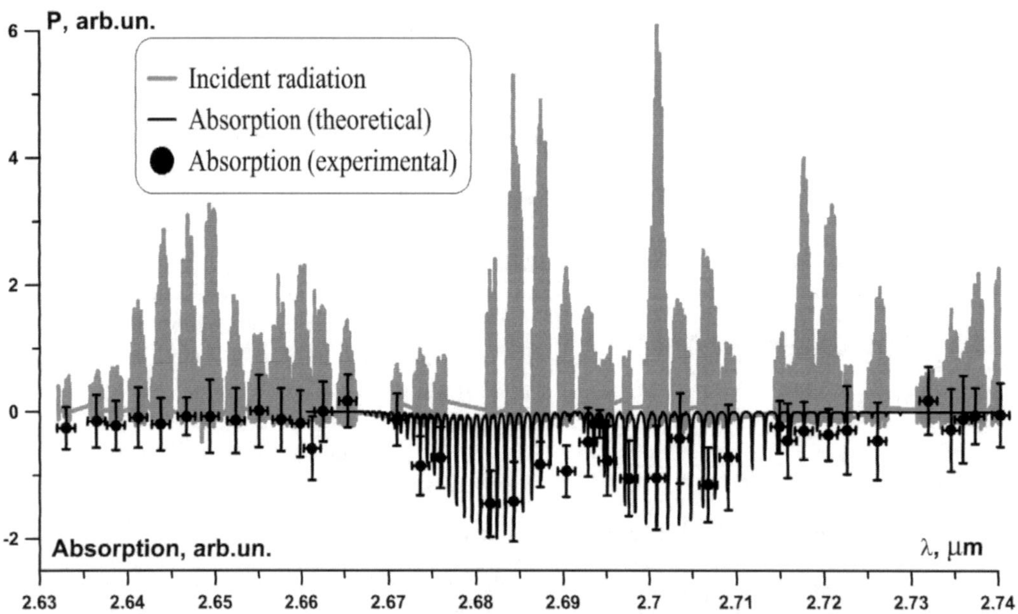

Figure 7. Spectra of incident and transmitted SFG radiation (gray and black graphs respectively; the last graph multiplied by 10) and theoretically calculated absorption spectrum in gas cell with CO_2 (black line, pending negative of ordinate axis). Experimental values of absorption (black dots) were calibrated on the lines, which are not absorbed by CO_2 molecules (λ~2.63 – 2.66 μm)

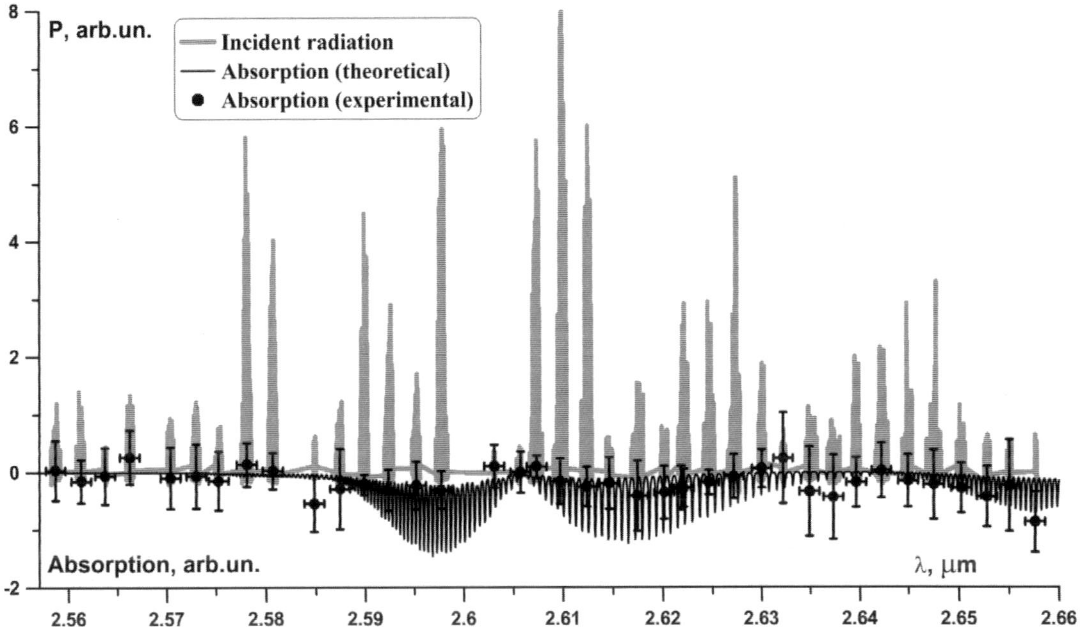

Figure 8. Spectra of incident and transmitted SFG radiation (gray and black graphs respectively; the last graph multiplied by 5) and theoretically calculated absorption spectrum in gas cell with N_2O (black line, pending negative of ordinate axis). Experimental values of absorption (black dots) were calibrated on the lines, which are not absorbed by N_2O molecules ($\lambda \sim 2.56 - 2.58$ μm)

From the absorption measurement data, we see that the experimental and calculated agree satisfactorily over the entire spectral range in which the measurements were made. We also need to take into account the uncertainty in measurement of the radiation energy by the OPHIR calorimeter, which is no greater than 5%. Some discrepancies between the calculation and the measurements can be explained by not sufficiently high spectral resolution of the spectrometer. It is not allowed complex SFG spectrum of the multi-line CO laser detects narrow absorption lines of the investigated gases.

5. Conclusions

As a result of experiments it was obtained SFG spectra of CO laser radiation. Absorption of SFG radiation in gas cell with a length of 10 cm, containing gaseous nitrous oxide N_2O and carbon dioxide CO_2 at different partial pressures with the addition buffer gas (nitrogen) of up to atmospheric pressure were measured. Analysis of experimental data shows that in those spectral intervals where the absorption of radiation is small, the experimental values of the absorption vary around zero. In other intervals, the measured absorption coefficients, on average, take the values close to the calculated values, but the spectral resolution of the spectrometer was not sufficiently high to allow complex SFG spectrum of the multi-line CO laser and detect narrow absorption lines of the investigated gases.

Acknowledgments

This research is partially supported by the Russian Foundation for Basic Research (Project No. 13-05-98074 and 15-02-08037) and Educational-Scientific Complex of P.N. Lebedev Physical Institute.

6. References

[1] R.J. Keyes, Springer Optical and Infrared Detectors (Topics in Applied Physics V. 19), (1980).
[2] O.G.Buzykin, A.A.Ionin, S.V.Ivanov, et al., Laser and Particle Beams, 18, 697 (2000).
[3] O.G.Buzykin, A.A.Ionin, S.V.Ivanov, et al. Bulletin of the Russian Academy of Sciences, Physics, 66 (7), 1055-1060 (2002).

[4] O. A. Romanovskii, O. V. Kharchenko Journal of Applied Spectroscopy, V. 66, Issue 6, pp 985-988 (1999).

[5] A.I. Karapuzikov, I.V Ptashnik, I.V Sherstov. et al. Infrared Physics and Technology, V.41, N 2, p. 87-96 (2000)

[6] S.V. Ivanov, V.I. Novoderezhkin, V.Ya. Panchenko, et al. Opt. Eng., V.33(10), P.3202, (1994).

[7] A.A. Ionin, Electric discharge CO laser, in: M. Endo and R. Walter (Eds.) Gas Lasers, CRC Press - Taylor and Francis Group, Boca Raton, Florida, USA, pp. 201-237, (2007).

[8] N.G. Basov, G.D. Hager, A.A. Ionin, et al. "Efficient pulsed first-overtone CO laser operating within the spectral range of 2.5-4.2 μm". IEEE J. Quantum Electronics, V.36. P.810, (2000).

[9] E.Yu. Bondarchuk; A.A. Ionin; A.Yu. Kozlov; et al. "Spectroscopic analysis of multicomponent gas mixtures with wide range carbon monoxide laser", Proceedings SPIE, V. 7994, p. 799413 (2011)

[10] A.A. Ionin, Yu.M. Klimachev, A. Yu. Kozlov et al. Journal of Applied Spectroscopy, Vol. 81, No. 2, p.309-312 (2014)

[11] H. Dahnke, D. Kleine, W. Urban et al., Appl. Phys. B 72, 121–125 (2001)

[12] A.A. Ionin, I.O.Kinyaevskiy, Yu.M. Klimachev, et al. Optics Letters 37 (14), p. 2838–2840 (2012).

[13] Yu.M. Andreev, A.A. Ionin, I.O. Kinyaevskiy, et al. Quant. Electron. 43 (2), p. 139-143 (2013).

[14] O.V. Budilova, A.A. Ionin, I.O. Kinyaevskiy, et al. "Broadband two-stage frequency conversion of CO laser radiation in $AgGaSe_2$ crystal," Optics Letters, 41 (4), p.777-780 (2016)

[15] Yu.M.Andreev, O.V.Budilova, I.O.Kinyaevskiy, et al. Optics Letters, 40 (13), p.2997-3000 (2015)

Radiative transitions in quasimolecules: H+H$^-$ collisions

A V Dadonova[1] **and A Z Devdariani**[1,2]

[1] Department of Theoretical Physics and Astronomy, Herzen University, St.- Petersburg 191186, Russia

[2] Department of Optics and Spectroscopy, St.-Petersburg State University, St.- Petersburg 198904, Russia

E-mail: alladadonova@mail.ru

Abstract. The $H + H^- \rightarrow H^- + H + \hbar\omega$, $H + H^- + \hbar\omega \rightarrow H + H + e$ reactions give examples of bound-bound and bound-free electron transitions in the $H + H^-$ quasi-molecules temporally formed during collisions. The present work is aimed at the calculations of spectral profiles produced in the reactions of resonant charge exchange $H + H^-$. The oscillator strength of photodetachment of quasimolecules has been calculated for the first time.

1. Introduction

The study of radiative transitions in single ions or atoms is a traditional approach in plasma spectroscopy. On the other hand, radiative transitions in collisional quasi-molecules, that is temporary molecules formed during ion or atom collisions, are also of importance. Such transitions are directly connected with the dynamics of ion or atom collisions. The study of the transitions produced in the $H + H^-$ collisions has been initiated in [1] where bound-bound transitions were analysed. The aim of the present work is to get down to the case of bound-free transitions. Thus, the main new result of the present work is the calculation of the oscillator strengths of bound-free transitions in collisions or, in other words, photodetachment during collisions.

In the frame of the adiabatic approach the process of charge exchange $H + H^-$ can be regarded as evolution of an atomic system along gerade and ungerade potential energy curves of the adiabatic states of a quasi-molecule formed during collisions. In the adiabatic approach emission is treated as the dipole transition between two quasi-molecular states involved [2].

The main characteristics of the radiative transitions involved in the description of optical transitions in quasi-molecules have been calculated in the frame of the zero-range potential model [3]. In the model the energy state difference and other characteristics of the process can be expressed in a closed form in terms of the Lambert W(z) function [4]. Another type of electron transitions can also be treated in the frame of the zero-range potential model, namely $H + H^- + \hbar\omega \rightarrow H + H + e$ as an example of bound-free transitions in quasi-molecules. For the first time the spectral profiles in charge exchange reaction $H + H^-$ are presented the generalization of calculations and the oscillatory strength calculated as a function of momentum the photodetachment in these collisions.

2. Bound-bound transitions

The radiation process is treated as a dipole transition between the gerade U_g and the ungerade U_u adiabatic quasi-molecular states produced by the interaction of the ion and the atom in their ground states that is as an optical transition between the two quasi-molecular states formed during the collision.

The spectral profile in the quasi-static approximation is

$$I(\omega) = 4\pi R_c^2 \left[\frac{A(R_c)}{d\Delta U/dR_c}\right] exp\left(\frac{-U_u(R_c)}{kT}\right), \tag{1}$$

here U_u - the adiabatic energy of the initial ungerade state, ΔU is the difference between gerade and ungerade quasimolecular states,

$$A(R_c) = \left(\frac{4}{3g}\right)\left(\frac{\omega^2}{137}\right) D\left(R_c\right)^2 \tag{2}$$

is the Einstein coefficient for spontaneous emission of the quasimolecule, g statistical factor, D is the electric dipole moment for transitions between the two states involved [1]. In our case and R_c is the position of the Condon point that is the root of the equation

$$\Delta U\left(R_c\right) = \omega \tag{3}$$

that has the solution at $\alpha R \geq 1$ as

$$R_c = \frac{1}{\alpha} LambertW\left(\frac{2\alpha^2}{\omega}\right) \tag{4}$$

The results of the profile calculations are shown in Figures 1 and 2 for two temperatures versus the frequency of the radiation and the interatomic distance, respectively.

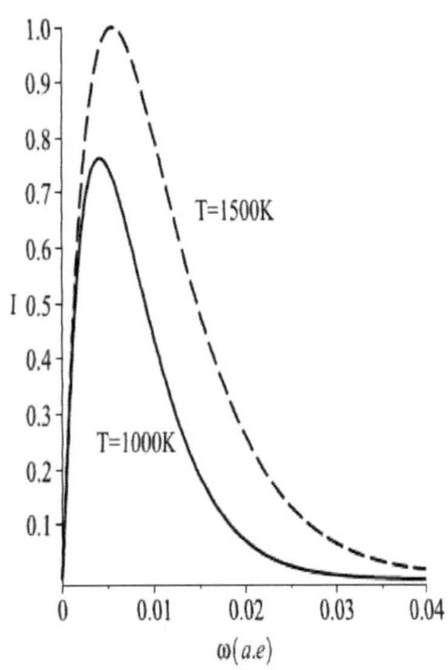

Figure 1. Spectral profile produced in the collisions versus frequency of the radiation for two temperatures.

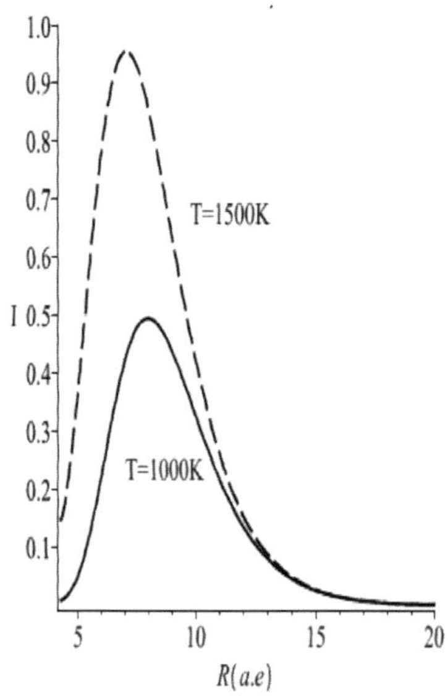

Figure 2. Spectral profile produced in the collisions versus the interatomic distance for two temperatures.

It should be stressed that the maxima of the intensities are formed by electron transitions at 7 - 9 a_0. It means that the radiative transitions are followed by the formation of the unstable H_2^- molecules in their high rotational states. The last conclusion seems to support the prediction Golser et al [5]. We emphasize that for the first time a summary of the results presented in this paper

3. Bound-free transitions

In the previous Section the radiative processes produced by charge exchange have been considered as an example of bound-bound transitions in quasi-molecules. Here, we consider bound-free transitions in quasi-molecules or, in other words, photodetachment taking $H - H^-$ collisions as a typical example.

The cross-section of photodetachment is proportional to the density of oscillator strengths which is the main characteristic of the dipole transitions

$$\sigma(\omega) = \frac{2\pi^2}{c}\frac{df_{bc}}{d\omega} = \frac{4\pi^2}{c}\omega_{bc}\left|X_{bc}\right|^2 \tag{5}$$

For simplicity, we discuss the one-dimensional case. Then the matrix element is

$$X_{bc} = \int \langle \Psi_b \left|x\right| \Psi_c \rangle \, dx \tag{6}$$

In the LCAO approximation quasimolecular gerade and ungerade wave functions in the frame of the zero-range potentials model are written as

$$\Psi_b^{gu} = \sqrt{\alpha_{gu}}\left[exp(-\alpha_{gu}\left|x + R/2\right|) \pm exp(-\alpha_{gu}\left|x - R/2\right|)\right] \tag{7}$$

$$\Psi_c^{gu} = \frac{1}{\sqrt{2\pi k}}\left[sin(k\left|x + R/2\right|) \pm sin(k\left|x - R/2\right|)\right] \tag{8}$$

Then the matrix element dipole moments can be expressed by the formula:

$$X_{b=g,c=u} = \sqrt{\frac{\alpha_g k}{\pi}}\frac{\alpha_g}{\omega^2}cos^2\left(\frac{kR}{2}\right) + \sqrt{\frac{\alpha_g}{\pi k}}\frac{\alpha_g R}{4\omega}sin\left(kR\right) \tag{9}$$

$$X_{b=u,c=g} = \sqrt{\frac{\alpha_u k}{\pi}}\frac{\alpha_u}{\omega^2}sin^2\left(\frac{kR}{2}\right) - \sqrt{\frac{\alpha_u}{\pi k}}\frac{\alpha_u R}{4\omega}sin\left(kR\right) \tag{10}$$

where $\omega = (\alpha_{gu}^2 + k^2)/2$

Although the quasi-molecular matrix elements increase with increasing R it does not lead to difficulties at large R. In this case $U_u = U_g$, and it is obligatory to use the sum of amplitudes of two different ways of detachment [2].

The position of the Condon point that is the root of the equation:

$$\omega = \frac{\alpha_u^2(R_c) + k^2}{2} \tag{11}$$

The Figure 3 shows the results for the oscillatory strength for the radiation transition between gerade and ungerade quasimolecular states as a function of momentum.

Collisions result in a lowering of the threshold of the electron detachment compered with the single ion H^- because ungerade quasi-molecular term is repulsive.

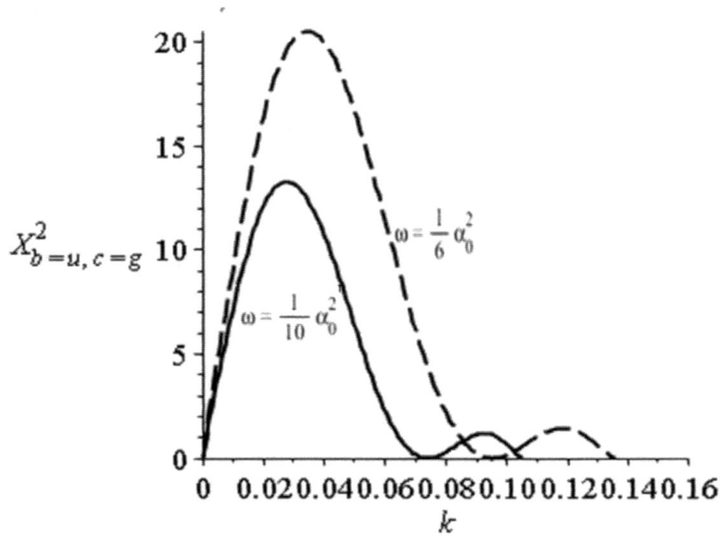

Figure 3. Oscillatory strength as a function of momentum.

4. Conclusion

The spectral profiles in charge exchange reaction $H + H^-$ are presented together with preliminary results related to photodetachment in these collisions. The main results obtained are as follows. First, the emission is produced in the collision $H + H^-$ in the absence of any initial electronic excitation in the colliding particles. Second, quasi-molecular radiative transitions lead to the formation of ro-vibrational states of the H_2^- molecules at comparatively large interatomic distances about 8 - 10 a.u. Third, collisions result in a lowering of the threshold of the electron detachment compared with the single ion because ungerade bound quasi-molecular term is repulsive. The results obtained may contribute to the cold and thermal plasma spectroscopy.

References

[1] Devdariani A Z, Dadonova A V 2015 *Russian Journal of Physical Chemistry B* **9(4)** 591-592.
[2] Devdariani A Z 2015 *Opt.spektr.* **119(3)** 333-337.
[3] Demkov Yu N, Ostrovskii V N 1988 *Plenum Pub Corp* 288.
[4] Corless R M, Connet G H, Hare D E G, Jeffrey D J, Knuth D E 1996 *Adv. Comput. Math.* **5(1)** 329-359.
[5] Golser R, Gnaser H, Kutschera W, et al 2005 *Phys. Rev. Lett.* **94** 223003.

Journal of Physics: Conference Series **769** (2016) 012044 doi:10.1088/1742-6596/769/1/012044

Quantum magnetometers as a base for atomic clock

E A Sagitov[1], S V Ermak[1], M V Petrenko[2] and V V Semenov[1]

[1] Institute of Physics, Nanotechnology and Telecommunications, Peter the Great Saint Petersburg Polytechnic University, 29 Politekhnicheskaya str., St. Petersburg, 195251, Russia
[2] Ioffe Physical Technical Institute, 26 Politekhnicheskaya str., St. Petersburg, 194021, Russia

E-mail: serge_ermak@mail.ru

Abstract. The possibility of the atomic clock based on two quantum magnetometer system development is demonstrated. One of the magnetometers is self-oscillation type, another is M_Z-type magnetometer based on end-state UHF resonance. The laser pumping of ^{87}Rb atoms placed into antirelaxation coated sell is provided. The magnetometers frequency difference fluctuations experimental results are represented and Allan deviation is determined. The role of the radio-optical resonance frequency light shift different components is denoted for the determination of the quantum magnetometers optimal operation mode. The effect of the light shift compensation is demonstrated.

1. Introduction

Atomic clocks implement the synchronization of telecommunication and precise time systems. They ensure correct operations of stationary and mobile systems, ensuring their reliability and accuracy. The basis for the atomic clock constructing is the optical pumping method, which have been developed since the beginning of the 50s of XX century [1]. The optical pumping leads to the redistribution of populations in the atom energy structure ground state, that creates the conditions to monitor the magneto-dipole transitions in the hyperfine structure (HFS), providing precise measurement of the resonance frequency, that determine the stability of an atomic clock [2]. The main measurement error source is the light frequency shift associated with the influence of non-resonant light components [3].

Theoretical and experimental researches have shown the ability to reduce the influence of the light frequency shift component compensation in the case of end-state magneto-dependent HFS transitions (end-state resonance), which is not possible in the case of magneto-independent 0-0 transition [4]. In the case of the atomic clock on end-state resonance, Zeeman transitions in self-oscillating magnetometer can be selected to compensate the magnetic fluctuations in an external magnetic field. Hereby, the compensation of the magneto-dependent HFS end-state resonance magnetic dependence and the additional frequency destabilizing factors of the atomic clock is provided [5].

Practically, this compensation is achieved by simultaneous signal registration of end-state HFS resonance using the technique of synchronous detection and self-oscillating quantum magnetometer. These oscillations are on the magneto-dipole transitions average frequency of the ^{87}Rb atoms ground state. The further processing of the received signals is provided.

Thus, the frequency light shifts compensation is very impotent for atomic clock long term stability achievement, and is learning in this article. The effect of the light shift component compensation is demonstrated. To achieve this objective it is necessary to solve following problems:

- the experimental research of the end-state HFS resonance frequency stability in laser pumped ^{87}Rb atoms, determination of Allan variance and its subsequent analysis;
- the development of the compensation destabilizing factors methods using the signals of two magnetometers;
- the elaboration of the recommendations for optical pumping modes optimization in alkaline atoms placed into the absorption cell with wall anti-relaxation coating and allowing to reduce frequency errors of the atomic clock.

2. The method of optical pumping

The main task is the creation of non-equilibrium population of atoms in the energetic structure. This can be achieved by method of optical pumping, which is widely used in modern devices of quantum electronics and primarily includes exemplary frequency measurements and a whole range of precision quantum magnetometers used in various technical applications, and in a study of fine physical phenomena. The basic substance of such devices is alkali metal atoms.

The object of optical pumping in this energy structure is the ground state, where in thermodynamic equilibrium conditions all magnetic sublevels have equilibrium populations. Optical pumping, i.e. the resonant light effect on the atoms, leads to the inversion of the populations in the ground state, that creates conditions for monitoring magneto-dipole transitions in ultra-fine structure.

Until recently, the technique of optical pumping has used spectral lamps filled with the same matter as the pumping atoms. With the progress in laser sources development there is a great interest to use laser pumping with various alkali metals to produce wide range of quantum devices based on various physical effects. The first experiments with laser pumped caesium atoms allowed one to reach almost 100% polarization of alkali metal atoms in the magneto-dipole hyperfine transition of ground state [6]. An interesting continuation of such research was the precision measurement of the resonance frequency variation on 0-0 transition caused by the dynamic Stark effect offsets the energy sublevels of the atoms in the pumping radiation field [7, 8].

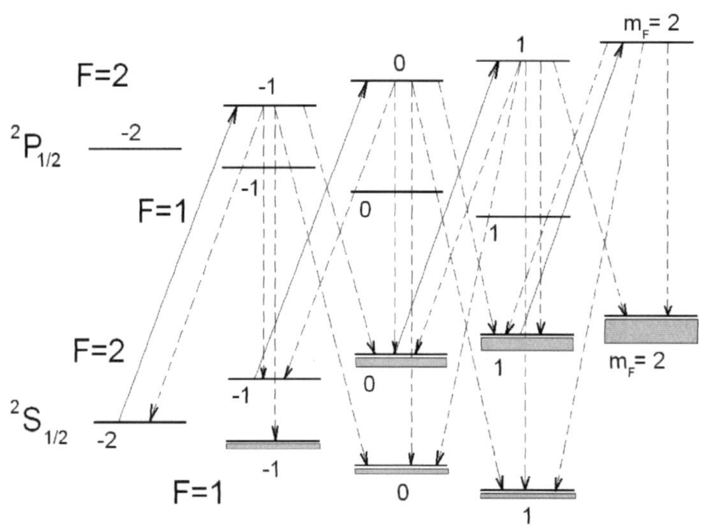

Figure 1. The scheme of optical orientation of 87Rb atoms in the ground state in terms of electro-dipole transitions by σ+ polarized laser light of D_1 line. Solid arrows indicate direct transitions, dashed arrows indicate spontaneous transitions.

Figure 2. Functional scheme of a two-channel system of quantum magnetometers with optical pumping.

Another method of optical pumping is the optical orientation [1]. The energetic structure of ^{87}Rb atoms ground state ($^2S_{1/2}$), first excited state ($^2P_{1/2}$) and electro-dipole transitions by circularly polarized light ($\sigma+$ or $\sigma-$ polarization) is presented in figure 1. Optical orientation by circularly polarized light, where, as a result, the implementation of the magnetic quantum state selection rules depletes the number of magnetic sublevels with extreme value projection of a full momentum. Non-equilibrium population is created on a magnetic quantum sublevels. This method is used in quantum magnetometers.

3. The system of two quantum magnetometers

The basis of the experimental research was the magnetic induction meter scheme based on the system of two quantum magnetometers with optical pumping (SQMOP), one of which was self-oscillating low frequency (LF) magnetometer, another – passive Z-magnetometer based on the UHF radio-optical resonance, which resonance frequency corresponds to magneto-dependent transitions between HFS sublevels with extreme value of a magnetic quantum number. The simplified functional scheme of such device is presented in figure 2.

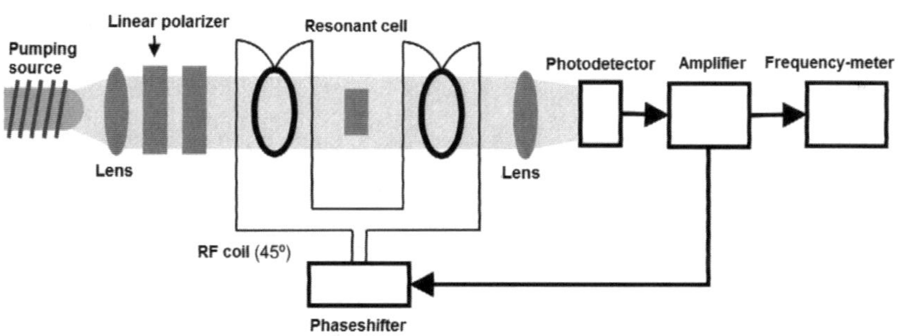

Figure 3. Self-oscillating magnetometer, LF channel.

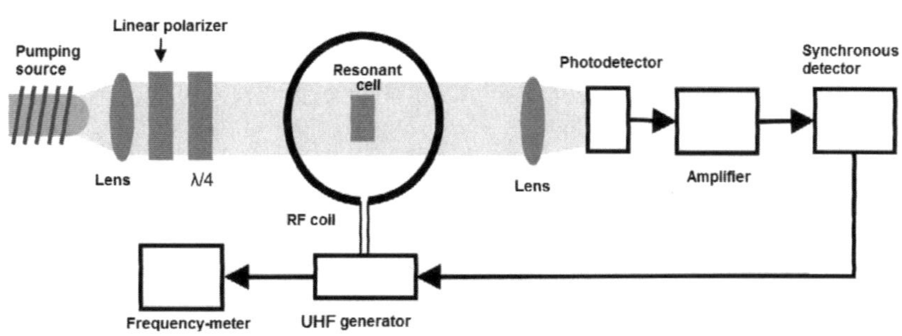

Figure 4. Z-magnetometer, UHF channel.

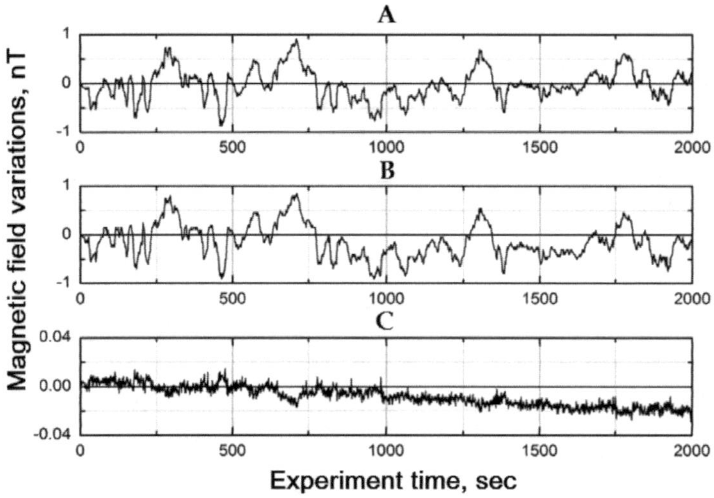

Figure 5. The compensation of magnetic fluctuations in the system of two quantum magnetometers (A – LF channel, B – UHF channel, C – difference between the magnetic variations registered by LF and UHF channels).

The functional schemes of LF self-oscillating and UHF Z-type magnetometers are presented in figure 3 and figure 4. The original result of this research is the use one cell with anti-relaxation coating, containing alkali atoms for two magnetometers, that provides additional possibilities in the light shift reduction.

Experimental verification of the light shift compensation effect was carried out in the context of the ^{87}Rb atoms with laser optical pumping of ground state F=2 by light component $S_{1/2} - P_{3/2}$ of D_2 line in SQMOP. The functional scheme of SQMOP contains well known elements used in gas-cell atomic clock. The scheme is operated on the UHF radio-optical resonance signals and on the weighted average resonance frequency of the Zeeman structure. This scheme provides simultaneous measurement of magnetic field fluctuations by registration of the UHF measuring channel error signal and by measurement of the LF measuring channel signal frequency, which react to changing of an external magnetic field and laser pumping parameters. Then, the signals from two channels are detected by the registration scheme where they are mutually subtracted (see figure 5). This method provides compensation of the magnetic fluctuations and light shift components [4]. Further, Allan variance for the differential signal of the two quantum magnetometers was calculated.

Figure 6 presented Allan variance for two directions of circular polarization of pumping light (σ+ and σ–) with light power of ~100 μW, that corresponds to the quality factor maximum mode (QFM) of LF and UHF radio optical resonance signals. The represented experiments are provided under other identical conditions. Thus, there is a principal possibility of scalar and vector light shift mutual compensation by choosing a suitable sign of circular polarization of pumping light, that allows realizing the reduced resonance frequency light shift of the atomic clock based on the two quantum magnetometers system.

The greatest interest in applications represents the area of averaging time 10^2–10^3 s. It was the cause to carry out a number of experiments for the quantum magnetometers operating mode optimization in time of averaging exceeding a time barrier in hundreds of seconds [9].

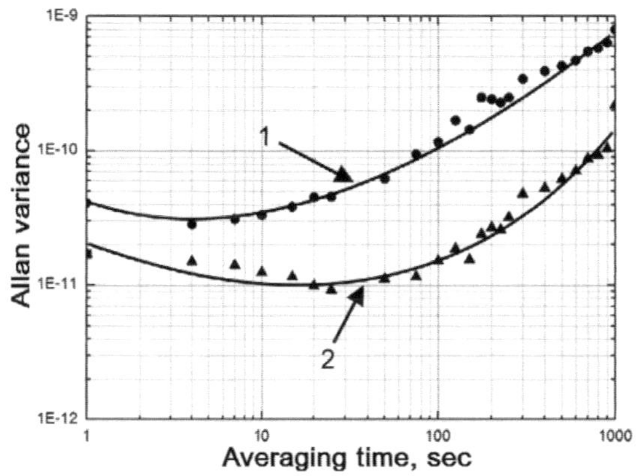

Figure 6. Allan variance for system of two quantum magnetometers. Graphs 1 and 2 corresponds the polarization of pumping light of σ– and σ+ types.

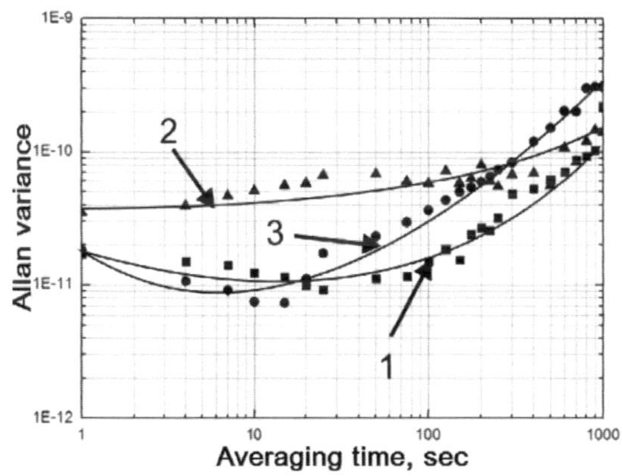

Figure 7. Allan variance for the system of two quantum magnetometers. Graphs 1 and 2 corresponds to laser power of ~100 μW (QFM mode) and ~25 μW (MLTS mode) under σ+ polarization of optical pumping, graph 3 – Allan variance for atomic clock on 0-0 transition.

It was found experimentally, that this variance limit corresponds to a certain intensity of the pumping light (see figure 7). The intensity in this case was ~25 μW under σ+ polarization of optical pumping. These conditions correspond to the mode of maximal long-term stability (MLTS).

In figure 7 similar dependences for SQMOP are given in QFM mode (dependence 1) and for atomic clock on 0-0 transition (dependence 3). Short-term stability of SQMOP in the MLTS mode (see figure 7) significantly concedes to similar parameter during the operation of the device in the MQF mode that is explained by decrease in a quality factor of observed signals because there is discrepancy in intensity of pumping to its optimum level within a flicker floor. However, at long operation of the device in MLTS mode is justified by more flat dependence of Allan variance with time growth that in the conditions of the light shift component compensation allows improving the SQMOP precision properties. In the MQF mode a flicker floor of SQMOP and atomic clock on 0-0 transition within several tens of seconds of supervision practically coincide, however, the advance in area of long-term stability in figure 7 shows undoubted advantage of the studied scheme in comparison with atomic clock on 0-0 transition.

Thus, the correct choice of the laser pumping light polarization allows light shift various component mutual compensation of the atomic clock resonant frequency and therefore, increases its long-term stability.

4. Summary

It is shown, experimentally, that the effect of the light shift components compensation allows reducing the frequency Allan variance of the atomic clock based on the system of two quantum magnetometers in comparison with the atomic clock on 0-0 transition.

Journal of Physics: Conference Series **769** (2016) 012044 doi:10.1088/1742-6596/769/1/012044

References

[1] Happer W 1972 *Rev. Mod. Phys.* **44** 170
[2] Riehle F 2004 *Frequency Standards: Basics and Applications* (Weinheim: WILEY-VCH Verlag GmbH & Co. KGaA)
[3] Happer W and Mathur B S 1967 *Phys. Rev.* **163** 12
[4] Baranov A A, Ermak S V, Sagitov E A, Smolin R V and Semenov V V 2015 *J. Exp. Theor. Phys.* **121** 393
[5] Jau Y Y, Post A B, Kuzma N N, Braun A M, Romalis M V and Happer W 2004 *Phys. Rev. Lett.* **92** 110801
[6] Siahatgor S and Hochuli V E 1969 *Quantum Electron.* **5** 295
[7] Arditi M and Picque J L 1975 *Optics Comm.* **15** 317
[8] Arditi M and Picque J L 1975 *J. Phys. B: Atom. Phys.* **88** 331
[9] Aleksandrov E B and Vershovskii A K 2009 *Phys. Uspekhi* **52** 573

Journal of Physics: Conference Series **769** (2016) 012045 doi:10.1088/1742-6596/769/1/012045

On a theory of light scattering from a Bose-Einstein condensate

V M Ezhova, L V Gerasimov and D V Kupriyanov

Department of Theoretical Physics, Peter the Great Saint-Petersburg Polytechnic University, St-Petersburg 195251, Russia

E-mail: kupr@dk11578.spb.edu

Abstract. We consider a quantum theory of elastic light scattering from a macroscopic atomic sample existing in the Bose-Einstein condensate (BEC) phase. Following the second quantized formalism we introduce a set of coupled and closed diagram equations for the polariton propagator contributing to the T -matrix and scattering amplitude. Our approach allows us to follow important density corrections to the quasi-energy structure caused by static interaction and radiation losses associated with incoherent scattering in the case of near resonance excitation.

1. Introduction

Light scattering on complex quantum systems is a potentially interesting process for developing various quantum interface protocols organized between the light and matter subsystems. The coherent joint propagation of light and matter wave through a degenerate quantum gas existing in the Bose-Einstein condensate (BEC) phase had been predicted in [1] even before such a matter state had been created in the laboratory. After the first successful experimental realization of BEC in alkali-metal systems in [2, 3], evident signatures of cooperative dynamics in light scattering on the condensate have been observed in experiments [4]-[5]. The strong coherent scattering of light on a sample led to the condensate fragmentation; this effect have been explained in [5] in terms of a Kapitza-Dirac diffraction phenomenon. These experiments have encouraged experimental and theoretical efforts towards deeper understanding and realistic description of light scattering process from degenerate bosonic gas [7]-[11]. In [9]-[11] a semiclassical approach for light scattering from a BEC considered as an ideal atomic gas have been applied to describe the basic results of experiment [4, 6].

In the present report we propose a theoretical approach based on a standard formalism of the quantum scattering theory. We consider BEC in the framework of the Gross-Pitaevskii model [12, 13] and describe the optical excitation of such a complex quantum system via introducing the self-consistent Green's function analysis.

2. Scattering problem in conditions of quantum degeneracy

The quantum-posed description of the photon scattering problem is based on the formalism of the T-matrix, which is defined by

$$\hat{T}(E) = \hat{V} + \hat{V}\frac{1}{E-\hat{H}}\hat{V} \tag{1}$$

where $\hat{H}$ is the total system Hamiltonian consisting of the nonperturbed part $\hat{H}_0$ and an interaction term $\hat{V}$ such that $\hat{H} = \hat{H}_0 + \hat{V}$. The energy argument E is an arbitrary complex parameter in Eq. (1). Then the scattering process, evolving from an initial state $|i\rangle$ to the final state $|f\rangle$, is expressed by the following relation between the differential cross section and the transition amplitude, given by the relevant T-matrix element considered as a function of the initial energy E_i:

$$d\sigma_{i\to f} = \frac{\mathcal{V}^2 \omega'^2}{\hbar^2 c^4 (2\pi)^2} \left| T_{g'\mathbf{e}'\mathbf{k}';\, g\mathbf{e}\mathbf{k}}(E_i + \mathrm{i}\,0) \right|^2 d\Omega \tag{2}$$

Here the initial state $|i\rangle = |g; \mathbf{e}, \mathbf{k}\rangle$ is specified by the incoming photon's wave vector $\mathbf{k}$, frequency $\omega \equiv \omega_k = ck$, and polarization vector $\mathbf{e}$, and the atomic system populates a particular collective ground state $|g\rangle$. In our case initially $|g\rangle = |\mathrm{BEC}\rangle^N$ performs a collective state of N atoms in the BEC phase described by the Gross-Pitaevskii model. The final state $|f\rangle = |g'; \mathbf{e}', \mathbf{k}'\rangle$ is specified by a similar set of quantum numbers, excepting that $|g'\rangle$ can be now a disturbed condensate state for inelastic channels, and the solid angle Ω is directed along the wave vector of the outgoing photon $\mathbf{k}'$. The presence of quantization volume $\mathcal{V}$ in this expression is caused by the second quantized structure of the interaction operators. The optical theorem links the total cross section with the diagonal T-matrix element

$$\sigma_{\mathrm{tot}} = -\frac{2\mathcal{V}}{\hbar c} \operatorname{Im} T_{ii}(E_i + \mathrm{i}\,0) \tag{3}$$

which gives a convenient tool for the cross-section evaluation via calculation of only one T-matrix element for the elastic forward scattering.

In the second quantized representation the interaction term $\hat{V}$ in Eq. (1), taken in the rotating wave approximation, is given by

$$\hat{V} = -\sum_n \int d^3r \left[\hat{\mathbf{d}}_{n0} \hat{\Psi}_n^\dagger(\mathbf{r}) \hat{\Psi}_0(\mathbf{r}) \hat{\mathbf{E}}^{(+)}(\mathbf{r}) + h.c. \right] \tag{4}$$

where $\hat{\mathbf{d}}_{nm}$ is a component of the dipole operator, and where $n = 0, \pm 1$ and $m = 0$ respectively specify the single atom angular momentum of the exited and non-degenerate ground states for the BEC consisting of atoms with a ${}^1S_0$ ground state. $\hat{\mathbf{E}}^{(+)}(\mathbf{r})$ is the positive frequency component of the electric field operator and the interaction term is described in the long wavelength dipole approximation, see [14]. The operators $\hat{\Psi}_0(\mathbf{r})$ and $\hat{\Psi}_n^\dagger(\mathbf{r})$ are the second quantized annihilation and creation operators of an atom at position $\mathbf{r}$ respectively in the non-degenerate ground and in a particular excited state and in accordance with the model we take

$$\hat{\Psi}_0(\mathbf{r}) |\mathrm{BEC}\rangle^N = \Xi(\mathbf{r}) |\mathrm{BEC}\rangle^{N-1} \tag{5}$$

where $\Xi(\mathbf{r})$ is the ordering parameter ("wavefunction") of the condensate.

The scattering amplitude expressed by "on-shell" T-matrix elements contributing to Eqs. (2) and (3) and considered for the elastic scattering channel can be expressed by the Green's function of a single particle excitation propagating in the BEC:

$$\begin{aligned} T_{fi}(E) \quad &= \quad \frac{2\pi\hbar (\omega_{\mathbf{k}'}\omega_{\mathbf{k}})^{1/2}}{\mathcal{V}} \iint d^3r'\, d^3r \sum_{n',n} (\mathbf{d}\mathbf{e}')^*_{n'0} (\mathbf{d}\mathbf{e})_{0n}\, \mathrm{e}^{-\mathrm{i}\mathbf{k}'\mathbf{r}' + \mathrm{i}\mathbf{k}\mathbf{r}}\, \Xi^*(\mathbf{r}')\, \Xi(\mathbf{r}) \\ &\times \left(-\frac{\mathrm{i}}{\hbar} \right) \int_0^\infty dt\, \mathrm{e}^{\frac{\mathrm{i}}{\hbar}(E - E_0^N + \mathrm{i}0)t}\, \mathrm{i}\, G_{n'n}(\mathbf{r}', t; \mathbf{r}, 0) \end{aligned} \tag{6}$$

where E_0^N is the initial energy of the condensate consisting of N particles and

$$\mathrm{i}G_{n'n}(\mathbf{r}', t'; \mathbf{r}, t) = \langle \mathrm{BEC} | T \Psi_{n'}(\mathbf{r}'; t') \Psi_n^\dagger(\mathbf{r}; t) | \mathrm{BEC} \rangle^{N-1} \tag{7}$$

is the time ordered (causal) Green's function (propagator) associated with the polariton-type (i.e. superposed between field and atom) quasi-particle excitation propagating through the condensate consisting of $N-1$ particles. The operators contributing to the polariton propagator are the original atomic operators transformed in the Heisenberg representation and dressed by the entire interaction process. We additionally assumed that the condensate itself is a stable system, which cannot be modified by the interaction (4) directly without perturbation by an incoming photon.

In accordance with the standard definitions of the diagram technique [15] the polariton propagator obeys the following

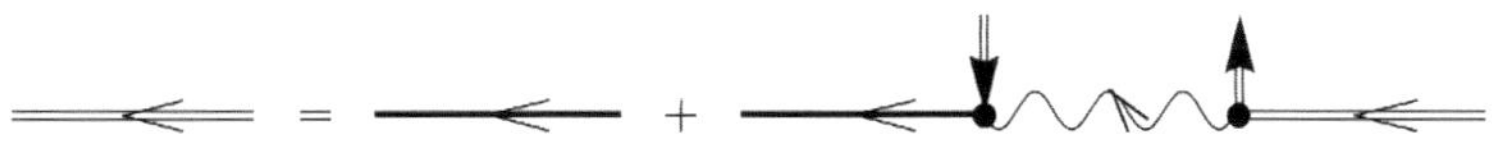

(8)

where it is visualized as doubly-straight line dressed by all the interaction processes. The inward and outward vertical arrows image the ordering parameters and form the self-energy part responsible for coherent conversion of the excitation between the field, in which free dynamics is expressed by undressed wavy line, and an atom subsequently recovering to the condensate phase. The degradation of this coherent process is caused by the interaction with the vacuum modes when the excited atom emits a photon spontaneously and escapes coherent dynamics with further drifting through the condensate as a spectator.

The latter process is described by an incomplete atomic propagator (imaged by straight solid line in the diagrams) obeying the following Dyson-type equation

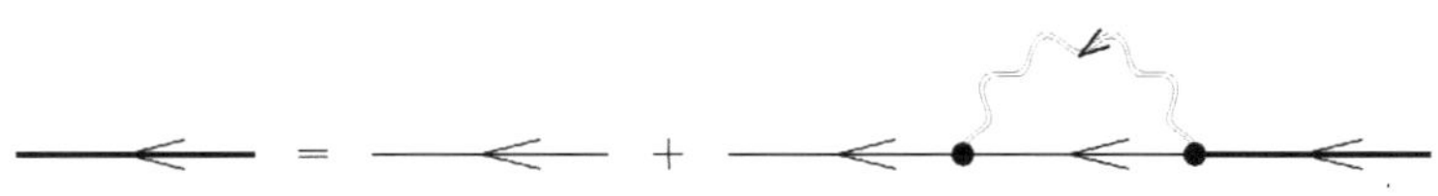

(9)

which has to be considered together with the equation for the field propagator

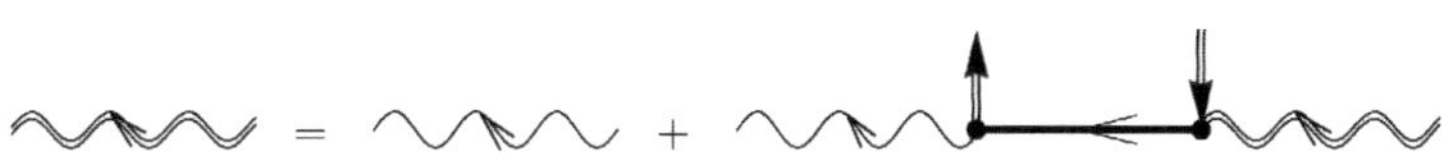

(10)

These two diagram equations reproduce the self-consistent dynamics of an atomic dipole interacting with its environment similar the situation in a disordered atomic gas. Any optical excitation created from the condensate has a chance to be incoherently re-emitted into the vacuum modes. As can be verified by precise analytical analysis, both the diagram equations lead to the same structure of polarization operator, dielectric constant and renormalization of the spontaneous decay rate as in disordered atomic gas of the same density. The relevant definitions and calculations have been done in [16], which we will further apply when solving the basic equation (8)

3. Polariton propagator

For an infinite, homogeneous and isotropic medium, which physically requires that the sample size would be much larger than the radiation wavelength, the solution of Eq. (8) can be found in the reciprocal space as linear combination of the transverse and the longitudinal components with respect to the wave vector argument. In our estimates below, with considering the internal binding energy in the condensate as weak, we will ignore the chemical potential of the condensate as a negligible quantity in comparison with the basic spectral parameters such as spontaneous decay rate, recoil energy etc. Then the Fourier components of the polariton propagator (7) can be expanded as follows

$$G_{nn'}(\mathbf{k},\omega) = G_{\parallel}(k,\omega)\,\frac{k_n k_{n'}}{k^2} + G_{\perp}(k,\omega)\left[\delta_{nn'} - \frac{k_n k_{n'}}{k^2}\right] \tag{11}$$

where we associated the vector indices in the quasi-particle wave vector $\mathbf{k}$ with the quantum numbers of the atomic excited state.[1]

Let us first consider and discuss the transverse component of the polariton propagator, which is given by

$$G_{\perp}(k,\omega) = \left[\omega - \omega_0 - \frac{\hbar k^2}{2\mathrm{m}} + \frac{4\pi n_0 d_0^2}{3\hbar} + \frac{\mathrm{i}\sqrt{\epsilon(\omega)}\gamma}{2} - \frac{4\pi n_0 d_0^2 \omega^2}{\hbar(\omega^2 - c^2k^2)}\right]^{-1} \tag{12}$$

where "m" is the atomic mass, $n_0 = |\Xi(\mathbf{r})|^2 = \text{const}$ is the density of atoms in the BEC, d_0 is the modulus of the dipole transition matrix element (same for all transitions), and ω_0 is the transition frequency for the atom considered in a two-level approximation, γ is its natural radiation decay rate. The dielectric susceptibility $\epsilon(\omega)$ can be analytically calculated for arbitrary atomic density, see [16] for derivation details.

In Fig. 1 the transverse component of the polariton propagator Eq.(9) is shown as a function of its frequency ω and wavenumber k for the parameters specified in the caption. As a basic wavelength we use rubidium $D2$-line with $\lambda = 780\,nm$. These plots show that for large frequency detunings the excitation mode approximately approaches the photon-type dispersion law $\omega \sim c|\mathbf{k}|$. The key feature of this result suggests fast propagation of the polariton through the sample and possible strong coherent scattering of light from the sample boundaries where the polariton is created. The coherent scattering could manifest itself in those conditions when a normal atomic gas of the same density would be practically transparent. The strong scattering of light from a BEC of alkali-metal atoms for off-resonant detunings of a few GHz have been observed in experiments [5, 6].

With approaching the point of atomic resonance $\omega \to \omega_0$ the optical excitation shows behavior similar to that of a non-condensed disordered atomic gas of the same density. The collective dipole polarization is driven by the propagating field and the environment of proximal dipoles induces the well known red static Lorentz-Lorentz shift $-4\pi n_0 d_0^2/3$ from the atomic resonance. However there is an extra frequency shift, induced by the polarization interaction with the condensate particle, which is given by the last term in the right-hand side of Eq.(12). Considering such an exciton-type quasi-particle as motionless with negligible momentum $p = \hbar k \ll \hbar\omega/c$ the dependence on ω vanishes and this part of the interaction also becomes static. In this limit the transverse component of the polariton propagator coincides with its longitudinal part, such that the excitation process becomes isotropic, and is given by

$$G_{\parallel}(k,\omega) = \left[\omega - \omega_0 - \frac{\hbar k^2}{2\mathrm{m}} - \frac{8\pi n_0 d_0^2}{3\hbar} + \frac{\mathrm{i}\sqrt{\epsilon(\omega)}\gamma}{2}\right]^{-1} \tag{13}$$

The spectral behavior of the longitudinal component of the polariton propagator is shown in Fig. 2 for the same parameters as in Fig. 1.

[1] The used notation for the momentum argument in the spatial Fourier transform has a signature but should not be directly identified as the wave vector of the scattered photon defined in the previous section.

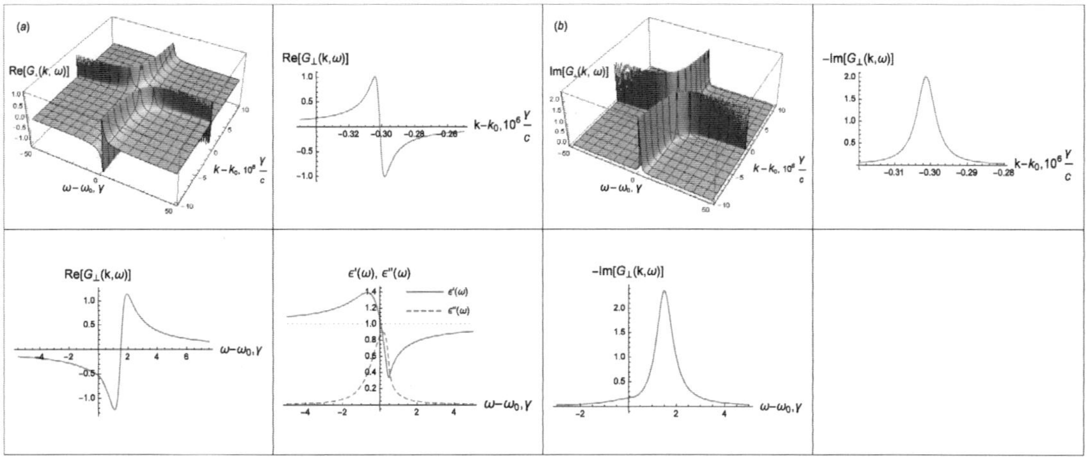

Figure 1. The transverse component of the polariton propagator $G_{\perp}(k,\omega)$ plotted for the density $n_0(\lambda_0/2\pi)^3 = 0.05$, where $\lambda_0 = 780\,\text{nm}$ is the radiation wavelength at atomic resonance: (a) Real part and (b) Imaginary part. Other graphs show the resolved spectral profiles, taken at the edges of three-dimensional plots (a) and (b), and the dielectric permittivity of the sample $\epsilon(\omega) = \epsilon'(\omega) + i\epsilon''(\omega)$ calculated in [16].

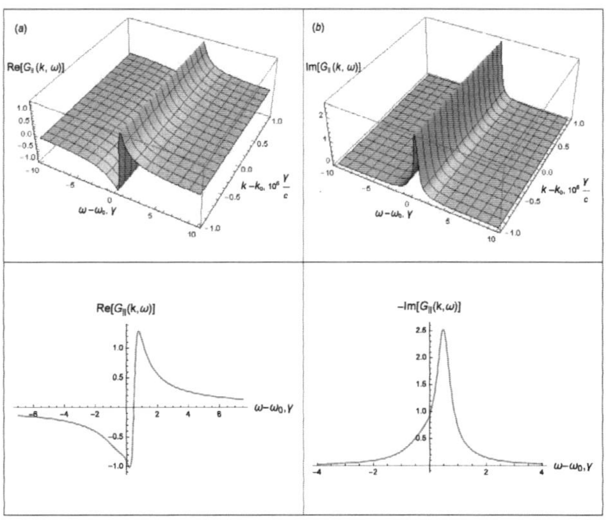

Figure 2. Same as in Fig. 1 but for longitudinal component of the polariton propagator $G_{\parallel}(k,\omega)$.

4. Conclusion

In the present paper we have shown that light propagates through a degenerate atomic gas existing in the BEC phase as a superposed state of field and matter - polariton wave. The polariton, created on the sample boundaries, can lead to strong coherent scattering of light from the sample edges in a broad frequency spectrum. This potentially valuable property could be applied for developing optomechanical quantum interface schemes between the non-classical light and condensate with using superfluidity as a main macroscopic quantum property of such a matter state.

Acknowledgements
This work was supported by the RFBR grant 15-02-01060, 16-32-50091, by the Foundation "Dynasty" and by the External Fellowship Program of RQC.

References
[1] H. D. Politzer Phys. Rev. A **43**, 6414(R) (1991)
[2] M.H. Anderson, J. R. Ensher, M. R. Matthews, C. E. Wieman, and E. A. Cornell, Science **269**, 198 (1995)
[3] K. B. Davis, M.-O. Mewes, M. R. Andrews, N. J. van Druten, D. S. Durfee, D. M. Kurn, and W. Ketterle, Phys. Rev. Lett. (PRL) **75** 3969 (1995)
[4] S. Inouye, A. P. Chikkatur, D. M. Stamper-Kurn, J. Stenger, D. E. Pritchard, W. Ketterle, Science **285**, 571 (1999)
[5] D. Schneble, Y. Torii, M. Boyd, E. W. Streed, D. E. Pritchard, W. Ketterle, Science **300**, 475 (2003)
[6] A. Hilliard, F. Kaminski, R. le Targat, C. Olausson, E. S. Polzik, and J. H. Müller, Phys. Rev. A **78**, 051403(R) (2008)
[7] M. G. Moore and P. Meystre, Phys. Rev. Lett. **83**, 5202 (1999)
[8] O. Zobay, G. M. Nikolopoulos,Phys. Rev. A.**73**, 013620 (2006)
[9] E. D. Trifonov, Laser Physics, **12**, No. 1, pp. 211–216 (2002)
[10] Ya. A. Avetisyan, E. D. Trifonov, Phys. Rev. A **88**, 025601 (2013)
[11] Y. A. Avetisyan and E. D. Trifonov, Phys.-Usp. **58** 286 (2015)
[12] E. P. Gross, Nuovo Cimento **20** 454 (1961); J. Math. Phys. **4** 195 (1963)
[13] L. P. Pitaevskii, J. Exptl. Theoret. Phys. (U.S.S.R.) **40**, 646-651 (1961)
[14] C. Cohen-Tannoudji, J. Dupont-Roc, G. Grynberg, *Atom-Photon Interactions. Basic Processes and Applications* (John Wiley & Sons, Inc., 1992)
[15] V. B. Beresteskii, E. M. Lifshits, L. P. Pitaevskii, *Course of Theoretical Physics: Quantum Electrodynamics* (Oxford: Pergamon Press, 1981); E.M. Lifshits, L.P. Pitaevskii, *Course of Theoretical Physics: Statistical Physics. Part II* (Oxford: Pergamon Press, 1980).
[16] I. M. Sokolov, M. D. Kupriyanova, D. V. Kupriyanov, M. D. Havey Phys. Rev. A **79** 053405 (2009).

Investigation of coherent population trapping signals in ^{87}Rb cells with buffer gas

M I Fedorov[1], S V Ermak[1], M V Petrenko[2], E N Pyatyshev[3] and V V Semenov[1]

[1] Institute of Physics, Nanotechnology and Telecommunications, Peter the Great Saint Petersburg Polytechnic University, 29 Politekhnicheskaya str., St. Petersburg, 195251, Russia
[2] Ioffe Physical Technical Institute, 26 Politekhnicheskaya str., St. Petersburg, 194021, Russia
[3] Laboratory of Nano- and Microsystem Techniques, Peter the Great Saint Petersburg Polytechnic University, 29 Politekhnicheskaya str., St. Petersburg, 195251, Russia

E-mail: qfs-lab@mail.ru, m.petrenko@mail.ioffe.ru

Abstract. Characteristics of CPT (coherent population trapping) signal are investigated in small-size glass vapor cells containing ^{87}Rb and Ne buffer gas with narrow linewidth laser pumping on D_2 line. Parameters of CPT absorption signals are measured using small-size vapor cells with Ne buffer gas pressure in range of 200–400 Torr, cell temperature in range of 65–120 °C and values of laser pumping power of 30–400 μW/cm^2. Optimum conditions, under which the minimal value of short-term instability of resonance line is achieved, are obtained in experiments. CPT signals using vapor cells based on integrated technologies containing ^{87}Rb in atmosphere of Ne are also investigated. The CPT signals with typical linewidths of 2–3 kHz and signal-to-noise ratio of 1500 in 1 Hz bandwidth are observed, which allows one to provide relative frequency instability of 10^{-11} at 100 s.

1. Introduction

The development of modern methods of resonance signals inducing in optically oriented alkali atoms, such as the effect of coherent population trapping (CPT), and the progress in the establishment of coherent light sources with high stability have made possible the creation of a new generation of miniature quantum devices with optical pumping [1]. Such devices include miniature M*x* magnetometers [2] with sensitivity of 5 pT/Hz$^{1/2}$ and small-size quantum frequency standards based on alkali vapor cells (^{87}Rb, ^{133}Cs) with physical package volume less than 20 cm^3 and relative frequency instability at the level of 10^{-11} s$^{-1/2}$ [3-5]. Metrological characteristics of such devices are largely determined by the choice of the working alkali atoms, by the technology and the fabrication quality of the working vapor cell and by the parameters of the pumping source.

A necessary condition for the development of structure of miniature quantum devices with optical pumping is the use of relatively high pressures of inert gas (up to several atmospheres), injected into alkali vapor cells with the aim of reducing the relaxation effects in the working medium [6]. This inevitably causes the problem of the shift and broadening of resonance absorption line, which requires careful selection and adjustment of all elements of the physical package of a quantum device, which allows one to implement the minimum measuring error.

2. Experimental setup

In this work we investigated characteristics of CPT signal in vapor of ^{87}Rb isotope, which possesses simple structure of energetic levels in comparison with ^{133}Cs atoms. The use of neon (Ne) as a buffer gas is allowed, according to preliminary calculations, to ensure relative frequency instability of miniature atomic clocks of 10^{-11}–10^{-12} $s^{-1/2}$. The research was carried out on the universal experimental setup with laser pumping source, which made it possible to record CPT signals and signals of double radio-optical resonance. In the setup we provided for the heating of the working vapor cells in the range of temperatures of 20–120 °C and the modulation of the pumping source frequency up to 10 GHz. The functionality of the experimental setup also allow the wide range of investigations, related to the so-called light shifts of the resonance line in quantum device, which are one of the destabilizing factors of frequency standards [7-9].

We used the external cavity laser as a pumping source, which allowed to provide high stability of irradiation characteristics (spectral width less than 1 MHz) and high intensity (30–150 mW) in a wide range of adjustment. Photoreceiver served as a silicon photodiode with working area size 1×1 mm^2 and responsivity 0.1 A/W [10, 11]. We used the photodiode in short circuit mode, which excluded the influence of the dark current on the output signal when the bandwidth of registration is 0–20 kHz. The laser was tuned to the D_2 line of rubidium atoms (the wavelength is λ=780 nm) with adjustable polarization of the pumping radiation.

The scheme of experimental setup is presented on figure 1. A laser beam (L) was fed to an electro-optical phase modulator (EOM) modulated by a sinewave signal from a microwave generator (MWFG). At the EOM output, the first order spectral −1 and +1 components separated by twice the frequency of the modulating microwave signal appeared. The carrier component "0" was suppressed by 70% as compared with its intensity at the EOM input.

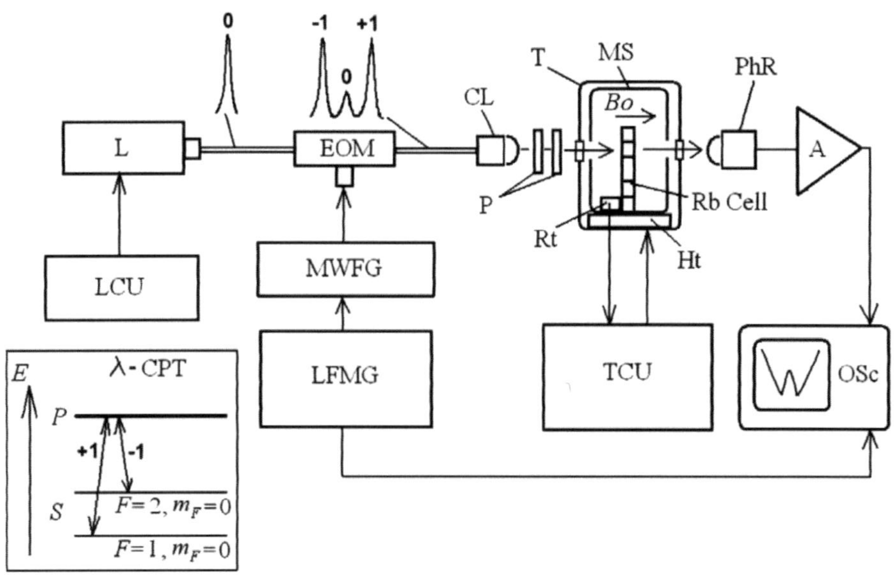

Figure 1. Experimental setup: L – laser, EOM – electrooptical modulator, MWFG – microwave generator, CL – collimator, P – polaroids, Rb Cell – cell with rubidium, PhR – photoreceiver, A – amplier, OSc – oscilloscope, T – thermostat, TCU – thermostat control unit, Ht – heater, Rt – thermistor, MS – magnetic shield, LCU – laser control unit, LFMG – low-frequency modulation generator, λ-CPT – a simplied structure of energy levels in rubidium in the case the CPT signal is produced according to the λ-scheme.

Then the light passed through a collimator (CL) and polaroids (P) (λ/2 and λ/4), was partially absorbed in the cell (Rb Cell) and was detected by a photoreceiver (PhR). After that the signal amplified by an amplifier (A) was fed to the oscilloscope (OSc) input.

When the difference between the frequencies of the first order spectral –1 and +1 components was equal to the resonant frequency of hyperfine transition of the ground state in ^{87}Rb atoms (6.834 GHz), so-called "dark" resonance appeared and the absorption of the laser light inside the vapor cell reduced. To select 0–0 magnetically independent transition, the cell was put into a uniform magnetic field B_0 (100 mG), which was parallel to the laser beam.

The cell temperature was stabilized with an accuracy of ±0.1 °C by a thermostat (T) that included a heating element (Ht) and a thermistor (Rt) that provided measurements of the cell temperature. The thermostat was controlled by a control unit (TCU). A magnetic shield (MS) screened the cell from external magnetic fields, the internal magnetic field B0 was produced by a built-in coil. A low-frequency modulation generator (LFMG) provided the modulation of the resonance conditions with frequency of 23 Hz for observation of CPT signals. A laser control unit (LCU) was used to stabilize laser current and temperature.

3. Coherent population trapping in small-size rubidium cells

In experiments we obtained the optimal values of the buffer gas pressure in the alkali vapor cell, the temperature of the cell and the intensity of the laser pumping radiation corresponding to maximal quality factor of CPT signal. We investigated the line shape of the resonance signals using small-size glass cylindrical cells with size of Ø3×3 mm, manufactured according to the conventional technology. The cells contained ^{87}Rb atoms with different pressures of buffer gas Ne (200, 300 and 400 Torr). Figure 2 shows the experimentally obtained dependence of the CPT signal from **f** – detuning the double modulation frequency of laser radiation relative the frequency of hyperfine transition, for the cells with temperature of T~80 °C, power density of circularly polarized laser radiation of 400 μW/cm^2 and signal registration bandwidth of 10 kHz. As can be seen from figure 2, increasing the buffer gas pressure more than 300 Torr leads to a decrease in the amplitude of the resonance signal. This, apparently, is connected with increase of collision frequency between alkali atoms and buffer gas atoms, which leads to depolarization effect. [12]

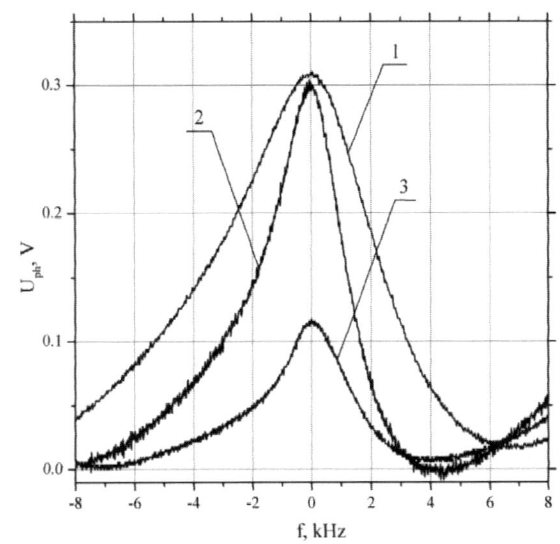

Figure 2. CPT signals at different neon pressures in the cell (1 – P_{Ne}=200 Torr, 2 – P_{Ne}=300 Torr, 3 – P_{Ne}=400 Torr).

Table 1. Parameters of CPT signal in dependence on pressure of Ne in the cell.

P_{Ne}, Torr	F, GHz	Δf, kHz	$(S/N)_{1\,Hz}$	Q	σ, $s^{-1/2}$
200	6.834733	5.90	3900	1.16×10^{6}	2.21×10^{-10}
300	6.834764	3.00	3800	2.28×10^{6}	1.16×10^{-10}
400	6.834780	2.89	2900	2.36×10^{6}	1.46×10^{-10}

The parameters of CPT signals for the vapor cells containing buffer gas Ne with pressure values of 200, 300 and 400 Torr are presented in table 1, where P_{Ne} – neon pressure, F – frequency of hyperfine transition, Δf – linewidth of the CPT resonance, $(S/N)_{1Hz}$ – the signal-to-noise ratio in the 1 Hz band, Q – the quality factor of the resonance equal to the ratio between the transition frequency and resonance linewidth and σ is the short-term instability of CPT signal. The minimal value of short-term instability was $\sigma=1.16\times10^{-10}$ $s^{-1/2}$. This value was observed using the cell with Ne pressure of 300 Torr with signal-to-noise ratio of 38 in registration bandwidth 10 kHz and with resonance linewidth of 3 kHz. The estimation of the short-term instability of resonance lines was performed according to the equation (1) [12].

$$\sigma=\left[\left(\frac{S}{N}\right)_{1Hz}\times Q\right]^{-1} \quad (1)$$

The increase in short-term instability at a pressure of neon of P_{Ne} = 400 Torr is caused by a decrease in the amplitude of the test signal, which reduces the signal-to-noise ratio. At the same time the large amount of instability when P_{Ne} = 200 Torr is due to the linewidth of the CPT resonance (see table 1).

In accordance with figure 2 the optimal cell with buffer gas by a factor of the linewidth is the cell with the pressure P_{Ne} = 400 Torr. Waveforms of the CPT signals at the intensity of the pumping radiation 180 μW/cm^2 and the cell temperature T = 65–85 °C, presented in figure 3, show a slight increase in linewidth with increasing temperature. At the same time the short-term frequency instability of the CPT signal (figure 4), corresponding to the so-called flicker limit of the quantum frequency standard, $\sigma = 4.69\times10^{-11}$ $s^{-1/2}$ is achieved at the temperature of the working cell T~75 °C.

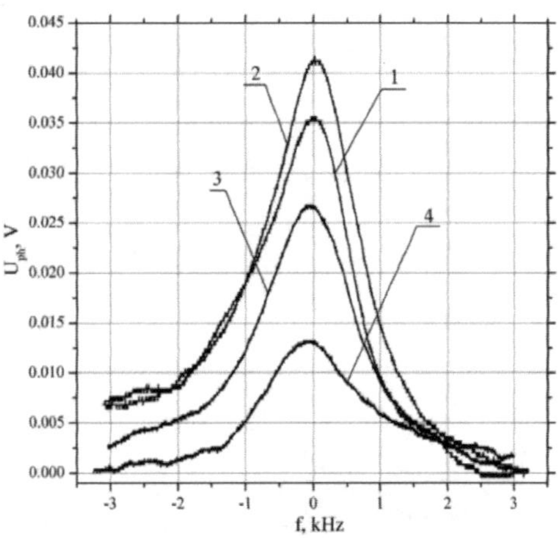

Figure 3. CPT signals at different temperatures of the cell with neon pressure P_{Ne}=400 Torr (1 – T=68.4°C, 2 – T=76.3°C, 3 – T=81.6°C, 4 – T=84.2°C).

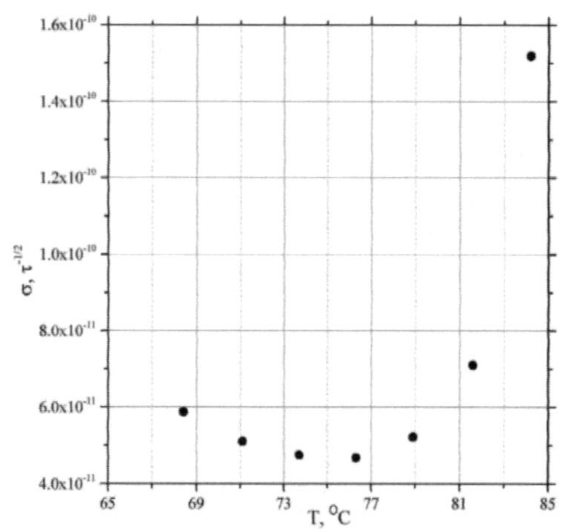

Figure 4. The dependence of a short-term instability estimation on the cell temperature at neon pressure in the cell P_{Ne}=400 Torr.

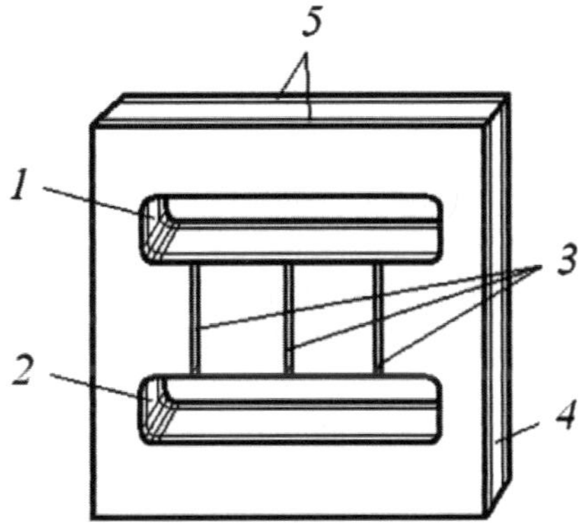

Figure 5. Microfabricated cell: 1, 2 – working and additional cavity, respectively, 3 – connecting channels, 4, 5 – silicon and glass plates, respectively.

From experimental data it follows that the intensity of the CPT signal is directly proportional to the intensity of pumping radiation, indicating the absence of saturation effects, while the linewidth decreases in proportion to the pumping intensity, reaching values of the order of 1 kHz when the laser power density is 30 $\mu W/cm^2$. When the pumping power density is 60 $\mu W/cm^2$, the optimal value of the parameter $\sigma = 4.49\times10^{-11}$ $s^{-1/2}$ is achieved.

4. Investigation of microfabricated cells

As it was shown the cell quality strongly depends on the amount of alkali metal, the buffer gas pressure and composition, and the presence of impurities in the cell and directly affects the stability and reproducibility of the characteristics of frequency standards and atomic magnetometers. The most appropriate technology that can ensure a high quality of cells for such devices is the micro electro-mechanical systems (MEMS) manufacturing technology. It combines the advantages offered by the techniques of microelectronic component fabrication and manufacturing and assembling miniature mechanical systems [13].

For the studies, cells with a universal construction (figure 5) were fabricated [5]. The additional cavity 2 was necessitated by the use of the method of recovery of an alkali metal from the rubidium dichromate salt including interaction of a material with an activating laser radiation [14]. Channels 3 provided, due to a small cross-section, a transfer of rubidium atoms into the working cavity without byproducts formed during recovery of alkali vapors.

The CPT signals with a typical linewidth of 2–3 kHz and a signal-to-noise ratio of 1500 in the 1 Hz bandwidth were observed in the experiments using chip-scale microfabricated rubidium cell, described above, with a working cavity volume of 1.2 mm^3 manufactured using the MEMS technique. By substituting the resonance signal parameters into expression (1), we obtain the estimate of the short-term frequency instability $\sigma = 1.4\times10^{-10}$ $s^{-1/2}$, which allows one to provide relative instability of 10^{-11} at 100 s.

5. Summary

The result of this work can be used in the development of miniature frequency standards and quantum magnetometers with laser pumping. The studies of CPT signals using small-size vapor cells with linear dimensions of the cavity on the order of 1 mm allowed us to obtain optimal parameters of the pumping intensity, buffer gas pressure and the cell temperature. The results showed that by adjusting the parameters of the physical package of a quantum device the values of frequency instability can be reached at the level of $\sigma = 5\times10^{-11}$ $s^{-1/2}$.

The results obtained in studies of parameters of resonance CPT signals from microfabricated cell manufactured using the MEMS technique suggest that this technique is efficient for the development of chip-scale quantum devices with laser pumping and frequency instability of $\sigma = 1.4\times10^{-10}$ $s^{-1/2}$.

Journal of Physics: Conference Series **769** (2016) 012046 doi:10.1088/1742-6596/769/1/012046

References

[1] Alzetta G, Gozzini A, Moi L and Orrios G 1976 *Nuovo Cimento* B **36** 5

[2] Schwindt P D, Lindseth B, Knappe S, Shah V Kitching J and Liew L A 2007 *Appl. Phys. Lett.* **90** 081102

[3] Kitching J, Knappe S and Hollberg L 2002 *Appl. Phys. Lett.* **81** 553

[4] Knappe S, Shah V, Schwindt P D, Hillberg L, Kitching J, Liew L and Moreland J 2004 *Appl. Phys. Lett.* **85** 1460

[5] Ermak S V Semenov V V, Piatyshev E N, Kazakin A N, Komarevtsev I M, Velichko E N, Davydov V V and Petrenko M V 2015 *St. Petersburg Polytechnical University Journal. Physics and Mathematics* **1** 37

[6] Riehle F 2004 *Frequency Standards: Basics and Applications* (Weinheim: WILEY-VCH Verlag GmbH & Co. KGaA)

[7] Semenov V V, Baranov A A and Ermak S V 2011 *Proc. of Joint Conf. of the IEEE IFCS & EFTF* (San Francisco: IEEE) p 580

[8] Baranov A A, Semenov V V and Ermak S V 2012 *St. Petersburg Polytechnical University Journal. Physics and Mathematics* **141** 152

[9] Baranov A A, Ermak S V and Semenov V V 2013 *Opt. Spectrosc.* **114** 337

[10] Zabrodsky V V, Belik V P, Aruev P N, Ber B Y, Bobashev S V, Petrenko M V and Sukhanov V L 2012 *Tech. Phys. Lett.* **38** 812

[11] Artyomov A P, Baksht E H, Tarasenko V F, Fedunin A V, Chaikovsky S A, Aruev P N, Zabrodskii V V, Petrenko M V, Sobolev N A and Sukhanov V L 2015 *Instrum. Exp. Tech.* **58** 102

[12] Knappe S 2007 *Comprehensive Microsystems* **3** 571

[13] Pyatyshev E N, Lurie M S, Popova I V and Kazakin A N 2001 *J. Nano Microsyst. Tech.* **6**, 32

[14] Hasegawa M, Chutani R, Gorecki C, Boudot R, Dziuban P, Giordano V, Clatot S and Mauri L 2011 *Sensors and Actuators* A **167** 594

Analysis of tooth tissues using Raman spectroscopy

E.V. Timchenko[1], P.E. Timchenko[1], L.T. Volova[2], A.Yu. Rosenbaum[2], A.Yu. Kulabukhova[1]

[1]Samara National Research University.
Samara

[2]Institute of Experimental Medicine and Biotechnologies. Samara State University.
Samara

E-mail: laser-optics.timchenko@mail.ru, timpavel@mail.ru, csrl.sam@mail.ru

Annotation. The results of experimental studies of healthy tooth tissue and tooth tissues during caries disease are presented. Features of Raman spectrum of tooth tissues during caries disease are obtained: the main changes are detected at wavenumbers 956 cm^{-1}.1069 cm^{-1}, corresponding to phosphates, and 1241 cm^{-1}, 1660 cm^{-1}, corresponding to collagen III and collagen I, respectively. Were introduced criteria allowing to detect caries and to identify weakening of tooth tissues, preceding the caries. The reliability of research results is confirmed by scanning electron microscopy.

1.Introduction

In dentistry a significant share occupy problems that associated with diseases and lesions of tooth tissues. The most common of them is caries, which is irreversible microbial disease that affects calcined tooth structure. [12]. Initially, it affects enamel and remains asymptomatic until it reaches the dentin and pulp. [3] At the same time it is known that the first signs of dental caries, which lead to a weakening of tooth tissue and can prevent its development, is very difficult to diagnose.

At present, are widely used non-destructive optical methods for assessing teeth. Among them is widespread Raman spectroscopy (RS) method, which is simple and requires no sample preparation. [4] So in [5-6] was assessed parallel-polarized and cross-polarized (polarization direction of return perpendicular to the polarization direction of transmission) Raman spectra at 959 cm^{-1} intensity. In this article concluded that these introduced ratios of polarization can be used to assess the degree of demineralization.

In the study of authors [7] was shown that using Raman spectroscopy it is possible to assess mineral component of tooth tissue. In this work the carious areas of enamel are compared with healthy areas of the tooth at wavenumbers 960 and 880 cm^{-1} corresponding to $(PO_4)^{3-}$, and $(CO_3)^2$, respectively, since these peaks clearly shows the difference between healthy and carious areas of tooth. It also was experimentally proved that the enamel has a high concentration of mineral substances.

The author of work [8] studied the structural features of hard tooth tissues during cariosity, during non-carious lesion at the example of a generalized pathological abrasion of teeth. Also were studied pulpless teeth using infrared Raman spectroscopy. Was studied ratio of CO_3/ PO_4, which shows a decrease of crystallinity by substituting phosphate-ion with carbonate-ion and mineral/organic matrix. It was established that the greatest ratio of CO_3/PO_4 observed in samples of enamel and dentin of pulpless teeth.

In dentistry, there are a number of devices for diagnosis of dental diseases. For example, the scanner ORTHOPHOS XG 3D, which has a high resolution (160/100μm.), which allows an accurate diagnosis

of caries sequelas and to determine effective methods and tactics of canals treatment, but when examining by device there is irradiation of both patient and physician. [9]
Apparatus KaVo DIAGNOdent, on the basis of fluorescence method allows to detect caries in the early stages, however, its significant disadvantage is a false triggering, for example, a reaction to dental plaque [10].
And device SIROInspect, also on the basis of fluorescence method can quickly and easily detect caries lesions during excavation process, but the accuracy depends on the doctor's professionalism [11].
The objective of this work is application of Raman spectroscopy method to diagnose caries and weakening of tooth tissues in its early stages.

2. Materials and methods

As objects of the study were used 11 samples of large and small molars of patients with a diagnosis of "caries" (K02).
Raman spectroscopy method was implemented using experimental stand, previously described in [12].
Measurements were carried out on the specified lines in Figure 1 at power in healthy area of 300 mW and 100 mW in carious area and exposure time of 20 seconds. The error of Raman spectroscopy method amounted to 3%.
For a comparative analysis of chemical composition was used a scanning electron microscope (SEM) JED – 2300 AnalysisStation(Japan).
As a studied object were different parts of tooth areas (enamel, dentin, cement), both in normal and in case of caries.

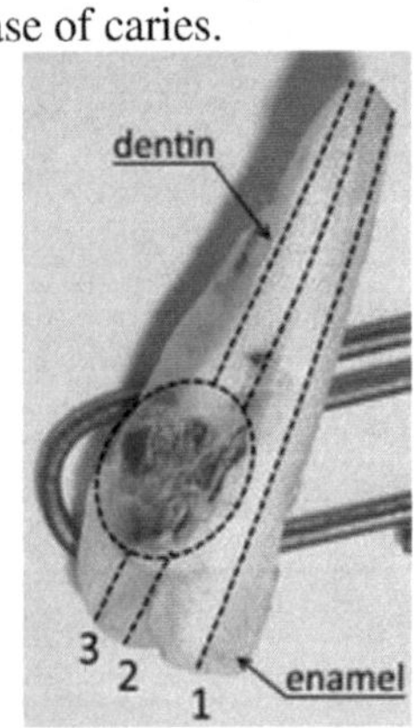

Figure 1. Studied tooth: a tooth-section - 1 line - dentin, enamel 2,3 line (see from the bottom upwards) – dentin, enamel. There is a carious lesion in dentin and enamel in the tooth's bulbous portion.

3.The results of research

Fig. 2 presents characteristic Raman spectra of tooth tissues obtained as a result of studies in lesioned and healthy areas of tooth.

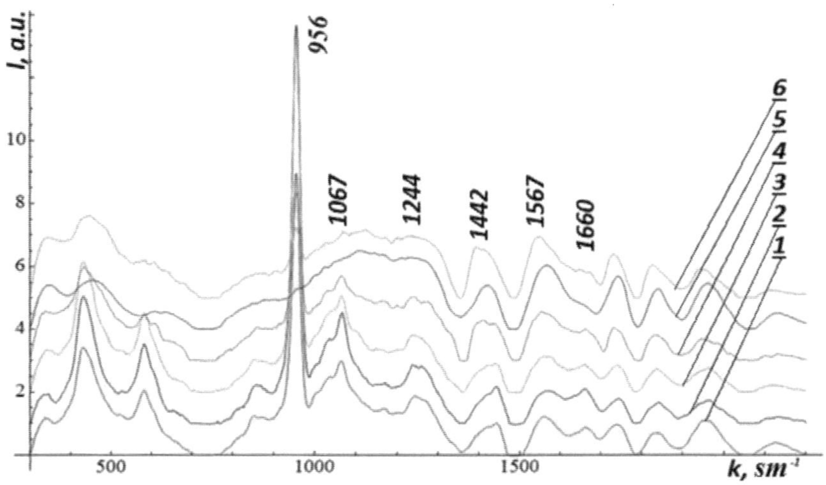

Figure 2. Raman spectra of healthy samples: cement (1), dentin (2), enamel (3) and lesioned by caries cement (4), dentin (5), enamel (6).

From Figure 2 we can observe spectral differences between various tissues of healthy teeth (1-3 notation), for example, dentin, compared with other tooth tissues, is characterized by gaining the line at wavenumber 1069 cm^{-1}, corresponding to oscillations of ν_1 $(CO_3)^{2-}$ substitution B-type where this group replaces phosphate and has a lower hardness [13]. Also in dentin are appear lines 1271 cm^{-1} and 1660 cm^{-1}, corresponding to collagen III and collagen I, respectively. The almost complete lack of these lines in enamel is explained by the fact that enamel is the the hardest tissue in the human body due to the high content of inorganic substances (to 97%), namely hydroxyapatite.

Mineral component of tooth is characterized by vibrational modes ν_1, ν_3 and ν_4 $(PO_3)^{4-}$ in hydroxyapatite molecule, Raman lines of which were registered at wavenumbers 581, 647, 955, 1038 and 1169 cm^{-1} [14,15], and by vibration modes of ν_1 $(CO_3)^{2-}$ substitutions B-type, where $(PO_4)^{3-}$ is substituted by $(CO_3)^{2-}$, and A-type [15], where OH^- is substituted by $(CO_3)^{2-}$ at 874, 1067, 1100 and 1435 cm^{-1} wavenumbers, respectively. The collagen component of Raman lines in addition to proline and hydroxyproline is represented by groups of amide III (in 1241-1271 cm^{-1} range) [14,16], amide II (in 1520-1580 cm^{-1} range) [14] and amide I (in 1630-1680 cm^{-1} range) [14,16].

As can be seen from the figure during lesions of tooh tissues with caries (designations 4-6) the main changes are observed in decrease of lines at 431, 581, 956, 1041 cm^{-1} wavenumbers, corresponding to phosphates and 1660 cm^{-1} corresponding to collagen I.

For samples with an initial stage of caries and a weakened mineral component spectra are transient. Initial changes in spectrum primarily associated with both the remineralization of hard tooth tissues and the loss of collagen I.

For diagnosis of caries development process and weakening of hard tooth tissues (enamel, dentin, cement, pulp) were introduced following coefficients, constructed two-dimensional dependences, shown in Figure 3:

$$D = I_{956}/I_{1069}$$
$$K = I_{1660}/I_{1241}$$

where Ii - intensity of line at the wavenumber;

D - coefficient proportional to the change of content of calcium carbonate-ions in tooth tissue (substitution of ions $(PO_4)^{3-}$ to $(CO_3)^{2-}$, B type);

K- coefficient characterizing collagen I loss compared to collagen III.

Journal of Physics: Conference Series **769** (2016) 012047 doi:10.1088/1742-6596/769/1/012047

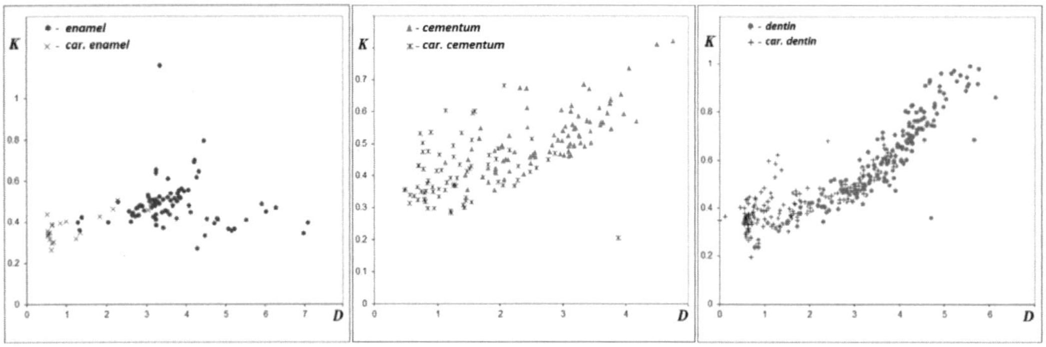

a) b) c)

Figure 3. Two-dimensional dependence of coefficients D and K for area of a) healthy enamel 1 and carious enamel 2; b) healthy cement 1; c) healthy dentin 1 and dentin with caries 2.

Fig. 3 shows a two-dimensional dependence of coefficients in different areas of tooth tissue. The figure shows that caries is characterized by a decrease of optical coefficients, due to a decrease of ions $(PO_4)^{3-}$ and collagen I in tooth tissues. In pathologic processes such as caries, there is a shift towards the replacement of B-type. Loss of collagen I is most clearly visible for dentin, as in cement and enamel it's initially small. Thus, Two-dimensional dependences in Fig. 3 allow to determine not only the type of tooth tissue, but also the presence of pathologic changes.

As a control method of analysis was used raster electron microscopy (REM). Fig. presents the microphotographs of the surface structure of healthy and carious tooth tissue samples.

According to REM data about element composition (Fig 4) the caries samples are characterised by the reduced content of phosphorus and calcium and increased content of carbon and nitrogen, which confirms the obtained results and the reliability of the introduced coefficients.

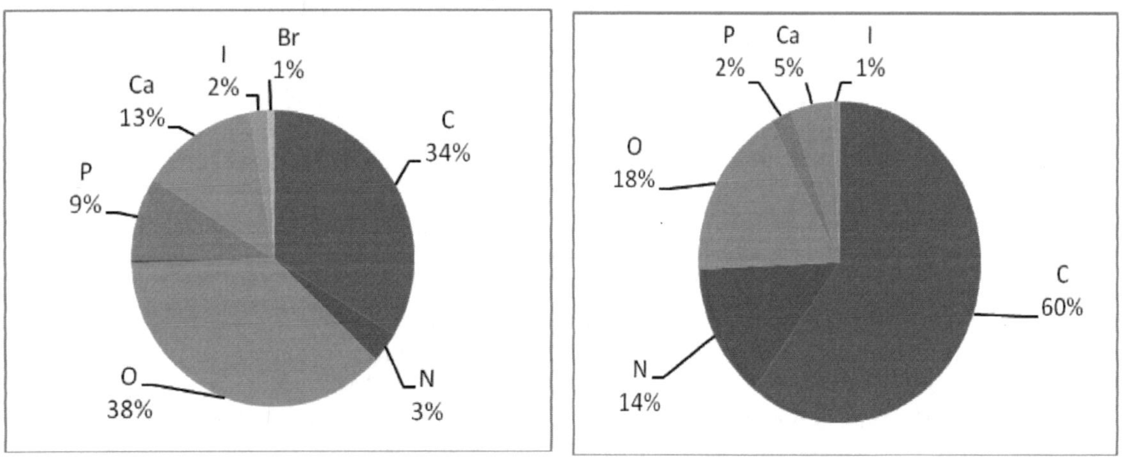

a) b)

Figure.4.The raster electron microscopy: a) - a healthy area of dentin; b) - a dentine surface with a diagnosis of caries.

4.Conclusion

Features of Raman spectrum for healthy and carious tooth tissues were obtained. It is established that caries is characterised by changing of spectrum at 956 cm^{-1},1069 cm^{-1} wavenumbers, corresponding to phosphate, and 1241 cm^{-1}, 1660 cm^{-1}, corresponding to collagen III and collagen I.

Two-dimensional analysis of introduced optical coefficients D and K, which allows to diagnose the beginning of development of caries in hard tooth tissues, were carried out. It is established that caries is characterised by decreasing of coefficients due to decreasing of $(PO_4)^{3-}$ions and collagen I in tooth tissue. Reliability of research results is confirmed by scanning electron microscopy.

The work was supported by the Ministry of Education and Science

References:

[1] B.A. Dye, S. Tan, V. Smith, B.G. Lewis, L.K. Barker, G. Thornton-Evans, P.I. Eke, E.D. Beltrán-Aguilar, A.M. Horowitz, C.H. Li N 2007Trends in oral health status : United States, 1988-1994 and 1999-2004 (Vital Health Stat;(248)) p 1-92

[2] R. Ramakrishnaiah, G. Rehman, S. Basavarajappa, A. A. Al Khuraif, B.H. Durgesh, A. S. Khan &IhteshamurRehman2015Applications of Raman Spectroscopy in Dentistry: Analysis of Tooth tissue (Applied Spectroscopy Reviews. –Volume 50, Issue 4)p 332-350

[3] RavikumarRamakrishnaiah, GhufranurRehman, SanthoshBasavarajappa, Abdulaziz Abdullah Al Khuraif, B. H. Durgesh, Abdul Samad Khan &IhteshamurRehmanApplications of Raman Spectroscopy in Dentistry: Analysis of Tooth tissue p 332-350

[4] M. Miyazaki, H. Onose, B. Moore V 2002Analysis of the dentin–resin interface by use of laser Raman spectroscopy(Dent. Mater. 18.) p 576–580

[5] A. C.-T. Ko, M. Hewko, M.G. Sowa, C. C. Dong, B. Cleghorn, and Lin-P'ingChoo-Smith A 2008Early dental caries detection using a fibre-optic coupled polarization-resolved Raman spectroscopic system(Opt Express; 16(9))p 6274–6284

[6] A. C.-T. Ko, Lin-P'ingChoo-Smith, M. Hewko, M. G. Sowa, C. C.S. Dong, and B. Cleghorn2006Detection of early dental caries using polarized Raman spectroscopy (Vol. 14) p 203-215

[7] S. Yang, B. Li, A. Akkus, O. Akkus, L. Lang Wide-Field Raman Imaging of Dental LesionsSchool of Dental Medicine(Case Western Reserve University, Cleveland, OH, 44106, USA. Raman Spectroscopy)

[8] A.S Ivanov Application of micro Raman spectroscopy to study the structural characteristics of hard tissue of teeth (Department propaedeutics and physiotherapy of dental diseases USMA)

[9] ORTHOPHOSXG 3D: самая популярная в мире рентгеновская система(SIRONE. The Dental Company)

[10] Laser Fluorescence and Quantitative Light Fluorescence(The WetpaintCariology Project)

[11] SIROInspect безопасный и надежный мониторинг кариеса(SIRONE.The Dental Company)

[12] E.V. Timchenko, P.E. Timchenko, L.T. Frolov Yu.V. Ponomareva, L.A. Taskina2014Issledovanie organomineralnoj structure of bone implants by Raman scattering. (Kvant.elektronika, 44 (7)) p 696-699

[13] AA Dolzhikov 2007 Histology of tooth tissues (Textbook for independent work of students of the Faculty of Dentistry, Belgorod State University, Faculty of Medicine, Department of Anatomy and Histology cheloveka.- Belgorod)

[14] R. Kumar, G. P. Singh, K. M. Grønhaug, N. K. Afseth, C. de Lange Davies, J. O. Drogset and M. B. LilledahlSingleCellConfocalRamanSpectroscopy of Human Osteoarthritic Chondrocytes: A Preliminary Study(InternationalJournalofMolecularSciences — OpenAccessJournal)

[15] I. Rehman, Z. Movasaghi, S. Rehman 2012Vibrational Spectroscopy for Tissue Analysis (Series in Medical Physics and Biomedical Engineering. CRC Press)p 271

[16] G. Mandair, M. Morris V 2015Contributions of Raman spectroscopy to the understanding of bone strength (BoneKEy Reports) p 620

Investigation of quantum efficiency of A3B5 laser heterostructure after FIB milling

M I Mitrofanov[1], **V P Evtikhiev**[2]

[1] St.Petersburg Polytechnic University, 29 Politekhnicheskaya, St Petersburg, 195251, Russian Federation
[2] Ioffe Institute, 26 Politekhnicheskaya, St Petersburg, 194021, Russian Federation

maxi.mitrofanov@gmail.com, evtikhiev@mail.ioffe.ru

Abstract. Al-containing layers oxidization was eliminated using UHV FIB system and thus, only ion implantation and amorphisation influence on quantum efficiency of A3B5 laser heterostructures should be investigated.

1. Introduction

Focused Ion Beam (FIB) nanofabrication is one of the fundamental techniques in research, and production development areas. FIB milling is widely used in integrated photonics as a rapid and flexible prototyping method for structures consisting of Si planar waveguides or Si photonic crystal based elements, because it is maskless direct etching on nanoscale [1].

The most important problem in silicon photonics is an integration of light emitters into planar photonic chip [2]. Hybrid technology like wafer bonding based on Si substrate is not a suitable solution for small and reliable active elements like laser or light emitting diodes [3]. Otherwise A3B5 laser heterostructures are used for small (less than 10 μm) integrated light emitters [4]. In application to A3B5 semiconductors FIB etching is used for manufacturing microresonators [5], subwavelength gratings [6], near-field couplers [7] and etc. However, FIB milling process is not suitable for fabrication of light emitters because of etching-induced defects such as, ion implantation, material amorphisation and deep oxidization of A3B5 semiconductor material. Oxidization appears because of low vacuum in a standard FIB chamber and high etching temperature. Both types of defects lead to crystal imperfections in the active layer of the structure, and therefore to nonradiative carrier recombination and low quantum efficiency of light emitters.

2. Experiment

2.1. Milling of laser heterostructure.

In order to investigate high energy ions influence on quantum efficiency we decided to conduct an experiment on molecular beam epitaxy grown $Al_{0.4}Ga_{0.6}As$/GaAs laser heterostructure with $In_{0.2}Ga_{0.8}As$ quantum well (fig. 1 a). To prevent oxidization of Al containing layers we used originally designed ultra-high vacuum (UHV) cross-beam FIB/SEM (scanning electron microscopy) system equipped with Orsay Physics Ga ion and electron columns. We produced trenches on the laser structure with different depths: 120, 250 and 550 nm with exposure time 2, 4 and 10 ms correspondingly. Measured trenches width is close to 100 nm. SEM image of the cross section for depth measuring is shown on figure 1, b. Two trenches are located over the quantum well, and the third trench crosses it (fig. 1 a). Beam parameters were 30 kV acceleration voltage and 50 pA probe current. Optical characteristics were investigated using micro photoluminescence (PL) technique with 671 nm pump lasing wavelength focused by 0.5A objective in 40 μm spot diameter.

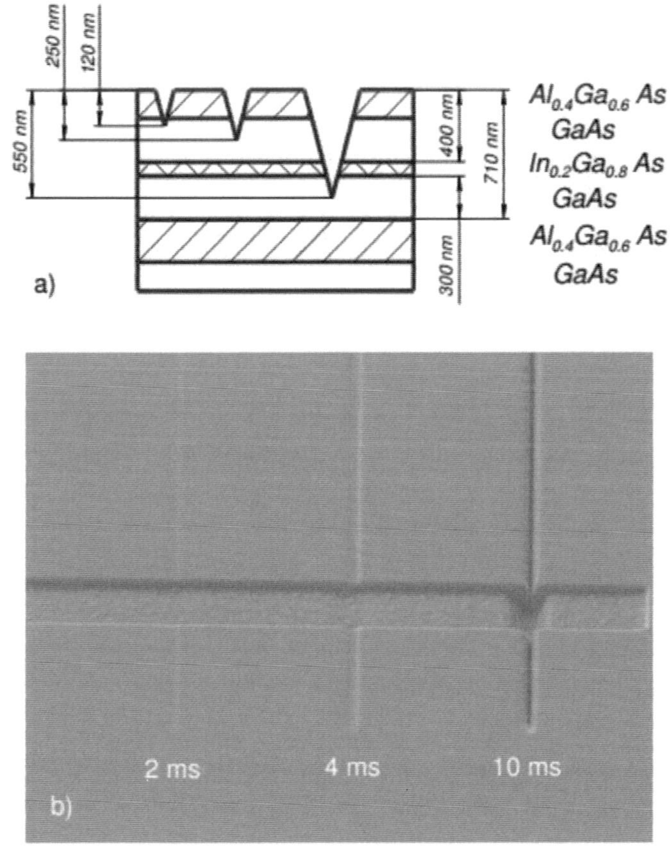

Figure 1. a – laser heterostructure scheme, b – etched trenches with 120, 250 and 550 nm depth, SEM image.

2.2. Measurement results and discussions

Figure 2 represents spatial distribution of micro photoluminescence (PL) intensity from the bulk structure and etched trenches. Experimental data is represented without mathematical processing and asymmetry of curves appears from aberrations of the optical system. The luminescence intensities from trenches 1 and 2 are close to each other and discern from trench 3. Assuming that luminescence from bulk structure is 100 % then trenches 1 and 2 have close to 97 % intensity and 3rd trench has 93 %. Spot of excitation is larger than trench width and radiation was collected from all layers, hence this graph is mostly qualitative. For quantitative examination the luminescence spectra were measured (figure 3). Low energy peak corresponds to luminescence from quantum well and high energy peak from waveguide layers. From this graph we clearly see that induced during FIB etching structural damages in quantum well lead to radical decrease of quantum efficiency, 60 % in comparison to the bulk structure. However, the light intensity decrease from trenches 1 and 2 (96 % and 93 %, correspondingly) is observed only in waveguide layer. Luminescence suppression by the shape of the etched surface due to the total internal reflection is not significant and hence not discussed in detail.

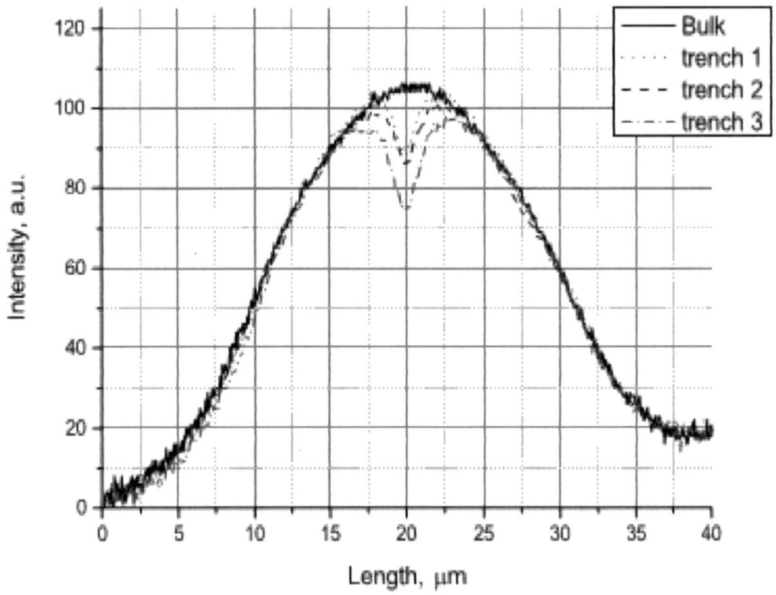

Figure 2. Spatial distribution of micro PL intensity. Bulk – 100 %, trench 1 and 2 – 97 %, trench 3 – 93 %.

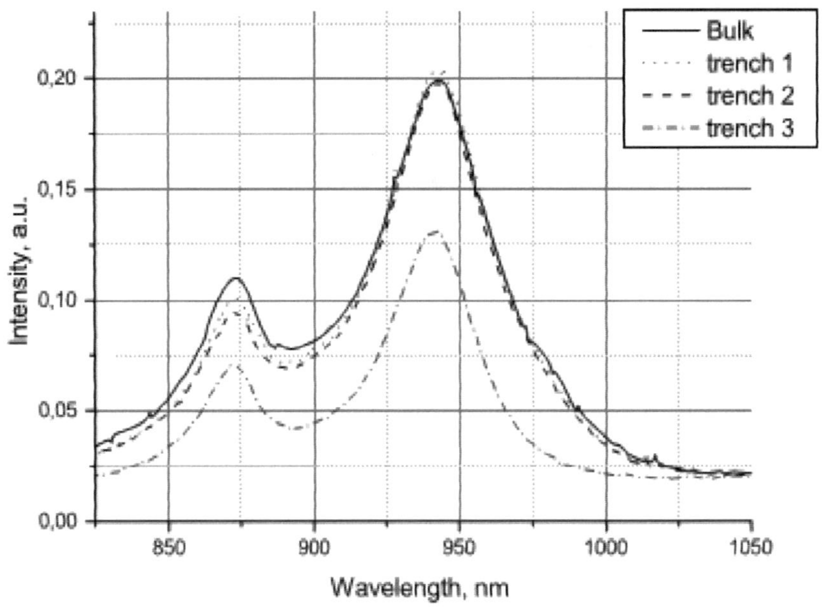

Figure 3. Micro PL spectra, intensity from bulk – 100 %, trenches: 1 – 96 %, 2 – 93 %, 3 – 60 %.

3. Conclusion

Several trenches on A3B5 laser heterostructure were etched using the UHV FIB system. Influence of radiation defects on luminescence efficiency of quantum well was investigated. As the result we can state that typical laser heterostructure can be etched in oxygen less atmosphere by 30 keV Ga FIB up to quantum well layer without decreasing of luminescence efficiency. Considering that width of FIB milling in A3B5 materials is less than 100 nm, this approach can serve as technological basic for light emitting elements in integrated nanophotonics.

References

[1] Ridder R M, Ay F, Kauppinen L J 2008 *Proc. of the First International Workshop on FIB for Photonics (Eindhoven, 2008)* University of Twente Enschede
[2] Liang D, Bowers J E 2010 *Nature photonics* **4** 511
[3] Fang A W et al 2008 *IEEE Photon. Tech. Lett.* **20** 1667
[4] Kryzhanovskaya N V et al 2012 *Semiconductors* **46** 1063
[5] Mialichi J R et al 2010 *ECS Trans.* **31** 289
[6] Pozo J et al 2006 *Proc. of SPIE* 6184
[7] Liu X et al 2009 *Proc. of NEMS 2009 (Shenzhen 2009) IEEE* 1000

Testing of Raman spectroscopy method for assessment of skin implants

E V Timchenko[1], P E Timchenko[1], L T Volova[2], S V Pershutkina[1], P Y Shalkovskaya[1]

[1] SamaraUniversity,
443086 Russia, Samara, Moscow highway, 34
[2] Institute of Experimental Medicine and Biotechnology (IEMB)
443079 Russia, Samara, ul. Gagarin, 20

E-mail: pshalkovskaya@mail.ru

Abstract. Results of studies of testing of Raman spectroscopy (RS) method for assessment of skin implants are presented. As objects of study were used samples of rat's skin material. The main spectral differences of implants using various types of their processing appear at wavenumbers 1062 cm^{-1}, 1645 cm^{-1}, 1553 cm^{-1}, 851 cm^{-1}, 863 cm^{-1}, 814 cm^{-1} and 1410 cm^{-1}. Optical coefficients for assessment of skin implants were introduced. The research results are confirmed by morphological analysis.

1. Introduction

Due to the high prevalence of thermal injuries and high mortality is still relevant problem of improving the results of treating patients with burn wounds [1]. The problem of restoring humans' defects of skin material is one of the most important problems of the modern world. [2]

For treatment of skin tissue burn defects are used different biological materials. Active development of biotechnologies resulted in the creation of entirely new regenerative technologies that allow to recover biological tissues, called biomatrix. For today, there is still no optimal method of their processing and control of effectiveness.

In studying the effectiveness of the method of manufacturing biomatrix is worth considering several major factors affecting the quality of the resulting implant. These include immunogenic properties of biomatrix caused by epidermal cells and endothelial cells in the dermis. The processed tissues must not contain cells, including cell components: cytoplasm and nucleuses. Their presence in biomatrix can promote a violation of cell biocompatibility of invitro and cause side effects in invivo conditions during subsequent processing. [3]

In the manufacture of skin biomatrix the treatment of samples is carried out by different methods [2]. However, the processing may be followed by a decrease of glycosaminoglycan's, proteoglycans and glycoproteins, which are part of biomatrix's intercellular substance that can lead to implant rejection [4].

At the same time, the loss of glycosaminoglycans leads to loss of turgor (elasticity) of material [5], which can also adversely affect the engraftment of implant.

Therefore, complex assessment of processing of skin biomatrixes is a highly urgent problem. There are number of methods to control the quality of biomatrixes, these include biochemical analysis (cytological [6, 7], histological [8]). However, this analysis is very time consuming and destructive [6]. Therefore, for assessment of skin biomatrixes it is necessary to use optical methods of control that are not destructive.

Raman spectroscopy [9, 10, 11] due to its simplicity, efficiency and non-invasiveness is suitable for quality control of results of tissue engineering [12, 13, 14].

In article [11] was used Raman spectroscopy method, as the optimal method for studying biological tissues. In the article were studied healthy and with tumor mass skin samples of rat mammary. Analysis of the principal components of matrix was carried out in 900 - 1790 cm^{-1} range of wavenumbers, typical for collagen, proteins, lipids, DNA, and amide I. The authors of work established that cancerous skin tissues have a higher protein concentration than the normal ones.

In article [14] this method was applied to assess rat's skin carcinoma and normal skin. Carried out analysis revealed that significant changes between tissues appear in the 1200-1600 cm^{-1} range, Amide III band including, δCH2, amide II, and in the range of wavenumbers above 2,800 cm^{-1} for CH2, CH3 and OH stretchings.

Main changes in spectra were observed in bands of collagen corresponding to line 1328 cm^{-1}, changing of which reveals the presence of pathology of skin.

Objective of the work - to carry out studies on the possibility of using Raman spectroscopy method for assessing skin implants.

2. Materials of research

As objects of study were used samples of rat's skin material subjected to 2 different types of processing: thermochemical treatment with incomplete epithelialization and chemical treatment of the skin tissue using "Lioplast"® (ТУ-9398-001-01963143-2004) technology.

Control of processing of skin matrix was carried out using an experimental stand shown in Fig. 1, which includes a high-resolution digital spectrometer Shamrock sr-303i with a built-in cooling chamber DV420A-OE, a fiber-optic probe for Raman spectroscopy (RS) RPB-785, combined with laser module LuxxMaster LML-785. 0RB-04 and coordinated with the laser Raman probe.

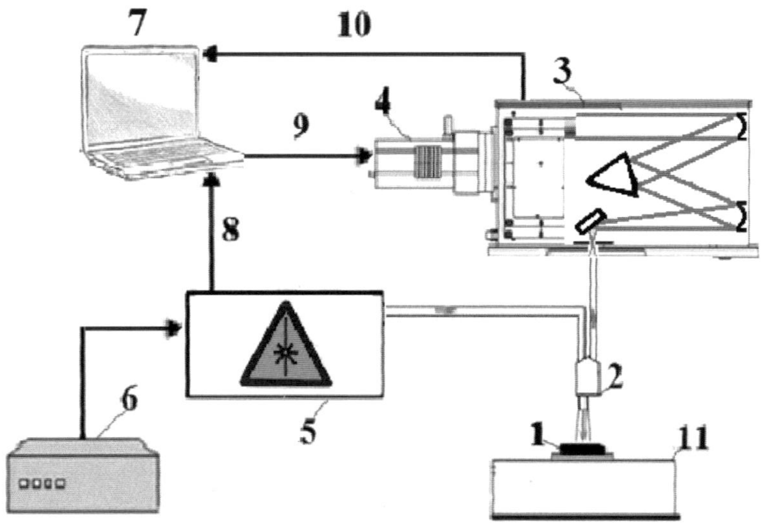

Figure 1. Experimental stand that implements RS method: 1 – the studied object; 2 - Raman Probe RPB785; 3 - spectrometer Shamrock sr-303i; 4 - Built-in cooling chamber DV420A-OE; 5 - Laser Module LuxxMasterRamanBoxx; 6 - power supply of the laser module; 7 - computer; 8, 9, 10 - Information electric cables; 11 - coordinate table

Built in broadband filter probe 2 is designed to isolate radiation in the studied spectral range, which is then sent via fiber to the spectrometer 3 with an integrated camera 4. The coordinate table 11 with stepper motor allows to scan the sample dimensionally with steps 1-1, 5 mm. Processing of the spectra was carried out in WolframMathematica 8 program. As a result of test experiments was selected laser power of 100 mW.
Also was calculated error of the method, which amounted 10, 1%

3. Results and discussion

Fig. 2 shows averaged Raman spectra of skin implants samples. Registered Raman bands correspond to vibration modes, which are represented in the table 1.

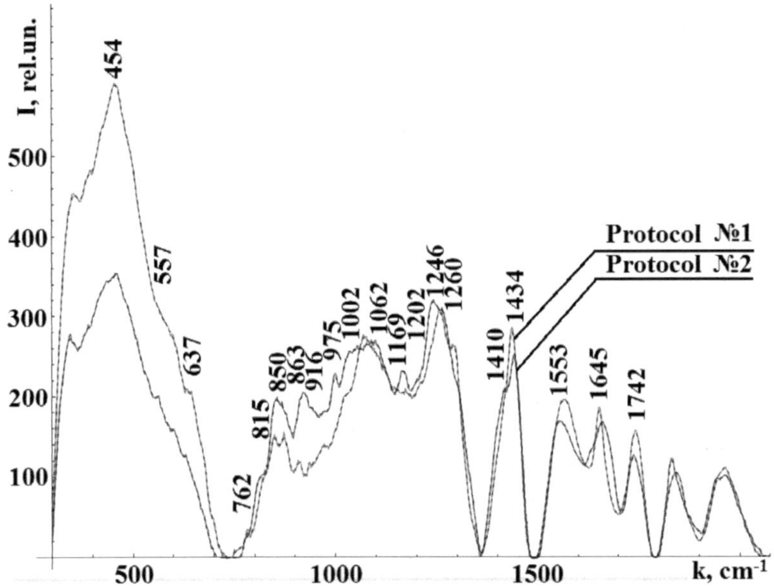

Figure2. Averaged Raman spectra of skin implants. №1, 2 - protocols of skin biomatrixes (1-chemical treatment, 2 –thermochemical treatment

Table 1. Explanation the Raman spectra of skin matrix

Wavenumber cm^{-1}	Fragment oscillation
863, 975	Vibration ribose (RNA)
1062	Glycosaminoglycan's (α-glycan's)
1202	Hydroxyproline, tyrosine
1260, 1268, 1271	Amide III (collagen)
1410, 1412, 1415	ν (C-O) from COO (amino acid glutamic acid)
1434, 1446	δCH2, δCH3, C-H vibrations (proteins and lipids)
1553, 1556- 1557	NH deformation Amide II
1645	Amide I

Fig. 2 shows that the Raman spectra of all samples are similar in nature.
However, at wavenumbers 1062 cm^{-1}, 1645 cm^{-1}, 1553 cm^{-1}, 863 cm^{-1} and 1410 cm^{-1} corresponding to glycosaminoglycan's, collagen type I, collagen type II, ribose (RNA) and epithelium [15, 16, 17], there are differences between the samples of skin implants. These components determine the quality of implants. Therefore, these wavenumbers can be used as control criteria of skin biomatrixes.

Relatively constant in skin implants is collagen III [14] corresponding to wavenumber 1260 cm^{-1}, so it was used as denominator in introduced optical coefficient:

$$k = \frac{I_i}{I_{1260}},$$

where I_i - intensity values at wavenumbers 1062 cm^{-1}, 1645 cm^{-1}, 1553 cm^{-1}, 850 cm^{-1}, 863 cm^{-1}, 1202 cm^{-1} and 1410 cm^{-1} corresponding to glycosaminoglycans, collagen type I, type II collagen, glycosaminoglycans, ribose (RNA) hydroxyproline and epithelium 1260 cm^{-1} - a line corresponding to collagen III.

Further, Fig. 3 shows the two-dimensional diagrams of introduced optical coefficients.

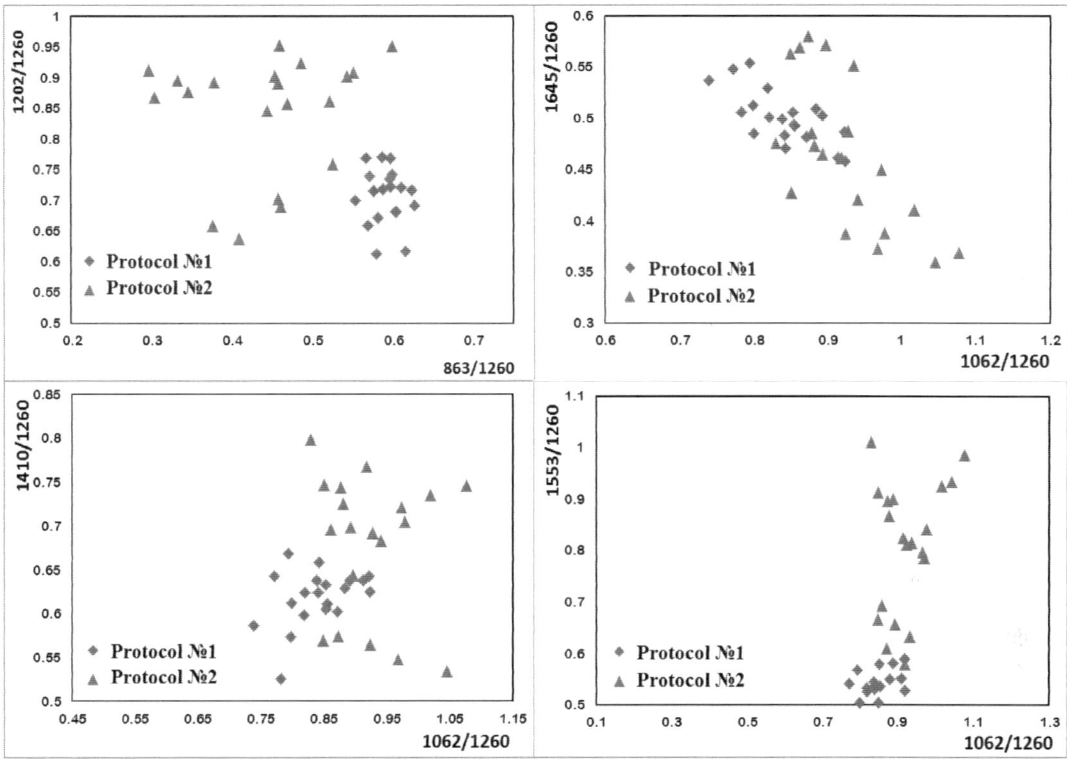

Figure3. Two-dimensional diagrams of introduced optical coefficients for rat's skin implants

Analysis of two-dimensional diagrams (Fig. 3) showed that the content of glycosaminoglycans (GAGs), corresponding to wavenumber 1062 cm^{-1}, more in implants made by protocol №2.This is due to the fact that when using this type of treatment, a chemical treatment is weaker, preserving GAG, compared to chemical treatment by protocol №1.However, when using the treatment protocol №2 in skin implant has a residual epithelium (optical coefficient I_{1410} / I_{1260} to 0, 8), the presence of which is expressed by the values of intensity of Raman lines on the wave of 1410 cm^{-1} (Fig. 2).This is due to the fact that with this type of processing, implants were not completely epithelialized.

Implants, made by protocol №1 contain cells, as evidenced by the value of optical coefficient I_{863} / I_{1260} to 0,63. Fig. 3 also shows that the optical coefficients vary depending on the types of implants manufacturing.

Based on the carried out analysis was constructed histogram comparing different types of treatment in the manufacture of skin biomatrixes and shown morphological analysis of skin implants made according to protocols (Fig. 4).

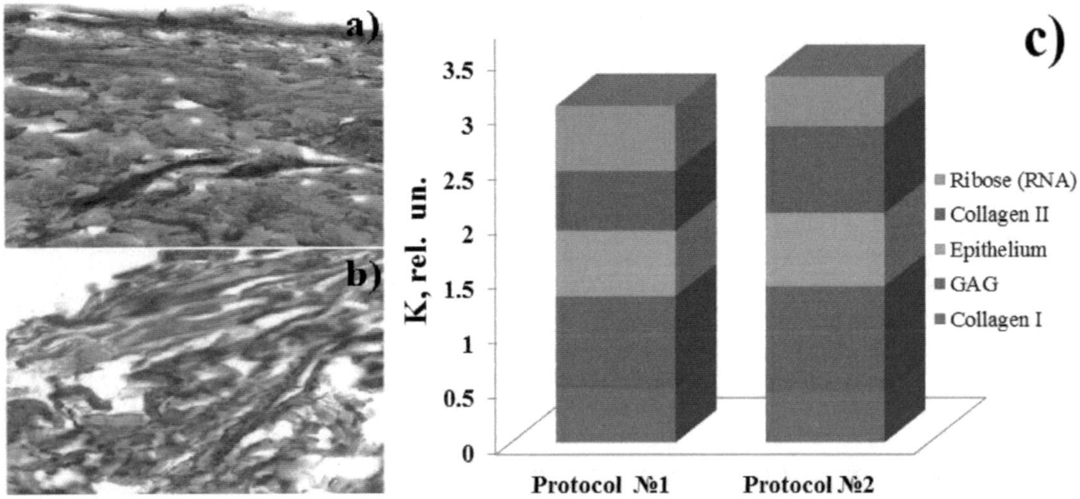

Figure4. I) Morphology of skin implants: a) rat skin made by protocol №1); b) rat skin made by protocol №2); II) Histogram of content of substances in rat implants, protocols (№1-2)

Using morphological analysis it was established that samples of skin tissue, made by protocol №2, have the structure of tissue with maintaining its main components, including hair follicles with the remnants of hair shaft. Implants which made by protocol №1 contain cellular components.

Also experimentally was found, that using a thermo-chemical treatment in the manufacture protocols of skin implants, their collagen structure is destroyed (collagen I), which is caused by a decrease of optical coefficient I_{1645} / I_{1260} to 0, 35.

Given analysis showed that introduced criteria allow to assess skin implants made by various types of treatment. Thus, Raman spectroscopy method can be used for assessing skin implants.

4. Conclusions

Features of Raman spectrum for skin biomatrix of rats with different protocols of manufacturing were obtained. It was established that main differences appear at wavenumbers 1062 cm^{-1}, 1645 cm^{-1}, 1553 cm^{-1}, 863 cm^{-1} and 1410 cm^{-1} corresponding to glycosaminoglycans, collagen type I, collagen type II, ribose (RNA) and epithelium. It is also experimentally found that using thermochemical treatment in protocols of manufacturing of skin implants their collagen structure is destroyed (primarily collagen I). Were introduced criteria allowing to assess skin implants made by various protocols, that allows to continue using Raman spectroscopy method for assessing skin implants.
The research results are confirmed by morphological analysis.

This work was supported by the Ministry of Education and Science.

5. References:

[1] Ledovskoy SN, Burda YE, Lazarenko VA 2008 Analysis of the clinical efficacy of fetal and adult allogeneic diploid fibroblasts in the treatment of burns border *Scientific J.: Advances modern science***9**92-94

[2] Gerasimov L 2000 The problem of burns on the threshold of XXI century *Scientific J. Kombustologiya***2** 36-42
[3] LiveseySA, HerndonDN, HollyoakMA, AtkinsonYH, NagA1995Transplanted acellular allograft dermal matrix. Potential as a template for the reconstruction of viable dermis*J. Transplantation***60**1-9
[4] Gubareva E A, Sotnichenko A, Gilewicz I, Macchiarini P 2012 Morphological evaluation of the quality decellularization heart and diaphragm of rats *J. Cell Implant and tissue engineering***3** 22-27
[5] Sergeyevich DS, PodhvatilinaNA, VasilyevaMB 2012Morphological and functional features of the aortic graft after detsellyulyarizatsii*J.Complex problems of cardiovascular disease***2** 3-6
[6] Kurapeev DI,Lavreshin AV, AnisimovSV2012Tissue engineering of heart valves: decellularizationallo- and xenografts*J.Cell Implant and tissue engineering***1** 34-39
[7] Isayev AD 2014 Skin cancer. Causes, symptoms and signs, diagnosis and treatment of cancer *J. Polismed***1** 33-39
[8] Afanasiev YI2012 Histology, embryology, cytology M.: Medicine p.800
[9] Bykov VL 2002Handbook on the Cytology and general histologyvol1,ed(St. Petersburg: SOTIS)p254
[10] Caspers PJ, Lucassen GW, PuppelsGJ2001 Combined In Vivo Confocal Raman Spectroscopy and Confocal Microscopy of Human *Skin J. of Investigative Dermatology***116**434-442.
[11] MatousekP, DraperER, GoodshipAE, ClarkIP, RonayneKL, ParkerAW2006 Non-invasive Raman spectroscopy of human tissue in *J. Central Laser Facility Annual Report***4**133-135
[12] RanieroL,CanevariRA, RamalhoLN, RamalhoFS, dos SantosEA, BitarRA, JalkanenKJ, MartinAA 2011 In and ex vivo Breast Disease Study by Raman Spectroscopy*J. of Theoretical Chemictry Accounts***130**1239-1247
[13] Khmaladze A, Ganguly A, Kuo S 2013 Tissue-Engineered Constructs of Human Oral Mucosa Examined by Raman Spectroscopy*J.Tissue Engineering Part***19**299-306
[14] Zhao JY, Chai JK, Song HF, Sun TJ, Li DJ, Liu LY, Gao QW, Liang LM 2011 Effects on collagen I and III after transplantation of porcine acellular dermal matrix with hyaluronic acid*J. Zhonghua Yi XueZaZhi***91**1276-80
[15] CiurleS, DeheleanCA, PinzaruS, FalamasA, MuresanA, LoghinF 2012 RAMAN SPECTROSCOPY INVESTIGATIONS OF THE CD1 NU/NU MOUSE SKIN PRE-CARCINOMA DAMAGES*J. Farmacia***3**448-456
[16] Zakharov VP,TimchenkoEV, TimchenkoPE, TaskinaLA, KozlovSV, MoryatovAA 2013 Backscattering spectroscopy to assess skin lesions ***J.*** *Proceedings of the Samara Scientific Center of the Russian Academy of Sciences***15** 126-130
[17] Tfaili S, Gobinet C, Josse G, Angiboust J, ManfaitM, Piot O 2012 Confocal Raman microspectroscopy for skin characterization: a comparative study between human skin and pig skin*J. Analyst***137**3673–3682
[18] Zanyar M, ShazzaR, IhteshamU2007Raman Spectroscopy of Biological Tissues *J. Applied Spectroscopy Reviews***42**493-541

Journal of Physics: Conference Series **769** (2016) 012050 doi:10.1088/1742-6596/769/1/012050

A novel type of quasi-phasematching for the second harmonic generation

E G Sokolovskaya[1,2], K A Fedorova[3], V V Dudelev[1,4], E U Rafailov[3], G S Sokolovskii[1]
[1] Ioffe Institute, St Petersburg, Russia
[2] Peter the Great Polytechnic University, St Petersburg, Russia
[3] Aston University, Birmingham, United Kingdom
[4] ITMO University, St Petersburg, Russia

E-mail: gs@mail.ioffe.ru

Abstract. We propose a novel type of quasi-phasematching for the second harmonic generation in periodically-poled nonlinear crystals. In contrast to the conventional quasi-phasematching where one (or few) quasi-wavevector(s) of periodical poling compensate for the momentum mismatch between a pair of the fundamental photons and the SHG one, with the proposed mechanism the momentum mismatch between several pairs of fundamental and SHG photons is compensated with one quasi-wavevector of periodical poling.

1. Introduction

Generation of the second harmonic (SHG) of infrared (IR) light offers one of the most preferred ways for realization of compact visible laser sources with a number of cutting-edge applications in microscopy, spectroscopy, biophotonics and photomedicine. Availability of compact, efficient and cost-effective IR laser diodes in the near-IR range of 0.8 - 1.5 μm supports this trend both from the market and technical sides [1]. However, the efficient SHG is possible only with simultaneous photon energy $E_\lambda=2E_{2\lambda}$ and momentum conservation $k_\lambda=2k_{2\lambda}$ (here and below the subscript 2λ denotes the fundamental and λ SHG wavelength). The last requirement of the photon momentum conservation (or "phase-matching") is difficult to achieve due to dispersion of the refractive index in the nonlinear crystal. Without phase-matching, the generated second harmonic grows and decays as the fundamental and SHG waves go in and out of phase over each coherence length $L_c=\frac{1}{2}\lambda/|n_\lambda-n_{2\lambda}|$, where n_λ and $n_{2\lambda}$ are the refractive indexes at SHG and fundamental wavelength correspondingly [2].

Since its first introduction in 1960s [2,3], the preferred approach for the phase-matching between interacting harmonic and fundamental waves is the periodical poling (or "quasi-phase-matching" - QPM) of the ferroelectric nonlinear crystals. This is normally achieved by periodically reversing the crystals polarization under large electric field. The proper phase relationship between the propagating waves is maintained with the poling period being double the coherence length $\Lambda=2L_c$. Under this condition, the SHG efficiency is maximized because of momentum conservation by the quasi-wave-vector of the periodical poling $k_\Lambda=2\pi/\Lambda=\Delta k=k_\lambda-2k_{2\lambda}$ as shown schematically in figures 1(a) and (b). Tuning of the conversion wavelength is possible by introduction of multiple gratings or variable poling period [4], Fibonacci or Fourier-constructed quasi-periodical poling [5,6], shifting dispersion in the nonlinear crystal with temperature [7] or by introduction of the multimode waveguide [8,9].

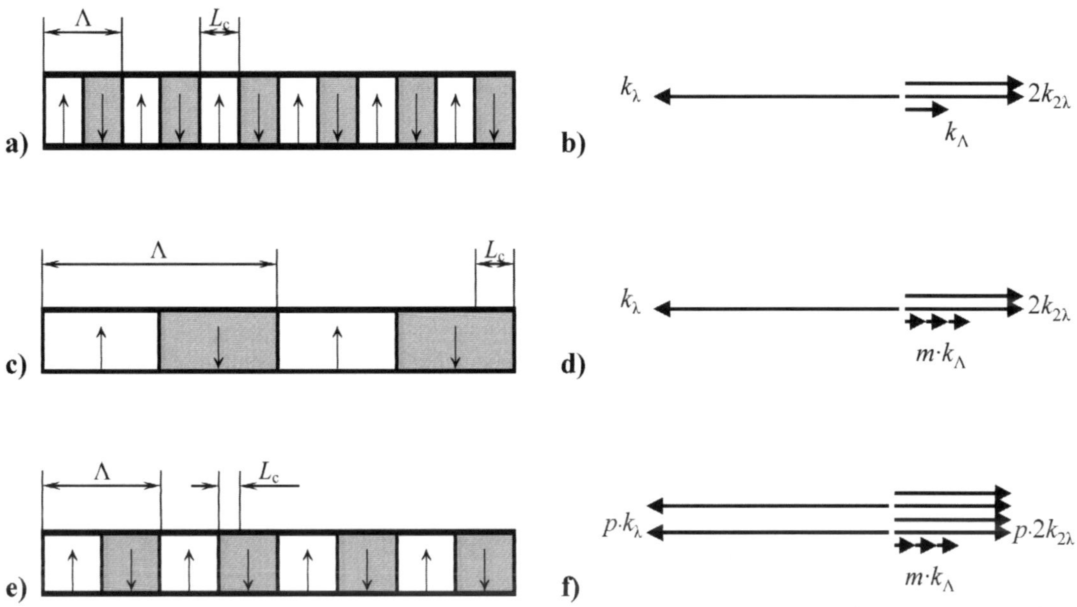

Figure 1 Schematic of periodical poling and momentum conservation for the first-order poling (a,b), high-order poling (c,d) and fractional-order poling (e,f) correspondingly. Note the reduction of the coherence length L_c for the fractional-order poling (e).

As shown by Fejer at al. [10], it is also possible to achieve QPM in the case of the high-order periodical poling that dramatically extends the spectral tuning range of SHG. With the high-order poling, the period is a multiple of the doubled coherence length $\Lambda = m \cdot 2L_c$, where the natural number m=2, 3, 4… is the order of poling. With the mth order poling, in contrast to the first-order poling, it is necessary to 'utilize' not one, but m quasi-wave-vectors of the periodical poling for compensation of the momentum mismatch $m \cdot k_\Lambda = \Delta k = k_\lambda - 2k_{2\lambda}$ as shown schematically in figures 1(c) and (d). Generally speaking, this process is very similar to the high-order diffraction with the conventional diffraction grating.

2. Fractional-order periodical poling for the second harmonic generation

In this paper, we propose a fractional order of poling period of nonlinear crystal. In contrast to the higher-order poling with m quasi-wave-vectors of the periodical poling compensating the momentum mismatch between a pair of fundamental photons and SHG one, fractional order poling enables momentum compensation for p pairs of fundamental and p SHG photons with one quasi-wave-vector of the periodical poling: $k_\Lambda = \Delta k = p \cdot (k_\lambda - 2k_{2\lambda})$. The effective coherence length in this case of the multi-photon momentum compensation reduces p-fold $L_c = \pi/\Delta k$ leading to the corresponding decrease of the fractional-order poling period $\Lambda = 2\pi/\Delta k = 2\pi/p \cdot (k_\lambda - 2k_{2\lambda})^{-1}$ being a 'fraction' of the first-order one. Also, it is possible to generalize this approach by merging both the high- and the fractional-order poling: $m \cdot k_\Lambda = \Delta k = p \cdot (k_\lambda - 2k_{2\lambda})$. In other words, momentum mismatch between p pairs of the fundamental and p SHG photons can be compensated with m quasi-wave-vectors of the m/p-order periodical poling as shown schematically in figures 1(e) and (f).

Now, it is important to understand the law for the spatial evolution of the SHG intensity with the fractional-order poling. With the conventional and higher-order poling, the rate of

Journal of Physics: Conference Series **769** (2016) 012050 doi:10.1088/1742-6596/769/1/012050

growth of the slowly varying SHG field amplitude in the one-dimensional periodically-poled nonlinear crystal is proportional to the square of the amplitude of the fundamental wave, spatially-modulated nonlinear coefficient and the phase term [10]:

$$\frac{dE_2}{dz} \sim E_1^2 d(z)\exp(-i\Delta kz) \quad (1)$$

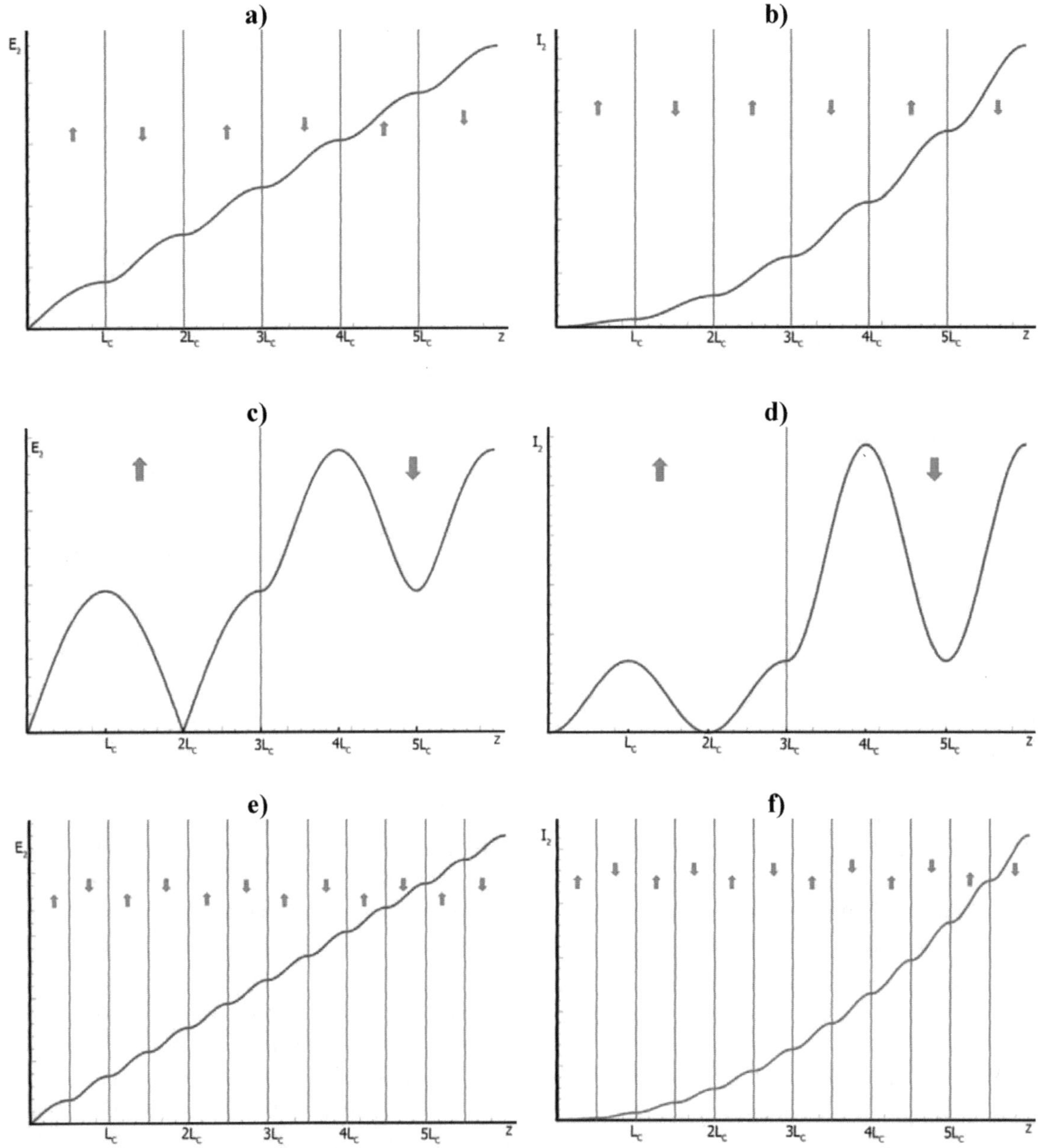

Figure 2 Spatial evolution of the SHG field amplitude and intensity for the first-order poling (a,b), third-order poling (c,d) and half-order poling (e,f).

This simple expression is valid under assumption of low conversion efficiency, loose focusing and absence of losses at fundamental and SHG wavelengths. Integration of (1) yields the spatial evolution of the SHG field amplitude:

$$E_2(z) \sim E_1^2 \int_0^z d(x)\exp(-i\Delta kx)dx \qquad (2)$$

where x is the integration variable and z is the coordinate along the direction of propagation of the fundamental and SHG waves in the periodically-poled nonlinear crystal. In the case of an 'ideal' first-order poling, with the period of poling corresponding to the momentum mismatch $\Lambda=2\pi/\Delta k=2\pi/(k_\lambda-2k_{2\lambda})$ and the module of material nonlinearity always having the maximal value $d(z)=\pm d_{eff}$ (corresponding to the meander-like modulation with negligibly thin inverted domain interfaces) integration of (2) yields the classical spatial evolution of the SHG field amplitude represented in figure 2(a). Figure 2(b) shows the spatial evolution of the SHG intensity being a square of the module of the field amplitude.

Taking the third-order poling as an example of the higher-order poling, one should increase the period of poling 3-fold: $\Lambda=6\pi/\Delta k$ and keep the momentum mismatch value $\Delta k=k_\lambda-2k_{2\lambda}$. With meander-like modulation and thin domain interfaces, integration of (2) and its square provides one with the spatial evolution of the SHG field amplitude and intensity as shown in figures 2(c) and (d). Comparing figures 2(a,b) and 2(c,d) correspondingly it is easy to note significant decrease of the conversion efficiency with the third-order poling due to uncompensated oscillations of the SHG field in the first two thirds of each half-period of poling.

In contrast to a higher-order poling, a fractional poling period becomes a 'fraction' of the first-order one $\Lambda=6\pi/\Delta k$ not due to 'artificial' denomination by the fraction value p but because of increase of the effective momentum mismatch $\Delta k=p\cdot(k_\lambda-2k_{2\lambda})$ which in turn is due to compensation of momentum mismatch between p pairs of fundamental and p SHG photons with one quasi-wave-vector of the periodical poling as stated in the first paragraph of this section. Figures 2(e) and (f) show the spatial evolution of the SHG field amplitude and intensity for the simplest case of the ½ period poling, with periodically-poled grating quasi-wave-vector compensating for the momentum mismatch of two pairs of fundamental and two SHG photons. One can note striking similarity between the figures 2(e,f) and 2(a,b) which differ only with the period of 'ripples' caused by the SHG and fundamental waves coming in and out of phase. This should be attributed to an 'ideal' compensation of the multi-photon momentum mismatch with the shortened period of poling.

3. Conclusion

In summary, we have discussed a novel mechanism of quasi-phasematching for the second harmonic generation in periodically-poled nonlinear crystals. The proposed mechanism is based on compensation of the momentum mismatch between several pairs of fundamental and SHG photons with one quasi-wavevector of periodical poling. This is the 'opposite' to the conventional quasi-phasematching where momentum mismatch between two fundamental and one SHG photon is compensated with the one (or few) quasi-wavevector(s) of periodical poling.

Acknowledgment
This work was partially supported by the Russian Ministry of Education and Science (Grant ID: RFMEFI60714X0101).

References

[1] E U Rafailov, M A Cataluna, and W Sibbett 2007 Mode-locked quantum-dot lasers *Nature Photon.* **1**, 395–401.

[2] J A Armstrong, N Bloembergen, J Ducuing, P S Pershan 1962 Interaction between light waves in nonlinear dielectric *Phys. Rev.* **127**, 1918-1939.

[3] P A Franken, J F Ward 1963 Optical harmonic and nonlinear phenomena *Rev. Mod. Physics* **35**, 23-39.

[4] D A Chestnut, S V Popov, J R Taylor, T D Roberts 2006 Second-harmonic generation to the green and yellow using picosecond fiber pump sources and periodically poled waveguides. *Appl. Phys. Lett.* **88,** 071113.

[5] A Bahabad, N Voloch, A Arie, R Lifshits 2007 Experimental confirmation of the general solution of the multiple-phase-matching problem *J. Opt. Soc. Am. B* **24**, 1916-1921.

[6] S N Zhu, Y Y Zhu, N B Ming 1997 Quasi-phase-matched third-harmonic generation in a quasi-periodic optical superlattice *Science* **278**, 843-846.

[7] K A Fedorova, M A Cataluna, P R Battle, C M Kaleva, I L Krestnikov, D A Livshits, E U Rafailov 2011 Orange light generation from a PPKTP waveguide end-pumped by a cw quantum-dot tunable laser diode *Appl. Phys. B* **103**, 41-43.

[8] K A Fedorova, G S Sokolovskii, P R Battle, D A Livshits, E U Rafailov 2012 Green-to-red tunable SHG of a quantum-dot laser in a PPKTP waveguide *Laser Phys. Lett.* **9**, 790-795.

[9] K A Fedorova, G S Sokolovskii, P R Battle, D A Livshits, E U Rafailov 2015 574 nm – 647 nm wavelength tuning by second-harmonic generation from diode-pumped PPKTP waveguides *Optics Letters*, **40**(5), 835-838.

[10] M M Fejer, G A Magel, D H Jundt, R L Byer 1992 Quasi-phase-matched second harmonic generation: tuning and tolerances *IEEE J. Quantum Electron.* **28**, 2631-2654.

Optical anisotropy of cubic photonic crystals under conditions of multiple-mode light propagation

T A Ukleev[1,2], D I Yurasova[3], N N Shevchenko[3], and A V Sel'kin[1,2]

[1] Solid State Physics Department, Saint Petersburg State University, Ulyanovskaya 1, Peterhof 198504, Russian Federation

[2] Spectroscopy of Solids Laboratory, Ioffe Physical Technical Institute, Polytekhnicheskaya 26, St Petersburg 194021, Russian Federation

[3] Synthesis of Peptides and Polymer Microspheres Laboratory, Institute of Macromolecular Compounds, RAS, Bolshoy pr. 31, St Petersburg 199004, Russian Federation

E-mail: t.ukleev@gmail.com, alexander.selkin@mail.ioffe.ru

Abstract. Bragg reflection spectra of light are studied for opal-like photonic crystals made of polystyrene spheres. A resonant enhancement of reflectivity is observed in cross-polarization configuration of the analyzer and polarizer when varying the azimuthal orientation of a sample in respect to the incidence plane. The cross-polarization effect takes place at oblique incidence of light on the lateral (111) crystal plane with the plane of incidence being non-perpendicular to the inclined (11-1) crystal plane. The effect is shown to be due to the multiple Bragg diffraction of light when the resonant Bragg conditions are fulfilled at a certain angle of incidence and azimuth for the lateral and inclined crystal planes simultaneously.

1. Introduction

Opal-like structures considered as a typical example of three-dimensional (3D) photonic crystals (PhC) have received much attention during last decades both due to their non-trivial optical properties and potential applications in devices [1, 2]. A number of interesting and still not adequately explored phenomena occurring in the frequency range of low energy stop-bands are associated with the multiple Bragg diffraction (MBD) effects [3-6]. Most prominently, these phenomena are manifested in reflection spectra at large angles of incidence as a doublet structure of the reflectance contour [6-8].

As a rule, the simplest geometry of reflection is chosen so that the incidence plane of light is perpendicular to both the lateral (111) and inclined (11-1) crystal planes, which corresponds to the incidence plane azimuth $\varphi = 0$ (see Fig.1). Such geometry allows one, in a rather simple way [5], to analyze separately the contribution to the reflection spectrum from the TE- and TM- electromagnetic modes propagating inside an opal-like PhC: these modes are excited in s- and p-polarization of incident light, respectively.

In this work, we discuss novel optical phenomena associated with resonant polarization mixing of electromagnetic modes in opal-like photonic crystals, which is the case, in spite of the high, fcc, symmetry of the sample, when the azimuth φ differs from $m\pi/3$ values (m=0,1,...,5).

2. Experimental

The opal-like PhCs under study were prepared from monodispersive polystyrene spheres self-assembled to the films of about 5 μm in thickness. The particles were synthesized by styrene and

Journal of Physics: Conference Series **769** (2016) 012051 doi:10.1088/1742-6596/769/1/012051

methacrylic acid emulsifier-free emulsion copolymerization. They were initiated by potassium persulfate, had a spherical shape, uniform surface and particle size distribution less than 2%.

The Bragg reflection spectra were measured in s-s, p-p and p-s configurations of polarized light when incoming s- or p-polarized radiation was incident on the lateral reflecting surface (111) of the PhC film.

The azimuth angle φ of the sample was measured from zero-value which corresponded to the perpendicular orientation of the icidence plane in respect to the inclined crystal planes (020) and (11-1). The sample was precisely oriented using measured Bragg reflection spectra under the MBD conditions [6].

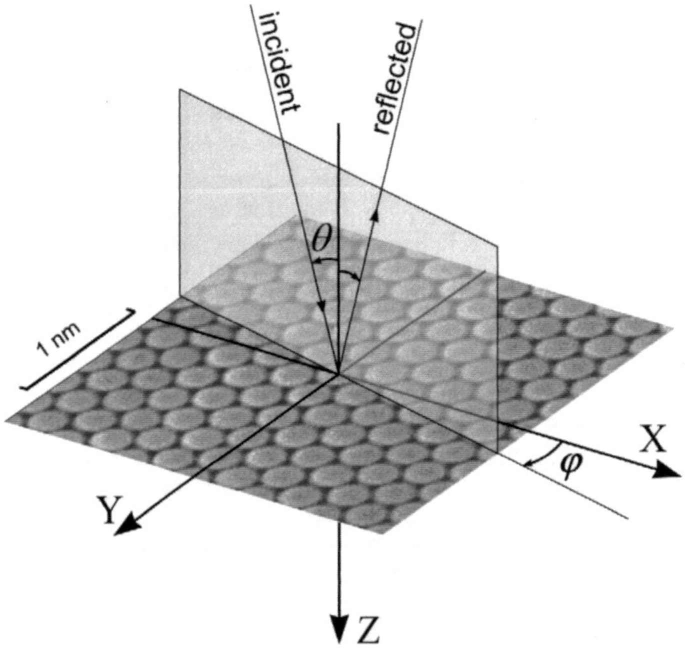

Figure 1. Geometry of Bragg reflection of light from an opal-like photonic crystal: the XY plane of the reference system coincides with the (111) lateral (reflecting) surface of the sample, the azimuth φ of the incidence plane is reckoned from the XZ plane which is perpendicular to the inclined crystal planes (11-1) and (002), θ is the angle of incidence.

Figure 2 shows the s-s reflection spectra obtained at the azimuthal orientation $\varphi = 15^0$ of the plane of incidence. In Fig. (a), the topogram of all the spectra measured is depicted where, together with the main reflection peak band, the reflectance deep is clearly seen crossing the peak band in the incidence angle range $\sim 50^0 < \theta < 62^0$. Some reflectance contours recorded at three angles of incidence are correlated with the topogram in Fig. (b).

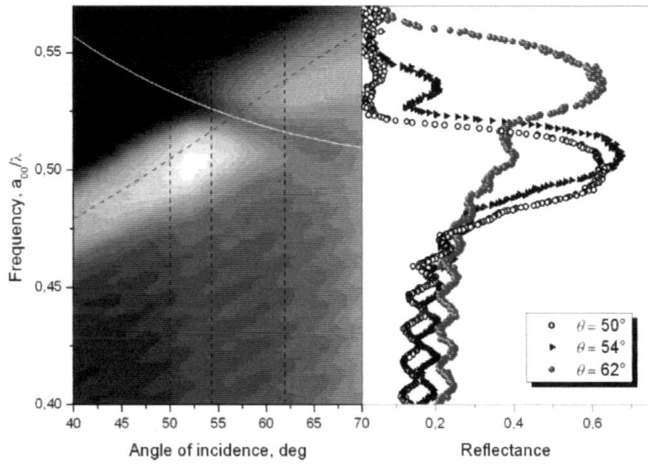

Figure 2. (a) The topogram of measured reflectance in s-s configuration of polarization at the azimuth $\varphi = 15^0$ of the incidence plane; superimposed lines are calculated with the formulae (1) and (2) giving Bragg conditions for light reflection from the crystal planes (111) (black dashed line 1) and (11-1) (white solid line 2). (b) s-s reflectance spectra obtained at three different angles of incidence θ, as indicated. a_{00} is the smallest distance between the particle centers in the (111) plane, λ is the wavelength of incident light.

A very surprising effect appears when measuring the same spectra in p-s cross-polarization configuration (see Fig. 3). Instead of expected absence of reflection signal, there exists a rather bright spot on the topogram (Fig.a). This spot is observed in the region where the deep in s-s reflection is registered (cf. Fig. 2), the maximum value of the p-s reflectance being about as high as 10%! It should be emphasized here that with decreasing azimuth φ (up to $\varphi = 0$) the p-s reflection intensity tends to zero.

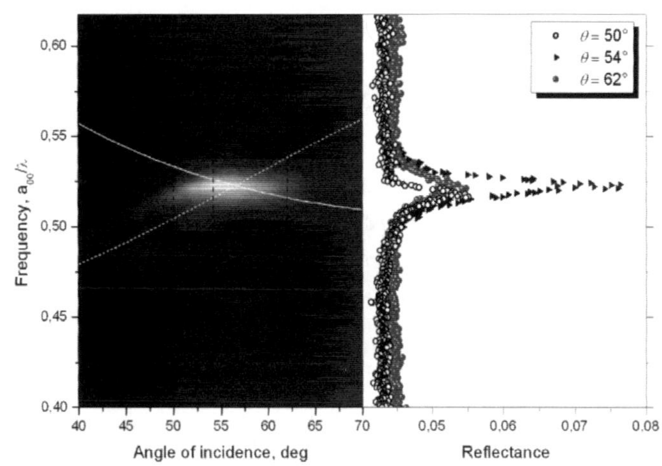

Figure 3. (a) The topogram of measured reflectance in p-s cross-polarization configuration at the azimuth $\varphi = 15^0$ of the incidence plane; superimposed lines are calculated with the formulae (1) and (2) giving Bragg conditions for light reflection from the crystal planes (111) (white dotted line 1) and (11-1) (white solid line 2). (b) p-s reflectance spectra obtained at three different angles of incidence θ, as indicated. a_{00} is the smallest distance between the particle centers in the (111) plane, λ is the wavelength of incident light.

3. Discussion

In accordance with the technique developed in [6], based on comprehensive analysis of Bragg reflection spectra at $\varphi = 0$, we carefully measured the parameter values of the PhC sample under study. The mean distance between neighboring particles placed in the lateral plane is a_{00} = 279 nm, the average dielectric constant is ε_0=2.185, the coefficient of uniaxial compression along [111] direction is η = 0.94. These values allow us to describe in a good approximation [7] the spectral positions λ_{111} and λ_{11-1} of the main features appearing in reflection spectra using the following Bragg-type formulae:

$$\lambda_{111} = \sqrt{8/3}a_{00}\eta\sqrt{\varepsilon_0 - \sin^2\theta}\,, \tag{1}$$

$$\lambda_{11-1} = \frac{\sqrt{6}a_{00}\eta}{1+2\eta^2}\left(\eta\sqrt{2}\sin\theta\cos\varphi + \sqrt{\varepsilon_0 - \sin^2\theta}\right), \tag{2}$$

where the indices 111 and 11-1 differing the wavelengths λ are referred to Bragg diffraction from the corresponding crystal planes. As follows from above formulae, only λ_{11-1} depends on the azimuth φ of the incidence plane. λ_{111} gives the s-s reflection peak band position as a function of the angle of incidence θ, whereas λ_{11-1} describes the behavior of the s-s reflection deep.

Figure 2 demonstrates clearly that the spectral positions of the reflectance maxima and deeps are fitted well by Eqs. (1) and (2), respectively. This allows us to consider unambiguously the spectral deep observed at the 15° azimuth in the vicinity of the interception point

$$\lambda_* = \frac{4\sqrt{3}a_{00}}{4-\eta^{-2}}\sin\theta_*\cos\varphi_* \tag{3}$$

for the theoretical curves (1) and (2) as being due to the MBD effect. Because the enhancement of reflectivity in cross-polarization configuration (see Fig. 3) takes place also in the vicinity of this interception point, we can ascribe such enhancement to the MBD effect, as well.

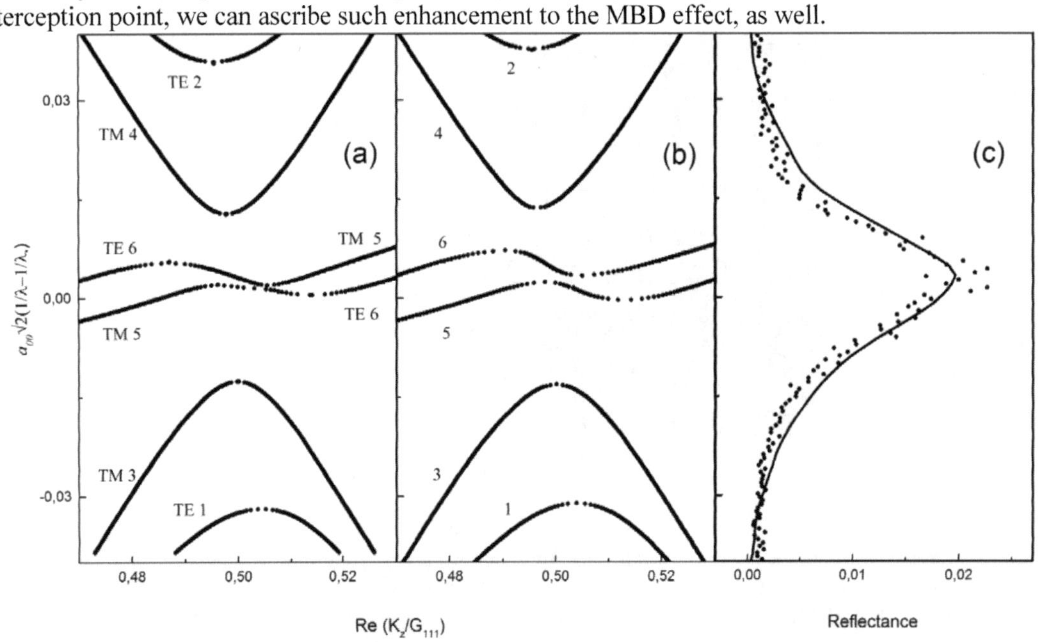

Figure 4. The eigenmode energy spectra (a,b) of the opal-like PhC as compared with the reflection spectrum for p-s cross-polarization at the angle of incidence θ= 54° (a), 56° (b,c) and at the azimuthal angle φ= 0° (a), 15° (b,c). At φ= 0° the p-s signal disappears practically. On the panel (c), the experimental and theoretical data are represented by points and the solid line, respectively.

When calculating energy and reflectance spectra we took into account 3 reciprocal lattice vectors G_{111}, G_{11-1} and G_{020} which form the Bloch mode amplitude in the vicinity of the λ_* point (3). Fig. 4(a) shows that at $\varphi = 0^o$ inside the crystal there exist the electromagnetic mode which can be classified as TE- and TM-modes. It worth to note that the TM5 and TE6 curves intersect at $\sim\lambda_*$. However when $\varphi \neq 0^o$ (in our case $\varphi = 15^o$, see Fig. 4(b)) the anti-crossing region arises around the point λ_*, and polarization mixing between all the modes 1-6 takes place. Such kind of mixing is more essential in the vicinity of the crossing point. Therefore the maximum of the cross-polarization signal should be located near this point. As shows the panel (c) the calculated curve is well fitted to the experimental data. As follows from our theoretical calculations, the FWHM of the p-s reflectance contour is determined by the lifetime $\tau = \Gamma^{-1}$ of the electromagnetic state. We have obtained the value $\Gamma = 50$ meV which allowed us to describe theoretically well all our experimental data including the reflectivity contours in p-p and s-s configurations.

It should be noted that in spite of a weak uniaxial anisotropy of the sample under study the observed cross-polarization effect does not associate with such kind of anisotropy and is, in fact, the intrinsic property of an ideally cubic photonic crystal with the opal-like structure. There exist a number of techniques for preparing artificial opals which allow one to make samples with certain optical and other properties (see, for example, [9] and references therein). Unfortunately, preparing opal-like PhCs with ideal cubic symmetry meets, as a rule, specific technological difficulties which can lead to a number of additional experimentally observed anisotropy effects (as it was pointed, in particular, in [9-11]). The cross-polarization effect analysed in our work is described adequately in the framework of the dynamical Bragg diffraction theory [6, 8]. Making use of this theory we present for the first time detailed comparison between the experimental and theoretical cross-polarization spectra and established the characteristic lifetime ~ 0.1 ps of the electromagnetic state in the vicinity of the stop-band, which corresponds to the measured Γ value.

4. Conclusion

The results presented in this work demonstrate that in spite of high (nearly cubic) symmetry of opal-like photonic crystals their optical properties exhibit strong anisotropy which becomes very much more pronounced when the multiple Bragg diffraction conditions are fulfilled. In such a case, the resonant polarization mixing of diffraction processes governed by two or more systems of crystal planes takes place. So, for some directions of propagation, the TE- or TM-classification of electromagnetic modes fails because other eigenmodes are excited with polarization states different from ones of strongly defined TE- and TM-modes. As a result, at p- or s-input polarization, reflected light becomes elliptically polarized, which appears experimentally as a resonant cross-polarization effect. It should be concluded that this effect can be considered as a new direct manifestation of the multiple Bragg diffraction in 3D photonic crystals.

References

[1] Galisteo-López J F, Ibisate M, Sapienza R, Froufe-Pérez L S, Blanco Á, López C. 2011. Self-Assembled Photonic Structures. Adv. Matter. 23(1):30-69.

[2] Joannopoulos J D, Johnson S G, Winn J N, Meade R D. 2008. Photonic Crystals: Molding the Flow of Light. 2nd edition. Princeton: Princeton University Press.

[3] van Driel H M, Vos W L. 2000. Multiple Bragg wave coupling in photonic band-gap crystals. Phys. Rev. B. 62:9872-9875.

[4] Romanov S G, Maka T, Sotomayor Torres C M, Müller M, Zentel R, Cassagne D, Manzanares-Martinez J, Jouanin C. 2001. Diffraction of light from thin-film polymethylmethacrylate opaline photonic crystals. Phys. Rev. E. 63:056603.

[5] Shishkin I I, Rybin M V, Samusev K B, Golubev V G, Limonov M F. 2014. Multiple Bragg diffraction in opal-based photonic crystals: Spectral and spatial dispersion. Phys. Rev. B. 89:035124.

[6] Sel'kin A V. 2004. Structural characterization of photonic crystals by Bragg reflection spectroscopy. Proceedings of 12th Symposium "Nanostructures: Physics and Technology" (St Petersburg, Russia, June 21-25, 2004), St Petersburg: Ioffe Istitute, p.111-112.

[7] Bazhenova A G, Sel'kin A V, Men'shikova A Yu, Shevchenko N N. 2007. Polarization-dependent suppression of Bragg reflections in light reflection from photonic crystals. Physics of the Solid State, 49(11):2109-2120.

[8] Fedotov V G, Sel'kin A V, Ukleev T A, Men'shikova A Yu, Shevchenko N N. 2011. Resonant multiple diffraction of light in 3D opal-like photonic crystals. Phys. Status Solidi B. 248(9):2175-2179.

[9] Sabataityte J, Simkiene I, Babonas G –J, Reza A, Baran M, Szymczak, Vaisnoras R, Rasteniene L, Golubev V, Kurdyukov D. 2007. Physical studies of porphyrin-infiltrated opal crystals. Material Sciences and Engeneering C, 27:985-989.

[10] Plekhanov A I, Chubakov V P, Chubakov P A. 2011. Polarization Anisotropy of Single Crystal Opal Films. Physics of the Solid State. 53(6):1145-1151.

[11] Romanov S G. 2007. Anisotropy of Light Propagation in Thin Opal Films. Physics of the Solid State. 49(3):536-546.

Influence of «defocusing» on the correlation function in the problem of particles suspension accounting

T A Vovk, T Yu Nikolaeva, N V Petrov
Department of Photonics and Optical Information Technology, *ITMO University* Kronverksky Av. 49, St. Petersburg 197101, Russia

E-mail: tatiana_vovk@corp.ifmo.ru

Abstract. This paper describes influence of such parameter as "defocusing" on the ratio of the peak of the correlation function to its pedestal, calculated for adjacent planes of the volume with suspended particles. This ratio is used in the method of statistical accounting of the particles in the volume. Numerical modelling and experimental validation are performed.

1. Introduction

Nowadays, digital image processing techniques find application in many fields of science and technology. They help to implement numerically different measurement methods of various characteristics of particle flows or distributions of suspended particles. Application of digital image processing of particles includes basic research: the study of the dynamics of particle fluxes (two-phase flow of gas, supersonic flow, etc.) [1, 2]; the study of transparent medium (to prevent the defects of glass, etc.) [3]. There are lots of methods of coherent processing of particle images, among which are laser anemometry of particle images (PIV, or particle image velocimetry) [4] and digital holography methods (DH) [5, 6]. The correlation technique has been used for the displacement [8] and deformation [9] measurement, as well as for roughness measurements [10]. Width [10] and position [8, 9] of correlation peak are analyzed usually to extract useful data, namely, information about roughness of the illuminated surface, simulated displacements on image and deformation field. This study, which is a continuation of [7], proposes calculation of the cross-correlation of images of two adjacent slices of a volume to estimate the characteristics of the distribution of the particles. We study particles in a volume of optically transparent medium. We examine the changing of peak to pedestal ratio when the concentration, size and other statistical parameters of the particle distribution are varied. It is considered to be more reliable, allowing to achieve higher speeds of calculations as compared to other methods.

2. Statement of the problem

We consider the case of scattering particles distributed in the volume of the optical medium. The purpose of this study is to estimate the efficiency and determinate the applicability of the particle image statistical description based on analysis of cross-correlation function parameters. We consider a simple case of the distribution of particles in a volume to identify and verify the basic properties and behavior of the correlation function. Such a property is correlation function peak separation in event of «defocusing» of the particles, peak broadening with an increasing size of repetitive elements, etc.

3. Algorithm based on cross-correlation calculation

We have developed a simulation model for the validation of the method of statistical accounting of the particles in the volume and evaluation of its effectiveness as a tool to determine the nature of the distribution of particles suspended in a volume. This simulation has been implemented using National

Instruments LabVIEW development environment. The modelling program consists of three consecutive blocks, which are examined and explained below.

3.1. Simulation of the distribution of particles suspended in the volume of optical medium
On the first stage, program simulates a suspension of particles in a volume with specified distribution, size and number of particles. The volume of the optical medium is defined as a set of segments with a specific distribution of scattering particles in each of them. We consider random distribution (equally probable in each segment) in order for the developed model of the volume of optical medium to be in accordance with the conditions that most frequently take place in the research problems of particles. The mathematical model of particle distribution in a volume is described in [11].

3.2. Modelling of the formation of particles volume images
Second stage is the image processing: program calculates two adjacent layers of the volume with the mathematical apparatus of the scalar diffraction theory. To calculate the electromagnetic field propagation from coherent light source through the volume of the optical medium, the method of propagation of the angular spectrum of plane waves implemented through two fast Fourier transform can be used [12]. We use an algorithm based on the edge-point linking and thresholding technique that was described in [13] to detect layers of volume with particles in it.

3.3. Cross-correlation analysis
Cross-correlation function is a method for assessing the degree of correlation between two sequences (or arrays in two-dimensional case). Consider two arrays *f* and *g*. Cross-correlation is defined as:

$$(f \star g)_{i,g} = \sum_k f_k^* \cdot g_{i,j+k} \tag{1}$$

Here *i* and *j* are indices that indicate coordinates of a point where correlation calculation is performed, *k* is a shift-indice and *f** indicates a complex conjugation on *f*. In case of more efficient and faster computation using fast Fourier transform algorithm [14] cross-correlation function can be calculated as:

$$f \star g = \hat{F}^{-1}\left[(\hat{F}[f])^* \cdot (\hat{F}[g])\right] \tag{2}$$

Here f and g are two-dimensional arrays, $\hat{F}[g]$ is the Fourier transform on g and $\hat{F}^{-1}[g]$ is inverse Fourier transform. Cross-correlation function of two focused layers of particle distribution depends on many parameters. Among them are the distance between the layers Δz, particles' diameter D, concentration C (depends on number of particles in each segment N_P and number of segments N), transparency of particles T, wavelength of the laser source λ, etc.

It is more convenient and faster to analyze two-dimensional case of correlation function. We can move from a three-dimensional correlation function to a two-dimensional one avoiding loss of information. Due to the symmetry of the particles themselves and of the distributions we study, cross-correlation turns to be symmetrical as well. The transition from 3D to 2D is illustrated in Fig. 1.

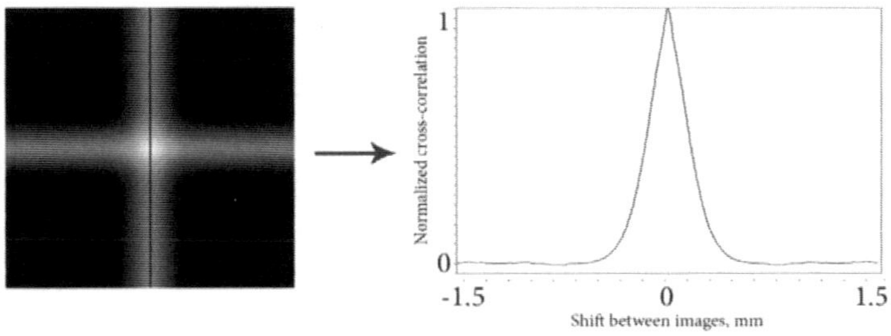

Figure 1. Transition from normalized 3D cross-correlation function to its 2D cross-section. The cut place is indicated

4. Study of the dependence of the correlation function form on parameters of the particles

4.1. Dependence of correlation on space between slices of volume

While Δz is not large enough to change the pattern of two images of different slices of volume, correlation peak is wide and has no inflections or borders. The difference between focused slices increases with a growth of Δz and the form of cross-correlation function changes: border between the peak and the pedestal of the cross-correlation function becomes distinguishable, videlicet, peak becomes narrower but pedestal remains the same. This effect of peak indication is caused by the nature of distinction between images, namely, the diffraction of scattered light. The larger is the distance between layers of particles Δz, the greater is the effect of diffraction, and defocused particles are surrounded by diffraction rings. We performed the experimental validation and computational simulation of the mutual change between Δz and correlation function of two adjacent layers. Dependence of cross-correlation on thickness between slices of volume for each case is shown in Fig. 2, 3.

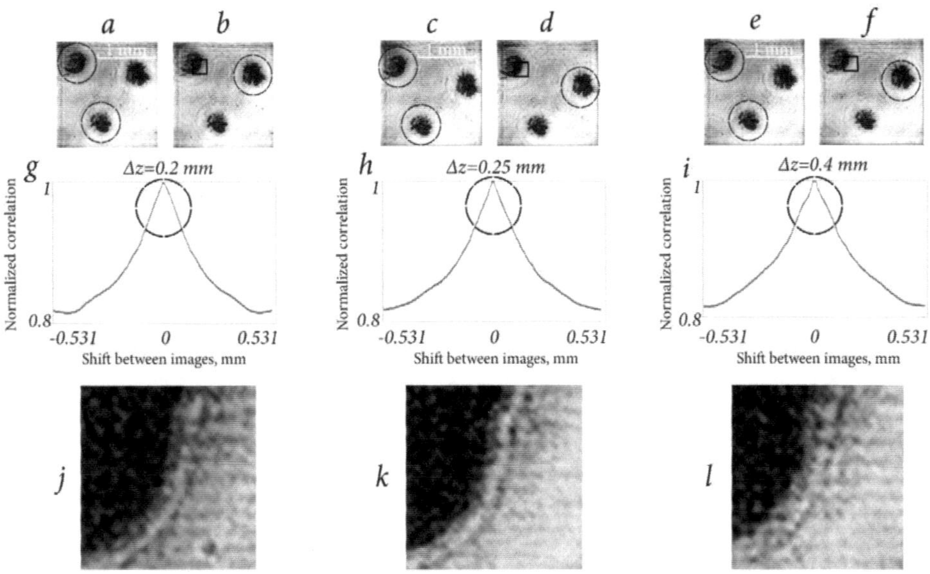

Figure 2. Normalized cross-correlation sections {g, h, i} between 2 slices ({a, c, e} – first for observer, {b, d, f} – second for observer) of volume of real particles where $D \approx 0.42$ mm, $N_P = 2$ on the first slice, $N_P = 1$ on second slice and Δz varies. Sizes of detected slices are 1,062 × 1,062 mm, $\lambda = 532$ nm. The distance from the object to the CCD equals to 6.7 mm. The difference in diffraction rings of defocused particles is shown in {j, k, l}

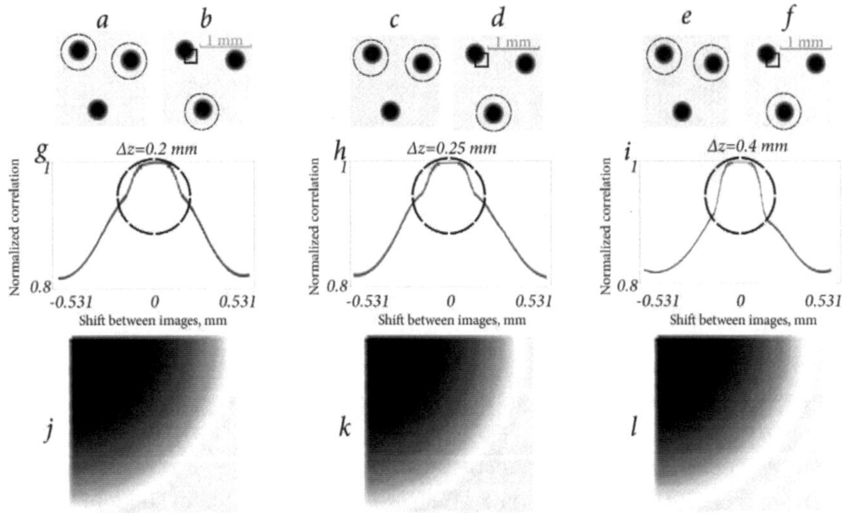

Figure 3. Normalized cross-correlation sections {g, h, i} between 2 slices ({a, c, e} – first for observer, {b, d, f} – second for observer) of volume of spherical particles. The difference in diffraction rings of defocused particles is shown in {j, k, l}. All parameters are similar to Fig. 2

The dashed circles are given to indicate focused particles, closer examination of the segments is marked with square frames. In the experimental and simulation cases we can see a separation of the peak of the correlation function from the pedestal due the diffraction of light.

4.2. Dependence of correlation on diameter of the particles

Correlation between two focused layers increases with increasing particles' size. It is understandable from the definition of cross-correlation function as a sum of different shifts between arrays in each pair of coordinates as shown in equation (1). The larger is the size of similar objects on images (focused and defocused particles), the greater is the shift between them will be to contribute to the peak of correlation function. We performed a simulation of particles with the same coordinates in each layer and increasing diameters D to reach maximum accuracy of the estimation of this dependence. The result of it is shown in Fig. 3. Peak broadening due to the increased size of the particles is observed.

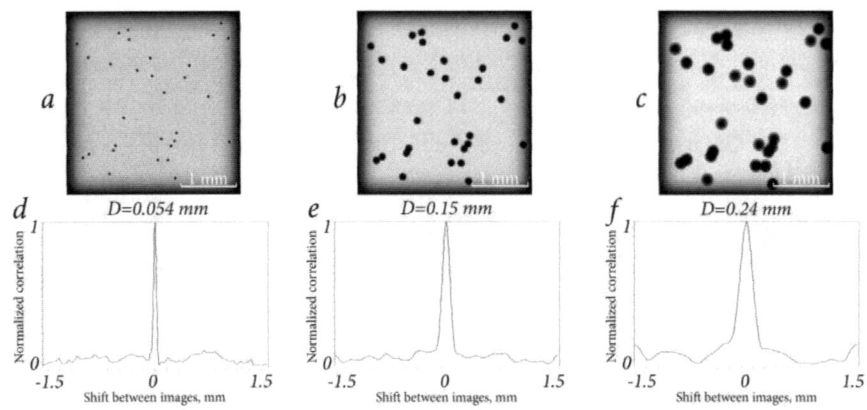

Figure 3. Random distribution particles {a, b, c} with $\Delta z = 0.15$ mm, $N = 3$, $N_P = 10$, $T = 0$ and their cross-correlations {d, e, f} calculated for 2nd and 3rd slices (from observer). Sizes of detected slices are 3.072 x 3.072 mm, $\lambda = 632.8$ nm

4.3. Dependence of correlation on concentration of the particles in a volume

Correlations between the pixels of adjacent segments of the volume are increasing with the increase of the concentration, or quantity of particles in each layer in volume. This effect can be explained by the increase of repetitive patterns with the increase of concentration in the case of equiprobable distribution of particles. This is evident in the reduction of the pedestal and the increase of the peak of the cross-correlation function (Fig. 4). Here a rapid decrease of the pedestal and it's detachability from the increased peak can be seen. Border between peak (p) and pedestal (b) are indicated.

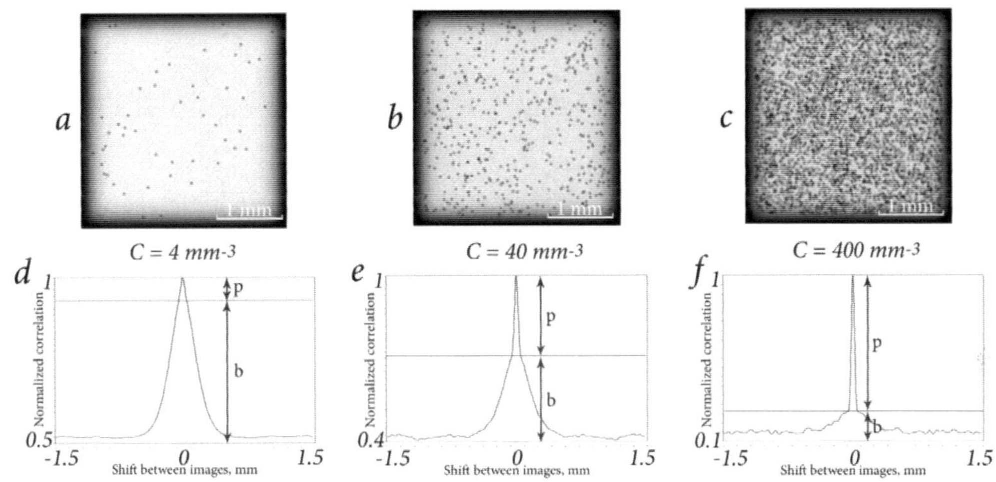

Figure 4. Random distribution of particles {a, b, c} with $\Delta z = 0.15$ mm, $D = 0.054$ mm, $N = 10$, $T = 0.01$ and increasing concentration C and cross-correlation {d, e, f} calculated for 2nd and 3rd slices (from observer). Sizes of detected slices are 3.072 x 3.072 mm, $\lambda = 632.8$ nm

Using the last stage of algorithm, namely, the correlation analysis of these images and the peak to the pedestal ratio calculation we can describe the distribution and concentration of particles in the volume. The dependence of the peak to the pedestal ratio on number of particles in each slice turns to be linear for random distribution.

5. Conclusion

In this study, we demonstrated an approach to the analysis of opaque particles suspended in a volume of the optical medium, which uses the method of digital holography and calculation of cross-correlation function. Numerically propagating the field that has passed through the investigated volume and comparing adjacent layers of it with the instrument of correlation analysis, we did a basic research of correlation function in addition to the study of particles. For validation of this method we investigated the basic characteristics of the cross-correlation function in the case of random distribution of the particles in a volume. Thus, we suggested and confirmed with the help of simulation and experiment that with the increase of the distance between the layers of particles, peak of correlation separates from the pedestal. Moreover, we have confirmed a consequence of the definition of the correlation function, namely, that an increase of the diameter of the particles results in the correlation peak broadening. Finally, we found that the ratio of the peak of the correlation function to its pedestal is linearly dependent on the concentration of particles in the volume. To obtain the results listed above a numerical simulation model has been developed. Identified characteristics and patterns can later be used to develop more accurate and efficient methods of characterization of opaque particles.

6. Acknowledgement
T. A. V. and T. Yu. N. acknowledge financial support from Government of Russian Federation, Grant No. 074U01. N. V. P. acknowledges financial support from Russian Ministry of Education and Science project within the state mission for institutions of higher education, agreement №2014/190.

7. References
[1] Voronetski A, et al. 2012 Measuring the spatiotemporal parameters of motion of self-luminous particles in high-temperature supersonic flow *J. Opt. Technol.* **79** 1
[2] Hesselink L 1988 Digital image processing in flow visualization *Ann. Rev. Fluid Mech.* **20** 1
[3] Johansson E, Benckert L, Sjodahl M 2004 Phase object data obtained from defocused laser speckle displacement *Appl. Optics* **43** 16
[4] Peterson K, Regaard B, Heinemann S, Sick V 2012 Single-camera, three-dimensional particle tracking velocimetry *Opt. Express* **20** 8
[5] Pitkäaho T, Niemelä M, Pitkäkangas V 2014 Partially coherent digital in-line holographic microscopy in characterization of a microscopic target *Appl. Optics* **53** 15
[6] Dyomin V, Kamenev D 2014 Methods of processing and retrieval of digital holograms particles and their practical application *Radiophysics and Quantum Electronics* **57** 8-9
[7] Nikolaeva T, Petrov N 2014 Statistical study of coherent images of particles in the volume of optical medium *Proc. SPIE* 921612-921612-8
[8] Luu L, Wang Zh, Vo M, Hoang Th, Ma J 2011 Accuracy enhancement of digital image correlation with B-spline interpolation *Opt. Lett.* **36** 16
[9] Fricke-Begemann T 2003 Three-dimensional deformation field measurement with digital speckle correlation *Appl. Opt.* **42** 34
[10] Malov A, Pavlov P 2012 Determination of surface roughness parameters optically opaque parts by speckles using spiral beams *Computer Optics* **36** 3
[11] Nikolaeva T, Petrov N 2015 Study of transparent particles in the volume of optical medium using digital holography and singular-optics approach *Proc. SPIE* 95080Y-95080Y-11
[12] Goodman J 1961 Introduction to Fourier Optics *McGraw-Hill Book Co.* 441
[13] Nikolaeva T, Petrov N 2015 Characterization of particles suspended in a volume of optical medium at high concentrations by coherent image processing *Opt. Eng.* **54** 8
[14] Sjödahl, M 1994 Electronic speckle photography: increased accuracy by non-integral pixel shifting *Appl. Opt.* **33**

Investigation of temperature characteristics of UV-LEDs with different GaN/AlGaN heterostructures

A S Evseenkov[1], S Yu Kurin[2], P O Tadtaev[1], S A Tarasov[1], A V Solomonov[1]
[1]St. Petersburg Electrotechnical University "LETI", 5 Prof. Popova Street, St. Petersburg 197376, Russia.
[2]GaN-Crystals Ltd., 27 Engels Ave, St. Petersburg 194156, Russia,

E-mail: as.evseenkov@gmail.com

Abstract. The quantum efficiency of AlInGaN-based LED structures with different active region thickness at different temperatures and biases was studied. The strong influence of the thickness of the active region on the characteristics of the samples was observed. Through the decrease of the active region thickness not only the emissive power dwindled but also thermodynamic properties tended to deteriorate and low temperature stability of the active region was observed. It was shown that the decrease in internal quantum efficiency through the increase of the current density was caused by charge leak from the quantum well. The heating of the active region due to the higher Auger recombination frequency led to the shift in the peak wavelength of the emission spectrum and lower emission power.

1. Introduction

At the present time, one of the main problems in development of LEDs and gallium nitride based lamps [1,2] is the decrease of the emission power of a single led in the matrix composition. Using a smaller number of LEDs in the lamp reduces its size, allows for easier electrical circuitry development and reduces the overall cost of the product. Watt-ampere characteristic is the most important characteristic of the emitters. It describes the change of radiation power with the increase in the values of direct current. Ideally, this dependence is linear. Unfortunately, in real emitters the increase of the power slows down and then goes into decline as the current grows. This limits the maximum achievable power. Such nonlinearity may be explained by self-heating of the crystal and the decrease in the efficiency of LEDs with increasing current density. The reasons for the low efficiency might be the lack of high-quality epitaxial heterostructures [3-6], low light extraction ratio [7], the influence of embedded piezoelectric fields in quantum well, charge leaks, structural defects and Auger recombination processes. A short wavelength ultraviolet emission requires an increase in the aluminum nitride concentration in a solid solution of the active region. This leads to additional technological complexity and increases the number of structural defects. In addition, in the ultraviolet range of wavelengths the absorption coefficient of epitaxial layers increases dramatically. This leads to the stronger influence of the thermal effects on the device's characteristics. What is more, the processes of self-heating depend on the device's emitting crystal properties and its package.

In order to study the processes of self-heating in the test-system [8], a new patented method of determining the temperature of the active region is used. It is based on a detailed analysis of the spectral characteristics of electroluminescence and Van-Rusbreka-Shockley theory, which considers the effect of impurities and defects on the structure's performance. The method is quick, non-contact and allows for investigation of the individual emitters included in the composition of the LED lamp or other LED-based products with no need for unsoldering. What is more, the technique allows to determine the temperature of the active region of the chip directly on the substrate surface. Thus, the method is applicable for carrying out quick temperature diagnostics of the structures at all stages of production of LEDs and LED-based products.

However, studies have shown that in case of light emitting structures with one substrate the difference in temperature parameters can sometimes reach considerable values, which is caused by the features of integrated grouped technique used in chips production. This leads to the necessity to control the quality of the structures within the substrate. Conventional methods based on determining the power and some other parameters of the structures do not always provide researchers with data for valid conclusions about structures'quality. Fast non-contact temperature measurements solve this problem. The diagnostic technique to evaluate the emitting chip temperature in the composition of the substrate was developed, followed by an automatic analysis of the obtained characteristics and valid conclusions about the quality of the structure. Time required for testing one single chip does not exceed 1 second. The technique is completely ready for the infusion to the production process.

In this paper an experimental investigation of the optical and electrical properties of ultraviolet light-emitting diodes based on GaN/AlGaN heterostructures was carried out, electroluminescence spectra, current-voltage characteristics and other important characteristics were investigated. Special attention is paid to the study of the influence of thermal processes on the spectral characteristics, optical power and emission efficiency of the structures. What is more, the temperature of the active region was determined and the effects of the composition on the values of these temperatures were studied.

2. Samples and Experimental techniques

In this paper UV-LEDs with emission wavelengths in the range from 355 to 385 nm were investigated. The method of chloride-hydride vapour phase epitaxy (CHVPE) was used to create light-emitting heterostructures. The means of reducing the mismatch of lattice parameters of semiconductor materials included growing a thick (over 100 nm) active region without the formation of dislocations and other defects.

The construction of UV LED heterostructure consisted of the following layers (starting from the substrate): AlN seed layer with a thickness of 200 nm; buffer area of alternating layers of AlGaN of variable composition with a total thickness of 3.5 μm; n-AlGaN hole-blocking layer (dopant - Si) with a thickness of 800 nm; AlGaN active region with a thickness of 80-150 nm; p-AlGaN electron-blocking layer (dopant - Mg) with a thickness of 200 nm; p-GaN upper contact layer 100 nm thick. The composition of the electron-blocking layer was selected so that, on the one hand, the electron leakage into the p region was suppressed and, on the other hand, the necessary hole injection level into the Al was retained. In this work the electron-blocking layer with a gradual decrease of Al concentration from 15% to 6% was used (counting from the active layer). The linearly decreasing gradient Al concentration in the electron-blocking layer is known to significantly improve structure's efficiency. Active zones contain 1-2% of aluminum. Two types of LED structures were made: ones with a crystal size of 1 × 1 mm without package and ones with the crystal size 0.31 × 0.31 mm packaged. Metal compositions of Ti/Al and Ni/Al were used for N-type and p-type electrodes respectively. Processing and packing of samples was performed using an external service.

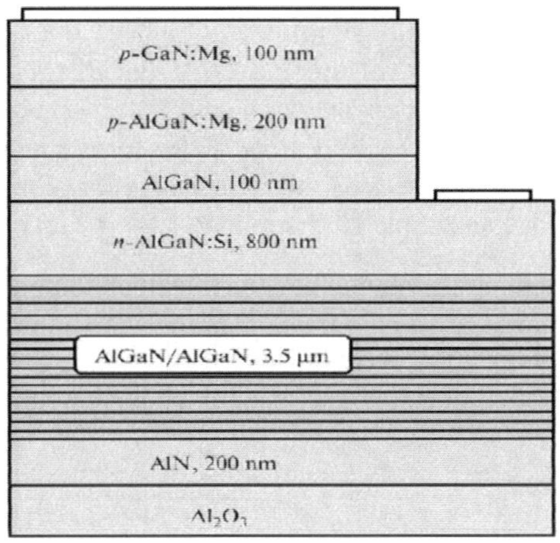

Figure 1. The structure of the ultraviolet light-emitting diode

3. Results and Discussion

In this paper an experimental study of the characteristics of a significant number of ultraviolet light emitting diodes based on different types of heterostructures and GaN/AlGaN is presented. The study of heterostructures was performed using the test set developed by the authors [2, 9, 10]. The main tasks were the study of the thermal processes which take place in these structures, identification of their ground and evaluation of the possibility of the application of thermal parameters analysis in estimating the structures' quality. For further analysis of the obtained data two samples were selected and, being characterized by their best performance parameters - power and efficiency, are hereinafter referred to as a sample of the first type (first sample) and a sample of the second type (second sample).

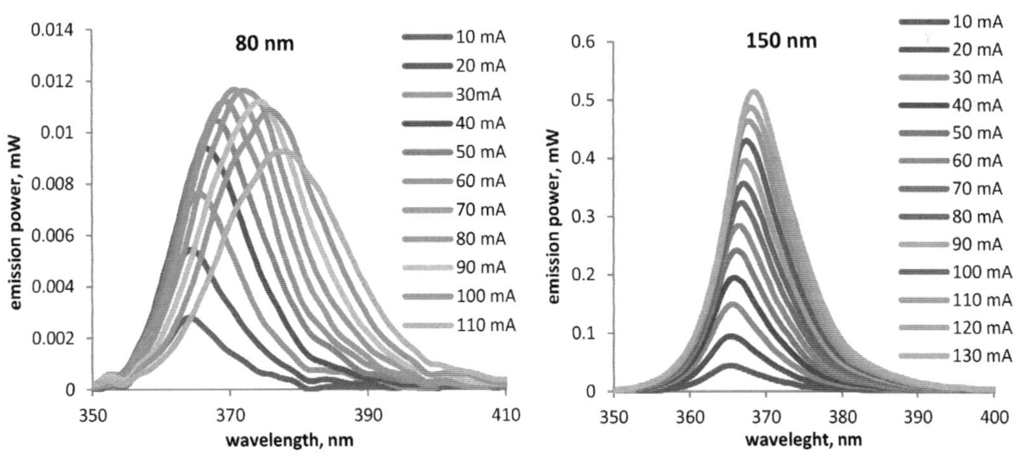

Figure 2. Electroluminescence spectra of UV LEDs with the active region thickness of 80 nm and 150 nm

The main characteristic of any light emitting diode is its emission spectrum. For a full investigation of the features of the samples, the spectral characteristics measurement must be performed within the whole operational range of currents. The emission spectra of the investigated samples are shown in Fig. 2. A very significant difference in all spectra parameters is observed, including the peak's shift and the change of the area under the curve through the increase of the direct current. The results may be also presented as the dependences of the emission power (Fig. 3), efficiency (Fig. 4) and the emission wavelength at the maximum (Fig. 5) on the direct current. The analysis of the characteristics shows that for the samples of the second type not only significantly higher values of the power (up to 5 mW) are achieved, but also that the emission power grows almost linearly. For a sample of the first type a significant deviation from the straight dependence occurs at 24 mA already; at 80 mA the power saturation is observed and it the curve starts to decline with a further increase in the current. This limits the achievable values of the emission power by more than an order of magnitude. Similar processes are observed for the efficiency - for the sample of the second type it exceeded 6% and saturated at higher currents (40 mA).

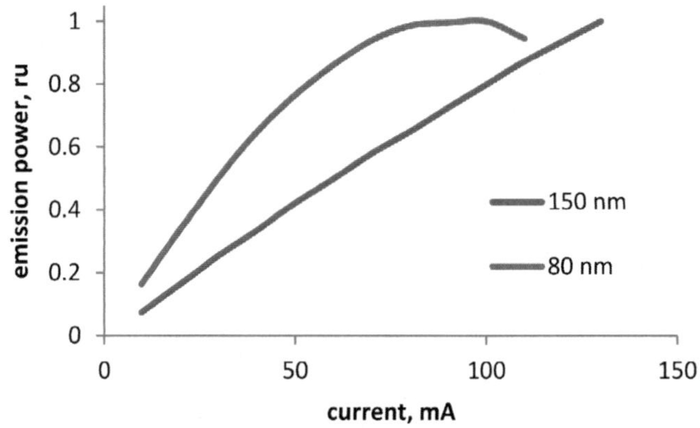

Figure 3. Power versus current for the samples with an active region thickness of 80 nm and 150 nm

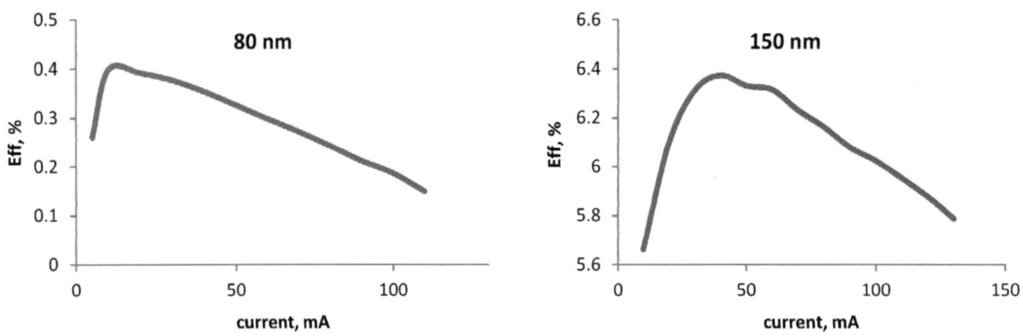

Figure 4. The dependence of the efficiency on the current for the samples with an active region thickness of 80 nm and 150 nm

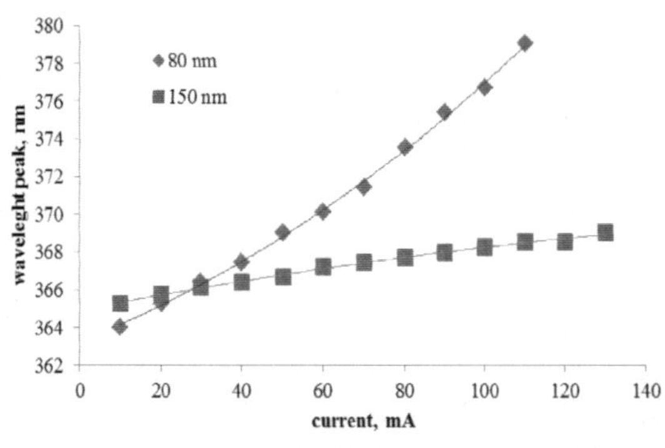

Figure 5. The dependence of the emission wavelength at the maximum on the current for the samples with an active region thickness of 80 nm and 150 nm

Preliminary information about the processes allows us to analyze the dependence of the peak wavelength of spectral characteristics on the direct current (Fig. 5). It is shown that in both cases there is a shift towards the longer wavelengths and for the first sample equals 14 nm while raising the current up to 110 mA, while for the second one the peak is shifted by only 3 nm. It is known that a shift towards longer wavelength is determined by self-heating processes in the structure, and that was further confirmed by the measurements of the spectral characteristics of the LED at a constant current and increasing the ambient temperature. In order to study these processes we analyzed the influence of the temperature on the structures' performance (Fig.6).

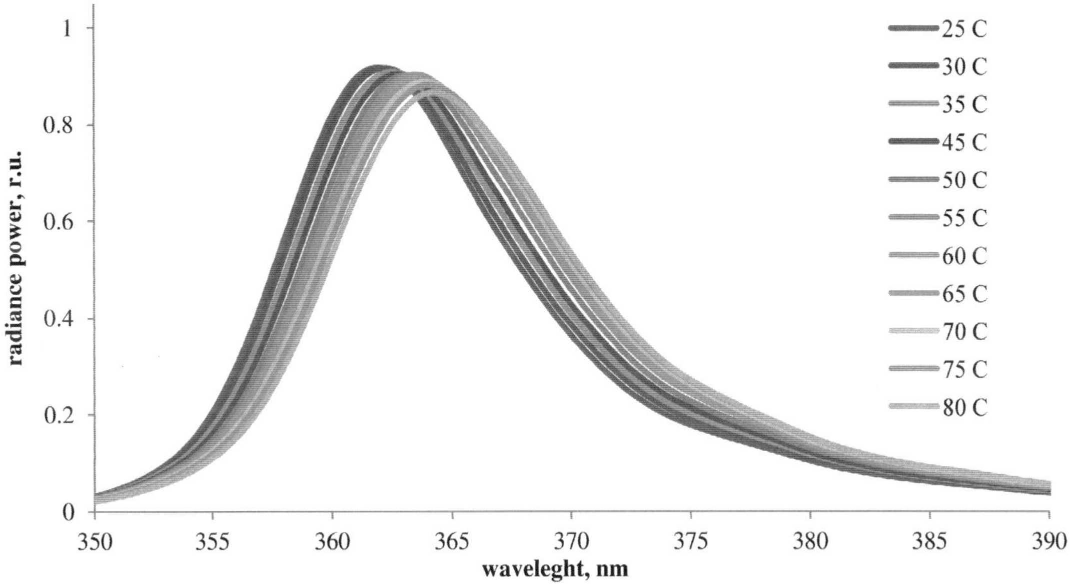

Figure 6. The emission spectra for the sample with an active region thickness of 80 nm at various temperatures

It is shown that the intensity of self-heating processes of the sample of the second type at 100 mA is more than 2.5 times lower than that of the samples of the first type for which the local temperature may exceed 200 degrees Celsius. This may be due to the following facts. Samples of the first type are characterized by a much higher density of the defects, which is associated with the non-optimized compositions of buffer layers and a smaller thickness of the active region. The features of these structures in terms of defects presence were investigated previously by the methods of secondary ion mass spectrometry, luminescence spectroscopy, atomic force microscopy and X-ray analysis; the research results confirmed these conclusions. Increasing the density of the defects increases the probability of non-radiative recombination, which leads to the additional heating of the structures. Therefore, temperature diagnostics allow for a high precision estimation of the presence of defects in the emitting region of the LEDs.

4. Conclusions

The study of the spectral and temperature characteristics of UV LEDs, created by CHVPE, was done. The experiment showed that the peak wavelength of the investigated samples was in the range of 363-367 nm at an operating current, the half-width of the emission spectrum didn't exceed 10-15 nm. At the operating current of 20 mA the temperature of the active region of the first sample was 36 degrees Celsius, but at 100 mA it raised up to 246 degrees Celsius; the optical output power and efficiency were equal to 0.29 mW and 0.4% respectively.

The second sample showed the best performance. At the operating current of 20 mA the temperature of the active region was equal to 36 degrees Celsius, at 100 mA it raised to 119 degrees Celsius. Optical output power and efficiency at an operating current were equal to 3.9 mW and 6.1% respectively.

The results can be explained by the main difference between the samples - the thickness of the active region. Small area dissipates heat worse and the study confirms that passing the current through such a small volume of the semiconductor leads to a sharp deterioration in its power characteristics. What is more, a smaller area of photon generation at the same current results in the lower intensity of the emission.

There is another phenomenon which influences the LEDs' parameters. Achieving ultraviolet emission requires higher concentration of the aluminum if the solid solution which, in turn, leads to the bigger lattice mismatch between the substrate and the grown layers. This complicates the production process significantly as the quality of the heterostructures becomes more crucial. Results obtained during this study may prove that production technique is not defectless enough. Thus, the method used in this study may be used for the analysis of the quality of the created structures.

Acknowledgments
The work was supported by the Ministry of Education and Science of the Russian Federation in the framework of the project of the state task in the field of scientific activity, project number 16.1307.2014K.

References
[1] Panfeng J, Naixin L, Tongboo W 2011 *Journal of Semiconductors* Vol. **32** 11
[2] Menkovich E A,Tarasov S A, Lamkin I A, Kurin S Yu, Antipov A A, Roenkov A D, Barash I S, Helava H I, Makarov Yu N 2013 *Journal of Physics: Conference Series* **461** 012027
[3] Kim J K Alhmed N, Frank W. 2008 *Appl Phys Lett* **93** 221111
[4] Evseenkov A S, Tarasov S A, Lamkin I A, Solomonov A V, Kurin S Y. 2015 IEEE North West Russia Section Young Researchers In Electrical And Electronic Engineering Conference, ELCONRUSNW 2015 27-29
[5] Kurin S, Antipov A, Barash I, Roenkov A, Makarov Y, Usikov A, Helava H, Solomonov A, Tarasov S, Evseenkov A, Lamkin I. 2015 *Physica Status Solidi (C) Current Topics in Solid State Physics* **12** № 4-5 369-371
[6] Evseenkov A S, Tarasov S A, et al 2015 *IOP Publishing Journal of Physics: Conference Series* **643** 012033
[7] Son J H, Jong U K, Yang Hee Song 2012 *Advanced Materials* **24** 2259-2262
[8] Gu Y, Narendran N. 2004 *Proceedings of SPIE* **5187**: 107-114
[9] Solomonov A V, Tarasov S A, Men'kovich E A, Lamkin I A, Kurin S Yu, Antipov A A, Barash I S, Roenkov A D, Helava H I, Makarov Yu N 2014 *Semiconductors* **48** 245–250
[10] Menkovich E A, Solomonov A V, Tarasov S A, Yurgin P A 2014 *Functional materials* **21** 186-189

Journal of Physics: Conference Series **769** (2016) 012054 doi:10.1088/1742-6596/769/1/012054

Selective Au-AlGaN photodetectors for the 350 - 370 nm wavelength range

S M Altimime[1], A S Evseenkov[1], I I Mikhailov[1], S A Tarasov[1], I A Lamkin[1], M Y Andreev[1], A V Solomonov[1], S Yu Kurin[2]
[1]Saint-Petersburg Electrotechnical University "LETI", Prof. Popova 5, St. Petersburg 197376, Russia
[2]GaN-Crystals Ltd., 27 Engels Ave, St. Petersburg 194156, Russia,

E-mail: ialamkin@mail.ru

Abstract. The selective Au-AlGaN-based Schottky barrier photodetectors operating in different UV ranges have been fabricated. The wide bandgap window effect and the over barrier emission are proposed as methods for control of the photosentivity spectrum. Selective photodetectors were fabricated with the following parameters: a full width at half maximum of 5-6 nm with a maximum at 355 nm, 362 nm, 366 nm, and a sensitivity up to 140 mA / W.

1. Introduction

Devices based on photodetectors are widely used in virtually all areas of human activity [1], [2]. Recently, the need to create selective photodetectors, that can reliably provide radiation level control in a given wavelength range under conditions of high background illumination, has become more evident [3]. The use of optical filters to achieve high selectivity is often unacceptable due to their high cost and high degradability. Development of selective photodetectors for the ultraviolet region of the spectrum is especially important [4], [5]. Wide bandgap Aluminum Gallium Nitride (AlGaN) solid solutions, recognized as one of the most promising materials, can be used to achieve this goal. To successfully implement such detectors into production, it is necessary to develop a fabrication method that provides high parameters as well as low cost of the devices [6]. The abovementioned requirements can be met with structures based on the Schottky barrier [7]. Such structures can be used for designing high-performance UV photodetectors by means of comparatively simple and inexpensive methods [8].

2. Samples and experimental technique

The AlGaN epitaxial layers were grown by hydride vapour phase epitaxy (HVPE) technique [9], [10]. The top *n*-doped AlGaN layer had a carrier concentration of 10^{17} cm^{-3}. Metal layers were deposited by vacuum thermal evaporation at a residual gas pressure of 10^{-5} Torr. Au was used as a rectifying contact material. Al layer with Ti underlayer were used for forming Ohmic contacts [11], [12]. Prior to metal deposition the structures were cleaned with potassium hydroxide (KOH) during 2 min and washed with distilled water. To achieve good metal adhesion the structures were heated during deposition to a temperature of 300 °C. Photodetector characteristics were measured using a set-up based on a monochromator with a diffraction grating and a xenon lamp as a light source.

3. Experimental results and discussions

Using the wide bandgap window effect and the over barrier emission in Au-AlGaN structure, it is possible to fabricate a selective photodetector with a narrow range of photosensitivity. Selectivity was achieved by using the wide bandgap window effect. Figure 1 shows the structure of the photodetector based on Au-AlGaN contact with a AlN content of 0.1. When the photodetector is illuminated from the metal side, two peaks are present in the photosensitivity spectrum, where the long-wavelength peak is caused by the over barrier emission. When increasing the energy of incident photons, interband transitions occur resulting in a short-wavelength peak at 346 nm. When the photodetector is illuminated from the side of the transparent sapphire substrate, light passes through the wide bandgap window, and in the spectral characteristics only the over barrier emission peak is observed. Thus, it is possible to fabricate a selective photodetector with the photosensitivity maximum at 355 nm.

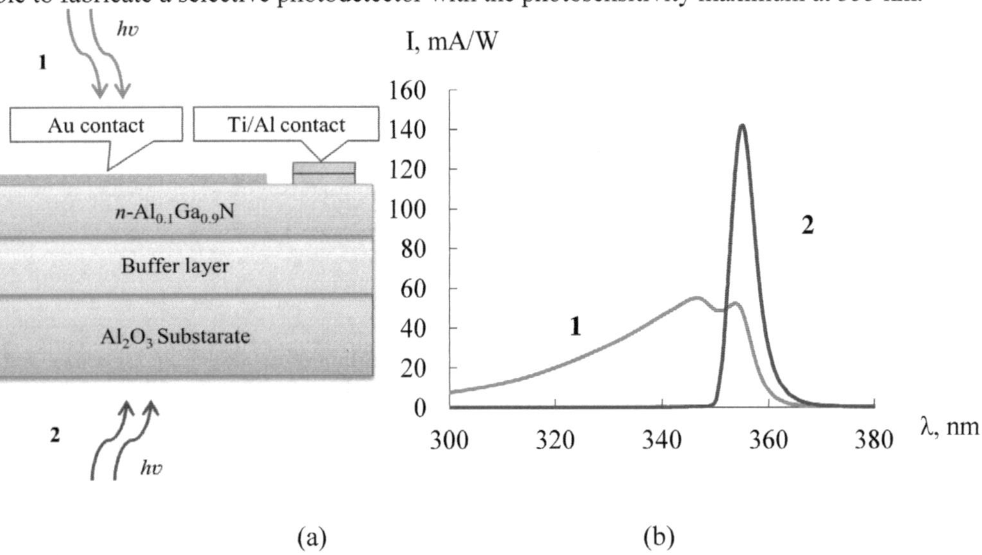

Figure 1. a) Au-$Al_{0.1}Ga_{0.9}N$ photodetector structure b) photosensitivity spectrum of: 1 - the top-illuminated photodetector, 2 - the back-illuminated photodetector.

Figure 2 shows processes that take place in the sample being illuminated from sides of metal layer and the substrate. When the sample is illuminated by monochromatic light with an energy hv > $q\varphi_B$ from the side of metal contact, excited electrons in the metal layer have sufficient energy to overcome the barrier (Process 1 in Figure 2). If hv > E_g, and the metal film is thin enough, then the light passes through the metal layer, which results in electron-hole pairs generation in the semiconductor (Process 2 in Fig 2).

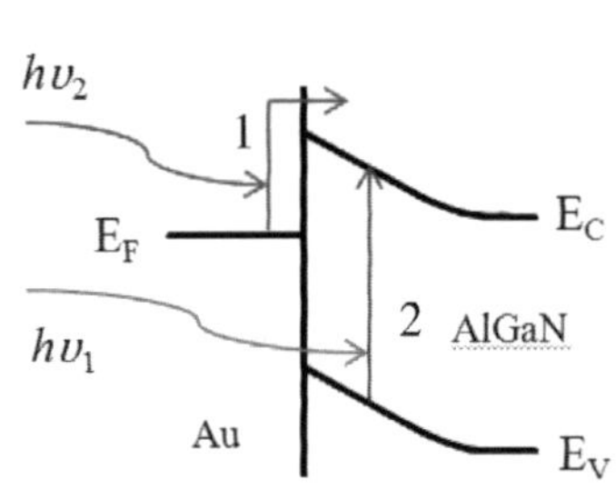

Figure 2. Energy diagram of photoexcitation processes

The excitation of electrons in the metal layer with energy sufficient to overcome the potential barrier (hv > $q\varphi_B$) is also possible when the sample is illuminated from the substrate side (Process 1 in Fig 2). However, when hv > E_g, the light is intensively absorbed on the back surface of the top semiconductor layer. Therefore, the probability for electron-hole pairs generated in this region to reach the metal-semiconductor interface is very small. The photosensitivity spectrum position along the wavelength axis can be controlled by varying the composition of the solid solution that forms a Schottky barrier with Au. For this purpose, a photodetector based on epitaxial layer AlGaN with an AlN content of 0.06 was fabricated (Fig. 3a). In this case, by illuminating the photodetector from the metal side, the over barrier emission can be recognized as an inflection in the long-wavelength region of the obtained spectrum. When this structure is illuminated from the substrate side, the top AlGaN layer with an AlN content of 0.06 acts as a wide bandgap window. Thus, it is possible to fabricate a selective high photosensitive photodetector with a peak sensitivity wavelength of 362 nm (Fig. 3b).

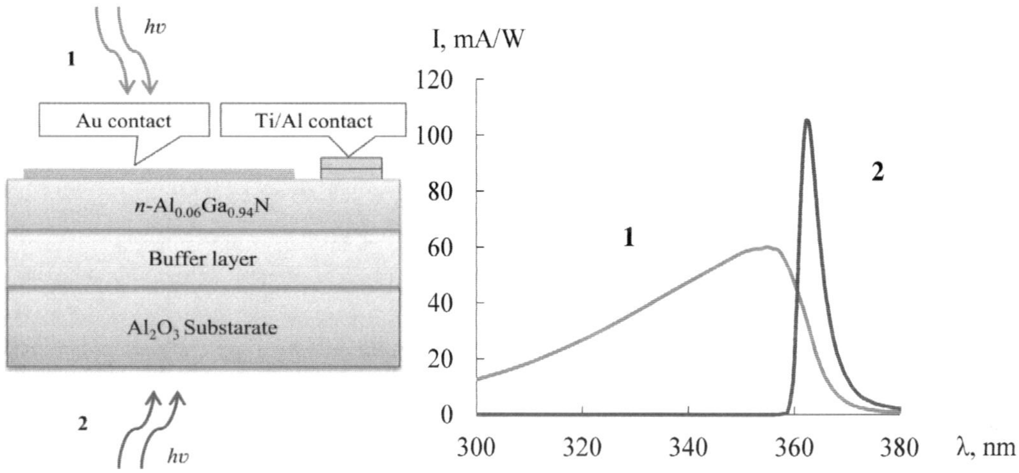

(a) (b)

Figure 3. a) Au-$Al_{0.06}Ga_{0.94}N$ photodetector structure b) photosensitivity spectrum of: 1 - the top-illuminated photodetector, 2 - the back-illuminated photodetector.

It is possible to control the sensitivity in both short- and long-wavelength spectrum regions if an extra layer is used in the photodetector structure (Fig. 4a). In this case GaN layer, acting as a wide bandgap window, sets the short-wavelength boundary of the photosensitivity spectrum to 360 nm (Fig. 4b).

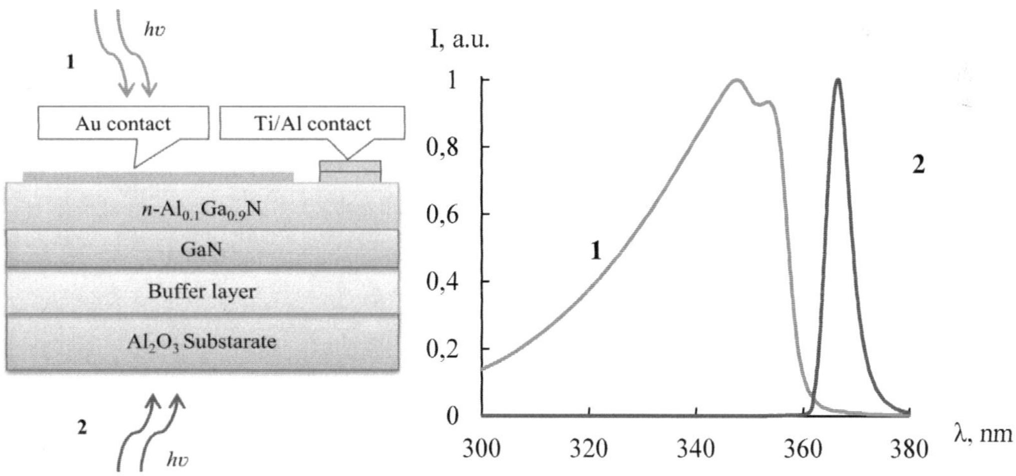

(a) (b)

Figure 4. a) Au-$Al_{0.1}Ga_{0.9}N$-GaN photodetector structure b) photosensitivity spectrum of: 1 - the top-illuminated photodetector, 2 - the back-illuminated photodetector.

The long-wavelength boundary is set by the potential barrier height of the Au-AlGaN contact with an AlN content of 0.06.

Although the maximum photosensitivity value of the structure with an extra GaN layer is lower in relation to the structures considered above, this structure is designed to show how the effects of wide bandgap window and over barrier emission when applied in combination can be used to control the photodetector sensitivity range. Thus, here we have demonstrated a wavelength selective Au-AlGaN photodetector with the possibility of setting its selectivity (see Table 1).

Table 1. Characteristics of Samples.

Parameter	Au-$Al_{0.06}Ga_{0.94}N$	Au-$Al_{0.1}Ga_{0.9}N$/ GaN	Au-$Al_{0.1}Ga_{0.9}N$
λ_{max} under illumination from the side of metal, nm	355	347 and 353	346 and 353
λ_{max} under illumination from the side of substrate, nm	362	366	355
FWHM under illumination from the side of metal, nm	40	30	30
FWHM under illumination from the side of the substrate, nm	5	5	6
The sensitivity range under illumination from the side of metal, nm	200-380	200-375	200-375
The sensitivity range under illumination from the side of substrate, nm	358-378	361-375	350-375

Selective photodetectors were fabricated with the following parameters obtained: a full width at half maximum (FWHM) of 5-6 nm, photosensitivity in the range of 351 to 373 nm with a maximum at 355 nm, 362 nm, 366 nm, and a sensitivity up to 140 mA / W.

4. Conclusion

Selective photodetectors based on solid solution AlGaN have been fabricated. On the basis of the structure with the top epitaxial AlGaN layer (with an AlN content of 0.1) that forms a Schottky barrier with Au, a selective photodetector with a photosensitivity maximum at 355 nm and a FWHN of 6 nm has been fabricated. With the AlN content reduced to 0.06 the photosensitivity maximum shifted to 362 nm, whereas the FWHN decreased to 5 nm. By using an extra GaN layer, which acts as a wide bandgap window in the Au-$Al_{0.1}Ga_{0.9}N$ structure, the sensitivity maximum was shifted to 366 nm. It has been shown how varying the solid solution composition, which acts as a wide bandgap window, can change the range of the photosensitivity in the short-wavelength region of the spectrum.

Acknowledgment
The study was supported by the Ministry of Education of the Russian Federation within the framework of the state project in the field of scientific activity, project number 16.1307.2014K.

References

[1] Kung P, Yasan A, McClintock R, Darvish S R, Mi K, Razeghi M 2002 *Proc. SPIE* **4650** 199.
[2] Helava H I, Menkovich E A, Kurin S Yu, Antipov A A, Roenkov A D, Barash I S, Tarasov S A, Shmidt N M, Makarov Yu N, Lamkin I A 2013 *Journal of Physics: Conference Series* **461** 012028.
[3] Petrov A A, Tarasov S A, Lamkin I A, Menkovich E A, Solomonov A V, Kurin S Yu 2014 *Journal of Physics: Conference Series* **572** 012063.
[4] Lamkin I, Tarasov S 2013 *Journal of Physics: Conference Series* **461** 012025.
[5] Ulmer M, Razeghi M, Bigan E 1995 *Proc. SPIE* **2397** 210.

[6] Menkovich E A, Tarasov S A, Lamkin I A 2012 *Functional Materials* **19** 233.
[7] Lamkin I A, Andreev M Y, Tarasov S A, Solomonov A V, Kurin S Y *Proceedings of the 2015 IEEE North West Russia Section Young Researchers in Electrical and Electronic Engineering Conference, ElConRusNW 2015* 2015 30.
[8] Rumyantsev S L, Pala N, Shur M S, Gaska R, Levinshtein M E 2001 *Applied Physics Letters* **79** 866.
[9] Kurin S, Antipov A, Barash I, Roenkov A, Usikov A, Helava H, Ratnikov V, Shmidt N, Sakharov A, Tarasov S, Menkovich E, Lamkin I, Papchenko B, Makarov Y 2014 *Physica status solidi (C)* **3** 813.
[10] Barash I, Roenkov A, Kurin S, Antipov A, Helava H, Tarasov S, Menkovich E, Lamkin I, Makarov Yu. 2013 *Physica Status Solidi (C) Current Topics in Solid State Physics* **10** 289.
[11] Lamkin I A, Tarasov S A, Petrov A A, Menkovich E A, Solomonov A V, Kurin S Yu 2014 *Journal of Physics: Conference Series* **572** 012062.
[12] Andreev M Y, Lamkin I A, Tarasov S A, Mikhailov I I, Solomonov A V *Proceedings of the 2015 IEEE North West Russia Section Young Researchers in Electrical and Electronic Engineering Conference, ElConRusNW 2015* 2015 13.

Mode selection control in microring resonators

I V Levitskii[1,*], V P Evtikhiev[2]

[1]Peter the Great St. Petersburg Polytechnic University, St. Petersburg, Russia
[2]Ioffe Institute, St. Petersburg, Russia

*Corresponding author: levitskyar@gmail.com

Abstract. We present numerical simulations of a novel microring resonator geometry, based on subwavelength hollow core defects inside the resonator cavity in order to study its influence on mode structure. The microring resonator has 9 µm outer diameter and 5 µm inner diameter, subwavelength defects are round shaped and have 100 nm diameter. Results allowed us to formulate basic approaches for controlling mode structure of ring microresonators.

1. Introduction

The rapid growth of information flows leads to the need of continuous improvement of instruments and devices for its processing, storage and transmission. The advent of WDM channels, allowing information packets pass simultaneously on dozens of carrier frequencies using a single path, principally increased competitive solutions based on optical transmission and processing of information. Thus, Photonic Integrated Circuits (PICs) have been subject to extensive research over the last years [1,2]. A key component in many PICs is the optical resonator. Various types of resonators were demonstrated, including, for example, microring resonators (MRR) [3,4], Photonic Crystals (PhCs) cavities [5], Bragg resonators, Fabry-Perot resonators, microdisk resonators, vertical-external-cavity surface-emitting-laser (VECSEL) [6] etc. The potential applications for such resonators include laser sources, modulators, filters, sensors, optical delay lines, nonlinear optics and more.

MRRs used in PIC applications typically have small ring thickness and therefore support only whispering gallery modes (WGMs), but in general microring resonators sustain higher order radial modes. The mode structure changes of microring lasers caused by focused ion beam (FIB) milling of pits and grooves were demonstrated [7]. However, the information about proper positioning and selecting proper size, shape and number of defects is absent. Thus, we conducted more detailed research on possibility of mode selection control via subwavelength defects.

2. Perforated microring resonators

Recently it was shown that microring resonators with couple of microns (ranging from less than 2 µm up to 12 µm) diameter containing active region can sustain laser generation [8,9]. We obtained a number of micro-photoluminescence spectra of GaAs microring resonators with active region based on InAs QD. We found that MRR with 9 µm outer diameter and 5 µm inner diameter exhibits the best performance (among our set). So in this paper we considered MRR of this size. The simulations were performed using finite element method (FEM) in COMSOL Multiphysics software. While the full three-dimensional simulations are necessary for determining parameters of the device, as intended for use in real world applications, two-dimensional simulations can also provide some insight into the structure of the resonator and its performance. Also as the geometry of the structure becomes complicated, with finer details, a smaller grid step size is required. Thus, simulations of a structure with fine details in 3D become time-consuming. Only a single microring resonator problem can be fully reduced from 3D to 2D calculations – this is possible due to axial symmetry of the problem. In this paper all results come from 2D simulations, except mode analysis in a single microring resonator stated above. The ring consists of material with refractive index $n_1 = 3.43$ (corresponding to refractive index of GaAs on ~1.2 µm wavelength). The ambient media has refractive index $n_0 = 1$.

For further investigation we obtained eigenmodes of the MRR and chose 3 modes with different radial order numbers – WGM, 2nd and 3rd radial modes. For example, the light intensity distribution of the 3rd order radial mode is shown in the figure 1.

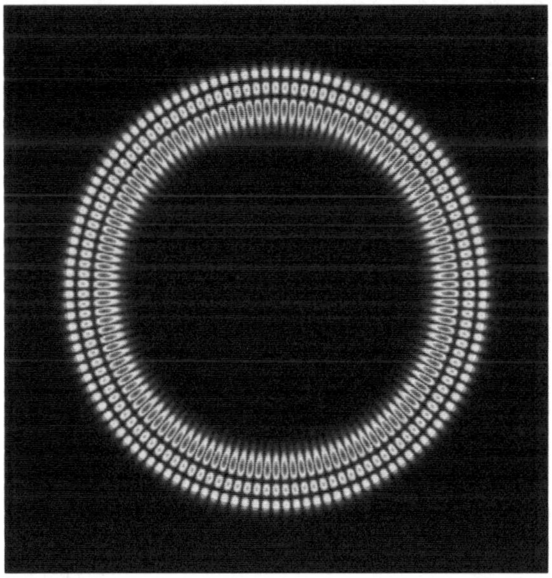

Figure 1. 3rd order radial mode in the MRR.

2.1. Microring resonator with single defect

The size of the light intensity maximums of the modes averaged in 100 nm, so one should expect that defects of this size would be critical for light propagation in the MRR. Indeed, bigger defects will cause diffraction and the anticipated effect will not be local, on the other hand smaller ones have neglectable influence. In comparison with the light wavelength in the material (~340 nm), the defects are subwavelength-sized. The simplest modification of the MRR is a single subwavelength hollow core defect of 100 nm diameter positioned in the maximum of the light intensity of the mode. The light intensity distribution of the 3rd order radial mode in the locally perforated MRR is shown in the figure 2. The defect affects the mode in such a way, that maximums of the light intensity redistribute from defect area to the edge of the MRR. But this influence is not enough to remove this mode from eigenmode spectrum of the MRR. Also this defect doesn't affect WGM and 2nd order mode because it's located relatively far from their maximums. Of course, radial mode has axial symmetry, and it can be rotated in such a way that defect falls in between maximums of the light intensity distribution, but it will be noted that in real world applications the symmetry is violated.

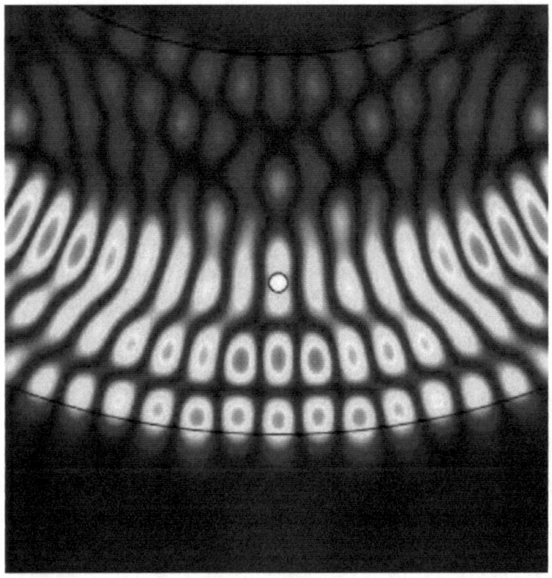

Figure 2. 3rd order mode in the MRR with single defect

2.2. Microring resonator with several defects

The next straightforward step is to add one more defect. But our simulations have shown that only equilateral triangle configuration of 3 subwavelength defects is sufficient to suppress 3rd radial mode and removes it from eigenmode spectrum. Figure 3 shows the light intensity distribution calculated on the frequency corresponding to the chosen 3rd radial mode.

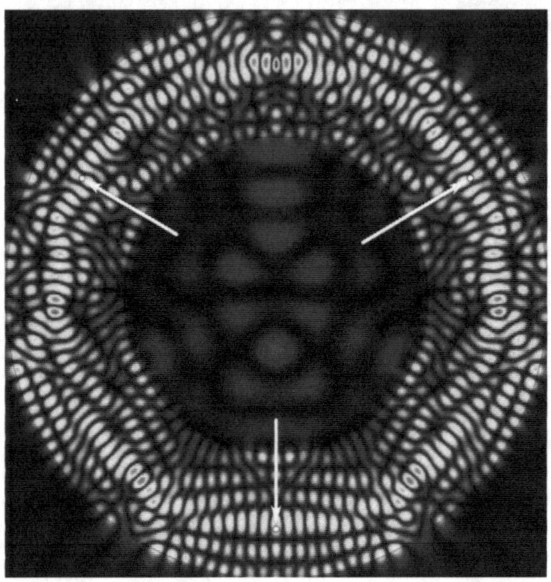

Figure 3. The light intensity distribution in the perforated MRR.

1st, all 3 defects fall into maximums of the light intensity distribution, 2nd huge number of the light intensity maximums filling the entire volume of the MRR.

Both these facts result in removing the 3rd radial mode from the eigenmode spectrum, and it is in the agreement with the results of solving eigenvalue problem. This approach can be tuned to remove any other mode just by placing defects in the proper position corresponding to maximum of the light intensity of mode.

2.3. Add-drop filter

All the previous results were obtained for MRR alone, but in real world applications input and output of light are required. It is achieved by placing planar waveguides in the vicinity of MRR, so the light can exhibit tunneling back and forth. By varying the tunneling gap one can tune the Q-factor of the system. In integrated optics this system is known as add-drop filter. It should be noted that the addition of waveguides to MRR brakes the axial symmetry of the system and "fixes" MRR modes, so the previous considerations of placing defects in intensity maximums were correct (in the system MRR-waveguide mode is bound to the gap).

We simulated add-drop filter consisting of the MRR (diameter of 9 μm) and 2 planar waveguides placed opposite from each other. Waveguides have 300 nm thickness and the same material properties as MRR, the gap between waveguide and MRR is 100 nm. The principal scheme is shown in the figure 4 (a).

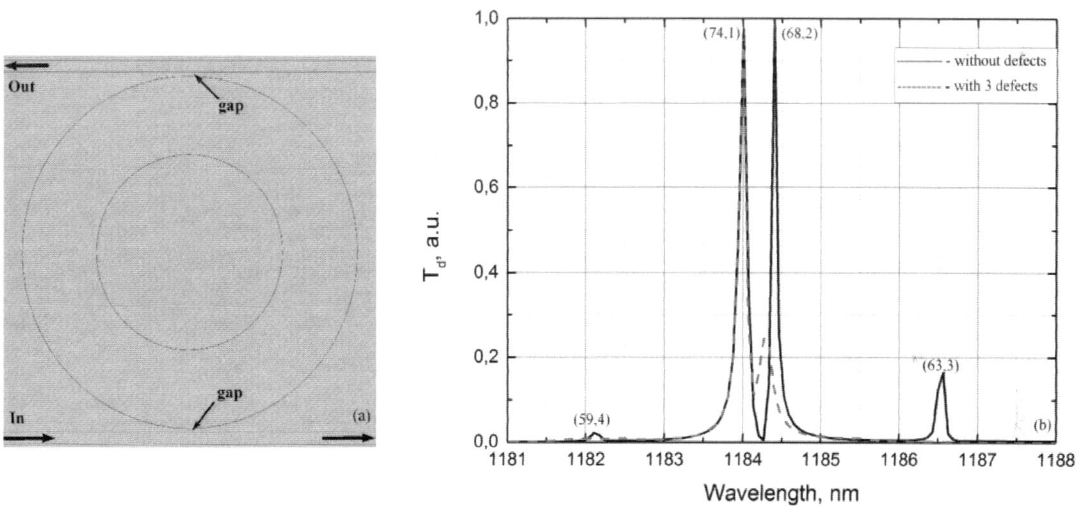

Figure 4. (a) Scheme of the add-drop filter. (b) Transmission spectrum of the add-drop filter. Numbers in brackets (*M*,*N*) indicate azymuthal *M* order and radial *N* order of a mode.

The calculated spectrum of the system is shown in the figure 4 (b). Black solid line corresponds to MRR without any defects. In this case 4 modes of different orders are excited. Red dashed line corresponds to MRR with 3 defects placed in equilateral triangle configuration. In that case the WGM still excites, but the 2nd order radial mode exhibits partial suppression, while the 3rd order mode is totally suppressed as was expected. Also 2nd order radial mode exhibits blue shift due to change in effective refractive index for this mode caused by defects. According to the principle of reciprocity, these results also can be used in the case of microring laser coupled with waveguides.

3. Conclusion

As a result we formulate basic principles of controlling mode spectrum of MRR via subwavelength hollow core defects: size of a defect should match spatial size of light intensity maximum of the chosen mode and not exceed $\lambda_{mat}/4$, defect should be positioned in light intensity maximum of the chosen mode for maximum influence. Also we state that configuration of 3 defects is sufficient for suppressing different MRR modes.

References

[1] Kirchain R and Kimerling L 2007 *Nature Photonics* **1** 303–5
[2] Rabus D G 2007 *Integrated Ring Resonators: The Compendium* (Berlin: Springer)
[3] Ilchenko V S and Matsko A B 2006 *IEEE Journal of Selected Topics in Quantum Electronics* **12** 1 3–14
[4] Mohammadi F, Ahmadi V and Gandomkar 2012 *Lightwave Technology* v.**30** 23 3720–6
[5] Akahane Y, Asano T, Song B and Noda S 2003 *Nature* **425** 944–7
[6] Kuznetsov M *et al.* 1997 *IEEE Photon. Technol. Lett.* **9** (8) 1063
[7] Zhukov A E *et al.* 2014 *Opt. Express* **22** 25782–7
[8] Zhukov A E *et al.* 2014 *Semiconductors* v.**48** 12 1626–30
[9] Zhukov A E *et al.* 2012 *Semiconductors* v.**46** 8 1027–31

High-temperature lasing in diode microdisk lasers with InAs/InGaAs quantum dots

E I Moiseev[1], N V Kryzhanovskaya[1,2], Yu S Polubavkina[1], M V Maximov[1], M M Kulagina[2], S I Troshkov[2], Yu M Zadiranov[2], A A Lipovskii[1,5], I S Mukhin[1,3], M Guina[4], T Niemi[4], A E Zhukov[1]
[1]St Petersburg Academic University, 8/3 Khlopina, 194021, St Petersburg, Russia
[2]Ioffe Institute, 26 Polytechnicheskaya, St Petersburg, 194021, Russia
[3]ITMO University, Kronverkskii 49, St Petersburg, 197101, Russia
[4] Optoelectronics Research Centre, Tampere University of Technology, Tampere, Finland
[5]Peter the Great St.Petersburg Polytechnic University, St. Petersburg, 195251 Russia

E-mail: moiseev@spbau.com

Abstract. We demonstrate that quantum dot microdisk lasers are able to operate under continuous wave current injection at 100 °C. We also present a novel method for increasing a side mode suppression ratio in microdisk lasers.

1. Introduction

High-temperature operating diode lasers are attractive for a variety of applications including on-chip communication. For this purpose, microlasers are more suitable owing to their small footprint. In(Ga)As quantum dots (QDs) are successfully used in edge-emitting (macro) lasers capable of operating at temperatures as high as 163-220 °C [1, 2]. However, the maximal temperature of CW lasing reported for QD micodisk laser is 50 °C [3]. It is strongly desired to have a single-frequency lasing spectrum for optical communication. However, QD-based microring and microdisk lasers typically demonstrate multi-frequency lasing on whispering gallery modes of different azimuthal and/or radial orders. In order to enhance single frequency lasing, various methods have been proposed (for example, [4, 5]), which are usually based on suppression of side modes, whereas intensity of a dominant mode remains unchanged or even degrade.

In this work, we present results on high-temperature operation of microdisk laser. The key modification, which results in successful lasing even at 100 °C, was the use of deep localized quantum dots in combination with their multiple stacking. We also present our preliminary results on characterization of QD microdisks with nano-antennas attached to the microdisk side wall. Owing to more effective light outcoupling, side mode suppression ratio in such microdisk increases owing to enhancement of the dominant mode intensity rather than introducing additional loss for some undesired modes.

2. Injection-pumped microdisk lasers

An epitaxial structure was grown by molecular beam epitaxy on an n+ GaAs(100) substrate. An active region comprises 10 layers of InAs/InGaAs quantum dots. We used a 0.44-µm-thick GaAs waveguiding layer with $Al_{0.25}Ga_{0.75}As$ *n*- and *p*-type doped claddings. A 31-µm – diameter microdisk was etched with chemical plasma process. Ohmic contacts were formed to p+ GaAs cap layer and n+ substrate. Microlaser was mounted on a copper heatsink and tested at room temperature and elevated temperatures under continuous wave injection. In-plane emitted light was collected with a piezoelectrically adjustable ×100 Olympus LMPlan IR objective and analysed with a Horiba FHR 1000 monochromator and a Horiba Symphony InGaAs CCD array (spectral resolution 30 pm).

Emission spectra taken at 100 °C under various bias current are shown in figure 1(a). The spectra contains a series of narrow lines, which correspond to $TE_{1,1,m}$ whispering gallery modes of different azimuthal number *m*. The dominant mode, which is $TE_{1,1,242}$, is centred near 1304 nm. The next most intensive modes are $TE_{1,1,243}$ and $TE_{1,1,241}$. Integrated intensities against bias current for these three modes are shown in figure 1(b). The light-current characteristic of the dominant mode is shown in figure 1(b). A pronounced threshold behaviour is observed with the threshold current of 13.8 mA (1.8 kA/cm^2). To our best knowledge, this is the highest CW lasing temperature among QD microdisk/microring lasers grown on GaAs substrates. The intensities of the side modes start decreasing as the injection current exceeds the lasing threshold.

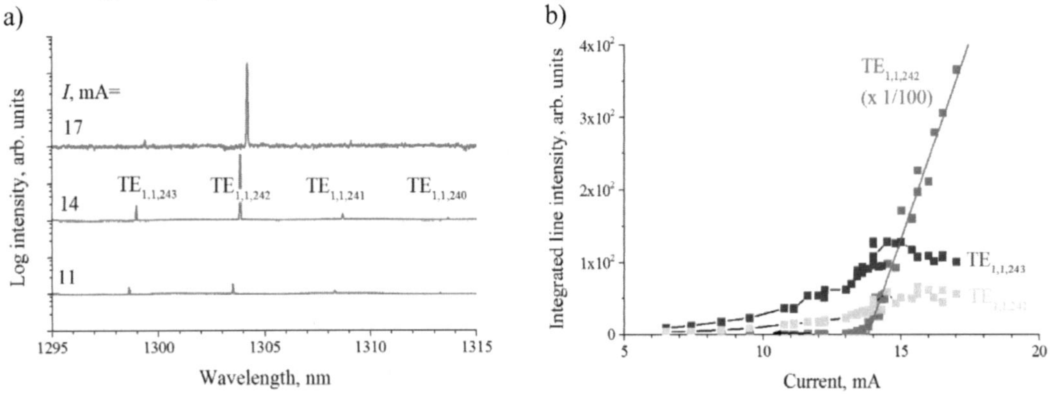

Figure 1. 100 °C emission spectra (a) and integrated intensity of WGMs as a function of bias current (b).

Threshold current is shown in figure 2 as a function of temperature. At 25 °C lasing starts at 4.4 mA. The characteristic temperature in the 25-100 °C interval was estimated to be 66 K. At 110 °C the microdisk does not lase.

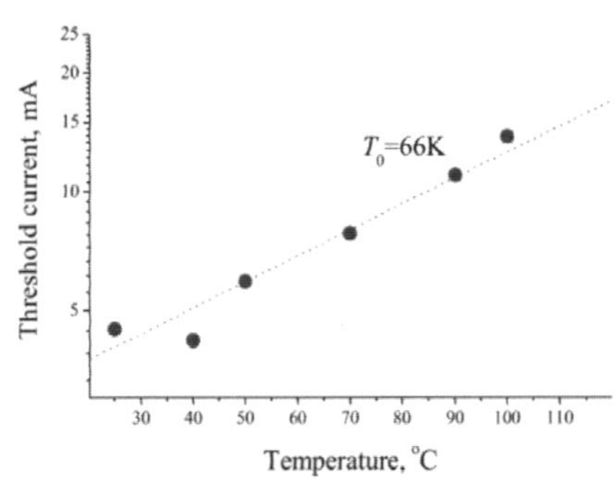

Figure 2. Temperature dependence of threshold current and its fit with characteristic temperature of 66 K.

3. Modification of micro-resonator spectral characteristics using nano-antennas

A far-field emission of the microdisk lasers is isotropic making coupling into waveguide inefficient. Another drawback of the microdisk geometry is multi-mode lasing due to the small free spectral range, whereas many applications require a single-mode emission. In order to increase the output without increasing the threshold and improve the side-mode suppression ratio we propose to introduce a nano-antenna to the edge of the resonator. Proposed method was tested using optically pumped QD micodisk laser. The epitaxial structure was synthesized by molecular-beam epitaxy on a GaAs(100) substrate with a Riber-49 MBE machine. An active region represents 5 layers of InAs/InGaAs quantum dots inserted into a 0.22-μm-thick GaAs waveguiding layer cladded with 400-nm-thick $Al_{0.98}Ga_{0.02}As$ layer from the substrate side. Spectral position of quantum dot ground-state transition was located around 1.28 μm at room temperature. Microdisk resonators were fabricated using photolithography and Ar+ ion beam etching. Outer diameter of the microdisks was 6 μm. The $Al_{0.98}Ga_{0.02}As$ bottom cladding layer was selectively oxidized to be transformed into an AlGaO oxide.

Nano-antennas were formed by means of electron-beam induced deposition with $C_9H_{16}Pt$ precursor gas using a Carl Zeiss CrossBeam 1540XB microscope. The precursor gas was injected trough a micro-nozzle into a spatial region of electron beam focus. An operation pressure in the microscope chamber was $2{*}10^{-5}$ mBarr, electron beam diameter and current was 2-3 nm and 50 pA, respectively, Pt-C nano-antenna growth rate was about 160 n/s. The nano-antennas formed under such growth conditions have a polycrystalline structure representing an array of Pt nano-crystallites of 2-3 nm inserted into an amorphous carbon matrix. Figure 3 depicts a microscopic image of the microdisk with a nano-antenna attached to the microdisk. The nano-antenna diameter is 150 nm, its height is 1.3 μm.

Figure 3. Scanning electron microscopy image of 6-μm-diameter microdisk with 150-nm-wide 1.3 μm-high nano-antenna.

Optical characteristics were studied under an YAG:Nd laser excitation (λ=532 nm) with laser beam focused on a sample surface with an Olympus LMPlan IR 100 NA0.8 objective lens. The same lens was used for luminescence signal collection. An FHR1000 monochromator and a Horiba Symphony multi-channel cooled InGaAs photodetector were used for signal detection (spectral resolution was 0.03 nm). Measurements were conducted at room temperature with various excitation power.

In order to clarify effect of nano-antennas on microdisk performance we compare properties of the microlaser before and after formation of the nano-antenna (as shown in figure 3). Luminescence spectra are shown in figure 4. In both cases the microdisk has the dominant mode wavelength around 1277 nm. The next most intense modes are located near 1290 and 1303 nm.

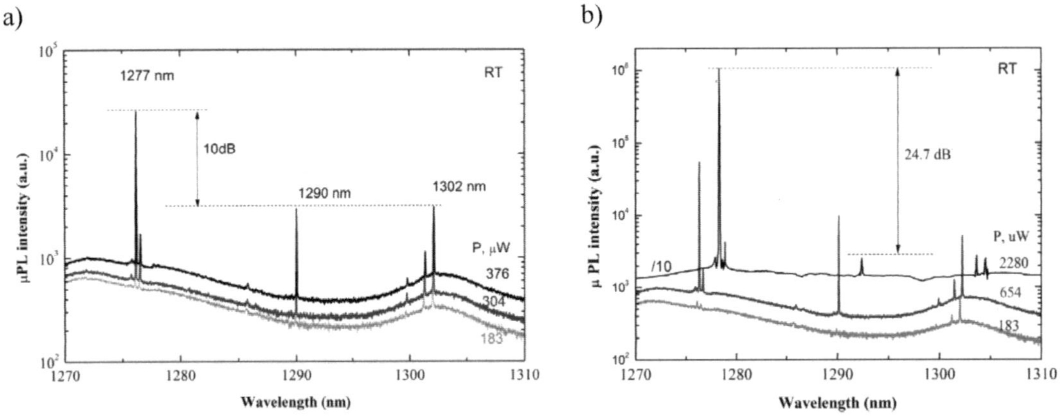

Figure 4. Lasing spectra of microdisks before (a) and after (b) nano-antenna deposition taken at room temperature at various pump power.

Figure 5(a) summarizes side mode suppression ratio (SMSR) for the microdisk lasers before and after nano-antenna deposition. It is seen that the maximal SMSR increases from only 10 dB in the microlaser without nano-antenna to 24.7 dB in the microlaser with antenna. The modified structure is also characterized by better light outcoupling from the resonator into a free space as it is demonstrated in figure 5(b), where the dominant mode intensity is shown as a function of pump power. It should be noted that the nano-antenna increases the maximal output light intensity by more than 20 times.

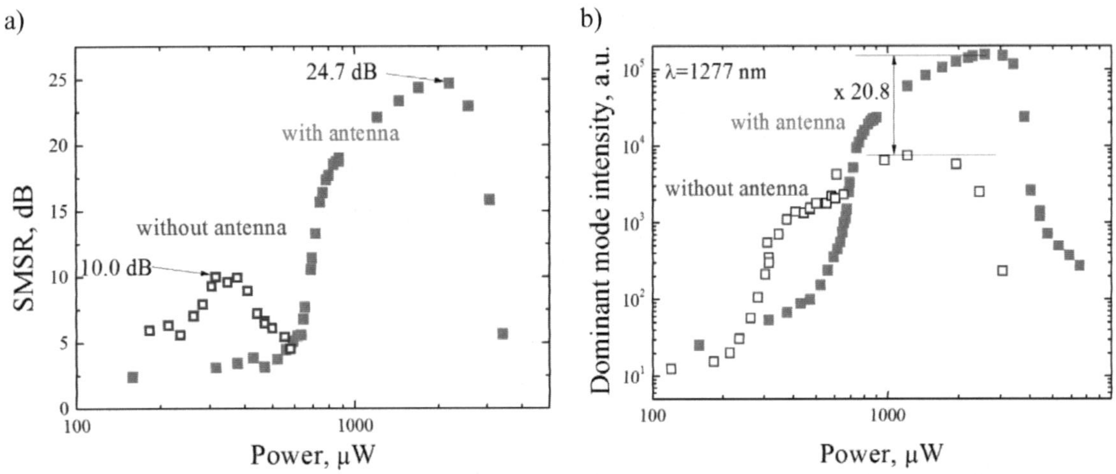

Figure 5. Side mode suppression ratio (a) and dominant mode intensity (b) as function of pump power for microdisks with and without nano-antenna.

4. Conclusions

In conclusion, lasing in a quantum dot microdisk laser at temperature as high as 100 °C has been for the first time. The threshold current density was 1.8 kA/cm^2; lasing wavelength 1.3 μm. We also proposed and realized a novel method of mode selection in microresonators with quantum dots. The method is based on an enhancement of a dominant mode intensity by means of a nano-antenna attached to the microdisk. Using Pt-C antennas formed with electron-beam assisted deposition, we achieved the maximal SMSR increment by 14 dB and the dominant mode enhancement by 20 times in room-temperature spectra of 6-μm-diameter QD microdisk lasers.

5. Acknowledgement
This work was supported by the project the Ministry of Education of the Russian Federation RFMEFI61615X0052 "Development of microlasers with disk geometry based on quantum-sized active region for integrated photonics" (agreement # 14.616.21.0052).

6. References

[1] Ledentsov N N, Shchukon V A, Maximov M V, Shernyakov Yu M, Payusov A S, Gordeev N Yu, Rouvimov S S 2015 *Semicond. Sci. Technol.* **30** 105005

[2] Kageyama T, Nishi K, Yamaguchi M, Mochida R, Maeda Y, Takemasa K, Tanaka Y, Yamamoto T, Sugawara M, Arakawa Y 2011 *Conf. Lasers and Electro-Optics Europe*, Munich, Germany.

[3] Munsch M, Claudon J, Malik N S, Gilbert K, Grosse P, Gérard J-M, Albert F, Langer F, Schlereth T, Pieczarka M M, Höfling S, Kamp M, Forchel A, Reitzenstein S 2012 *Appl. Phys. Lett.* **100** 031111

[4] Backes S A, Cleaver J R A, Heberle A P, Köhler K 1998 *J. Vac. Sci. Technol. B* **16** 3817

[5] Kryzhanovskaya N V, Mukhin I S, Moiseev E I, Shostak I I, Bogdanov A A, Nadtochiy A M, Maximov M V, Zhukov A E, Kulagina M M, Vashanova K A, Zadiranov Yu M, Troshkov S I, Lipovskii A A, Mintairov A 2014 *Opt. Exp.* **22** 25782

Light-emitting structures based on colloidal quantum dots of cadmium sulphide having a high color rendering index

E M Stepanov[1], P O Tadtaev[1], O S Vatalev[1], I I Mikhailov[1], S A Tarasov[1], A V Solomonov[1]

Department of Microelectronics,[1]Saint-Petersburg Electrotechnical University "LETI",5 Prof. Popov Str., St. Petersburg197376, RUSSIA

E-mail: stepanovem2015@mail.ru

Abstract. Quantum dots created by colloidal synthesis (CQD`s) were investigated. The paper shows how by changing the size and concentration of CQDs the emission spectrum of the structure with a color rendering index greater than 90 was achieved. Creation of the light-emitting structures involved spin-coating of CdS CQDs onto the glass substrate and consequent high-vacuum annealing. Methods allowing the increase of the emission intensity to achieve better approximation of the radiation spectra shape of white LEDs' spectra.

1. Introduction

Creation and investigation of structures containing quantum dots (QDs) are some of the most important tasks of modern Photonics as soon as the QDs-based devices exert substantially enhanced performance. The widespread application of such structures is limited to relatively high cost of their creation. From this point of view, the methods of colloidal synthesis are very promising, allowing simplification of the production technology of such particles [1, 2]. Quantum dots created by colloidal synthesis (CQDs) are highly efficient and enable the emission wavelength's adjustment by the synthesis parameters' variation [3]. It is essential that CQDs absorb light in a wide range of wavelength (Figure 1) [4] which makes excitation of their photoluminescence possible by the same short-wave emitter.

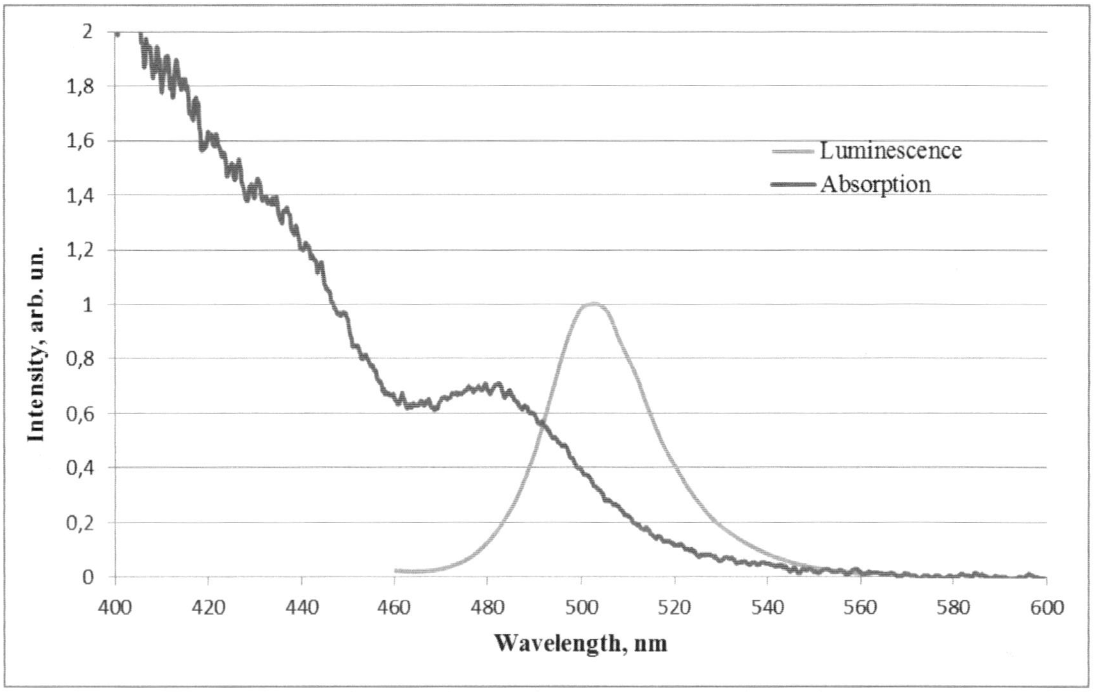

Fig. 1.The absorption and photoluminescence spectra of the CdSe CQD's.

A colloidal solution of QDs consists of a large number of nanocrystals with different sizes. Consequently, emission from the sample will occur in a wide range of wavelengths. Therefore, the peak of the photoluminescence (PL) spectrum will have a certain width. The greater the dispersion of particle size is, the greater the width of the peak will be. So far, the technology of the QDs' colloidal synthesis allows to obtain samples with a high degree of monodispersity, which, in this case, results in the half-width of the PL peak of several tens of nanometers [5].

Another feature of colloidal synthesis is the presence of the second PL peak besides the primary one [6]. It is considered to happen due to the presence of the defects such as surface states and others. However, the character of such defects remains poorly studied which complicates the prediction of their influence on the formation of the structure's spectrum.

The application of CQDs in the light-emitting structures' and, consequently, thereon based devices' creation offers some advantages [7]. First of all, it is a high spectral purity of the color, due to a narrow half-width of the emission line. Additional advantages can be obtained if CQDs are deposited on porous matrices [8-10] which increases the control over the shape of the spectra. Another important feature is that QDs are not an organic material which makes them more photostable than any organic structure, providing the possibility of increasing power of the exciting pump and structure's durability.

Color rendering index is one of the main characteristics of emission sources in the visible range of the spectrum. The importance of obtaining the high color rendering index is due to the high sensitivity of live forms to this parameter. Therefore, light emitting devices that have a higher value of color rendering index are more commercially successful.

2. Sample preparation

The cadmium sulfide CQDs were used to create light-emitting structures, since the dots of this type have a high value of quantum yield in the visible range of spectrum [11-13]. The toluene was used as a solvent. To create the structures, CQDs of various sizes were spin-coated onto the glass substrate (15 seconds, 2000 rpm). These parameters define the film thickness on the surface. Further annealing was performed in a high vacuum in order to evaporate the solvent and to form an ordered layer from the amorphous structure. This procedure is necessary for creating structures with electrical pump, since it requires the transmission of charge carriers. Annealing time was 30 minutes at the temperature of 90 degrees Celsius. The resulting thickness of the active region was several nanometers. Wavelengths of the main luminescence peaks were in the range from 380 nm to 480 nm (Figure 2).

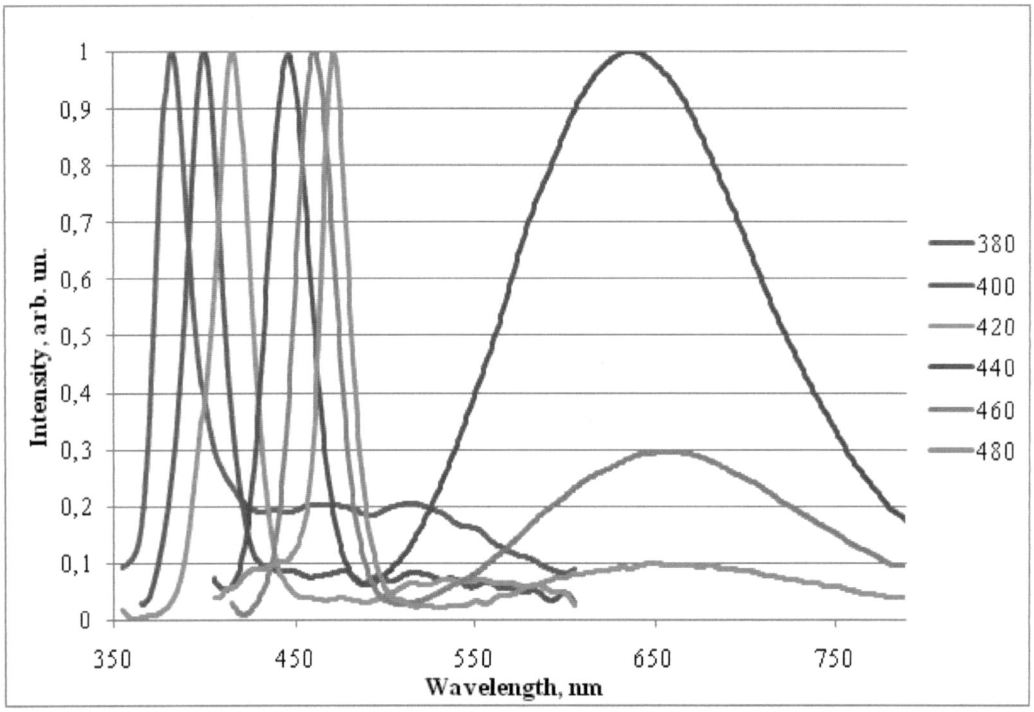

Fig. 2.The photoluminescence spectra of the CdS CQDs.

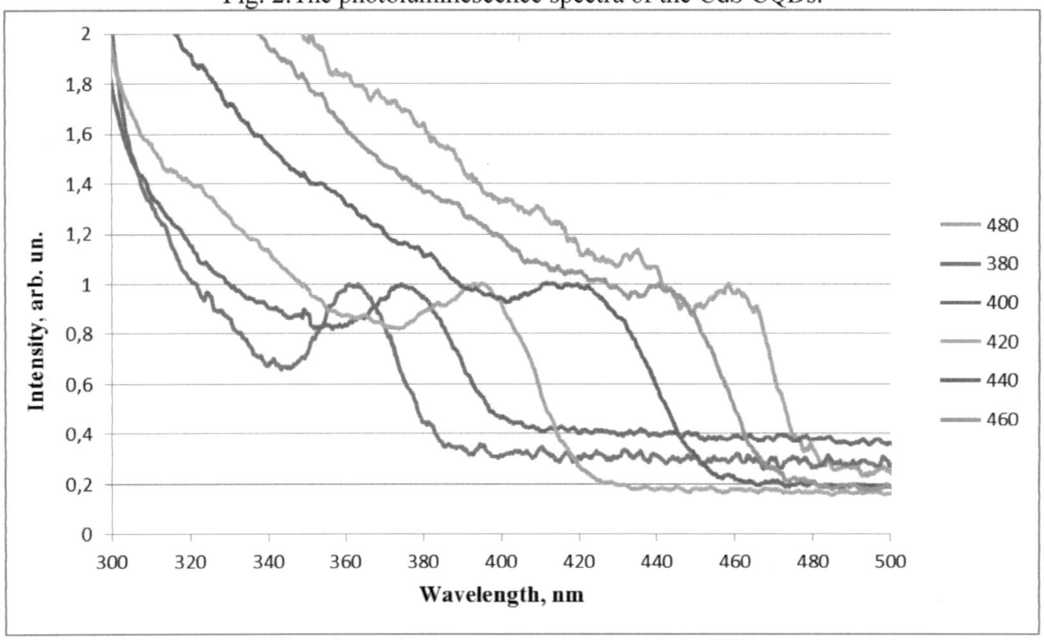

Fig. 3.The absorption spectra of the CdS CQDs.

Knowing the position of the first excitonic peak, we can estimate the mean size of the nanoparticles in ensemble. The estimation is based on empirical data (1) and (2), obtained from transmission electron microscopy [13-15]. (1) and (2) stand for CdSe and CdS CQDs respectively. The results of calculations are summarized in the Table 1.

$$a = 1{,}6122 \cdot 10^{-9} \lambda^4 - 2{,}6575 \cdot 10^{-6} \lambda^3 + 1{,}6242 \cdot 10^{-3} \lambda^2 - 0{,}4277 \cdot \lambda + 41{,}57 \quad (1)$$

$$a = -6{,}6521 \cdot 10^{-8} \lambda^3 + 1{,}9557 \cdot 10^{-4} \lambda^2 - 9{,}2352 \cdot 10^{-2} \lambda + 13{,}29 \quad (2)$$

where a – particle size, λ – wavelength of first excitonic peak.

Table 1. The sizes of the CdS and CdSe CQDs.

				CdS			CdSe
The emission wavelength, nm	380	400	420	440	460	480	500
Size, nm	2.3	2.6	3.1	4.1	4.8	5.7	2.3

The presence of the second peaks on the PL spectra (Fig. 2), as mentioned, is due to the defects such as surface states. These peaks' presence appears to be random in the solutions of CQDs consisting only of cores. In the solutions of "core/shell" CQDs density of these defects seem to decrease. The CQDs with significant intensity of the second peak were selected from a set in order to provide better correlation between the spectra`s shape of our structures and the spectra`s shape of white LEDs.

3. Results and discussion

The CQDs mixture with increased color rendering index containing a certain composition of different sized QDs was designed. Shortwave part of such structures' visible luminescence spectrum consists of major CQDs luminescence peaks; the long-wave part of the spectrum is caused by the surface states emission. A spectrum of a black body at 6504 K was chosen as an ideal one because of its similarity to an object CRI equal to 100. The modeling aim is to create a mixture of different sized CQDs which spectrum is identical to ideal (black body) one so the intensity of CQDs' peaks fit the black body's spectrum at the corresponding wavelengths. The resulting CQDs spectrum is shown on Figure 4 and in this case the mixture's CRI is 99.9.

Above described CdS CQDs samples allowed us to obtain an emission spectrum with a color rendering index of 90.6 (Figure 5). Each QDs ensemble's contribution could be controlled by mass concentration and quantum yield accounting. The concentration of CdS nanoparticles in the original solution was 5 mg/mL (100 mg/mL for CdSe). Nanoparticles' quantum yield in the CdS samples was about 10% (40% for CdSe sample). The original concentration (5 mg/mL) was used to prepare a mixture of CdS CQDs. CdSe CQDs were diluted by toluene solvent to 1.25 mg/mL concentration with the aim of intensity decrease.

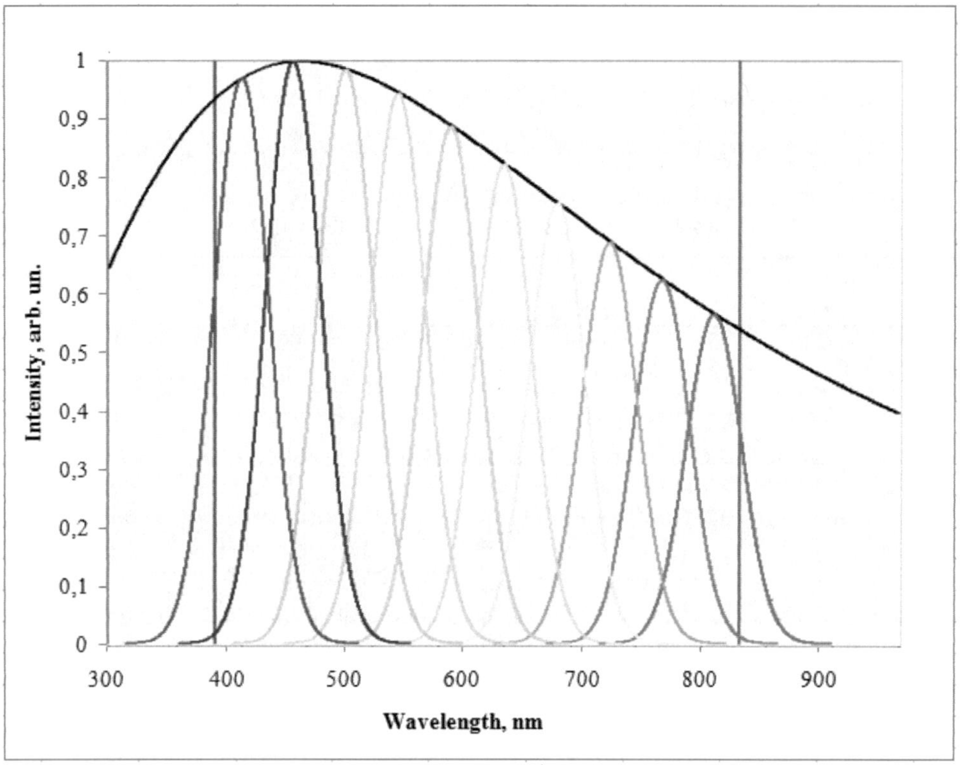

Fig. 4.The Black Body's (6504 K) spectrum and the modeled CQDs spectra

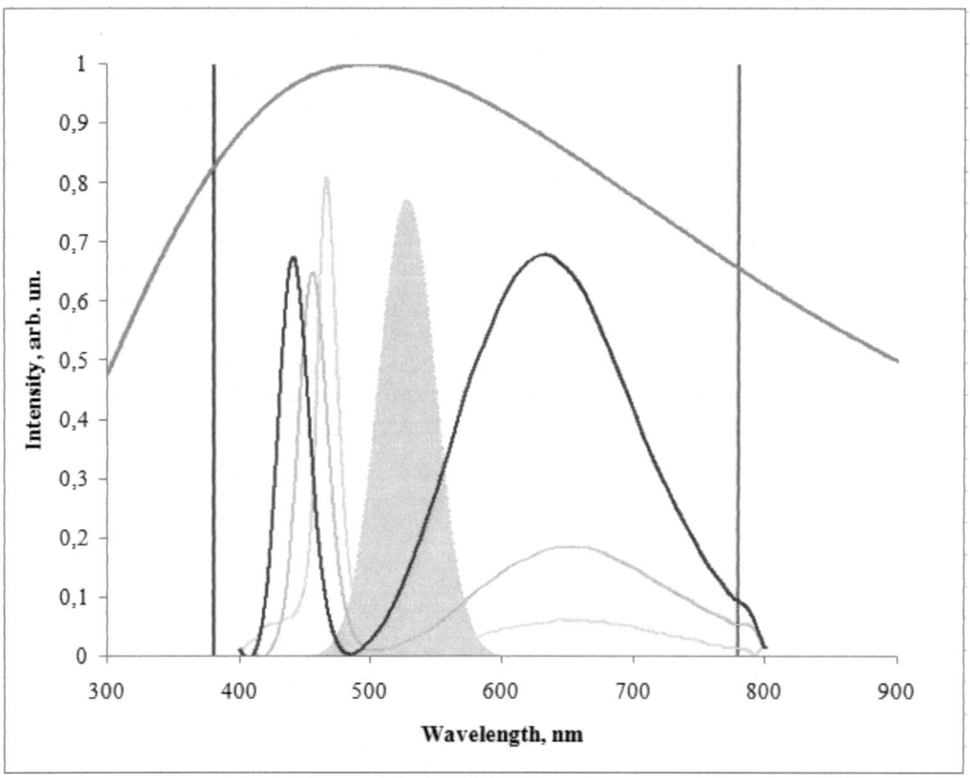

Fig. 5.The modeled CdS CQDs mixture spectrum

CdS CQDs with 440, 460, 480 nm emission wavelengths and CdSe CQDs with 500 nm emission wavelength were used for CRI calculation (Fig. 5) although actual wavelengths are slightly different.

The measured spectrum of designed light emitting structure containing CQDs is shown on Figure 6 where the red curve is a black body's spectrum. The summary spectrum covers almost the whole visible range and has three typical peaks. The work principle of this structure is based on photoluminescence mechanism, i.e. it is an optical pump structure. An optical pump is provided by UV gallium-aluminum nitride LED with the maximum of emission at 209 nm [16].

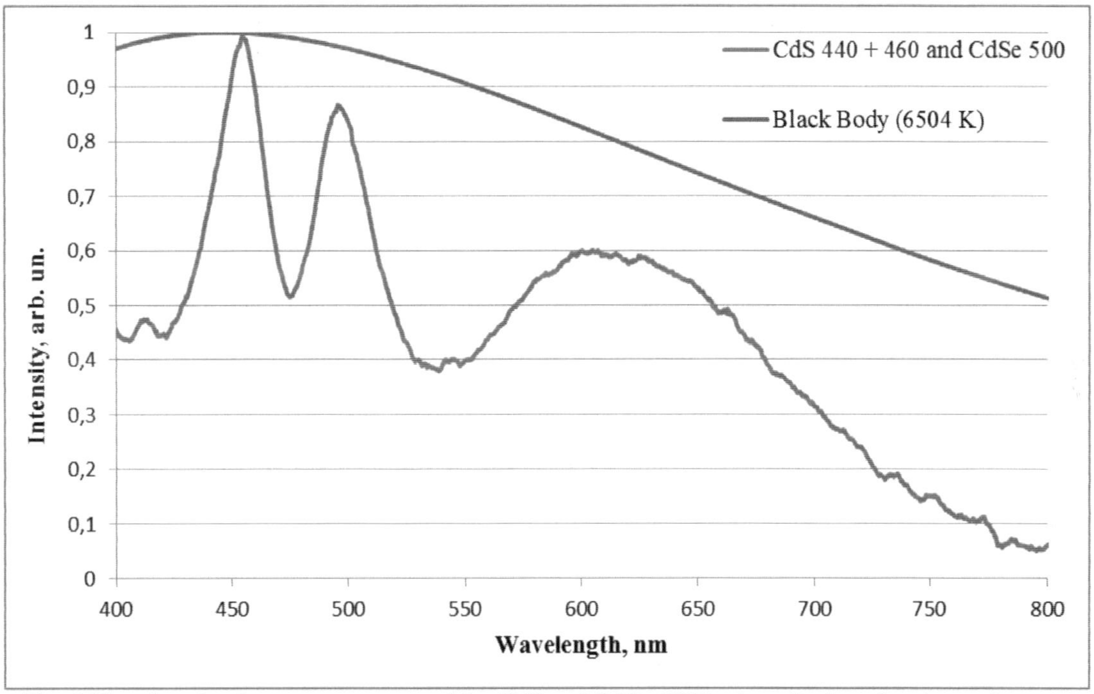

Fig. 6.The photoluminescence spectra of the CdS and CdSe CQDs deposited on the glass substrate.

The density of QDs film deposited on the glass substrate was especially high after the evaporation the solvent. Theoretically, this could cause an energy transfer from smaller dots to larger ones. This mechanism of energy transfer is called Förster mechanism and the main required condition for this transfer is a small distance between the dots (5-10 nm). Förster mechanism affects PL spectra by decreasing the intensity of short-wave peaks and increasing the intensity of long-wave ones. In our case, the intensities of all peaks were proportional to calculated ones. It might be concluded that the Förster energy transfer impact on the structure's spectra is extremely small. Perhaps, it is due to the dots ligand shell which is about 5 nm thick.

Details of the electrical pump structures creation are also determining factors of the final result. It is important to use inorganic layers in the structure because of organic materials' rapid degradation. For the purpose of creating high color rendering index LEDs a structure with optical pumping should be used because of its high efficiency and characteristics' stability.

4. Conclusion

The modeling of luminescence spectral characteristics of various sized CdS CQDs mixture was carried out and the possibility of obtaining CRI greater than 95 was shown. The structure with a CRI greater than 90 was successfully created. A simple and inexpensive method of creating light-emitting nanostructures was used. It was shown how the selection of the different sized CQDs' concentration can be used to effectively control the light emitting structure's spectral characteristics. The creation of a structure with color rendering index exceeding 95 by presented method is a work in progress.

5. Acknowledgments
The work was supported by the Ministry of Education of the Russian Federation in the framework of scientific activity state task project number 16.1307.2014K.

References
[1] I Mikhailov, S Tarasov, A Solomonov, O Aleksandrova, L Matyushkin, D Mazing, Journal of Physics: Conference Series **572**, 012029 (2014).
[2] C Murray,S Sun, W Gaschler, H Doyle, T Bentley, C Kagan, 2001, IBM J.Res.Dev. **45** (2001).
[3] D Mazing, L Matyushkin, O Aleksandrova, I Mikhailov, V Moshnikov, S Tarasov, Journal of Physics: Conference Series **572** , 012028 (2014).
[4] Y Shirasaki, G Supran, M Bawendi and V Bulovic, 2012, Nature Photonics, **7** (2012).
[5] C Burda, S Link, M Mohamed and M El-Sayed, 2001, J. Phes. Chem. B **105**, 12286 (2001).
[6] D Romanovskiy, S Tarasov, G Galiev, S Pushkarev, Journal of Physics: Conference Series **541**,012080 (2014).
[7] I Mikhailov, S Tarasov, A Solomonov, L Matyushkin, D Mazing, Functional materials **21**, 142 (2014).
[8] S Tarasov, O Aleksandrova, I Lamkin, A Maksimov, E Maraeva, I Mikhailov, V Moshnikov, S Musikhin, S Nalimova,NPermyakov, Yu Spivak and P Travkin, Semiconductors **49**, 1710 (2015).
[9] S Tarasov, I Gracheva, K Gareev, O Gordyushenkov, I Lamkin, E Men'kovich,V Moshnikov and A Presnyakova, Semiconductors **46**, 1584 (2012).
[10] VMoshnikov, I Gracheva, V Kuznezov, A Maximov, S Karpova, A Ponomareva, Journal of Non-Crystalline Solids **356**, 2020 (2010).
[11] S Tarasov, O Aleksandrova, A Maksimov, E Maraeva, L Matyushkin, E Men'kovich, V Moshnikov, S Musikhin, Semiconductors **48**, 1729 (2014).
[12] I Mikhailov, S Tarasov, I Lamkin, M Andreev, A Solomonov, Journal of Physics: Conference Series **643** , 012080 (2015).
[13] C Shen and W Tseng,Inorg. Chem. **48**, 8689 (2009).
[14] A Rogach, A Kornowski, M Gao, A Eychmuller and H Weller, J. Phys. Chem. B**103**, 3065 (1999).
[15] W Yu, LQu, WGuo and XPeng, 2003, Chem. Mater.**15**, 2854 (2003).
[16] E Menkovich, S Tarasov, I Lamkin, Functional Materials**19**, 233 (2012).

ECV investigation of ion-implanted photosensitive silicon structures for backside illuminated CCDs

G Yakovlev, D Frolov, A Zubkova and V Zubkov
Department of micro- and nanoelectronics, St. Petersburg State Electrotechnical University "LETI", Prof. Popov str. 5, St. Petersburg, 197376, Russia

E-mail: gy@etu.ru

Abstract. Boron implanted Si structures used in back-side illuminated and electron bombarded charge coupled devices (CCD) were investigated by means of electrochemical capacitance-voltage (ECV) profiling. A set of test structures were specially fabricated using various energies and doses of boron implantation, as well as different materials were used as their coating layers: Al and silicon oxide. The concentration profiles of free charge carriers across the sample were experimentally obtained. Further, using Poisson equation and Fredholm equation of the first kind the distribution of free charge carrier concentration and electric field intensity in depth of the samples were calculated. By analyzing and comparison of simulated and experimentally obtained concentration profiles, the recommendations for optimization of sample parameters were proposed aiming at increase of sweeping field and decrease of surface potential impact on charge carrier transport.

1. Introduction

Fabrication of photon- and electron-sensitive devices with charge transfer implies a thinned from the back side CCD matrix with full removal of highly-doped epi-layer [1-3]. Untreated surface in such CCD does limited sensitivity to the UV radiation and low-energy electrons with short penetration depth because of the surface depletion [4, 5]. To overcome this a p^+-area in the epi-layer is created, which provides both the bends over the surface potential caused by thinning processes, and also the sweeping field (to the front side contacts) for minority charge carriers generated by electron bombarding during device operation [6].

Ion implantation together with anneal is typically used to form highly doped area in the vicinity of the backside surface. For optimization of ion implantation it is very important to have a feedback and to investigate the obtained free charge carrier distribution profiles. One of the most useful measurement techniques for such task is the electrochemical capacitance-voltage (ECV) profiling. It shows up by wide measurement range of free charge carrier concentration (from 10^{14} to 10^{20} cm^{-3}) and ability to provide measurement results on depth from few nanometers up to few microns [7, 8]. The significant feature of the technique is simplified Schottky contact preparation.

The aims of this investigation are to measure the boron implanted Si samples by ECV technique, to simulate free charge carrier distribution and distribution of electric field in the samples. The results of experiments and simulation will allow making recommendations for optimization of the ion implantation process to increase the sweeping field and decrease the surface potential influence on charge carrier transport.

2. Samples and experimental technique

The test structures under investigation were low-doped 20 μm thick Si epi-layers with resistance of 12 Ohm·cm, grown on (100) Si substrate. Boron implantation was performed directly in the epi-layer or through coating layers. After implantation at 450°C the annealing for 60 min was used. The structure specification is shown in Table 1.

Table 1. Test structures specification

Sample number	Implantation energy, keV	Implantation dose, ion/cm^2	Coating layer material	Coating layer thickness, nm
1	13	$2\cdot10^{14}$	-	-
2	13	$5\cdot10^{14}$	-	-
3	13	$8\cdot10^{14}$	-	-
4	10	$2\cdot10^{14}$	Al	45
5	15	$2\cdot10^{14}$	Al	45
6	20	$2\cdot10^{14}$	SiO_x	60

The measurements were performed at room temperature using ECVPro profiler (Nanometrics). The aqueous solution of 0.2M NH_4HF_2 with addition of isopropyl alcohol (IPA) (up to 20% of total volume) was used to form electrolyte barrier with nominal contact area of 0.1 cm^2. During ECV profiling the samples were etched gradually with 5 nm step. To verify the etching depth and the surface quality the atomic force microscopy (AFM Solver NEXT) was used. The frequency of ac test signal was changed gradually from 0.1 to 5 kHz during the experiment (to eliminate the influence of series resistance on fast dropping capacitance).

In figure 1 the etching pit profile of sample #6 is shown. As it could be seen, the etching depth according to AFM measurements is 0.95 μm, which is in good agreement with ECV results (1.0 μm). The difference could be assisted, for instance, with a quite small AFM measurement area (20x20 μm) in contrast to the electrolyte contact area (0.1 cm^2). The etched pit has a very sharp edge, the average surface roughness after etching is 5 nm, even better than before the etching (15 nm). It confirms the right choice of the electrolyte.

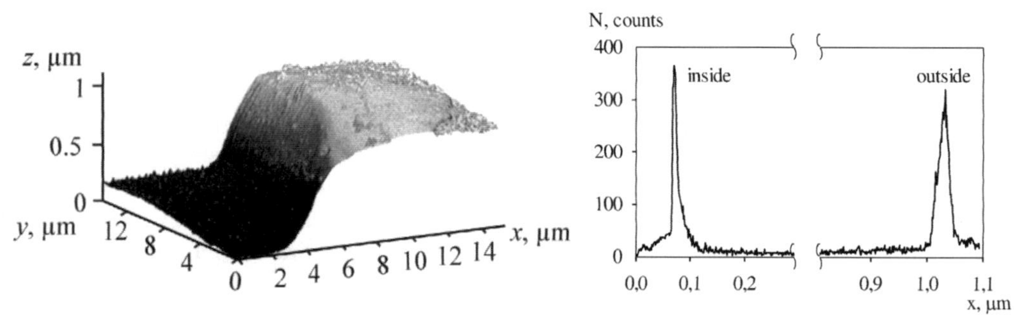

Figure 1. 3D AFM plot and height histogram of the sample's #6 etched hole.

3. The experimental results and discussion

The main feature of the investigated ion-implanted structures is high doping level in the vicinity of the surface and shallow ion distribution profile, which needs keeping in mind the surface conductivity and sharp alteration of capacitance during the experiment.

3.1. Implantation in the samples without coating layers

The obtained free charge carrier profiles for samples #1-3 (no coating layers, various implantation doses) are shown in figure 2. With the increase of implantation dose the value of concentration in peak does increase, while the location of the peak shifts to the surface. The asymmetry of the profiles in the vicinity of the maximum and their shift could be associated with the dopant diffusion. The obtained parameters characterizing the carrier distribution in all measured samples are summarized in Table 2.

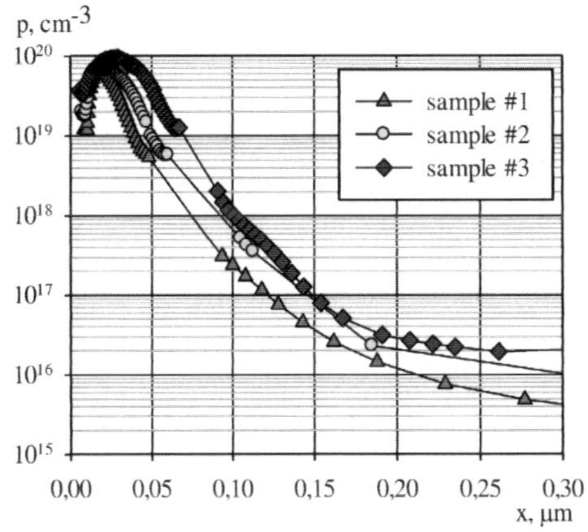

Figure 2. Apparent free charge carrier profiles of samples #1-3 (without coating layers) with various implantation doses.

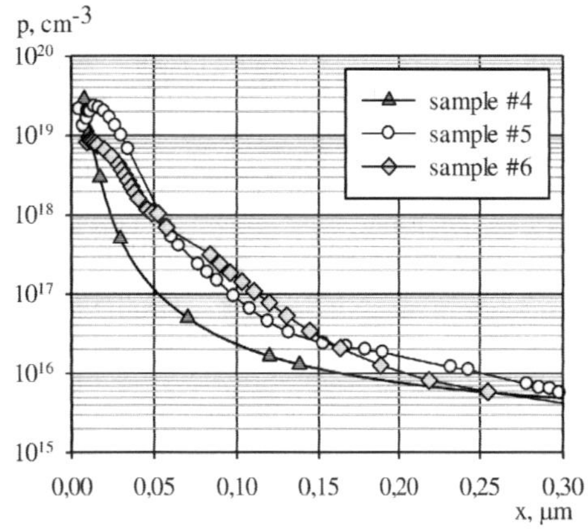

Figure 3. Apparent free charge carrier profiles of samples #4-6 implanted through various coating layers.

Table 2. Parameters of the measured free carrier concentration profiles and simulated electric field profiles for the investigated samples

Sample number	Carrier concentration in peak, cm^{-3}	Distance from the peak to the surface, nm	FWHM for the carrier concentration profile, nm	Distance from the electric field maximum to the surface, nm	Electric field peak value, mV/nm
1	$6.5 \cdot 10^{19}$	18	18	31	2.7
2	$6.7 \cdot 10^{19}$	21	27	44	2.5
3	$9.5 \cdot 10^{19}$	29	37	67	1.8
4	$2.9 \cdot 10^{19}$	8	-	16	5.1
5	$2.3 \cdot 10^{19}$	14	28	38	2.7
6	$8.6 \cdot 10^{18}$	12	32	36	2.2

Note that all concentration profiles in figures 2 and 3 are smeared on Debye length, which is determined by the intensity of exchange coupling between dopants and free charge carriers:

$$L_D = \sqrt{\frac{\varepsilon \varepsilon_0 kT}{q^2 N}}, \qquad (1)$$

wherein the Debye length itself depends on the concentration.

3.2. Implantation through coating layers

The experimental free charge carrier concentration profiles of the samples implanted through Al (#4, 5) and SiO_x (#6) layers are shown in figure 3. As it could be seen, the concentration profiles shift to the surface with the decrease of implantation energy (samples #4 and #5). As it was expected, these samples have smaller peak concentration than samples #1-3 because of smaller implantation dose.

4. Analysis

4.1. The electric field simulation

The use of the coating layer allows one to shift free charge carrier concentration profile closer to the surface. Wherein, it is very important to control the maximum of the gradient of the free charge carrier distribution, which in turn determines the position of the electric field maximum [9]. Usually, the material of the coating layer and its thickness are fixed for specific device class. For example, for visible-blind devices Al layer acts as a filter, which cut the undesirable portion of the spectra. Since the photocathode material of the electron sensitive back-side illuminated devices usually is transparent for visible range of the spectrum, the use of the Al coating layer allows one to avoid the device blooming in this spectral range. In this case only photo-generated electrons (accelerated by the bias enough to overcome the coating layer) are detected.

The main factor, which influences the sensitivity of the photodetector element, is charge collection efficiency (CCE) [10]. The necessity of the photo-generated electrons to overcome the coating layer decreases the CCE (because of the high acceleration biases). But even this condition is not enough. To obtain the highest CCE the electron should get into the sweeping electric field area. It means that it must enter the area behind the electric field intensity maximum, which is determined by the implanted concentration profile. It leads to the necessity of the acceleration bias increasing. However, if the implantation through a coating layer took place, the doping profile shifts to the surface with the consequent shift of the electric field maximum. Thus, by varying the implantation energy and dose, the electric field maximum could be shifted nearly to the interface. And now, the applied bias value is only limited by the thickness of the coating layer needed to be overcome. These conditions contribute to the sweeping field formation without "dead layer" [11, 12], which is a summarized thickness of the low CCE area. This area is limited, on one hand, by the coating layer surface, and the electric field maximum on the other hand.

The analytical expression for the electric field intensity distribution could be obtained from the Poisson equation

$$\frac{\partial^2 \varphi(x)}{\partial x^2} = -\frac{\rho(x)}{\varepsilon\varepsilon_0}. \tag{2}$$

Using the single donor assumption, after the math operations one could get

$$E(x) = \frac{kT}{q}\frac{1}{n(x)}\frac{\partial n(x)}{\partial x}, \tag{3}$$

where $n(x)$ is the free charge carrier distribution. We used this formula to obtain the electric field distribution in the investigated samples.

The electric field distribution for the sample #4, calculated from the experimental free charge carrier profile, is shown in figure 4. The peak value of the electric field intensity is 5.1 mV/nm ($5.1\cdot10^4$ V/cm). Sample #4 has the highest (of all measured samples) the electric field peak intensity. The electric field maximum is shifted to the surface by the coating layer and is located at depth 16 nm from the surface. The peak electric field intensity values and its locations are summarized in Table 2. To our knowledge, based on the experience of numerous measurements (most are not presented in this paper), for getting high values of sweeping electric field the surface carrier concentration must be greater than 10^{19} cm^{-3}.

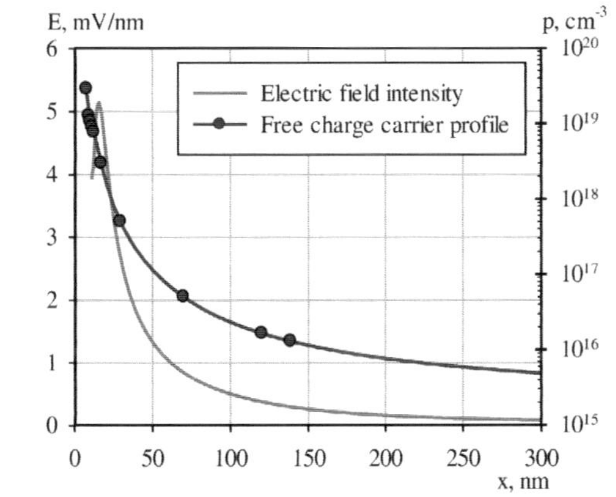

Figure 4. Depth distribution of the electric field intensity (red) and free charge carriers (blue) of sample #4 with Al coating layer. Experimental data – circles.

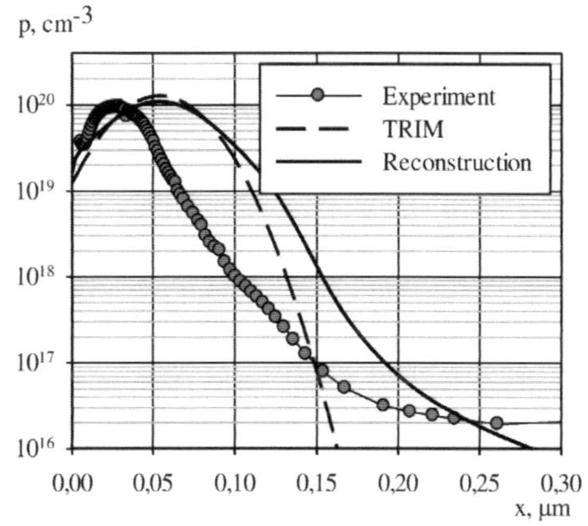

Figure 5. The comparison of experimental and simulated data for sample #3: experimental charge carrier profile (circles); TRIM simulation for dopant distribution (dashed line); free charge carrier profile, derived from TRIM data (solid line).

4.2. Ion implantation simulation

It is useful to compare the results obtained in our investigation with the simulation by TRIM software, which is frequently used for estimation of implanted profiles. There are lots of works, where free charge carriers concentration profiles are obtained by CV-technique, which is widely used for semiconductor characterization. However, the concentration profiles obtained by conventional CV are limited by space charge area penetration (~100 nm for $5\cdot10^{17}$cm^{-3} and for similar samples as in this paper it will be ~25 nm [13, 14]), while ECV (due to possibility of etching) allows one to obtain full profile information, even for the samples with very broad range of noticed concentrations.

The TRIM output is the doping profile, so to provide the comparison it is necessary to recalculate the doping profile to free charge carrier distribution. The latter is not identical to the doping profile because of the Debye tailing. When calculate, we do assume that all impurity is ionized (this is the case for room temperature).

The standard math processing for this is the convolution with known broadening function. However, in our case the broadening function isn't constant and is dependent on concentration through the Debye length (1). The impurity ions and free charge carriers are charged oppositely, therefore the higher the dopant concentration is, the higher the attractive force between ions and free carriers and the less the Debye length become.

Our recalculation was based on the Fredholm equation of the first kind:

$$y(x) = \int K(x,x')f(x')dx' \quad (4)$$

where $f(x')$ is the doping distribution, $K(x,x')$ is the equation core (the broadening function), determined by the Debye length:

$$K(x,x') = e^{-|x|/L_D(x')} \quad (5)$$

We show in figure 5 two free charge carrier profiles for sample #3: (1) obtained from the experiment and (2) simulated by TRIM and processed by Fredholm equation. As one can see, there is a small (not more than 30 nm) mismatch between experimental and simulated position of extremum. We explain this mismatch by the extremely low annealing temperature in contrast to the standard Si implantation technology. Note that TRIM does not take into account the annealing temperature and its duration. The default TRIM settings are based on the suggestion that all impurity is built into the lattice and is electrically active, while the interaction between charged ions isn't taking into account.

Moreover, in our case ion implantation took place with quite low energies (10-45 keV). The use of TRIM for simulation in this energy range could lead to the error around 30% [15], so generally TRIM is used for energies around 0.2-1 MeV. We conclude that these features are the main reason of the experimental and simulation mismatch.

5. Conclusion

In this work the specially prepared Si samples, implanted by boron with various energies and doses directly and through coating layers (Al and SiO_x), were investigated. Free charge carrier profiles of the samples were obtained by electrochemical CV technique. From the experimental data the electrical field distributions in the surface vicinity were obtained. It was shown that for the formation of high sweeping field under the illuminated surface, the concentration profile must be sharp with the surface value not less than 10^{19} cm^{-3}. To obtain this it is necessary to provide ion implantation through the coating layer, wherein the doping distribution maximum must be located inside this layer. The optimal conditions, at which the electric field maximum is located in the vicinity of the surface, could be obtained by varying of implantation dose and energy for the given coating layer thickness. So, the experiment and simulation gave for 45 nm thick Al coating layer the following optimal implantation parameters: energy – 10 keV, dose – $2 \cdot 10^{14}$ ion/cm. For the sample with these parameters the electric field maximum ($5.1 \cdot 10^{4}$ V/cm) is located at depth around 16 nm from the surface. The use of the ion implantation with the optimal parameters allows one to compensate decreasing of CCE after sample thinning, and also allows one to decrease accelerated bias with no drop of the device sensitivity. The comparison between experimental carrier concentration profiles and TRIM simulations was made. The mismatch between the simulation and experimental data could be attributed to neglecting of TRIM for the technology process features, particularly the annealing temperature.

References

[1] Hiroven L, Jiggins S, Sergent N, Zanda G and Suhling K 2014 *Rev. Sci. Instrum.* **85** 123102
[2] Gruwe M 1997 *Nucl. Instr. Meth. Phys. Res.* A **387** 282
[3] Richard J and Vittot M 1992 *Nucl. Instr. Meth. Phys. Res.* A **315** 368
[4] Nikzad S, Jones T, Elliott S, Cunningham T, Deelman P, Walker A and Oluseyi H 2000 *Proc. Conf. Space Astrophysics Detectors and Detector Technologies* (Baltimore: STScI Press)
[5] Yakovlev G, Frolov D, Zubkova A, Levina E, Zubkov V, Solomonov A, Sterlyadkin O and Sorokin S 2016 *Semiconductors* **50** (3) 324-330
[6] Stern R, Catura R, Kimble R, Davidsen A, Winzenread M, Blouke M, Hayes R, Walton D and Culhane J 1987 *Opt. Eng.* **26** 875
[7] Ambridge T and Faktor M 1975 *J. Appl. Electrochem.* **5** 319
[8] Zubkov V, Kucherova O, Frolov D and Zubkova A 2013 *Phys. stat. sol.* C **10** 342
[9] Blouke M, Delamere W and Womack G 1991 *Proc. IEEE CCD Workshop* (Waterloo: University of Waterloo Press) p 14
[10] Bertoluzzi L and Shuai M 2013 *Phys. Chem. Chem. Phys.* **15** 4283
[11] Lee C M, Chang S P, Chang S J and Wu S I 2013 *Int. J. Electrochem. Sci.* **8** 7634
[12] Wall B et al *Nucl. Instrum. Meth.* A **744** 73-79
[13] Kapila D, Kulkarni M, Fernando C, Davis J, Vasanth K and Pollack G 1998 *Tech. Proc.Int. Conf. Modeling and Simulation of Microsystems* (Santa Clara: Computational Pubns) p 199
[14] Oosterhoff S 1998 *Nucl. Instrum. Meth.* B **30** 1
[15] Ziegler J 2006 *Ion Implantation – Science and Technology* (Chicago: IIT Press)

Application of strongly focused pulsed electron beam for the reaction wheels balancing

A O Borduleva[1], G A Bleykher[1], V V Solovev[2], V P Krivobokov[1] and M N Babihina[3]

[1] Department of Experimental Physics, National Research Tomsk Polytechnic University, Tomsk, 634050, Russian Federation

[2] Research and Development Centre "POLUS", Tomsk, 634050, Russian Federation

[3] Department of General Physics, National Research Tomsk Polytechnic University, Tomsk, 634050, Russian Federation

E-mail: tcybenkoao@gmail.com

Abstract. In the given work the material removing possibility by the strongly focused pulsed electron beam was investigated. The optimal mode of flywheels balancing was found. At this mode the power density is 1.6 MW/cm^2 and pulse duration is 0.65 s. At such parameters the evaporation rate is equal to 11 g/s·cm^2. It is possible to vary the amount of remote material from 1 to 100 mg, that is sufficient to balance flywheel. It is found that treatment by an electron beam does not change the material structure.

1. Introduction

One of the satellite main systems is orientation system which lets stabilizing the position of its axes relative to the specified destinations. Attitude control system includes reaction wheel which allows controlling satellite movement around the center of mass. Wheel rotates on the high speed; thereby creates internal moments that allow changing the angular position of the satellite relative to the base reference system, without changing the center of mass. Usually three wheels are established in satellite. Their axes are aligned with the satellite major axis of inertia. Fig. 1 shows the structure of the reaction wheels.

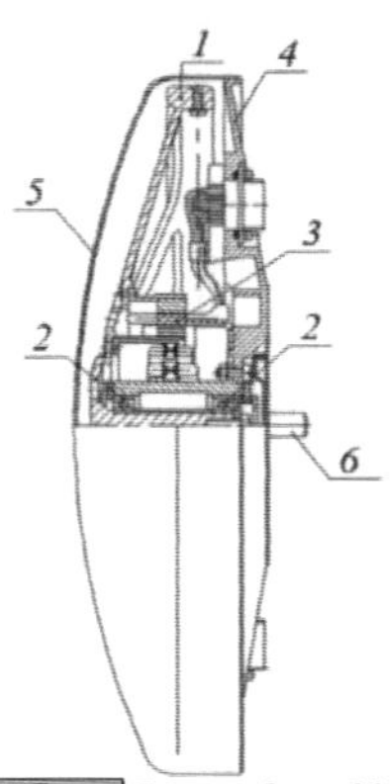

Figure 1. The construction of reaction wheels: 1 - ring of wheel, 2 – supports, 3 - electromotor, 4 - a base on which the rotor is fixed, 5 - protective cover, 6 – spindle.

In the reaction wheel manufacturing process there is a number of problems, one of which is its mass unbalance. This problem can lead to the satellite avoidance from its orbit. Currently, there are several techniques for flywheel unbalance correction: mechanical technique and treatment by concentrated energy fluxes (precision work).

Figure 2. The mechanical method of the flywheel balancing.

The first method involves holing in the flywheel, and, if necessary, balancing it by placing the particular weights in the form of small bolts inside the holes (Fig. 2). Such laborious technique requires heavy expenses of time and money. Therefore there is a need to develop other flywheel balancing methods, for example treatment by concentrated energy fluxes that includes laser and electron beam technologies. In this case the main mechanism of material removing is evaporation, because metal temperature riches a high magnitude under the influence of high-energy particles. This technique is more precise and can be more effective than the mechanical method.

Nowadays, laser technology of balancing is developing, but there are some disadvantages such as high energy reflection coefficient (~80%) and liquid melt ejection. Electron beam application allows to avoid these problems, because the energy reflection coefficient of the electron beam is much lower (~15 %) than for the laser beam, consequently the process efficiency is higher. Varying the electron beam parameters the liquid melt ejection can be avoided.

In the present study the electron beam treatment is suggested as the method for flywheel balancing. Application of strongly focused pulsed electron beam allows local and short-time metal treatment.

2. Experimental

2.1. Electron beam treatment

Electron-beam evaporation was carried out on the EBU - 0,5 - 6 electron-beam unit with plasma cathode (Fig.3). This unit is fitted out by a two-lens focusing system which gives an opportunity to reduce the beam divergence angle. The maximum diameter of the beam can reach the value of 0.8 mm, considering that the distance between the focusing system and the sample can be ~0.5 m. Such system allows producing a beam with the diameter less than 300 microns [1].

The investigated samples were made from steel AISI 321. Their thickness was 5 mm and diameter was 20 mm.

During the experiment the acceleration voltage was 28 kV; working pressure was 10^{-1} Pa. The distance to the focusing coil was 200 mm that corresponds to the beam diameter equal to 0.32 mm [1]. Such parameters as beam current, duration and number of pulses were variated to determine the optimal mode of evaporation.

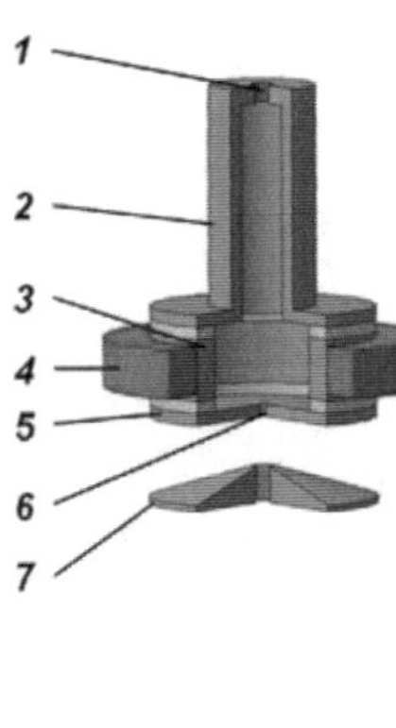

Figure 3. The electrode system of plasma cathode gun: 1 – the gas flow channel, 2 - the hollow cathode, 3 - the anode, 4 - the permanent magnet, 5 - the emitter cathode; 6 - the emission channel, 7 - the extractor (accelerating electrode) [2].

2.2. Methodology

The amount of removed material was measured using the PA413 laboratory analytical weighing scales OHAUS Pioneer. The macrostructure was investigated by the TRIO -1044 microscope. The samples hardness was measured by Vickers method with the load of 50 grams by the KB 30S Pruftechnik GmbH Hardness tester.

3. Results and discussion

Experiments show that the minimum of the beam current while the evaporation process was 35 mA; maximum magnitude was 52 mA. Above this value there are some negative effects, such as liquid melt ejection and energy beam expenditure on the thermal conductivity, i.e. overheating of samples (Fig.4). The evaporation rate at different current (35, 40, 52 mA) was investigated (Fig.5). Almost all energy was used to heat the sample in case when the initial energy was equal to 28 kV (the secondary radiation yield is very small), and the power density was 10^6 W / cm^2. Fig. 5 shows that the removal rate of material reaches the maximum at the pulse duration of ~ 0.15...0.3 s. However, at these pulse duration magnitudes, the evaporation process is non-stationary and the curve nature is periodic (Fig.4). It can be explained by the auto oscillations, which are presented during the electron beam treatment.

When the pulse duration is 1 s a large amount of energy is coming to the sample consequently heats it. There is vapor formation near samples surface, which leads to loss of beam energy, and evaporation process becomes energy intense. From these observations it can be seen that the optimal pulse duration is 0.5…0.65 s.

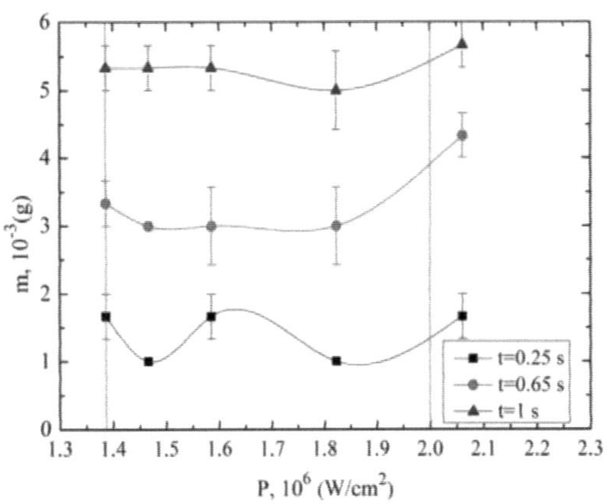

Figure 4. The removed material mass versus the electron beam power density.

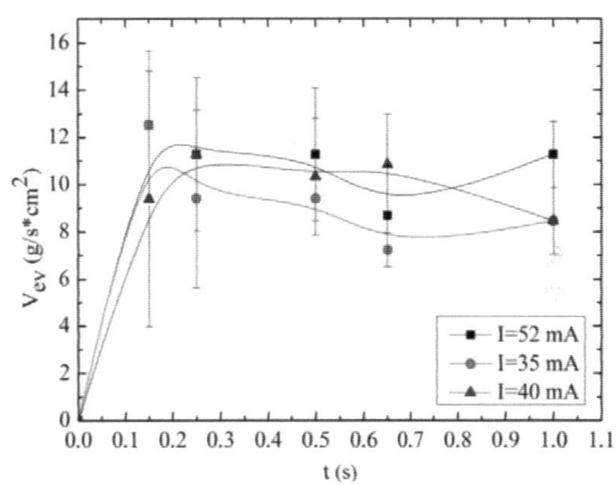

Figure 5. The dependence of the evaporation rate on the electron beam pulse duration.

Besides, it can be seen from Fig.5 that the evaporation rate at 52 mA almost equals the rate at 40 mA, therefore it can be noted that the more energetically effective current beam is 40 mA.

In addition, in this work it was necessary to investigate the macrostructure and mechanical properties of treated samples, because the presence of defects can lead to the reaction wheel damage in the process of operation. Hardness investigation shows that electron beam treatment does not affect the mechanical properties of material (Fig.6). It was shown that the depth from the beam thermal effect significantly depends on the pulse duration. When it was 1 s, the depth of molten zone was 5 mm (sample thickness), but at the value of 0.65 s the maximum width of the molten zone was 6 mm, and the depth was 4.4 mm (Fig.7). Such dimensions are negligible for the flywheels treatment, because its diameter varies from 30 to 35 cm. Also it can be seen that there are not any bulk defects (pores and cracks).

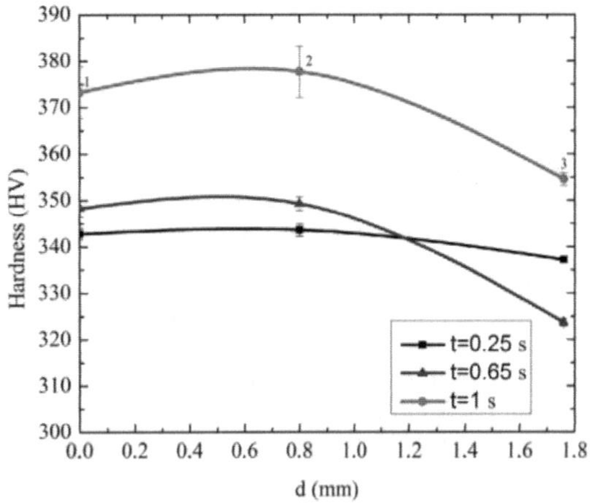

Figure 6. Samples hardness distribution: 1 – area which was treated by electron beam 2 - between the treatment zone and steel, 3 – steel.

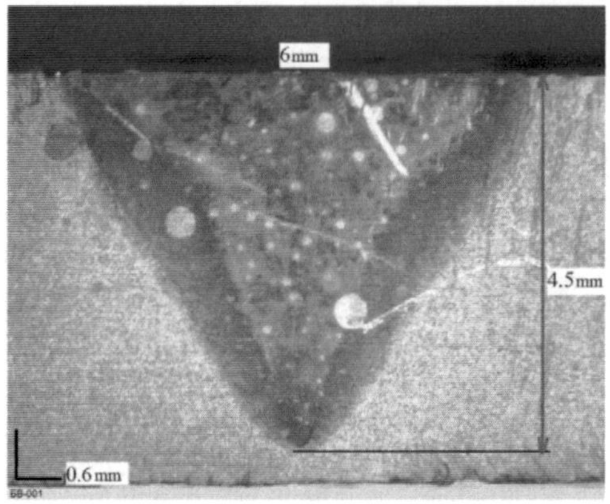

Figure 7. The macrostructure of the molten zone at t = 0.65 s.

From these results it may be concluded that the treatment by strongly focused pulsed electron beam does not lead to the defects appearance.

4. Conclusions

It was shown that strongly focused pulsed electron beam can be apply as the method of gyroscope flywheel balancing. The optimal mode for material removing from the steel surface includes following parameters: beam power density of $1.6 \cdot 10^6$ W/cm^2, pulse duration of 0.65 s, evaporation rate of 11 g/s$\cdot$cm^2. The amount of removed metal may be varied from 1 to 100 mg per a cycle.

The investigation of the samples macrostructure and hardness showed that the electron beam treatment does not change the metal structure significantly.

References

[1] Kornilov S Yu, Osipov I V and Rempe N G 2009 *Instruments and Experimental Techniques* **52** 409

[2] Rempe N, Kornilov S, Beniyash A and Hassel T 2012 *Welding and Cutting* **11** 123

Journal of Physics: Conference Series **769** (2016) 012060 doi:10.1088/1742-6596/769/1/012060

Simulation of femtosecond pulsed laser ablation of metals

R V Davydov, V I Antonov

Peter the Great St.Petersburg Polytechnic University, Saint Petersburg, Russia

E-mail: davydovroman@outlook.com

Abstract. In this paper a mathematical model for femtosecond laser ablation of metals is proposed, based on standard two-temperature model connected with 1D hydrodynamic equations. Wide-range equation of state has been developed. The simulation results are compared with experimental data for aluminium and copper. A good agreement for both metals with numerical results and experiment shows that this model can be employed for choosing laser parameters to better accuracy in nanoparticles production by ablation of metals.

1. Introduction

There is a growing interest in the nanofabrication of materials and their applications in various fields of life and technology, such as electronics, energy generation, health care and storage. A great deal of progress in this field has relied on the use of lasers. Production of nanoparticles can be done in several ways, one of them is laser ablation [1].

Despite extensive research work, accurate prediction of the ablation process is still lacking, because it significantly depending on laser parameters, surrounding medium and target material characteristics. To analyze the physical processes at high energy densities, when laser is used, an adequate description of the thermodynamic properties of matter over a broad region of states including the normal conditions and plasma at high pressures and temperatures is required.

Nowadays a two-temperature model (TTM) has been widely employed for solving ultrashort laser processing of metals [2-5]. This continuous model describes the energy transfer inside a metal with two coupled generalized heat conduction equations for the temperatures of the electrons and the lattice. To describe the material removal processes TTM is often inserted into a hydrodynamic code. But the choice of an equation of state, required for solving hydrodynamic equations, can significantly affect the results.

The article is organized as follows. It begins with describing the mathematical model, then we construct the appropriate semi-empirical two-temperature equation of state for model. Finally, we demonstrate the results for aluminium and copper and compare it with experimental data.

2. Mathematical model of laser ablation

In this work we describe the evolution of material parameters using the conservation of mass, momentum and energy of electron, and ion subsystems in a two-temperature single-fluid 1D Lagrangian form:

$$\frac{\partial}{\partial t}\left(\frac{1}{\rho}\right) + \frac{\partial v}{\partial m} = 0 \qquad (1)$$

$$\frac{\partial v}{\partial t} + \frac{\partial P}{\partial m} = 0, v = \frac{\partial x}{\partial t} \qquad (2)$$

$$\frac{\partial \varepsilon_e}{\partial t} + P_e \frac{\partial v}{\partial m} = \frac{\partial}{\partial m}\left(k\rho \frac{\partial T_e}{\partial m}\right) - \frac{\alpha_{ei}}{\rho}(T_e - T_i) + J_L \quad (3)$$

$$\frac{\partial \varepsilon_i}{\partial t} + P_i \frac{\partial v}{\partial m} = \frac{\alpha_{ei}}{\rho}(T_e - T_i) \quad (4)$$

where m is the mass coordinate, $dm = pdx, m = \int_{x_0}^{x} pdx$, direction of x-axis is chosen perpendicular to the irradiated surface of the metal, ρ is the density, ρ_0 is the initial density, v is the velocity, t is the time, P_e and P_i are the pressures of electrons and ions, ε_e and ε_i are the internal energies of electrons and ions, $P = P_e + P_i$ and $\varepsilon = \varepsilon_e + \varepsilon_i$ are the full pressure and the internal energy, T_e and T_i are the temperatures of electrons and ions.

For the coefficient of electron-ion relaxation α_{ei} we use formula from work [6] and energy of the absorbed laser radiation J_L, described by:

$$J_L = \frac{F_{abs}}{\tau_L \delta \sqrt{\pi} \rho} \exp\left(-\frac{t^2}{\tau_L^2}\right) \exp\left(-\frac{x(m,t) - x(m_0,t)}{\delta}\right) \quad (5)$$

where τ_L is the laser pulse duration, F_{abs} is the laser radiation energy, δ is the skin depth of metal.

The electron wide-range thermal conductivity coefficient k is constructed as an interpolation between limiting metal and hot plasma cases [7]. The viscosity and thermal conductivity of the ions are neglected in this model. To calculate the ablation depth, we perform an integration of the mass flux through the surface using the expression:

$$d = \frac{1}{\rho_0} \int_0^t (\rho v)|_{x=x_0} \, dt' \quad (6)$$

3. Equation of state

To solve the system of hydrodynamic equations (1) – (4) we construct the wide-range semi-empirical two-temperature equation of state. The metal is expected to consist of the same electrically neutral atomic cells with atomic weight A and charge Z. The Helmholtz free energy of a single atomic cell is the sum of three terms, describing the electronic and ionic components, and the interaction between them:

$$F = F_e + F_i + F_{ei} \quad (7)$$

As the main thermodynamic parameters we use atomic cell volume V and temperatures T_e, T_i. The number of free electrons in one atomic cell we call the degree of ionization and denote by the letter y. The pressure of the electrons is assumed to be the pressure of an ideal Fermi gas of electrons with density y/V and temperature T_e, where y is the solution of the modified equation of ionization:

$$\mu_F\left(\frac{y}{V}, T_e\right) + I(y) - B(V, T_e) = 0 \quad (8)$$

Here μ_F - the chemical potential of an ideal Fermi gas, $I(y)$ - ionization potential, which is a smooth function of y and is constructed as a smoothing spline on the known experimental values of the successive stages of ionization $I(1)$, $I(2)$ etc.

$B(V, T_e)$ - negative correction, which purpose is the description of cold ionization of highly compressed matter:

$$B(V, T_e) = b\left(\frac{Z}{V}\right)^{\beta} (1 + \mu T_e V^{\sigma})^{-1} \quad (9)$$

where b, β, σ, μ - parameters, determined from shock compression data of metals.

For the free energy of an ideal Fermi gas we use approximation:

$$F_e = yT_e[\frac{3}{5}\varphi - \frac{3}{2}\ln\left(1 + \frac{5}{2\varphi}\right)] \quad (10)$$

where

$$\varphi = \frac{1}{2}(3\pi^2)^{\frac{2}{3}}(\frac{y}{V})^{\frac{2}{3}}/T_e \quad (11)$$

From (10) we get formulas for the chemical potential, pressure and internal energy of the free electrons:

$$\mu_F = \frac{1}{2}(3\pi^2)^{\frac{2}{3}}(\frac{y}{V})^{\frac{2}{3}} + T_e[\left(1+\frac{2}{5}\varphi\right)^{-1} - \frac{3}{2}\ln(1+\frac{5}{2\varphi})] \quad (12)$$

$$P_e = \frac{1}{5}(3\pi^2)^{\frac{2}{3}}(\frac{y}{V})^{\frac{5}{3}} + T_e(\frac{y}{V})\left(1+\frac{2}{5}\varphi\right)^{-1} \quad (13)$$

$$E_e = \frac{3}{2}VP \quad (14)$$

F_i is describing the transition from the crystal state at low temperatures to the state of an ideal ion gas at high temperatures:

$$F_i = \frac{3}{2}T_i ln\frac{1+\lambda G}{V^{\frac{2}{3}}T} \quad (15)$$

where

$$G = (\frac{4\pi}{3})^{\frac{1}{3}}\frac{y_c{}^2(V)}{V^{\frac{1}{3}}T_i} \quad (16)$$

and $y_c(V)$ assumed to be a known function of the atomic volume V, defined by the equation of ionization, λ - parameter of interpolation.

From (15) we get:

$$P_i = \frac{T_i}{V}\frac{1+3\lambda(\frac{1}{2}-\frac{dln y_c(V)}{dlnV})G}{1+\lambda G} \quad (17)$$

$$E_i = \frac{3}{2}T_i\frac{1+2\lambda G}{1+\lambda G} \quad (18)$$

And finally, for P_{ei} we take into account that it must not affect the equation of ionization, full pressure should be equal zero at normal density and zero temperature, component of the internal energy must take into account the Coulomb attraction between the free electrons, at low densities P_{ei} must decrease rapidly, electron-ion interaction at a fixed density should decrease as the temperature increases.

$$P_{ei} = -\frac{1}{5}(3\pi^2)^{\frac{2}{3}}(\frac{y_0}{V_0})^{\frac{5}{3}}(\frac{V_0}{V})^{\frac{4}{3}}\frac{1+\delta}{1+[\delta+(1+\delta)T_e/T^*](\frac{V}{V_0})^{\gamma}} \quad (19)$$

Here V_0, γ, δ, T^* are the parameters, determined from the known values of density, sublimation energy, isothermal compressibility and thermal expansion coefficient.

4. Results

To verify the model we determine the dependence of ablation depth on the laser pulse duration after a single laser shot and compare with experimental data for aluminium and copper. Ultrashort laser pulses are generated by an amplified all solid-state Ti:Saphire laser chain [8].

Low energy pulses are extracted from a mode-locked oscillator (1.6 nJ/pulse, 80 MHz, 800 nm, 120 fs). The pulses are then injected into an amplifying chain including: an optical pulse stretcher, a regenerative amplifier associated with a two-pass amplifier using a 20 W Nd:YLF laser as pumping source, and a pulse compressor. Linearly polarized pulses with wavelength centered on 800 nm, an energy of 1.5 mJ at 1 kHz repetition rate and typical duration of 170 fs were delivered.

The samples are mounted on a three-motorized-axes system with 0.5 μm accuracy. Experiments are performed in the image plane of an aperture placed before the objective. Focusing objectives of 25 mm or 10 mm focal lengths to obtain fluences in a range of 0.5 to 35 J/cm^2 with the same beam size, 16 μm in diameter, in the image plane.

For ablation depth measurements, grooves are machined by moving the sample, which is adapted such that each point on the groove axis undergoes 8 consecutive irradiations at each target pass. The number of times the sample passes in front of the fixed beam can be adjusted. Figure 1 shows a scanning electron microscopy (SEM) picture of the machined grooves on copper for 2, 4, 6, 8 and 10 passes.

The groove depth is measured by an optical profilometer with a 10 nm depth resolution. The ablation rates for each groove are deduced taking into account the sample speed, the repetition rate of the laser, the beam size and the number of passes. For each energy density an ablation depth averaged over a few tens of laser shots.

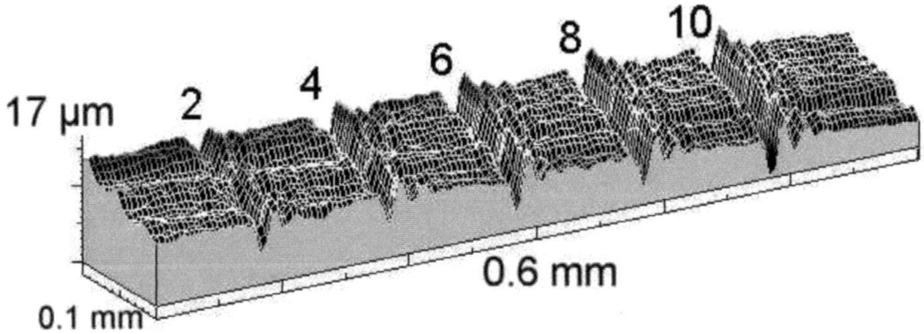

Figure 1. SEM picture of machined-groove profiles, from 2 to 10 passes, on a copper sample.

As we can see in figure 2 and in figure 3 results of the simulation are pretty close to experimental data both for copper and aluminium in a range of 0.5 to 35 J/cm^2.

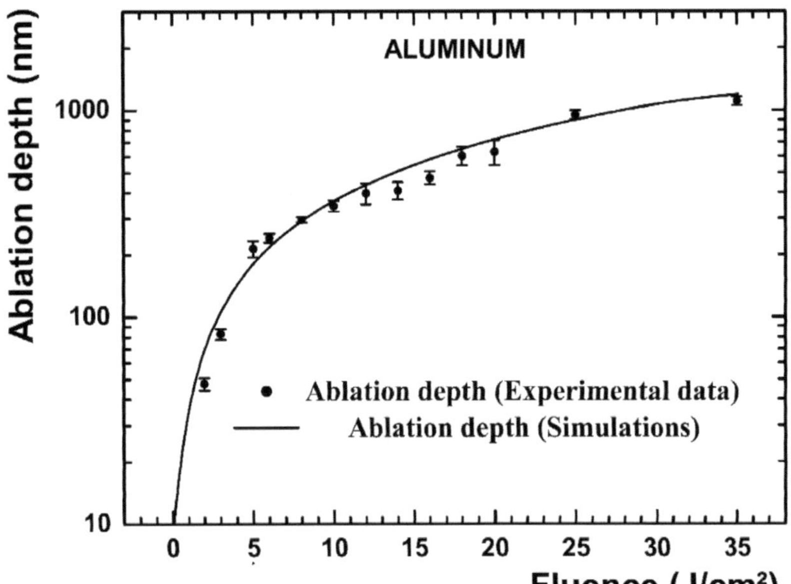

Figure 2. Experimental and numerical ablation depth as a function of the laser fluence on an aluminum target obtained with a 170 fs laser pulse.

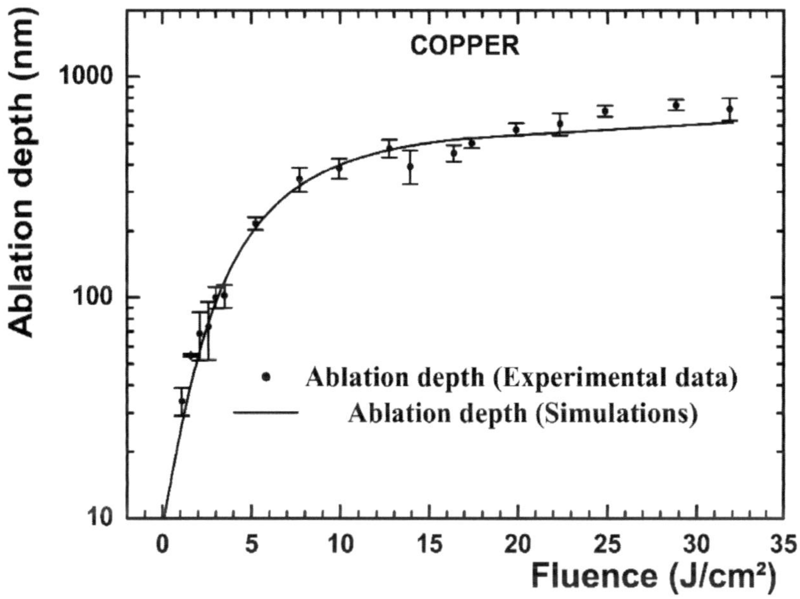

Figure 3. Experimental and numerical ablation depth as a function of the laser fluence on a copper target obtained with a 170 fs laser pulse.

5. Conclusion

In this paper a mathematical model for femtosecond laser ablation of metals is proposed, based on standard two-temperature model connected with 1D hydrodynamic equations. Wide-range equation of state has been developed. The simulation results are compared with experimental data for aluminium and copper. A good agreement for both metals with numerical results and experiment shows that this model can be employed for choosing laser parameters to better accuracy in nanoparticles production by ablation of metals.

References

[1] Makarov G N 2013 *Phys. Usp.* **56** 643–682
[2] Li Q, Lao H, Lin J, Chen Y and Chen X 2011 *Appl. Phys. A.* **105** 125-129
[3] Ren Y, Chen J and Zhang Y 2012 *J. Heat Mass Transfer* **55** 1620-1627
[4] Mishra S and Yadava V 2013 *Opt. Laser Technol.* **48** 461-474
[5] Zhang J, Chen Y, Hu M and Chen X 2015 *Appl. Phys.* **117** 063104
[6] Petrov Yu V 2005 *J. Laser and Particle Beams* **23** 283-289
[7] Povarnitsyn M E, Andreev N E, Levashov P R, Khishchenko K V and Rosmej O N 2012 *Phys. Plasmas* **19** 023110
[8] Colombier J P, Combis P, Bonneau F, Harzic R Le and Audouard E 2005 *Phys. Rev. B* **71** 165406

Low-frequency ultrasound attenuation in sodium chlorate crystals

AL Igisheva[1,a], EG Soboleva[2,a] and TB Krit[3,b]

[1,2]Yurga Institute of Technology Tomsk Polytechnic University,
Leningradskaya st., 26, 652055, Yurga, Russia
[3]Moscow State University,
Leninskie Gory 1/2, 119991, Moscow, Russia

[a]email: sobolevaeno@mail.ru,
[b]email: timofey@acs366.phys.msu.ru

Abstract. The authors study reduction of longitudinal ultrasonic oscillations in sodium chlorate crystals within the temperature range of 78 – 525 K and relative amplitudes of oscillatory deformation of 10^{-7} – 10^{-5} according to the method of compound piezoelectric vibrator at the frequencies of ≈ 10^{5} Hz.

1. Introduction

As a rule, various mechanisms causing decay of vibrations are divided into relaxation, resonance and hysteresis [1]. Under the periodic deformation of the solid body all mechanisms reveal themselves to some extent but their contribution to the process of vibration decay is different. When the frequency of oscillations changes from infrasound to hypersound we obtain a number of peaks (maximums) on the curve of internal friction of the real crystal. This kind of Q^{-1} curve with maximums, by analogy with the theory of light, is sometimes called the "mechanical" absorption spectrum. It is impossible to create a method which would allow measuring the internal friction continuously in the interval of frequencies from 10^{-4} to 10^{11} Hz. That is why the experiments were carried out with the application of a number of methods where the frequencies overlapped each other. The experimental difficulties of Q^{-1} measuring in a wide interval of frequencies are often compensated by an indirect technique, namely – by obtaining the dependence of the internal friction upon the temperature. Curve Q^{-1} (T) is often called "thermal spectrum" for short [1].

At the present moment it is estimated that in ion monocrystals for the frequency range ~ 10^{5} Hz the basic contribution to the internal friction is made by dislocation mechanisms (for example, decay determined by the thermoelastic effect, for metals makes about 50% of general decay, and for semiconductors and non-conductors its contribution to general decay does not exceed several per cent [2]). Internal friction can be determined by two different phenomena: inelasticity and plastic deformation of the system. Inelasticity is characterized by amplitude-independent parameters of a part of deformation energy input into the system. Plastic deformation, as distinguished from inelasticity, is a function of amplitude of deformation of the studied macroscopic system.

Systematic study of internal friction in the kilohertz range of frequencies in the crystals of A_1B_7 type (alkali halide crystals with ionic bond) was completed by S.P. Nikanorov and B.K. Kardashev [3]. In the course of theoretical analysis of experimental data the given authors studied the role of Peierls barrier and the role of fields of force of point defects of different origin under thermally activated oscillating motion of dislocations near the equilibrium state. According to the results of measurements of the amplitude-dependent absorption of ultrasonic oscillations the authors unambiguously proved that in the stress field of the ultrasonic waves the dislocations dissipate the energy of oscillations at the same obstacles which play an important role in formation of the stress level of the flow. It is of interest to obtain the data on internal friction in crystals with significant covalent component of the interatomic bond. The crystal of sodium chlorate ($NaClO_3$) is one of such objects [4].

2. Calculation ratios

Study of inelastic properties of $NaClO_3$ monocrystal was completed in two variants. Amplitude-independent internal friction within the temperature interval 77 – 525 K at the frequencies of ≈ 100 kHz was studied by the "half-width" of the resonance curve

$$Q^{-1} = \frac{\Delta f}{f_p}, \tag{1}$$

where Δf – interval of frequencies between the resonance curve points, corresponding oscillation amplitudes equaling to 0.707 of the maximum amplitude, and f_p – resonance frequency. Description of the unit and details of the experimental method for this case are provided in monograph [5].

Amplitude-dependent internal friction was studied under the same frequencies and discrete temperatures of 300, 400 and 500 K with application of an electric acustic unit with load resistance of R = 1,2 kilohm, similar to that described in [3]. Logarithmic decrement δ was taken as a measure of internal friction, and it was found as

$$\delta = \left(\frac{2bd_{12}}{s_{22}}\right)^2 \frac{R_э}{f_p m_b}, \quad R_э = R\left(\frac{U}{U_R} - 1\right), \tag{2}$$

and amplitude of relative oscillative deformation of the sample was calculated according to the formula

$$\varepsilon = \frac{s_{22}}{2bd_{12}\ell_0 f_p} \cdot \frac{U_R}{R}. \tag{3}$$

Here b, d_{12}, s_{22} and f_p – accordingly thickness, piezoelectric modulus, compliancy constant and resonance frequency of piezoelectric quartz, m_b – mass of vibrator (piezoelectric quartz + sample), R_e – equivalent resistance of vibrator at the moment of resonance, U and U_R accordingly – decrease in current at the section "piezoelectric quartz – resistance R" and on the resistance at the moment of resonance, ℓ_0 – length of the sample.

$NaClO_3$ monocrystals were grown from aqueous liquids of a CP preparation by slow evaporation under room temperatures. Samples in the form of cores were cut out according to the wet thread technique and their length along the crystallographic direction <100> corresponded to the condition of resonance excitation of the double compound vibrator at the basic frequency of 100 kHz for the longitudinal standing ultrasonic wave. Maximum amplitude of the relative oscillative deformation in the crest of the standing wave ε changed within two orders (from 10^{-7} to 10^{-5}). Internal friction measurements were completed with an average error of about 10%.

3. Results and their discussion

Thermal dependence of internal friction in $NaClO_3$ samples practically in the whole region of solid phase existence is presented in Figure 1. Here we also present the changes of Young modulus in direction <100> of sodium chlorate monocrystal. The fact that at once catches the researcher's eye that the mechanical spectrum of $NaClO_3$ sample contains plenty of internal friction maximums (five completely resolved ones under the temperatures 110, 225, 260, 400 and 450 K) and an ascending branch of Q^{-1} maximum under T > 500 K. Function $E_{<100>}$, first of all, is considerably non-linear and, second, it has a number of anomalies. The obtained mechanical spectrum is even more surprising as under similar conditions in terms of temperature and frequency Q^{-1} (T) and $E_{<100>}$ (T) in NaCl crystal have a common linear form – with the growth of the sample temperature internal friction and elastic modulus change monotonously (Q^{-1} increases, $E_{<100>}$ decreases).

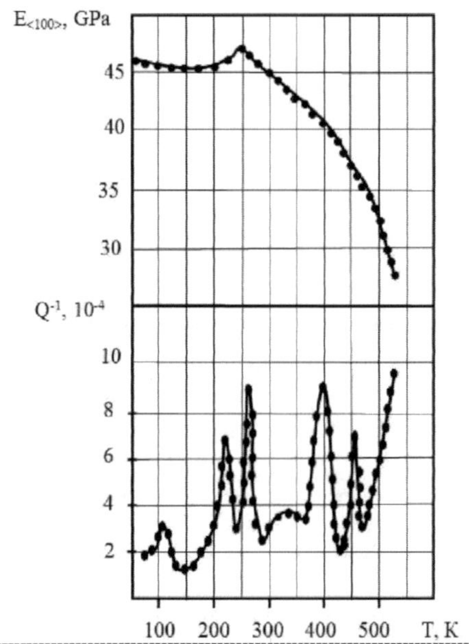

Figure 1. Thermal changes of internal friction and Young modulus of $NaClO_3$ monocrystal in the cube plane.

The observed peculiarities in the thermal behavior of internal friction and elastic modulus in sodium chlorate crystal are obviously related to the order-disorder processes in its lattice. In simplified form the interpretation of results can be presented as follows. When the crystal of $NaClO_3$ is heated, its anion sublattice, consisting of pyramidic groups of ClO_3^- with atom of chlorine in the vertex and atoms of oxygen in the base, is destabilized first. Near the temperature of 110 K in the local areas of anion sublattice atoms of chlorine start transitions between two equivalent sites and, by convention, this can be considered the trigger mechanism to the so called isostructural (isomorphous) phase transition in the crystal of sodium chlorate under low temperatures. Maximum Q^{-1} is followed by changing the sign of elastic modulus thermal dependence – $E_{<100>}$ increases in the interval of 170 – 260 K. A certain culmination of the described process is achieved under the temperatures of 225 – 260 K when the intensity of chlorine atoms transitions becomes enough to impart the quasi-molecular group ClO_3 additional degree of freedom – hindered rotation (libration). Separate anions randomly ("in a Brownian way") change their orientation for two directions. An indirect proof to this is the fact that under the temperature of 260 K $NaClO_3$ crystal becomes an auxetic [6, 7] (its Poisson's ratios $\sigma_{<100>}$ and $\sigma_{<110,001>}$ become negative). As the temperature of the crystal grows more and more anions become involved into the process of reorientation which finally leads (T ≈ 400 K) to cation sublattice destabilization. Activation energy of cation vacancy is 0.46 electron-volt and energy of Frenkel pair formation 1.98 electron-volt [5]. Q^{-1} peak under T = 400 K in $NaClO_3$, this way, is associated with elastic relaxation in the lattice of this crystal due to sodium ions "jumping" between the sites. When the temperature approaches 500 K ($T_{melting}$= 536 K) the crystal becomes strongly ionized (chlorate ions separate from sodium ions) and, as Curie point in $NaClO_3$ is taken over the melting temperature (T_c = 593 K), then, it is supposed that spontaneous polarization can be caused only by dipoles appearing due to free rotation of ClO_3 group.

Our studies of internal friction in $NaClO_3$ monocrystal under the temperature of T>300 K correlate to anomalous behavior of many physical properties of the given material in this temperature region (Fig. 2) [5].

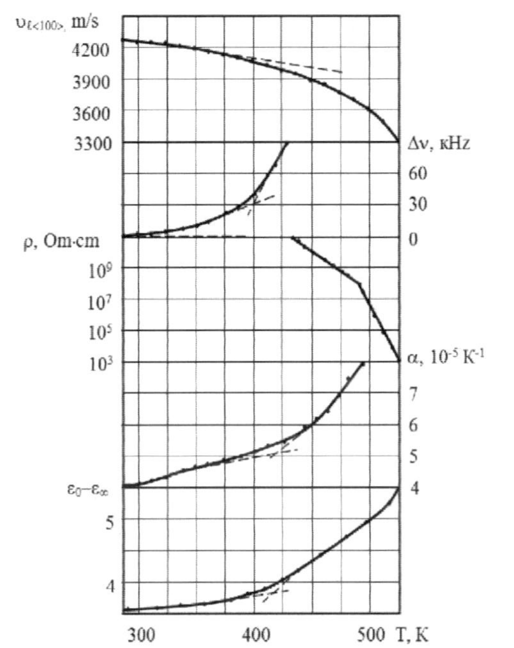

Figure 2. High-temperature changes of longitudinal waves velocity $\upsilon_{\ell<100>}$, frequency displacement of nuclear quadruple resonance ^{35}Cl $\Delta\nu$, specific electrical resistance ρ, thermal expansion coefficient α and difference of low- (ε_0) and high-frequency (ε_∞) transmittivities of $NaClO_3$ monocrystal.

An unexpected and surprising result was obtained in the experiments on amplitude dependence of internal friction in the studied object (Fig. 3). We mean peak δ under remarkably small amplitudes of oscillative deformation (near 10^{-7}). The plastic properties of $NaClO_3$ crystal in the ultrasound field of such low intensity are not quite understandable. Results presented in Fig. 3 clearly demonstrate that the view of δ (ε) curves significantly depends upon the temperature of the samples. The most marked peak of internal friction corresponds to the temperature of sodium chlorate sample of 400 K under which many non-linear properties start (Fig. 2). Let us notice that on dynamic curves of deformation for some materials which have absolutely different chemical nature than our object of study, for example, metals [3, 8], we observe similar "sharp plasticity point".

In the piezoelectric (sodium chlorate) and in the ferromagnetic (steel) identical "structural characteristics" caused by such insignificant mechanical effects of elastic pulse advancing, indicate a single mechanism of quasi-microplasticity effect characteristic for various solids being significantly different in nature and physical and mechanical properties [9]. The developed microplasticity of $NaClO_3$ under the temperature of 400 K is associated with the beginning of the given crystal transition into the superionic state due to cation sublattice "melting". Elastic wave modulates the depths of potential wells of quasi-free sodium ions in the conducting sublattice leading to the redistribution of Na^+ due to thermally activated motion. This phenomenon is the phenomenon of inelastic relaxation [10].

The test experiment with NaCl crystal (curve 4 in Fig. 3) demonstrates the typical view of function δ (ε) with amplitude dependence of internal friction under $\varepsilon \geq 10^{-5}$ determined by motion of dislocations. In $NaClO_3$ crystal under T = 500 K amplitude independent component of internal friction is practically absent (curve 3 Fig. 3).

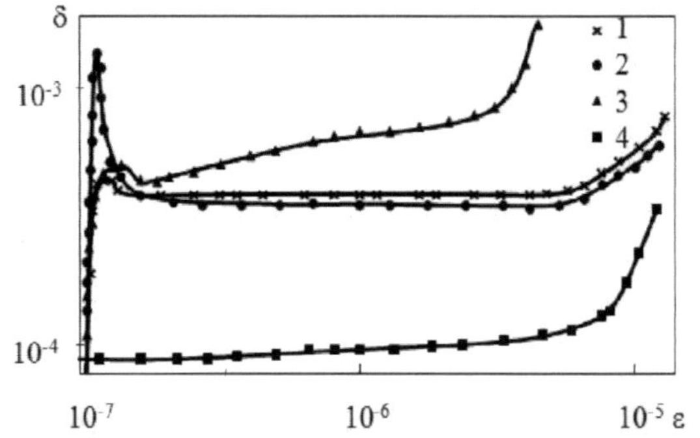

Figure 3. Amplitude dependence of internal friction of $NaClO_3$ (1, 2, 3) and NaCl (4) crystals. 1, 4 – 300 K, 2 – 400 K, 3 – 500 K.

4. Findings

1. It is established that mechanical spectrum of $NaClO_3$ sample contains maximum of internal friction (five completely solved ones under the temperatures of 110, 225, 260, 400 and 450 K) and an ascending branch of Q^{-1} maximum under $T > 500$ K.
2. A marked peak of internal friction is revealed under $T = 400$ K when many non-linear properties of sodium chlorate start.
3. The study revealed that the test experiment with NaCl crystal demonstrates the typical view of $\delta\ (\varepsilon)$ function with amplitude dependence of internal friction under $\varepsilon \geq 10^{-5}$ determined by motion of dislocations. In $NaClO_3$ crystal under $T = 500$ K amplitude independent component of internal friction is practically absent.

References

[1] Postnikov VS 1974 *Internal friction in metals* Moscow 352p.
[2] Truell R, Elbaum C, Chick B 1972 *Ultrasonic methods in solid physics* Moskow 307p.
[3] Nikanorov SP, Kardashev BK 1985 *Elasticity and dislocation inelasticity of crystals* Moskow 250p.
[4] Soboleva EG, Igisheva AL and Krit TB 2015 *IOP Conf. Series: Materials Science and Engineering* **91** 1-7.
[5] Belomestnykh VN, Soboleva EG 2009 *Acoustic, elastic and inelastic properties of sodium halogenate crystals* Tomsk 276p.
[6] Alderson KL, Alderson A, Grima JN, Wojciechowski KW 2014 *Phys. Status Solidi (b)* **250** 263.
[7] Goldstein RV, Gorodtsov VA, Lisovenko DS 2013 *Phys. Status Solidi (b)* **250** 2038.
[8] Kolarov D, Boltov A, Boncheva N 1979 *Mechanics of plastic medium* Moskow 302p.
[9] Klimenov V, Borozna V, Zaitsev K 2009 *METAL 2009 - 18th International Conference on Metallurgy and Materials* 273-278.
[10] Aliyev AE, Saidov AA, Khabibullayev PK, Shinder II 1995 *Acoustic Journal* **42** 322 – 333.

Journal of Physics: Conference Series **769** (2016) 012062 doi:10.1088/1742-6596/769/1/012062

Investigation of nanomechanical oscillators based on amorphous carbon whiskers in vacuum and at ambient pressure

S Lukashenko[1,3], I Mukhin[1,2], A Veniaminov[4], I Sapozhnikov[3], A Mozharov[2], D Kupriyanov[5], A Golubok[1,3]

1 Department of material science and nanotechnology, ITMO University, 197101 Kronverkskiy pr., 49, Saint-Petersburg, Russian Federation

2 Laboratory of renewable energy sources, St Petersburg Academic University RAS, 194021 Khlopina, 8/3, Saint-Petersburg, Russian Federation

3 Nanotechnology department, Institute for Analytical Instrumentation RAS, 190103 Rizhsky pr. 26, Saint-Petersburg, Russian Federation

4 Center for Information Optical Technologies, ITMO University, 199034 Birzhevaya ln. 14-16, Saint-Petersburg, Russian Federation

5 Mechatronics department, ITMO University, 197101 Kronverkskiy pr. 49, Saint Petersburg, Russian Federation

E-mail: lukashenko13@mail.ru

Abstract. Nanomechanical oscillators (NMO) on the base of amorphous C nanowhiskers, localized on the tops of W needles have been created and studied. Trajectories of resonant oscillations were visualized using a scanning electron microscope and a confocal laser scanning microscope. Resonant frequencies and the quality factor of NMO were determined at low pressure and in air. Reduction of the nanomechanical oscillators quality factor after the transition from vacuum condition to ambient pressure was not observed.

1. Introduction

The development of nanotechnology stimulates research in the field of nanomechanics. One of the important areas of nanomechanics is the creation and study of various types of nanomechanical oscillators (NMO), which are made from various materials and have various geometry. From a fundamental point of view, the interest to NMO is caused by the fact that due to high resonant frequencies, such mechanical systems can pass into quantum mechanical ground state at low temperatures [1]. From the applied point of view, the NMO is perspective sensitive elements for precision sensors for ultralow physical values measurement [2-7]. The switching at decreased air pressure between gas viscosity (low Q-factor) and free molecular (high Q-factor) regimes, and peculiarities in behavior of oscillators were analyzed [4,8-12]. In earlier researches NMO used to be investigated in the form of the tubes, beams or plates fixed on one or two ends on the bases with the bigger dimensions in comparison with NMO. In this work we studied the NMOs in the form of single amorphous carbon nanowhiskers (CNW) grown on the tops of the tungsten needles.

2. Experimental setup

In this paper, we investigated the NMO in form of amorphous carbon nano-beam attached to its base by one side. NMO was grown on the top of sharp tungsten probe (W-probe) with ~100 nm tip radius. Amorphous CNW was created using the technology of focused electron beam (FEB) in the chamber of a scanning electron microscope (SEM) Inspect S50 (FEI, USA), according to the method described in work [3]. Depending on the growth parameters CNW with diameter (50-200) nm and a length of (0.5 to 3) μm were created.

W probes were fabricated from polycrystalline W wire with base diameter of 150 μm using the electrochemical etching method in 5% KOH solution. FEB technology allows to grow NMO with different sizes and shapes and to measure the amplitude-frequency characteristics (AFC) of as-grown structure using the SEM trajectory visualization method [8]. It should be noted that the creation of structures and the study of their resonance characteristics at the low pressure were conducted in a single cycle in the chamber of a scanning electron microscope with the procedure described in [9]. The registration of the NMO oscillations at atmospheric pressure was conducted using a confocal laser scanning microscope (CLSM) LSM 710 based on AXIO Imager Z1 (Carl Zeiss, Germany) and the 405 nm laser. In this case, the basement with piezoactuator and W needle was replaced from SEM to CLSM 'Figure 1'. Laser beam was focused on the top of W-probe with whisker perpendicularly to plane of oscillation. Measurements were taken in reflection and transmission regimes.

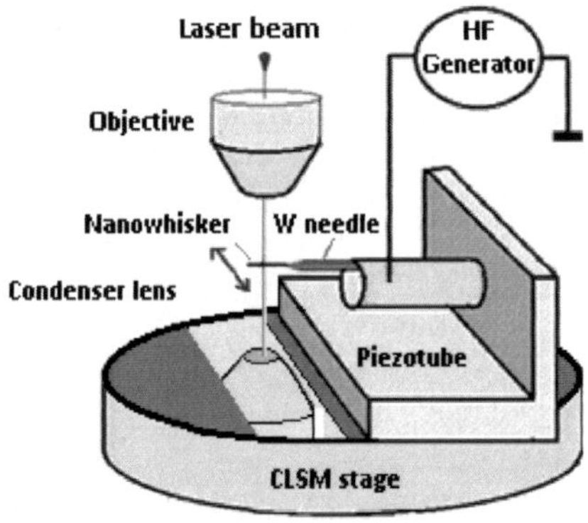

Figure 1. Sketch of setup for excitation and oscillation detection at atmospheric pressure.

'Figure 2a' shows SEM image of the static whisker, grown on the top of the W-probe. 'Figure 2b' presents SEM image of the resonant oscillation trajectory at low pressure (LP) of $p = 2 * 10^{-3} Pa$. 'Figure 2c,d' shows CLSM optical images of the same whisker at atmospheric pressure (AP).

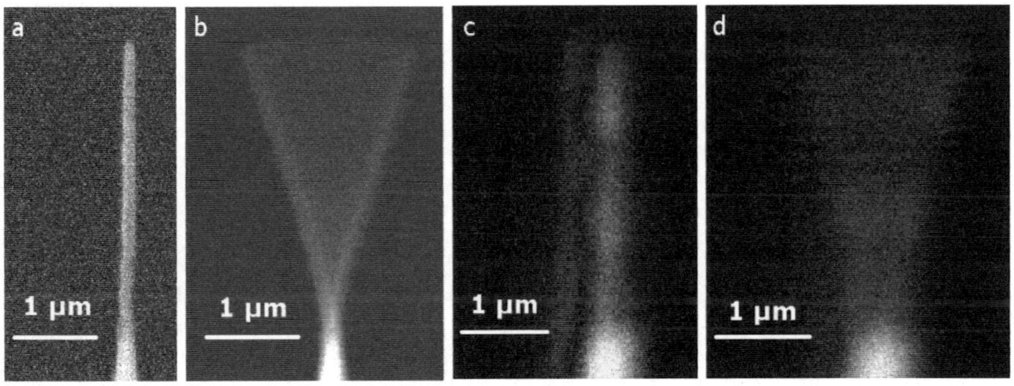

Figure 2. Whisker images at the top of the W-probe. (a, b) - SEM images in the secondary electron regime at LP for static (a) and oscillated at the resonance (b) whisker; (c, d) - CLSM optical images in reflection regime at the AP for static (c) and oscillated at the resonance (d) whisker.

According to 'Figure 2a', the lateral whisker dimension is ~100 nm and length ~2.3 μm. 'Figure 2b' depicts the resonance of the whisker. Left and right borders of the trajectory are brighter than main image due to longer time staying on boundary positions. As result, more secondary electrons are detected 'Figure 1a' that increases the brightness of the oscillating whisker. Images presented on 'Figure 2a', b reflect a real whisker dimensions and amplitude of resonant oscillations, because the lateral resolution of used SEM has a value of ~7 nm. CLSM lateral resolution is determined by laser wavelength equal ~400 nm that is 4 times larger than whisker lateral dimension. It determines the blurring of whisker optical image and dimension enhancing in comparison with the real size 'Figure 2c'. However, it is still possible to measure the AFC at AP, because the blurring amplitude of the CLSM optical image related to whisker oscillations 'Figure 2d' is larger than optical resolution.

3. Results and discussions

'Figure 3' shows AFC of the CNW measured at LP 'figure 3a' and AP 'figure 3b' and repeated low pressure (RLP, 'figure 3c'). Measurement data was approximated using Lorentzian function that allows to calculate Q-factor with relative error of 25% and 15% for CLSM and SEM measurements, respectively. The errors of measurement of Q-factor are caused by an error of measurement of amplitude of CNW oscillations. These errors emerge at determining the borders of "fan" on SEM or CLSM images. Larger tolerance at CLSM measurement related to larger blurring of the optical image 'figure 2d'. Fig.3d shows linear dependence of CNW oscillation amplitude versus voltage applied to the actuator, that proves the linear oscillation regime.

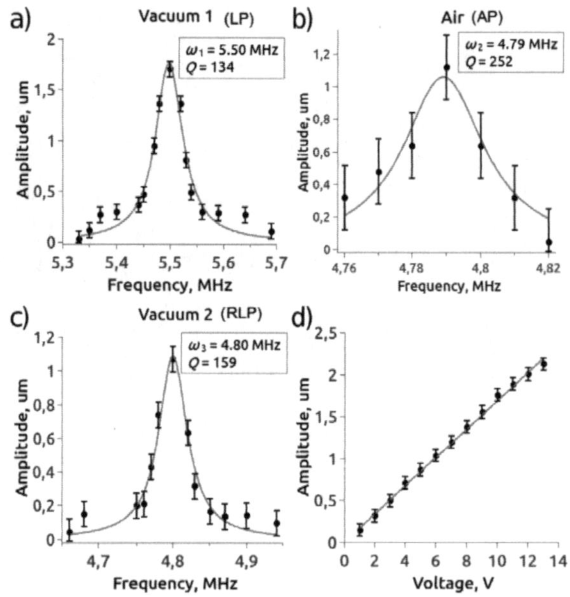

Figure 3. The amplitude frequency characteristics (AFC) of the CNW measured at low pressure (a), ambient pressure (b) and repeated low pressure (c). CNW amplitude versus actuator voltage (d).

Whisker resonant frequency at LP $\omega_{LP} = 5{,}49MHz$ 'Figure 3', line 1). Meanwhile, W-probe has several resonant peaks at resonant frequency$\omega_0 = 3{,}56MHz$ of the W needle. Q-factor of the whisker at LP calculated as $Q_{LP} = \omega_{LP}/\Delta\omega_{LP} = 134$, where $\Delta\omega_{LP}$ is the AFC width measured at the $1/\sqrt{2}$ of the peak level. Resonant frequency of the whisker measured at AP using CLSM is ('Figure 3' $\omega_{AP} = 4{,}79$MHz line 2) with $Q_{AP} = 252$. Repeating the measurement at RLP, the resonant frequency decreases on 700 kHz to $\omega_{RLP} = 4{,}80$MHz comparing to the first measurement at LP and increases on 10 kHz comparing to the measurement on AP 'figure 3c'. Obtained Q-factor at RLP is $Q_{RLP} = 159$. Measurements presented on 'fiigure 3' were repeated for different samples. Obtained resonant frequencies were slightly different, but the tendency of results for the measurements at LP and AP was similar 'table 1'.

No	Dimensions	LP regime		AP regime		RLP regime	
	a x b x L , nm x nm x um	Q ± 15%	ω , MHz	Q ± 25%	ω , MHz	Q ± 15%	ω , MHz
1	100 x 109 x 2,33	134	5,5	252	4,79	159	4,8
2	110 x 170 x 1,8	103	11,34	185	9,85	120	9,85
3	110 x 135 x 3,11	140	5,46	160	4,67	182	4,79

Table 1. Experimental data of sizes, resonant frequencies and Q-factor for different samples obtained at low and ambient pressures.

The internal frictions in the amorphous carbon and losses at the connection point of the CNW with the W tip mainly cause the energy losses of the CNW oscillations in LP regime. In case of AP extra losses are linked with environment gas emerge. These losses are determined by the gas flowing regime (gas viscosity or free molecular). As shown in [13, 18] the gas flowing regime for oscillator with resonant frequency$\omega$and lateral size D is determined by Weisenberg parameter $W_{Wi} = \omega\tau$, where$\tau$is the relaxation time of gas media and Knudsen number, $K_n = l_p/D$ where l_p is a mean free path of the gas molecule. If $W_{Wi} < 1$ and $K_n = 1$ ($W_{Wi} > 1$ and $K_n > 1$) the gas viscosity (free molecular) regime is realized. According to [13], our measurements at AP were carried out at gas viscosity regime ($W_{Wi} < 1$ and $K_n = 1$) and Q_{AP} has to be several times lower than Q_{LP} [14].

The single oscillator motion can be described by equation [19]:

$$\ddot{x} + 2\delta\dot{x} + \omega_0^2 = A_0\cos\Omega t \quad (1),$$

where $\omega_0$is the resonant frequency without damping, δ - damping factor, A_0 and Ω - amplitude and frequency of external periodic driving force. The excited resonance frequency ω of the single oscillator in the case of friction with the low damping coefficient is determinated by the Q-factor: [19]

$$\omega_{LP,AP} = \sqrt{\omega_0^2 - 2\delta^2} = \omega_0\sqrt{1 - \frac{1}{2Q_{LP,AP}^2}} \quad (2),$$

where index *LP* and *AP* consider oscillator parameters in LP and AP regimes, respectively.
According to measured frequencies ω_{AP} decreases on 710 kHz comparing to a ω_{LP} and due to equation (2) Q_{AP} should be two order lower than Q_{LP} that does not match the experimental data.
It is also known, that several water monolayers can condense on the solid surface at ambient pressure [20-23]. Such adsorbed water can increase the NMO mass and decrease the resonant frequencyω. In order to estimate the water mass, we use simple equation that describes relation of the oscillator mass increasing and frequency shift Δm [18]:

$$\frac{\Delta\omega}{\omega} = \frac{\Delta m}{2m_{eff}} \quad (3).$$

To shift resonant frequency of 710 kHz water thickness should be ~7 nm, that is thicker than water layer which can be condensed at AP regime.
Thus, difference between resonant frequencies in LP and AP regimes could not be explained neither by gas viscosity losses nor with effective mass increasing by condensed water on the CNW surface.
The effective mass of NMO oscillated in gas environment also could be increased due to mass of gas layer moved on the phase with NMO. In accordance with [24] the thickness of gas layer is in the range (1-5) μm, that corresponds to ~10 kHz resonant frequency shift in AP regime. This shift can't explain the obtained differences between ω_{AP} and ω_{LP} too.
Also in our measurements in the air the laser beam could have an impact on the oscillator as in [13]. But in measurements in ambient pressure we used 1% laser output (power). We did not observe any visible changes in resonant frequencies and q-factor of oscillators after output was increased to 15%. That is why we consider that the bolometric effect and radiation pressure effects should not have primary influence in our case.
Therefore, the obtained results can be explained with taking into account that NMO system "W tip - CNW" is considered as the system of two coupled oscillators. This coupled NMO system is described as [16]:

$$\ddot{x} + \gamma_1\dot{x}_1 + w_1^2X_1 + hX_2 = A_1e^{iwt}$$
$$\ddot{x} + \gamma_2\dot{x}_2 + w_2^2X_2 + hX_1 = 0 \quad (4),$$

Where γ_1 and γ_2 are oscillators damping parameters of the W tip and CNW, respectively, h is coupling coefficient of the oscillators, $A_1e^{i\omega t}$ the external actuator force. If *f* is proportional only to oscillator's shift, power coupling is realized, if *f* is proportional only to velocity, then the coupling is inertial. In case of simultaneous dependency of *f* between shift and velocity, there is mixed coupling. Power coupling analysis shows strong dependence of the resonant frequencies and Q-factors on the coupling coefficient due to interference and effective energy transfer between oscillators. In accordance with (4) there is a regime when CNW resonant frequency decreases and Q-factor of CNW increases with decrease of the coupling coefficient.
Considering the large resonant frequency shift (710kHz) and the Q-factor increasing at AP regime, we suggest that NMO system "W tip – CNW" is the system of coupled oscillators, while coupling of oscillators is decreased via thin adsorbed water layer on their surfaces. It is well known that for removing adsorbed water, solid surface has to be heated with temperatures higher than 100 °C [21]. That is why resonance frequency and Q factor of CNW at RLP regime don't return to the initial LP values.
For detailed explanation of the experimental data extended knowledge about oscillator coupling regimes (power, inertial, mixed) is needed.

4. Conclusions

In this paper, we experimentally demonstrated that Q-factor of the nano-mechanical oscillator in the form of the carbon nanowire fixed on the top of the sharp tungsten tip does not decrease (and even slightly increases) at transition from low pressure ($p = 2 * 10^{-3} Pa$) to ambient pressure. Such effect is connected with special behavior of NMO system of the coupling oscillators 'W tip – CNW". There are special conditions when the energy effectively transfers from the W tip to CNW that compensate the friction losses in the gas viscosity regime. The achieved results show perspective ways of the nanomechanical oscillators application as highly sensitive sensors with enhanced Q-factor for ambient pressure measurements.

5. Acknowledgment

This work is supported by Goverment of the Russian Federation (074-U01) through ITMO Early Carrier Fellowship scheme, Russian Foundation for the Basic Research (14-02-31703 mol_a, 16-32-60094 mol_a_dk), ITMO University grant 615893 and Program UMNIK.

6. References

[1] O'Connell A Hofheinz M 2010 *Nature* vol 464 pp 697-703
[2] Li M Tang H Roukes M 2007 *Nature nanotechnology vol 2 pp* 114-120
[3] Dikin D Chen X Ding W Wagner G Ruoffa R 2003 *Journal of applied physics* vol 93 pp 226-230
[4] Yum K Wang Z Suryavanshi A 2004 *Journal of applied physics* vol 96 pp 3933-3938
[5] Treacy M Ebbesen T Gibson J 1996 *Nature vol 381* pp 678-680
[6] Jensen K Kwanpyo K 2008 *Nature Nanotech.* vol 3 pp 533-537
[7] Durand C Casset F Renaux P Abele N Legrand B Renaud D Ollier E Ancey P Ionescu A Buchaillot L 2008 *IEEE Electron device letters* vol *29* pp 494-496
[8] Lissandrello C Yakhot V Ekinci K 2012 PRL vol 108 p 084501
[9] Karabacak D Yakhot V Ekinci K 2007 PRL vol 98 p 254505
[10] Ekinci K Karabacal D Yakhot V 2008 PRL vol 101 p 264501
[11] Biedermann L Tung R Raman A 2010 Nanotechnology vol 21 p 305701
[12] Belov M Quitoriano N Sharma S 2008 Journal of Applied Physics vol 103 p 074304
[13] Gil-Santos E Ramos D Pini V 2013 New Journal of Physics vol 15 p 035001
[14] Mukhin I Fadeev I Zhukov M Dubrovskii V Golubok A 2015 *Ultramicroscopy* vol *148 pp* 151-157
[15] Lukashenko S Lysak V Sapozhnikov I Mukhin I Golubok A 2015 *Journal of Physics: Conference Series* vol 643 12114
[16] Bhiladvala R Wang Z 2004 *Physical review E* vol 69 p 036307
[17] Rekhviashvili S Rozenberg B Dremov V 2008 *JETP Letters* vol *88* pp 887-891
[18] Meirovitch L 1986 *McGraw-Hill Science/Engineering/Math*
[19] Judge A Photiadis M Vignola J Houston B Jarzynski J 2007 *Journal of Applied Physics* vol 101 p 013521
[20] Rhoads J Kumar V Shaw S Turner K 2013 *Int. J. of Non-Linear Mechanics* vol 55 pp 79–89

Normal-state Nernst coefficient in high-temperature superconductors of the Y-Ba-Cu-O system

O A Martynova and V E Gasumyants

Peter the Great St. Petersburg Polytechnic University, Polytechnicheskaya str., 29, St. Petersburg, 195251 Russia

E-mail: olya218@yandex.ru

Abstract. A comparative experimental study of the normal-state Nernst coefficient, Q, in high-temperature superconductors of the $YBa_2Cu_3O_y$ system with different types of deviations from the stoichiometry was performed. The peculiarities of the $Q(T)$ dependences characteristic of Y-based HTSC, as well as the specific features of the influence of different types of doping on the absolute Q value were revealed. The obtained $Q(T)$ dependences were analyzed on the basis of a narrow-band model together with the thermopower temperature dependences measured on the same samples. It is shown that the developed approach allows one to describe quantitatively all the experimental Nernst results and to determine from their analysis the values of the charge-carrier mobility and the degree of the dispersion law asymmetry. The data on the variation of these parameters under influence of different types of doping are obtained and discussed with respect to the modification of the energy spectrum structure induced by studied dopants.

1. Introduction

Besides anomalously high values of the critical temperature, T_c, high-temperature superconductors (HTSC) are well-known to be characterized by unusual properties in the normal (non-superconducting) state. One of the examples is the behavior of the transport coefficients, whose temperature and concentration dependences demonstrate some peculiarities in comparison with those of the standard objects of the solid state physics, such as metals and semiconductors. Obviously, this is related to the specific features of the normal-state band spectrum structure, but questions on its main parameters and the mechanisms of its modification under doping, as well as on parameters of the charge-carrier system are still under discussion. This is why a lot of models based of different and even mutually exclusive assumptions were proposed to describe the electron transport phenomena in HTSC [1-7].

Note that despite a huge number of the experimental results devoted to an analysis of the temperature dependences of the resistivity, ρ, thermopower, S, and Hall coefficient collected already in the earlier years of studying the HTSC-materials (see, for example, reviews [8-10] and references therein), the data on the behavior of the Nernst coefficient, Q, at $T > T_c$ are limited in number, although this coefficient, according to the classical theory of the electron transport, is quite informative, especially in case of polycrystalline samples. The main reason for this is the fact that the quantitative analysis of the $Q(T)$ dependences cannot be performed when one uses a model-independent approach. Besides, to obtain the reliable information from such an analysis it is necessary to discuss the experimental Nernst data together with the results on other transport coefficients in the framework of a unified model concept on the energy spectrum structure in a studied material. Such an approach can be realized on the basis of the narrow-band model [7] allowing one to perform the joint quantitative analysis of the $S(T)$ and $Q(T)$ dependences measured on the same samples [11, 12].

For the above reasons, this paper is devoted to the comparative experimental study of the Nernst coefficient in yttrium-based HTSC samples with different types of doping and quantitative analysis of the obtained results together with the thermopower data.

2. Samples and Experimental Details

The samples of $YBa_2Cu_3O_{7-x}$ ($x = 0.09$-0.52), $YBa_2Cu_{3-x}Zn_xO_y$ ($x = 0.05$-0.12), $Y_{1-x}Ca_xBa_2Cu_3O_y$ ($x = 0.05$-0.2), $Y_{1-x}Ca_xBa_{1.5}La_{0.5}Cu_3O_y$ ($x = 0.2$-0.4), and $Y_{1-x}Ca_xBa_{2-x}La_xCu_3O_y$ ($x = 0.0$-0.5) were used for investigations. Such compositions were chosen for a comparative analysis because in these cases increasing doping level influences the critical temperature and transport coefficients values in different ways that should make it possible to trace the Nernst coefficient variations under different types of changes in the band spectrum and charge-carrier system parameters. Ceramic samples were prepared by the standard solid-state processing technique from the high-purity oxides and carbonates mixed in the required proportions. The oxygen deficit in $YBa_2Cu_3O_{7-x}$ was created by an annealing of the initial sample in vacuum at different temperatures for 2 h, for samples with cation doping the final oxygen saturation was not performed. In case of the $Y_{1-x}Ca_xBa_2Cu_3O_y$ system we used the samples characterized by an oxygen deficit created by an annealing in vacuum at 450 °C for 2 h. X-ray diffraction analysis has shown all the samples to be almost of single phase with amount of foreign impurities not exceeding 1-2%. The sample homogeneity was controlled by measuring the local values of the thermopower in various points on the sample surface at room temperature.

The $\rho(T)$ dependences were measured by the standard four-probe low-frequency ac ($f = 20$ Hz) method. The thermopower was measured by a differential method relative to copper electrodes at the temperature difference between the two ends of a sample of about 2 K and then calculated by correcting for the absolute thermopower of copper. The Nernst coefficient was measured in a constant magnetic field of 1.8 T. We used thin samples of about 1 mm in the ∇T direction to increase the Nernst signal and a reversible magnetic field to exclude a contribution of even magnetic effects to the measured Nernst voltage. Details of the Nernst coefficient measurements can be found elsewhere [12]. The resistivity and thermopower were measured in the temperature range $T = T_c$–300 K, the Nernst coefficient was measured at $T = 77$–310 K. The absolute error in determining the Nernst coefficient values when measuring the $Q(T)$ dependences did not exceed 10 %, the minimal securely fixed Q value at $T = 300$ K in the mobility units was about $5\cdot10^{-3}$ cm^2/(V·s).

3. Experimental Results

The critical temperature values for all the studied samples determined as a midpoint of the resistive superconducting transition are shown in figure 1. One can see that the T_c value changes with doping in various ways. In the $YBa_2Cu_3O_{7-x}$, $YBa_2Cu_{3-x}Zn_xO_y$, and $Y_{1-x}Ca_xBa_{2-x}La_xCu_3O_y$ systems, T_c decreases, but at different rates, whereas in the $Y_{1-x}Ca_xBa_2Cu_3O_y$ and $Y_{1-x}Ca_xBa_{1.5}La_{0.5}Cu_3O_y$ systems increasing calcium content leads to a T_c rise. The physical reasons for a T_c restoration in two last cases were earlier discussed elsewhere [13, 14].

The thermopower temperature dependences demonstrate all the well-known peculiarities typical for yttrium-based HTSC including some specific features characteristic of calcium-containing samples [13]. A detailed description and analysis of the thermopower results can be found in our earlier papers [13-16]. For this reason we will focus here on the experimental Nernst results.

For all the studied samples a character of the $Q(T)$ dependence remains qualitatively unchanged. As examples, figure 2 presents the experimental results for two of these systems, namely, $YBa_2Cu_{3-x}Zn_xO_y$ and $Y_{1-x}Ca_xBa_{1.5}La_{0.5}Cu_3O_y$. One can see that the Nernst coefficient is positive at $T = 300$ K and its values are extremely small (they do not exceed 1 cm^2/(V·s) in the mobility units for all the doping types and ranges, see figure 3). As the temperature decreases, the Q value increases slightly, then the $Q(T)$ dependence demonstrates a broad maximum followed by a sharp fall of the Q value.

Additionally, in the $Y_{1-x}Ca_xBa_{2-x}La_xCu_3O_y$ samples with $x = 0.2$ and $x = 0.3$, a transition to negative Q values is observed at $T < 150$ K.

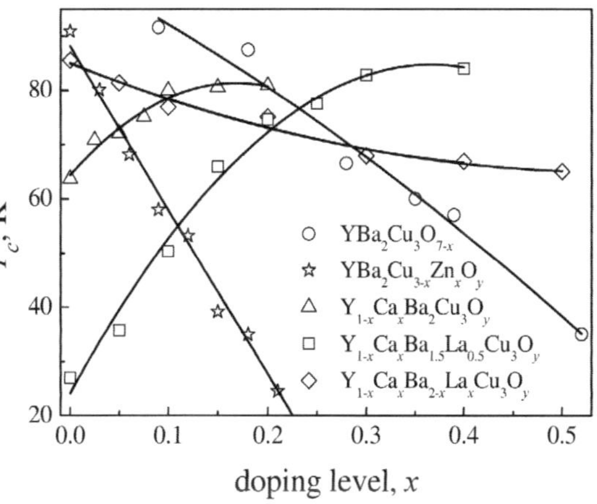

Figure 1. Critical temperature vs doping level

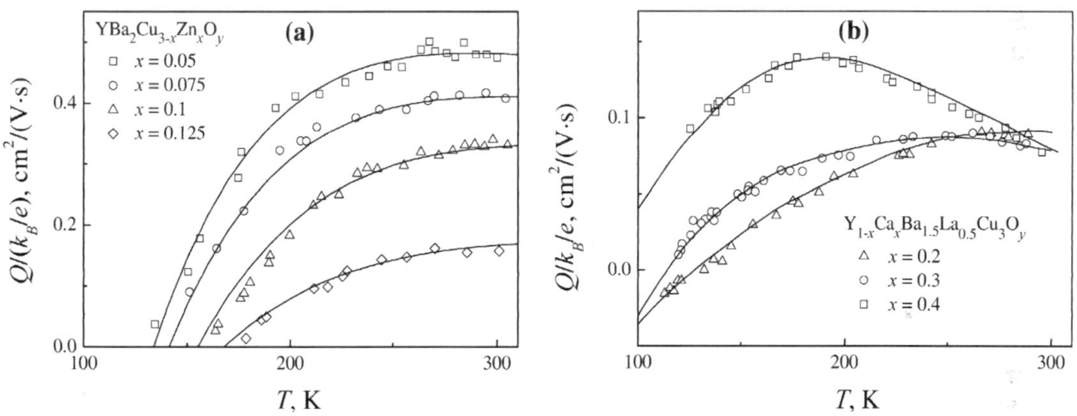

Figure 2. Nernst coefficient vs temperature for $YBa_2Cu_{3-x}Zn_xO_y$ (a) and $Y_{1-x}Ca_xBa_{1.5}La_{0.5}Cu_3O_y$ (b) samples. Symbols present the experimental data, lines are the calculation results

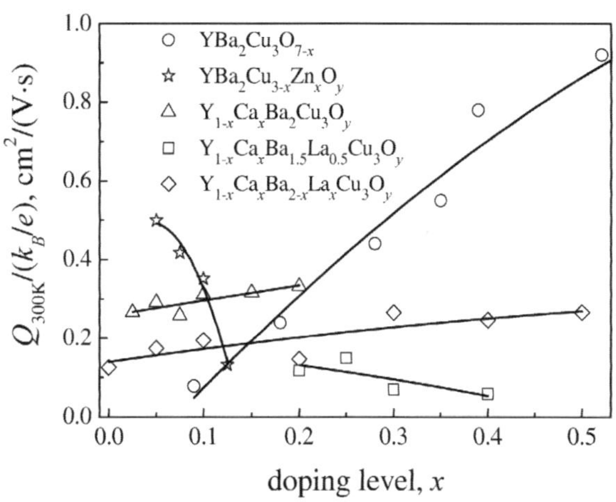

Figure 3. Nernst coefficient at $T = 300$ K vs doping level

The effect of doping level on the Nernst coefficient value at $T = 300$ K for all the systems studied is shown in figure 3. For $YBa_2Cu_3O_{7-x}$ with different oxygen content the $Q(T = 300$ K) value increases almost linearly and quite significantly with increasing doping level (from 0.078 to 0.92 $cm^2/(V·s)$ as x increases from 0.09 to 0.52). In the $Y_{1-x}Ca_xBa_2Cu_3O_y$ and $Y_{1-x}Ca_xBa_{2-x}La_xCu_3O_y$ systems, the Nernst coefficient value increases with x very slightly, while in the $YBa_2Cu_{3-x}Zn_xO_y$ and $Y_{1-x}Ca_xBa_{1,5}La_{0,5}Cu_3O_y$ systems the $Q(T = 300$ K) value decreases with increasing doping level, although to varying degrees. Thus, in general there is no correlation between the doping-induced variations of the Q values, on the one hand, and values of both T_c (see figure 1) and other transport coefficients (see [13-17]), on the other hand.

3. Analysis of the experimental data and discussion

The experimental results on the Nernst coefficient were quantitatively analyzed on the basis of a narrow-band model [7, 13] together with the data on the thermopower temperature dependences obtained earlier for the same samples [13-17]. According to this model, if the band responsible for the conduction process is narrow (which means that its half-width does not exceed the values of k_BT by more than an order of magnitude, where k_B is the Boltzmann constant), the analytic expression for the transport coefficients can be obtained by using the simplest rectangular approximation for the density-of-states, $D(E)$, differential conductivity, $\sigma(E)$, and Hall conductivity, $\sigma_H(E)$ functions (see figure 4).

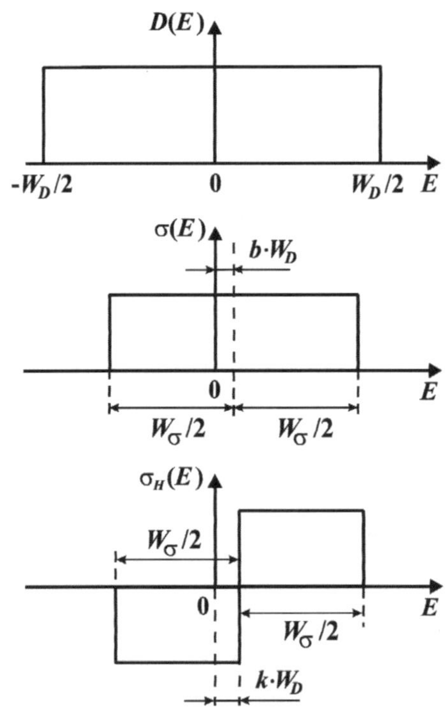

Figure 4. Approximations for the $D(E)$, $\sigma(E)$, $\sigma_H(E)$ functions used in the narrow-band model

It is necessary to stress that widths of rectangles approximating these functions can be different due to both a character of the energy dependence of the corresponding functions and a localization of the charge-carriers at the band edges [7, 13]. If so, the W_D value is the total effective bandwidth, W_σ is the effective width of the delocalized states interval, and the W_σ/W_D ratio characterizes the degree of state localization. Besides, these rectangles can be shifted relative to each other by the $b·W_D$ value, where b is the band asymmetry degree, in case the $D(E)$ function is asymmetric [7, 13, 14]. At last, as shown by the theoretical analysis of the Nernst coefficient behavior in case of a narrow conduction band [11], for the correct description of the $Q(T)$ dependences in HTSC one should take into account a possible asymmetry of the dispersion law. This can be modeled by shifting the point of change in the sign of the $\sigma_H(E)$ function by the value of $k\cdot W_D$ relative to the middle of the band (see figure 4). The k parameter value thus characterizes the degree of the dispersion law asymmetry.

In doing so, using the standard kinetic integrals and the above approximations for the $D(E)$, $\sigma(E)$, $\sigma_H(E)$ functions (see figure 4), we have obtained the following expressions quantitatively describing the temperature dependences of the thermopower and Nernst coefficient [7, 11, 13]:

$$S=-\frac{k_B}{e}\left\{\frac{W_\sigma^*}{\sinh W_\sigma^*}\left[\exp(-\mu^*)+\cosh W_\sigma^*-\frac{1}{W_\sigma^*}(\cosh\mu^*+\cosh W_\sigma^*)\ln\frac{\exp(\mu^*)+\exp(W_\sigma^*)}{\exp(\mu^*)+\exp(-W_\sigma^*)}\right]-\mu^*\right\},$$

$$Q=\frac{1}{eT}u\left(\frac{I_{H_1}}{I_0}-\frac{I_{H_0}I_1}{I_0^2}\right),$$

where $\mu^*\equiv\mu/k_BT=\ln\frac{\sinh(FW_D^*)}{\sinh\left((1-F)W_D^*\right)}-2bW_D^*$, $W_D^*\equiv W_D/2k_BT$, $W_\sigma^*\equiv W_\sigma/2k_BT$, μ is the electrochemical potential, e is the elementary charge, u is the band-averaged electron mobility, $I_0=\int_{-W_\sigma/2+bW_D}^{W_\sigma/2+bW_D}\left(-\frac{\partial f_0}{\partial E}\right)dE$, $I_1=\int_{-W_\sigma/2+bW_D}^{W_\sigma/2+bW_D}\left(-\frac{\partial f_0}{\partial E}\right)EdE$, $I_{H_0}=\int_{-W_\sigma/2+kW_D}^{W_\sigma/2+kW_D}\left(-\frac{\partial f_0}{\partial E}\right)\mathrm{sign}(E-kW_D)dE$, $I_{H_1}=\int_{-W_\sigma/2+kW_D}^{W_\sigma/2+kW_D}\left(-\frac{\partial f_0}{\partial E}\right)\mathrm{sign}(E-kW_D)EdE$, f_0 is the Fermi-Dirac function, F is the degree of band filling with electrons, which is equal to the ratio of the number of electrons to the total number of states in the band.

Thus, the calculated $S(T)$ and $Q(T)$ dependences are fully determined by the values of the six model parameters (W_D, W_σ, F, b, u, k), all of which have a clear physical meaning. It is important that the first four of them can be unambiguously (or, at least, with a small error) determined from the quantitative analysis of the experimental temperature dependence of the thermopower for a sample of the specific composition that was repeatedly demonstrated in our earlier publications [7, 13, 14, 16, 17]. Analyzing the experimental $Q(T)$ dependence measured for the same sample one can use these values, so that in such a case the calculated $Q(T)$ dependence is determined by the values of only two additional model parameters (u, k). Moreover, as clearly seen from the above formulae, only one parameter, k, influences a character of the temperature dependence of the Nernst coefficient, while the mobility value, u, determines the absolute Q values. All these facts essentially simplify the analysis of the experimental results making it possible to determine the u and k values from the $Q(T)$ analysis and to simultaneously specify the values of four parameters preliminary obtained from the $S(T)$ analysis. Thus, one can definitely determine the values of all six model parameters from the joint analysis of the experimental $S(T)$ and $Q(T)$ dependences.

We have used the described approach for analyzing the experimental results obtained. Results of the thermopower analysis, values of the main model parameters (W_D, W_σ, F, b) determined from this analysis, and discussion of their variations under doping can be found in [13-17]. Here we present only the summary of the results obtained for the effective bandwidth variations (see figure 5). The W_D value is seen to change with increasing doping level in different ways. The oxygen deficit in the $YBa_2Cu_3O_{7-x}$ system leads to a strong band broadening, in the $Y_{1-x}Ca_xBa_2Cu_3O_y$ and $Y_{1-x}Ca_xBa_{2-x}La_xCu_3O_y$ systems the W_D value increases rather insignificantly, in the $YBa_2Cu_{3-x}Zn_xO_y$ system it remains almost unchanged, and in the $Y_{1-x}Ca_xBa_{1.5}La_{0.5}Cu_3O_y$ system a strong narrowing of the band is observed. Thus, we have a possibility to analyze the Nernst coefficient behavior in cases of all the possible types of the energy spectrum modifications under doping.

The calculated $Q(T)$ curves for samples of the $YBa_2Cu_{3-x}Zn_xO_y$ and $Y_{1-x}Ca_xBa_{1.5}La_{0.5}Cu_3O_y$ systems are shown in figure 2 together with the experimental data. It is seen that for all the samples a good agreement between the experimental and calculated results is achieved. This fact can be considered as a powerful argument in favor of using the narrow-band model as an effective tool for the complex quantitative analysis of the electron transport phenomena in HTSC in the normal state.

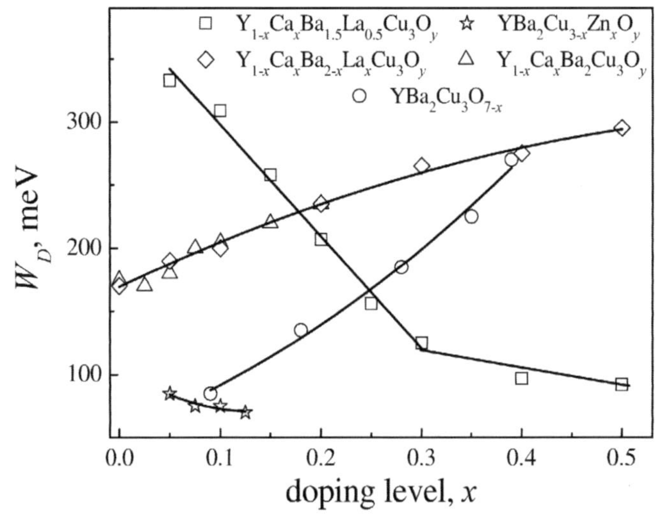

Figure 5. Total effective bandwidth vs doping level

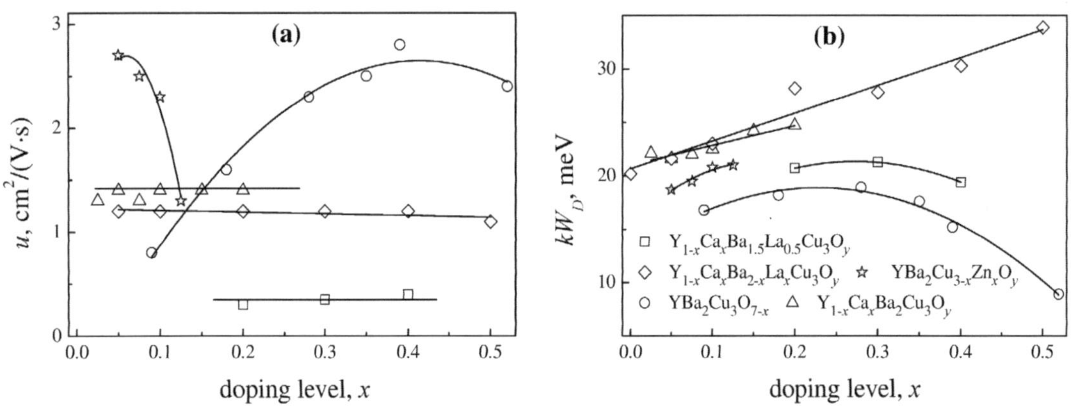

Figure 6. Variation of the electron mobility (a) and the absolute value of the energy shift of the Hall conductivity function (b) with doping level

The results obtained from the Nernst coefficient analysis are presented in figure 6. One can see that for all the studied samples the electron mobility is extremely low, its values do not exceed 3 cm^2/(V·s). In our opinion, this is resulted from a large effective mass of electrons that is usually characteristic of narrow-band systems. A character of the mobility variation with doping depends strongly on a type of deviation from the stoichiometry. In the $Y_{1-x}Ca_xBa_{1.5}La_{0.5}Cu_3O_y$, $Y_{1-x}Ca_xBa_{2-x}La_xCu_3O_y$, and $Y_{1-x}Ca_xBa_2Cu_3O_y$ systems, the u value remains almost unchanged, in the $YBa_2Cu_3O_{7-x}$ system the mobility increases with oxygen deficit, in the $YBa_2Cu_{3-x}Zn_xO_y$ system its value falls sharply with increasing Zn content. To explain such a different mobility behaviour it is necessary to take into account a character of the energy spectrum modification under doping in the studied systems.

As shown in our publications devoted to the analysis of the thermopower results for different doped Y-based HTSC-systems [7, 13, 14, 16, 17] on the basis of the narrow-band model, the modification of the energy spectrum under doping is mainly related to the Anderson localization [18] induced by a rising system disordering when introducing an impurity into the lattice or changing the oxygen content. If so, a single non-isovalent doping usually leads to an increase in both the W_σ and W_D values and simultaneously to a decrease in their ratio [7, 13]. However, in co-doped and calcium-containing systems, as well as in cases of some specific dopants, increasing impurity content can lead to additional peculiarities in a character of the conductive band modification [13, 14, 16, 17]. As already mentioned, a detailed description of the variation of the energy spectrum parameters under doping in all the studied systems is presented in our earlier publications [13-17]. Here we use these results to explain the observed changes in the charge-carrier mobility.

The mobility value is well-know to be proportional to the averaged relaxation time of carriers, but inversely proportional to their effective mass. For this reason, a broadening of the conductive band (an increase in W_σ and W_D values), leading to a decrease of the effective mass, should result in a mobility increase. On the other hand, increasing doping level induces a rise of the system disordering (a decrease in the W_σ/W_D ratio) that enhances the probability of the charge-carrier scattering and thus decreases the relaxation time value resulting in a mobility reduction. As a result, these two factors, characterizing the band spectrum modification under doping, influence the charge-carrier mobility in an opposite way. A character of the effect of a particular dopant on the u value is determined by which of these factors is prevailing in case of the studied doped system. In the $Y_{1-x}Ca_xBa_{2-x}La_xCu_3O_y$ and $Y_{1-x}Ca_xBa_2Cu_3O_y$ systems, a slight band broadening with x is accompanied by a small decrease in the W_σ/W_D ratio [13, 14, 16], i.e., a slight increase in the system disordering. In the $Y_{1-x}Ca_xBa_{1.5}La_{0.5}Cu_3O_y$ system the conductive band narrows strongly and simultaneously the degree of system disordering decreases (the W_σ/W_D ratio increases), both due to the effect of charge compensation realized in co-doped systems [13, 17]. As a result, in all these three systems, the two above effects compensate the influence of each other on the mobility value and it remains unchanged with doping. In the $YBa_2Cu_3O_{7-x}$ system the band broadens much stronger than the W_σ/W_D ratio falls [7, 13] that results in an increase in the charge-carrier mobility. In the $YBa_2Cu_{3-x}Zn_xO_y$ system, increasing Zn content leads to a slight band narrowing, but at the same time the degree of the charge localization increases strongly [13] because of destroying the conductive band [19], i.e., two factors considered above are summarized and the u value falls quite strongly. Thus, for all the studied systems the observed variation of the charge-carrier mobility can be explained by taking into account the character of the energy spectrum modification induced by the specific type of doping.

When analyzing the modification of the dispersion law (characterized in our model by the k parameter value), it is more reasonable to consider the absolute value of the energy shift of the point of changing sign of the $\sigma_H(E)$ function, $k \cdot W_D$. Dependences of this value on the doping level are presented in figure 6 (b). It is seen that, despite qualitatively different and in some cases very strong changes in the W_D value with increasing doping level (see figure 4), the $k \cdot W_D$ value varies rather insignificantly in a range from 10 to 30 meV for all the types and contents of dopants. Taking into account the small $k \cdot W_D$ values, a limited range of their variations, and a simplicity of the model used for transport coefficients analysis, one can conclude that the $k \cdot W_D$ value remains approximately unchanged in Y-based HTSC samples of varying compositions. It is also necessary to note that, in contrast to the asymmetry of the density-of-state function (characterized in our model by the b parameter), which is, in case of Y-based HTSC, characteristic of only calcium-containing samples [7, 13, 14, 16, 17], the dispersion law asymmetry is observed in all the studied samples including the calcium-free ones. Both these facts make is possible to conclude that the asymmetry of the dispersion law is not related to an effect of a specific impurity, but rather the fundamental property of the energy spectrum structure in Y-based HTSC.

4. Conclusions

Thus, this paper presents the results of the comparative study of the Nernst coefficient behaviour in the normal state for Y-based high-temperature superconductors with different types of deviations from the stoichiometry. The main peculiarities of the temperature and concentration dependences of this coefficient are revealed and discussed. It is shown that the narrow-band model can be used for the joint quantitative analysis of the thermopower and Nernst coefficient allowing one to describe all the specific features of their temperature dependences and to determine the values of all the model parameters characterizing the energy spectrum structure and charge-carrier system properties in samples of different compositions. The values of the charge-carrier mobility for all the studied samples were determined. The mobility was found to be extremely low, its values do not exceed 3 $cm^2/(V \cdot s)$. The observed character of the mobility change with increasing doping level for all the studied systems can be explained by the energy spectrum modification occurring due to an influence of the specific type of doping. The energy spectrum structure in all the studied samples is shown to be characterized by an asymmetry of the dispersion law. The degree of this asymmetry changes with doping so that the absolute value of the energy shift of the point of changing sign of the Hall conductivity function varies slightly, being the fundamental characteristic of Y-based high-temperature superconductors.

Acknowledgments

The study was supported by the grant of the St. Petersburg government for young scientists.

References

[1] Forro L, Lukatela J, Keszei B 1990 *Solid State Commun.* **73** 501
[2] Nagaosa N, Lee P A 1990 *Physical Review Letters* **64** 2450
[3] Cohn J L, Skelton E F, Wolf S A, Liu J Z 1992 *Phys. Rev. B* **45** 13140
[4] Xin Y, Wong K W, Fan C X, Sheng Z Z, Chan F T 1993 *Phys. Rev. B* **48** 557
[5] Alexandrov A S, Bratkovsky A M, Mott N F 1994 *Phys. Rev. Lett.* **72** 1734
[6] Newns D M, Tsuei C C, Huebener R P, van Bentum P J M, Pattnaik P C, Chi C C 1994 *Phys. Rev. Lett.* **73** 1695
[7] Gasumyants V E, Kaidanov V I, Vladimirskaya E V 1995 *Physica C* **248** 255
[8] Ong N P In *Physical Properties of High Temperature Superconductors II*, ed by Ginsberg D M, World Scientific, pp 459-507 (1990)
[9] Iye Y In *Physical Properties of High Temperature Superconductors III*, ed by Ginsberg D M, World Scientific, pp 285-361 (1992)
[10] Kaiser A B, Ucher C In *Studies of High Temperature Superconductors, vol. 7*, ed by Narlikar A V, Nova Science Publishers, pp 353-392 (1991)
[11] Ageev N V, Gasumyants V E 2001 *Phys. Solid State* **43** 1834
[12] Gasumyants V E, Ageev N V, Elizarova M V 2005 *Phys. Solid State* **47** 202
[13] Gasumyants V E In *Advances in Condensed Matter and Materials Research, Vol. 1*, ed by Gerard F, Nova Science Publishers, pp 135-200 (2001)
[14] Gasumyants V E, Vladimirskaya E V, Elizarova M V, Ageev N V 1998 *Phys. Solid State* **40** 1943
[15] Gasumyants V E, Elizarova M V, Patrina I B 1999 *Phys. Rev. B* **59** 6550
[16] Gasumyants V E, Elizarova M V, Patrina I B 2000 *Supercond. Sci. Technol.* **13** 1600
[17] Vladimirskaya E V, Gasumyants V E, Patrina I B 1995 *Phys. Solid State* **37** 1084
[18] Anderson P W 1958 *Phys. Rev.* **109** 1492
[19] Gupta R, Gupta M 1999 *Phys. Rev. B* **59** 3381

The accelerator neutron source for boron neutron capture therapy

D Kasatov[1], A Koshkarev[1], A Kuznetsov[1], A Makarov[1], Yu Ostreinov[1,2*], I Shchudlo[1], I Sorokin[1], T Sycheva[1], S Taskaev[1,3], L Zaidi[4]

[1]Budker Institute of Nuclear physics, Novosibirsk, Russian Federation
[2]Novosibirsk State Technical University, Novosibirsk, Russian Federation
[3]Novosibirsk State University, Novosibirsk, Russian Federation
[4]University of science and technology HOUARI BOUMEDIENE, Algeria

[*]E-mail: wtfsnoo@gmail.com

Abstract. The accelerator based epithermal neutron source for Boron Neutron Capture Therapy (BNCT) is proposed, created and used in the Budker Institute of Nuclear Physics. In 2014, with the support of the Russian Science Foundation created the BNCT laboratory for the purpose to the end of 2016 get the neutron flux, suitable for BNCT. For getting 3 mA 2.3 MeV proton beam, was created a new type accelerator - tandem accelerator with vacuum isolation. On this moment, we have a stationary proton beam with 2.3 MeV and current 1.75 mA. Generation of neutrons is carried out by dropping proton beam on to lithium target as a result of threshold reaction $^{7}Li(p,n)^{7}Be$. Established facility is a unique scientific installation. It provides a generating of neutron flux, including a monochromatic energy neutrons, gamma radiation, alpha-particles and positrons, and may be used by other research groups for carrying out scientific researches. The article describes an accelerator neutron source, presents and discusses the result of experiments and declares future plans.

1. Introduction

Currently, Boron Neutron Capture Therapy (BNCT) is considered as a promising technique for treatment of malignant tumors [1]. It provides selective destruction of tumor cells by prior accumulation inside them a stable boron-10 isotope and subsequent irradiation with epithermal neutrons (Figure. 1). Because of the absorption of a neutron by boron, a nuclear reaction takes place with a large release of energy in the cell, leading to its death. Clinical trials on nuclear reactors showed that BNCT could treat glioblastoma, brain metastases of melanoma and several other tumors. For the widespread introduction of this technique in practice, compact sources of epithermal neutrons based on charged particle accelerators are required.

Currently in the Budker Institute of Nuclear Physics (BINP), at the accelerator obtained a stationary proton beam with energy 2.3 MeV and a current of 1.75 mA with high monochromatic energy 0.1% and stability of current efficiency of 0.5%, carried out neutron generation [2] and research of radiation influence on cells structure [3]. For therapy should increase the proton beam current of at least to 3 mA. 16-20 November 2015 we conducted a series of experiments of irradiation of cells at our facility together with the staff and PhD students from the Hospital of Tsukuba University (Japan).

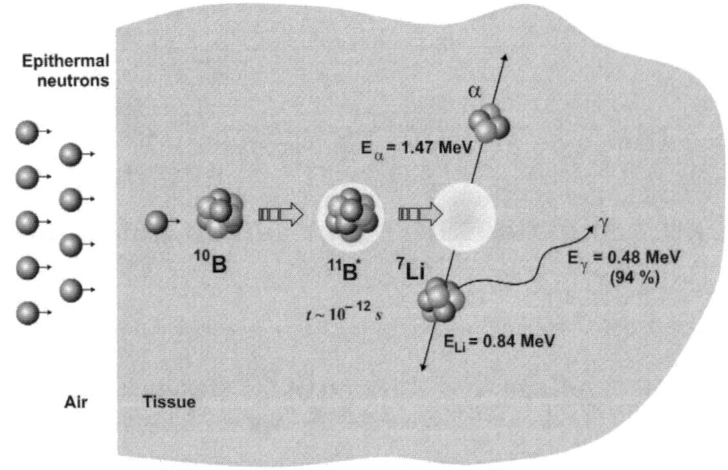

Figure. 1. Scheme of $^{10}B(n,\alpha)^{7}Li$ reaction

2. Accelerator based on epithermal neutron source

It is world-recognized [1] that the best reaction to form the epithermal neutron beam for BNCT is the $^{7}Li(p,n)^{7}Be$ reaction: neutron production is high, and the neutron spectrum is relatively soft. The source of epithermal neutrons based on the original tandem accelerator with vacuum insulation and the lithium neutron producing target was proposed in BINP [3]. The scheme of the proposed neutron source is shown at Figure 2. The facility consists of a source of negative hydrogen ions, a tandem accelerator generating a proton beam and a neutron producing target. Because of the complexity of creating such an accelerator neutron source for producing the proton beam it a new type of accelerator has been proposed, the tandem accelerator with vacuum insulation. It was also proposed to implement $^{7}Li(p,n)^{7}Be$ reaction, despite of the complexity of manufacturing the target: low melting temperature, low thermal conductivity and high chemical activity of lithium.

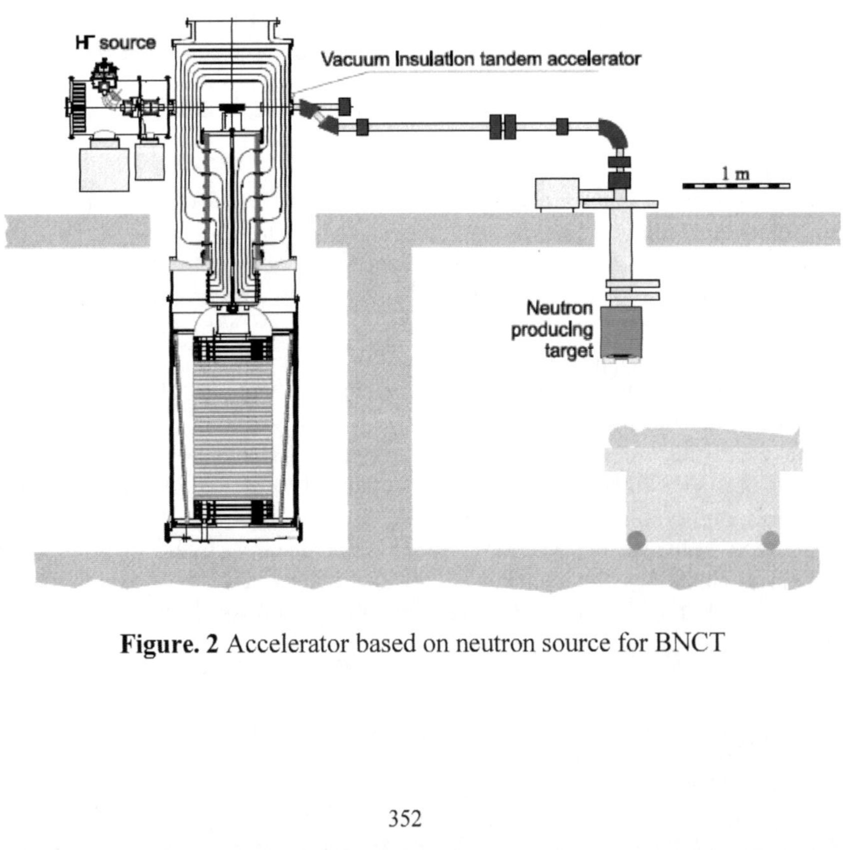

Figure. 2 Accelerator based on neutron source for BNCT

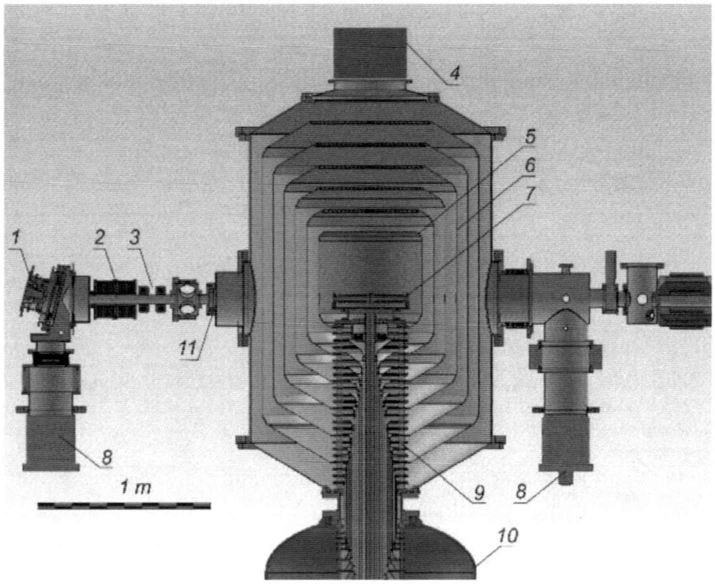

Figure.3 Tandem accelerator with vacuum insulation: 1 – source of negative hydrogen ions, 2 – magnetic lenses, 3 – correctors, 4 – cryogenic pump, 5 – high voltage electrode, 6 – intermediate electrodes, 7 – gas stripping cell, 8 – turbomolecular pump, 9 – insulator, 10 – high-voltage power supply, 11 – inlet diaphragm.

Figure. 3 shows the accelerator. Coming from the source (1) negative hydrogen ion beam with 23 keV energy is rotated in a magnetic field at an angle of 15 degrees, focused by a pair of magnetic lenses (2), injected into the accelerator and accelerated up to 1 MeV. In the gas (argon) stripping cell (7), which is installed inside the high-voltage electrode (5), negative hydrogen ions are converted into protons. Then protons by the same 1 MV potential are accelerated to 2 MeV energy. The potential for the high-voltage (5) and five intermediate electrodes (6) of the accelerator is supplied by a high-voltage source 10 (most of the source is not shown) through the insulator (9), wherein the resistive divider is set. Evacuation of gas is performed by turbomolecular pumps (8) mounted at the ion source and at the exit of the accelerator, and a cryogenic pump (4) via jalousies in the electrodes.

3. Additional flux of charge particle in accelerating channel

The main problem of achieving the objective of the Laboratory is double increase of the proton beam current. To clarify the reasons of breakdowns preventing an increase in current the process of acceleration of the ion beam in the accelerator has been studied. It was found that the acceleration of the ion beam is accompanied by additional flux of charged particles in the accelerating channel. It was found [4] that the appearance of charged particles is due to ionization of stripping gas and residual gas by the injected beam and due to penetration of positive argon ions of the stripping cell into the accelerating channel. The electrons caused by ionization of the gas are accelerated mainly to 1-MeV and absorbed by the elements of high-voltage electrode or stripping cell leading to a significant bremsstrahlung. Penetration of positive argon ions from the stripping cell into the accelerating channel is accompanied by their further acceleration to 1 MeV, which can lead to a significant secondary emission of electrons due to absorption of accelerated argon ions in the wall of the vacuum chamber.

4. Experiment for suppressing accompanying flux of charged particles

To suppress the flux of accompanying particles it was proposed and made two modernizations. Firstly, we improved vacuum conditions particularly in the beginning of the acceleration of the ion beam. For

Journal of Physics: Conference Series **769** (2016) 012064 doi:10.1088/1742-6596/769/1/012064

this purpose we have made a new input node, wherein a vacuum resistance formed as a cooled diaphragm was set and a cryopump On-Board 10 (CTI-Cryogenics, USA) with a pumping speed 2500 l/s for argon and 5000 l/s for hydrogen was mounted. At Figure.4 is shown a cryopump on input node, that place is pointed 11 at Figure.3

Figure.4 Input node. 1– Cryopump

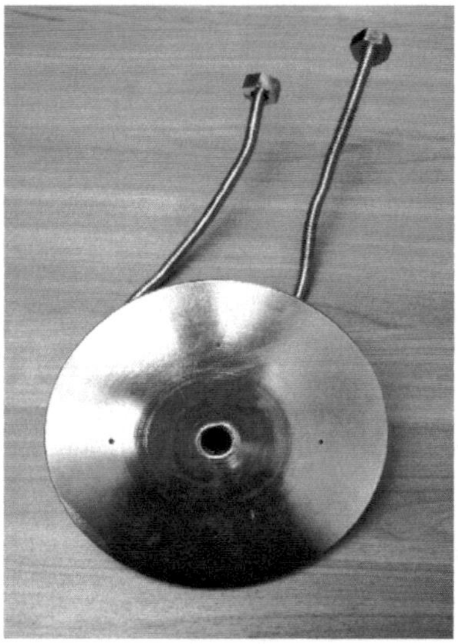

Figure.5 Water cooled diaphragm

Secondly, the wall of the vacuum chamber bombarded by positive argon ions and ultraviolet radiation from the stripping tube. We set an argon detector with grid at the inlet of accelerator on the water-cooled diaphragm. 300 Volt was applied on the grid to block a secondary electron emission. Also, metal circle was set under same voltage behind the water-cooled diaphragm to block electrons which may permeate with H^- beam. A photograph of modification describe above is shown in Figure. 6a, 6b.

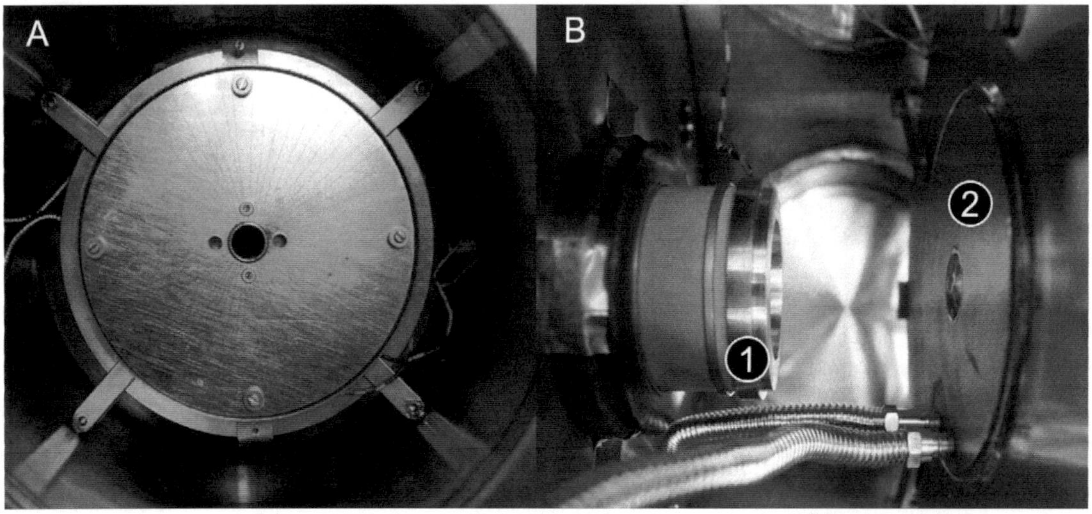

(a) (b)

Figure.6. (a) Argon detector with grid under voltage. (b) 1 – Metal circle under voltage, 2 – Water-cooled diaphragm.

5. The results

During the experiment, the voltage have being changing at the grid and ring. The obtained dependences are presented in Figure.7, 8.

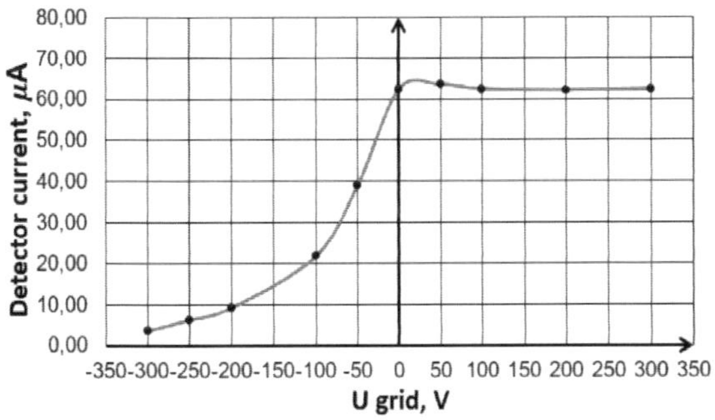

Figure.7 Counter-current of argon at detector

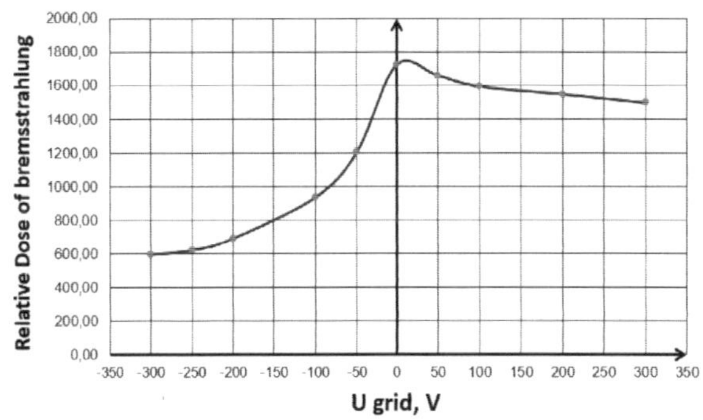

Figure.8 Relative dose of bremsstrahlung

The experiment produced the following results:

- Inclusion of additional cryopump on inlet of accelerator reduces bremsstrahlung dose 1.4 times.
- When voltage is applied on a grid on inlet of accelerator, dose bremsstrahlung is decreased in 2.5 times and the current at the detector in 10 times (Figure.7, Figure.8).

The main result is the additional flux of electrons accelerated up to 1 MeV was reduced in 3.2 times, up to 0,5% from injected beam, through inclusion additional cryopump and applied voltage on grid. The total flux of charged particles was reduced up to 5% from current of injected ion beam.

6. Future prospect

Figure.9 Stripping target

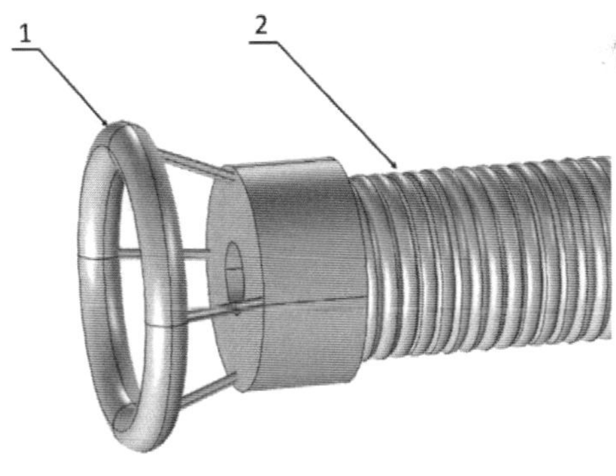

Figure.10 1– Electrostatic ring. 2–Stripping target

There are 2 modifications of stripping target (Figure.9) for suppression of the ion flux from it. It was proposed to use a stripping target with electrostatic rings. The design of modified stripping target is shown at Figure.10, idea of electrostatic ring is the following, voltage about 100V is supplied on the rings placed at the both end of tube, and subsequently argon ions are held in potential barrier inside tube. The design of target with magnetic suppressing is described in [5].

7. Conclusion

At Budker Institute of Nuclear Physics an accelerator based epithermal neutron source for BNCT is created and the Laboratory for preparation for the therapy is formed. It is necessary to increase the proton beam current from 1.75 to 3 mA. Additional flux of charged particles in accelerating channel doesn't allow to increase a beam current and produce a high dose of radiation. Experiments were conducted for improving the vacuum conditions which allowed to significantly reduce the accompanying currents in the accelerator and the dose of bremsstrahlung. The next step will be to conduct BNCT at the facility.

8. Acknowledgments

The study was performed by a grant from the Russian Science Foundation (Project No.14-32-00006) with support from the Budker Institute of Nuclear Physics.

References

[1] Sauerwein W, Wittig A, Moss R, and Nakagawa Y 2012 Neutron capture therapy. Principles and applications. *Springer*

[2] Kuznetsov A, Malyshkin G, Makarov A, Sorokin I, Sulyaev Yu, and Taskaev S 2009 First experiments on neutron detection on the acceleratorbased source for boron neutron capture therapy , *Techn. Phys. Letters* **35** 346-348.

[3] B. Bayanov, V. Belov, E. Bender, M. Bokhovko, G. Dimov, V. Kononov, O. Kononov, N. Kuksanov, V. Palchikov, V. Pivovarov, R. Salimov, G. Silvestrov, A. Skrinsky, and S. Taskaev 1998 Accelerator based neutron source for the neutron-capture and fast neutron therapy at hospital *Nucl. Instrum. Meth. A.* **413/2-3** 397-426

[4] Kasatov D, Makarov A, Taskaev S, Shchudlo I, Recording of current accompanying an ion beam in a tandem accelerator with vacuum insulation 2015 *Tech. Phys. Lett.* **41** 139-141

[5] Makarov A, Ostreinov Yu, Taskaev S, Vobly P 2015 Modification of the argon stripping target of the tandem accelerator. *Applied Radiation and Isotopes* **106** 53-56.

New scheme of the microwave signal formation for quantum frequency standard on the atoms of caesium-133

A A Petrov and V V Davydov

Peter the Great Saint-Petersburg Polytechnic University, Saint-Petersburg, Russia

Alexandrpetrov.spb@yandex.ru

Abstract. In present work several directions of quantum frequency standard modernization are considered. A new implementation of a frequency synthesizer and a magnetic field control unit are presented. Experimental study of a frequency synthesizer showed improvement parameters of a microwave-excitation signal, such as step frequency tuning, time frequency tuning, range of generating frequencies and spectral characteristics. Magnetic field control unit eliminates one of the most important perturbing factors affecting the long-term frequency stability. Daily frequency stability of quantum frequency standard improved on 15%.

1. Introduction

The accurate measurement of time and frequency is vital to the success of many fields of science and technology. Examples from atomic physics are atom-photon interactions, atomic collisions, and atomic interactions with static and dynamic electromagnetic fields. Geodesy, radio-astronomy (very long baseline interferometry) and pulsar astronomy rely on the availability of stable local frequency standards and uniform timescales. The same is valid for operation of satellite-based navigation systems.

Quantum frequency standards on the atoms of caesium are used as a clock generators in the communications equipment and in a data transmission devices, applied in the satellite navigation systems GLONASS and GPS as a clock generators and in the various metrological services. Also these standards perform a role of the reference signals with high precision and stability in radio equipment.

Considering the high importance of precision atomic clocks in science and technology and a vast area of their application, modernization of existing and development new quantum frequency standards are the urgent tasks.

The process of frequency standards modernization includes various directions: reducing energy consumption, weight and dimensions, improving metrological characteristics. And for frequency standards characterized by the fact that modernization may not be for the whole construction and may be for individual units or blocks. And in present work one of the directions of modernization of the caesium atomic clock is considered.

The article begins with the theoretical foundation of a caesium atomic clock operation and the role of frequency synthesizer and magnetic field control unit. Then we calculate range of generated frequency, step frequency tuning, frequency hopping and suppression of lateral amplitude components. Then we present a new implementation of a digital frequency synthesizer. After that, we describe how new synthesizer construction can be used for elimination one of the contributions to the long-term frequency instability. Finally, several results of actual experiments are given, and advantages of the new schemes which lead to improvement metrological characteristic of frequency standard are discussed.

2. Principles of quantum frequency standard on atoms of caesium-133 operation.

The work of a caesium atomic clock is based on the principle of adjustment a highly stable crystal oscillator frequency to quantum transition frequency of caesium-133 atoms [1-3]. Figure 1 shows a block diagram of a caesium atomic clock.

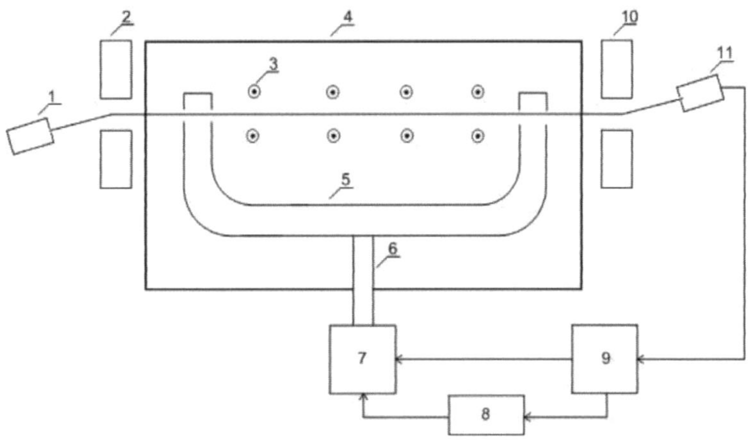

Figure 1. Block diagram of a caesium atomic clock. 1 - a source of caesium atoms, 2 - magnet polarizer, 3 - magnetic field, 4 - magnetic shield, 5 - Ramsey resonator, 6 - the waveguide, 7 - frequency converter, 8 - crystal oscillator, 9 - automatic frequency control system, 10 - magnet analyzer, 11 – detector.

The output signal frequency 5MHz of the crystal oscillator 8 is supplied to the frequency converter 7. Frequency converter consists of the frequency synthesizer, the mixer signals and the multiplier signals. In the frequency synthesizer input signal frequency 5MHz is converted to the signal frequency 12,631772 MHz and supplied to the input of mixer signals. In the multiplier signals input signal frequency 5 MHz is multiplied to the frequency 270 MHz and then to frequency 9180 MHz. This signal frequency 9180 MHz is also supplied to the input of mixer signals. As a result, the output signal of the frequency converter is the signal of ultrahigh frequency 9192,631772 MHz. This signal is supplied in the waveguide 6 and then on the input of atomic beam tube.

In caesium atomic clock with the help of magnet polarizer 2 the atoms are prepared such that they are either in the $|F=4, m_F=0>$ or in $|F=3, m_F=0>$ state. Afterwards the atoms interact with an electromagnetic field that induces transitions into the former unoccupied state.

A magnetic field is used to separate energetically the otherwise degenerate magnetic sub-levels in order to allow the excitation of the clock transition $|F = 3, m_F = 0> \rightarrow |F = 4, m_F = 0>$ isolated from the other transitions. By convention such a field is referred to as the C field as it is applied between the fields of the polariser and the analyser which historically were called the A field and the B field, respectively. The magnitude of the C field is chosen as a compromise between two conflicting requirements. First, it has to be large enough to separate the otherwise overlapping resonances. Second, the C field shifts the resonance frequency quadratically which has to be corrected. However, in a larger field the frequency of the clock is influenced to a larger extent by fluctuations of the magnetic field. In the scheme of a commercial Cs clock the C field is often generated by a coil with windings in the paper plane wound around the Ramsey resonator and hence, points perpendicularly to that plane. Owing to the dependence of the frequency of the clock transition from the magnetic field, efficient magnetic shielding has to be provided in order to attenuate the ambient magnetic field and the magnitude of the associated fluctuations.

The atoms in the former unoccupied state are detected and allow one to determine the frequency of the interrogating field where the transition probability has a maximum. The observed transition frequency is corrected for all known frequency offsets that would shift the transition frequency from the unperturbed transition and is used to produce a standard frequency or pulse per second every 9192631772 cycles [1-3].

Scanning the frequency ν of the atomic resonance leads to a detector current. The signal shows the Ramsey resonance structure on a broader, so-called, Rabi pedestal.

The central feature with the maximum at the transition frequency ν_0 is used to stabilize the frequency of the crystal oscillator to the atomic transition frequency. To this end, the frequency from the synthesizer is modulated across the central peak. The signal from the detector is phase-sensitively detected in the automatic frequency control system, integrated and this signal is used for stabilising the frequency of the crystal oscillator.

Improvement the accuracy of the signal frequency 12,631770 MHz results in better accuracy of the resonant frequency of the atomic transition.

The main characteristic of the frequency synthesizer is its ability to impact the characteristic of frequency stability of the quantum frequency standard output signal. Frequency instability introduced by the synthesizer is determined by the lateral discrete spectrum components of the signal that occurs in dividing, multiplying, mixing frequency signals, the accuracy of the generated frequency, and the impact on the signal of natural and technical noise.

Experimental study showed up that the present method of generating the output signal of the frequency synthesizer needs to increase the accuracy. The large resolution of step frequency is necessary. New scheme of the frequency synthesizer is designed using direct digital synthesis (DDS - Direct Digital Synthesis) [4]. This method allows to generate the output signal of the synthesizer with accuracy more than 10^{-6} Hz. Step frequency tuning ΔF_{out} calculated by the formula below:

$$\Delta F_{out} = \frac{F_{clk}}{2^N}, \quad (1)$$

where F_{clk} is the clock frequency, N is the capacity of accumulator.

In our scheme the clock frequency is equal F_{clk}=20 MHz, the capacity of accumulator is equal N=40. Step frequency tuning is equal $\Delta F_{out} = 1{,}8\,10^{-5}$ Hz.

The application of direct digital synthesis gave the possibility of obtaining the generated frequencies in a wide range (0 - 3 MHz), in contrast to previous schemes, where this feature was absent. Range of generated output frequencies may be calculated by the formula below:

$$F_{out} = \frac{M * F_{clk}}{2^N}, \quad (2)$$

where M is the frequency code in decimal, F_{clk} is the clock frequency, N is the capacity of accumulator.

The implementation of our proposed method enables us to control the frequency of the output signal frequency synthesizer in real time. Time hopping t_h is the sum of the time adjustment the digital part of the system and the delay time of the low-pass filter (LPF). Time adjustment digital part in synthesizer is three clock cycles. The delay time of the LPF is inversely proportional to the low-pass filter bandwidth. If the cut-off frequency is equal $F_{cut} = F_{clk}/4$ then time delay is about four clock periods. It turns out that time hopping is equal $t_h = 7/F_{clk}$. The clock frequency F_{clk} is equal 20 MHz, then the time hopping is equal $t_h = 0.35$ μs. This ensures a high rate of frequency tuning, which makes it possible to more efficiently adjust the crystal oscillator frequency in contrast the previously used scheme.

To meet the requirements for spectral purity of output signal 10 bit DAC was used. It is possible to obtain the suppression of lateral amplitude components in the spectrum of the output signal is not worse than -60 dB. In figure 2 new design of the frequency converter is presented.

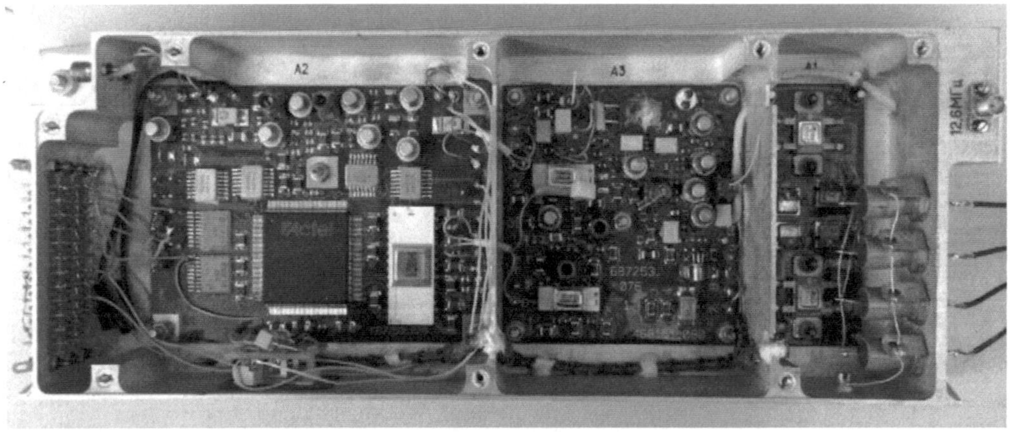

Figure 2. A new block of the microwave generator circuit.

3. Experimental research of synthesizer's output signal.

As experimental researches showed, new scheme of the frequency synthesizer has great advantages in comparison with previously used scheme. In figure 3, 4, 5 and 6, as an example, spectrums of the output signal measured in the different bands for new scheme and previously used scheme respectively are presented.

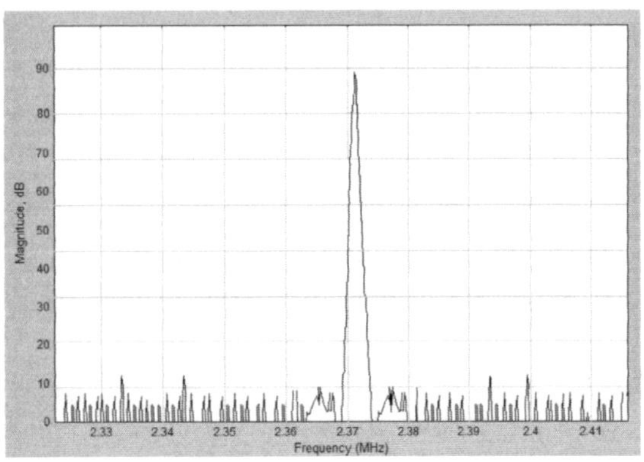

Figure 3. Spectrum of the new synthesizer scheme measured in the band of 8 kHz.

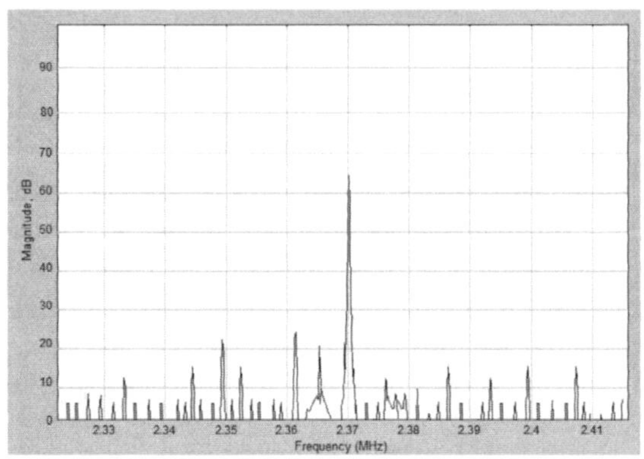

Figure 4. Spectrum of the previously used scheme measured in the band of 8 kHz.

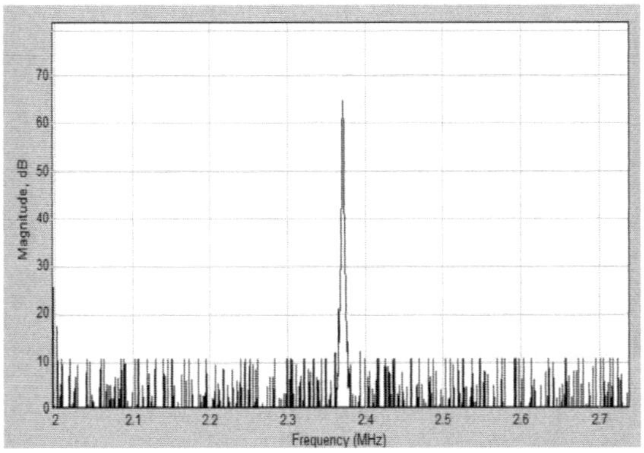

Figure 5. Spectrum of the new synthesizer scheme measured in the band of 700 kHz.

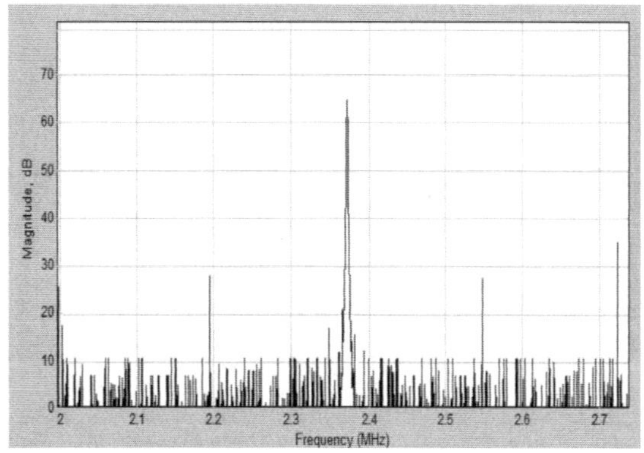

Figure 6. Spectrum of the previously used scheme measured in the band of 700 kHz.

Suppression of lateral discrete components in the spectrum of microwave-excitation is improved on 25 dB in the both bands [5, 6].

The shift of the peak of the line resonance atomic transition Δf is arising due to there are lateral discrete components in the microwave-excitation signal spectrum. The shift is defined by the following expression:

$$\frac{\Delta f}{f} = \frac{A}{I}\frac{(\delta f)^2}{f(f-f_s)}, \quad (3)$$

where A is the amplitude of the lateral components, I is the amplitude of the carrier, δf is the width of the spectral line, $(f - f_s)$ is the detuning of the lateral components relative to peaks of the spectral line.

In the developed scheme of the frequency synthesizer the suppression of lateral components is not less than -88 dB, spectral line width is equal 500 Hz and the band of the lateral components is equal 1 kHz. Then relative frequency shift of the atomic transition is equal $\frac{\Delta f}{f} = 1.08 \cdot 10^{-14}$. Previously obtained result is equal $\frac{\Delta f}{f} = 4.83 \cdot 10^{-13}$ Hz.

By reducing the frequency shift of the atomic transition Δf the more fine-tuning on the center of the resonance line is occur. This leads to a more accurate determination of the value of the nominal output frequency of frequency standard and, consequently, improves the long-term frequency stability of a cesium atomic clock [5].

4. Eliminate frequency shift.

How we noted earlier, improved spectral characteristics directly affect on long-term frequency stability. In addition the new design of the frequency synthesizer allows to eliminate one of the most important perturbing factors affecting the long-term frequency stability too.

The stable isotope ^{133}Cs has a two hyperfine states F = 4 and F = 3 which are split in the magnetic field into 16 components. In accordance with the selection rules seven transitions between the components of hyperfine sublevels are possible. These are represented in figure 7.

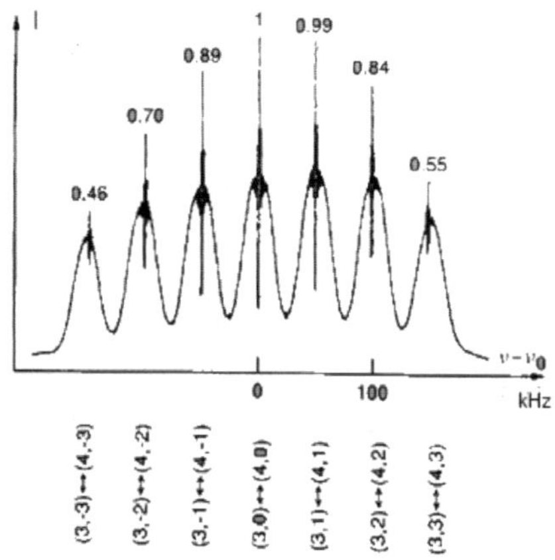

Figure 7. Microwave resonances in atomic beam tube.

The central resonance |F =3, m_F=0> → |F =4, m_F=0> due to the Zeeman effect expose a quadratic frequency shift. With the help of formula (5) we can calculate a frequency shift.

$$\Delta f_{B^2} \approx 4.2745 * 10^{-2}\ \mathrm{Hz} * \left(\frac{6*10^{-6}\mathrm{T}}{\mathrm{uT}}\right)^2 = 1{,}5388\ \mathrm{Hz}. \quad (4)$$

For a typical value of the C field near 8 uT the frequency shift is 2.7 Hz corresponding to a relative frequency shift of $3\cdot 10^{-10}$.

The accuracy of the output signal quantum frequency standard is dependent on the shift of the central resonance. It should be noted that not only the central resonance is exposed the frequency shift, but also all six transitions (3, m_F) ↔ (4, m_F), which $\Delta m_F = 0$. To express these changes as a function of B and atomic constants use the equation Bright Rabi:

$$E(F, m_F) = -\frac{hv}{2(2I+1)} - g_I \mu_B B m_F + \varepsilon \frac{hv}{2}\left(1 + \frac{4m_F}{2I+1}x + x^2\right)^{\frac{1}{2}}, \quad (5)$$

where E(F,m_F) is the change energy of atoms in the ground state; I is the quantum number of nuclear spin; g_I is the factor Lande for electron; μ_B is the Bohr magneton; B is the magnetic field; m_F is the magnetic quantum number; $x = \frac{(g_j+g_I)\mu_B B}{hv}$.

This formula can be used for calculation the frequency shift of any transition between two hyperfine sublevels, depending on the magnetic field. Revealing this expression, we find that the first member is proportional to the magnetic field B. For cesium beam primary frequency standards we must consider the quadratic member of this expression [1, 5].

In modern quantum frequency standards the magnetic field is maintained by the active stabilization system. For this purpose the neighboring transition |F =3, m_F=1> ↔ |F =4, m_F=1> is used. The method of C-field adjustment is similar to the method of frequency adjustment to the main maximum. For this purpose, the average value of the sampling frequency v_i is changed from the value of the v_{Cs} to $v_{Cs} + \Delta v$, where v_{Cs} is the frequency of the main transition of the cesium atom, Δv - difference between the transitions for a preset value of the magnetic field. Then the value of magnetic field is adjusted in a such way that the frequency of transition |F =3, m_F=1> ↔ |F =4, m_F=1> match the preset frequency. This adjustment set up automatically several times per minute. The value of the applied field is automatically maintained at a predetermined level. In this case, excluded effects associated with changes in the magnetic field (for example, long-term drift of the current source, temperature dependence, effect of external magnetic field, etc.)

This feature gave us the opportunities to close the ring-locked loop frequency oscillator on the neighboring resonance transition.

Journal of Physics: Conference Series **769** (2016) 012065 doi:10.1088/1742-6596/769/1/012065

5. Experimental research of quantum frequency standard's output signal.

Experimental research of quantum frequency standard's output signal consisted of measuring the output frequency and calculation the Allan variance, which allows to evaluate the frequency instability.

The formula for calculation the Allan variance is as follow:

$$\sigma_y^2 = \frac{\sum_i^n \sigma_{0i}^2}{n-1}, \tag{6}$$

where $\sigma_{0i} = \frac{f_{i+1} - f_i}{f}$ i-th relative frequency variation, n- number of variations.

After processing the obtained values of relative frequency variations were plotted Allan variance via time for the new and the previous design of frequency synthesizer. (Figure 8).

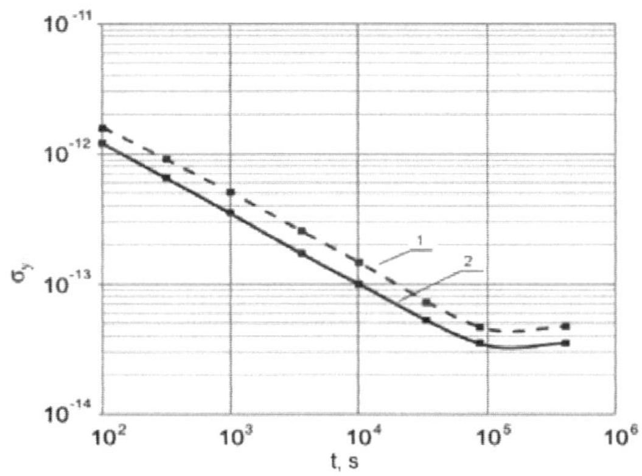

Figure 8. The Allan variance via time for new (2) and the previous (1) design of frequency synthesizer.

From these results it is clear that the use of new design of the frequency synthesizer makes it possible to obtain a better frequency stability of quantum frequency standard output signal. It is achieved through an improved spectrum of the output signal of the frequency synthesizer and by compensating the frequency offset central transition.

6. Conclusion.

Due to a new algorithm of formation microwave - excitation output signal, parameters such as the step frequency tuning, time frequency tuning, range of generated frequencies and spectral characteristics have been improved.

The range of generated output frequencies is expanded, the possibility of detuning the neighboring resonance frequency of spectral line that makes it possible to adjust the C-field in quantum frequency standard is implemented.

Metrological characteristics (daily frequency stability) of the quantum frequency standard on the atoms of caesium - 133 with new scheme of the microwave – excitation signal improved on 15%.

References

[1] Riechle F 2004 *Frequency Standards. Basics and Applications* (Weinheim: WILEY-VCH Verlag GmbH Co.KGaA)

[2] Oduan K and Gino B 2002 *Chronometry and Basics of GPS* (Moscow: Technosphere)

[3] Dudkin V I and Pachomov L N 2002 *Quantum Electronics. Devices and Their Applications* (Moscow: Technosphere)

[4] Ridiko L I 2005 *Components and Technology* vol 7 p 83

[5] Claude A, Noel D, Vincent D and Jacques V 2003 Physical origin of the frequency shifts in cesium beam frequency standards: related environmental sensitivity *IEEE Trans. on UFFC* vol 39 pp 412–421
[6] Petrov A A and Davydov V V 2015 Improvement frequency stability of caesium atomic clock for satellite communication system *Proc. 15th International Conference NEW2AN 2015 and 8th Conference ruSMART 2015* pp 739-744

Dual-tunable ferrite-ferroelectric dynamic magnonic crystal

I A Ustinova[1], A A Nikitin[1,2], A B Ustinov[1,2], B A Kalinikos[1] and E Lähderanta[2]

[1] Department of Physical Electronics and Technology, St.Petersburg Electrotechnical University, 197376 St.Petersburg, Russia
[2] Department of Mathematics and Physics, Lappeenranta University of Technology, Lappeenranta 53850, Finland

E-mail: ustinovairin@yahoo.com

Abstract. A dynamic magnonic crystal based on a planar multiferroic structure consisting of a ferrite and a ferroelectric layers has been designed and theoretically investigated for the first time. Spatially periodic electric field applied to the ferroelectric layer turns the regular multiferroic waveguide into periodic one. An electrodynamic model of the dynamic magnonic crystal and theoretical analysis of its performance is presented.

Recent years an increased interest for investigation of spin waves propagating in spatially periodic magnetic film waveguide (the so-called magnonic crystals) takes place [1, 2]. A feature of the spin wave spectrum of such structures is the presence of band gaps which appear due to Bragg reflection. Dynamic magnonic crystals (DMC) cause a special interest [3, 4], because of the ability to "turn on" and "turn off" its periodic waveguiding properties. For example, a magnonic crystal made of a ferromagnetic film and a meander conductor located near its surface was described in [3]. The current flow caused a sine-like variation of the film's internal magnetic field, which could be modulated by changing the amount of current. In the work [4] DMC was realized as a periodic structure with a variable width. The edges of the structure were screened by conductors with electric current.

Another topical trend is investigation of artificial materials which demonstrate multiferroic properties [5-9]. Among them are ferrite-ferroelectric layered structures, in which hybrid spin-electromagnetic waves (SEWs) may propagate [10]. Until now, the main attention was given to the investigation of spatially homogeneous ferrite-ferroelectric waveguiding structures, while spatially periodic structures received a small attention. For example, a propagation of SEWs in a multiferroic periodic structure fabricated with a one-dimensional magnonic crystal and a ferroelectric slab had been experimentally investigated for the first time in [11]. The dispersion and frequency-selective properties of periodic ferrite-ferroelectric structures had been theoretically studied in the works [12, 13].

This work is devoted to theoretical investigation of microwave properties of multiferroic dynamic magnonic crystal. Investigated structure is shown in figure 1. It consists of several layers and metal electrodes. The layers are a ferrite film (1), a dielectric substrate (2), and a ferroelectric slab (3). The ferroelectric slab is covered by thin metal electrodes. The electrode on the bottom side is continuous. The electrode on the top side is a grid consisting of metal strips. Both electrodes are assumed to be transparent for the microwave fields in the structure due to small thickness which is much smaller than the skin depth [14]. Ferroelectric slab is assumed to be in a paraelectric phase with a high dielectric constant. The dielectric constant is controlled by the bias voltage applied to the top and bottom electrodes. This assumption is correct for barium strontium titanate ceramic of the composition $Ba_{0.5}Sr_{0.5}TiO_3$ at the room temperature.

The ferrite-ferroelectric structure serves as a waveguide for the SEWs. The electrodes do not affect to the propagation of SEWs in the structure. The electrodes produce periodic polarization of the ferroelectric layer, as is shown in figure 1. It is well known that the polarization of the ferroelectrics decreases their dielectric permittivity ε_d. Therefore, application of the voltage to the electrodes produces periodic modulation of ε_d in the investigated structure. This periodic modulation can be "turned on" and "turned off" by the bias voltage. Thus, the described multiferroic structure is a dynamic magnonic crystal for the SEWs.

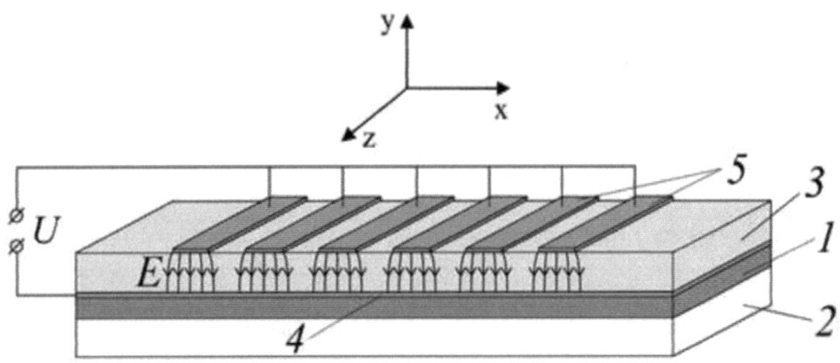

Figure 1. Schematic representation of the dynamic magnonic crystal based on ferrite-ferroelectric structure.

Consider the development of the electrodynamic model of the structure. We assume that the SEWs propagate perpendicular to the grid electrode in the direction opposite to x axis in a tangentially magnetized (along z axis) ferrite layer surrounded by free space (Figure 1). In this configuration maximum of the SEW energy is located at the interface between the ferroelectric layer and the ferrite film. We do not take into account existence of the dielectric substrate (2) because it has a low dielectric constant (about 10).

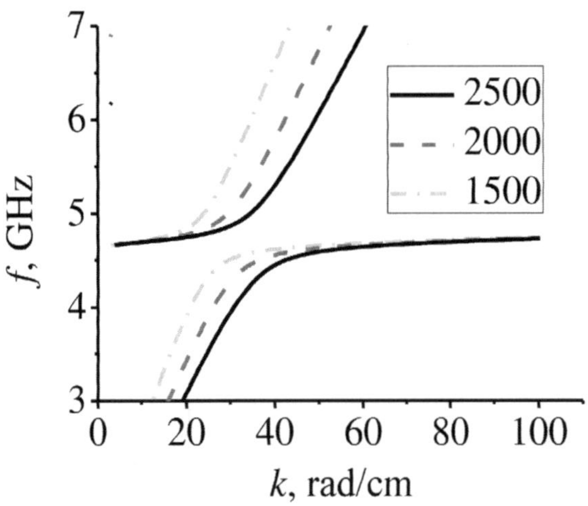

Figure 2. The spectrum of SEWs in the ferrite-ferroelectric structure.

The theoretical model was developed in two stages. In the first stage we derive the dispersion equation and calculate the dispersion characteristics of the SEWs in the homogeneous ferrite-ferroelectric waveguiding structure. In order to derive the dispersion equation, we use a method described in [15]. In this method four Helmholtz equations for the variable component of the electric field E_z of TE wave are obtained from Maxwell's equations. The first and fourth of them correspond to free half-spaces above and below the structure, the second equation describes the propagation of electromagnetic waves (EMWs) in the ferroelectric layer, and the third one describes the propagation of the SEWs in the ferrite film taking into account the microwave magnetic permeability tensor. The solutions of the Helmholtz equations give expressions for the components E_z in each layer. Further H_x components of magnetic field are obtained from E_z via the Maxwell's equations. Substitution of E_z and H_x into the electrodynamic boundary conditions at the layers' interfaces gives us a system of six linear homogeneous algebraic equations with respect to six variables. The vanishing of the system's determinant results in the following dispersion relation:

$$\tan(\kappa_d L_d)\left[tan(\kappa_f L_f)A + (\kappa_0^2 - \kappa_d^2)\mu_\perp\right] + \kappa_d\left[tan(\kappa_f L_f)B + 2\kappa_0\mu_\perp\right] = 0, \quad (1)$$

where $A = \left(\kappa_d^2\mu_\perp - \frac{\kappa_0 k\alpha}{\mu}\right)\left(\frac{k\alpha}{\kappa_f\mu} - \frac{\kappa_0\mu_\perp}{\kappa_f}\right) + 2\kappa_0\kappa_f$, $B = \frac{\kappa_0^2\mu_\perp^2}{\kappa_f} - \frac{k^2\alpha^2}{\kappa_f\mu^2} - \kappa_f$, k is a SEW longitudinal wave number, L_d and L_f are the thicknesses of the ferroelectric and ferrite layers, respectively, $\kappa_0 = \sqrt{k^2 - \omega^2\mu_0\varepsilon_0}$, $\kappa_d = \sqrt{\omega^2\mu_0\varepsilon_0\varepsilon_d - k^2}$, and $\kappa_f = \sqrt{\omega^2\mu_0\mu_\perp\varepsilon_0\varepsilon_f - k^2}$ are transverse wave numbers in free space, in the ferroelectric layer and in ferrite layer, respectively, $\mu_0 = 4\pi \cdot 10^{-7}$ H·m^{-1}, $\varepsilon_0 = 8.85 \cdot 10^{-12}$ F·m^{-1}, $\mu_\perp = \frac{\mu^2 - \alpha^2}{\mu}$, $\mu = [\omega_H(\omega_H + \omega_M) - \omega^2]/(\omega_H^2 - \omega^2)$, $\alpha = \omega\omega_M/(\omega_H^2 - \omega^2)$, $\omega_H = 2\pi\gamma H$, $\omega_M = 2\pi\gamma \cdot 4\pi M_s$, $\gamma = 2.8$ MHz/Oe, H is the magnetic field, M_s is the saturation magnetization.

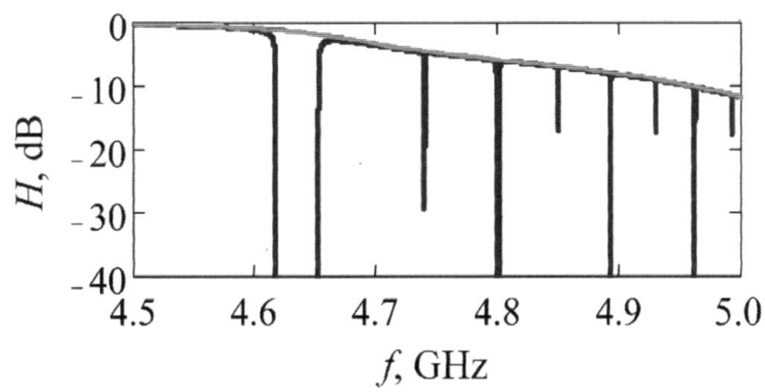

Figure 3. Amplitude-frequency characteristics of unpolarized (grey line) and periodically polarized (black line) ferrite-ferroelectric structure.

As a second step we calculate the dispersion relation of the SEWs in periodically polarized ferrite-ferroelectric structure. We use following formula obtained by the method of coupled waves [16]:

$$\cos K\Lambda = \cos k_1 l_1 \cos k_2 l_2 - \frac{k_1^2 + k_2^2}{2k_1 k_2} \sin k_1 l_1 \sin k_2 l_2, \quad (2)$$

where K is a Bloch wave vector, Λ is a period of the structure, k_1 и k_2 are wave numbers of SEWs propagating in polarized and unpolarized sections of the structure with ε_{d1} и ε_{d2}, respectively, l_1 and l_2 are the distances passed by the wave in the polarized and unpolarized sections of the structure (see

Figure 1). Finally, the amplitude-frequency characteristics (AFC) of the multiferroic periodic structure are calculated by following formula:

$$H = 20\lg e^{-\alpha x}, \quad (3)$$

where $\alpha = \left|\frac{dk(\omega)}{dH}\right| \Delta H + \varepsilon_d \left|\frac{dk(\omega)}{d\varepsilon_d}\right| \tan\delta$ is a spatial damping decrement of the SEWs [17], ΔH is the half-width of the curve of the ferromagnetic resonance, tanδ is a tangent of dielectric losses of the ferroelectric layer.

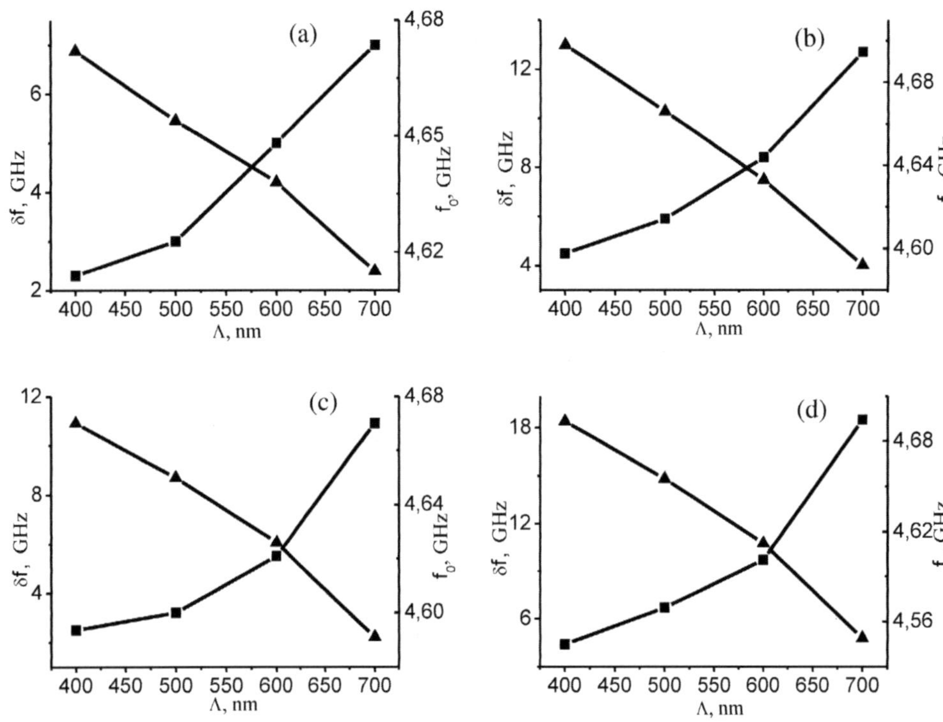

Figure 4. Width of the first band gap (squares) and its centre frequency (triangles) as a function of period for l_1=*100* μm and l_2=*300* μm (a, b) and for l_1=*300* μm and l_2=*100* μm (c, d) as well as for L_f=*5* μm (left graphs) and L_f=*10* μm (right graphs).

The dispersion characteristics of the SEWs propagating in the homogeneous ferrite-ferroelectric structure were calculated with the use of the dispersion equation (1). An example of the calculation results is shown in figure 2. The calculation was carried out for the ferrite layer having parameters corresponding to single-crystal yttrium iron garnet (YIG) film: $4\pi M_s = 1750$ G, $L_f = 10$ μm, $H = 1000$ Oe, $\varepsilon_f = 14$, $\Delta H = 0.5$ Oe. The calculation was also carried out for the ferroelectric layer having parameters corresponding to the ceramic solid-state solution of $Ba_{0.5}Sr_{0.5}TiO_3$: $\varepsilon_d = 1500 - 2500$, that depends on the applied electric field, $L_d = 500$ μm, $\tan\delta = 10^{-2}$. As can be seen from the figure, a region of a strong dispersion of the SEWs is located at the wave number region from 25 rad/cm to 50 rad/cm. This area corresponds to the maximum hybridization between the spin waves in the ferrite film and the EMWs in the ferroelectric slab. The most significant modulation of the dispersion of the SEWs under the application of the electric field to the ferroelectric slab occurs in this area.

The dispersion and transmission characteristics of the SEWs in the DMC for different ε_d are also calculated. Application of the voltage leads to the appearance of band gaps in the SEWs spectrum and, consequently, to the appearance of characteristic dips in the AFC of the waveguiding structure (see figure 3). The first band gap is relatively wide while the other are relatively narrow. This is due to the fact that the first band gap is located in the area of the SEWs strong dispersion. The centre frequency of the first band gap, as well as its width increases with increase of the bias voltage.

The calculations were made for a variety of the structure parameters. Figure 4 shows the dependences of the centre frequency of the first band gap and its width on the period of the structure with $l_1 \neq l_2$. The analysis of the results shows that the increase of the period Λ of the structure increases the width of the first band gap. This is due to the reduction of the Bloch wave vector $K_b=\pi/\Lambda$ with an increase in Λ. Therefore the frequency of the first band gap shifts, and it is getting closer to the area of strong dispersion. According to the above discussion, the application of voltage leads to a significant change of the dispersion in that area. It leads to the broadening of the band gap.

Similar calculations were made for the structure with $l_1=l_2$. For example, the calculations for $\varepsilon_{d1} = 2000$ and $\varepsilon_{d2} = 2500$ and the ferrite film thickness $L_f=5$ μm show that the increase of the structure's period leads to increase of the first band gap, which is equal to 21 MHz, and the decrease of its centre frequency, which is equal to 70 MHz. As well as for the film thickness $L_f=10$ μm the increase of the first band gap is equal to 37 MHz and the decrease of its centre frequency is equal to 130 MHz. Thus, it is seen from the results that it is more preferably to use relatively thick ferrite films to increase the electric tunability of the ferrite-ferroelectric DMC.

In conclusion, it is easy to achieve an efficient electrical tuning of one or more band gaps that are located at the frequencies of the strong dispersion of SEWs by choosing the geometry of the DMC and physical parameters of the multiferroic layered structure. Thus, dynamic magnonic crystals based on the planar multiferroic structure consisting of ferrite and ferroelectric layers can find a variety of applications.

Acknowledgement
The work on the development of the theory of SEW spectra at SPbETU was supported by RFBR (Grant № 15-32-20357 mol_a_ved). The work on the development of the model of periodic structure at SPbETU was supported by Ministry of Education and Science of Russian Federation (Project "Goszadanie"). The work on numerical simulation of DMC characteristics at SPbETU was supported by the Russian Science Foundation (Grant № 14-12-01296). The work at LUT was supported by Academy of Finland.

References

[1] Krawczyk M and Grundler D 2014 *J. Phys.: Condens. Matter* **26** 123202
[2] Nikitov S A, Kalyabin D V, Lisenkov I V, Slavin A N, Barabanenkov Yu N, Osokin S A, Sadovnikov A V, Beginin E N, Morozova M A, Sharaevsky Yu P, Filimonov Yu A, Khivintsev Yu V, Vysotsky S L, Sakharov V K and Pavlov E S 2015 *Physics-Uspekhi* **58**
[3] Chumak A V, Nuemann T, Serga A A, Hillebrands B and Kostylev M P 2009 *J. Phys.* D*: Appl. Phys.* **42** 205005/1-8
[4] Nikitin A A, Ustinov A B, Semenov A A, Chumac A V, Serga A A, Vasyuchka V I, Lähderanta E, Kalinikos B A and Hillebrands H 2015 *Appl. Phys. Lett.* **106**. 102405
[5] Vopson M M 2015 *Critical Reviews in Solid State and Materials Sciences*. **40** 223-50
[6] Vaz C A F 2012 *J. Phys.: Condens. Matter*. **24** 333201
[7] Lu C, Hu W, Tian Y and Wu T 2015 *Appl. Phys. Rev.* **2** 021304
[8] Ahmed A, Goldthorpe I A and Khandani A K 2015 *Appl. Phys. Rev.* **2** 011302
[9] Sun N X and Srinivasan G 2012 *Spin* **2** 1240004
[10] Demidov V E, Kalinikos B A and Edenhofer E 2002 *J. Appl. Phys.* **91** 10007
[11] Ustinov A B and Kalinikos B A 2014 *Tech. Phys. Lett.* **40**. 568-70
[12] Drozdovskii A V, Nikitin A A, Ustinov A B and Kalinikos B A 2014 *Tech. Phys.* **59** 1032-35

[13] Morozova M A, Grishin S V, Sadovnikov A V, Romanenko D V, Sharaevskii Y P and Nikitov S A 2015 *Magnetics, IEEE Transactions on* **51**
[14] Ustinov A B, Srinivasan G and Kalinikos B A 2007 *Appl. Phys. Lett.* **90** 031913
[15] Gerson T J and Nadan J S 1974 *IEEE Trans. MTT.* **22** 757-63
[16] Yariv A and Yeh P 1984 *Optical waves in crystals* (New York: Wiley)
[17] Ustinov A B, Kalinikos B A, Tiberkevich V S, Slavin A N and Srinivasan G 2008 *J. Appl. Phys.* **103** 063908

Aspects of symmetry of Electromechanical Coupling Factors in Piezoelectric Single Crystals

A Zamkovskaya and E Maksimova
Crimean Federal V.I. Vernadsky University, Simferopol, Crimea, Russian Federation

E-mail: trabem.z@gmail.com

Abstract. This paper presents the method for the calculation of anisotropic piezoelectric properties of single crystals and the graphical display of the results in 3 D. Crystallographic preferred orientations were determined for piezoelectric modules and electromechanical coupling factor, which measures the ability of a material to interconvert electrical and mechanical energy.

1. Introduction

Certain materials produce electric charges on their surfaces as a consequence of applying mechanical stress. The induced charges are proportional to the mechanical stress. This is called the direct piezoelectric effect and was discovered in quartz by Piere and Jacques Curie in 1880.

The direct effect can be written as the relationship between the 2nd rank stress tensor x_{jk} and 1st rank electric polarization vector P_i, linked by the 3rd piezoelectric tensor d_{ijk} as follows:

$$P_i = \sum_{j=1}^{3} \sum_{k=1}^{3} d_{ij} X_{jk},\ i = 1,2,3 \quad (1)$$

In the literature the tensor d_{ijk} is reported for single crystals in the practical and compact Voigt matrix notation. The conversion from Voigt notation d_{il} to tensor d_{ijk} notation $d_{ijk} = d_{il}$ when l = 1; 2; 3 and $d_{ijk} = \frac{1}{2} d_{il}$ when l = 4; 5; 6. The direct effects can also be written in Voigt matrix notation as:

$$P_i = \sum_{l=1}^{6} d_{il} \sigma_j ,\ i = 1,2,3 \quad (2)$$

All crystals belong the 21 non-centrosymmetric points groups are piezoelectrically active, with the exception of the cubic 432, [1].

However, the description of the properties of piezoelectric materials is impossible without the involvement of the elastic and dielectric coefficient as the piezoelectric property by definition implies the connection between the elastic and dielectric characteristics. From this point of view, for comparison of their piezo properties useful so-called electromechanical coupling factor k (ECF):

$$k^2 = (\text{Stored mechanical energy / Input electrical energy}) \quad (3)$$

or

$$k^2 = (\text{Stored electrical energy / Input mechanical energy}) \quad (4)$$

A system of the ECFs is introduced to describe the conversion and takes into account the symmetry of a piezoelectric material, orientations of its crystallographic axes, etc. A larger absolute value of the

ECF |k| implies a more efficient performance of a piezoelectric material at a specific mode of oscillation and k^2 is used to characterize the magnitude of the transducer bandwidth. Knowledge of the ECF values of piezoelectrics is important for energy-harvesting applications in the context ofrelationships between the amount of stored and supplied energy.

The single crystals (SCs) are anisotropic, exhibiting different material properties depending on the cut of the materials. The single crystal piezoelectric tensor can be visualized by indicatory surfaces, which radius vector is proportional to the value of the property in arbitrary direction n, [2]. For the construction of three-dimensional models of surfaces of piezoelectric coefficients was used MTEX open-source package, [3].

2. Research method

The work presents the aspects of anisotropy and symmetry of longitudinal piezoelectric effect. The piezoelectric effect is called longitudinal when the polarization and mechanical stress directions coincide.

Since the electromechanical coupling factor depends on the number of anisotropic properties: electrical ε (2nd rank tensor), piezoelectric d (3rd rank tensor) and elastic properties s (4th rank tensor), it is also anisotropic and can be described via the indicatory surface.

The base of visualizations of surface orientation dependence of ECF is the value of the k in a direction n:

$$r = k_n = d/(\mathcal{E}_0 \mathcal{E} S)^{1/2} \tag{5}$$

The analysis of the received indicatory surfaces allows to define the symmetry and the anisotropy of the properties, and, if necessary, establish the directions of its extreme value.

3. Analysis and discussion

Although piezoelectric ceramics are widely used for a large number of applications, single crystal materials retain their utility, being essential for applications such as frequency stabilized oscillators and surface acoustic devices.

The most popular single-crystal piezoelectric materials are quartz (SiO_2), lithium niobate ($LiNbO_3$), lithium tantalite ($LiTaO_3$), barium titanate ($BaTiO_3$) and etc.

Quartz has been the first and widespread piezoelectric material. Below T = 847 K quartz is related to the 32 symmetry class. According to room-temperature experimental data, these piezoelectric coefficients are d_{11} = 2.30 pC/N and d_{14} = –0.693 pC/N, table 1.

Taking into account elastic compliances of the quartz and its dielectric permittivity, we obtain the ECFs as follows, figure 1.

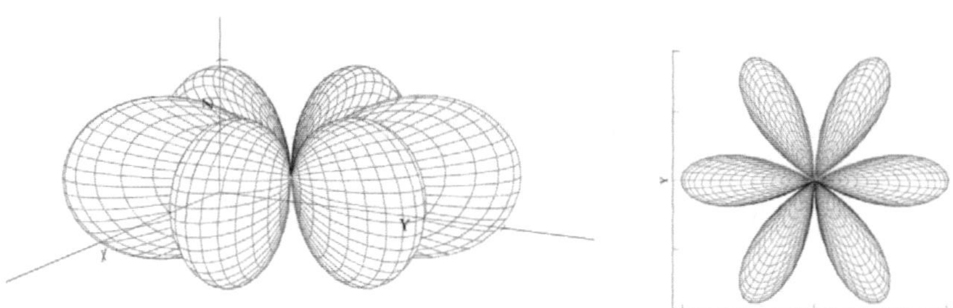

Figure 1. The indicatory surface of longitudinal electromechanical coupling factor k of quartz (SiO_2) and its projection onto the plane (XOY).

This indicatory surface consisting of alternating positive and negative parts is related to the $\bar{6}$m2 symmetry class. Maximum longitudinal piezoelectric effect (as well as the maximum value of k) is observed along the polar axis of symmetry of the second order. Along the Z-axis of the longitudinal piezoelectric effect is not observed.

Table 1. Experimental values of elastic compliances s (in 10^{-12} Pa^{-1}), piezoelectric coefficients d (in pC/N), dielectric permittivities $\varepsilon/\varepsilon_0$ of some piezoelectric SCs at room temperature, [4].

	$BaTiO_3$	CdS	SiO_2
s^E_{11}	7,92	20,69	12,79
s^E_{12}	-3,80	-9,93	-1,54
s^E_{13}	-1,28	-5,81	-1,10
s^E_{14}			4,46
s^E_{22}	8,05		
s^E_{23}	-3,80		
s^E_{33}	7,92	16,97	9,56
s^E_{44}	11,9	66,49	19,78
s^E_{55}	30,2		
s^E_{66}	13,6	61,36	
d_{11}			2,3
d_{14}			-0,693
d_{15}	126	-13,98	
d_{24}	269		
d_{31}	-189	-5,18	
d_{32}	-24,5		
d_{33}	225	10,32	
$\varepsilon^\sigma_{11}/\varepsilon_0$	265	9,35	4,520
$\varepsilon^\sigma_{33}/\varepsilon_0$	2130	10,38	4,6

To the best of our knowledge, the important group of piezoelectric SCs consist of compounds such as α-ZnS, CdS, CdSe, ZnO, etc. from the 6mm symmetry class. The ECFs for the CdS have the form on figure 2.

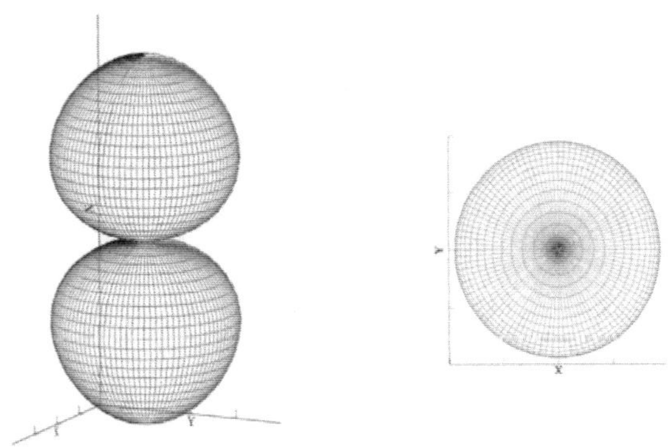

Figure 2. The indicatory surface of longitudinal electromechanical coupling factor k of CdS and its projection onto the plane (XOY).

The symmetry of this surface identical with the symmetry of the indicatory surface of the piezoelectric modules in CdS and describes the class ∞mm. Perpendicular to the Z-axis there is anti-symmetry plane, since its upper and lower parts have different signs.

Barium titanate $BaTiO_3$ (the perovskite-type ferroelectric crystals) is one of the most thoroughly studied and most widely used piezoelectric materials, [1, 4]. Just below the Curie temperature (120°C), the vector of the spontaneous polarization points in the [001] direction (tetragonal phase), below 5 °C it reorients in the [011] (orthrhombic phase) and below - 90 °C in the [111] direction (rhombohedral phase). Example of electromechanical and piezoelectric constants surfaces of the $BaTiO_3$ single crystals (4mm symmetry class) are given in figure 3.

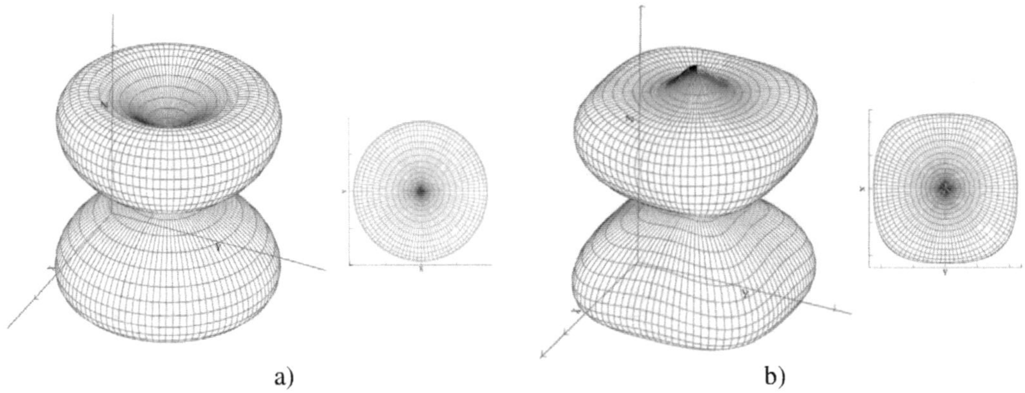

a) b)

Figure 3. The indicatory surface of longitudinal piezoelectric effect in $BaTiO_3$ and its projection onto the plane (XOY) (a); the indicatory surface of longitudinal electromechanical coupling factor k in $BaTiO_3$ and its projection onto the plane (XOY) (b).

In contrast to the previous cases, symmetry of surfaces are different, and respectively, the directions with extreme values of properties are also different.

The indicatory surface of piezoelectric modules is the rotational surface, i.e. it possesses the symmetry axis of infinite order, directed along the coordinal Z axis, chosen along the fourth order axis in SCs. The symmetry class of this surface is ∞mm. The piezoelectric effect is zero perpendicular to the fourth order symmetry axis (Z axis). The maximum piezoelectric effect in the SCs of barium titanate can be obtained by cutting out the rod along the directions cone with the half of its internal angle of 51°5′ to the Z axis.

The symmetry of electromechanical coupling factor orientation dependence is defined by the indicatory surfaces superposition of values describing it. Herewith the determining role is played by those surfaces which symmetry is lower. Thus, in the barium titanate electromechanical coupling factor indicatory surface the fourth order symmetry axis appears which caused by the aspects of its elastic properties. The symmetry of this surface 4mm is equal to the symmetry of the crystal itself. The maximum electromechanical coupling factor in this crystal can be obtained by deforming barium titanate along directions being at the angle 49°23′ to the Z axis.

4. Conclusion

We calculated efficiency of mechanical-to-electrical energy conversion, depending on the direction of applied load with respect to the crystallographic axes for a wide range of single piezoelectric crystals.

References

[1] Sirotin Y I, Shaskol'skaya M P 1979 *Fundamentals of crystal physics* (Moscow: Science)
[2] Kholkin A L, Pertsev N A, Goltsev A V 2008 *Piezoelectric and Acoustic Materials for Transducer Applications* (Springer)
[3] Mainprice D, Hielscher R & Schaeben H 2011 Calculating anisotropic physical properties from texture data using the MTEX open-source package (Geological Society, London, Special Publications)
[4] Bowen Christopher R, Topolov Vitaly Yu, Hyunsun Alicia Kim 2016 *Modern Piezoelectric Energy-Harvesting Materials* (Springer Series in Materials Science)

Journal of Physics: Conference Series **769** (2016) 012068 doi:10.1088/1742-6596/769/1/012068

Effect of nitride chemical passivation of the surface of GaAs photodiodes on their characteristics

E V Kontrosh[1], N M Lebedeva[1], V S Kalinovskiy[1], F Yu Soldatenkov[1], V P Ulin[1]

[1]Ioffe Institute, 26 Polytekhnicheskaya, St. Petersburg, Russian Federation, 194021

E-mail: kontrosh@mail.ru

Abstract. Characteristics of GaAs photodiodes have been studied before and after the chemical nitridation of their surface in hydrazine sulfide solutions, which leads to substitution of surface As atoms with N atoms to give a GaN monolayer. The resulting nitride coatings hinder the oxidation of GaAs in air and provide a decrease in the density of surface states involved in recombination processes. The device characteristics improved by nitridation are preserved during a long time.

1. Introduction

Monolayer gallium nitride (GaN) layers formed on the surface of GaAs crystals can serve for an effective chemical and electronic passivation of the surface of semiconductor devices based on these crystals. Owing to the high strength of Ga – N bonds, coherent nitride coatings exhibit a chemical stability sufficient for protecting the GaAs surface from being oxidized in air. This precludes the appearance of intraband surface states characteristic of the oxidized GaAs surface. These states create parasitic conductance channels on the surface of structures with p – n junctions, such as photodiodes. In addition, when surface arsenic atoms are replaced with the substantially more electronegative nitrogen, newly formed surface states fall deeper into the valence band of GaAs and become electrically inactive. Thus, the surface recombination rate substantially decreases in a nitridized crystal [1], which, in turn, markedly improves the surface – dependent electronic and optical characteristics of the given semiconductors. The present study is concerned with the influence exerted by the chemical surface passivation on the efficiency of GaAs photodiodes, associated with the minimization of the leakage current due to the tunnel – trap (excess) and recombination charge transport mechanisms.

2. Sample fabrication

GaAs photodiode (PD) p – i – n structures were fabricated in two stages by liquid – phase epitaxy and Zn diffusion. Lightly doped n^0 – GaAs layers were epitaxially grown from a Ga – As melt in the hydrogen atmosphere at temperatures of 750 to 700 °C. To reduce the content of residual impurities in the n^0 – GaAs layers being grown, minor amounts (up to 0, 05 at %) of rare – earth elements (Yb, Sc, Er, Y) were added to the Ga – As melt [2, 3]. The free electron concentration in the n^0 – GaAs layers was in the range from 1×10^{14} to 5×10^{15} cm^{-3}. The diffusion of zinc was performed on the surface of an epitaxial layer to form a p^+ layer with a hole concentration of about 5×10^{19} cm^{-3} [4]. PD chips had the form of mesa structures with ring-shaped ohmic contacts fabricated by optical lithography and wet etching down to the substrate.

The GaAs surface was nitridized by its treatment in a buffer solution of hydrazine hydrochloride in hydrazine hydrate (pH = 8,5) containing sodium sulfide [5, 6] in a concentration of 0,01 M.

After the natural oxide layer is removed in the alkaline medium, arsenic atoms on the GaAs surface selectively interact with SH^- anions, which results in that molecules of thioarsenic acid are formed and electrophilic adsorption centers appear on exposed gallium atoms. Then, mutually competing processes of adsorption of univalent OH^- and SH^- anions and nucleophilic hydrazine molecules present in the alkaline sulfide solutions develop on these centers. Although the primary coverage of gallium atoms is predominantly formed via chemisorption of univalent anions, the ability of nucleophilic nitrogen atoms in hydrazine (N_2H_4) molecules to successively form two or three bonds with electrophilic gallium atoms provides to these atoms the possibility of an irreversible substitution of the chemisorbed anions. Depending on the coordination of the incorporated nitrogen atoms, determined by the initial crystallographic orientation of the GaAs surface, doubly coordinated amino groups NH or triply coordinated nitrogen atoms remain on the surface after the final dissociation of the adsorbed hydrazine molecules. Thus, the GaAs (111) B, (100), and (110) surfaces and those with intermediate crystallographic orientations are terminated with electronegative nitrogen atoms, which form substantially stronger chemical bonds with gallium atoms, compared with arsenic atoms. In this case, it can be said that a solid GaN layer coherent with the crystal is formed. As a result, the nitrogen –related surface states are localized within the valence band of GaAs and become electrically inactive, and the surface of the nitridized crystal becomes resistant against oxidation by atmospheric oxygen. At the same time, in the case of the gallium (111) A surface, treatment in a hydrazine sulfide solution yields an adsorption layer composed of sulfur dimers that pairwise bind gallium atoms [6]. This coating also provides an electronic passivation and chemical stability of the GaAs (111) A surface. Consequently, irrespective of the set of orientations of the facets forming the mesa structures, their treatment in hydrazine sulfide solutions leads to a stable electronic and chemical passivation of the intricate surface of a gallium arsenide device structure.

3. Results and discussion

Forward dark current – voltage $J-U$ characteristics and external quantum efficiency spectra were measured for GaAs PDs before and after the surface nitridation. The forward dark $J-U$ characteristics were measured at room temperature at current densities $J = (10^{-9} - 1)$ A/cm^2 and voltages $U = (0 - 1, 5)$ V. The external quantum efficiency spectra were measured at wavelengths $\lambda = (350 - 900)$ nm.

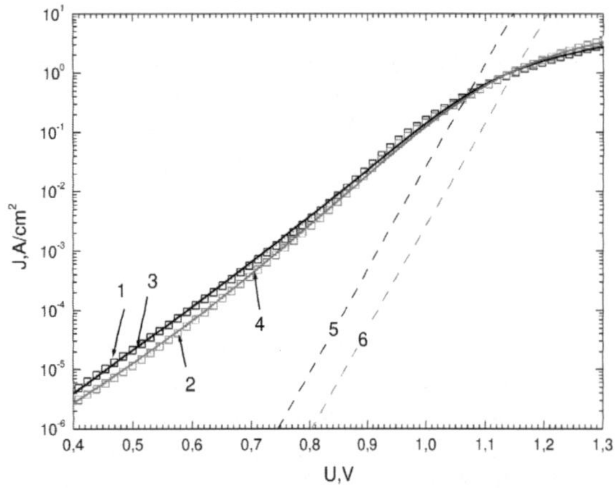

Figure 1. Forward dark current – voltage characteristics of GaAs photodiodes: (1, 2) experimental; (3, 4) calculated before and after the deposition of GaN, respectively; (5, 6) calculated exponential curves corresponding to the diffusion charge transport mechanism, $A = 1$ (Shockley), $J_{0d} = 1{,}1 \times 10^{-19}$ and $1{,}1 \times 10^{-20}$ A/cm^2 before and after the deposition of GaN, respectively.

Figure 1 shows calculated and experimental forward dark $J-U$ characteristics before and after the surface nitridation of GaAs PDs. Analysis of the forward dark $J-U$ characteristics, carried out using method described in [7, 8], shows that the pre-exponential factor J_{0d} corresponding to the diffusion charge transport mechanism with $A = 1$ (Shockley) decreased by an order of magnitude, from 10^{-19} to 10^{-20} A/cm^2, after the experiment in which a GaN monolayer was formed.

It follows from the results of measurements of the external quantum efficiency spectra (in figure 2) that the sensitivity of the photodiode increases by more than 8% at wavelengths of 400 – 870 nm after its surface is nitridized.

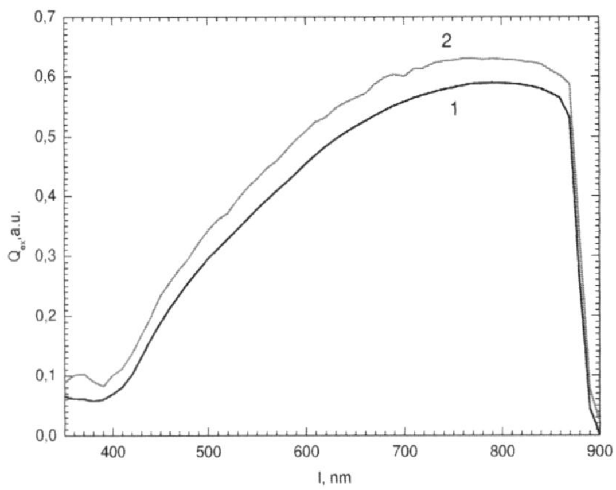

Figure 2. Spectra of the external quantum efficiency Q_{ext} of the GaAs PDs under study: (1) before and (2) after the deposition of a GaN monolayer.

4. Conclusion

A method was suggested for producing a gallium nitride monolayer on the side and front surfaces of GaAs PDs via chemical nitridation in hydrazine sulfide solutions.

Forward dark current – voltage characteristics and external quantum efficiency spectra were measured for GaAs PDs before and after the nitridation. An analysis of the characteristics before and after the nitridation of the front and side surfaces of GaAs PDs in hydrazine sulfide solutions demonstrated that the method suggested for producing a gallium nitride monolayer provides the following:

- The pre-exponential factor of the dark $J-U$ characteristic, which corresponds to the diffusion charge transport mechanism with $A_d = 1$ (Shockley), decreases by an order of magnitude, from 10^{-19} to 10^{-20} A/cm^2.
- The external quantum efficiency increases by more than 8% at wavelengths $\lambda = 400 - 870$ nm.

Acknowledgments

This work was supported by Ministry of Education and Science of the Russian Federation (Grant Agreement № 14.604.21.0088 dated June 27, 2014, unique identifier of applied research (project) RFMEFI60414X0088).

References

[1] Berkovits V L Ulin V P Losurdo M Capezzuto P Bruno G Perna G Capozzi V Wet chemical nitridation of GaAs(100) by hydrazine solution for surface passivation 2002 *Appl. Phys. Lett* **80** 3739

[2] Danilchenko V G Korolkov V I and Soldatenkov F Yu 2009 *Semiconductors* **43** 1055 – 7

[3] Soldatenkov F Yu 2007 *Russian Federation patent* No 2297690

[4] Soldatenkov F Yu Sorokina S V Timoshina N Kh Khvostikov V P Zadiranov Y M Rastegaeva M G and Usikova A A 2011 *Semiconductors* **45** 1219-26

[5] Paget D Berkovits V L Ulin V P Ozanam F Dumas P Kubsky S Lahlil K Bideux L Monier G Real time infra-red absorption analysis of nitridation of GaAs(001) by hydrazine solutions 2013 *J. Electrochem. Soc* **160** H229 – H236

[6] Berkovits V L Ulin V P Tereshchenko O E Paget D Rowe A C Chiaradia P Doyle B P Nannarone S. Chemistry of wet treatment of GaAs(111)B and GaAs(111)A in hydrazine-sulfide solutions 2011 *J. Electrochem. Soc* **158** D127 – D135

[7] Kalinovsky V S Kontrosh E V Dmitriev P A Pokrovsky P V Chekalin A V Andreev V M Current flow and efficiencies of concentrator InGaP/GaAs/Ge solar cells at temperatures below 300K 2014 *AIP Conf. Proc* **1616** 8

[8] Kontrosh E V Malevskaya A V Lebedeva N M Grebenshchikova E A Kontrosh L V Ill`inskaya N D Kalinovsky V S Stady of the photovoltaic characteristics of multi-junction solar cells fabricated by one-step and two-step separation etching 2015 *ISJAEE* **19 (183)**

Grain boundary scattering contribution to the thermopower in Cr-Si nanocrystalline films

S V Novikov[1], **A T Burkov**[1,2]

[1] Ioffe Institute, 194021, Polytechnicheskaya 26, Saint-Petersburg, Russia
[2] St.Petersburg National Research University of Information Technologies, Mechanics and Optics, 197101, Kronverkskiy pr, 49, Saint-Petersburg, Russia

E-mail: S.Novikov@mail.ioffe.ru

Abstract. We study transport properties of $Cr_{1-x}Si_x$ films (x = 0.65, 0.67, 0.72, 0.85), produced by magnetron sputtering of composite targets on unheated Si/SiO_2 substrates. The as-deposited films have amorphous structure, which can be converted by isothermal annealing into nanocrystalline state with the average crystal grain size of about 15 nm. We found that in the nanocrystalline composites there is an additional large contribution to the thermopower due to energy dependent scattering of the charge carriers on the nanocrystal interfaces. This contribution to the thermopower is strongly dependent on the film composition. The maximum contribution have the films of stoichiometric $CrSi_2$ composition.

1. Introduction

Thermoelectric energy conversion has a large potential to increase power system efficiency by utilization of the waste heat. Recently, considerable efforts were directed to investigate the prospects of the thermoelectric converters for utilization waste heat from the internal combustion engines, garbage incinerators, metallurgical furnaces, and so on. In order to be economically viable the thermoelectric converters should be efficient and cheap. The efficiency of the converters to a large part is determined by the properties of the thermoelectric materials from which the active parts of the converters are made. The thermoelectric efficiency of a material is characterized by parameter Z:

$$Z = \frac{S^2}{\rho\kappa}, \tag{1}$$

where S - Seebeck coefficient or thermopower, ρ - resistivity and κ - thermal conductivity. The nanostructurization is considered as a promising way to increase the efficiency [1, 2]. One of the practically important types of the nanostructured materials is the nanocrystalline (NC) composites. It is commonly agreed that the NC materials are the polycrystalline substances with average crystal grain size below about 100 nm [3]. Two main effects are expected to contribute to the enhancement of the efficiency in the NC thermoelectrics: a strong reduction of thermal conductivity, and increase of the thermopower due to energy-dependent scattering (energy filtering) of the charge carriers at the boundaries of nanoparticles.

The preparation of NC materials with well defined properties is quite a challenging task. The main difficulty is to get the NC state with clean nano-crystal interfaces. In this work, the

NC state is prepared by crystallization of amorphous material in the course of annealing at a certain temperature, i.e. via solid state reaction. The NC state can be further transformed into conventional microcrystalline state (MC) by long annealing at sufficiently high temperatures. In this way it was possible to compare thermoelectric properties of amorphous, NC and MC states of the same sample.

2. Experimental procedures

The amorphous films were deposited by magnetron sputtering of composite targets onto unheated Si/SiO_2 substrates. The deposition procedures were carried out in a high-vacuum chamber equipped with a turbo-molecular pump and with a mass flow controller for maintaining Ar working pressure at $4{\cdot}10^{-3}$ mbar. The deposition rate was in the range 20-60 nm/min.

$Cr_{1-x}Si_x$ films with composition x=0.65; 0.67; 0.72; 0.85 with thickness from 20 to 300 nm were prepared. The film composition was determined with Rutherford Backscattering Spectroscopy (RBS) and Energy Dispersive X-ray Spectroscopy (EDXS). The films were further transformed into nanocrystalline state by annealing with in-situ thermopower and electrical resistivity measurements. The resistivity and thermopower were measured simultaneously. 4-point, DC procedure was used in the resistivity measurements, while a differential method was employed for measurement of thermopower. The measurement set-up was described elsewhere [4]. The measurements were done in the inert helium gas atmosphere.

3. Results and discussion

Amorphous films were transformed into nanocrystalline state during thermal annealing. The crystallization of $Cr_{1-x}Si_x$ films into the NC state was confirmed by structural analysis [5]. At isothermal annealing the crystallization rate is strongly dependent on the annealing temperature and can be experimentally observed above about 450 K. Below this temperature the crystallization rate is too low for experimental observation. The crystallization temperature increases with increasing silicon content in the $Cr_{1-x}Si_x$ composite films [6].

The evolution of the resistivity and thermopower during the isothermal annealing at T_{ann}= 495 K is shown in Fig. 1. Both, the resistivity and the thermopower have a characteristic sharp increase at the onset of the crystallization.

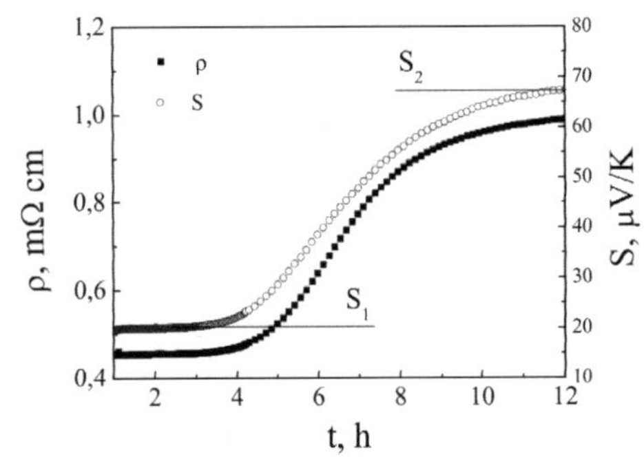

Figure 1. Dependencies of thermopower and resistivity on annealing time at T_{ann} = 495 K of $CrSi_2$ film.

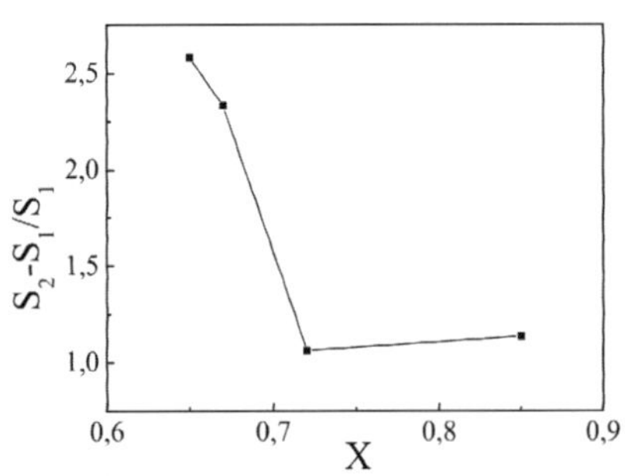

Figure 2. Change of the thermopower due to the crystallization versus film composition. Thickness of the films is 100 nm.

According to the film structure investigation [5] the crystallization starts with the nucleation of nanocrystals of $CrSi_2$ with the average size of 15 nm. In the course of the crystallization the number of the nanocrystals increases, while the average grain size remains in the range

10–20 nm. At the early stages of the crystallization process the film consists mainly of the amorphous phase with diluted inclusions of the $CrSi_2$ nanocrystals. With annealing time the amount of the NC phase increases.

Interaction between charge carries and nanocrystals depends on ratio of crystalline size and the de Broglie wavelength λ. λ is given by

$$\lambda = \frac{h}{m^* V},$$

where h - Planck's constant, m^* - effective mass, V - thermal mean velocity. The effective mass of holes for this material $m_h^* = 3.2m_0$ [7].

The thermal mean velocity can be calculated from:

$$\frac{3}{2}kT = m^* V^2/2,$$

where k - Boltzmann constant, T - absolute temperature. Thus $\lambda = 5$ nm, so the nanocrystal can be regarded as a point defect or point-like scattering center because the average grain size comparable to de Broglie wavelength.

Accordingly, the increase of the resistivity and thermopower (Fig.1) is related to the increasing number of the scattering centres with selective scattering of the charge carriers at the boundary of the nanocrystals. This scattering gives essential contribution to both, to the resistivity and to the thermopower Figs.1 and 2. Of course, at later annealing stages the NC grains can not be considered as independent point-like scattering centres, they even can form a percolating cluster of NC phase, nevertheless the scattering on the grain boundaries remains the dominating scattering mechanism for charge carries as long as the grain size is comparable to the mean free path of the carriers. In accord with this our earlier experimental observation demonstrated that the power factor S^2/ρ of the Cr-Si and $MnSi_{2.2}$ compounds in the nanocrystalline state exceeds the power factor in the microcrystalline state due to enhancement of thermopower [8].

More detailed analysis of the resistivity and thermopower during crystallization can be based on the Matthiessen's rule for resistivity Eq. 2, and Nordheim-Gorter rule for thermopower [9]:

$$\rho = \rho_a + \rho_b, \tag{2}$$

$$S = (S_a - S_b)\frac{\rho_a}{\rho_a + \rho_b} + S_b, \tag{3}$$

where S and ρ are the total thermopower and resistivity of the sample, $S_a \equiv S_1$ and ρ_a - thermopower and resistivity arising from charge carriers scattering in amorphous phase, S_b and ρ_b - thermopower and resistivity arising from charge carriers scattering at the NC grain interfaces.

The Matthiessen's rule Eq. 2 and the Nordheim-Gorter rule Eq. 3 are valid for diluted alloys, i.e. under the condition that the interaction of charge carriers with the alloy matrix and with the impurity scattering centres are independent processes. In the context of the presents work this corresponds to the initial stages of the annealing, Fig.3. According to the Eq. 3 the total thermopower, measured during annealing should be a linear function of the reciprocal total resistance. Fig.3 shows that, indeed, the linear function law is realized on the first stage of annealing. S_b was found from slope of Fig.3. Fig.4 shows value of S_b for different film compositions.

It was found that grain boundary scattering contribution to the thermopower of the nanocrystalline state depends on compositions. S_b is comparable to thermopower of bulk crystalline $CrSi_2$ [10]. The maximum thermopower of selective scattering have films of stoichiometric composition.

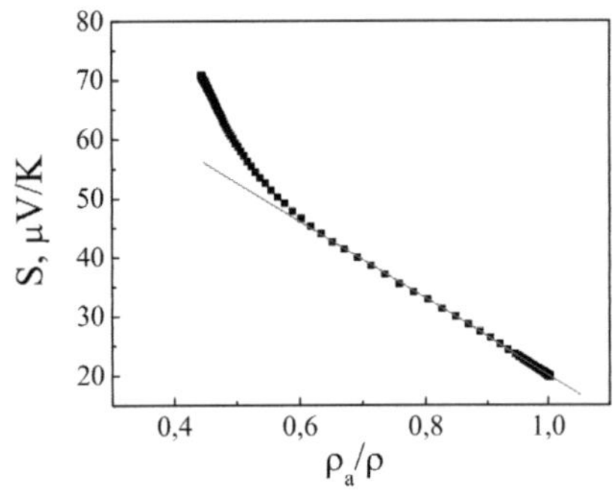

Figure 3. Dependence of the total thermopower on inverse total resistivity for $CrSi_2$ film.

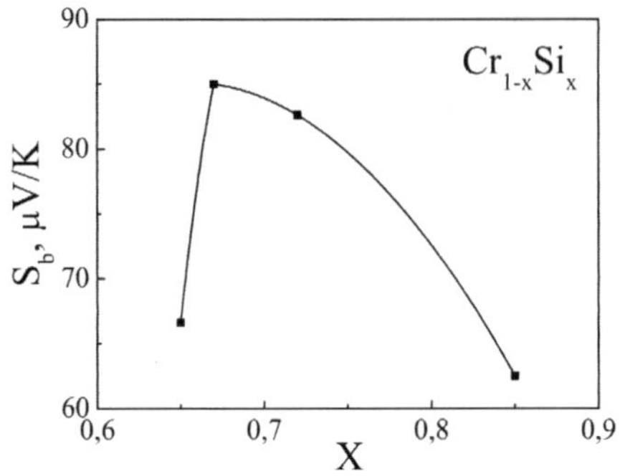

Figure 4. Dependence of the thermopower arising from charge carriers scattering at the boundaries on composition. Thickness of the films is 100 nm.

4. Conclusions

Grain boundary scattering contribution to the thermopower due to selective scattering is large. For this reason, the power factor of the nanocrystalline state is larger than power factor of microcrystalline state. Stoichiometric composition of Cr-Si has maximum grain boundary scattering contribution to thermopower.

Acknowledgments

This work was supported by Russian Foundation for Basic Research under Grant No. 14-08-31177 mol_a and 16-38-60084 mol_a_dk. SNV gratefully acknowledges financial support by Stipend of the President of the Russian Federation SP-1795.2015.1.

References

[1] Hicks L and Dresselhaus M 1993 *Phys. Rev. B* **47**(19) 12727–12731
[2] Kanatzidis M 2010 *Chemistry of Materials* **22** 648–659
[3] Surynarayana C 1995 *International Materials Reviews* **40** 41–64
[4] Burkov A, Heinrich A, Konstantinov P P, Nakama T and Yagasaki K 2001 *Measurement Science and Technology* **12** 264–272
[5] Schumann J, Gladun C, Monch J I, Heinrich A, Thomas J and Pitschke W 1994 *Thin Solid Films* **246** 24–29
[6] Novikov S, Burkov A and Schumann J 2014 *Journal of Electronic Materials* **43** 2420–2424
[7] Nikitin E, Tarasov V and Zayzev V 1973 *Physics of the Solid State* **15**(4) 1254–1256
[8] Novikov S, Burkov A and Schumann J 2013 *Journal of Alloys and Compounds* **557** 239 – 243
[9] Guenault A M 1972 *Journal of Physics F: Metal Physics* **2** 316
[10] Solomkin F, Zaitsev V, Novikov S, Samunin Y and Isachenko G 2013 *Technical Physics* **58** 289–293

Pre-transitional evolution of central peaks and transverse acoustic phonon branch in single crystal lead zirconate titanate with Ti concentration 0.7%

D A Andronikova[1,2], I A Bronwald[1,2], R G Burkovsky[2], I N Leontiev[3], N G Leontiev[4], A A Bosak[5], A V Filimonov[2] and S B Vakhrushev[1,2,6]

[1] Ioffe Institute, 26 Politekhnicheskaya, St Petersburg 194021, Russian Federation
[2] Peter the Great St.Petersburg Polytechnic University, 29 Polytechnicheskaya, St.Petersburg, 195251, Russian Federation
[3] Southern Federal University, ul. Bolshaya Sadovaya 105/42, Rostov-on-Don, 344006, Russian Federation
[4] AzovBlack Sea Engineering Institute, Don State Agrarian University, ul. Lenina 21, Zernograd,Rostov oblast, 347740, Russian Federation
[5] European Synchrotron Radiation Facility, BP 220, F-38043 Grenoble Cedex, France
[6] Faculty of Physics, Saint-Petersburg State University, 198504 Petrodvorets, Ulyanovskaya str., 1, Saint-Petersburg, Russia

E-mail: `andronikova.daria@gmail.com`

Abstract. Inelastic X-ray scattering measurements have been done to study the lattice dynamics in lead zirconate titanate solid solution with 0.7% of $PbTiO_3$. The temperature evolution of central peak and low-energy transverse phonon branches has been traced. Temperature dependent in-plane transverse polarized acoustic phonon branch in <1 1 0> direction has been revealed. The central peaks of two types have been found. The central peak at small wave vectors can be attributed to the relaxational-type soft ferroelectric mode, while the latter at $\mathbf{Q}$ = (1.5 -0.5 0) could be linked to the formation of M-superstructure in the intermediate ferroelectric phase.

1. Introduction

Lead zirconate titanate (PZT) is well known and widely used ferroelectric material. PZT attracts a lot of attention because of its exceptionally good piezoelectric properties. Another motivation to study PZT is a variety of properties and structures, demonstrated by PZT with different composition. Pure lead zirconate ($PbZr_{1-x}Ti_xO_3$ at x=0) is the prototypical antiferroelectric material. According to a set of reports [1–4] between the cubic perovskite paraelectric phase and the antiferroelectric phase, in the narrow temperature range, the intermediate ferroelectric phase exists. Addition of titanium increases the temperature range of stability of the intermediate phase [5]. Cubic-to-intermediate phase transition is accompanied by doubling of the cell parameters of the paraelectric cubic lattice along two directions [6] and results in the appearance of M-superstructure with coordinate (H$\pm\frac{1}{2}$ K$\pm\frac{1}{2}$ 0) in the diffraction pattern.

While antiferroelectricity is known for a long time, the understanding of this phenomenon is still incomplete. It has been shown that the antiferroelectricity in lead zirconate originates from the anomalously high impact of flexoelectric coupling on the transverse acoustic spectra [7]. Due to that the lattice becomes unstable against incommensurate modulations over a large volume in the Brillouin zone along <1 1 0> directions, but the transition to the incommensurate phase is avoided by lock-in phase transition to the observed antiferroelectric phase due to sufficiently strong Umklapp interaction. This mechanism was confirmed for pure $PbZrO_3$ by the analysis of inelastic and diffuse X-ray scattering [7]. However, this description does not account for the appearance of the intermediate ferroelectric phase.

Diffuse scattering could give a hint in the understanding of the lattice dynamics. Our measurements revealed the appearance of strong anisotropic diffuse scattering around integer-value Bragg spots in paraelectric phase of PZT with 0.7 % Ti concentration [8]. The shape of diffuse scattering in the PZT is a similar to that observed in pure lead zirconate. Intensity of diffuse scattering demonstrates fast growth on approaching the transition temperature from the high temperature. Such behavior could be explained by the softening of the phonon modes close to the center of the Brillouin zone.

To study the details of the lattice dynamics of lead zirconate-titanate with low Ti concentration we have performed inelastic X-ray scattering at single crystal $PbZr_{0.993}Ti_{0.007}O_3$.

2. Experimental details

The single crystal of PZT with x = 0.007 has been studied. The synthesis process was described earlier [9]. The crystal preparation included polishing and etching with hydrochloric acid. The resulting crystal was $100 \times 100 \times 500$ μm. The lattice dynamics in PZT was studied using the backscattering spectrometer ID28 at the European Synchrotron Radiation Facility (France). It was operated at X-ray energy E = 21.75 keV (Si (11 11 11) monochromator) providing the energy resolution of 1.5 meV FWHM.

According to [10] temperature range of the intermediate phase is about 500 - 455 K on heating. Our measurements were performed at 4 temperatures above and at one temperature below the temperature of transition in the intermediate phase.

Previous studies of lead zirconate lattice dynamics [7] have demonstrated that phonons along <1 1 0> directions have anomalously low and temperature dependent energy. Because of that in this experiment we concentrated on wave vectors **q** along the [-1 -1 0] direction. The measurements were performed in the vicinity of the (2 0 0) reciprocal lattice point. Such choice of the direction and the reciprocal lattice point gives the dominant contribution of the transverse phonon to the experimentally measured spectra. The ID28 spectrometer multianalyzer construction allows us to register 9 reciprocal lattice points simultaneously. As the result the region around (2 0 0) reciprocal lattice point surrounding [-1 -1 0] direction has been studied.

3. Results

Fig.1 shows spectra for a set of reciprocal points and different temperatures. All spectra contain clearly visible symmetric phonon resonances and a central peak (at zero energy transfer). The two low-energy (about 3 meV) Stokes and anti-Stokes peaks are associated with the transverse acoustic (TA) mode with "in-plane" polarization vector. The high-energy (about 8 meV) Stokes and anti-Stokes peaks can be attributed to the transverse optic (TO) phonons. Phonon resonances were fitted using damped harmonic oscillator line shape convoluted with the experimental resolution [11, 12]. The central peak was fitted by Lorentzian function [12]. In Fig.1 the fitting results are shown by lines while experimental data are shown by points.

The obtained dispersion curves for the TA phonons along [-1 -1 0] direction are shown in Fig.2. In the paraelectric phase the TA phonon dispersion branch demonstrates low energy which does not exceed 4 meV. Notable feature of the dispersion curve is the change of its slope at q > 0.3.

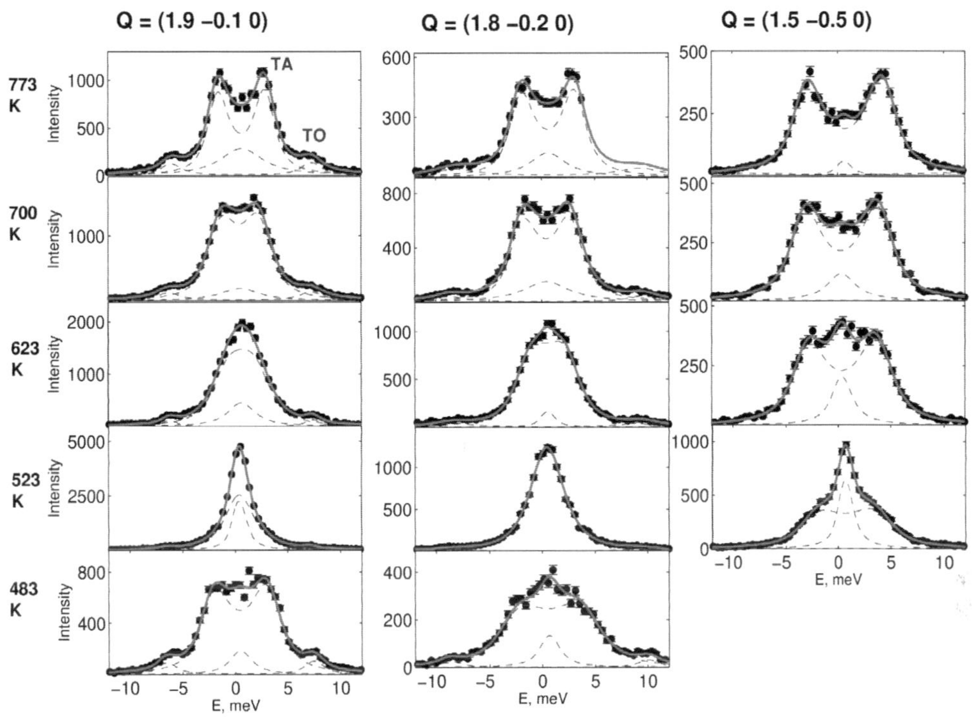

Figure 1. Inelastic X-ray scattering spectra, measured at **Q** = (1.9 -0.1 0), **Q** = (1.8 -0.2 0) and **Q** = (1.5 -0.5 0) reciprocal lattice points for temperatures 773K, 700K, 623K, 523K and 483K. The experimental data are shown by points, the fit results are shown by lines.

The in-plane polarized TA phonons demonstrate a distinct temperature dependence. In the paraelectric phase the temperature decrease results in the lowering of the TA phonon branch energy. In the intermediate phase (at 483K) the energy of the TA in-plane phonon notably increases. No pronounced temperature dependence of the TO in-plane phonon energy has been found. This is in a good agreement with the results obtained for pure lead zirconate.

The anisotropy of TA in-plane phonon was studied. In Fig.3 the TA in-plane phonon energy distribution in reciprocal space is shown for temperature 700K. Along $< 110 >$ directions minima in phonon energy are clearly observed, a similar anisotropy of the TA phonon was found in pure lead zirconate earlier [7].

As in lead zirconate [7] in the PZT all spectra contain central peaks. To analyze the central peak features we plotted the contour maps of scattering intensity in the q-E coordinates (Fig.4). The energy of the TA and TO phonons, determined from the fits of the experimental spectra, are indicated by points. At small wave vector **q** the high intensity central peak exists at 523 K, close to the transition temperature. The intensity of this central peak decreases with wave vector increase, at **Q** = (1.6 -0.4 0) the central peak intensity becomes minimal, as compared to the surrounding q-points along the [-1 -1 0] direction. At Brillouin zone boundary (**Q** = (1.5 -0.5 0)) another central peak is clearly observable. Both central peaks demonstrate an increase of intensity on approaching the paraelectric-intermediate phase transition temperature.

Different distributions of central peak intensity in reciprocal space allows to suggest that the central peak at small wavevector **q** and central peak at the M-point are the results of different

Journal of Physics: Conference Series **769** (2016) 012070 doi:10.1088/1742-6596/769/1/012070

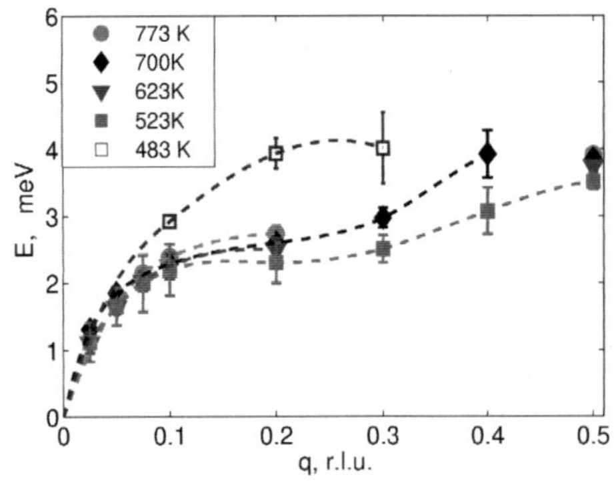

Figure 2. Temperature evolution of TA in-plane phonon dispersion along [-1 -1 0] direction near (2 0 0) reciprocal lattice point (**Q** = (2-q -q 0)). Points are fitting results, lines are guide for eyes.

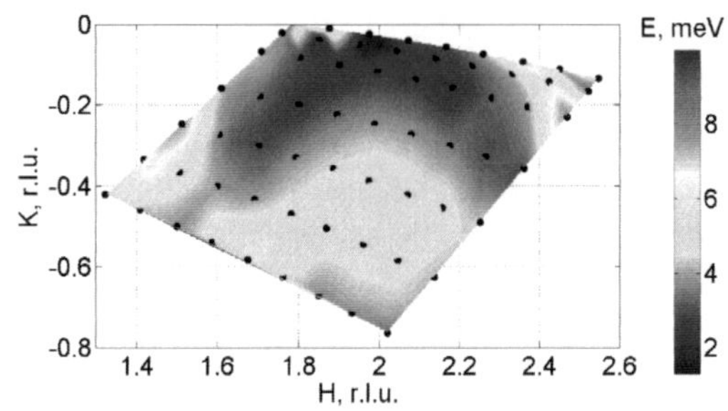

Figure 3. TA in-plane phonon energy distribution in reciprocal space coordinate around (2 0 0) reciprocal lattice point for temperature 700K.

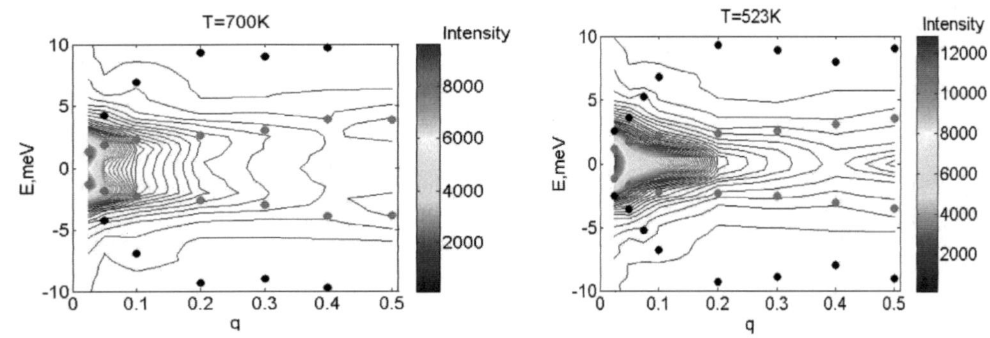

Figure 4. Contour maps of inelastic X-ray spectra for [-1-10] direction near Bragg reflection (2 0 0) for temperatures 700K and 523K, red points indicates the TA phonon energy, black points – the TO phonon energy

processes: the central peak at small **q** could be linked with some critical excitation, related to the dielectric anomaly. The central peak at the M-point could be relevant to the arising of the M-superstructure, which appears below the paraelectric-intermediate phase transition.

4. Conclusions

Using inelastic X-ray scattering the lattice dynamics of the lead zirconate titanate single crystal with Ti concentrations 0.7% has been studied. Anomalously low energy TA in-plane polarized phonon has been found in $< 110 >$ directions. On approaching to the phase transition temperature in the paraelectric phase the energy of the TA phonon decreases. The anisotropy of the phonon surface dispersions is similar to the anisotropy of the diffuse scattering published in [8] for the same material.

The central peaks of two types have been found. The central peak at small **q** ($q < 0.4$) could be associated with the ferroelectric soft mode, which is responsible for the observed dielectric anomaly. It has been shown for lead zirconate [13] that soft ferroelectric mode manifests itself as a combination of central peak and optic phonon. As the result while the temperature decreases towards to transition temperature the intensity of this central peak sharply increases and optic phonon energy demonstrates just a little energy change in lead zirconate. This behavior could originate from coupling of the soft optic phonon and some noncritical relaxation process. In this case the critical increase of the central peak intensity in PZT with 0.7% of Ti could be a manifestation of the optic mode softening close to the Brillouin zone center. The central peak at the Brillouin zone boundary could be linked with M-superstructure formation, which appears below the paraelectric-intermediate phase transition.

5. Acknowledgments

R.G. Burkovsky, D.A. Andronikova and I.A. Bronwald acknowledge the support by Grant of the President of Russian Federation for state support of Russian young scientists MK-5685.2016.2. The work of A.V. Filimonov was performed under the government order of the Ministry of Education and Science of RF.

References

[1] Tanaka M, Saito R and Tsuzuki K 1982 Electron microscopic studies on domain structure of $PbZrO_3$ *Japanese Journal of Applied Physics* **21** 291
[2] Fujishita H and Hoshino S 1984 A study of structural phase transitions in antiferroelectric $PbZrO_3$ by neutron diffraction *J. Phys. Soc. Jpn.* **53** 226-234
[3] Dec J and Kwapulinski J 1989 Crystallogeometry of phase transitions in $PbZrO_3$, single crystals *J. Phys.: Condens. Matter* **1** 3389-3396
[4] Xu Z, Dai X, Viehland D, Payne D, Li Z and Jiang Y 1995 Ferroelectric domains and incommensuration in the intermediate phase region of lead zirconate *Journal of the American Ceramic Society* **78** 2220-2224
[5] Shirane G, Sawaguchi E and Takagi Y 1951 Dielectric properties of lead zirconate *Phys. Rev.* **84** 476-481
[6] Ricote J, Corker D, Whatmore R, Impey S, Glazer A, Dec J and Roleder K A tem and neutron diffraction study of the local structure in the rhombohedral phase of lead zirconate titanate *Journal of Physics: Condensed Matter* **10** 1767
[7] Tagantsev A *et al.* 2013 The origin of antiferroelectricity in $PbZrO_3$ *Nature Communications* **4** 2229
[8] Andronikova D *et al.* 2015 Critical scattering of synchrotron radiation in lead zirconate-titanate with low titanium concentrations *Physics of the Solid State* **57** 2441-2446
[9] Leontiev N, Smotrakov V and Fesenko E 1982 *Izv. Akad. Nauk SSSR, Neorg. Mater.* **18**
[10] Fesenko O, Smotrakov V and Leontev N 1983 *Physics of the Solid State* **25** 1958-1963
[11] Dorner B 1993 *Coherent Inelastic Neutron Scattering in Lattice Dynamics* Springer Tracts in Modern Physics
[12] Burkovsky R, Andronikova D, Bronwald Y, Krisch M, Roleder K, Majchrowski A, Filimonov A, Rudskoy A and Vakhrushev S 2015 Lattice dynamics in the paraelectric phase of $PbHfO_3$ studied by inelastic x-ray scattering *Journal of Physics: Condensed Matter* **27** 335901
[13] Burkovsky R *et al.* 2014 Lattice dynamics and antiferroelectricity in $PbZrO_3$ tested by x-ray and brillouin light scattering *Phys. Rev. B* **90** 144301

Journal of Physics: Conference Series **769** (2016) 012071 doi:10.1088/1742-6596/769/1/012071

Spin-wave self-modulation instability in a perpendicularly magnetized magnonic crystal

A V Drozdovskii[1,2], B A Kalinikos[1], A B Ustinov[1], and A Stashkevich[2,3]

[1]Department of Physical Electronics and Technology, St. Petersburg Electrotechnical University, St. Petersburg, 197376 Russia
[2]International laboratory "MultiferrLab", ITMO University, St. Petersburg, 197101 Russia
[3]LSPM (CNRS-UPR 3407), Universite Paris 13, Sorbonne Paris Cite, 93430 Villetaneuse, France

E-mail: drozdovskyiav@gmail.com

Abstract. Self-modulation instability of microwave spin waves monochromatically excited in a periodic magnetic film structure, a magnonic crystal, has been observed. The phenomena occurs only at the frequencies corresponding to the band gaps of the spin-wave spectrum of the magnonic crystal. Depending on the frequency within the band gap the nonlinear waveforms have a shape of bright or dark pulses.

1. Introduction

During last years a great attention has been given to the study of tunable materials [1] both for their properties and for the high number of potential applications. Magnonic crystals (MCs) are new class of tunable materials with the periodically modulated magnetic properties where spin waves (SW) can propagate [2-5]. Reviving interest to such structures is stimulated by fundamental and applied researches. In particular, the periodic magnetic waveguides were used for development of the tunable resonators and filters, signal-to-noise-enhances, logic elements, sensors and other [6-13]. Microwave properties of such structures can be tuned in a wide frequency range with magnetic field. Most attention was given to study the linear properties of MCs and possibilities of their reconfiguration [2, 14]. However, the magnonic crystals produced from the ferrite films may demonstrate strong nonlinearity for high spin wave power [7].

The phenomenon of modulation instability that appears due to four-wave parametric process is typical of many nonlinear media (see, e.g.,[15]). The modulation instability is commonly classified by two types, spontaneous and induced instability [15, 16]. Induced instability arise from the interaction of two initial monochromatic signals owing to nonlinearity of the medium. Spontaneous modulation instability (called also "self-modulation instability") appears due to the decay of an initial monochromatic signal into waves with close wave numbers. As is known, both types of modulation instability can lead to the formation of envelope solitons.

In this work, we report the first experimental study of self-modulation spin-wave instability in perpendicularly magnetized magnonic crystals.

2. Experimental investigation

The magnonic crystal was fabricated from an yttrium-iron garnet (YIG) film with chemical etching. The fabricated sample of the magnonic crystal had a width of 2 mm and a length of 30 mm. The YIG film with thickness of 9 μm and saturation magnetization of 1750 Gs was grown by liquid-phase epitaxy on a gadolinium-gallium garnet substrate. The half-width ΔH of the ferromagnetic resonance curve of the YIG film was 0.55 Oe measured for the frequency of 4500 MHz. The film had free surface spins. It was confirmed with preliminary experimental research using the methods of magnetic well and spin-wave spectroscopy [17-20].

To avoid possible unwanted reflected spin waves, the edges of the MC were cut at an angle less than 45^0.The experimental periodic structure had a period Λ of 400 μm, a groove length of 50 μm, and a groove depth of 1.5 μm.

A delay line structure was used for the measurements. Spin waves were excited and received using short-circuited input and output microstrip antennas having a length of 2 mm and a width of 50 μm. The distance between the antennas was 5.2 mm. Thus, there were 13 periods of the structure between the transmitting and receiving antennas. The experimental device was placed between poles of an electric magnet generating a uniform magnetic field with strength of 3050 Oe. The field was directed perpendicular to the plane of the magnonic crystal. Such mutual orientation of the field and the propagation direction of spin waves corresponds to the excitation of forward volume spin waves in the magnonic crystal. For this magnetic field direction, the nonlinear self-interaction coefficient N is positive.

The experiment was carried out in several stages. First, the transmission characteristics of spin waves in a knowingly linear regime of their excitation and propagation were studied. The measured amplitude–frequency characteristic is shown by solid lines in figure 1. As is seen, the spin-wave spectrum of the magnonic crystal has pass bands and stop bands (band gaps) with strong group velocity dispersion and high damping of spin waves. The stop bands are formed due to Bragg resonances.

The numerical calculation of the amplitude–frequency characteristic was performed with the use of the method similar to that described in [17, 18], i.e., in terms of the transmission matrix. The calculation did not take into account the losses for the excitation and reception of the spin waves. Nevertheless, the plots indicate good quantitative agreement between measured and calculated frequencies of the stop bands.

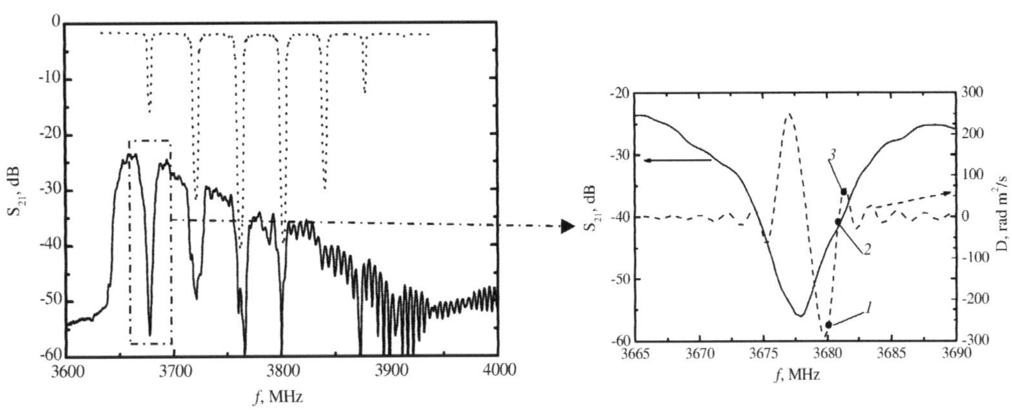

Figure 1. The experimental (solid line) and theoretical (dotted line) transmission characteristics of magnonic crystal. Dashed line shows the frequency dependence of the spin-wave dispersion coefficient near the frequency of the first Bragg resonance.

Journal of Physics: Conference Series **769** (2016) 012071 doi:10.1088/1742-6596/769/1/012071

At the second stage of the experiments we studied the possibility of the emergence of self-modulation instability of the spin-waves propagating in the magnonic crystal. We measured the spectra and envelopes of the output microwave signal under the systematic variation of the power and frequency of the input monochromatic signal. The results show that the spin-wave instability in the magnonic crystal emerged in a threshold manner at P_{in} = 18 dBm for frequencies within high-frequency slope of the stop band. For these parameters, low-amplitude satellites emerged in the spectrum of the output signal. The amplitude of these frequency harmonics increased and the spectrum was enriched with an increase in the input power. The frequency region of the the self-modulation instability was also broadened.

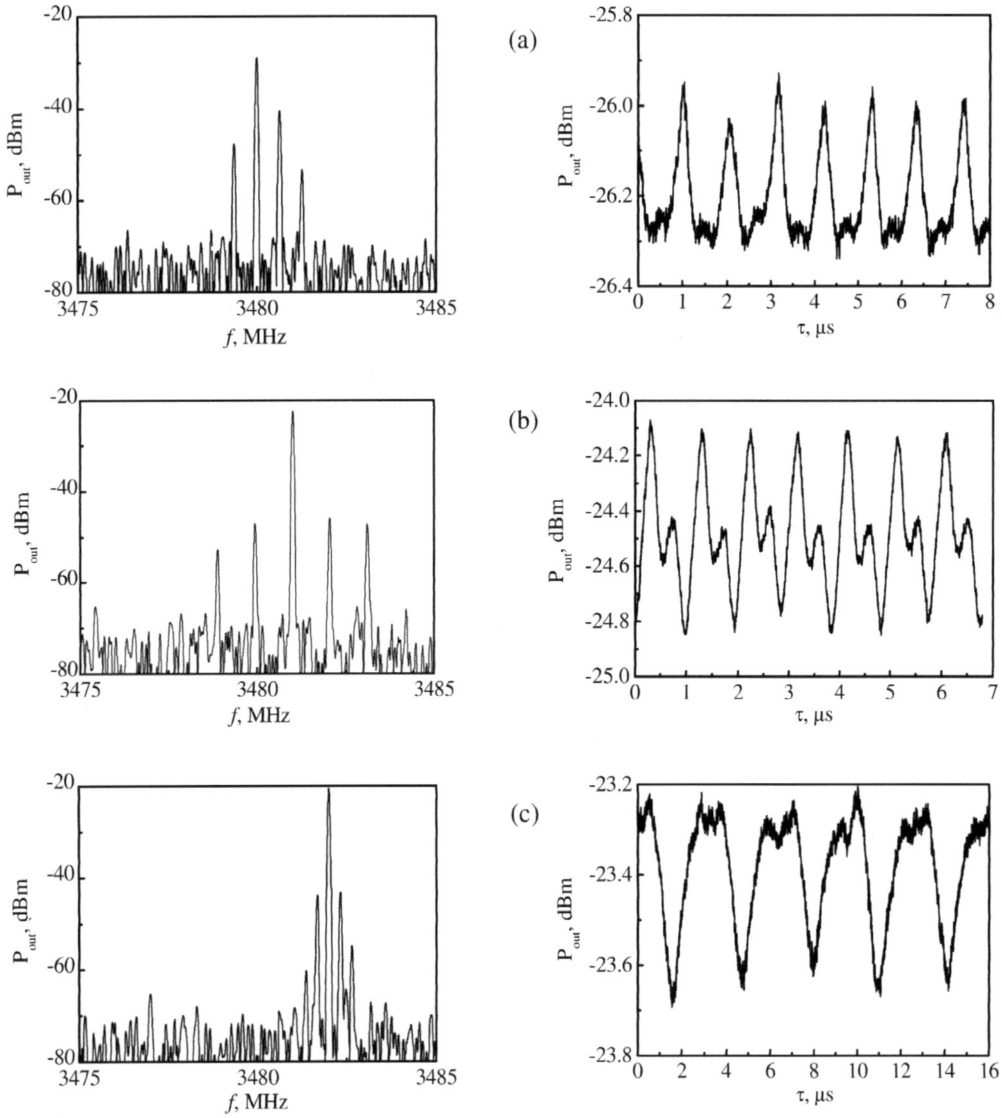

Figure 2. Frequency spectra (left) and the respective envelopes (right) of the output signal measured for various frequencies of input microwave signal: (a) f_{in} = 3680 MHz, (b) f_{in} = 3681 MHz, (c) f_{in} = 3682 MHz.

The shape of output waveform depends not only on the power, but also on the frequency of the input microwave signal. Figure 2 exemplifies the frequency spectra (left panel) and the corresponding envelopes of the output microwave signal (right panel) measured for frequencies near the first Bragg resonance for P_{in} = 20 dBm.

Figure 2,a shows the measured results for the frequency of the input signal of f_{in}=3680 MHz and negative dispersion coefficient (see point *1* on figure 1). In this case the output bright pulses were observed. This agrees with the theoretical concepts of the Lighthill criterion for solitons ($D \cdot N<0$) [14]. In the region where the dispersion coefficient changes its sign (see point *2* on figure 1) the pairs of bright and dark pulses were observed (see figure 2,b). For areas with a positive dispersion coefficient (see point *3* on figure 1) and $D \cdot N>0$ the formation of dark pulses was observed as is shown in figure 2,c.

It should be mentioned that the self-modulation instability was observed also near the second and the third stop bands of the magnonic crystal. Outside the stop bands, i.e., in the regions where the dispersion coefficient is low, the inherent modulation instability was not observed in the entire power range of the experimental setup.

3. Conclusions

In conclusion, self-modulation spin-wave instability in a magnonic crystal has been studied for the first time. The formation of periodic bright and dark spin-wave envelope soliton trains has been detected at the edges of stop bands.

Acknowledgments

The experimantal work at SPbETU was supported in part by Ministry of Education and Science of Russian Federation (Project "Goszadanie"). The fabrication and characterization of the magnetic samples at SPbETU was supported in part by Russian Science Foundation (Grant# 14-12-01296). The work at ITMO University was supported by the Government of the Russian Federation (Grant 074-U01).

References

[1] Ye Z. G. 2008 *Handbook of advanced dielectric, piezoelectric and ferroelectric materials: Synthesis, properties and applications* (Elsevier).

[2] Vogel M, Chumak A V, Waller E H, Langner T, Vasyuchka V I, Hillebrands B, Freymann G 2015 *Nature Physics* **11** 487–491

[3] Kruglyak V V, Demokritov S O, Grundler D 2010 *J. Phys. D: Appl. Phys.* **43** 264001

[4] Nikitov S A, Tailhades Ph, Tsai C S 2001 *JMMM* **236** 320-330

[5] Ustinov A B, Grigorieva N Y, Kalinikos B A 2008 *JETP Lett.* **88(1)** 31–35

[6] Gouzerha J, Stashkevich A A, Kovshikov N G, Matyushev V V, Desvignesa J M 1991 *JMMM* **101(1)** 189-190

[7] Ustinov A B, Drozdovskii A V, Kalinikos B A 2010 *Appl. Phys. Lett.* **96** 142513

[8] Grishin S V, Beginin E N, Dulin Y V, Nikitov S A, Sharaevskii Y P 2012 *Tech. Phys.Lett.* **38(7)** 638-641

[9] Zhu Y, Chi K H, Tsai C S 2014 *Appl. Phys. Lett.* **105** 022411

[10] Grishin S V, Beginin E N, Sharaevskii Yu P, Nikitov S A 2013 *Appl. Phys. Lett.* **103** 022408

[11] Grishin S V, Beginin E N, Morozova M A, Sharaevskii Yu P, Nikitov S A 2014 *J. Appl. Phys.* **115** 053908

[12] Inoue M, Baryshev A, Takagi H, Lim P B, Hatafuku K 2011 *Appl. Phys. Lett.* **98** 132511

[13] Kryshtal R G, Medved A V 2012 *Appl. Phys. Lett.* **100** 192410

[14] Nikitin A A, Ustinov A B, Semenov A A, Chumak A V, Serga A A, Vasyuchka V I, Hillebrands B 2015 *Appl. Phys. Lett.* **106(10)** 102405

[15] Remoissenet M 1996 *Waves Called Solitons: Concepts and experiments* (Springer-Verlag, Berlin).
[16] Kivshar Y S, Agrawal G 2003 *Optical solitons: from fibers to photonic crystals* (Academic press).
[17] Soohoo R F, 1965 *Thin Magnetic Films* (Harper and Row).
[18] Gurevich A G, Melkov G A, 1996 *Magnetization oscillations and waves* (CRC press).
[19] Kalinikos B A, Kovshikov N G, Kolodin P A, and Panchurin I P 1985 *Electronn. Tekn. Ser. Elektr. SVCh* **10** 53
[20] Kalinikos B A, Slavin A N, 1986 *J. Phys. C: Solid State Phys.* **19** 7013
[21] Seshadri S R 1986 *J. Appl. Phys.* **60** 1758
[22] Chumak A V, Serga A A, Wolff S, Hillebrands B, Kostylev M P 2009 *Appl. Phys. Lett.* **94** 172511

Theoretical Investigation of Dual Tuning of Solitonic Processes in Multiferroic Structures

M A Cherkasskii[1], A A Nikitin[1,2], A B Ustinov[1,3], A Stashkevich[2,3], B A Kalinikos[1]

[1]Department of Physical Electronics and Technology, St. Petersburg Electrotechnical University, St. Petersburg, 197376 Russia
[2]International laboratory "MultiferrLab", ITMO University,197101, St. Petersburg, Russia
[3]LSPM (CNRS-UPR 3407), Universite Paris 13, Sorbonne Paris Cite, 93430 Villetaneuse, France

e-mail: macherkasskii@hotmail.com

Abstract. The solitonic wave processes in a multiferroic structure based on ferroelectric and ferrite layers are studied. The influence of external electric and magnetic fields on frequency and wave-number ranges, where bright and dark solitons can exist, are analysed. The investigation was carried out with the nonlinear Schrödinger equation. Results show that an increase of the electric field shifts the boundary between bright and dark solitons to long-wave region. An increase in magnetic field results in the opposite effect.

The term "soliton" usually refers to a propagating nonlinear pulse or wave packet which preserves its shape without dispersive spreading. Study of envelope solitons is one of the fundamental problem in the modern physics. Investigation is carried out in many areas of physics, for example, the physics of wave phenomena in ferromagnetic films, nonlinear optics [1, 2], low-temperature physics [3] etc.

It is well known that the formation and propagation of the envelope solitons are usually described using the nonlinear Schrödinger equation (NLSE) [4]. This equation and its solution were intensively investigated for different waveguiding media [5-7]. It is clear that the equation coefficients describe characteristic features of waveguide medium where solitons can exist. Therefore, tuning the medium properties leads to a change in the coefficients and consequently, in the end results in a modification of the soliton parameters. From the fundamental and applied point of view, it is interesting to analyze a relation between the properties of the medium and the soliton parameters.

Media, where waves with different nature can propagate simultaneously, attracts special scientific interest. One of the most striking examples is multiferroics. Active study of artificial multiferroic media in a form of layered structures composed of ferromagnetic and ferroelectric layers has recently begun [8-10]. Their linear wave properties were well studied. At the same time, there are only few papers devoted to the investigation of nonlinear wave properties of the multiferroics [11-14]. Thus, dependences of the soliton parameters on multiferroic properties remain unexplored.

A main advantage of the multiferroic structures in comparison with other waveguide media is dual tunability of their properties by external electric and magnetic fields. This behavior is due to electrodynamic interaction between the microwave electromagnetic and spin waves in the layered ferrite-ferroelectric structures. This interaction leads to formation of hybrid spin-electromagnetic waves (SEW). Note, that electric tuning of the SEW spectrum is possible due to an explicit dependence of dielectric permittivity in the ferroelectric layer on the bias electric field whereas magnetic tuning is provided by a dependence of magnetic permeability of the ferrite layer on the bias magnetic field.

The aim of this work is to study the influence of external electric and magnetic fields on peculiarities of solitonic excitations. For that purpose, we have investigated the influence of the electric and the magnetic fields on the dispersion and nonlinear coefficients of NLSE. This investigation have enabled us to study the influence of fields changing on frequencies and wave-number ranges where bright and dark solitons can exist. It is worth noting that the bright solitons are bell-shaped and they can be described with the hyperbolic secant function. The dark solitons have the form of localized "dark" holes created on the CW background. The last one are described with hyperbolic tangent.

In contrast with the previous work [11], where our attention was focused on an influence of *AC* electric and magnetic fields on the solitonic processes, this work is devoted to dual tunability by the external *DC* electric and magnetic fields.

In the generalized coordinates the nonlinear Schrödinger equation has the following form

$$i\frac{\partial u}{\partial t}+\frac{D}{2}\frac{\partial^2 u}{\partial x^2}-N|u|^2 u=-i\Gamma u, \tag{1}$$

where $D=\partial^2\omega/\partial k_x^2$ is the dispersion coefficient, $N=\partial\omega/\partial|u|^2$ is the nonlinear coefficient, and Γ is the relaxation frequency.

It is important to note that the envelope solitons are formed due to a balance between two opposite wave processes. The dispersion spreading and the self-phase modulation compete with each other during the propagation of waves. The first effect is described by the dispersion coefficient *D*; the second one is described by the nonlinear coefficient N. It is well known that if both coefficients have the same signs $DN>0$, then dark solitons can be formed. If the signs are opposite $DN<0$, then bright solitons can be excited [1]. It was recently shown that multiferroics have double wave nonlinearity [11]. This nonlinearity can lead to a change of the coefficient signs in different ranges of wave numbers. Hence, if we tune the coefficients, we obtain a modification of the ranges of the bright and dark solitons modify.

The investigation was carried out in several steps. The linear dispersion relation has been derived and solved numerically. The double wave nonlinearity in the dispersion relation was considered. The dispersion and nonlinear coefficients were calculated by taking the derivatives of the relation. Then we have investigated the influence of the external electric and magnetic fields on the coefficients. Finally, we have analyzed the dual tunability of frequency and wave-number ranges, where bright and dark solitons can exist.

The investigated multiferroic structure is shown in figure 1. The bilayer structure was considered. A substrate for the ferrite layer is negligible in view of its relatively small dielectric permittivity and absence of magnetic properties.

We assume that the SEW propagate along the x-axis in the tangentially magnetized structure. The z-axis is parallel to the direction of the bias magnetic field. The structure is infinite in the x-z-plane and is surrounded by free space. Therefore, in the considered case we can use the linear dispersion equation of the SEW from the [15].

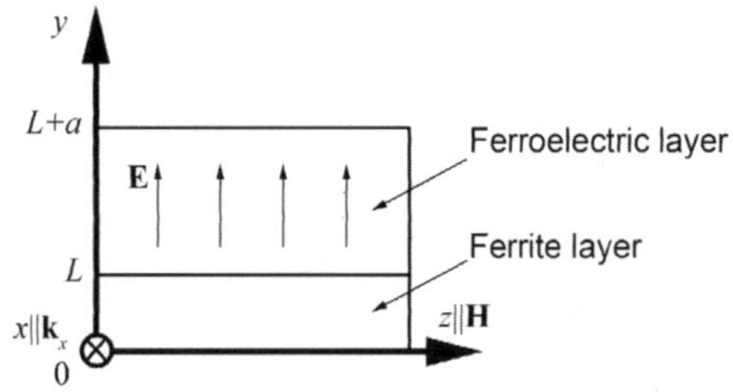

Figure 1. Geometry of the investigated structure

Solving the linear dispersion equation, we have used parameters which are typical for an yttrium iron garnet (YIG), namely, the saturation magnetization was $M_0 = 1750$ G, the permittivity was $\varepsilon_f = 14$ and the thickness was $L = 20\ \mu$m. The value of the external magnetic field was $H = 1500$ Oe. The ferroelectric layer thickness was $a = 500\ \mu$m, dielectric permittivity was equal to 660 at zero electric field. Note that these parameters correspond to the barium strontium titanate (BSTO) ceramic in paraelectric state at room temperature [16].

In addition, we have varied the external electric and magnetic fields. It is necessary to point out that the parameter, which describes the influence of the magnetic field, has been included in the dispersion equation. However, this equation does not involve the electric filed itself. To take into account the electric field we used the following dependence of the BSTO dielectric permittivity:

$$\varepsilon_a = \varepsilon_{a0} - rE^2, \tag{2}$$

where the coefficient r was calculated from the experimental work [16] and its value was equal to 7.7×10^{-8} m^2 / V^2.

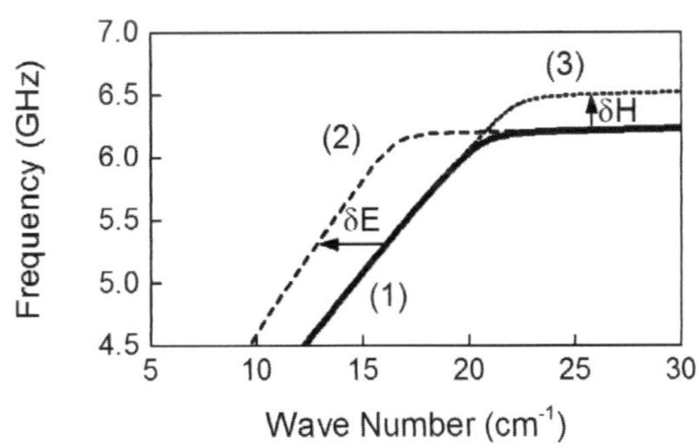

Figure 2. Dispersion characteristic of the SEW, where $\delta E = 20$ V/cm, $\delta H = 100$ Oe.

The numerical solutions of the dispersion equation for different values of the fields are shown in figure 2. The curves on the figure are numbered. These numbers are corresponded to the different parameter sets. The parameters from the first set are mentioned above. The other sets differ only by values of external electric and magnetic fields. The difference is given below.

It is well known that the SEW spectrum consists of the fast and the slow dispersion branches, but only slow branches are plotted for better visibility. Note that the slow SEW are usually excited in experiments. For that reason, our main attention was focused on these waves.

The spectrum of the SEW is formed because of hybridization of spin waves in the ferrite layer and electromagnetic waves in the ferroelectric layer. There is the point of the phase synchronism between these two types of waves. This point has the following coordinates $f = 6.214$ GHz, $k_x = 20$ cm^{-1} at $\delta E = 0$ and $\delta H = 0$. Due to the hybridization, the waves before and after the phase synchronism point have different nature. This statement has been corroborated by the investigation of the electric and magnetic fields variation.

In our investigation we have found that the increase of the electric field leads to a shift of that part of spectrum which corresponds to the electromagnetic-like waves. The dashed curve in the figure 2 demonstrates this. The curve was calculated using $\delta E = 20$ V/cm and it corresponds to the second parameter set. The increase of the magnetic field shifts the other part of the dispersion curve. This part demonstrates spin-wave-like behavior. In this case we used $\delta H = 100$ Oe. This case corresponds to the third parameter set. The point of the phase synchronism moves to a short-wave region with the magnetic field increase and vice versa to a long-wave region with the electric field increase.

Therefore, the variations of electric and magnetic fields result in different spectrum transformation. This difference is defined by hybrid nature of the SEW. It is the striking example that SEW are highly tuned and as a consequence of that devices based on the SEW will have high tunability, respectively.

The nonlinear dispersion relation has been derived in order to obtain the NLSE coefficients. This relation includes the double wave nonlinearity. Previously it was shown that the double nonlinearity should be taken into account for the multiferroic structures [11]. In accordance with this paper, we have transformed the *linear* dispersion equation into *nonlinear*. Subsequently, the formulae describing the dispersion and nonlinear coefficients were obtained.

In spite of the nonlinear coefficient of spin-waves in a free ferrite film, which can be either positive or negative in the whole wave-number range, the SEW nonlinear coefficient changes the sign in the different ranges of the wave numbers. The curves in figure 3(a) demonstrate this statement. Changing in signs is conditioned by the double wave nonlinearity of the SEW.

The dispersion coefficient of the SEW does not change the sign and it remains negative in the whole wave-number range. The values of the expression $|DN| / \max(|DN|)$ are shown in figure 3(b). As it was mentioned above the sign of DN defines the type of the envelope solitons that can propagate in media described by the NLSE.

It is clearly seen from the figure 3 that the electric and magnetic variation have different influence on the nonlinear and dispersion coefficients. The increase of the electric filed leads to the shift of the nonlinear coefficient curve to a long-wave region, while the magnetic field variation acts oppositely. The ranges of wave numbers, where either bright or dark solitons can exist, are shifted synchronously with the nonlinear coefficient. Hence, these ranges can be tuned by the electric and magnetic field. These fields variation impacts oppositely.

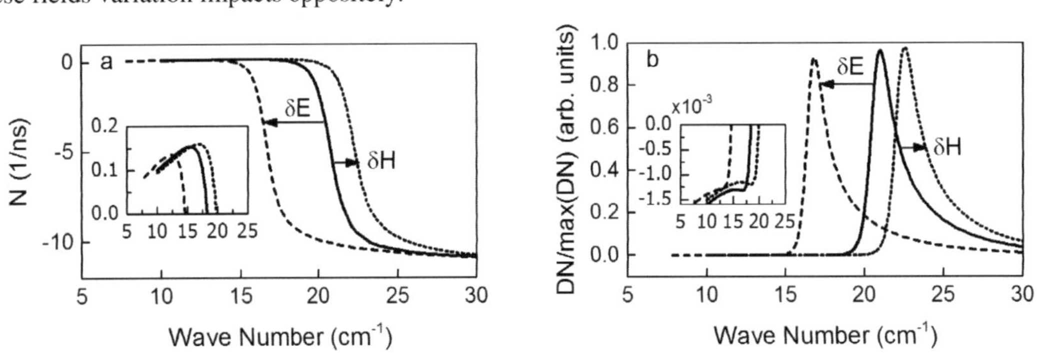

Figure 3. The nonlinear coefficient (a). The product of the dispersion and the nonlinear coefficients normalized on maximum value (b).

Table 1. Summary table of the parameter sets.

No	H, Oe	E, V/cm	ε_a	k_b, cm^{-1}	f_b, GHz
1	1500	0	660	18.5	5.770
2	1500	20	456	14.6	5.720
3	1600	0	660	20	6.070

The table 1 provides a summary of the results. In this table k_b and f_b designate the boundary wave number and the boundary frequency respectively. This boundary values separate the ranges, where bright and dark solitons can exist. The bright solitons exist if a carrier wave number and frequency are less than k_b and f_b respectively and contrariwise if a carrier wave number and frequency are greater than k_b and f_b respectively, then the dark solitons can exist.

In summary, as a result of the investigation it has been obtained that, the nonlinear solitonic excitations can be tuned by variation of the external electric and magnetic fields. As it was shown the tuning is possible due to hybrid nature of the SEW. The magnetic field influences spin-wave nature of the SEW whereas electromagnetic nature of the SEW can be controlled by the electric field. The effect of these fields is the opposite. The increase of the electric field shifts the phase synchronism point and the boundary between bright and dark solitons to long-wave region. On the other hand, the increase of the magnetic field can lead to the reverse effect.

Acknowledgments

The work on the development of the theory of SEW spectra at ITMO University was supported by the Government of the Russian Federation (Grant 074-U01). The work on the development of the theory of nonlinear SEWs at SPbETU was supported by Ministry of Education and Science of Russian Federation (Project "Goszadanie"). The work on numerical simulation of solitonic phenomena at SPbETU was supported by the Russian Science Foundation (Grant# 14-12-01296).

References

[1] Kivshar Y S, Agrawal G P, *Optical solitons. From Fibers to Photonic Crystals* (Academic Press, 2003)
[2] Akhmediev N N, Ankiewicz A, *Solitons: nonlinear pulses and beams* (Chapman & Hall, 1997)
[3] Kevrekidis P G, Frantzeskakis D J, Carretero-González R, *Emergent Nonlinear Phenomena in Bose-Einstein Condensates* (Springer-Verlag, Berlin, 2008)
[4] Remoissenet M, *Waves called solitons: Concepts and Experiments* (Springer-Verlag, Berlin, 1999)
[5] Sulem C, Sulem P L, *The nonlinear Schrödinger equation: self-focusing and wave collapse* (Springer Science & Business Media, 1999)
[6] Taha T R, Ablowitz M I 1984 *Journal of Computational Physics* **55** 203
[7] Serkin V N, Hasegawa A 2000 *Phys. Rev. Lett.* **85** 4502
[8] Nan C-W, Bichurin M I, Dong S, Viehland D, Srinivasan G 2008 *J. Appl. Phys.* **103** 031101
[9] Sun N X, Srinivasan G 2012 *Spin* **3** 1240004
[10] Tagantsev A K, Sherman V O, Astafiev K F, Venkatesh J, Setter N 2003 *Journal of Electroceramics* **11** 5
[11] Cherkasskii M A, Kalinikos B A 2013 *JETP Letters* **97** 611
[12] Cherkasskii M A, Kalinikos B A 2013 *Technical physics letters* **39** 182
[13] Ustinov A B, Kondrashov A V, Nikitin A A, Kalinikos B A 2014 *Appl. Phys. Lett.* **104** 234101
[14] Ustinov A B, Kondrashov A V, Nikitin A A, Cherkasskii M A, Kalinikos B A 2014 *JETP Letters* **100** 835
[15] Nikitin A A, Ustinov A B, Vitko V V, Semenov A A, Belyavskiy P Y, Mironenko I G, Stashkevich A A, Kalinikos B A, Lähderanta E 2015 *J. Appl. Phys.* **118** 183901
[16] Cao L Z, Cheng B L, Wang S Y, Zhou Y L, Jin K J, Lu H B, Chen Z H, Yang G Z 2005 *J. Appl. Phys.* **98** 034106

Investigation of frequency response of microwave active ring resonator based on ferrite film

M I Martynov, A A Nikitin, A B Ustinov, and B A Kalinikos

Department of Physical Electronics and Technology, St. Petersburg Electrotechnical University, St. Petersburg, 197376 Russia

E-Mail: ustinov_rus@yahoo.com

Abstract. The complex transmission coefficient of active ring resonators based on ferrite-film delay lines was investigated both theoretically and experimentally. Influence of the parameters of the delay line on the transmission coefficients was investigated. It was shown that the resonant frequencies of the ring depend on the ferrite film thickness and the distance between spin-wave antennae. These dependences give possibility to control the shape of the transmission coefficient that in combination with magnetic tuning provide flexibility for microwave applications.

During the past two decades the increased research interest to active ring systems is observed [1, 2]. Nowadays they are widely used for study nonlinear physical phenomena such as solitons, modulation instability, and chaos, as well as for signal processing and generation [3]. Tunable microwave active ring resonators or oscillators can be easily fabricated with the use of delay lines based on the thin ferrite films such as epitaxial yttrium iron garnet (YIG) films.

Distinguished features of the YIG-film devices are low insertion losses and magnetic field tunability. Performance characteristics of such devises are determined by the features of the spin wave (SW) dispersion. It makes the YIG films attractive for applications in miniature tunable microwave devices. In particular, a narrow band filters [4, 5], matched filters [6], stationary and chaotic signal generators [7, 8, 9, 10] based on active rings have been developed and studied experimentally.

The aim of this work is theoretical and experimental investigation of microwave properties of the spin-wave active ring resonators based on the YIG films.

Schematic diagram of the resonator is shown in the Fig. 1.a. This circuity consists of spin-wave delay line, variable attenuator, microwave amplifier and two directional couplers for input and output signals. All elements of the scheme are connected consequently in order to form the closed loop. The spin-wave delay line is used to introduce time delay in the ring. The wideband microwave amplifier and the variable attenuator provide adjustable amplification of the microwave signal to compensate the losses in the delay line. We will call this amplification as a ring gain and denote it by letter *G*. We assume that the gain has no frequency dependence.

Consider now the formation of transmission coefficient of the active ring resonator. The transmission coefficient has multiresonance behavior at frequencies determined by the dispersion law in the delay line. For resonance frequencies phase shift of the microwave signal circulating in the ring is equal to $\Delta\varphi = 2\pi n$, where n is a number of the signal circulations. Therefore, the resonance condition for SW wave numbers is $k_n\left(f_{res}\right) = 2\pi n / d$, where $k\left(f\right)$ is a dispersion relation of the SW, d is the length of the delay line. Note that in the case of the spin wave delay line the delay of all other electrical interconnections is negligibly small. The formation of the transmission coefficient of active ring resonator illustrated in the Fig. 1.b. As is shown, resonant wave numbers $k_n(f_{res})$ define the resonance frequencies through the dispersion characteristic. Therefore, one of the advantages of the active ring resonators is the magnetic tunability.

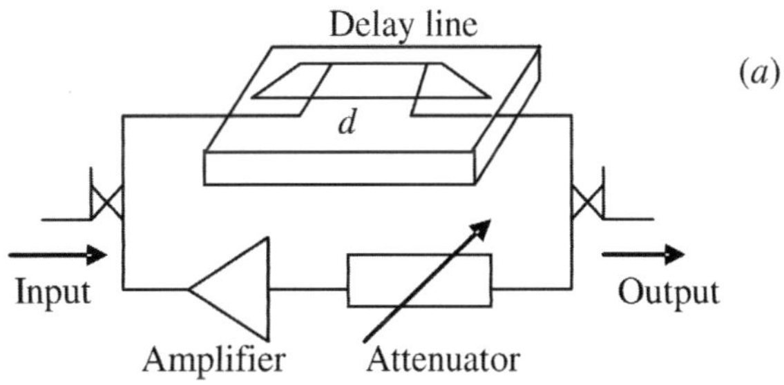

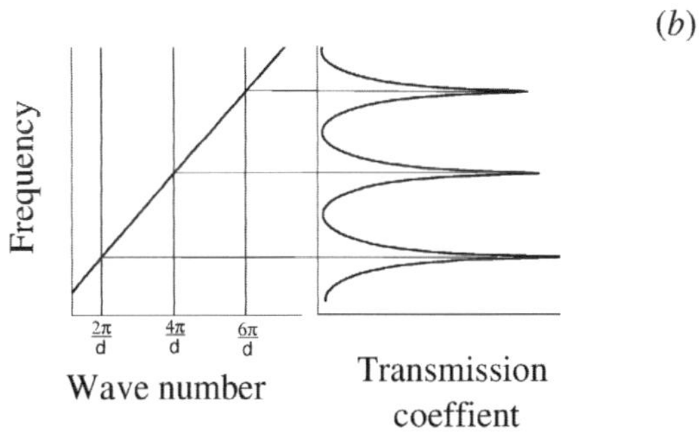

Figure 1. Block-diagram of an active ring resonator (*a*). Spin wave dispersion characteristic and AFC of the resonator (*b*)

For development of the theoretical model we assume that a monochromatic signal

$$\dot{A}_{in}(\omega) = A_0 e^{i\omega t} \tag{1}$$

is supplied to the input port. This signal circulates in the ring, as was described above. The output signal is a superposition of an infinite number of the circulating waves. As a result, an expression for the output signal after *m* circulations has the following form

$$\dot{A}_{out}(\omega) = A_0 \left[\sum_{m=1}^{\infty} e^{-i(k(\omega)-i\alpha(\omega))md} \cdot e^{mg} \right] e^{i\omega t} \tag{2}$$

where $k(\omega)$ is the wave number, $\alpha(\omega)$ is the damping decrement, and g is the amplifier gain coefficient. In order to derive generalized theory the SW dispersion relation $k_{sw}(\omega)$ and the damping decrement $\alpha_{sw}(\omega)$ will be used on the final step.

A complex transmission coefficient of the resonator circuit was found as a ratio of the output and input complex amplitudes, i.e.

$$\dot{H}(\omega) = \dot{A}_{out}(\omega) / \dot{A}_{in}(\omega), \tag{3}$$

In general case it can be written in the following form

$$\dot{H}(\omega) = \sqrt{H_p(\omega)} \exp[i\phi(\omega)], \quad (4)$$

where $H_p(\omega)$ is the power transmission coefficient and $\phi(\omega)$ is the phase-frequency characteristic of the active ring resonator. After summation and simple mathematical manipulations we obtain

$$H_p = \frac{1}{2} \frac{e^{g-\alpha(\omega)d}}{\cosh(g - \alpha(\omega)d) - \cos(k(\omega)d)}, \quad (5)$$

Expression for the phase-frequency characteristic is

$$\phi = \arctan\left[\frac{\sin(k(\omega)d)}{e^{g-\alpha(\omega)d} - \cos(k(\omega)d)}\right] \pm R\pi, \quad (6)$$

where $R = 0,1,2...$

It is important to note, that derived relations for the power transmission coefficient and the phase-frequency characteristic can be used for investigation of active ring resonators based on various delay lines. In the case of spin wave delay line the SW dispersion relation $k_{sw}(\omega)$ and the damping decrement $\alpha_{sw}(\omega)$ should be substituted in the Eq. 5 and 6, respectively.

The SW wave-number $k_{sw}(\omega)$ is calculated using an appropriate dispersion equation. For the surface spin waves used in this work the wave-number is calculated as [12]

$$k_{sw}(\omega) = -\ln\left(1 - 4\left(\omega^2 - \omega_H\left(\omega_H + \omega_M\right)\right)/\omega_M^2\right)/2L, \quad (7)$$

where L is a thickness of the YIG film, $\omega_H = |g|\mu_0 H_0$, $\omega_M = |g|\mu_0 M_0$, $|g| = 1.76 \cdot 10^{11}$ rad·s^{-1}·T^{-1} is a gyromagnetic ratio for an electron spin, $\mu_0 = 4\pi 10^{-7}$ H·m^{-1} is a vacuum permeability, H_0 is an external magnetic field, and M_0 is a saturation magnetization. The spatial damping decrement for the surface spin waves is calculated as

$$\alpha_{sw}(\omega) = 2\pi|g|\omega\Delta H \cdot L^{-1}\left(\left(\omega_H + \omega_M/2\right)^2 - \omega^2\right)^{-1}, \quad (8)$$

where ΔH is a half of the ferromagnetic resonance line-width.

Let us turn now to investigation of an experimental prototype of the active ring resonator. Its main elements were a microwave amplifier, the original microwave spin-wave delay line, and variable attenuator in order to adjust the gain in the ring.

The spin-wave delay lines are utilized a single-crystal YIG films grown on a gadolinium gallium garnet (GGG) substrate by liquid phase epitaxy. The films had a width of 2 mm, a length of 40 mm, various saturation magnetization M_0 of 1750, 1660, and 1780 G, and various thickness L of 5.7, 9.64, and 13.6 μm, respectively. Two 50 μm wide and 2 mm long short circuited microstrip antennae were used for excitation and reception of spin waves in the delay line. Original experimental prototype allows to investigate an influence of the distance d between the antennae on the transmission coefficient. The measurements were carried out at d equal 3, 6, and 9 mm. Antennae were fed by microstrip transmission lines of 50 Ohm characteristic impedance. The microstrip lines were fabricated with conventional photolithography on grounded alumina substrates with thickness of 500 μm. The YIG film stripes were positioned over the antennae with YIG film side in contact with the antennae.

The SW delay lines are placed between the poles of permanent magnet. The bias magnetic field H_0 in a range from 1226 Oe to 1330 Oe is applied in the plane of the YIG film and parallel to the antennae. This orientation provides excitation of surface spin waves in the YIG films.

Consider now the principle of operation of the SW active ring resonator. A microwave signal supplied to the input directional coupler begins to circulate in the ring. The signal losses Λ in the YIG delay line are compensated by the microwave amplifier. The signal amplification is controlled by a variable attenuator and is characterized by the ring gain G. The active ring resonator could operate as a passive microwave filter for G less than Λ or as a microwave oscillator, otherwise. A value of G, for which the ring breaks up to the auto-oscillation regime, may be used as a reference point. In the work, this G value is taken as a reference zero point.

Fig. 2 shows representative data on the frequency response of the active ring resonator measured for $G = -1.5$ dB, $L = 9.64$ μm, and $d = 6$ mm by the symbols. The well-pronounced spikes correspond to the resonant frequencies of the ring. Theoretical data for frequency response, shown by solid line in Fig. 2, well coincides with experimental characteristic.

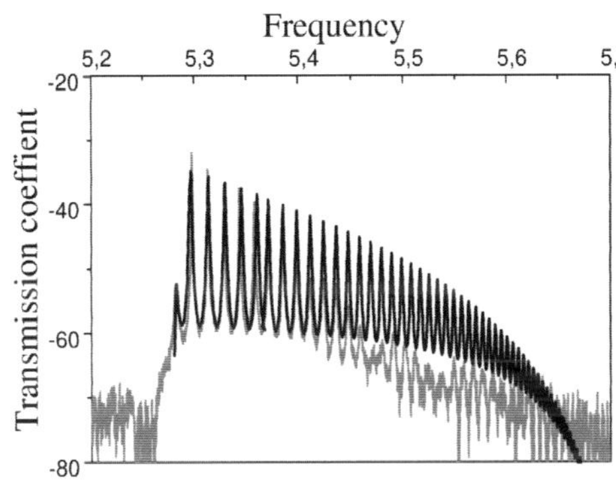

Figure 2. Power transmission coefficient for the active ring resonator based on the ferrite film with thickness of 9.64 μm and distance between antennae $d = 6$ mm.

From general point of view it is clear that transmission coefficient is defined by the phase shift of the signal in the ring $\varphi = k(\omega)d$. It can be considered also through delay time $\tau = d\varphi / d\omega$. Further, we will use the time delay in order to describe properties of the active ring resonators. As is clear only two parameters of the SW delay line define the time delay that are the distance between antennae d and YIG film thickness L. Influence of the both parameters are analyzed below.

Solid lines in Fig. 3 show theoretical dependences of the resonant frequencies from the distance between the antennae. Symbols show resonant frequencies measured for distance d of 3, 6, and 9 mm. As is seen increasing of the distance, otherwise increasing of the time delay, leads to decrease of a distance between adjacent resonant frequencies Δf (Fig. 4). This feature is one of the main advantage of the spin wave active ring resonator that allows to change a shape of the transmission characteristic.

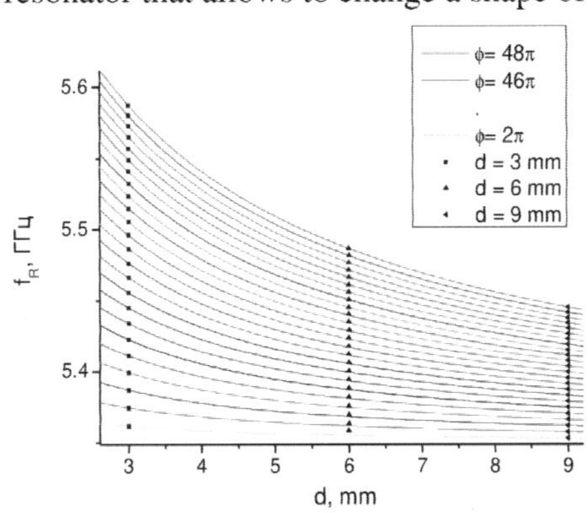

Figure 3. Dependences of the resonant frequencies of the active ring resonator from the distance between SW antennae d for the ferrite film thickness of 5.7 μm.

Second factor defining the resonance frequency position is the film thickness L. As is seen from the dispersion equation (7), the spin-waves in the ferrite films with various thickness has the different dispersion laws. As a result, positions of the resonant frequencies and range between adjacent resonance peaks Δf are different. Theoretical dependences of Δf values on frequency in ferrite films with different thicknesses calculated with Eq. 5 are shown by solid lines in Fig. 5. Experimental points are shown by symbols. The distance between antennae d for Fig. 5 equals 6 mm. Fig. 4 and 5 show a constructive flexibility of SW active ring resonators for creation of matched filters.

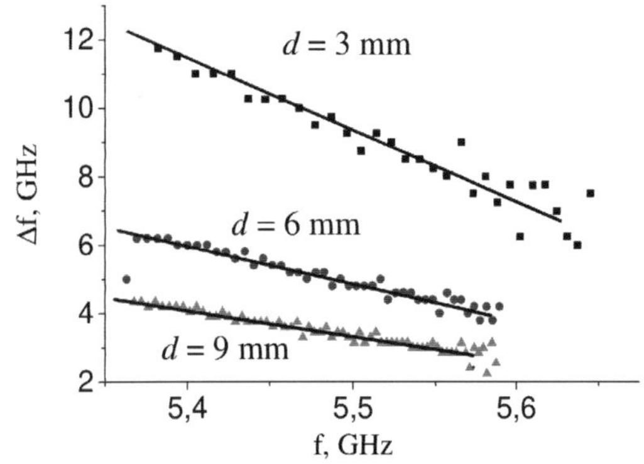

Figure 4. Dependences of the Δf from the frequency for the ferrite film with 5.7 μm thickness.

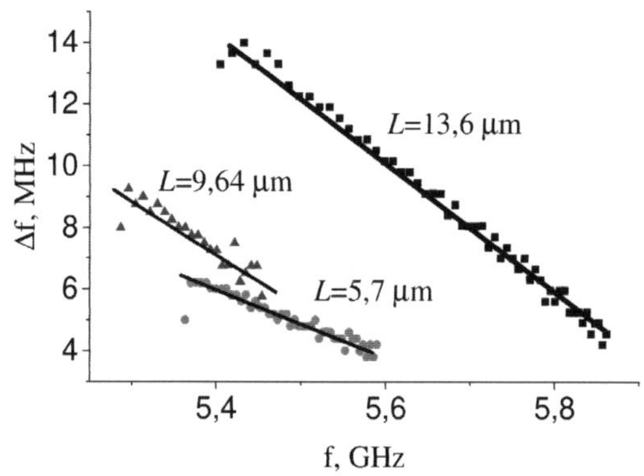

Figure 5. Dependences of the Δf from the frequency for distance between antennae d = 6 mm.

Variation of the bias magnetic field from 1226 Oe to 1330 Oe provided the tuning range for lowest resonant frequency of 5.4-5.7 GHz. In this range shift of the resonant frequencies are described by the dependences of dispersion relation versus magnetic field. Quality factor of the single resonant peak depends on the ring gain and it reaches maximum at $G = 0$.

Experimental dependences of the quality factor of the most pronounced resonance peak on the distance for three values of the film thickness are shown in the Fig. 6. The maximum quality factor measured for 13.6 μm film at 9 mm between antennae reaches 25 000. A distance between antennae d makes a primary contribution to the delay time. Increase of the delay time provides higher quality factor. However, due to higher values of the ferromagnetic resonance line-width in the thinner film quality factor decreases with the thickness.

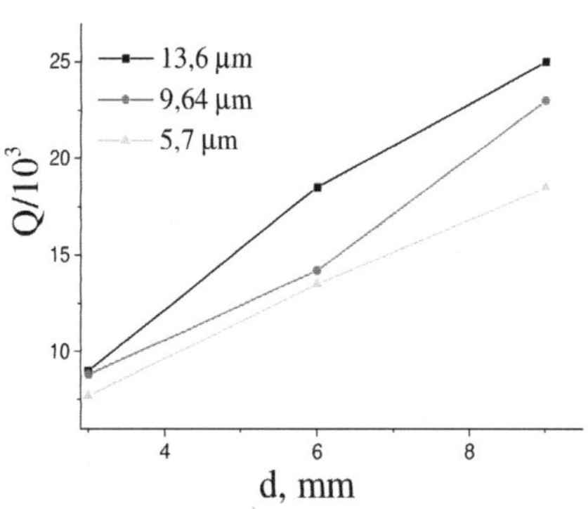

Figure 6. Dependences of the quality factor on the distance between antennae.

In conclusion, the theoretical and experimental investigation of the active ring resonator based on SW delay line was carried out. Transmission coefficient and phase-frequency characteristic of the active ring resonator were derived. Theoretical dependences for frequency response well coincided with experimental results. The quality factor of the resonator was about $2 \cdot 10^4$. Therefore the active ring resonators are promising for microwave applications, in particular for comb-frequency filtering.

Acknowledgments

The theoretical work was supported in part by Ministry of Education and Science of Russian Federation (Project "Goszadanie"). The experimental work was supported by Russian Science Foundation (Grant# 14-12-01296).

References

[1] Kivshar Yu and Agraval G 2003 Optical solitons: From fibers to photonic crystals *Academic Press, Amsterdam*

[2] Wu M 2011 *Solid State Physics Vol. 62*

[3] Grishin S V, Sharaevskii Y P, Nikitov S A, Beginin E N, Sheshukova S E, 2011 *IEEE Trans. Mag*. **47** 3716

[4] Demidov V E , Kalinikos B A , Kovshikov N G , Edenhofer P 1999 *Electron. Lett.* Vol. **35** 1856-1857

[5] Ustinov A B, Srinivasan G, Kalinikos B A 2008 *Appl. Phys. Lett.* Vol. **92** 193512

[6] Porokhnyk A A, Ustinov A B, Kovshikov N G and Kalinikos B A 2009 *Technical Physics Letters*, ***35*(9)** 843-846

[7] Demidov V E and Kovshikov N. G. 1997 *Journal of Experimental and Theoretical Physics Letters*, ***66*(4)** 261-265

[8] Scott M M, Kalinikos B A and Patton C E 2001 *Applied Physics Letters* ***78*(7)** 970-972

[9] Grishin S V, Beginin E N, Morozova M A, Sharaevskii Y P and Nikitov S. A. 2014 *Journal of Applied Physics* ***115*(5)** 053908

[10] Kondrashov A V, Ustinov A B and Kalinikos B A 2010 *Technical Physics Letters*, ***36*(3)** 224-227

[11] Kalinikos B A, Scott M M and Patton C E 2000 *Phys. Rev. Lett.* **84**, 4697

[12] Stancil D D 2009 and Prabhakar A 2009 Spin waves *Springer*

Journal of Physics: Conference Series **769** (2016) 012074 doi:10.1088/1742-6596/769/1/012074

Electronic Structure of PbSe Nanowires

I D Avdeev and M O Nestoklon
Ioffe Institute, 194021 St. Petersburg, Russia
E-mail: ivanAvdeev505@gmail.com

Abstract.
We present the tight binding calculations of the lead selenide nanowires: energy spectra of quantum confined states as a function of nanowire radius, dispersion in the full Brillouin zone, and the radial part of local electronic state density, which helps us to recognise valley splitting in the spectra. Also, we compare our results to KP perturbation theory predictions.
We show that the value of the valley splitting is comparable with the distance between two levels of size quantization and that it strongly depends on the arrangement of the atoms in the wire.

1. Introduction
The electronic structure of Lead Selenide nanocrystals (NC) was studied a lot by a variety of different methods, including KP perturbation theory, Density Functional Theory (DFT), and semi-emperical methods like a tight-binding or pseudopotential. However, there are few works dedicated to study intertervalley energy splitting. In our work we will show that the contribution of an intervalley spliting is large and one can not ignore it in the calculations.

The revival of interest for these material is explained by the fact that wide range of nanostructures can be synthesized by modern chemistry methods. The stusy of quantum confinement effects opens an access to new industrial applications, despite the fact that PbSe is widely used material in optoelectronic devices such as lasers, detectors, etc. Strong quantum confinement effects allows for a wide range of operating wavelength $[1 \div 5]\mu$.

Unfortunately, the intervalley interaction is not well studied, it depends on NC symmetry, number of atomic layers, NC size, etc., and is strongly strongly suppressed in structures with reduced symmetry [1]. To analyse how symmetry affect on the spectra we studied different types of wire. Since there is no exact theory that can describe the puzzling behaviour and predict the valley-mixing constant. To study the intervalley splitting in more details, we use a tight-binding (TB) calculations of the electronic structure of a PbSe nanowires.

2. Structure, and PbSe details
PbSe has the rocksalt crystal structure with the lattice constant about 6.1Å. The bulk crystal has the O_h^5 symmetry group. It is a narrow direct band gap semiconductor ($E_g \approx 0.28eV$ at $300K$) with the band extrema at the L point of the Brillouine zone. Group of the wave vector at L point is D_{3d}. All lead chalcogenides have a similiar bulk dispersion and lattice structure, in our study we consider only PbSe, but qualitatively the results are similar for other materrials from this group.

In the work we study cylindrical wires. The point group of a periodical structure is defined by the point group of the elementary cell of the structure. We define the wire as an array of atoms which lay inside an ideal cylinder. The cylinder is determined by the radius, the axis direction and by the position of its center in the bulk semiconductor.

In general case of an arbitrary axis directoin and the position of cylinder center, the resulting symmetry of the wire would be low. However, we are able to construct highly symmetric wires. We set the wire axis direction along the [111] L valley. Then we consider three wire types with different microscopic structure. The first type is obtained by placing the center at an atom, the second is by placing the center between atoms and the third is by placing the center at the middle of a chemical bond, see Fig. 1. The surface of a wire, which is very close to a circle for large radius, is discretized on an atomic level. As a result, the arrangement of atoms at the wire surface is different and the microscopic symmetry will be different in three cases.

We notice that the position of the center does not always coincide with inversion center of a wire. Moreover, the second type of wire has no inversion center. This fact is really interesting since the lack of inversion may produce k-dependent spin-splitting, which we are able to calculate.

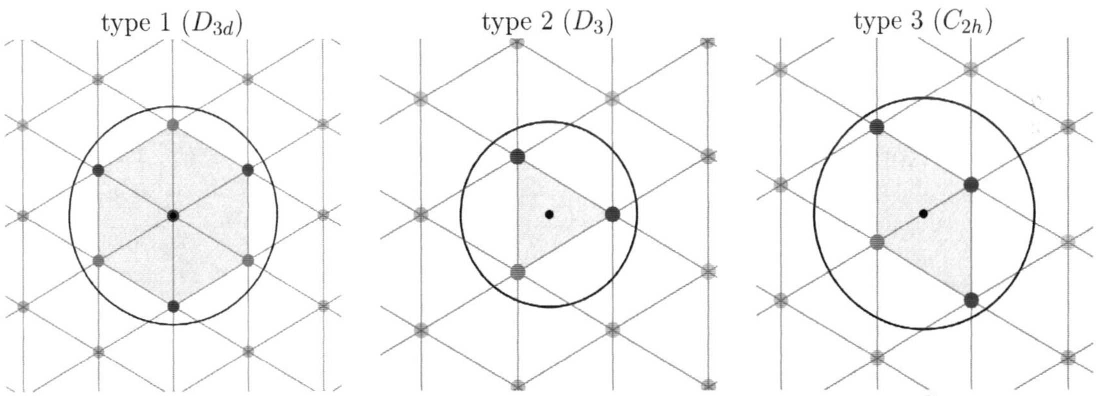

Figure 1. (Color online) Atomic arrangement for the three different types of wire. The red dots are Pb atoms, the blue ones are Se, solid circle is the border of an ideal cylinder, small black dot is the wire axis. The lattice in the figures is formed by three atomic planes (000), (111), (222) of the bulk crystal, as seen from [111] direction. Real atomistic structure within the ideal cylinder is highlighted.

The choice of the growth direction of the wire along the L valley leads to valley splitting of four equivalent L valleys in one longitudinal and three equivalent transverse due to mass anisotropy, which are quite easy to recognise. Because there are no irreducible representations in these groups with dimension greater then two, it is natural to expect futher splitting of the valleys by the wire interfaces with expected degeneracy $1 + 2 + 1$ for the ground state.

3. Details of the calculations

In our work we use the tight-binding method with the KP calculations for comparison.

3.1. TB method

Tight-binding method is a widely used semi-empirical method. It gives good results in a reasonable computational time and it is quite simple to implement (compared to *ab initio* methods). We use this advantage to calculate the wires with a wide radius range.

The method is based on decomposition of the electron Shrödinger equation in the local basis of Löwdin molecular orbitals [2] – orthogonalized atomic-like orbitals. The Hamiltonian for periodical structures in tight-binding method writes as:

$$H_{ml,m'l'}(\mathbf{k}) = \sum_j e^{i(\mathbf{R}_j+\mathbf{r}_l-\mathbf{r}_{l'})\mathbf{k}} \langle \phi_{ml}(\mathbf{r}-\mathbf{r}_{jl})|H|\phi_{m'l'}(\mathbf{r}-\mathbf{r}_{jl'})\rangle \ , \qquad (1)$$

where $\mathbf{k}$ is the wavevector, m is the orbital index, j is the primitive cell index, l, l' are atom indexes, $\phi_{ml}(\mathbf{r})$ is a Löwdin orbital.

To calculate the energy spectrum we have to solve an eigenproblem for a Hermitian sparse matrix of size $[N_{atoms}N_{orbitals} \times N_{atoms}N_{orbitals}]$ at each k point of interest. Tight-binding approximation restrict the amount of neighbors in hamiltonian matrix with the coordination number (6 for cubic lattice) and the hamiltonian matrix takes a band form, hence computational difficulty is only about $\mathcal{O}(N_{eigenvectors} \cdot N_{atoms} \cdot N_{orbitals})$. For instance, elementary cell of 20 Å has about 500 atoms, and H_{tb} size is about $[10000 \times 10000]$.

For our work we use $sp^3d^5s^*$ parametrization with spin-orbital interaction applied on p orbitals. Parameters were fitted [1] to accurately reproduce experimental effective masses at the L point and energies of special points in Brillouin Zone obtained in GW approximation. Then this set of TB parameters was also used in the KP approximation to compare two different approach.

3.2. KP method

This is the most popular method for calculating an energy spectra of nanocrystals. The method is in a reasonable agreement with experimental data for large NCs sizes. The main advantage of the method is the possibility to obtain the analytical solution in most cases.

All latest KP models describing PbSe [3, 4] are based on four-band envelop function model [5] with the following Hamiltonian for each L valley:

$$\hat{H}(\hat{\mathbf{p}}) = \begin{pmatrix} \left(\frac{E_g}{2} + \frac{\hat{p}_z^2}{2m_l^-} + \frac{\hat{p}_\perp^2}{2m_t^-}\right)\hat{U}_2 & \frac{P_l}{m_0}\hat{p}_z\hat{\sigma}_z + \frac{P_t}{m_0}\left(\hat{p}_\perp\hat{\sigma}_\perp\right) \\ \frac{P_l}{m_0}\hat{p}_z\hat{\sigma}_z + \frac{P_t}{m_0}\left(\hat{p}_\perp\hat{\sigma}_\perp\right) & -\left(\frac{E_g}{2} + \frac{\hat{p}_z^2}{2m_l^+} + \frac{\hat{p}_\perp^2}{2m_t^+}\right)\hat{U}_2 \end{pmatrix},$$

where $\hat{U}_2$ is 2×2 unit matrix, m_0 is a free electron mass, σ are Pauli matrices, and $P_{t(l)}$ is the transverse (longitudinal) momentum matrix component. KP parameters (momentum matrix, masses, E_g) were fitted to the dispersion near L point (namely, its second and fourth derivatives) calculated in the TB approximation. Usually, the KP hamiltonian is solved in perturbation theory $H = H_{symm} + H_{an}$, $|H_{an}| \ll H_{symm}$ by first finding a quantum confined solutions of spherically symmetric part of the KP hamiltonain H_{symm} and treating anisotopic part as a perturbation. It was shown that the asymmetric part is significantly smaller that the main term [6, 4] for dots and wires, especially for ground states. Despite that wires have the cylindrical symmetry, but they can be considered with the spherically symmetric hamiltonian [3]. The spherically symmetric hamiltonian produces fourfold degenerated states for all types of wire. More advanced cylindrical KP model allows to estimate valley splitting with respect to a mass anisotropy [4].

However, there is an additional intervalley splitting due to mixing at the interfaces, which can not be reproduced without significant modifications of KP model [1].

4. Results

Below we present the results of tight-binding calculations: the dependence of the quantum confined levels on a wire radius compared with KP results, the spin splitting in the wires without spacial inversion center and the structure of local density of states for different quantum confined levels.

4.1. Quamtum confimenet in the wires

One of the main result is a energy of quantum confined in the wires, calculated for different arrangement of atoms within the wires (1, 2, 3). We show only two of them, because the third type is very similar to the second. Tight-binding calculations are mapped on KP resuts with the same parameters. A thin line at the middle of the energy scale for both of those Figures 2, 3 shows $\pm\frac{E_g}{2}$, assuming zero energy corresponds to middle of the band gap.

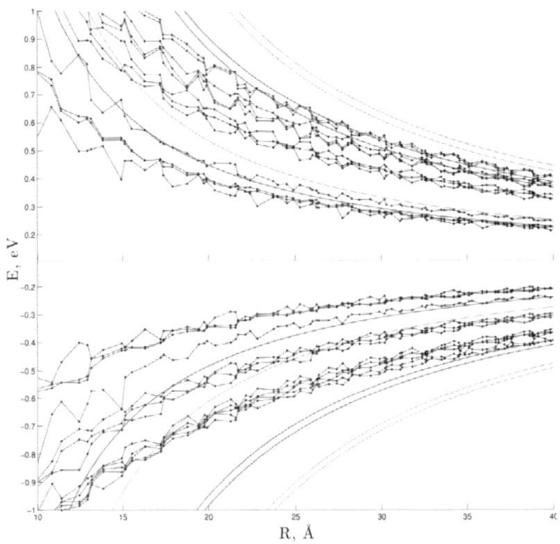

Figure 2. (Color online) Energy of quantum confimed states as a function of wire radius, the D_{3d} wire ("type 1"),

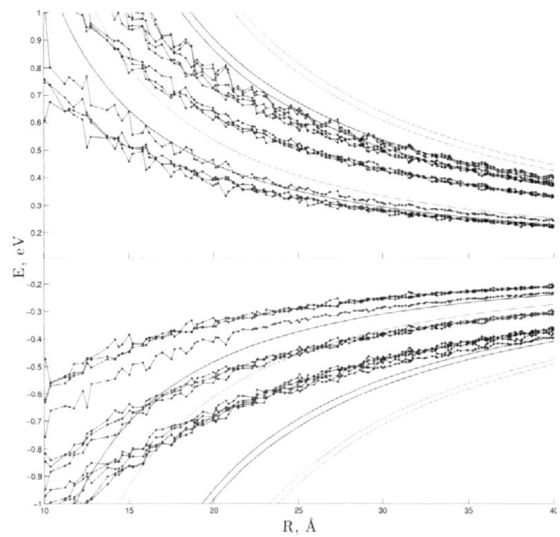

Figure 3. (Color online) Energy of quantum confimed states as a function of wire radius, the D_3 wire ("type 2").

TB calculations, KP singlet states, KP triplet states

There are 12 first subbands in conduction and valence bands calculated at the L point of the one-dimensional Brillouin zone, defined as follows:

$$k_L = \mathbf{k}_L \cdot \mathbf{n}$$

All the states in the KP theory have two quantum number n, m. Wavefuncitons of electrons localized in the wire is bessel functions $J_m(r\lambda_n)$, with the quantum number m corresponding to orbital momentum and n is the number of Bessel function roots in the interval $(0, R]$. Following [7, 3] there are three bulk subbands corresponded to $\{m_{\frac{1}{2}}, n_1\}$ (ground state), $\{m_{\frac{3}{2}}, n_1\}$ (first excited), $\{m_{\frac{1}{2}}, n_2\}$ (second excited).

Fourfold degenerated bulk states are splitted by mass anisotropy into triplet and singlet, which can be clearly seen for the ground state energies in Figs. 2,3. For the first and second excited states there are no such obvious observations. Because of strong intervalley interaction these states can be shuffled. As a general rule, which has been explained in [1] for the quantum dots, there is strong symmetry-induced suppression of intervalley splitting in the wires of lower symmetry.

The analyzis of the valley splittings is hindered by the levels mixing and the fact that the value of valley splittings is compared with the distance between levels with different quantum numbers n, m. We were unable to find a simple criterion to extract KP quantum numbers from TB calculations, the only reliable approach is to analyze the spacial structure of the wavefunctions computed in TB method, see below.

Journal of Physics: Conference Series **769** (2016) 012074 doi:10.1088/1742-6596/769/1/012074

4.2. Dispersion

Dispersion in the half-Brillouine zone, calculated in the TB method for 20Å wires is shown in the Figures 4, 5.

It is clear that for the ground states of the "type 1" wire effective masses can be calculated just as second derivative. However, the excited states are repulsed by k-dependent mixing and a special procedure should be used to remove all interaction effects. Another striking effect is that for the second wire type there is no Kramers degeneracy. It is well known that it may be lifted in the structures without inversion center. This effect is linear:

$$\Delta E_{spin} = \alpha(k - k_L), \tag{2}$$

with the spin-splitting constant $\alpha(R)$ that is highly oscillating similar to intervalley splitting. Due to large spin-orbit interaction in PbSe (larger than the band gap in this material), the values of the spin splitting is also large.

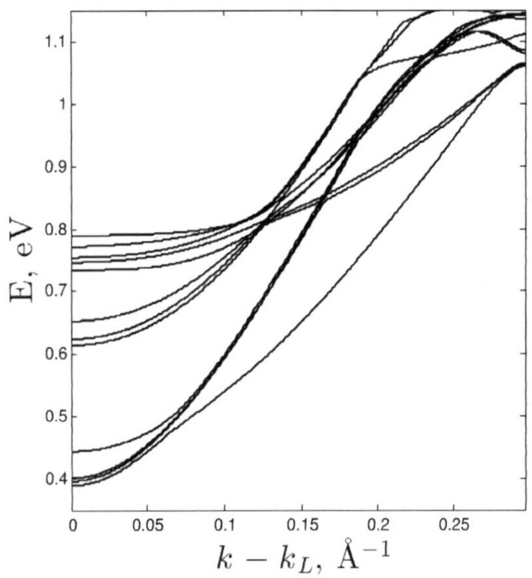

Figure 4. Energy dispersion in the half-Brillouine zone in the 20Å wire of "type 1",

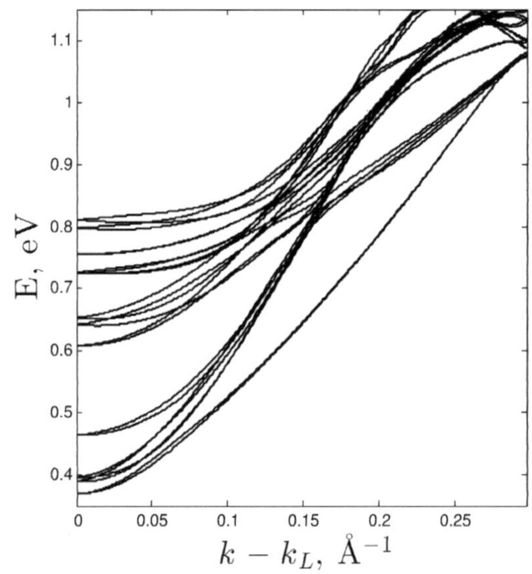

Figure 5. Energy dispersion in the half-Brillouine zone in the 20Å wire of "type 2".

4.3. Wavefunctions

To compare the states calculated in TB model with KP results, we used the analysis of local density of states (LDOS). Electron density (especially for big wire radiuses) is in a qualitative agreement with the density of confined states calcualted in the KP approximation. It allows us to classify the states in accordance with their valley structure and quantum numbers. Electron density can be calculated as follows:

$$|\Psi(j)|^2 = \sum_l |V_c(j,l)|^2 + |V_a(j,l)|^2, \tag{3}$$

Where j is the atomic index in (111) plane (i.e. cartesian coordinates $\{x, y\}$), l is orbital index, $V_{c(a)}$ is an eigenvector value at a cation or an anion.

There is a qualitative difference between the first and the second excited states. The lowest momentum number $m_{\frac{1}{2}}$ has a non-zero LDOS at the center of a wire, which allows us simpy distinguish such states.

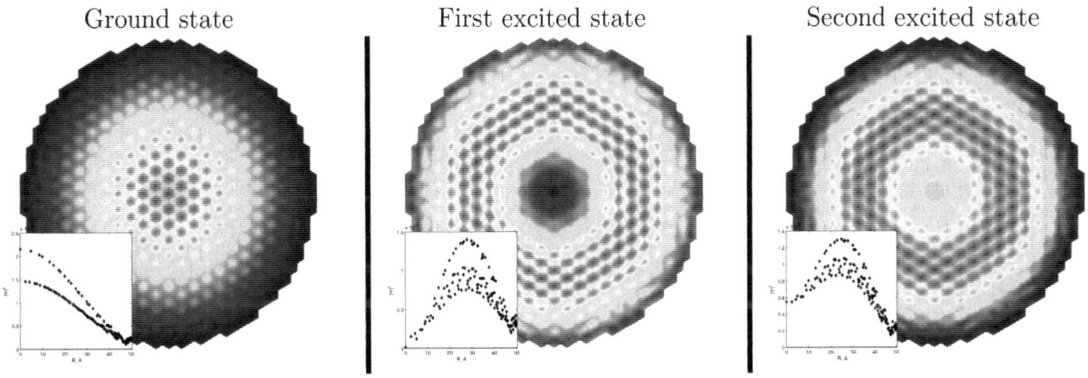

Figure 6. (Color online) LDOS of first three confined states in the 50Å wire of "type 1".

5. Conclusion

Our TB calculations are in good agreement with KP theory for big wire radiuses, but the splitting of the valley triplet is the result of an surface-induced intervalley mixing and can't be predicted by a KP theory. The value of intervalley splitting is large which menas that KP theory may not be used for the quantitative analysis of electron states in PbSe wires.

The effect of an intervalley splitting on excited states is even more pronounced than for the ground level as the order of the levels is changed in comparison with KP predictions. We stress that this effect is important for the analysis of optical spectra in such structures: as shown in [8], optical transitions from a ground valence to excited conduction states are about 30% of transitions between ground states.

Also, we show that in the wires without spacial inversion the linear spin splitting may be observed and its value is giant in comparison with other semiconductors.

References

[1] Poddubny A N, Nestoklon M O and Goupalov S V 2012 *Phys. Rev. B* **86** 035324
[2] YU P and Cardona M 2010 *Fundamentals of Semiconductors: Physics and Materials Properties* 4th ed (Springer-Verlag Berlin Heidelberg)
[3] Goupalov S V 2011 *Phys. Rev. B* **84** 037303
[4] Bartnik A C, Efros A L, Koh W K, Murray C B and Wise F W 2010 *Phys. Rev. B* **82** 195313
[5] Dimmock J O and Wright G B 1964 *Phys. Rev.* **135** A821–30
[6] Kang I and Wise F W 1997 *J. Opt. Soc. Am. B* **14** 1632–46
[7] Goupalov S V 2013 *J. Phys. Chem. C* **117** 6476
[8] Goupalov S V 2009 *Phys. Rev. B* **79** 233305

Microwave radiation absorption by 2D-electrons in the type II composite InAs/GaSb/AlSb quantum wells in a magnetic field

I V Kochman[1], A I Veinger[1], M P Mikhailova[1], P V Semenikhin[1], K V Kalinina[1], R V Parfeniev[1], V A Berezovetz[1], A Hospodková[2], J Pangrác[2] and E Hulicius[2]

[1] Ioffe Physical-Technical Institute, 194021, Politkhnicheskaya, 26, St. Petersburg, Russia

[2] Institute of Physics CAS CR, v.v.i., Cukrovarnická 10, 162 00, Prague, Czech Republic

E-mail: kochman@mail.ioffe.ru

Abstract. Strong Shubnikov – de Haas (SdH) oscillations were observed in the derivative of microwave absorption (f = 10 GHz) in the InAs/GaSb/AlSb composite quantum wells (CQWs) using electron-paramagnetic-resonance spectroscopy at low temperatures (2.7–20 K) and in the magnetic field up to 14 kOe. CQWs were grown on the n-GaSb:Te(100) and n-InAs:Mn(100) substrates with various width of QWs by MOVPE. Predominance contribution of the bulk n-GaSb substrate in SdH oscillations was manifested. Two frequencies of the SdH oscillations were found from Fourier analysis, which is connected to warping of the Fermi surface of GaSb. Unusual angular indicatrix was observed in dependence on orientation of the samples in the magnetic field. Obtained results can be explained by inversion asymmetry, which is a feature of the substances with lack of inversion centres. For CQWs grown on n-InAs:Mn ($n_s = 1.1 \times 10^{17}$ cm^{-3}) substrate, only several SdH oscillations with higher period were observed. Taking into account isotropic Fermi surface of bulk InAs, we succeeded to extract a contribution of the 2D carriers of InAs QW $\sim H_{\perp}$, where $H_{\perp} = H_{const} cos\Theta$, from bulk substrate oscillations using special spline interpolation from angular dependence of SdH oscillatory amplitudes in the angle range 0–90°. 2D electron concentration in the InAs QW $n_s \approx 1 - 3 \times 10^{11}$ cm^{-2} was evaluated from oscillatory period.

1. Introduction

Magnetotransport, spin-dependent and topological properties of the composite quantum wells (CQWs), based on the type II broken-gap InAs/GaSb/AlSb heterojunctions are actively studied in the last decades [1-6]. These structures are promising for spintronics and nanoelectronics side by side with HgTe [7] and graphene [8].

The unique properties of the type II broken-gap InAs/GaSb heterojunctions are due to the conduction band of InAs lying 150 meV lower than the valence band of GaSb [9, 10]. Electrons and holes are spatially separated, but they can tunnel between the wells. This allows the coexistence of closely separated electron (InAs) and hole (GaSb) two-dimensional gases. It gives the possibility to observe mixing conduction and valence band states and their hybridization under applied electric or magnetic fields [6]. In the narrow QWs (a_e < 85 Å), electron level E_1 is always higher than heavy hole level H_1 and we obtain normal semiconductor regime [2]. In wider well electron level E_1 will be lower than hole level H_1, so the band energy diagram of the QW will be inverted, i.e. E_{g0} is negative [1, 2]. If the QWs are undoped, there will be only electrons in such semimetal InAs QW due to their transfer from GaSb valence band which accounts for Coulomb interaction.

Previously magnetotransport, optical and spin-dependent phenomena were studied on the type II broken-gap GaInAsSb/p-InAs heterojunctions with self-consistent QWs grown by liquid-phase epitaxy (LPE) with sharp heterointerface and high mobility electron channel at the interface. At low temperatures, hybridization of the energy bands has been observed in the wide range of magnetic fields. For the first time the quantum Hall effect was observed and semimetal-insulator transition was obtained at $H_{const} > 16$ T in such broken-gap structures [10-13].

We report here the absorption of microwave radiation power and the observation of Shubnikov-de Haas (SdH) oscillations using electron-paramagnetic resonance spectroscopy (EPR) on InAs/GaAs/AlSb CQWs grown by metalorganic vapor phase epitaxy (MOVPE) on the lattice-matched GaSb and InAs substrates.

2. Experimental

It is known that InAs, GaSb and AlSb belong to 6.1 Å family with near equal crystal lattices [14]. It allows growing up lattice-matched structures. The InAs/GaSb/AlSb CQWs were grown in Laytec EPIRAS 200TT equipped AIXTRON 200 machine by low-pressure MOVPE at 560 °C on n-GaSb:Te(100) ($n_s = 3 \times 10^{17}$ cm^{-3}) and on n-InAs:Mn(100) ($n_s = 1.1 \times 10^{17}$ cm^{-3}) substrates. Sb-rich interfaces between InAs and GaSb QWs were realized by special switching sequence of metalorganics [15, 16]. QW materials and thicknesses were chosen to create semimetal structures, according to [2]. Several structures with different QW widths were prepared. A buffer layer (30 nm) was grown on the substrate followed by undoped InAs QW ($a_e = 15$ nm or 12.5 nm) and GaSb QW ($a_h = 10$ nm or 8 nm) surrounded on both sides by AlSb barriers (30 nm). Thin GaSb cap layer (3 nm) was grown on the top of the structure to prevent AlSb oxidation. The growing processes were in-situ controlled by reflectance anisotropy spectroscopy (RAS) [16]. The InAs QWs widths were chosen to get an inverted band structures (figure 1).

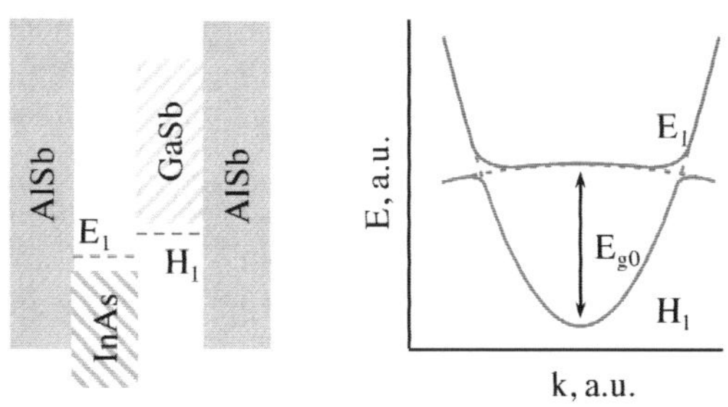

Figure 1. Energy band diagram of the inverted InAs/GaSb/AlSb structure. InAs well width $a_e = 125$ Å, GaSb well width $a_h = 80$ Å, band gap $E_{g0} = 150$ meV.

The calculated positions of E_1 and H_1 quantum levels in the InAs/GaSb/AlSb QW for the InAs well width $a_e = 12.5$ nm and GaSb $a_h = 8$ nm were the following: $E_1 = h^2/8m_e a_e^2 = 100$ meV, $E_{H1} = E_{g0} - h/8m_h a_h^2 = 136$ meV. Here $E_{g0} = 150$ meV is inverted band gap, $m_e = 0.023m_0$ and $m_h = 0.41m_0$. So, H_1 level lies higher than E_1 level in QW. Figure 2 demonstrates schematic structures of both CQWs grown on n-GaSb and n-InAs substrates, respectively.

Microwave radiation power absorption with frequency ~10 GHz and SdH oscillations were studied in the QW grown on n-GaSb and n-InAs substrates by EPR technique at low temperatures ($T = 2.7$-20 K) in the magnetic fields up to 14 kOe. The EPR-spectrometer E-112 VARIAN with helium gas flow cryostat Oxford Instruments E-910 was used [17]. The EPR technique is contactless and the deposition of Ohmic contacts on samples is not required [18]. The samples of 5 × 3 × 0.3 mm^3

were placed in the cavity and the outside magnetic field was applied. The microwave absorption power derivative (*dP/dH)* was registered in the experiments.

Magnetic field dependence of the cyclotron frequency (ω_c), amplitudes of the SdH oscillations and their temperature dependence were studied. Angle indicatrix was determined by rotation of the samples in the magnetic field. Figure 3 demonstrates the position of the samples with regard to constant magnetic field (H_{const}) and alternate current (I_{alter}) which arises in the induction from microwave radiation field (H_{alter}). Microwave power $P \sim 1$ mW (for H_{alter}).

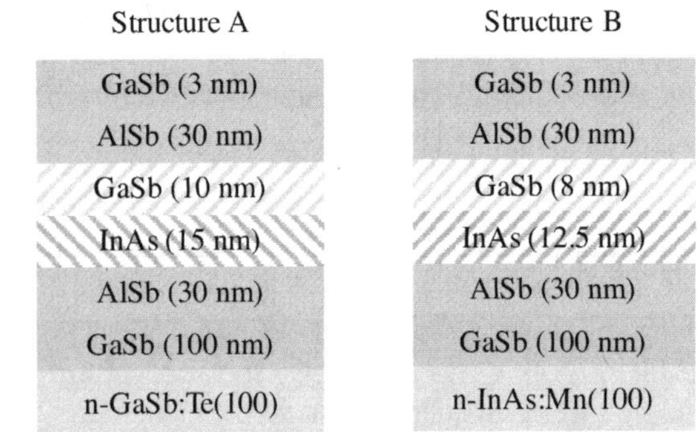

Figure 2. Schematic structure of the CQWs grown on n-GaSb:Te(100) substrate ($n_s = 3 \times 10^{17}$ cm^{-3}) and n-InAs:Mn(100) substrate ($n_s = 1.1 \times 10^{17}$ cm^{-3}).

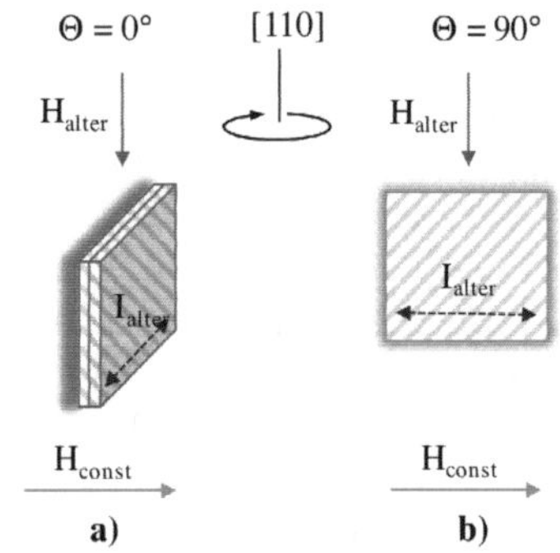

Figure 3. Position of the QW plane relative to magnetic field H_{const} at various angles Θ of the sample rotation. a — $\Theta = 0°$, b — $\Theta = 90°$.

3. Results and discussion

3.1. InAs/GaSb/AlSb CQWs grown on n-GaSb substrate

At first we are going to describe experimental results obtained for InAs/GaSb/AlSb CQW grown on n-GaSb:Te(100) substrate with carrier concentration $n_s = 3 \times 10^{17}$ cm^{-3}. We observed strong SdH oscillations with small period at $T = 2.7$ K. In the range of 4-14 kOe, there were more than 20-50 of them which were attenuated at T near 15-20 K (figure 4). Microwave current induces in the the microwave field. Magnetoresistance changes from traverse one to longitudinal one. Usually, the last of them is much less than traverse magnetoresistance. Oscillations of the microwave power derivative correspond to the transverse magnetoresistance $\Delta\rho/\rho_x$. Longitudinal magnetoresistance at H_{const}|| of the sample plain was not registered. The field of the cyclotron resonance H_c was evaluated from the dependence of filling factor on the reverse magnetic field. In our case we observed n = 25 to 60 Landau levels and $H_c = 39$ T was obtained.

Fourier analysis of the SdH oscillations allows us to find 2 frequencies which can belong to the warping of the Fermi surface of the n-GaSb [19]. Rotation of the sample A (CQW on the n-GaSb:Te(100) substrate) in the magnetic field shows unusual angular dependence (figure 5). It is not corresponding to expected $H_{const}cos\Theta$ dependence which is a feature of 2D-carriers. Amplitude oscillations decay with the angle change from 0 to 60° (relates to [111] orientation) and then increase up to 90°. Then we can see new decay or nodes of the amplitude intensity at $\Theta = 130°$ (orientation [110]) and amplitude increase at $\Theta = 180°$. We can explain these results by the main contribution of n-GaSb substrate to the SdH oscillations. This effect may be connected with the bulk inversion asymmetry (BIA) which is inherent to III-V semiconductors with zinc-blende structure with the lack of the inversion centers [19-21]. Recently the effect of the inversion asymmetry was observed experimentally in n-GaSb with concentration of $n_s > 10^{18}$ cm^{-3}, in n-HgSe [22] and n-InSb [23].

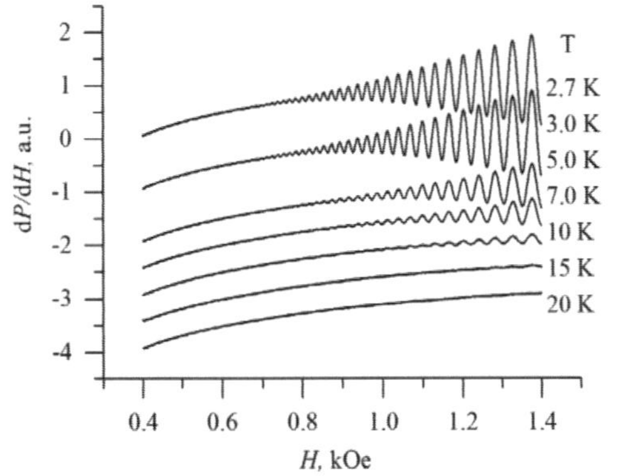

Figure 4. SdH oscillations amplitudes of the derivative of microwave radiation power dP/dH for the sample A grown on n-GaSb substrate, $T = 2.7$ K. $\Theta = 60°$

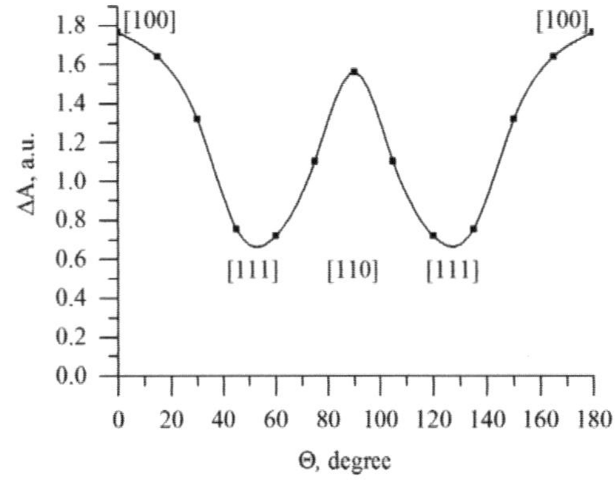

Figure 5. Angular indicatrix of the SdH oscillation amplitude by rotation of the sample A in the magnetic field. $H_{const} = 14$ kOe, $T = 2.7$ K. Orientation of the crystal is presented in the figure.

Inversion asymmetry obeys the spin-splitting of the conduction band in semiconductors with warping of Fermi surfaces and was theoretically calculated in [20-22]. A value of the spin-splitting of the conduction band is very small, less than ~1% and it is about $C_2E_F = 0.005$ meV (as it was found in [24]) for n-GaSb with concentration of $n_s > 10^{18}$cm^{-3} and $E_F = 0.095$ eV, where parameter C_2 is equal to 0.005. For n-GaSb with $n_s \approx 3 \times 10^{17}$ cm^{-3} and $E_F = 125$ meV, the value of the conduction band spin-splitting can be evaluated as $C_2E_F = 0.006$ meV.

3.2 InAs/GaSb/AlSb CQWs grown on n-InAs substrate

The microwave radiation absorption and SdH oscillations were measured for the sample B with semimetal InAs/GaSb/AlSb QW (figure 1, 2), grown on the n-InAs:Mn(100) substrate. In this case, only several wide SdH oscillations were observed (about seven) which corresponded to this sample in the magnetic field range of 4-14 kOe (figure 6).

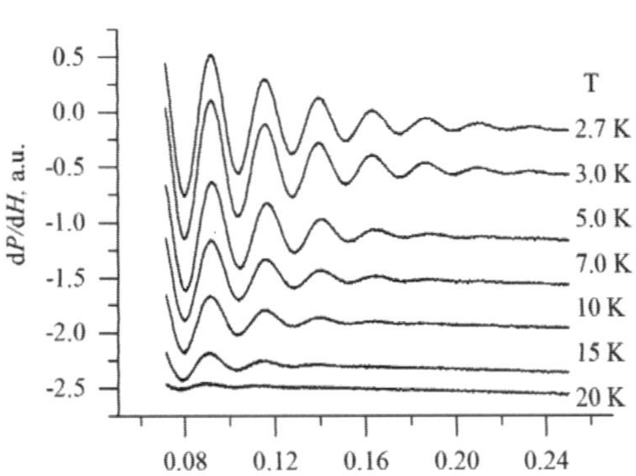

Figure 6. SdH oscillations of the microwave power absorption derivative versus reverse magnetic field at various temperatures for the sample B (n-InAs:Mn(100) substrate). $\Theta = 90°$

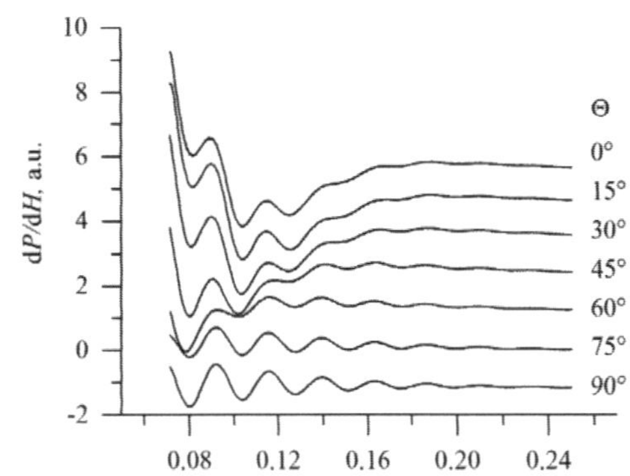

Figure 7. Angular dependence of SdH oscillations of the microwave power derivative by various orientations of the sample B in the reverse magnetic field. $T = 2.7$ K.

The angular dependence of the SdH oscillations amplitude was obtained by rotation of the sample in magnetic field in the angle range of 0-90°. As the Fermi surface of n-InAs is isotropic [22], we tried to extract the contribution of 2D-InAs QW from complete pictures of the SdH oscillations with high period at $T = 2.7$ K and at H_{const} = 4-14 kOe, with taken into account a typical angular dependence for 2D-carriers which has to be proportional to $H_{const}cos\Theta$.

We calculated angular indicatrices of the amplitude oscillations that belong to 2D-InAs QW by the subtraction the oscillation pictures at 90° from the whole picture oscillations at the other angles 0° – 75° (figure 7). End-point result is presented on figure 8, where angular dependences of the SdH oscillations for microwave absorption of the inverse magnetic fields is shown. As one can see, a position of the amplitude oscillation minimum at $1/H \approx 0.12$ kOe^{-1} shift to higher fields according to $\sim H_{const}cos\Theta$. 2D electron concentration in the InAs QW n_s = 3 × 10^{11} cm^{-2} was evaluated from oscillation period [25].

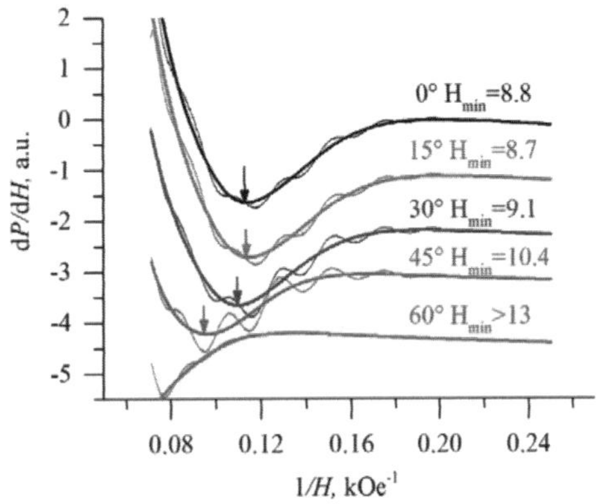

Figure 8. The spline interpolation of the oscillatory function for various angles Θ obtained by Savitsky - Golay method. Position of trend minima detected by arrows. Thin curves are the SdH oscillations amplitudes, solid curves are splines.

4. Conclusion

The type II broken-gap InAs/GaSb/AlSb CQWs heterojunctions were for the first time grown by MOVPE on the lattice-matched n-GaSb:Te(100) (structure A) and n-InAs:Mn(100) (structure B) substrates with QWs width InAs 12.5 nm, for GaSb 8 nm to get inverted band structure.

Strong SdH oscillations at derivate microwave radiation absorption (f = 10 GHz) were observed at low temperatures $T = 2.7$-20 K by using the electron-paramagnetic resonance spectrometer in the magnetic fields up to 14 kOe for both types of the CQWs grown on n-GaSb:Te(100) and on n-InAs:Mn(100).

Predominance contribution to SdD oscillations amplitudes relates to bulk electrons of n-GaSb:Te substrate as was established by rotation of sample A in the magnetic field. That was demonstrated by Fourier analysis and by measurement of the angular dependence on rotation of the substrate in the magnetic field. Two frequencies of SdH oscillations were found.

Observed unusual angular indicatrix of the amplitude for different orientation of H_{const} in dependence on the crystal orientation can be explained by inversion asymmetry, which is inherent to III-V semiconductors with zinc-blende structure without the inversion center. This effect obeys spin-orbit splitting of the conduction band spectrum and agrees to warping of the n-GaSb Fermi surface at high concentration. A small value of the spin-splitting parameter was evaluated according to Seiler, Becker and Roth and may be about $C_2E_F = 0.006$ meV for n-GaSb:Te(100) substrate with the concentration of $n = 3 \times 10^{17}$cm^{-3}.

Taking into account isotropic Fermi surface of n-InAs, we succeeded to extract a contribution of the 2D-carriers of InAs QW to SdH oscillation amplitudes. Angle dependence of SdH oscillation positions of amplitudes $\sim H_{const}cos\Theta$ was observed in the magnetic field range of 10-14 kOe at T = 2.7 K.

Further growing of the InAs/GaSb/AlSb QW structures on the semi-insulating substrates (GaAs, p-GaSb or p-InAs:Mn with lower concentration) allows us to study quantum phenomena in these unique and interesting QW structures more in detail. Investigation of the magnetotransport at higher magnetic fields and studying a conductivity and hybridization of electron-hole bands by the gating CQW samples under applied electric and magnetic fields at low temperatures are also in the project.

5. Acknowledgement

This work was supported in part by RBRF grant N 15-02-031151 and the Czech Science Foundation project number 13-15286S. Authors thank M.O. Safonchik for technical assistance.

References

[1] Naveh Y and Laikhtman B 1995 *Appl. Phys. Lett.* **66** 1980
[2] Liu C, Hughes T L, Qi X L, Wang K and Zhang S C 2008 *Phys. Rev. Lett.* **100** 236601
[3] Knez I, Du R R and Sullivan G 2011 *Phys. Rev. Lett.* **107** 136603
[4] Knez I, Rettner C T, Yang S H, Parkin S S P, Du L, Du R R and Sullivan G 2014 *Phys. Rev. Lett.* **112** 026602
[5] Nichele F, Pal A N, Pietsch P, Ihn T, Ensslin K, Charpentier C and Wegscheider W 2014 *Phys. Rev. Lett.* **112** 036802
[6] Yang M J, Yang C H, Bennet B R and Shanabrook B V 1997 *Phys. Rev. Lett.* **78** 4613
[7] Hasan M Z and Kane C L 2010 *Rev. Mod. Phys* **82** 3045
[8] Kane C L and Mele E J 2005 *Phys. Rev. Lett.* **95** 226801
[9] Chang L L and Esaki L 1980 *Surf. Sci.* **98** 70
[10] Mikhailova M P and Titkov A N 1994 *Semicond. Sci. Technol.* **9** 1279
[11] Mikhailova M P, Moiseev K D and Yakovlev Yu P 2004 *Semicond. Sci. Technol.* **19** R109
[12] Parfeniev R V, Moiseev K D, Berezovets V A, Averkiev N S, Mikhailova M P, Nizhankovskii V I and Kaczorovski D2009 *J. Magn. Magn. Mater.* **321** 712
[13] Averkiev N S, Berezovets V A, Mikhailova M P, Moiseev K D, Nizhankovskii V I, Parfeniev R V and Romanov K S 2004 *Phys. Solid State* **46** 2153
[14] Kroemer H 2004 *Physica E* **20** 196
[15] Kindl D, Toušková J, Hulicius E, Pangrác J, Šimeček T, Jurka V, Hubik P, Mareš J J and Krištofik J 2004 *J. Appl. Phys.* **95** 1811
[16] Mikhailova M P, Veinger A I, Kochman I V, Semenikhin P V, Kalinina K V, Parfeniev R V, Berezovets V A, Hospodkova A, Pangrác J and Hulicius E NATO Advanced Research Workshop on THz Diagnostics of CBRN effects and Detection of Explosives & CBRN, Izmir, Turkey, November 3-6, 2015, *Abstract Book* **1** 46
[17] Veinger A I, Zabrodskii A G, Tisnek T V and Biskupski G 1998 *Semiconductors* **32** 497
[18] von Bardeleben H J, Jia Y Q, Manasreh M O and Stuz C E 1993 *Appl. Phys. Lett.* **62** 90
[19] Matveenko A V and Parfenev R V 1971 *Sov. Phys. Semicond.* **5** 667
[20] Silva EAA e 1992 *Phys. Rev.* B **46** 1921
[21] Averkiev N S and Golub L E 1999 *Phys. Rev.* B **60** 15582
[22] Seiler D G and Stephens A E 1991 *Landau Level Spectroscopy* vol. 27 ed G Landwehr and E I Rashba (Amsterdam: North-Holland) , p 1035
[23] Dobrowolska M, Chen Y, Furdyna J K and Rodriguez S 1983 *Phys. Rev. Lett.* **51** 134
[24] Seiler D G, Becker W M and Roth L M 1970 *Phys. Rev.* B **1** 764
[25] Altarelli M, Maan J C, Chang L L and Esaki L 1987 *Phys. Rev.* B **35** 9867

Intersubband light absorption by holes in InAsSb/AlSb quantum well heterostructures

N V Pavlov[1], **G G Zegrya**[1]

[1] Ioffe Institute of RAS, 194021, St. Petersburg, Politekhnicheskaya st., 26, Russian Federation

E-mail: pavlovnv@mail.ru

Abstract. The absorption coefficients of intersubband optical transitions in the valence band of the AlSb/InAs$_{0.84}$Sb$_{0.16}$/AlSb quantum wells in framework of the four-band Kane model. It is established that the light absorption by holes may lead to the laser generation breakdown. It is shown that we need to decrease the quantum well width to range a<6 nm to neutralize the negative influence of this effect.

1. Introduction

One of the most important trends in modern semiconductor physics is creation of new optoelectronic devices in mid-infrared range (2-10 μm). This is due to the fact that there are the region of absorption of many toxic industrial gases, one of the windows of atmospheric transparency and maximum absorption tissues of the human body in this wavelength range. That can be applied to create a telecommunications and security systems, location and night vision equipment, gas analyzers, diagnostic and laser surgery devices. One of the most promising materials for solving this problem are deep quantum wells (QWs) based on InAsSb semiconductor compounds as the active region [1, 2]. These compounds are characterized by minimum values of the bandgap width E_g and effective electron mass m_c among A_3B_5 semiconductors [3, 4]. There are significant nonparabolicity of light (LH) and spin-orbit split-off (SH) holes energy spectrum in these heterostructures. This leads to significant corrections to the dimensional quantization levels energy and wave functions as compared to those in the case of the parabolic dispersion law even for the ground state in the valence band.

One of the basic mechanisms of internal light losses in semiconductor heterostructures is the light scattering by free charge carriers. The light scattering by free electrons in heterostructures based on A_3B_5 narrow-gap semiconductors is well studied. The light scattering by holes in quantum wells is studied within the Luttinger model for light and heavy holes (HH) but in order to take into account the spin- orbit split-off holes it is necessary to use the four-model Kane.

The aim of the present investigation is calculating of the optical absorption coefficient for HH-LH and HH-SH intersubband transitions in the AlSb/InAs$_{0.84}$Sb$_{0.16}$/AlSb heterostructure with a deep QW. The nonparabolicity is described with the four-band Kane model.

2. Basic Equations

We used the charge carriers wave functions of type $\psi = \Psi_s|s\rangle + \mathbf{\Psi}|\mathbf{p}\rangle$, where $|s\rangle$ and $|\mathbf{p}\rangle$ are the s- and p- type Bloch functions, respectively, and Ψ_s and $\mathbf{\Psi}$ are spinors. The Kane equations near the Γ-point in the spherical approximation are given by the following expressions [5]:

$$(E_C - E)\Psi_s - iP\nabla\Psi = 0$$

$$(E_V - \delta - E)\Psi - iP\nabla\Psi_s + \frac{\hbar^2}{2m_0}(\gamma_1 + 4\gamma_2)\nabla(\nabla\Psi)$$

$$-\frac{\hbar^2}{2m_0}(\gamma_1 - 2\gamma_2)[\nabla\times[\nabla\times\Psi]] + i\delta[\sigma\times\Psi] = 0 \quad (1)$$

where E_C and E_V are the energy of the conduction band and the valence band edges, respectively, P is the Kane matrix element, $\delta=\Delta_{so}/3$ is the spin-orbit split off constant, m_0 is the free electron mass, γ_1 and γ_2 are the generalized Lattinger parameters, $\boldsymbol{\sigma}=(\sigma_x, \sigma_y, \sigma_z)$ are the Pauli matrices.
The charge carrier energy spectra in the valence band in the four-band Kane model are:

$$E_h = \delta - \frac{\hbar^2(k_h^2+q^2)}{2m_h},$$

$$E_{lh,so} = -\frac{\delta}{2} - \frac{\hbar^2(k_{l,so}^2+q^2)}{4}(m_l^{-1}+m_h^{-1}) \pm \left\{2\delta^2 + [\frac{\delta}{2} - \frac{\hbar^2(k_{l,so}^2+q^2)}{4}(m_l^{-1}-m_h^{-1})]^2\right\}^{\frac{1}{2}}, \quad (2)$$

where $m_l^{-1} = -\frac{2P^2}{\hbar^2 E_{lh}} + m_0^{-1}(\gamma_1+4\gamma_2)$ and $m_h^{-1} = m_0^{-1}(\gamma_1-2\gamma_2)$, E_h, E_{lh} and E_{so} are the HH, LH and SH energies, respectively, k_h, k_l and k_{so} are the quantized HH, LH and SH wave vector components, respectively, q is the wave vector component in QW plane.
The number of transitions is given by formula:

$$Q = Q_{\parallel}\cos^2\beta + Q_{\perp}\sin^2\beta, \quad (3)$$

where β is the angle between the vector potential amplitude $\boldsymbol{A_0}$ and QW plane and

$$Q_{\perp,\parallel} = \frac{4\pi^2P^2e^2A_0^2}{\hbar^3c^2}\sum_{k_h}\int_0^{\infty}\frac{qdq}{(2\pi)^2}\sum_{k(q)}\sigma_{\perp,\parallel}f(E_h)(1-f(E))\delta(E_h-E-\hbar\omega), \quad (4)$$

$$\sigma_{\parallel} = \pi(\left|I_y^s\right|^2 + \left|I_z^s\right|^2),$$

$$\sigma_{\perp} = 2\pi\left|I_x^a\right|^2,$$

I_i are the overlap integrals between s-component of final state wave function and i-component of initial state wave function, $i=x,y,z$. index a denotes transitions with spin conversion, index s denotes transitions without spin conversion, ω is the optical transition frequency, $k(q)$ is the wave vector of final state, f is the Fermi function.
We considered the optical transitions where $\boldsymbol{A_0}$ is parallel to the QW plane. The absorption coefficient of intersubband optical transitions can be calculated using the following equation:

$$\alpha_{\parallel} = \frac{2\pi^2e^2}{nc\hbar^3}\frac{1}{a}\frac{1}{\hbar\omega}\sum_{k_h}m_r(E)\sigma_{\parallel}f(E_h)(1-f(E)), \quad (5)$$

Where $m_r(E) = \frac{1}{m_{l,so}^{-1}(E) - m_h^{-1}}$, m_{so} is the SH effective mass.

3. Results

In our calculations we used the following parameters: QW width a=8 nm, holes concentration $p=10^{12}$ cm^{-2} and temperature T=300 K.
In Figures 1-3 the absorption coefficient of various intersubband optical transitions and total absorption coefficient by holes are presented.

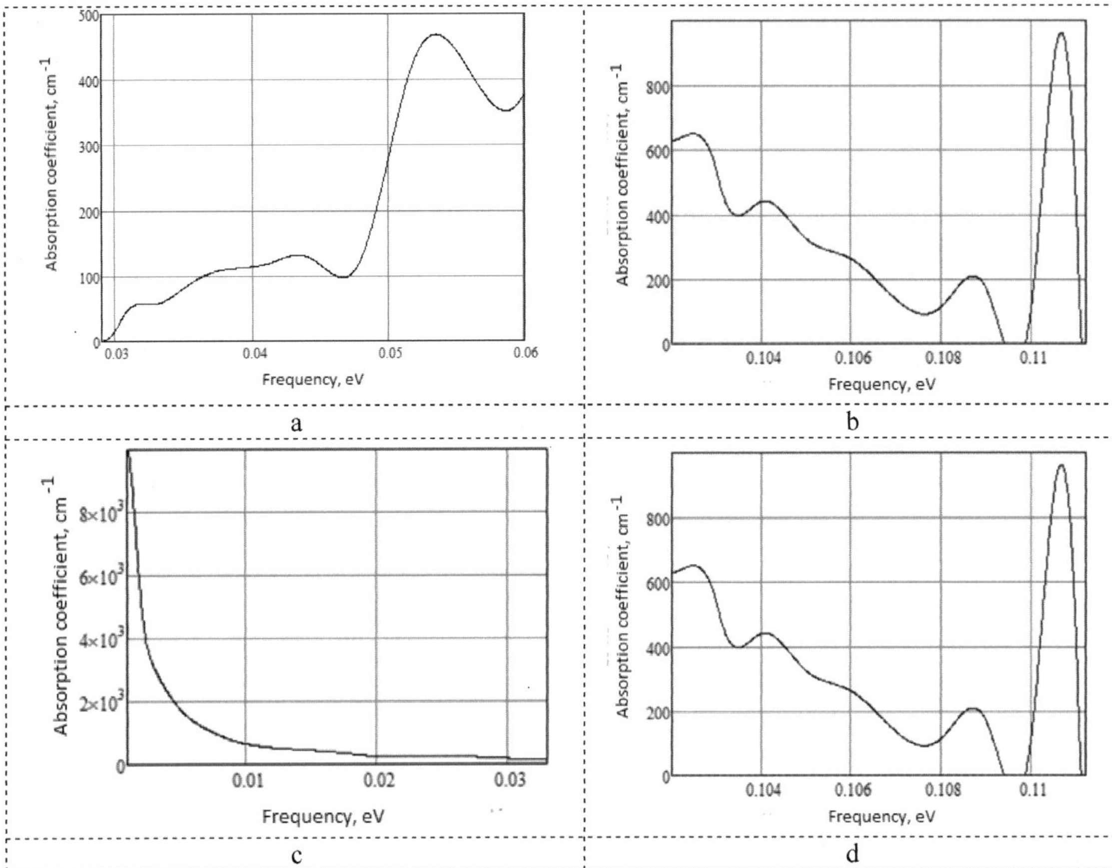

Figure 1. The absorption coefficient of HH1-LH1 (a), HH1-LH2 (b), HH2-LH1 (c), HH2-LH2 (d) optical transitions.

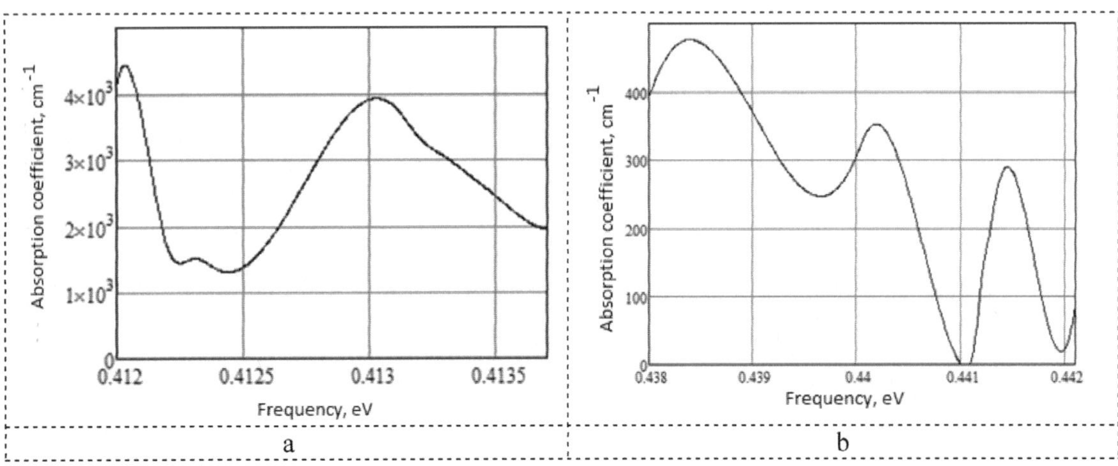

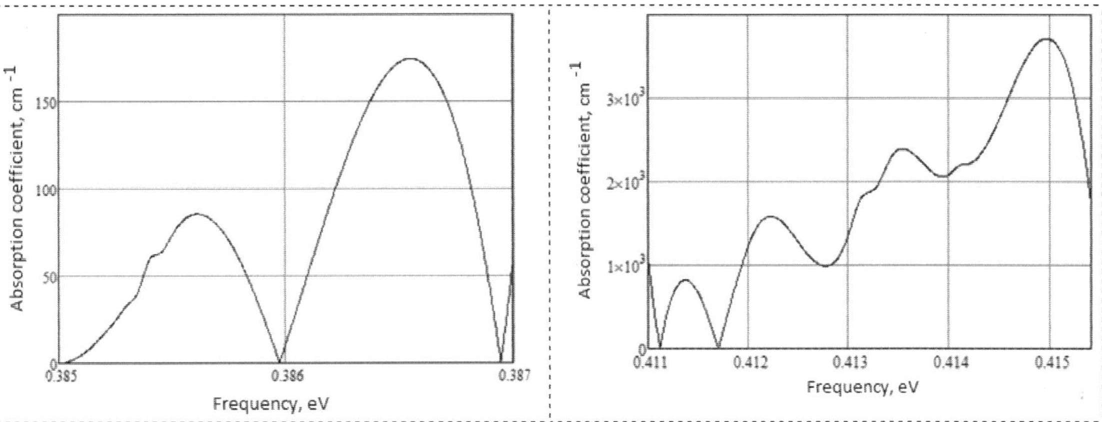

c d

Figure 2. The absorption coefficient of HH1-SH1 (a), HH1-SH2 (b), HH2-SH1 (c), HH2-SH2 (d) optical transitions.

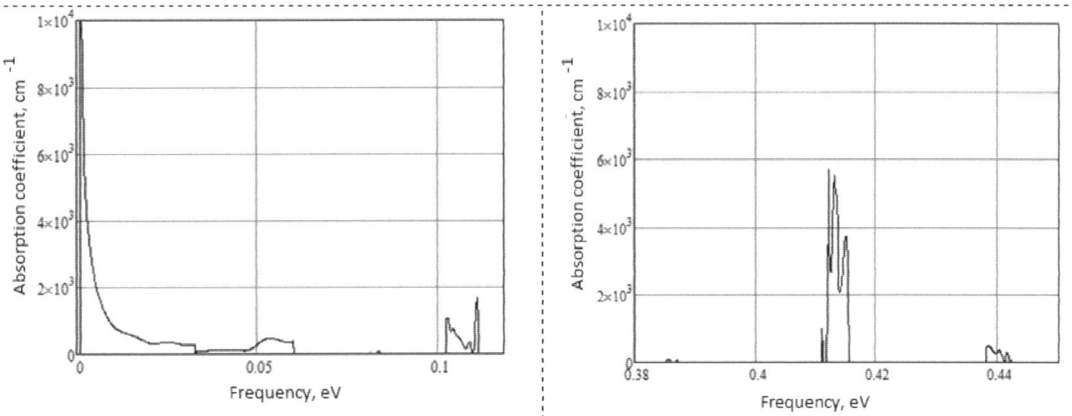

a b

Figure 3. Total absorption coefficient of HH-LH (a) and HH-SH (b) optical transitions.

As can be seen, transition HH2-LH1 has the maximum absorption coefficient. It causes by HH2 and LH1 energy levels resonance. Transition HH1-SH1 has large absorption coefficient that causes the fact that $m_h \approx m_{so}$.

Transitions HH1-SH1 and HH2-SH2 have the absorption coefficient value much higher than allowable value of internal optical losses in laser heterostructures. Thus we need to increase the effective energy gap value decreasing the QW width to achieve the conditions of the laser generation. The calculated value of QW width according to laser generation is a<6 nm.

4. Conclusions

Thus, it has been demonstrated that light absorption by holes is the very efficient way of internal optical losses and may cause the laser generation breakdown in $AlSb/InAs_{0.84}Sb_{0.16}/AlSb$ quantum wells with QW width a>6 nm.

This work was partly supported by Grant of the Science School #5062.2014.2.

References

[1] Zegrya G G 1997 *Mid-infrared strained diode lasers* (Antimonide-Related Strained Layer Heterostructures) ed Manasreh M O (Gordon & Breach Science Publishers, Amsterdam) p 273

[2] Mikhailova M P, Danilov L V *et al.* 2013 *Superlinear luminescence and enhancement of optical power in GaSb-based heterostructures with high conduction band offsets and nanostructures with deep quantum wells* (The Wonder of Nanotechnology: Quantum Optoelectronic Devices and Applications) ed Razeghi M *et al.* (SPIE Press, Bellingham, WA) p 105

[3] Vurgaftman I, Meyer J R, Ram-Mohan L R 2001 *Appl. Phys.* **89** 5815

[4] Cripps S A, Hosea T J C, Krier A *et al.* 2007 *Appl. Phys. Lett.* **90** 172106

[5] Zegrya G G and Polkovnikov A S 1998 *J. Exp. Theor. Phys.* **86** 815

Journal of Physics: Conference Series **769** (2016) 012077 doi:10.1088/1742-6596/769/1/012077

Mn-assisted molecular-beam epitaxy growth (Ga,Mn)As nanowires

R R Reznik[1,2,5], Yu B Samsonenko[1,4], A I Khrebtov[1], A D Bouravleuv[1,3], P Werner[6] and G E Cirlin[1,4,5]

[1]St-Petersburg Academic University – Nanotechnology Research and Education Centre RAS, Khlopina 8/3, 194021, St-Petersburg, Russia
[2]Peter the Great St.Petersburg Polytechnic University, Polytechnicheskaya 29, 195251, St-Petersburg, Russia
[3]Ioffe Physical Technical Institute RAS, Politekhnicheskaya 26, 194021, St-Petersburg, Russia
[4]Institute for Analytical Instrumentation RAS, Rizhsky 26, 190103, St-Petersburg, Russia
[5] ITMO University, Kronverkskiy pr. 49, 197101 St-Petersburg, Russia
[6]Max Planck Institute of Microstructure Physics, Weinberg 2, D-06120, Halle, Germany

E-mail: moment92@mail.ru

Abstract. Arrays of (Ga,Mn)As crystal nanowires on a GaAs (100) substrate were obtained using molecular-beam epitaxy at the substrate temperature 485°C. From the high energy electron diffraction patterns, the crystallographic phase of the nanowires is detected to be cubic which is supporting by *ex situ* microscopy study.

1. Introduction

Recently, interest has continued in one-dimensional nanostructures—crystal nanowires (NWs) grown using different technological methods, including molecular-beam epitaxy. This interest is due to the unique properties of these structures associated with a small number of structural defects, as well as the possibility of the NWs growing on substrates with large mismatch in lattice parameters. As a rule, the NWs growth corresponds to the so-called vapor–liquid–crystal method with a catalyst based on different metals. To activate such growth, the most frequently used metal is Au [1–3]. Self-catalyzed growth of the crystal nanowires is also possible, i.e., growth using its own material. In the case of GaAs, for example, gallium may serve as the catalyst for NW growth [4, 5]. Also, successful synthesis of (Ga,Mn)As NWs on GaAs(111)B substrates when using Mn as the catalyst has recently been reported [6].

It is worth noting that semiconductor compounds based on (Ga,Mn)As are very promising materials as applied to spintronics, since they have properties that are inherent to semiconductors as well as magnetic materials. Such semiconductors are called diluted magnetic semiconductors. One of the main problems in such materials use is that their Curie temperature is far from room temperature [7].

Transition to synthesis of nanostructures with reduced dimensionality, such as two-dimensional layers or quasi-structures like NWs, may be one of the ways to solve this problem. In contrast to the films, synthesis of the NWs opens more opportunities to create defect-free structures on a mismatched substrate.

In this paper, we showed the principal possibility to synthesize (Ga, Mn)As NWs on GaAs(100) substrates using Mn as a catalyst.

2. Experiments

The growth experiments were carried out using an EP1203 MBE system equipped with Ga, Mn atomic and As_4 molecular cells. First, epi-ready GaAs(100) substrates were thermally cleaned to eliminate native oxide layer at 620°C. The deoxidation process was monitored by the reflection high-energy electron diffraction (RHEED) method. The temperatures of the cells were equal to 1000°C, 373°C, and 720°C, for Ga, As, and Mn correspondingly. After completion of the oxide removing process, the temperature of the substrate was lowered by 40–50°C, and the growth of the GaAs buffer layer was carried out over 15 minutes to smooth out the surface. Then the temperature was decreased by another 10°C and Mn shutter was opened (As shutter was closed during Mn deposition stage) for 2 minutes to form Mn terminated nanoclusters used as a catalyst for the self-catalyzed growth of NWs later on. As the working temperature (485°C) was reached, the Mn, Ga, and As cells were simultaneously opened, and the (GaMn)As NWs started to grow. The time of NWs nucleation evaluated from RHEED pattern was 30 seconds. The total growth time was equal to 25 minutes at the GaAs growth rate equals to 1 ML/s. It should be noted, that the growth of NWs occurred under slightly metal stabilized growth conditions, i.e. the As_4/Ga flux ratio was 0.4.

3. Results

As it was noticed in [4], the predominant crystallographic phase of NWs was cubic, when GaAs NWs on a Si(111) surface were formed in Ga-stabilized conditions. The same type is characteristic for the obtained NWs array, as we detect from the RHEED patterns (Figure. 1), since the diffraction picture shows the presence of the cubic phase, only. The lack of characteristic reflections related to the hexagonal phase (in particular, 0001) indicates the absence of a wurtzite phase in the obtained structures. As was shown in [8, 9], in the process of GaAs NW growth on Si(100), Si(111), and GaAs(111)B substrates with Au as the catalyst, the predominant crystallographic phase is wurtzite. However, under definite growth conditions, a transition from a cubic structure to a hexagonal one and vice versus is possible. For our samples, such a transition is not characteristic, which is exclusively due to the influence of Mn. It should be noted that the NW formation on a substrate with orientation (100) can occur in four equivalent directions <111>. As follows from the RHEED pictures in our case, only two preferential directions of growth, which is likely caused by features of NW formation in this system one can observe.

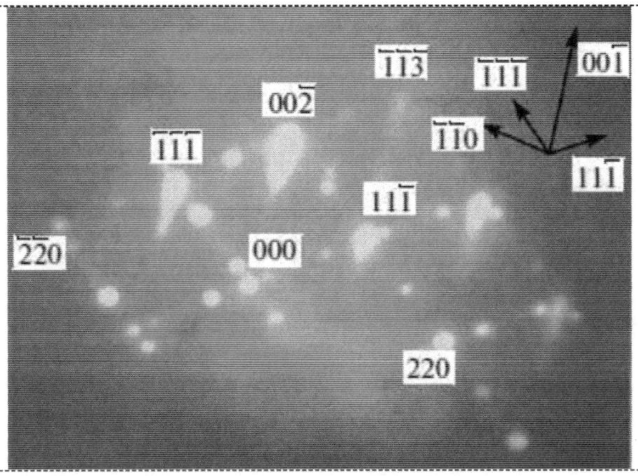

Figure 1. RHEED patterns of GaMnAs NWs on GaAs(100) substrate in (110) projection after termination of growth.

The morphology of the NWs obtained was studied using a scanning electron microscope (SEM) Zeiss Supra 25. SEM images of (Ga,Mn)As NWs shown in Figure 2 (a, b) demonstrate that NWs have typical lengths between 0.8 μm and 4 μm and diameters 50–90 nm. The nanowire density is about 10^7 cm^{-2}. Most of NWs have a preferential growth direction along <111> and <110>. Some of them are oriented along <310> crystallographic directions. Remarkably, (Ga,Mn)As NWs do not have any branches. Many of them are slightly tapered, which can be caused by the relatively low growth temperature resulting in limitations on Ga adatom diffusion lengths [10, 11].

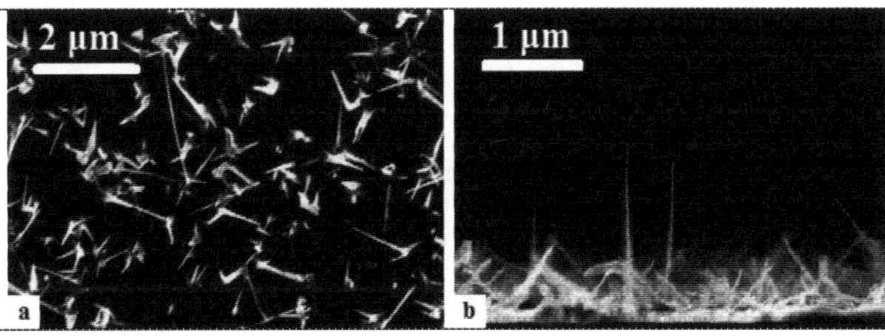

Figure 2. SEM micrograph of GaMnAs NWs on Ga(100) substrate: (a) sample surface, (b) image of the cleavage sample surface..

The structural and chemical characterization of the grown samples was carried out by transmission electron microscopy (TEM). For this purpose, the samples were prepared in two different ways. First, the NWs were chopped from the substrate and suspended on a carbon grid. The growth direction of the NWs and their morphology were analyzed in a JEOL JEM 4010 microscope, which allowed a high-resolution TEM investigation (HRTEM) of the crystal structure, including lattice defects detection. Second, the chemical nature of precipitates found on the substrate surface as well as the chemical composition were investigated by high-angle angular dark- field scanning transmission electron microscopy (HAADFSTEM) combined with energy-dispersive X-ray analysis (EDX). For this analysis, we used a FEI TITAN 80/300 equipped with a probe corrector, which allowed a lateral resolution of 0.1 nm. Figure 3 shows a typical example of a <110> oriented NW obtained as a result of TEM measurements. The NW has a length of 2 μm, diameter is lower 100 nm, which is changed due to slight tapering. Even in such long NWs, we could not find crystal defects, such as dislocations or precipitates. However, some of them may have planar features as stacking faults lying parallel to the growth direction (see Figure 3). The NWs have pure zincblende crystal structure. The EDX analysis of (Ga,Mn)As also yields no local precipitation of the Mn within the detection limit of 0.5%. Thus, Mn is homogeneously distributed over the NWs with the concentration less the 10^{20} cm^{-3}.

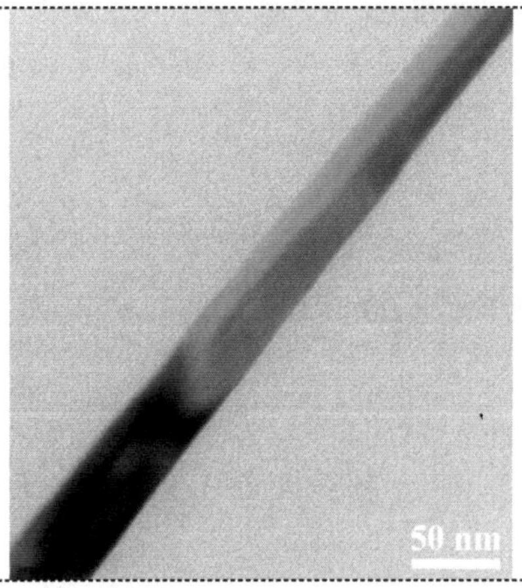

Figure 3. TEM image of a <110> oriented NW. The tip at the upper right contains the growth front and the catalytic droplet, which contains a high concentration of Mn. The dark features inside the NWs are caused by bending of the crystal lattice, known as "bending contours." In the center, a stacking fault (SF) is located lying parallel to the growth direction.

4. Conclusions

In summary, (Ga,Mn)As NWs have been grown by MBE on GaAs(100) substrates using Mn as a catalyst. The majority of the NWs have preferential growth direction along <111> and <110>. The NWs do not have any extended defects, such as dislocations or precipitates. The EDX analysis of (Ga,Mn)As also yielded no local precipitation of Mn within the detection limit of 0.5%. Some of the NWs contain planar features such as stacking faults lying parallel to the growth direction.

Acknowledgments

The nanowire samples were grown under the support of Russian Science Foundation (Project No 14-12-00393). This work was partially supported by RFBR (grant 14-02-00154 A) and Skolkovo Foundation (6 dd. 30.12.2015).

References

[1] Dubrovskii V G, Cirlin G E and Ustinov V M Semiconductor nanowhiskers: Synthesis, properties, and applications *Semiconductors* **43** no 12 1539-1584 (2009)

[2] Soshnikov I P Cirlin G E Tonkikh A A Samsonenko Yu B Dubrovskii V G Ustinov V M Gorbenko O M Litvinov D and Gerthsen D Formation of GaAs nanowhisker arrays by magnetron sputtering deposition *Phys.Solid State* **48** no 4 786-791 (2006)

[3] Tchernycheva M Cirlin G E, Patriarche G Travers L Zwiller L Perinetti U and Harmand J C Growth and characterization of InP nanowires with InAsP insertions *Nano Lett.* **7** no 6 1500–1504 (2007)

[4] Cirlin G E Dubrovskii V G Samsonenko Yu B Bouravleuv A D Durose K Proskuryakov Y Y Budhikar M Bowen L Kaliteevski M A and Abram R A Self-catalyzed, pure zincblende

GaAs nanowires grown on Si(111) by molecular beam epitaxy *Phys. Rev. B* **82** no 3 035302 (2010).

[5] Dubrovskii V G Cirlin G E Sibirev N V Jabeen F Harmand J C and Werner P New mode of vapor−liquid−Sslid nanowire growth *Nano Lett.* **11** no 3 1247 (2011)

[6] Bouravleuv A D Cirlin G E Romanov V V Bagraev N T Brilinskaya E S Lebedev S A Novikov S V LipsanenH and Dubrovskii V G Formation of (Ga,Mn)As nanowires and study of their magnetic properties *Semiconductors* **46** no 2 179-183 (2012)

[7] Chen L Yan S Xu P F Lu J Wang W Z. Deng J J Qian X Ji Y and Zhao J H Low-temperature magnetotransport behaviors of heavily Mn-doped (Ga,Mn)As films with high ferromagnetic transition temperature *Appl. Phys. Lett.* **95** no 18 182505 (2009)

[8] Soshnikov I P Cirlin G E Tonkikh A A Nevedomskii V V Samsonenko Yu B and Ustinov V M Electron diffraction on GaAs nanowhiskers grown on Si(100) and Si(111) substrates by molecular-beam epitaxy *Solid State Phys.* **49** no 8 1440-1445 (2007)

[9] Cirlin G E Dubrovskii V G Soshnikov I P Sibirev N V Samsonenko Yu B Bouravleuv A D Harmand J C and Glas F Critical diameters and temperature domains for MBE growth of III–V nanowires on lattice mismatched substrates Phys. *Status Solidi RRL* **3** no 4 112-114 (2009)

[10] Dubrovskii V G Sibirev N V Cirlin G E Tchernycheva M Harmand J C and Ustinov V M *Phys. Rev. E* **77** no 3 031606 (2008)

[11] Boeck J De Oesterholt R Van Esch A Bender H Bruynseraede C Van Hoof C and Borghs G Nanometer-scale magnetic MnAs particles in GaAs grown by molecular beam epitaxy *Appl. Phys. Lett.* **68** no 19 2744–2746 (1996)

Journal of Physics: Conference Series **769** (2016) 012078 doi:10.1088/1742-6596/769/1/012078

Resonant electronic excitation energy transfer by Dexter mechanism in the quantum dot system

D M Samosvat[1], O P Chikalova-Luzina[1], V M Vyatkin[2] and G G Zegrya[1]

[1]Ioffe institute, St.-Petersburg, Politekhnicheskaya st., 26, 194021,Russia
[2]St. Petersburg State Electrotechnical University, St. Petersburg, 197376, Russia

E-mail: samosvat@yandex.ru

Abstract. In present work the energy transfer between quantum dots by the exchange (Dexter) mechanism is analysed. The interdot Coulomb interaction is taken into consideration. It is assumed that the quantum dot-donor and the quantum dot-acceptor are made from the same compound A3B5 and embedded in the matrix of other material creating potential barriers for electron and holes. The dependences of the energy transfer rate on the quantum-dot system parameters are found using the Kane model that provides the most adequate description spectra of semiconductors A3B5. Numerical calculations show that the rate of the energy transfer by Dexter mechanism is comparable to the rate of the energy transfer by electrostatic mechanism at the distances approaching to the contact ones.

1. Introduction

Methods based on nonradiative energy transfer between quantum dots are in considerable current use in biology and medicine. The strong dependence of the energy transfer rate on the distance between the energy donor and the energy acceptor enables to detect the formation of antigen-antibody complexes, ensyme-substrate complexes, the DNA hybridization and cleavage [1] and also to study the structure and dynamics of biomolecules, where measurements of small distance within the molecule are necessary [2–4]. Results of these investigations are of the great importance for diagnostics and therapy of diseases, among which are oncological ones [5], (see also the literature in [1]). A correct interpretation of experimental data requires the adequate theory development. Resonant electronic excitation energy transfer occurs due to electric-dipole-dipole, electric-dipole-quadrupole and higher multipole interactions and the exchange interactions between the energy donor and the energy acceptor. The first interactions are of an electrostatic nature, the last one has its origin in the antisymmetry of the electronic wave function of the system including an energy donor and an energy acceptor. An equality of electron excitation energies in donor and acceptor is a necessary condition for the resonant energy transfer which follows from the energy conservation law. The theory descriptive of the energy transfer between molecules due dipole-dipole interaction has been developed by Forster [6]. Subsequently, Dexter developed the theory to describe the electrostatic and exchange mechanisms of the energy transfer between impurity atoms or ions in an insulating crystal [7]. It was indicated that the exchange contribution in the energy transfer can be important in two situations: first, distance between donor and acceptor is short and the donor and acceptor wave functions are not strong localized; second, a dipole-dipole transitions in the acceptor are forbidden. The use of the semiconductor quantum dots as the energy donor and the energy acceptor allows room for a wider range of biological and medicine experiments [8]. Theoretical description for the energy transfer between quantum dots by electrostatic mechanism was elaborated in [9, 10] assuming that the interactions between the quantum dot-donor electron and the quantum dot-acceptor electron is described by a Coulomb potential including all multipole terms. In present work the energy transfer between two quantum dots by the exchange mechanism is analyzed. The interdot Coulomb interaction is taken into consideration. It is assumed that the quantum dot-donor and the quantum dot-acceptor are made from the same compound A3B5 and placed in a matrix from other material creating potential barriers of finite heights for electrons (V_{cD}, V_{cA}) and holes (V_{hD},V_{hA}). The quantum dots are separated by a finite distance d. The system of two quantum dots is represented schematically in fig. 1. The energy transfer process is sketched in fig. 2.

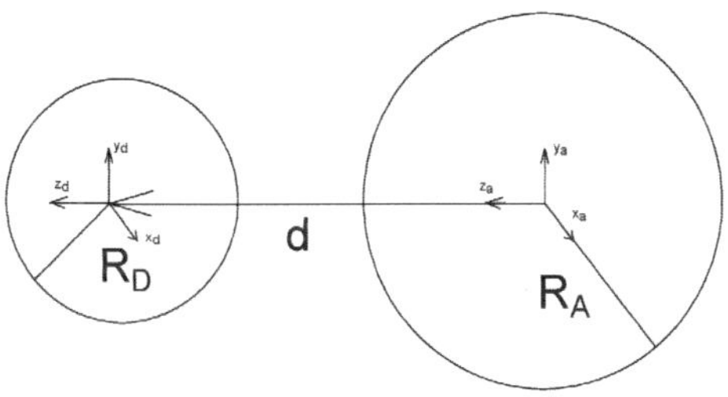

Figure 1. Diagram of two quantum dots: a donor of radius R_D and an acceptor of radius R_A.

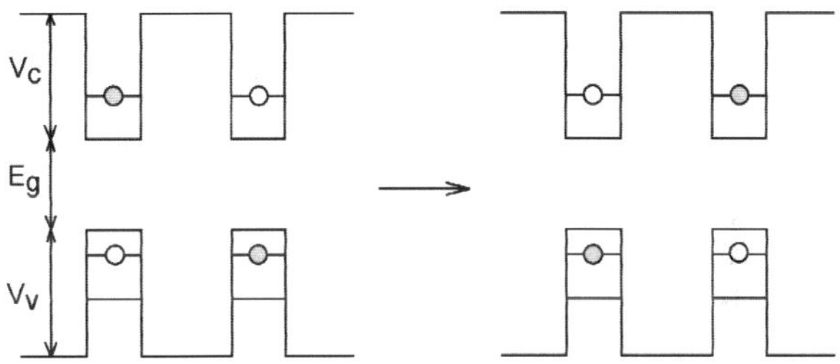

Figure 2. Schematic drawing of the energy transfer process between identical quantum dots ($R_D = R_A$).

In the initial state of the system the electron of the donor is in an excited state (in conduction band), the electron of the acceptor is in the ground state (in the valence band). As a result of the Coulomb interaction of the donor and acceptor electrons, the system passes in the final state where the electron of the acceptor is in a conduction band of the acceptor, in the donor the electron is in valence band.

2. Energy transfer matrix element

To obtain the rate of the nonradiative energy transfer between two quantum dots we need to calculate the energy transfer matrix element (the Coulomb interaction matrix element) for the system transition from the initial state to final state (see fig. 2). It can be expressed as

$$M_{if} = \sum_{\sigma_1,\sigma_2} \int d^3r_1 d^3r_2 \Psi_f(\xi_1,\xi_2) \frac{e^2}{\varepsilon|\mathbf{d}+\mathbf{r_1}-\mathbf{r_2}|} \Psi_i^*(\xi_1,\xi_2) \tag{1}$$

where $\xi_i = (\mathbf{r_i}, \sigma_\mathbf{i})$, $\sigma_\mathbf{i}$ are spin variables, $\mathbf{r_1}$ and $\mathbf{r_2}$ are the radius-vectors of electrons in the donor and acceptor, respectively, measured from the center of the corresponding quantum dot. The antisymmetric wave functions of the initial and final states of the system under consideration are given by

$$\Psi_i(\xi_1,\xi_2) = \frac{1}{\sqrt{2}}(\Psi_{cD}(\mathbf{r_1})\chi_{cD}(\sigma_1)\Psi_{hA}(\mathbf{r_2})\chi_{hA}(\sigma_2) - \Psi_{cD}(\mathbf{r_2})\chi_{cD}(\sigma_2)\Psi_{hA}(\mathbf{r_1})\chi_{hA}(\sigma_1)) \tag{2}$$

$$\Psi_f(\xi_1,\xi_2) = \frac{1}{\sqrt{2}}(\Psi_{hD}(\mathbf{r_1})\chi_{hD}(\sigma_1)\Psi_{cA}(\mathbf{r_2})\chi_{cA}(\sigma_2) - \Psi_{hD}(\mathbf{r_2})\chi_{hD}(\sigma_2)\Psi_{cA}(\mathbf{r_1})\chi_{cA}(\sigma_1)) \tag{3}$$

Here $\Psi_{cD}(r_i)$ and $\Psi_{hD}(r_i)$ are the electron and hole spatial wave functions of the donor (similarly of the acceptor); $\chi(\sigma_i)$ are the spin wave functions. Substituting wave functions (2) and (3) in the expression for the Coulomb interaction matrix element (1) results in dividing it into two terms:

$$M_{if} = M_{coul} - M_{ex} \tag{4}$$

where M_{coul} is the matrix element of the direct Coulomb interaction and M_{ex} is the matrix element of the exchange interaction. The matrix element of the direct Coulomb interaction for the system of two quantum dots is considered in detail in our previous work [10]. As present study focuses on the energy transfer by the exchange mechanism here we treat the matrix element of the exchange interaction that is of the form:

$$\begin{aligned} M_{ex} = \int d^3r_1 d^3r_2 \Psi^*_{cD}(\mathbf{r_1})\Psi_{cA}(\mathbf{r_1})\frac{e^2}{\varepsilon|\mathbf{d}+\mathbf{r_1}-\mathbf{r_2}|}\Psi_{hD}(\mathbf{r_2})\Psi^*_{hA}(\mathbf{r_2})\times \\ \times \sum_{\sigma_1,\sigma_2} \chi_{hD}(\sigma_2)\chi_{cA}(\sigma_1)\chi_{cD}(\sigma_1)\chi_{hA}(\sigma_2) \end{aligned} \tag{5}$$

Equation (5) determines the following selection rules in the exchange matrix element: M_{ex} is not equals to 0 when $\chi_{cD} = \chi_{cA}$ and $\chi_{hD} = \chi_{hA}$. However $\chi_{c(D,A)}$ is not necessary equal to $\chi_{h(D,A)}$ so that the spin function on both quantum dots may change simultaneously. Ignoring the χ's, one can see that this integral represents the electrostatic interaction between two charge clouds $Q_c(\mathbf{r_1}) = e\Psi^*_{cD}(\mathbf{r_1})\Psi_{cA}(\mathbf{r_1})$ and $Q_h(\mathbf{r_2}) = e\Psi^*_{hA}(\mathbf{r_2})\Psi_{hD}(\mathbf{r_2})$.

3. Charge carrier wave functions

As it is known the Kane model provides the most adequate description of the energy spectra and wave functions of semiconductors A3B5 [11]. In this model the electron and hole wave functions can be written as

$$\Psi = \Psi_s + \mathbf{\Psi}\left|p\right\rangle, \tag{6}$$

where $|s\rangle$ and $|\mathbf{p}\rangle$ are Bloch wave functions. The s-type functions describe the states in the conduction band; the p-type functions describe the states in the valence band. Kane's equations for the envelope functions and their solutions within the quantum dot and under the barrier are obtained in spherical approximation in [12].In the consideration of the exchange interaction contribution in the energy transfer we have restricted ourselves to the calculations of the matrix element M_{ex} with the electron wave function $\Psi_s\,|s\rangle$ that does not contain an admixture of the p-states and one of two heavy hole wave functions $\mathbf{\Psi_{h1}}\,|\mathbf{p}\rangle$ where

$$\Psi_s = Aj_j(k_c r)Y_{jm}(\theta,\phi) \tag{7}$$

$$\mathbf{\Psi_{h1}} = A_1 j_j(k_h r)\mathbf{Y}^j_{jm}(\theta,\phi) \tag{8}$$

Here $Y_{jm}(\theta,\phi)$ are the spherical functions, $\mathbf{Y}^j_{jm}(\theta,\phi)$ are the vector spherical harmonics,j and m are the angular moment and its projection on z axis, respectively, $j_j(k_c r)$ are the spherical Bessel functions, where k_c and k_h are the wave numbers for the electrons and holes, respectively; and A_1 are normalizing constants. The corresponding wave functions under the barrier are used in the calculations.

4. Exchange interaction matrix element

Restricting ourselves to the calculations of the matrix element M_{ex} with the electron wave function $\Psi_s\,|s\rangle$ (7) and the heavy hole wave function $\mathbf{\Psi_h}\,|\mathbf{p}\rangle$ (8), we obtain the estimation of the contribution of the exchange interaction in the energy transfer rate from below.

Considering only the coordinate part in the Eq. (6) and denoting Ψ_s and $\mathbf{\Psi}_{h1}$ functions for the donor and the acceptor by Ψ_{csD},Ψ_{csA} and by $\mathbf{\Psi}_{hD}$,$\mathbf{\Psi}_{hA}$ respectively, we can write the matrix element as follows:

$$M_{ex} = \int d^3r_1 d^3r_2 \Psi^*_{csD}(\mathbf{r_1})\Psi_{csA}(\mathbf{r_1})\frac{e^2}{\varepsilon|\mathbf{d}+\mathbf{r_1}-\mathbf{r_2}|}\mathbf{\Psi}_{hD}(\mathbf{r_2})\mathbf{\Psi}_{hA}(\mathbf{r_2}). \quad (9)$$

In order to calculate the matrix element we place the origin of the coordinates in the center of the quantum-dot acceptor. In this coordinate system Eq. (9) may be rewritten as

$$M_{ex} = \int d^3r'_1 d^3r_2 \Psi^*_{csD}(\mathbf{r'_1}-\mathbf{d})\Psi_{csA}(\mathbf{r'_1})\frac{e^2}{\varepsilon|\mathbf{r'_1}-\mathbf{r_2}|}\Psi_{hD}(\mathbf{r_2}-\mathbf{d})\Psi^*_{hA}(\mathbf{r_2}). \quad (10)$$

For the sake of simplicity of the calculations, we set values of the spherical functions and the vector spherical harmonics to be equal to their values at $\theta_1 = 0$, $\theta_2 = 0$, $\theta'_1 = \pi$ and $\theta'_2 = \pi$ that are determined by the angular moments and their projections on z axis as follows:

$$\begin{aligned}
&Y_{jm}(0,\phi) = \delta_{m0}\sqrt{\frac{2j+1}{4\pi}},\\
&Y_{jm}(\pi,\phi) = (-1)^j\delta_{m0}\sqrt{\frac{2j+1}{4\pi}},\\
&\mathbf{Y}^j_{jm}(0,\phi) = \begin{cases} -m\sqrt{\dfrac{2j+1}{8\pi}}\mathbf{e_m}, & m=\pm 1\\ 0, & \text{in the other cases.}\end{cases}\\
&\mathbf{Y}^j_{jm}(\pi,\phi) = (-1)^j\mathbf{Y}^j_{jm}(0,\phi).
\end{aligned} \quad (11)$$

It should be noted that this approximation allows the calculations only for $m_{cD} = m_{cA} = 0$ and $m_{hD} = m_{hA} = \pm 1$. At other values of the full angular moment projections on z axis, the more exact $Y_{jm}(\theta,\phi)$ and $\mathbf{Y}^j_{jm}$ should be taken into account. The problem may be conveniently solved in the cylindrical coordinate system, as the system of two quantum dots treated has the axis connecting their centers. In this case the z-depending part of the donor electron wave functions in the donor range and under the barrier can be written as

$$\begin{aligned}
\phi_{csD}(d-z'_1) &= A_{cD}j_{j_{cD}}(k_{cD}(d-z'_1)), & d-R_D \le z'_1 \le d;\\
\phi_{csD}(d-z'_1) &= B_{cD}k_{j_{cD}}(\kappa_{cD}(d-z'_1)), & 0 \le z'_1 \le d-R_D.
\end{aligned} \quad (12)$$

Boundary conditions for the functions are of the form:

$$A_{cD}j_{j_{cD}}(k_{cD}R_D) = B_{cD}k_{j_{cD}}(\kappa_{cD}R_D) \quad (13)$$

The z-depending parts of the acceptor electron wave functions, of the donor and acceptor hole wave functions and the boundary conditions for the functions can be written similarly. When calculating M_{ex} we do not take into account the z-depending part of the electron and hole wave functions at $z'_1 < 0$ and $z'_1 > d$ because the modified spherical Bessel function $k_j(x)$ decays rapidly with increasing x. In the cylindrical coordinate system the Coulomb potential of the system under study is given by

$$\frac{e^2}{\varepsilon r} = \frac{e^2}{\varepsilon\sqrt{p^2+z^2}}, \quad (14)$$

where $r = |\mathbf{r_1} - \mathbf{r_2}|$, $z = z_1' - z_2$, $p^2 = \rho_1^2 + \rho_2^2 - 2\rho_1\rho_2\cos(\varphi_1 - \varphi_2)$. Using the integral formula $\frac{1}{\sqrt{p^2+z^2}} = \int\limits_0^\infty \exp(-q|z|)J_0(qp)dq$ and the Grafs summation theorem for one can obtain on the simple but cumbersome manipulations the exchange matrix element (17) in the following form:

$$M_{ex} \approx \frac{e^2}{\varepsilon}4\pi^2 S\rho_{max}^3 \left[\int\limits_0^{R_A} \phi_{csD}(d - z_1')\phi_{csA}^*(z_1')dz_1' + \int\limits_{R_A}^{d-R_D} \phi_{csD}(d - z_1')\phi_{csA}^*(z_1')dz_1' + \int\limits_{d-R_D}^{d} \phi_{csD}(d - z_1')\phi_{csA}^*(z_1')dz_1'\right]\left[\int\limits_0^{z_1'} \phi_{hD}^*(d - z_2)\phi_{hA}(z_2)(P(\eta_1^2) - \eta_1)dz_2 + \int\limits_{z_1'}^{d} \phi_{hD}^*(d - z_2)\phi_{hA}(z_2)(P(\eta_1^2) - \eta_1)dz_2\right] \tag{15}$$

Here ρ_{max} is determined by the used approximation for $Y_{jm}(\theta,\phi)$ and $\mathbf{Y}_{jm}^{j}$, that is, condition $\rho_{max} \ll d$ should be met;

$$S = (-1)^{j_{cD}+j_{hD}+m_{hD}+1}\delta_{m_{cD},0}\sqrt{\frac{2j_{cD}+1}{4\pi}}\delta_{m_{cA},0}\sqrt{\frac{2j_{cD}+1}{4\pi}}\times \times(\delta_{m_{hD},1} + \delta_{m_{hD},-1})\sqrt{\frac{2j_{hD}+1}{8\pi}}\sqrt{\frac{2j_{hA}+1}{8\pi}}\delta_{m_{hD},m_{hA}}. \tag{16}$$

The quantities

$$P(\eta_{1(2)}^2) = \frac{4}{3\pi}\sqrt{\eta_{1(2)}^2 + 1}\left[\eta_{1(2)}^2 K\left(\left(\eta_{1(2)}^2 + 1\right)^{-1/2}\right) + \left(1 - \eta_{1(2)}^2\right)E\left(\left(\eta_{1(2)}^2 + 1\right)^{-1/2}\right)\right] \tag{17}$$

where $\eta_1 = (z_1' - z_2)/2\rho_{max}$ for $z_1' > z_2$ and $\eta_1 = (z_2 - z_1')/2\rho_{max}$ for $z_2 > z_1'$; $K(\xi)$ and $E(\xi)$ are the complete elliptic integrals of the first and second kind.

5. Exchange energy transfer rate

The nonradiative energy transfer rate is determined by the Fermi golden rule as

$$W_{i\to f} = \frac{2\pi}{\hbar}|M_{if}|^2\rho \tag{18}$$

where the density of final states $\rho = \frac{1}{\pi\Gamma}$, half-width of the energy level $\Gamma = \frac{\hbar}{\tau}$, τ - the lifetime of the electron-hole pair. The equation (18) is applicable only when interaction between charge carriers in QDs is weak. This means that energy level width is much larger than the transfer rate. In our calculations we took the radiative lifetime as $\tau_l = 10^{-9}s$. In further calculations (see fig. 3) it is shown that the exchange transfer rate is smaller than inverse charge carrier lifetime. In case of the strong interaction, it is necessary to use density matrix formalism [13]. Because the contributions of the direct Coulomb and exchange interactions between the quantum-dot donor and the quantum-dot acceptor are independent the total energy transfer rate is given by $W_{i\to f} = W_{coul} + W_{ex}$, where W_{coul} is the rate of direct Coulomb transfer obtained in [10], W_{ex} is the rate of the exchange energy transfer

$$W_{ex} = \frac{2\pi}{\hbar}|M_{ex}|^2\rho \tag{19}$$

Numerical calculations of exchange energy transfer rate (that is, the rate of energy transfer by Dexter mechanism) are carried out using expressions (19), (15), (12) and (13) for the case $R_D = R_A = R$ and taking into account the similar expressions for the z-depending parts of the acceptor electron wave functions, of the donor and acceptor hole wave functions and the boundary conditions for the functions. The basic transition in the quantum dots is considered. The dependencies of W_{ex} on the distance d are found for different values of the quantum dot radius R. The obtained dependences of W_{ex} on the distance d indicate that they are dominated

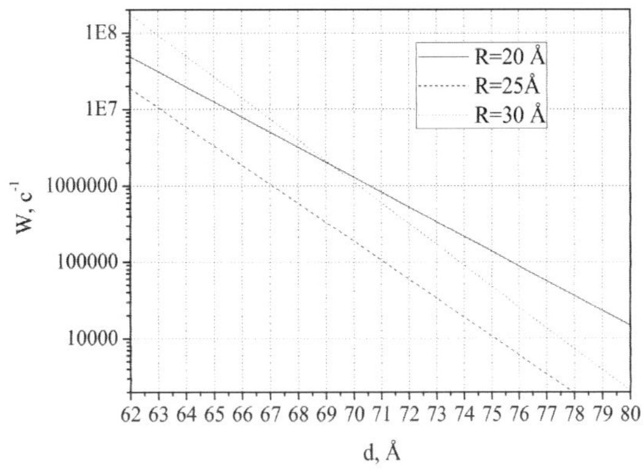

Figure 3. Calculus results of distance dependence of the energy transfer rate at the different quantum dot radii, radiative lifetime is $\tau = 10^{-9}$ s.

by the electron and hole wave functions in the range under the barriers, which are the modified Bessel functions. The different decay rates of W_{ex} are related to changing barrier height with radius of the quantum dot. A comparison between the values of W_{ex} and W_{coul} [10] at distances approaching to the contact ones shows that the contribution of the exchange interaction in the energy transfer should be taken into consideration in numerical calculations.

In conclusions, the results are useful in the correct interpretation of experimental data taking into account both the direct Coulomb interaction of the donor and acceptor electrons and their exchange interaction. This is of particular impotence for measurements of small distance within the biomolecule [2–4].

References

[1] Frasco M and Chaniotakis N 2009 *Sensors* **9** 7266
[2] Yabing L, Qiang M, Xinyan W and Xingguang S *Luminescence* **22**(N1)
[3] Xin-Yan W, Qiang M, Li Ya-Bing L B, Xing-Guang S and Qui-Han J *Canadian Journal of Analytical Sciences and Spectroscopy* **50**(N3)
[4] Schobel U, Egelhaaf H J, Brecht A, Oelkrug D and Gauglitz G 1999 *Bioconjugate Chem.* **10** 1107
[5] Tian-Cai L, Hai-Li Z and Jian-Hao W 2008 *Anal. Bioanal. Chem.* **391** 2819
[6] Forster T 1948 *Ann. Phys.* **2** 55
[7] Dexter D 1953 *J. Chem. Phys.* **21** 836
[8] Clapp A R, Medintz L and Mattoussi H 2006 *Chem. Phys. Chem.* **7** 47
[9] Kruchinin S Y, Fedorov A, Baranov A, Perova T and Berwick K 2008 *Phys. Rev. B* **78** 125311
[10] Samosvat D, Chikalova-Luzina O and Zegrya G *J. Exp. Theor. Phys.* **121**(1)
[11] Kane E 1957 *J. Phys. Chem. Solids* **1** 249
[12] Samosvat D and Zegrya G 2007 *J. Exp, Theor. Phys* **104** 951
[13] Agranovich V M and Galanin M D 1983 *Electronic excitation energy transfer in condensed matter* (Amsterdam, The Netherlands: North-Holland)

Waves in relativistic electron beam in low-density plasma

I Sheinman[1] and J Sheinman (Chernenco)[1]

[1]Physics Department, Saint-Petersburg ElectroTechnical University "LETI", 5 Prof. Popova Street, Saint-Petersburg, Russia

E-mail: isheinman@yandex.ru

Abstract. Waves in electron beam in low-density plasma are analyzed. The analysis is based on complete electrodynamics consideration. Dependencies of dispersion laws from system parameters are investigated. It is shown that when relativistic electron beam is passed through low-density plasma surface waves of two types may exist. The first type is a high frequency wave on a boundary between the beam and neutralization area and the second type wave is on the boundary between neutralization area and stationary plasma.

Key words: plasma, low-density plasma, ion focused regime, electron beam, wave generation, plasma instability.

Tasks of energy transformation of a high current relativistic electron beam to electromagnetic radiation are in the field of active investigation during recent years. Relativistic plasma RF electronics is focused on creation of high power RF amplifiers and generators [1, 2]. Plasma waveguides play role of decelerating systems. Low speed waves are used in linear particle wakefield accelerators where the correspondence between beam and electromagnetic wave speeds is necessary [3-5].

Waves in the beam – plasma system were investigated on the basis of hydro-electrodynamics approach. This method does not demand of additional defining of dielectric permittivity of a moving medium on the basis of particle dynamics equations. It also allows defining conditions of macroscopic balance of the loaded electron bunch in plasma. Similar problems were solved in [6], [7] within electrostatic approach.

In [8] steady and unstable surface electromagnetic waves on boundaries of relativistic plasma streams in the frequency range corresponding to positive values of dielectric permittivity were considered. In the same place existence of critical parameter was shown: the lowest mode number in cylindrical geometry corresponding to transition from slow waves to the fast. For an initial phase of development of hose instability of an electronic stream one more critical parameter – the cross wave size of the stream has been determined. In [9] surface electromagnetic waves on border of a relativistic electron beam in dense plasma in the frequency range corresponding to positive values of dielectric permeability were considered. As well as in case of the plasma stream, for the bunch in dense plasma critical parameters for instability development are the lowest mode number corresponding to transition from slow waves to fast and the cross wave size of the bunch.

As it will be shown below, unlike the bunch in dense plasma and from the plasma stream, in the rarefied plasma the dense bunch completely forces out electrons of plasma both from the beam volume, and from area of neutralization bigger than the beam radius. It leads to complication of dispersive properties of the beam-plasma system.

The hydrodynamic description of the cold multi component charged plasma [7] is based on the joint solution of Maxwell equations

$$\begin{cases} \nabla\times\mathbf{E}_\Sigma = -(1/c)(\partial\mathbf{B}_\Sigma/\partial t); \\ \nabla\times\mathbf{H}_\Sigma = (1/c)(\partial\mathbf{D}_\Sigma/\partial t)+(4\pi/c)\sum_j e_j n_{j\Sigma}\mathbf{v}_{j\Sigma}; \\ \nabla\mathbf{B}_\Sigma = 0; \\ \nabla\mathbf{D}_\Sigma = 4\pi\sum_j n_{j\Sigma}e_j \end{cases} \tag{1}$$

and hydrodynamics

$$\begin{cases} (\partial/\partial t+\mathbf{v}_{j\Sigma}\nabla)\mathbf{p}_{j\Sigma} = e_j\left[\mathbf{E}_\Sigma+(1/c)\mathbf{v}_{j\Sigma}\times\mathbf{B}_\Sigma\right]; \\ \partial n_{j\Sigma}/\partial t+\nabla(n_{j\Sigma}\mathbf{v}_{j\Sigma}) = 0, \end{cases} \tag{2}$$

where $\mathbf{E}_\Sigma$, $\mathbf{H}_\Sigma$ are intensities of electric and magnetic fields respectively; c is the speed of light; $\mathbf{D}_\Sigma$, $\mathbf{B}_\Sigma$ are inductions of electric and magnetic fields respectively; e_j, $n_{j\Sigma}$, $\mathbf{v}_{j\Sigma}$ are the charge, the concentration and the speed of a j-component of plasma respectively; $\mathbf{p}_{j\Sigma} = m_j\gamma_{j\Sigma}\mathbf{v}_{j\Sigma}$ is the impulse of the j-component of plasma, m_j is the mass of the particle of this component, $\gamma_{j\Sigma} = \left(1-|\mathbf{v}_{j\Sigma}|^2/c^2\right)^{-1/2}$ is its relativistic factor.

Systems of the equations of Maxwell and hydrodynamics (1), (2) have to be added with material relationships: $\mathbf{B}_\Sigma = \mathbf{H}_\Sigma$; $\mathbf{D}_\Sigma = \mathbf{E}_\Sigma$.

Let us consider the relativistic bunch of electrons in the motionless plasma environment with equilibrium concentration of particles n_1. We will consider that the bunch represents a cylinder with radius R_b containing electrons moving along its axes (axis z) with the speed $v_0 = \beta_0 c$ and the concentration n_0. Density of the bunch exceeds plasma density. It is possible to consider that more density electron beam moves in a compensating ionic background, completely forcing out plasma electrons. The electrostatic force of interaction of electrons with the ionic background is counterbalanced by ponderomotive force from own magnetic field of the bunch: $n_0 = n_i\gamma_0^2$, where $\gamma_0 = \left(1-\beta_0^2\right)^{-1/2}$. Out of the bunch the system (2) leads to a condition of an electric neutrality of the environment, from where follows that the bunch completely forces out plasma electrons from cylindrical area with the neutralization radius R_n. Neutralization radius R_n is connected with bunch radius R_b by expression:

$$R_n^2 = \frac{n_0}{n_i}R_b^2 = \gamma_0^2 R_b^2.$$

Out of neutralization area concentration of plasma electrons coincides with concentration of ions: $n_1 = n_i$.

In case the current of I_b and the speed of electrons of v_0 are set, and the density of the bunch exceeds the plasma density, there is the bunch self-focusing because of interaction of electrons to the ionic background. The established radius of the bunch can be determined as $R_b^2 = I_b\left(\pi n_1 e v\gamma_0^2\right)^{-1}$, where e is the elementary charge.

In relation to high-frequency electronic fluctuation the area of neutralization behaves as vacuum. Let the dielectric permittivity of the bunch is ε_2 and of external plasma is ε_1 (Figure 1). In works [10-

14] stability of the bunch in relation to its bend concerning the axis of the cylindrical ionic channel was investigated. We will analyze the bunch behavior at small disturbances of particle concentration in the cylinder.

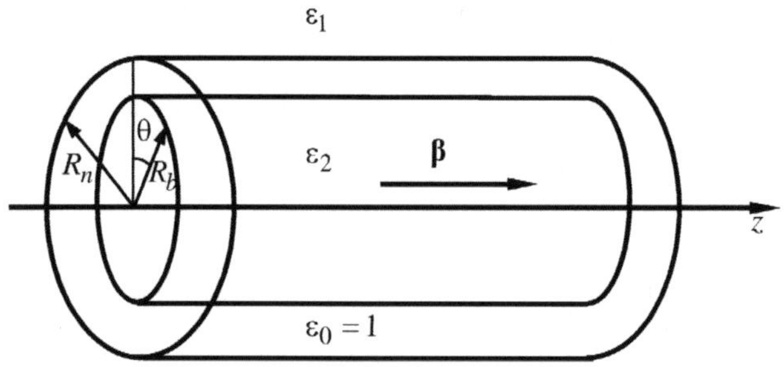

Figure 1. The beam – rarefied plasma system geometry.

Having entered designations

$$k_{p0}^2 = 4\pi e^2 n_0 \left(mc^2\right)^{-1};\ k_{p1}^2 = 4\pi e^2 n_1 \left(mc^2\right)^{-1};\ k_{p2}^2 = 4\pi e^2 n_2 \left(mc^2\right)^{-1},$$

we will receive for $n_0 > n_1$: $k_{p0}^2 = k_{p1}^2\gamma_0^2$, $R_n = \gamma_0 R_b$, $R_b = \left(\dfrac{4eI_b}{mk_{p1}^2 V\gamma_0^2}\right)^{1/2}$.

Solutions of linearized Maxwell's equations (1) and hydrodynamics (2) for the unlimited cylindrical bunch in the longitudinal direction we will look for in the cylindrical system of coordinates (r, θ, z) in the form:

$$\psi(r,\theta,z,t) = \psi(r)e^{i(\nu\theta + k_z z - \omega t)}$$

where ω is a circular frequency, k_z and ν are longitudinal and azimuthal wave numbers.

For receiving the dispersion equation we will use the boundary conditions following from Maxwell's equations:

$$\mathbf{n}\times\left(\mathbf{E}^{(2)} - \mathbf{E}^{(1)}\right) = 0,$$

$$\mathbf{n}\times\left(\mathbf{H}^{(2)} - \mathbf{H}^{(1)}\right) = \frac{4\pi}{c}\lim_{\Delta r\to 0}\int_{-\Delta r}^{\Delta r}\left(\sum_j e_j\left(\tilde{n}_j\mathbf{V}_j + n_j\mathbf{V}_j\right)dr\right)$$

where $\mathbf{n}$ is a normal to the bunch surface, $\mathbf{E}^{(2)}$, $\mathbf{E}^{(1)}$, $\mathbf{H}^{(2)}$, $\mathbf{H}^{(1)}$ are intensity of electric and magnetic fields on both sides from the medium boundary, Δr is counted from the boundary in the direction of the normal.

Solving in systems of equations of Maxwell (1) and hydrodynamics (2), with boundary conditions on boundaries of the bunch and the area of neutralization we will receive the dispersion equation for electromagnetic waves in the system:

$$P^2 A_{IIb}A_{KKn} - P(A_{IKb}A_{KIn} + A_{KIb}A_{IKn}) + A_{KKb}A_{IIn} - 2PQ_bQ_n(S_{0Kb} - S_{0Ib})(S_{0Kn} - S_{0In}) = 0$$

where $S_1 = \dfrac{K'_\nu(kR_n\eta_1)}{\eta_1 K_\nu(kR_n\eta_1)}$, $S_2 = \dfrac{I'_\nu(kR_b\eta_2)}{\eta_2 I_\nu(kR_b\eta_2)}$, $S_{0Kn} = \dfrac{K'_\nu(kR_n\eta_0)}{\eta_0 K_\nu(kR_n\eta_0)}$, $S_{0In} = \dfrac{I'_\nu(kR_n\eta_0)}{\eta_0 I_\nu(kR_n\eta_0)}$,

$S_{0Kb} = \dfrac{K'_\nu(kR_b\eta_0)}{\eta_0 K_\nu(kR_b\eta_0)}$, $S_{0Ib} = \dfrac{I'_\nu(kR_b\eta_0)}{\eta_0 I_\nu(kR_b\eta_0)}$, $P = \dfrac{I_\nu(kR_b\eta_0)}{I_\nu(kR_n\eta_0)}\dfrac{K_\nu(kR_n\eta_0)}{K_\nu(kR_b\eta_0)}$, $Q_n = \dfrac{\nu\eta}{kR_n}\left(\dfrac{1}{\eta_1^2} - \dfrac{1}{\eta_0^2}\right)$,

$$Q_b = \frac{\nu}{kR_b}\left(\frac{\eta}{\eta_0^2} - \frac{\tilde{\eta}}{\eta_2^2}\right), \quad A_{ijb} = \left(S_{0ib} - \varepsilon_2 S_2\right)\left(S_{0jb} - S_2\right) - Q_b^2, \quad A_{ijn} = \left(S_{0in} - \varepsilon_1 S_1\right)\left(S_{0jn} - S_1\right) - Q_n^2,$$

$$\{i,j\} = \{I,K\}, \quad \tilde{\eta} = \gamma^2\left((1-\varepsilon_2\beta^2)\eta + (\varepsilon_2 - 1)\beta\right), \quad \eta_0^2 = \eta^2 - 1, \quad \eta_2^2 = \eta^2 - 1 - (\varepsilon_2 - 1)\gamma^2(1-\eta\beta)^2,$$

$\eta_1^2 = \eta^2 - \varepsilon_1$, $\eta = k_z/k$, $I_\nu(x)$ is the modified Bessel function, $K_\nu(x)$ is the Macdonald function.

In case $R_n = R_b = R$ the equation coincides with the dispersion equation for waves on the relativistic plasma stream in motionless plasma [8]. In ultra-relativistic limit $\gamma \to +\infty$ two variants are possible. The first is $R_n \to +\infty$, in this case the dispersion equation describes surface waves on borders between the relativistic plasma stream and vacuum. The second is $R_b \to 0$, thus the equation describes surface waves on border of vacuum and motionless plasma.

The following algorithm was developed for search of roots of the dispersive equation. On the first step of algorithm intervals of equation root isolation in the complex plane were determined. Two methods were applied: search of intervals of transition through zero of imaginary and real parts of the function value and search of minima of the absolute value of function. The found coordinates were substituted as initial approach for search of roots by the method of secants.

After search of possible roots their check by substitution of the received values in initial expression was made. If thus initial expression accepted the value which is not exceeding $1 \cdot 10^{-8}$, the decision that the root is found was made.

Search of roots was run at various initial values of parameters of the wave number k_z and the bunch radius R_b. If the received value of the root slightly differed from the found values at other close k_z and R_b, we have considered the decision steady. Therefore not separate points, but curves on figures below are of interest.

Dependences of roots of the dispersion equation on the bunch radius for three first modes of the system are presented in tables 1 ($n_1 = 10^{12}\,\text{cm}^{-3}$) and 2 ($n_1 = 10^{14}\,\text{cm}^{-3}$) at $k = 1\,\text{cm}^{-1}$ ($f = 4.77$ GHz), $\beta = 0.9$. In the first and second lines of the tables real and imaginary parts of roots are shown in the dependence on bunch radius. Positions of roots on the complex plane are given in the third line of tables.

At $n_1 = 10^{11}\,\text{cm}^{-3}$ and lower there are no roots with $\text{Im}\,k_z \neq 0$ on zero mode. But on the first and the second modes there is a resonant convective instability of the bunch ($\text{Re}\,k_z \approx k$, $\text{Im}\,k_z \neq 0$) which takes place for all range of the bunch radius values (0.01 – 2.5 cm). Besides, there are slow modes of beam-plasma system with $\text{Re}\,k_z < k$, which are excited if the radius of the bunch exceeds some critical value.

At $n_1 = 10^{12}\,\text{cm}^{-3}$ plasma frequency is less than the chosen frequency $k > k_p$ ($\omega > \omega_p$). In this case there is the resonant convective instability of the bunch on zero mode for which it is fair $\text{Re}\,k_z \approx k$, $\text{Im}\,k_z \neq 0$. At $\nu > 0$ slow modes of beam-plasma system are accruing with $\text{Re}\,k_z < k$, which are excited if the radius of the bunch exceeds some critical value.

Roots with $\text{Re}\,k_z \approx k$, $\text{Im}\,k_z \neq 0$ were obtained at $n_1 = 10^{13}\,\text{cm}^{-3}$ on zero mode if the bunch radius is more than some critical value. As in the previous case there is the resonant convective instability of the bunch for those radius values and equilibrium concentration of particles of plasma.

At $n_1 = 10^{14}\,\text{cm}^{-3}$ $k < k_p$ ($\omega < \omega_p$) on zero mode the resonant instability of the bunch practically does not arise. However accruing are waves with negative real part of k_z, that is the waves moving towards the bunch. At $\nu > 0$ slow modes are accruing at excess by the radius of the critical value and with real part of wave number, close to zero, that means emergence of absolute instability.

Journal of Physics: Conference Series **769** (2016) 012079 doi:10.1088/1742-6596/769/1/012079

At $n_1 = 10^{14} - 10^{15}$ cm^{-3} there is the resonant convective instability of the bunch ($\mathrm{Re}\, k_z \approx k$, $\mathrm{Im}\, k_z \neq 0$) on the first and the second modes which is excited if the radius of the bunch exceeds some critical value.

Table 1. Roots of dispersion equation k_z vs bunch radius R_b at $n_1 = 10^{12}$ cm^{-3}.

$\nu = 0$	$\nu = 1$	$\nu = 2$
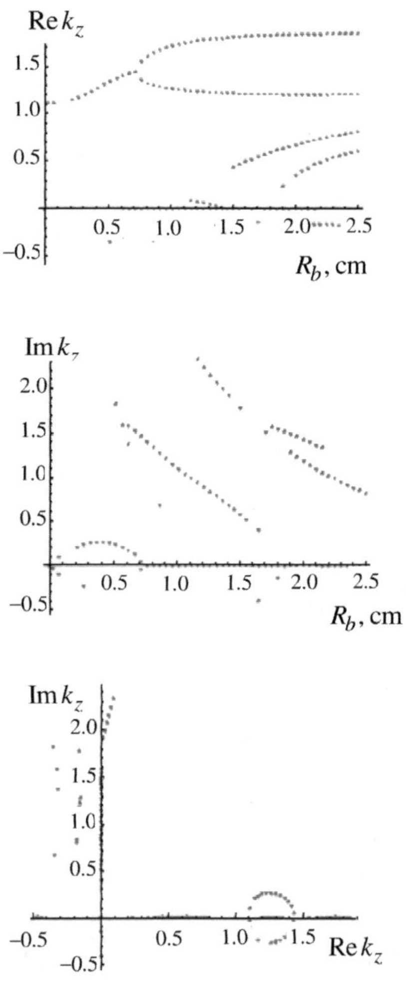	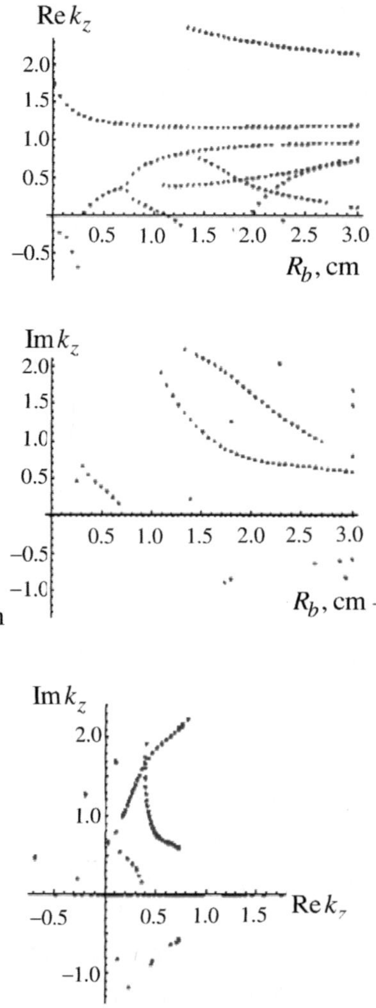	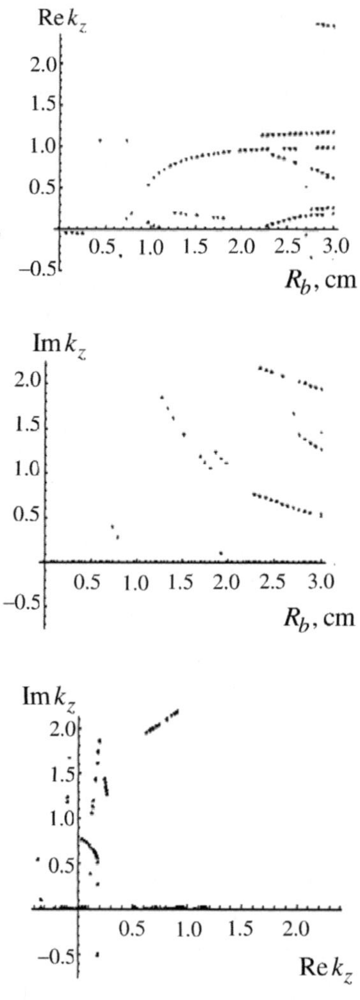

Journal of Physics: Conference Series **769** (2016) 012079 doi:10.1088/1742-6596/769/1/012079

Table 2. Roots of dispersion equation k_z vs bunch radius R_b at $n_1 = 10^{14}$ cm^{-3}.

$\nu = 0$	$\nu = 1$	$\nu = 2$

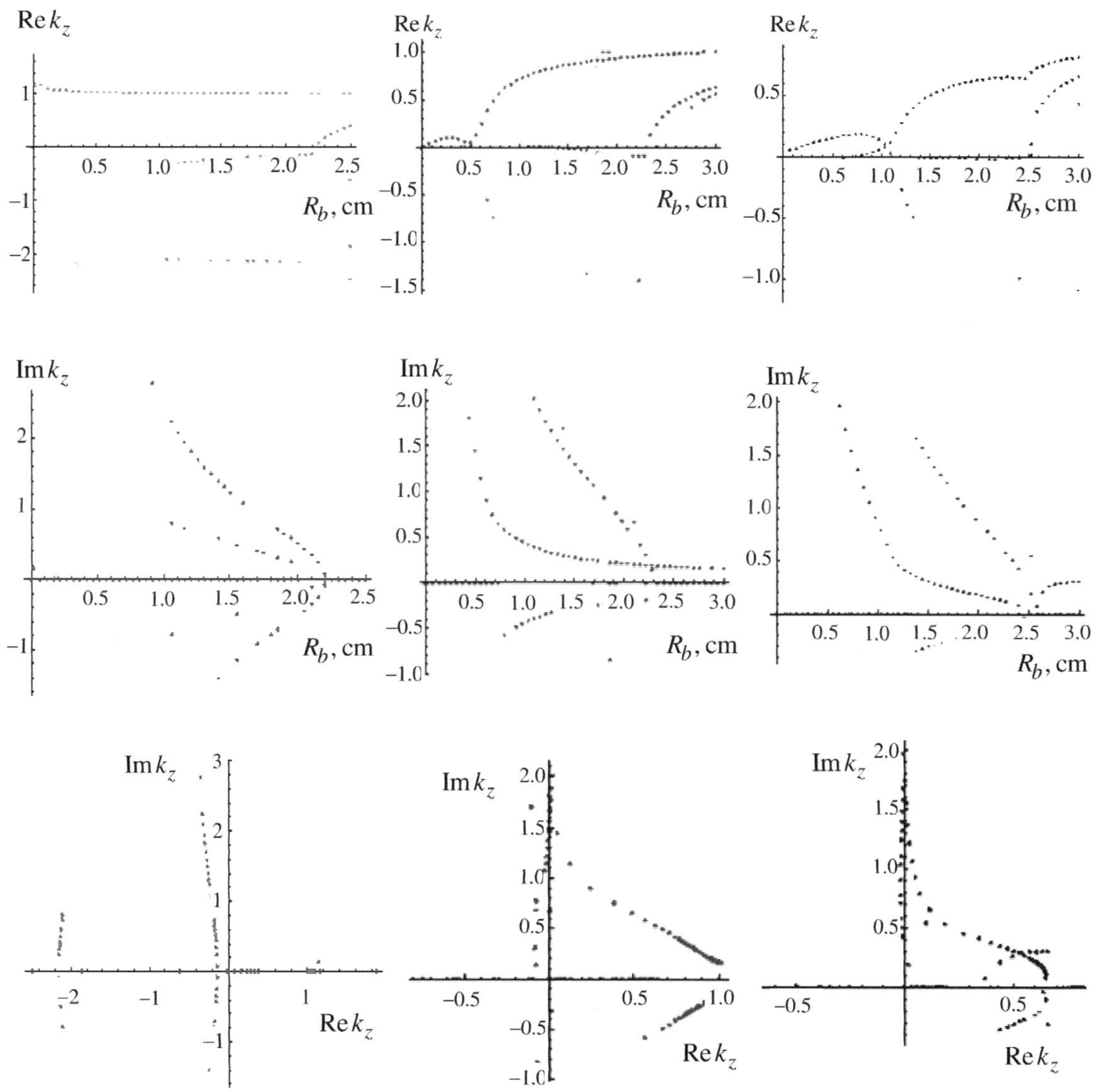

The analysis of the received dispersive equation shows that when passing the ultra relativistic bunch of electrons ($\beta = 0.999$) through the rarefied plasma there can be surface waves of two types: high-frequency on border of the relativistic bunch with area of neutralization and low-frequency on border of area of neutralization and motionless plasma. The first type corresponds to surface waves on border of the relativistic plasma stream with vacuum, the second – on border of the vacuum cylinder and motionless plasma.

References

[1] Kondratenko A N and Cooklin V M Osnovy plazmennoi elektroniki. M.: Energoatomizdat, 1988. 320 p

[2] Eksperimentalnaia reliativistskaia plazmennaia SVCh-elektronika. //Reliativistskaia plazmennaia SVCh-elektronika. M.: Nauka, 1994. 192 p. (Trudy IOFAN. V. 45)
[3] Plasma Particle Accelerators. John M Dawson in Scientific American, Vol. 260, No. 3, pages 54-61; March 1989
[4] Plasma Accelerators at the Energy Frontier and on Tabletops. Chandrashekhar Joshi and Thomas Katsouleas in Physics Today, Vol. 56, No. 6; pages 47-53; June 2003
[5] Accelerator Physics: Electrons Hang Ten on Laser Wake. Thomas Katsouleas in Nature, Vol. 431, pages 515-516; September 30, 2004
[6] Ronald C Davidson, "Theory of Nonneutral Plasmas (Advanced Book Classics)" Addison-Wesley. 1989. 199 p
[7] Sheinman I L Surface waves at the boundaries of relativistic plasma streams // Technical Physics, Volume 46, Number 5. May, 2001, pp. 532-539
[8] Sheinman I L Surface waves on a relativistic electronic beam in the dense plasma. (Poverkhnostnye volny na reliativistskom elektronnom puchke v plotnoi plazme). Izvestiia Vuzov Rossii. Radioelektronika. № 4. 2013. P. 76-83
[9] Lee E P Resistive hose instability of a beam with a Bennett profile. Phys. Fluids, Vol 21, No 8, c. 1327-1343; 1978
[10] Lampe M, Joyce G, Slinker S P, Wittum D H, "Electron-hose instability of a relativistic electron beam in an ion-focusing channel," *Phys. Fluids B*, Vol 5, No 6, 1993. — c. 1888-1901
[11] Wittum D H "Cumulative hose instabilities of a magnetically self-focused slab electron beam," *Phys. Fluids B*, Vol 5, No 12, 1993. — c. 4432-4455
[12] H. Lee Buchanan, "Electron beam propagation in the ion-focused regime," *Phys. Fluids*, Vol 30, No 1, 1987. — c. 221-231
[13] Wittum D H "Electromagnetic instability of the ion-focused regime," *Phys. Fluids B*, Vol 4, No 3, 1992. — c. 730-739

Investigation of the passage between LES and RANS subdomains in the framework of zonal RANS-LES approaches

M S Gritskevich[1], A V Garbaruk[1], F R Menter[2]

[1]Peter the Great St. Petersburg Polytechnic University, 195251, St. Petersburg, Russia
[2]Software Development Department, ANSYS Germany GmbH, 83714, Otterfing, Germany

E-mail: gritckevich@gmail.com

Abstract. A detailed investigation of the passage between LES and RANS subdomains for zonal RANS-LES simulations of wall-bounded flows is performed with the main focus on the turbulent quantities reconstruction at LES-RANS interfaces. It is shown, that most accurate results are obtained if the turbulent quantities are specified from a precursor RANS simulation, while employing the actual solution from the LES subdomain is slightly less precise.

1. Introduction

Zonal RANS-LES approaches are now routinely applied to industrial flow simulations in order to balance computational and solution accuracy requirements. Within such approaches LES (Large Eddy Simulation) is used only in areas where RANS (Reynolds Averaged Navier-Stokes) models are not suitable, while RANS is used elsewhere. The main issue of such approaches is to provide a consistent blending when the mean flow crosses the interfaces between LES and RANS subdomains [1–3]. Depending on the flow direction, two types of interfaces can be distinguished, namely the RANS-LES interface (the flow from RANS to LES), and the LES-RANS interface (the flow from LES to RANS).

From the physical point of view, specification of the conditions at RANS-LES interfaces is the most difficult case, since the information required for the LES modeling is not readily available within the RANS solution. It should be mentioned that a number of elaborated methods for generating of the resolved turbulence from time-averaged RANS profiles are available in the literature [3–5], the best of which are highly accurate and compatible with general-purpose CFD codes.

At the same time, despite the apparent physical simplicity of the conditions at LES-RANS interfaces, their implementation into general purpose CFD codes is far from being straightforward. Particularly, the instantaneous LES velocity should be converted into the corresponding mean fields [3], which in principle, could be achieved with the use of an averaging procedure. However, for practical reasons, such methods could be only rarely applied for complex problems since the reconstruction of the turbulence quantities requires second order statistics, which is known to be very sensitive to the averaging sample and, therefore, requires a high level of computational time.

A possible alternative to the averaging is proposed in [6], where it is suggested to combine an algebraic sub-grid model in LES subdomains with a differential model in RANS subdomains. Within such an approach, the differential equations for the turbulent quantities are solved in the entire domain and, therefore, could be used at LES-RANS interfaces, while the resolved structures are suppressed due to the high eddy viscosity and dissipative upwind scheme in RANS subdomains.

It should be mentioned that solving of the RANS equations in LES subdomains on the basis of the resolved flow field yields excessive levels of the 'RANS' turbulent kinetic energy and eddy viscosity due to the higher velocity gradients of the LES flow field. This problem could be potentially avoided by utilizing the turbulent quantities at the LES-RANS interface from a precursor RANS simulation, however, some additional problems with such methods are anticipated if the precursor and the actual solutions are substantially different. In addition, another issue, which is known as turbulence kinetic energy "double accounting" [3], originates from the presence of both modeled and resolved turbulence at the LES-RANS interface resulting in a strong overestimation of the total kinetic energy and total shear stresses. This problem could, in principle, be reduced if the underlying RANS transport equations have a mechanism for adaptation of the modeled turbulence to the resolved scales (such models are sometimes termed as 2nd generation RANS models [1–3]), which decreases the modeled turbulent kinetic energy and results in more consistent conditions at LES-RANS interfaces.

Thus, the objective of the paper is to investigate different methods for the computation of turbulent quantities at the LES-RANS interface. The tests are carried out for fully developed channel flow [7] in the framework of the general purpose code ANSYS Fluent [8].

2. Test case description

Simulations of the channel [7] are conducted at Reynolds numbers of $Re_\tau = u_\tau \cdot h/\nu = 395$ and $Re_\tau = 18000$ (u_τ is the friction velocity, $h = 0.5 \cdot H$ is the channel half-height, and ν is the kinematic viscosity).

The computational domain for this flow (Figure 1) is decomposed into three subdomains referred to as AUX-LES, LES, and RANS with the dimensions of $4H \times H \times 1.5H$, $2H \times H \times 1.5H$, and $16H \times H \times 1.5H$ respectively. The AUX-LES subdomain is used for the LES inflow turbulent content generation, thus eliminating the additional uncertainties due to synthetic turbulence generators.

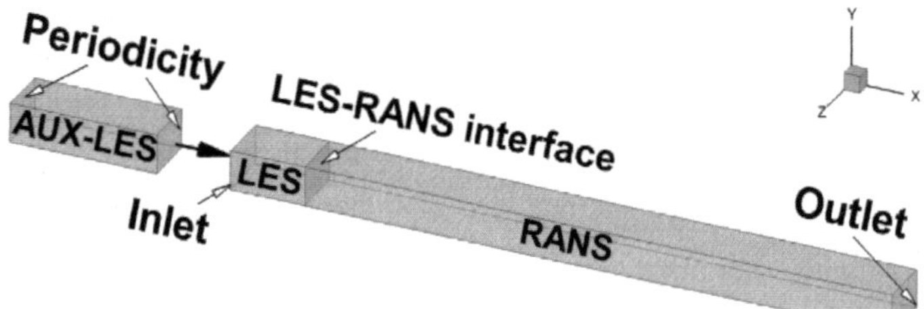

Figure 1. Computational setup for the developed channel flow

The computational grid, consists of about $2 \cdot 10^6$ hexahedral cells, has the grid-steps of $\Delta x/H = 1/10$ and $\Delta z/H = 1/20$ in the streamwise and spanwise directions, while in the wall-normal direction a dense clustering is employed to satisfy $\Delta y^+_w < 1$ for both Reynolds numbers.

The applied boundary conditions are shown in Figure 1. For all the considered subdomains, a y^+-insensitive wall treatment [9] is utilized at the top and bottom walls and the periodicity condition is specified in the spanwise direction. In the AUX-LES subdomain the periodicity condition is also specified in the streamwise direction and the flow is driven with a constant pressure gradient of $dp/dx = -2 \cdot \rho \cdot u_\tau^2/H$ (p is the pressure and ρ is the density), which is taken into account in the governing equations via the source term in the momentum equations. At the LES inlet, the velocity and turbulence quantities are specified from the profiles extracted at each time step from the periodic boundary of AUX-LES. At the LES-RANS interface the turbulent quantities are either specified from the actual LES solution (one-stage or 1S approach) or from the precursor RANS simulation (two-stage or 2S approach). Finally, the constant pressure is specified at the RANS outlet.

It is worth mentioning that the convective fluxes for the momentum equations are computed with the use of the second order central differencing scheme in the AUX-LES and LES subdomains, while in the RANS subdomain the second order upwind scheme is employed.

3. Assessment of two-stage approaches

For the 2S approach, the flow is firstly computed entirely with the underlying RANS model. In the second stage, this RANS solution for k and ω is frozen in the LES subdomain and kept in the background while the LES model is active. At the LES-RANS interface, the frozen solution from the two-equation model is re-activated, while resolved turbulence structures from the LES domain are allowed to pass into the RANS zone. It is then observed how quickly these structures are damped and how realistic the time averaged solution is compared to the known channel flow solution. Two underlining RANS models are considered for assessment of the two-stage approach [6], namely the

Shear Stress Transport model (2S-SST) [9] and the baseline version of the SST model (2S-BSL) [10], while in the AUX-LES and LES subdomains the eddy viscosity is calculated from the algebraic version of the Improved Delayed Detached Eddy Simulation (IDDES) [11].

As seen from the instantaneous vorticity iso-lines (Figure 2 and Figure 3) the resolved turbulent structures are observed downstream of the LES-RANS interface (x/H=0) for both models, as expected. Particularly, for the near-wall location, large spanwise-elongated vortices can be distinguished up to x/H=6, while at the channel center, the structures are more isotropic and decay slightly more rapidly regardless of the considered Reynolds number (the results for Re_τ=18000 are not shown).

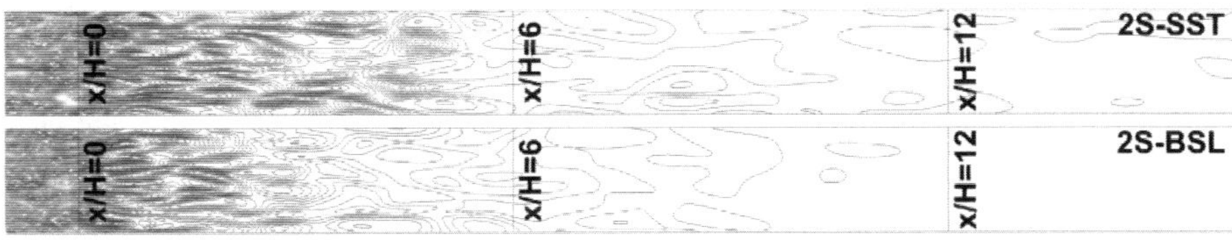

Figure 2. Vorticity iso-lines at the near wall section (y/H=0.05) for Re_τ=395.

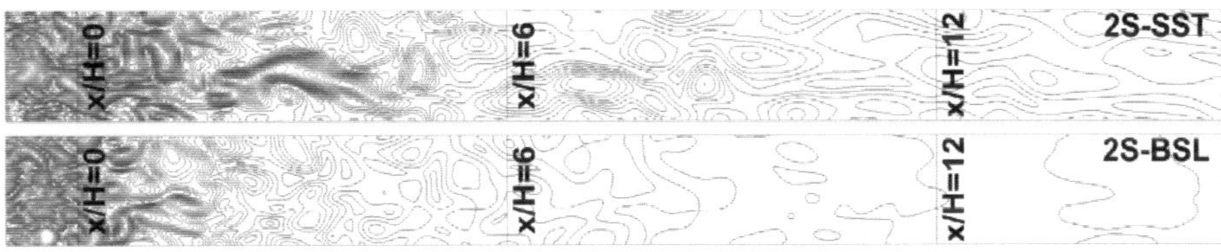

Figure 3. Vorticity iso-lines at the channel center (y/H=0.5) for Re_τ=395

The skin friction coefficient distributions (Figure 4) show a noticeable discrepancy near the LES-RANS interface for 2S-BSL, while for 2S-SST only a small deviation is observed near the LES-RANS interface, while further downstream almost perfect agreement is achieved. It should be mentioned that the difference between 2S-SST and 2S-BSL is mainly attributed to the SST limiter [9] of the former, which decreases the eddy viscosity due to the noticeably larger velocity gradients of the LES solution resulting in a slower decay of the resolved turbulence in the RANS subdomain.

For the mean velocity (Figure 5), 2S-SST yields almost identical solutions to those of the reference RANS solution independently of the Reynolds number, while 2S-BSL profile has a slightly smaller slope at the logarithmic part due to the higher value of the skin friction coefficient.

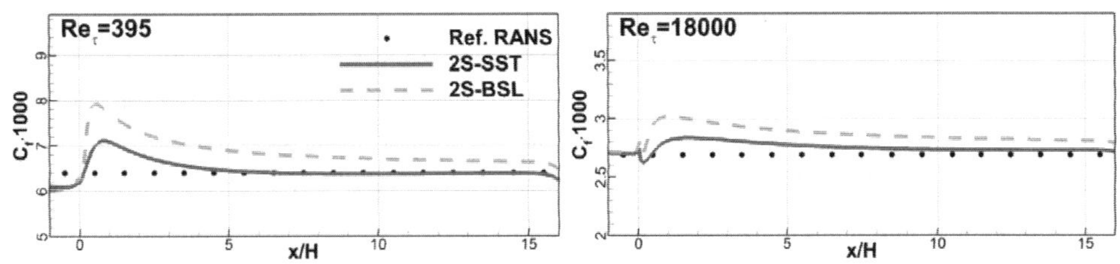

Figure 4. Distributions of the skin friction coefficient along the wall

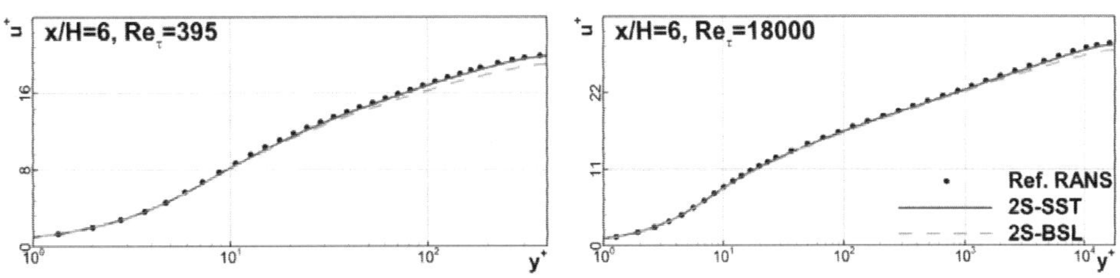

Figure 5. Profiles of the mean velocity at different Re numbers at x/H=6

Finally, the profiles of the total kinetic energy (sum of resolved and modelled) in Figure 6 show that 2S-SST and 2S-BSL provide almost identical solutions at the LES-RANS interface (x/H=0), which are noticeably higher than the reference RANS due to the "double accounting" effect. Further downstream the resolved turbulence is suppressed by the RANS model and the total kinetic energy is recovering back to the reference solution.

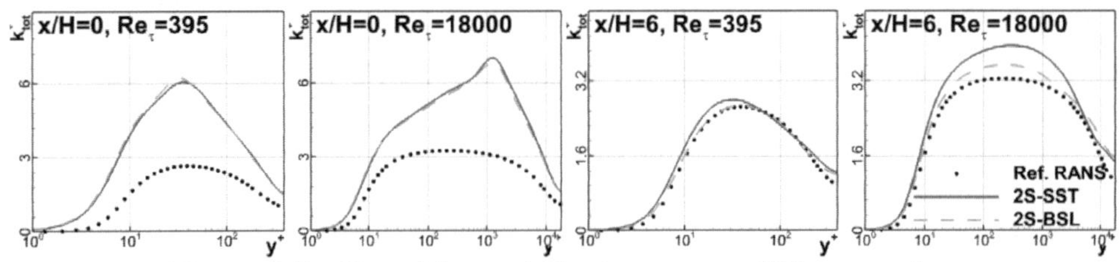

Figure 6. Profiles of the total kinetic energy at different sections

4. Comparison of one-stage and two-stage approaches

For the 1S approach, the two-equation RANS turbulence model is run passively in the background of the LES simulation domain based on the unsteady LES flow field and the solution of the RANS model is then re-activated downstream of the RAN-LES interface. As it has been shown in the previous subsection, 2S-SST provides slightly better results than 2S-BSL for the skin friction and therefore, this model is used as the reference solution for the 1S approaches. For that purpose two turbulence models are considered namely the SST model (1S-SST) [9] and the SST based Scale-Adaptive Simulation (1S-SAS) [12]. It should be mentioned, that 1S-SAS, which is a 2nd generation RANS model, has the mechanism for the specific dissipation rate amplification in the presence of the resolved turbulent structures and, therefore, it could be anticipated that the resolved turbulence of 1S-SAS would be observed for a relatively large distance downstream of the LES-RANS interface. Indeed, as seen from the vorticity iso-lines (Figure 7 and Figure 8) 1S-SAS sustains the turbulent structures in the entire RANS subdomain. In contrast, 1S-SST provides almost immediate decay of the resolved turbulence downstream of the LES-RANS interface, which is even more rapid than those of 2S-SST.

Figure 7. Vorticity iso-lines for at the near wall section (y/H=0.05) for Re_τ=395.

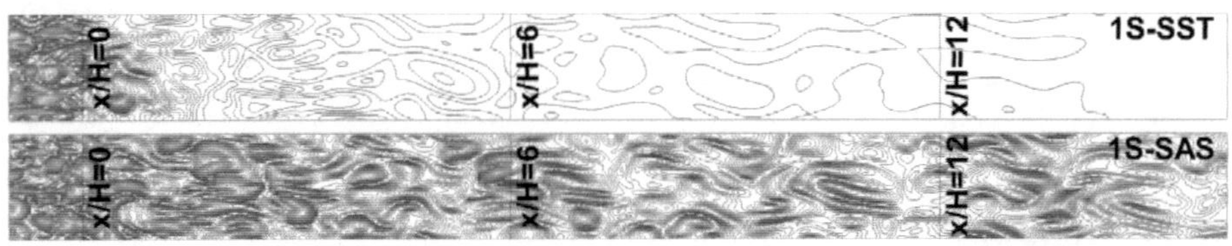

Figure 8. Vorticity iso-lines at the channel center (y/H=0.5) for Re_τ=395.

However, as seen from the skin friction coefficient (Figure 9), both 1S models are less accurate than 2S-SST for both considered Reynolds numbers. Particularly, 1S-SST shows a substantial deviation near the LES-RANS interface, which is relatively slowly decaying downstream, while the skin friction of 1S-SAS is lower than the reference solution in the entire RANS subdomain. Consistently, 1S-SST yields slightly lower values of the mean velocity at the channel center (Figure 10), while 1S-SAS shows the opposite trend. Particularly, such behavior of 1S-SAS for the mean velocity profile is similar to the logarithmic-layer mismatch [11].

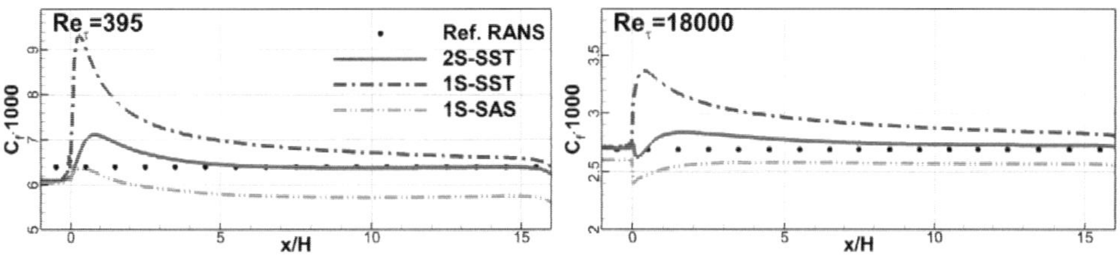

Figure 9. Comparison of the skin friction coefficient for 2S and 1S approaches

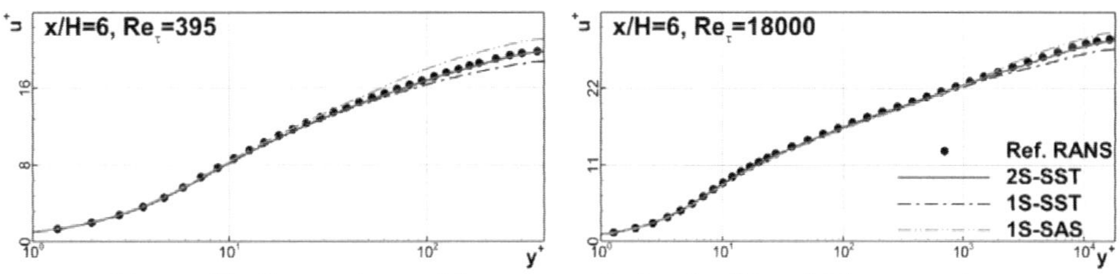

Figure 10. Comparison of the mean velocity for 2S and 1S approaches

As seen from the modeled kinetic energy (Figure 11) both 1S-SST and 1S-SAS are noticeably different from the reference RANS solution at the LES-RANS interface. Particularly, 1S-SST noticeably overestimates the modeled kinetic energy near the LES-RANS interface due to the excessive turbulence generation in the LES subdomain and the "double accounting" effect when the flow re-enters the RANS subdomain, while 1S-SAS yields lower modeled kinetic energy due to its inherent sensitivity to the resolved turbulent structures. Finally, the total kinetic energy (Figure 12) for both 1S-SAS and 1S-SST is substantially higher at the LES-RANS interface (*x/H*=0) than those of the reference RANS solution. Further downstream, 1S-SAS yields similar to 2S-SST profiles, while 1S-SST still noticeably overestimates the total kinetic energy at the channel center.

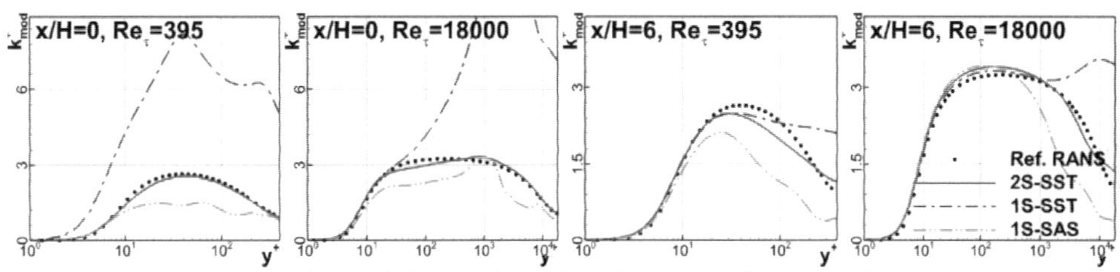

Figure 11. Comparison of the modeled kinetic energy for 2S and 1S approaches.

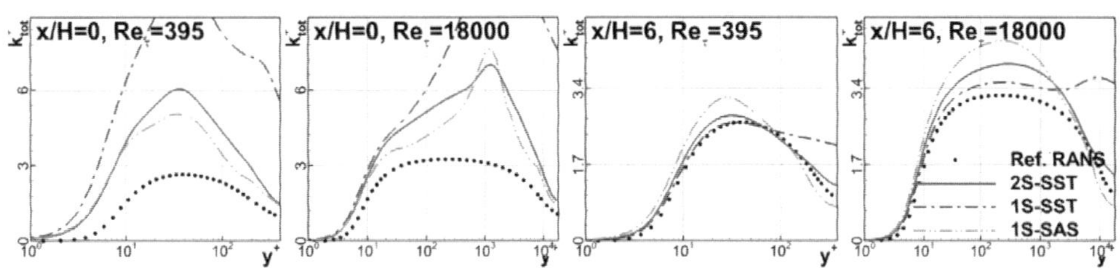

Figure 12. Comparison of the total kinetic energy for 2S and 1S approaches

5. Summary

Several approaches for modeling of the passage between LES and RANS subdomains have been investigated with the main focus on recovering proper turbulence quantities.

It has been shown that the considered two-stage approaches (2S-SST and 2S-BSL) yield more accurate results than the one-stage approach (1S-SST and 1S-SAS), however, the mean velocity and the skin friction coefficient of the latter are still in a reasonable agreement with the reference RANS solutions.

At the same time, the total kinetic energy of the two-stage approaches is strongly overestimated due to the "double accounting" effect at the LES-RANS interface, which is more pronounced for 2S-BSL than for 2S-SST due to the SST correction [9] of the latter. Further downstream, both models provide a relatively quick recovering of the total kinetic energy. Nevertheless, it is worth mentioning that the two-stage approaches are inappropriate when the RANS and LES subdomains are strongly coupled or if the flow has an inherent unsteadiness (e.g. due to unsteady body motions).

With regards to one-stage approaches, due to an excessive turbulence generation in the LES subdomain and the "double accounting" effect, 1S-SST noticeably overestimates the total kinetic energy with a relatively slow recovery downstream. At the same time, 1S-SAS reasonably agrees with the reference RANS solution due to its inherent mechanism for resolved turbulence detection ensuring a physically consistent solution in RANS subdomains. On the other hand, 1S-SAS requires a relatively long distance in the RANS subdomain to completely suppress the resolved turbulence.

Acknowledgement
The authors from St. Petersburg are partially supported by Russian Foundation for Basic Research (grant No. 14-08-31121) and Russian Science Foundation (grant No. 14-11-00060). The computations are conducted with the support of Saint-Petersburg Branch of the Joint Supercomputer Center of the Russian academy of sciences.

References
[1] Sagaut P., Deck S. and Terracol M. 2006 *Multiscale and Multiresolution Approaches in Turbulence*
[2] Menter F. R., Schütze J. and Gritskevich M. S 2012 Global vs. Zonal Approaches in Hybrid RANS-LES Turbulence Modelling *Progress in Hybrid RANS-LES Modelling, Notes on Numerical Fluid Mechanics and Multidisciplinary Design Volume 117* pp 15–28
[3] Frohlich J. and Von Terzi D. 2008 Hybrid LES/RANS Methods for the Simulation of Turbulent Flows *Progress in Aerospace Sciences* **44** 349–77
[4] Nolin G., Mary I., Ta-Phuoc L., Hinterberger C. and J. F 2006 Coupling from LES to RANS using eddy viscosity models *Direct and Large-Eddy Simulation VI* vol 6 pp 679–86
[5] von Terzi D, Mary I and Frohlich J 2009 Segregated LES/RANS Coupling Conditions for the Simulation of Complex Turbulent Flows *Numerical Simulation of Turbulent Flows and Noise Generation Volume 104* vol 104, ed C Brun, D Juvé, M Manhart and C-D Munz pp 231–52
[6] Cokljat D, Caridi D, Link G, Lechner R and Menter F 2009 Embedded LES Methodology for General-Purpose CFD Solvers *Proceedings of 6th International Symposium on Turbulence and Shear Flow Phenomena* pp 1191–6
[7] Moin P and Kim J 1982 Numerical investigation of turbulent channel flow *Journal of Fluid Mechanics* **118** 341–77
[8] Mathur S R and Murthy J Y 1997 A pressure-based method for unstructured meshes *Numerical Heat Transfer* **32** 195–215
[9] Menter F R, Kuntz M and Langtry R 2003 Ten Years of Experience with the SST Turbulence Model *Proceedings of 4th International Symposium on Turbulence, Heat and Mass Transfer* pp 625–32
[10] Menter F R 1993 Zonal two equation k-w turbulence models for aerodynamic flows *AIAA Paper* **1993-2906**
[11] Shur M L, Spalart P R, Strelets M K and Travin A K 2008 A hybrid RANS-LES approach with delayed-DES and wall-modeled LES capabilities *International Journal of Heat and Fluid Flow* **29** 1638–49
[12] Menter F R and Egorov Y 2010 The Scale-Adaptive Simulation Method for Unsteady Turbulent Flow Predictions. Part 1: Theory and Model Description *Flow Turbulence and Combustion* **85** 113–38

Journal of Physics: Conference Series **769** (2016) 012081 doi:10.1088/1742-6596/769/1/012081

Application of DDES and IDDES with shear layer adapted subgrid length-scale to separated flows

E K Guseva, A V Garbaruk and M Kh Strelets

Peter the Great St. Petersburg Polytechnic University, 29 Polytechnicheskaya Str., 195251, St. Petersburg, Russia

E-mail: katia.guseva@inbox.ru

Abstract. A comparative study is conducted of the original versions of Delayed Detached-Eddy Simulation (DDES) and Improved DDES (IDDES) and these approaches combined with "shear-layer-adapted" (SLA) subgrid length-scale proposed recently for resolving the issue of delayed RANS-to-LES transition in separated shear layers in global hybrid RANS-LES approaches. Computations were carried out of two separated flows: a transonic flow past M 219 cavity and a subsonic flow over NASA wall mounted hump. Results of the computations suggest that the use of the SLA subgrid length-scale considerably accelerates transition to resolved three-dimensional turbulence in the separated shear layers and substantially improves agreement with the experimental data.

1. Introduction

Hybrid RANS-LES approaches to turbulence modeling combine high quality of Large Eddy Simulation (LES) in separated flow regions and low computational costs of the methods based on the Reynolds Averaged Navier-Stokes equations (RANS) in the near wall regions of the flow. However near the boundary between RANS and LES regions (so-called "gray area"), where the solution is not either pure RANS or LES, accuracy of the hybrid approaches decreases. One of the examples is a tangible delay of formation of developed three-dimensional turbulent structures in separated shear layers typical of global hybrid methods like DES [1], which decreases quality of the solution not only in the shear layer but in the whole flow as well.

A remedy for this flaw has been suggested in a recent publication of Shur *et al.* [2], who proposed a new "shear-layer-adapted" (SLA) definition of the subgrid length-scale for global hybrid RANS-LES methods. Based on a set of numerical examples, it has been shown that replacing the standard DES definition of the subgrid length-scale (maximum local grid spacing) by the newly proposed one, Δ_{SLA}, ensures a rapid unlocking of the Kelvin – Helmholtz (KH) instability and significant acceleration of transition to the developed turbulence in free and separated shear layers. The present work is aimed at evaluating efficiency of the DDES [3] and IDDES [4] approaches combined with the SLA subgrid length-scale and at finding out whether any undesirable side effects caused by interaction of this length-scale with the empirical functions involved in formulations of DDES and IDDES takes place.

2. Shear-layer adapted subgrid length-scale and its incorporation into the DDES and IDDES approaches

A detailed outline of the physical background of the new subgrid length-scale definition is presented in [2]. The definition reads as:

$$\Delta_{SLA} = \widetilde{\Delta}_{\omega} F_{KH}(<\mathrm{VTM}>) \tag{1}$$

Here $\widetilde{\Delta}_{\omega} = \frac{1}{\sqrt{3}} \max_{n,m=1,8} |(\mathbf{I}_n - \mathbf{I}_m)|$, where $\mathbf{I}_n = \mathbf{n}_{\omega} \times \mathbf{r}_n$, $\mathbf{n}_{\omega}$ is the unit vector aligned with the vorticity vector, and $\mathbf{r}_n$ is the radius-vector of the n-th vertex of the considered grid cell. This quantity takes into account high anisotropy of the grids typically used in initial regions of the shear layers and, in particular, excludes a control of both maximum and minimum grid steps in these regions.

The second ingredient of the new length-scale (1) is the empirical function F_{KH} aimed at additional reducing of the subgrid length-scale in the initial regions of the shear layers. The argument of this function, a kinematic parameter VTM, is used to identify quasi-2D regions of the flow. It is defined as follows:

$$\mathrm{VTM} \equiv \frac{\sqrt{6}\,|(\hat{\mathbf{S}} \cdot \boldsymbol{\omega}) \times \boldsymbol{\omega}|}{\omega^2 \sqrt{3tr(\hat{\mathbf{S}}^2) - [tr(\hat{\mathbf{S}})]^2}}, \tag{2}$$

where $\hat{\mathbf{S}}$ is the strain rate tensor, $tr(\,)$ denotes the trace operation, and $\boldsymbol{\omega}$ is the vorticity vector.

The quantity denoted by <VTM> in (1) is the VTM averaged over the current and closest neighboring cells. It is close to zero in quasi-2D regions of the flow, whereas in regions with fully developed turbulence it is about 0.3 and higher. This property of <VTM> allows constructing the following non-dimensional empirical function:

$$F_{KH}(<\mathrm{VTM}>) = \max\{F_{KH}^{\min}, \min\{F_{KH}^{\max}, F_{KH}^{\min} + \frac{F_{KH}^{\max} - F_{KH}^{\min}}{a_2 - a_1}(<\mathrm{VTM}> - a_1)\}\}, \tag{3}$$

where $F_{KH}^{\max} = 1.0$ and the empirical constants are: $F_{KH}^{\min} = 0.1$, $a_1 = 0.15$, $a_2 = 0.3$.

As seen from (3), in the flow regions where <VTM> is less than 0.15, the function F_{KH} is equal to 0.1, thus ten times reducing the length-scale Δ_{SLA} compared to $\widetilde{\Delta}_{\omega}$. On the other hand, if <VTM> is higher than 0.3, the function equals 1.0, i. e., is not active.

In the inviscid flow areas and inside the attached boundary layers with no turbulent content the F_{KH} function should be deactivated as well.

For the inviscid regions, this is achieved by multiplying <VTM> by the quantity $\max\left\{1, \frac{0.2\nu}{\max\{(\nu_t - \nu_{t,\infty}), 10^{-6}\nu_{t,\infty}\}}\right\}$.

As far as the attached boundary layer regions are concerned, in the framework of DDES the function F_{KH} is modified as [2]:

$$F_{KH}^{\lim} = \begin{cases} 1.0, if\ f_d < (1-\varepsilon) \\ F_{KH}, if\ f_d \geq (1-\varepsilon) \end{cases}, \tag{4}$$

where f_d is the delay function of the DDES approach and $\varepsilon = 0.01$ is the empirical constant.

For the IDDES, based on the same ideas, we define the function $F_{KH}^{\lim}$ as follows:

$$F_{KH}^{\lim} = \begin{cases} 1.0,\ if\ \widetilde{f}_d > \varepsilon \\ F_{KH}, if\ \widetilde{f}_d \leq \varepsilon \end{cases}, \tag{5}$$

where $\widetilde{f}_d$ is the analog of the delay function in IDDES.

Within the DDES approach, the new length-scale Δ_{SLA} simply replaces the original definition $\Delta = \Delta_{\max}$. The modification of the IDDES approach is not as straightforward and implies only a partial replacement of $\Delta_{\max}$ with Δ_{SLA} in the original IDDES definition of the subgrid length-scale. Namely, the original definition $\Delta = \min\{\max[C_w d_w, C_w \Delta_{\max}, \Delta_{wn}], \Delta_{SLA}\}$ should be replaced with

$\Delta = \min\{\max[C_w d_w, C_w \Delta_{\max}, \Delta_{wn}], \Delta_{\max}\}$, where $C_w = 0.15$ is an empirical constant and Δ_{wn} is the wall-normal grid step.

3. Results and discussion

In this section we present computational set-ups and results of simulations of two test-cases chosen for evaluation of performance of modified (based on SLA subgrid length-scale) version of the DDES and IDDES approaches with Spalart-Allmaras background RANS model. All the simulations are carried out with the use of the NTS code which details are outlined in [5].

3.1. Transonic M 219 cavity flow

This flow studied in the experiments [6] has the Mach number M = U_0/a =0.85 and Reynolds number Re = HU_0/ν = $3.4 \cdot 10^5$, where U_0 is a free stream velocity and H is a quarter of the cavity depth. A quality of resolution of the shear layer separated from the leading edge of the cavity has a strong impact on the flow, in general, and on its unsteady characteristics, in particular, the latter being of crucial importance for an accurate prediction of the noise generated by the cavity. Therefore, this flow seems to be a good test case for evaluation of performance of the SLA subgrid length-scale.

The cavity is made in a flat plate and has the length of 20H, height of 4H. Its drawing and computational domain and grid with 3 million nodes used in the simulations are presented in Fig. 1. The grid clusters in the vicinity of the cavity walls and the shear layer which separates from the cavity leading edge. The time step in the simulations was equal to $2.5 \cdot 10^{-2} H/U_0$.

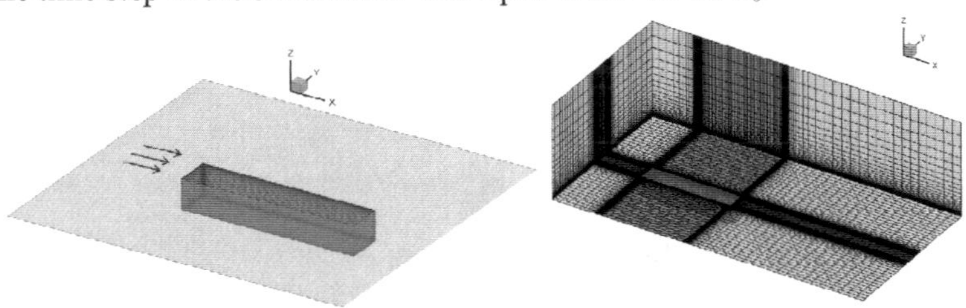

Figure 1. Drawings of the cavity and computational domain and grid used in simulations

Figures 2, 3 present instantaneous eddy viscosity and vorticity fields from the DDES computations with the use of the original and SLA subgrid length-scales. They clearly show that consistently with the design of the SLA length-scale, it leads to a tangible reducing of the turbulent viscosity compared to the original definition (see Fig. 2) which, in turn, leads not only to unlocking of the KH instability and accelerating transition to 3D turbulence in the shear layer but also to somewhat finer resolution of the turbulent structures inside the cavity (see Fig. 3). As a result, predicted sound pressure level and power spectra of the pressure fluctuations on the cavity floor turn out to be considerably closer to the experimental data than those of the standard DDES, the spectra improvement being quite pronounced in terms of both the broadband part level and frequencies of the tones (Fig.4).

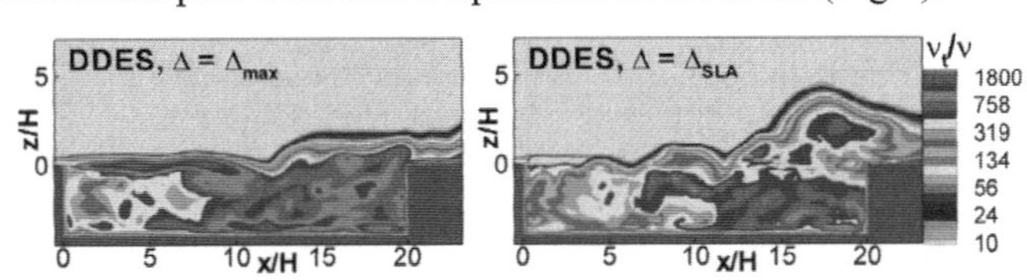

Figure 2. Snapshots of eddy viscosity obtained with the original and modified DDES versions

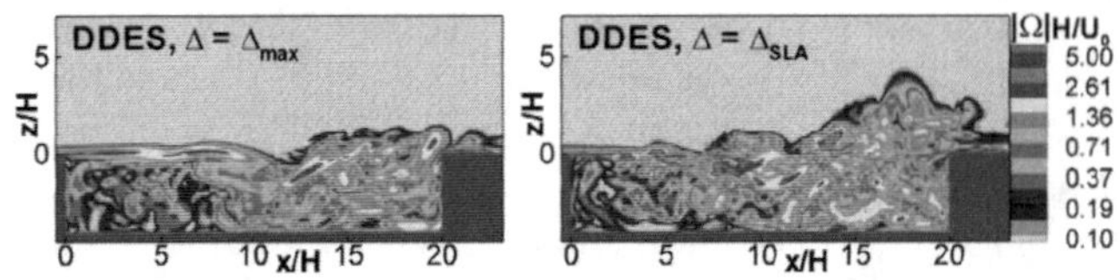

Figure 3. Snapshots of vorticity magnitude obtained with the original and modified DDES versions

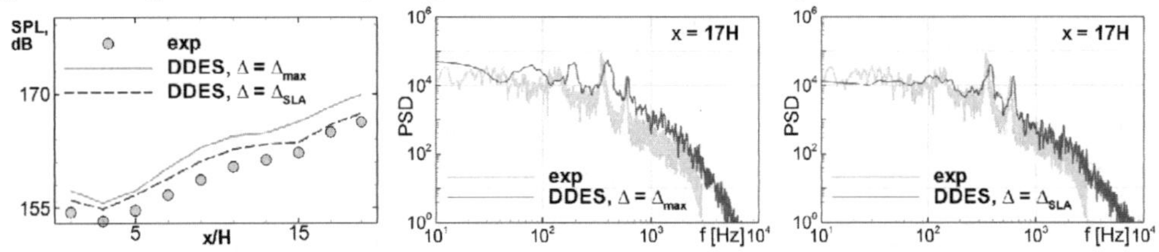

Figure 4. Sound pressure level and power spectra of pressure fluctuations on the cavity floor predicted by original and modified DDES versions

Similar but even more pronounced tendencies are observed in the results of IDDES of the considered flow. Namely, the modified IDDES version predicts more rapid roll-up of the separated shear layer (Fig.5) which leads to a more accurate prediction of the unsteady pressure characteristics on the cavity flow (Fig.6).

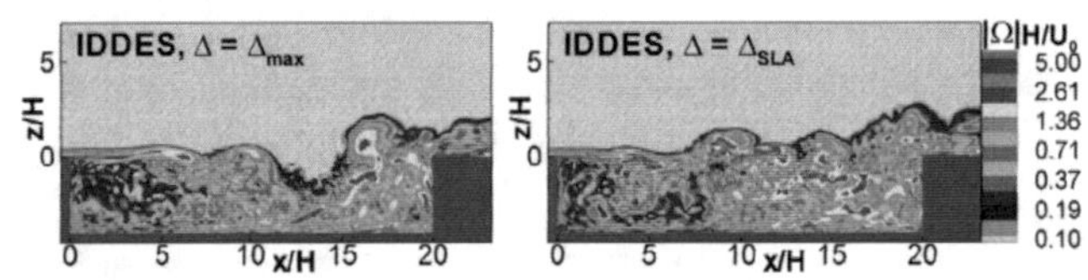

Figure 5. Snapshots of vorticity magnitude predicted by original and modified IDDES versions

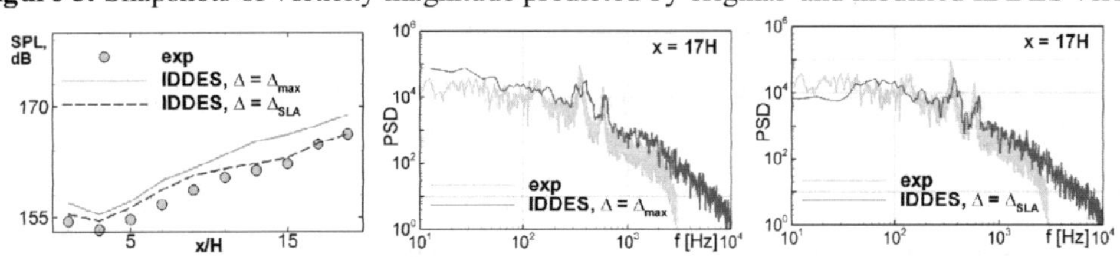

Figure 6. Sound pressure level and power spectra of pressure fluctuations on the cavity floor predicted by original and modified IDDES versions

3.2. Flow over the wall mounted hump

This is a flow over a so called 2D NASA wall-mounted hump which was studied in the experiments [7]. The Reynolds number based on the hump length, c, and reference velocity, U_0 (maximum free stream velocity at the inlet of the domain) in the experiment was equal to $9.36 \cdot 10^5$, and the height of the hump, h, was equal to $0.128c$.

A computational domain and a grid in XY-plane used in the simulations are shown in Fig.7. The grid has 510 and 126 cells in the x- and y-directions, respectively. A size of the domain in the homogeneous, z, direction is $0.4c$, and the grid-step, Δz, is uniform and equal to $5 \cdot 10^{-3}c$. The total grid size is about 5 Million. A time step in the simulations is equal to $2 \cdot 10^{-3}c/U_\infty$.

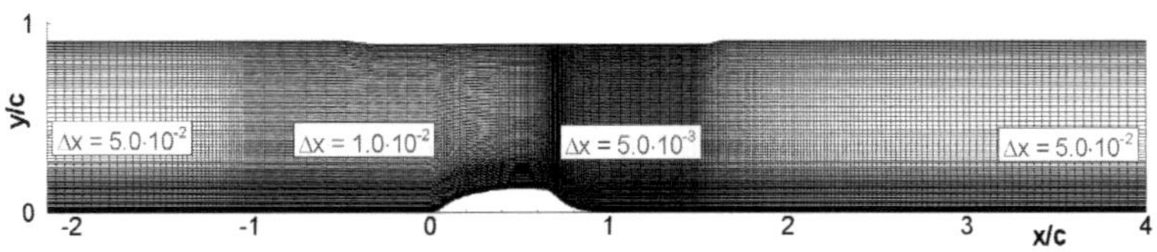

Figure 7. Computational domain and grid in XY-plane used for NASA hump simulations

Figures 8-10 present primary results of the DDES computations of this flow. As seen in Figs.8, 9 where instantaneous fields of turbulent viscosity and vorticity magnitude are depicted, the standard DDES suffers from a strong delay of transition to resolved 3D turbulence in the separated shear layer, whereas the use of the SLA subgrid length-scale reduces the turbulent viscosity, unlocks the KH instability in the shear layer, and accelerates the transition. As a result, mean characteristics of the flow, e.g., skin friction predicted by the modified DDES version better agree with the experimental data than those of the standard DDES (Fig. 10).

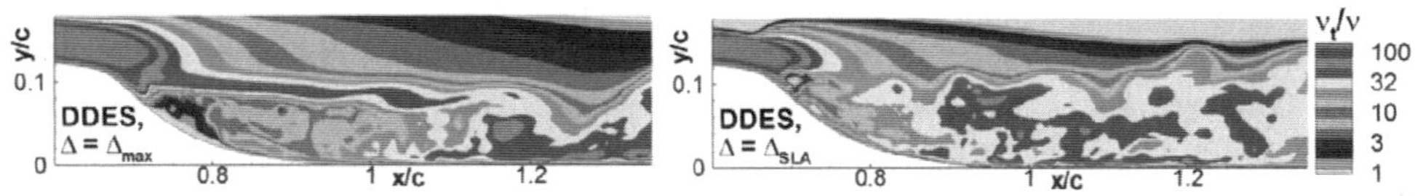

Figure 8. Snapshots of eddy viscosity obtained with the original and modified DDES versions

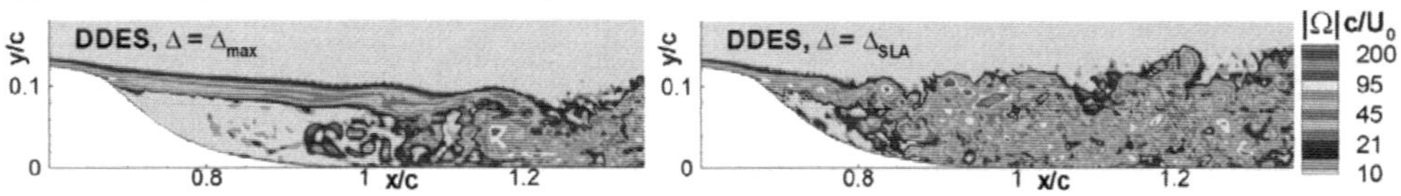

Figure 9. Snapshots of vorticity magnitude obtained with the original and modified DDES versions

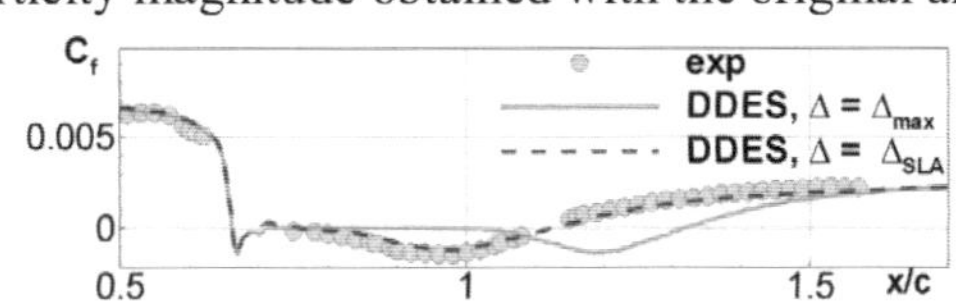

Figure 10. Mean friction coefficient distributions. Comparison of results obtained with the original and modified DDES versions and experimental data

In general, similar results are obtained with the use of IDDES approach. However, as seen from the comparison of vorticity snapshots obtained with the use of the original versions of DDES and IDDES (left frames in Figs. 9 and 11), the latter suffers from the delay of the RANS-to-LES transition in the separated shear layer not as severely as the former. Not surprisingly, the effect of the modification of the subgrid length-scale in the original IDDES also is not as pronounced as in the original DDES. For instance, the skin friction coefficients predicted by the original and modified version of IDDES are virtually identical and very well agree with the experimental data (Fig. 12). Nevertheless, the use of the SLA subgrid length-scale in the framework of IDDES still turns out to be helpful: it somewhat speeds the shear layer roll up (Fig. 11) which results in a corresponding improvement of the agreement with the experiment on the normal Reynolds stresses in the vicinity of separation (Fig. 13).

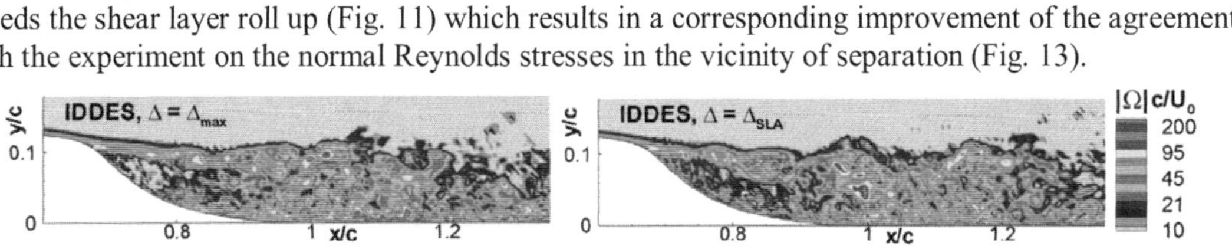

Figure 11. Snapshots of vorticity magnitude obtained with the original and modified IDDES versions

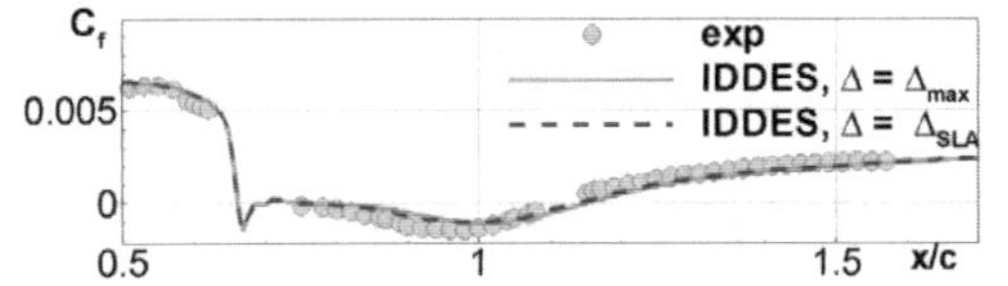

Figure 12. Distributions of skin friction predicted by original and modified IDDES versions

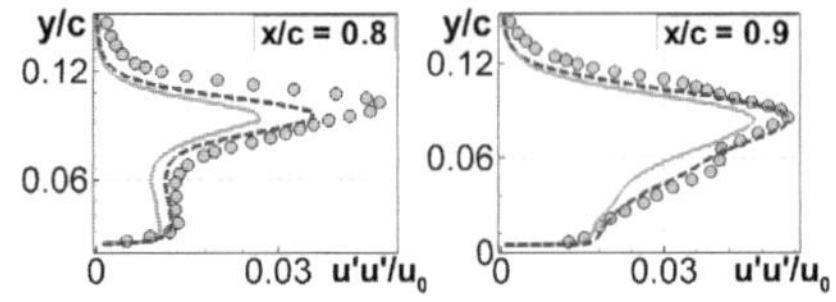

Figure 13. Resolved normal stress profiles at 2 flow cross-sections predicted by original and modified IDDES versions

Conclusions

A numerical study was conducted focused on testing of DDES and IDDES approaches combined with a recently proposed shear-layer-adapted subgrid length-scale. Two separated flows were considered: a flow with a fixed separation (transonic flow past M 219 cavity) and a flow with smooth-surface separation and reattachment (NASA wall-mounted hump). Results of the simulations show that shear-layer-adapted versions of both approaches surpass the original ones in terms of agreement with the experiments which is achieved thanks to a more rapid transition from RANS to LES in the separated shear layers. No negative side effects caused by a possible interaction of the solution-dependent length-scale with the empirical functions involved into the original DDES and IDDES formulations were observed

Acknowledgements

The study was carried out with the financial support of the Russian Science Foundation (Grant 14-11-00060). Computational resources have been provided by Saint-Petersburg Branch of the Joint Supercomputer Center of the Russian academy of sciences (http://scc.ioffe.ru/).

References

[1] Spalart P R, Jou W-H, Strelets M, and Allmaras S R 1997 Comments on the feasibility of LES for wings, and on a hybrid RANS/LES approach *Proceedings of first AFOSR international conference on DNS/LES*, pp. 137–147.

[2] Shur M L, Spalart P R, Strelets M K and Travin A K 2015 An enhanced version of DES with rapid transition from RANS to LES in separated flows *Flow Turbul. Combust.* vol. **95** pp 709-737

[3] Spalart P R, Deck S, Shur M L, Squires K D, Strelets M K and Travin A 2006 A New Version of Detached-eddy Simulation, Resistant to Ambiguous Grid Densities *Theor. Comput. Fluid Dyn.* vol. **20**, no. 3 pp. 181–195

[4] Shur M L, Spalart P R, Strelets M K and Travin A K 2008 A hybrid RANS-LES approach with delayed-DES and wall-modelled LES capabilities *Int. J. Heat Fluid Flow* vol. **29** no. 6 pp. 1638–1649

[5] Shur M L, Strelets M K and Travin A K 2004 High-Order Implicit Multi-Block Navier-Stokes Code: Ten-Year Experience of Application to RANS/DES/LES/DNS of Turbulence *7th Symposium on Overset Grids & Solution Technology October 5-7, 2004, Huntington Beach, CA, USA*

[6] Swalwell K, Sheridan J and Melbourne W 2003 Frequency Analysis of Surface Pressures on an Airfoil After Stall *21st AIAA Applied Aerodynamics Conference* no. June, pp. 1–8.

[7] Henshaw M J 2002 M219 cavity case: Verification and validation data for computational unsteady aerodynamics *Tech. Rep. RTO-TR-26, AC/323(AVT)TP/19*, QinetiQ, UK, pp. 453–472.

[8] Greenblatt D, Paschal K B, Yao C-S, Harris J, Schaeffler N W and Washburn A E 2004 A Separation Control CFD Validation Test Case *43rd AIAA Aerosp. Sci. Meet. Exhib.* 485

Adjustment of the k-ω SST turbulence model for prediction of airfoil characteristics near stall

A A Matyushenko, A V Garbaruk
Peter the Great St.Petersburg Polytechnic University
Russia, Saint-Petersburg, 195251, Politechnicheskaya, 29\

E-mail: alexey.matyushenko@gmail.com

Abstract. A version of k-ω SST turbulence model adjusted for flow around airfoils at high Reynolds numbers is presented. The modified version decreases eddy viscosity and significantly improves the accuracy of prediction of aerodynamic characteristics in a wide range of angles of attack. However, considered reduction of eddy viscosity destroys calibration of the model, which leads to decreasing accuracy of skin-friction coefficient prediction even for relatively simple wall-bounded turbulent flows. Therefore, the area of applicability of the suggested modification is limited to flows around airfoils.

1. Introduction

Prediction of airfoil characteristics in regimes with maximum lift and near stall is an important task for aviation and wind power, as well as for turbomachinery. Unfortunately, maximum lift coefficient and corresponding angle of attack are systematically overpredicted by the Reynolds Averaged Navier-Stokes (RANS) approach in combination with different semi-empirical turbulence models (see, for example, [1-3]). This disagreement is caused by delay of turbulent boundary layer separation under adverse pressure gradient on the suction side of the airfoil (Fig. 1). Since in the frame of RANS approach the separation position is controlled by the turbulence model, one of the ways to improve accuracy of the airfoil characteristics prediction is a special tuning of the models for such class of flows. This paper presents such adjustment for k-ω SST (2003) turbulence model [4].

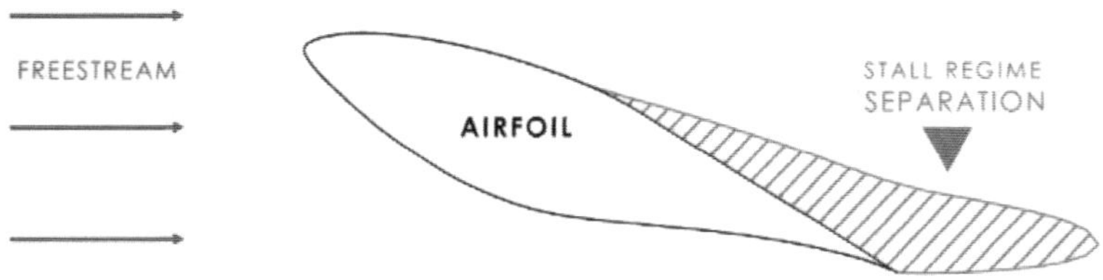

Figure 1. Scheme of flow around an airfoil at the near-stall regime.

2. Model formulation

The equations of the turbulent kinetic energy (k) and specific dissipation rate (ω) for the SST turbulence model are obtained from a combination of k-ε and k-ω turbulence models and read as:

$$\begin{cases} \frac{D(\rho k)}{Dt} = \nabla \cdot [(\mu + \sigma_k \mu_T)\nabla k] + P_k - \beta^* \rho\omega k \\ \frac{D(\rho\omega)}{Dt} = \nabla \cdot [(\mu + \sigma_\omega \mu_T)\nabla \omega] + \gamma \frac{\rho}{\mu_T} P_k - \beta\rho\omega^2 + 2(1 - F_1)\rho\sigma_{\omega 2} \frac{(\nabla k)\cdot(\nabla \omega)}{\omega} \end{cases} \quad (1)$$

where $P_k = \min(\mu_t S^2, 10\beta^* \rho k\omega)$ is the production term,
d_w – wall distance,
ρ – density,
μ and ν – dynamic and kinematic viscosity correspondingly,
S – strain rate.

Functions and constants of the model are as follows:

$$F_1 = \tanh(\arg_1^4), \quad \arg_1 = \min\left[\max\left(\frac{\sqrt{k}}{0.09\omega d_w}, \frac{500\nu}{d_w^2\omega}\right), \frac{4\rho\sigma_{\omega 2} k}{CD_{k\omega} d_w^2}\right], \quad CD_{k\omega} = \max\left(2\rho\sigma_{\omega 2}\frac{1}{\omega}\frac{\partial k}{\partial x_j}\frac{\partial \omega}{\partial x_j}, 10^{-10}\right)$$

$$F_2 = \tanh(\arg_2^2), \quad \arg_2 = \max\left(\frac{2\sqrt{k}}{0.09\omega d_w}, \frac{500\nu}{d_w^2\omega}\right) \quad (2)$$

$$\sigma_k = F_1\sigma_{k1} + (1 - F_1)\sigma_{k2}, \ \sigma_\omega = F_1\sigma_{\omega 1} + (1 - F_1)\sigma_{\omega 2}, \ \beta = F_1\beta_1 + (1 - F_1)\beta_2$$

$$\gamma = \frac{\beta}{\beta^*} - \frac{\sigma_\omega \kappa^2}{\sqrt{\beta^*}}$$

Coefficients of the model read as:

$\beta^* = 0.09, \ \kappa = 0.41, \ a_1 = 0.31$,

$\sigma_{k1} = 0.85, \ \sigma_{k2} = 1.0; \quad \sigma_{\omega 1} = 0.5, \ \sigma_{\omega 2} = 0.856; \quad \beta_1 = 0.075, \ \beta_2 = 0.0828.$

Eddy viscosity definition $\mu_T = \frac{\rho a_1 k}{\max(a_1\omega, SF_2)}$ includes the so called SST limiter, which is introduced into the model in order to prevent overprediction of the shear stress in the boundary layers under adverse pressure gradient. Therefore, behaviour of the model in such boundary layers as well as separation on the suction side of the airfoil is controlled by the a_1 constant. In the present work the SST model modification and adjustment for flows around airfoils is carried out by modification of a_1.

3. Problem definition

Four aerodynamic airfoils with different shapes and thicknesses (from 15% to 21%) were considered (Fig. 2). Experimental investigations [5 - 8] were carried out in low turbulence wind tunnels (I <1%). at relatively high Reynolds number (Re > 10^6) based on airfoils chord and freestream velocity.
A tunnel wall correction was applied to the airfoil characteristics and the angle of attack for comparison with freestream setup [9]. Since the experimental Mach number did not exceed 0.15, incompressible flow was considered.

Figure 2. Considered airfoils.

Numerical simulations of two-dimensional RANS equations in combination with the SST turbulence model were carried out using the double precision version of ANSYS Fluent 15.0. The pressure-based coupled solver was employed with the Second Order Upwind discretization scheme for the convective terms in all transport equations.
Fine structured C-type meshes were generated in 20Cx20C (C – airfoil chord) computational domain using ICEM CFD (Fig. 3). The meshes were refined normal to the wall in order to resolve the viscous sublayer (Δy_1^+<1), near the leading edge in the streamwise direction for a proper resolution of thin

boundary layer, and near the trailing edge. This results in about 400 points along the airfoil and a total mesh size of about 40,000 cells.
A constant velocity is specified at the inlet section of the computational domain. Inlet turbulent kinetic energy corresponds to experimental turbulence intensity and the specific dissipation rate is specified as $\omega = 10 \cdot U_\infty / C$ [10]. No-slip conditions are used on the airfoil surface and constant pressure is specified on the outlet.

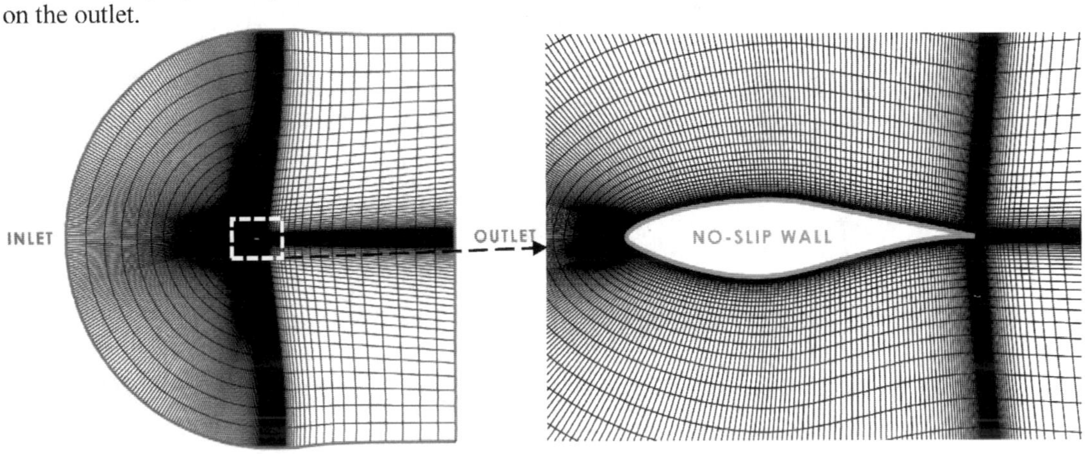

Figure 3. Computational domain, boundary conditions, and mesh.

4. Results

4.1. Model calibration

Preliminary results of the original SST model indicated that the size of the separation zone and position of the separation point on the suction side of the airfoil was substantially underestimated, especially for angles of attack near stall, which leads to an overestimation of the maximum lift coefficient. The calibration of a_1 was carried out for the S809 airfoil at $Re = 2 \cdot 10^6$ and angle of attack equals to 10°. One can see that decrease of a_1 leads to increase of the size of the recirculation zone and improves agreement with experimental data (Fig 4a, 5). The best agreement is achieved at $a_1 = 0.28$, so this value was used for further computations (Fig. 4b).

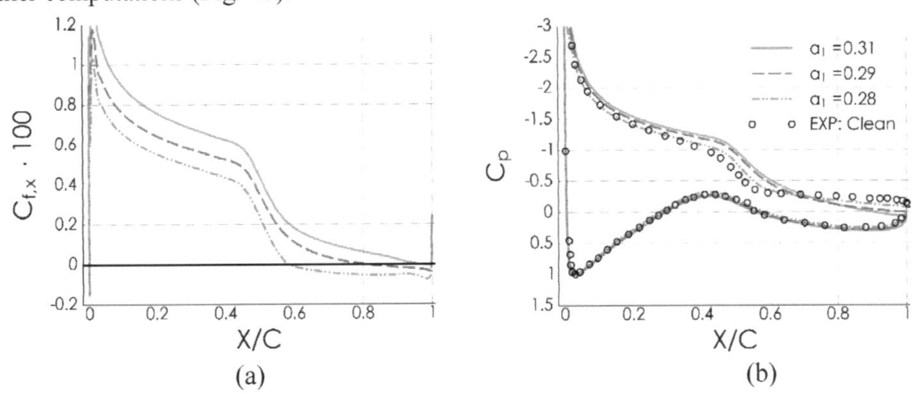

Figure 4. Effect of a_1 constant on the prediction of the skin-friction (a) and pressure (b) coefficients for the S809 airfoil at $\alpha = 10°$.

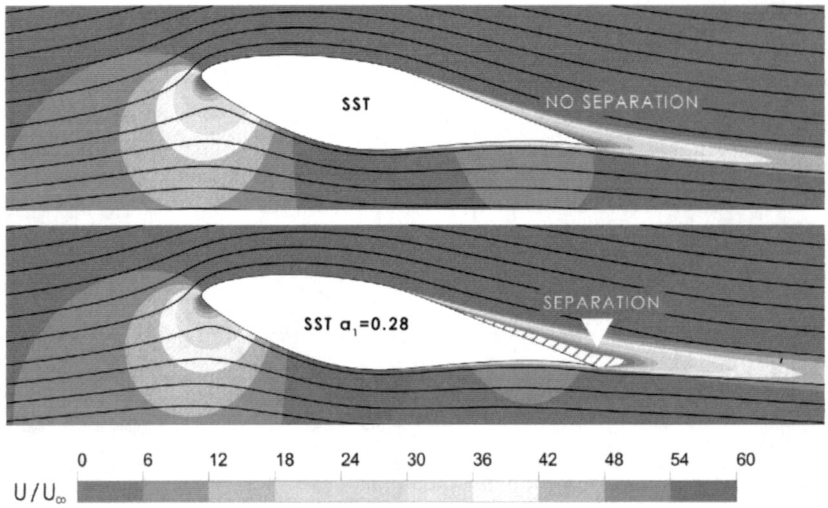

Figure 5. Contours of streamwise velocity component and streamlines near the S809 airfoil at $\alpha = 10^{\circ}$.

4.2. Prediction of flow around airfoils

Computations using the original and modified versions of the SST model show that both models predict virtually the same lift coefficient for low angles of attack when the flow is attached (Fig. 6). At higher angle of attack the modified SST model predicts lower value of the lift coefficient due to the larger size of the recirculation zone.

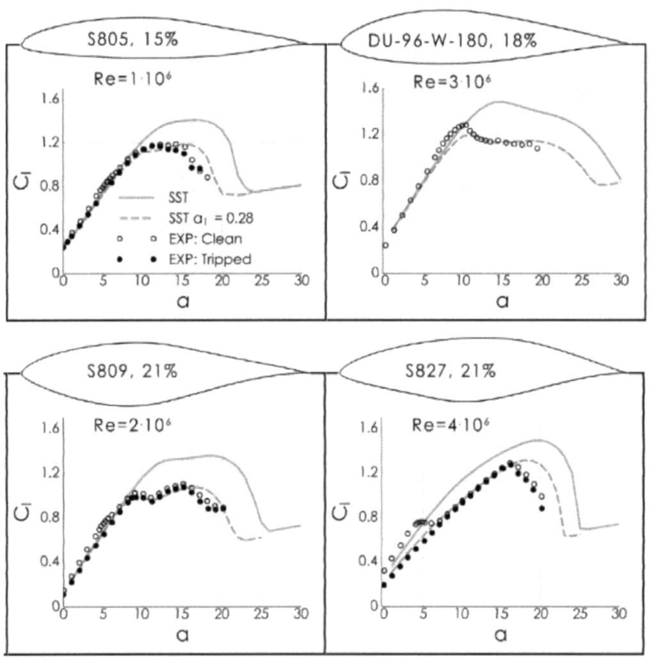

Figure 6. Comparison of experimental and computational lift coefficient at different angles of attack. Experiments were carried out for clean airfoil model (EXP: Clean) and model with rough leading edge (EXP: Tripped)

One can see (Fig. 7) that the same separation point is obtained at 4°-5° lower angles of attack. Lift coefficient distribution for modified version as well as pressure coefficient (Fig. 8) is in good agreement with the experimental data for all the considered airfoils over a wide range of angles of attack, while the original SST model strongly overpredicts lift value (error is about 25%). However even the modified model predicts somewhat higher deep stall angle. One of the possible reasons is elimination of laminar-turbulent transition effect [11].

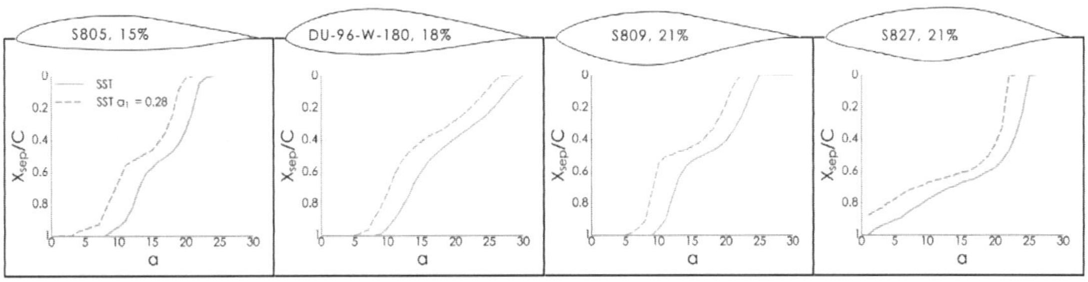

Figure 7. Position of computational separation point on the suction side of the airfoils at different angles of attack.

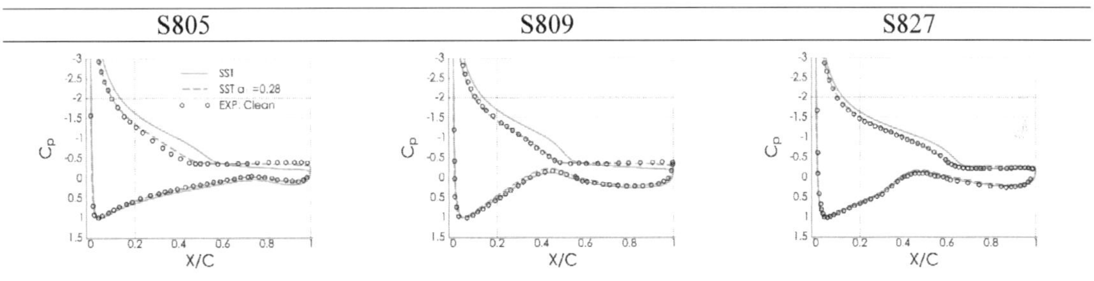

Figure 8. Computational and experimental pressure coefficient for different airfoils at angle of attack near stall (α = 15°).

4.3. Basic turbulence flows.

The SST model was calibrated by the authors on a set of basic turbulent flows and decrease of the a_1 constant could destroy this calibration. The effect of reduction of a_1 was checked with prediction of flow on the flat plate [12] and in CS0 diffuser [13]. Results of computations show that the modified model systematically underpredicts the value of the skin-friction coefficient for even relatively simple wall-bounded turbulent flows. In particular, the error in prediction of skin friction coefficient on the flat plate reaches 25% (Fig. 9a), which confirms once again the imperfection of semi-empirical turbulence models for all types of flows.

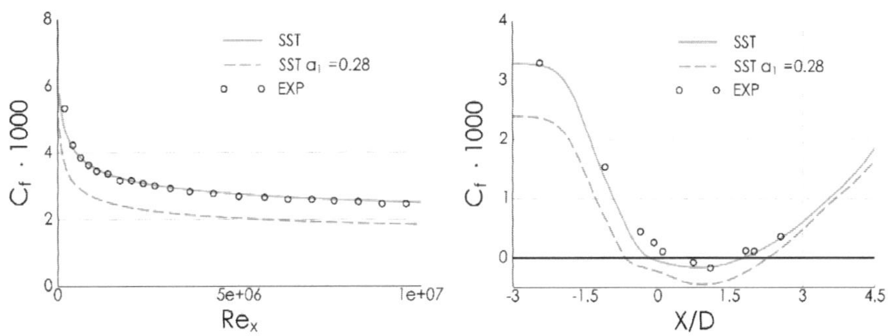

(a) (b)

Figure 9. Comparison of experimental and computational skin friction coefficient distribution for the flat plate (a) and CS0 diffuser (b).

5. Conclusion

Presented modification of the k-ω SST turbulence model with reduced a_1 constant (a_1 = 0.28) significantly improves the accuracy of prediction of aerodynamic characteristics over a wide range of angles of attack ($\alpha = 0^o \div 30^o$). Error in the prediction of maximal lift coefficient does not exceed 5% with the modified model (this error is about 25% for the original model). However, reduction of the a_1 constant destroys calibration of the model, which leads to decreased accuracy of the skin-friction coefficient even for relatively simple wall-bounded turbulent flows. In particular, the error of skin friction coefficient for the flat plate reaches 25%, which confirms once again the imperfection of semi-empirical turbulence models for all types of flows. As a result, the area of applicability of suggested modification is limited to flows around airfoils.

References

[1] Bertagnolio F., Sørensen N.N., Johansen J. 2006. Profile Catalogue for Airfoil Sections Based on 3D Computations. *Risø Technical Report Risø-R-1581, Risø National Laboratory, 72 pp.*

[2] Villalpando F., Reggio M., Ilinca, A. 2011. Assessment of Turbulence Models for Flow Simulation around a Wind Turbine Airfoil. *Modelling and Simulation in Engineering, V. Article ID 714146, 8 pp.*

[3] Bak C., Fuglsang P., Johansen J., Antoniou I. 2000. Wind tunnel tests of the NACA 63-415 and a modified NACA 63-415 airfoil. *Risø Technical Report Risø-R-1193, Risø National Laboratory, 108 pp.*

[4] Menter F. R., Kuntz M. & Langtry R.B. 2003. Ten Years of Industrial Experience with the SST Turbulence Model

[5] Somers D. M. 1997. Design and Experimental Results for the S805 airfoil

[6] Somers D. M. 2005. Design and Experimental Results for the S827 Airfoil Period of Performance 1998 – 1999

[7] Somers, D. M. 1997. Design and Experimental Results for the S809 Airfoil, *NRELSR-440-6918.*

[8] Timmer W. A. 2003. Summary of the Delft University Wind Turbine Dedicated Airfoils

[9] Allen H. J. and Vincenti W. G. 1944. Wall interference in a two-dimensional-flow wind tunnel, with consideration of the effect of compressibility. *NACA Rep. 782.*

[10] Menter F. R. Zonal Two Equation k-omega Turbulence Models for Aerodynamic Flows, AIAA Paper 93-2906, July 1993

[11] Матюшенко А.А., Гарбарук А.В., Смирнов П.Е. Ментер Ф.Р. 2015. Численное исследование влияния ламинарно-турбулентного перехода на характеристики аэродинамических профилей. *Тепловые процессы в технике. № 8. С. 338-343*

[12] Wieghardt, K., & Tillmann, W. 1951. On the turbulent friction layer for rising pressure. *NACA TM 1314.*

[13] Driver D.M. 1991. Reynolds shear stress measurements in a separated boundary layer flow. *AIAA-91-1787.*

Journal of Physics: Conference Series **769** (2016) 012083 doi:10.1088/1742-6596/769/1/012083

Renormgroup algorithm for the theory of the relativistic plasma resonance

I I Metelskii[1,2], V F Kovalev[1,2,3], V Yu Bychenkov[1,2]

[1] P.N. Lebedev Physical Institute of the Russian Academy of Sciences 53 Leninskiy Prospekt, Moscow, 119991, Russia

[2] All-Russia Research Institute of Automatics (VNIIA), ROSATOM 22, ul. Sushchevskaya Moscow, 127055, Russia

[3] Keldysh Institute of Applied Mathematics of the Russian Academy of Sciences Miusskaya sq., 4 Moscow, 125047, Russia

E-mail: metelski@lebedev.ru

Abstract. Renormgroup algorithm is used to reveal the nonlinear space-time stationary structure of the potential electric field in the vicinity of a plasma resonance. A two parameter transformation group with two renormgroup parameters related to non-relativistic and relativistic nonlinear effects is constructed and then used to transform the perturbation theory approximation to the desired electric field structure. Further applications of the obtained solution to problems of laser-plasma physics are discussed.

1. Introduction

The renormalization group (RG) algorithm was introduced in mathematical physics at the end of XX-th century [1–3] with the same aim of finding an improved solution (in comparison with the initial approximate solution) as the classical algorithm of Bogoliubov's RG method [4–7], but its implementation in mathematical physics is based on techniques of the modern group analysis. Since the appearance and the first successful application of RG algorithm in plasma physics [3] the approach proved its efficiency for many nonlinear problems of mathematical physics [8]. Here this approach is used to find the nonlinear space-time stationary structure of the potential electric field, excited by a powerful laser radiation in the vicinity of a relativistic plasma resonance. Nonlinear effects in the vicinity of the plasma resonance are of considerable importance for analyzing nonlinear absorption and harmonics generation in the laser plasma [9], and the component of the electric field, polarized along a plasma inhomogeneity gradient, plays the dominant role in these processes. The stationary structure of this field was discussed in detail in [10, 11] in the linear, perturbation theory (PT) approximation. Increasing the laser intensity makes it topical to involve nonlinear effects of the laser-plasma interaction into consideration even for non-relativistic electron velocities. At even higher laser intensities relativistic effects for plasma electrons near the plasma resonance becomes essential. Here we present an analytical approach for finding the stationary solution for the equations that describe nonlinear relativistic plasma oscillations in the vicinity of the plasma resonance. The key point in constructing this solution is the use of the RG algorithm.

2. Solution of basic equations via RG algorithm

Basic equations for the potential electric field, excited by the p-polarized laser wave in the inhomogeneous plasma with a density scale L in the vicinity of a plasma resonance $x = 0$, where the laser frequency ω_0 coincides with the Langmuir frequency of plasma electrons ω_L, are presented by a system of two nonlinear partial differential equations for electron dynamics and electric field structure,

$$\partial_t v + av\partial_x v = P\left(1 - \beta v^2\right)^{3/2}, \quad \partial_t P + av\partial_x P = -{\omega_0}^2 v \, . \tag{1}$$

Here a is the dimensionless constant proportional to the magnetic field amplitude in the plasma resonance point $x = 0$ in the linear approximation, which is defined by the laser wave amplitude; $\beta = a^2/c^2$, c is the speed of light; v and P are values of the electron velocity and the potential electric field along the gradient of the plasma inhomogeneity, which are normalized to the parameter a.

Notice that equations (1) contain two parameters, a and β, which make contributions to the nonrelativistic and the relativistic nonlinearities, respectively. Generally speaking, a measure of relativistic nonlinearity is the parameter $\beta \sim 1/c^2$ which tends to zero in the nonrelativistic limit. However, the parameter selection in the form $\beta = a^2/c^2$ simplifies further calculations. In the limiting case of $a \to 0, \beta \to 0$ equations (1) have the well-known solution [10] that describes the plasma resonance in the linear approximation. In non-relativistic case, $\beta = 0$, equations (1) by the use of the RG algorithm were solved and analyzed in [3]. We extend now this result to the relativistic case with $\beta \neq 0$.

Following the general scheme of the RG algorithm [8] we first define the *basic RG-manifold* as equations (1) in the extended space of independent variables that now include parameters a and β and construct the PT solution of (1) in the form of truncated perturbation series in powers of a and β. Next, using the group analysis technique we calculate the symmetry group admitted by the RG-manifold. This infinite-dimensional Lie point transformation group is defined by the infinitesimal operator

$$X = \xi^1\partial_t + \xi^2\partial_x + \xi^3\partial_a + \xi^4\partial_\beta + \eta^1\partial_v + \eta^2\partial_P \tag{2}$$

with coordinates ξ^i, $i = 1, ..4$ and η^j, $j = 1, 2$, which are functions of the variables t, x, v, P, parameters a, β. The coordinates of (2) can be defined from the solution of an overdetermined system of differential equations, named as *determining equations*,

$$\begin{aligned} &X_{(1)}\left(\partial_t v + av\partial_x v - \left(1 - \beta v^2\right)^{3/2} P\right)|_{(1)} = 0\,, \\ &X_{(1)}\left(\partial_t P + av\partial_x P + \omega_0^2 v\right)|_{(1)} = 0\,, \end{aligned} \tag{3}$$

that follow from the invariance test for the basic RG-manifold (1) with respect to the extended group with the group generator $X_{(1)}$. Here $X_{(1)}$ defines the prolongation of the generator X to the first derivatives and $|_{(1)}$ means that the result of the action of the operator $X_{(1)}$ is taken on the manifold defined by the equations (1) and all their differential consequences,

$$\begin{aligned} &X_{(1)} = X + \zeta_1^1\partial_{v_t} + \zeta_1^2\partial_{P_t} + \zeta_2^1\partial_{v_x} + \zeta_2^2\partial_{P_x}, \\ &\zeta_1^1 = D_t\eta^1 - \partial_t v D_t\xi^1 - \partial_x v D_t\xi^2 - \partial_a v D_t\xi^3 - \partial_\beta v D_t\xi^4, \\ &\zeta_2^1 = D_x\eta^1 - \partial_t v D_x\xi^1 - \partial_x v D_x\xi^2 - \partial_a v D_x\xi^3 - \partial_\beta v D_x\xi^4, \\ &\zeta_1^2 = D_t\eta^2 - \partial_t P D_t\xi^1 - \partial_x P D_t\xi^2 - \partial_a P D_t\xi^3 - \partial_\beta P D_t\xi^4, \\ &\zeta_2^2 = D_x\eta^2 - \partial_t P D_x\xi^1 - \partial_x P D_x\xi^2 - \partial_a P D_x\xi^3 - \partial_\beta P D_x\xi^4, \\ &D_t \equiv \partial_t + \partial_t v\partial_v + \partial_t P\partial_P, \ D_x \equiv \partial_x + \partial_x v\partial_v + \partial_x P\partial_P. \end{aligned} \tag{4}$$

In view of (1), (2) and (4), the system of determining equations for coordinates ξ^i and η^j is reduced to the following form,

$$\begin{aligned}
&a\eta^1 + avY\xi^1 - Y\xi^2 + v\xi^3 = 0\,,\\
&Y\eta^1 - z^{3/2}\left(PY\xi^1 + \eta^2\right) + \frac{3}{2}(1-z)Pz^{1/2}\left[\frac{\xi^4}{\beta} + 2\frac{\eta^1}{v}\right] = 0\ ,\\
&Y\left[a\eta^2 + \xi^2{\omega_0}^2 + \xi^3 P\right] = 0,\ Y\xi^3 = 0,\ Y\xi^4 = 0\,,
\end{aligned} \tag{5}$$

where we introduce the operator

$$Y \equiv \partial_t + av\partial_x + Pz^{3/2}\partial_v - \omega_0^2 v\partial_P\,,\quad z \equiv 1 - \beta v^2. \tag{6}$$

Omitting the calculation details we present the final expressions for the coordinates ξ^i and η^j of the group generator (2),

$$\begin{aligned}
\xi^1 &= -\eta^2/\omega_0^2 v + \xi_1(I_1, I_2, I_3) + \frac{1}{\beta}\left[\frac{6+\beta I_1}{2\sqrt{4+\beta I_1}}E(\rho,\sigma) - \right.\\
&\left. -\frac{1}{\sqrt{4+\beta I_1}}F(\rho,\sigma) + \frac{\sqrt{\beta}P/\omega_0}{4+\beta I_1}\sqrt{\frac{1/\sqrt{z}-1}{1+1/\sqrt{z}}}\right]\ ,\\
\xi^2 &= -a\eta^2/\omega_0^2 - \xi^3 P/\omega_0^2 + \xi_2(I_1, I_2, I_3)\ ,\\
\eta^1 &= \frac{z^{3/2}}{v}\left[\eta_1(I_1, I_2, I_3) - (P/\omega_0^2)\eta^2 - \right.\\
&\left. -\frac{\xi^4}{2\beta^2}\left(2 - 3/\sqrt{z} + 1/z^{3/2}\right)\right]\ ,\\
\eta^2 &= \eta^2(t, x, v, P)\ ,\quad \xi^{3,4} = \xi^{3,4}(a, \beta, I_1, I_2, I_3)\ .
\end{aligned} \tag{7}$$

Here $\xi_{1,2}$, $\xi^{3,4}$, and η^2 are arbitrary functions of their arguments; I_1, I_2, and I_3 are invariants of the operator Y, and are defined as follows:

$$\begin{aligned}
I_1 &= \frac{2}{\beta}\left(1/\sqrt{z} - 1\right) + P^2/\omega_0^2\,,\quad I_2 = x + aP/\omega_0^2\ ,\\
I_3 &= t + \sqrt{4+\beta I_1}E(\rho,\sigma) - \frac{2}{\sqrt{4+\beta I_1}}F(\rho,\sigma)\,.
\end{aligned} \tag{8}$$

In (7),(8) $F(\rho,\sigma)$ and $E(\rho,\sigma)$ are incomplete elliptic integrals of the first and second kind, respectively:

$$F(\lambda,k) = \int_0^\lambda \frac{d\mu}{\sqrt{(1-k\mu^2)(1-\mu^2)}}\ ,\quad E(\lambda,k) = \int_0^\lambda \frac{\sqrt{1-k\mu^2}}{\sqrt{1-\mu^2}}\,d\mu\ . \tag{9}$$

The operator (2) with coordinates (7) determines the most general symmetry group of continuous point transformations admitted by (1). Following the scheme of the RG algorithm now we perform the restriction procedure of this group on a particular PT solution of our problem. This procedure appears as a "combining" of different coordinates of the group generator (2) admitted by the RG-manifold. The vanishing condition for this combination on a particular solution leads to algebraic equalities that couple different coordinates ξ^i and η^j and give rise to desired RG generators, which are then used to construct the nonlinear solution to (1) from the PT solution. These two RG generators are given by

$$R_1 = \xi^2_{RG}\partial_x + \partial_a \ , \quad R_2 = \xi^1_{RG}\partial_t + \partial_\beta + \eta^1_{RG}\partial_v \ , \tag{10}$$

with coordinates ξ^1_{RG}, ξ^2_{RG}, and η^1_{RG},

$$\begin{aligned} \xi^1_{RG} &= \frac{1}{\beta}\left[\frac{6+\beta I_1}{2\sqrt{4+\beta I_1}}E(\rho,\sigma) - \frac{1}{\sqrt{4+\beta I_1}}F(\rho,\sigma) + \frac{\sqrt{\beta}P/\omega_0}{4+\beta I_1}\sqrt{\frac{1/\sqrt{z}-1}{1+1/\sqrt{z}}}\right], \\ \xi^2_{RG} &= -\frac{P}{\omega_0^2}, \ \eta^1_{RG} = -\frac{z^{3/2}}{2v\beta^2}\left(2+\frac{1}{z^{3/2}}-\frac{3}{\sqrt{z}}\right), \ \rho = \arcsin\frac{P/\omega_0}{\sqrt{I_1}}, \quad \sigma = \sqrt{\frac{\beta I_1}{4+\beta I_1}} \ . \end{aligned} \tag{11}$$

The completion phase of the RG algorithm implies utilization of RG-operators to find analytical expressions for the solution of our problem. This is performed by solving the related Lie equations and finding finite group transformations that enable to restore the nonlinear solution with finite values of a and β from the PT solution. It is easy to verify that generators R_1, R_2 form a commutative group. Therefore, finite group transformations, defined by these generators, can be viewed as two independent transformations with group parameters a and β, respectively. First, consider finite transformations generated by R_1. Since R_1 does not contain differentiation with respect to t, v and P, then these variables are invariant. Hence, the generator R_1 leads to the following finite transformations for the time t, coordinate x, field P and velocity v:

$$t_{\mathrm{I}} = t_{linear}, \ x_{\mathrm{I}} = \eta - aP\omega_0^{-2}, \ P_{\mathrm{I}} = P_{linear}(\chi,\eta), \ v_{\mathrm{I}} = v_{linear}(\chi,\eta) \, . \tag{12}$$

Here $\chi = \omega_0 t_{linear}$, and $\eta = x(0,\eta)$ denotes the value of the coordinate $x = x(a,\eta)$ at $a = 0$. Formulas (12) define nonlinear solution which takes nonrelativistic nonlinearity into account. The velocity and the electric field here are defined by the linear theory solution (indicated by the index *linear*), because they are invariant under transformation with the group parameter a. Thus, the nonlinear structure of the electric field and the electron velocity depend on the form of the electric field P_{linear}, which arises from the solution of the linearized system of equations (1). For definiteness sake we select for P_{linear} the linear solution obtained for cold electron plasma, characterized by the dielectric permittivity $\varepsilon = 1 - \omega_L^2/\omega_0^2$, with a linear ion density profile [1]. In this case electric field has a singularity of the order of $1/\varepsilon$ at the critical point $x = 0$, however it is eliminated in view of either finite values of the electron plasma temperature, or a finite electron collision frequency.

For thus specified choice of the linear theory field structure, the formulas (12) for the electric field and the electron velocity are rewritten as follows:

$$\begin{aligned} P \equiv P_{\mathrm{I}} &= -\frac{\omega_0^2 L^2}{\Delta^2+\eta^2}\left(\eta\cos\chi + \Delta\sin\chi\right) , \\ v_{\mathrm{I}} &= -\frac{\omega_0 L^2}{\Delta^2+\eta^2}\left(\eta\sin\chi - \Delta\cos\chi\right) , \\ x \equiv x_{\mathrm{I}} &= \eta + \frac{aL^2}{\Delta^2+\eta^2}\left(\eta\cos\chi + \Delta\sin\chi\right) . \end{aligned} \tag{13}$$

Here the plasma resonance width Δ is determined either by the thermal motion of electrons with the thermal velocity V_T, or by a finite value of the collision frequency ν:

$$\Delta = \max\{\nu L/\omega_0 \, ; \, \left(3V_T^2 L/\omega_0^2\right)^{1/3}\}. \tag{14}$$

Now we consider finite transformations generated by R_2. These transformations involve the group parameter β, and relate non-relativistic nonlinear solutions (13) with $\beta = 0$ to relativistic solutions with $\beta \neq 0$. As coordinate x and the electric field P are invariants of the group transformations with the generator R_2 that is $x_{\rm II}{=}x_{\rm I} \equiv x$ and $P_{\rm II}{=}P_{\rm I} \equiv P$, they are defined by the corresponding expressions (14). The finite transformations for the remaining variables, namely the velocity $v \equiv v_{\rm II}$ and "time" $\tau \equiv \omega_0 t_{\rm II}$ are given by the following formulas:

$$v = \pm\frac{1}{\sqrt{\beta}}\left[1 - \frac{1}{\left(1+\beta v_{\rm I}^2/2\right)^2}\right]^{1/2}, \quad \tau = \chi - \left(\sqrt{4+\beta I_1}E(\varphi;k) - \frac{2F(\varphi;k)}{\sqrt{4+\beta I_1}} - \varphi\right), \tag{15}$$

where

$$\varphi = \arcsin\frac{P/\omega_0}{\sqrt{I_1}}, \quad k = \sqrt{\frac{\beta I_1}{4+\beta I_1}}, \quad I_1 = \frac{2}{\beta}\left(1/\sqrt{z} - 1\right) + P^2/\omega_0^2. \tag{16}$$

The final formulas (13), (15), generated by RG algorithm, describe the sought-for nonlinear structure of the electric field P and the plasma electron velocity v that takes into account both relativistic, $\beta \neq 0$, and nonrelativistic, $a \neq 0$, plasma nonlinearities. The dependence of P and v upon coordinate x and time τ is defined implicitly through the parametric variables η and χ.

3. Conclusions

Using the RG algorithm we have constructed an analytical solution to the system of equations that describe relativistic plasma oscillations in the vicinity of the plasma critical density. The distinctive feature of the method is that we have employed the two-parameter algebra of RG generators to obtain exact nonlinear solutions in a wide range of nonlinear and relativistic plasma parameters. The two-parameter renormgroup gives the possibility to separate inputs from nonrelativistic and relativistic nonlinearities. This fact is an added reason for utilizing multi-dimensional renormgroups for systems with several parameters [12]. Notice that although to obtain the solution we used the hydrodynamic-type equations for the cold collisionless plasma, the RG algorithm looks attractive for pursuing further research in this area with different plasma models. The detailed analysis of the space-time nonlinear structure of the potential plasma field in the vicinity of a plasma resonance opens a possibility to calculate the power-law spectra of stationary plasma oscillations, to estimate the threshold for plasma wave-breaking, to analyze the nonlinear absorption and harmonics generation in a laser irradiated low-density plasma corona. These results will be published elsewhere.

Acknowlegements

This work was supported by the RSCF grant No. 14-12-00194.

References

[1] Shirkov D V 1991 Several topics on renorm-group theory in "Renormalization group '91", *Proc. of Second Intern. Conf., Dubna, USSR* ed. Shirkov D V and Priezzhev V B 1992 (WS, Singapore) pp 1-10; Kovalev V F, Krivenko S V and Pustovalov V V, *The Renormalization group method based on group analysis ibid.*, pp 300-314

[2] Kovalev V F, Pustovalov V V and Shirkov D V 1998 Group analysis and renormgroup symmetries, *J. Math. Phys.* **39** 1170–88 (*Preprint* hep-th/9706056)

[3] Kovalev V F, Pustovalov V V 1989 Functional self-similarity in a problem of plasma theory with electron nonlinearity. *Theoretical and mathematical physics* **81** 1060-71

[4] Bogoliubov N N and Shirkov D V 1955 On Renormalization Group in Quantum Electrodynamics *Dokl. Akad. Nauk. SSSR* **103** 203-06 ; *ibid.*, (in Russian) **103** 391-94

[5] Bogoliubov N N and Shirkov D V 1956 Charge Renormalization Group in Quantum Field Theory *Nuovo Cim.* **3** 845–63

[6] Bogoliubov N N and Shirkov D V 1956 The multiplicative renormalization group in the quantum theory of fields *Sov.Phys.JETP* **3** 57

[7] Bogoliubov N and Shirkov D 1949 and 1980 *Introduction to the Theory of Quantized Fields* (New York: Wiley-Interscience)

[8] Kovalev V F, Shirkov D V 2008 Renormgroup symmetries for solutions of nonlinear boundary value problems *Physics-Uspechi* **178** 849-65

[9] Erokhin N S, Moiseev S S, Mukhin V V 1974 Theory of second-harmonic generation in an inhomogeneous hot plasma *Nuclear Fusion* **14** 333-39

[10] Gershman B N, Denisov N G, Ginsburg V L 1957 The propagation of electromagnetic waves in a plasma (ionosphere) *Physics-Uspechi* **61** 561-12

[11] Davydova T A 1981 The dynamics of surface waves excitation, transformation and decay processes *Plasma Physics* **4** 921

[12] Stephens C R 1998 Why two renormalization groups are better than one. *International Journal of Modern Physics B* **12** 1379-96 (hep-th/9611062 8 Nov 1996)

Drag coefficient of solid micro-sphere between parallel plates

S A Poniaev[1,2], K K Soboleva[3], A I Sobolev[3], V V Dudelev[4], G S Sokolovskii[1]

[1]Ioffe Institute, St.Petersburg, Russia

[2]Mozhaiskiy Military Space Academy, St.Petersburg, Russia

[3]Peter the Great St.Petersburg Polytechnic University, St.Petersburg, Russia

[4]ITMO University, St.Petersburg, Russia

Email: serguei.poniaev@mail.ioffe.ru

Abstract. We present results of numerical simulation of the drag force acting on a spherical micro-particle in a two-wall channel formed by parallel plates. Dependence of the drag force on the distance between the particle and the channel wall is studied for different channel heights. We show the drag force in a two-wall channel with the height of the order of the particle diameter to be significantly higher than that in the single-wall counterpart.

1. Introduction

At present investigations of non-contact manipulation of micrometer-sized objects attract a considerable attention of researchers. The method is based on the optical trapping of the micrometer-sized particles as suggested in the 1970s [1]. The initial idea led to the development of a single-beam gradient force optical traps or "optical tweezers" capable of manipulation with the particles sized 1s-10s micrometers. Nowadays optical tweezers are widely used in many areas of physics and biology [2-5]. One of the emerging and fast-developing applications of the optical tweezers is lab-on-a-chip devices for manipulation and analysis of biological specimens. In such a prospective devices, micrometer-sized particles may be trapped and manipulated by the beams of semiconductor lasers which are the most compact and effective light emitters. However, application of the laser diodes to the optical trapping is hindered by the low spatial quality of their beams [6] resulting in the high beam propagation parameter M^2. The impact of the high M^2 values on the optical trap must be taken into account even for super-focused [7], non-diverging [8-10], and self-focused [11] laser diode beams and results in considerable reduction of the trapping force. On the other hand, if a particle is located in a wall-bounded flow, the particle drag coefficient can change in a wide range and the drag force can exceed the trapping force that leads to loss of trapping and escape of the particle from the laser beam. Therefore, it is highly important to have a proper understanding of the drag force acting on a micrometer-sized particle being moved near a wall of a channel.

The spherical solid particle motion in an unbounded flow and near a wall in a single-wall channel configuration has been extensively studied. These problems were considered in many publications, both experimental and theoretical, where the hydrodynamic forces acting on a spherical particle were investigated [12-14] and where several sphere drag coefficient approximations were suggested [15].

However, there are few investigations of the channel configuration with two walls in the form of parallel plates, and only a small part of them are experimental. For example, the experimental data on the drag force given in [16] allow one to compare the numerical simulation and experimental results only for the centre line between two parallel plates. There are no data for other areas of the channel.

2. Numerical model

Figure 1 shows schematically a sphere in a channel and characteristics of the computational zone. The sphere with diameter D moves with velocity V_s parallel to the channel wall. The channel height is H and the distance between the channel wall and the sphere surface is d.

The flow near the sphere moving with velocity V_s in a quiescent fluid can be described in the dimensionless form as

$$\nabla \cdot U = 0$$
$$\frac{\partial U}{\partial t} + (U \cdot \nabla)U = -\nabla p + \frac{1}{\mathrm{Re}}\nabla^2 U \tag{1}$$

where U is the dimensionless flow velocity, p is the dimensionless pressure, and the Reynolds number (the ratio of the inertial to viscous forces) is $\mathrm{Re} = V_s D/\nu$. The coordinates, velocity, time, and pressure in the governing equations are normalized by using the characteristic values D, V_s, D/V_s, ρV_s^2, respectively, where ρ is the fluid density, and ν is the fluid kinematic viscosity. When the sphere moves in a fluid, the drag force F_D acting on the sphere arises. This force is due to the pressure and viscous effects. The drag coefficient C_D can be expressed as the ratio between the drag force, unit kinetic energy, and sphere cross-section area $C_D = \frac{2F_D}{\rho V_s^2(1/4\pi D^2)}$.

To carry out numerical simulation, a reference frame was attached to the sphere. In this case the boundary conditions on the channel walls and inlet (the right-hand side of the computational domain) are given by u = -U and v = w = 0, the boundary conditions on the outlet (left-hand side of the computational domain) are $\frac{\partial U}{\partial x} = 0$, and the boundary conditions on the sphere surface are u = v = w = 0. In this paper a motion of micro-particles in a small-size channel is considered, so the typical Reynolds number is 10^{-3}.

A modified incompressible solver based on the OpenFOAM package [17] in a full 3D zone with a spherical particle was used for the numerical simulation. In the package, a finite-volume method was employed. For simulation, second-order spatial schemes with a minimal dissipation were chosen.

A grid and domain size independence study was performed by using an unbounded flow and single-wall channel flow as a benchmark for validation. The mesh convergence study was carried out by the 2x mesh refinement till the difference in the drag coefficient between the calculated and analytical values became as low as 1%.

There is a simple relation between the drag coefficient and Reynolds number for an unbounded flow, i.e., the flow without any walls [18]

$$C_{Dfree} = \frac{24}{\mathrm{Re}}(1 + 0.15\mathrm{Re}^{0.687}) \tag{2}$$

In the text below, C_{Dfree} will be used as a characteristic value for the drag coefficient normalization.

Table 1 shows results of the grid and domain size independence study for a unbounded flow configuration. The mesh size is denoted by MxN, where M is the number of cells in the radial direction and N is the number of cells around the central cross-section of the sphere, the domain size D_d is normalized by using D as a characteristic value. Figure 2 shows an example of the mesh and computational region in case of unbounded flow. All simulations were made for $\mathrm{Re}=10^{-3}$.

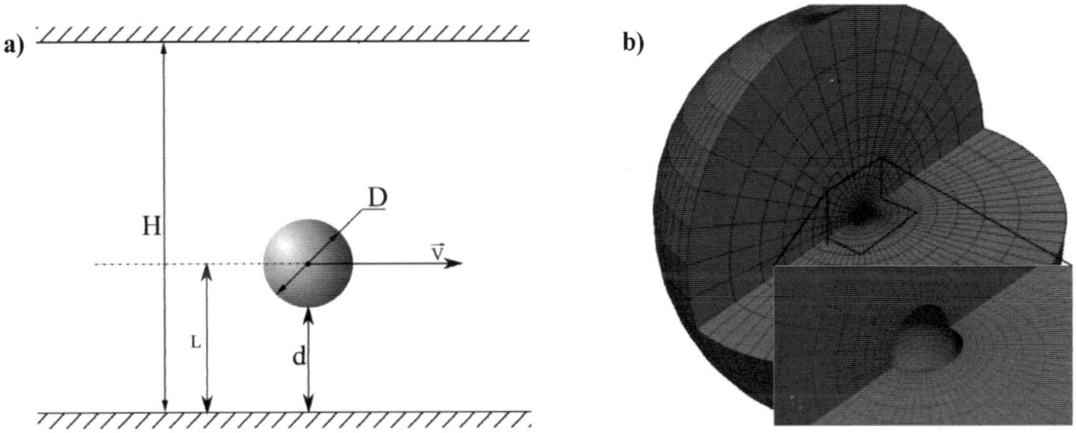

Figure 1. Schematic of sphere motion in the channel (a) and characteristics of the computational domain and mesh near the sphere (b).

Table 1. Drag coefficients for different computational domain and mesh sizes.

No.	Mesh size RxS	Domain size, D_d/D	C_{D_ns} numerical simulation	C_{Dfree} calculated from (2)	$\Delta = (C_{D\_ns}-C_{Dfree})/C_{D\_ns}*100$ %
1	10x20	10	24649	24031	2.5
2	20x40	10	25283	24031	5.0
3	10x20	20	24700	24031	2.7
4	20x20	20	24350	24031	1.3
5	20x40	20	24190	24031	0.7

Thus, the domain size equal to 20 sphere diameters and the mesh size equal to 20x40 was chosen for the numerical simulation.

Changes in the drag coefficient in a single-wall channel configuration have been considered by many authors. For the case when a sphere nearly touches the channel wall, i.e. d → 0, a lubrication theory [19] gives an approximation for the drag coefficient as:

$$C_D = \frac{24}{\mathrm{Re}}\left(0.9588 - \frac{8}{15} ln\left(2\frac{d}{D}\right)\right) \quad (3)$$

So it is evident that there is logarithmic singularity when the sphere approaches the channel wall. The approximation valid for any values of d was suggested in [15]:

$$\begin{gathered} C_D = C_{D0}\left(1 + \alpha \mathrm{Re}^{\beta}\right) \\ C_{D0} = \frac{24}{\mathrm{Re}}\left(1.028 - \frac{0.07}{(1+4(d/D)^2)} - \frac{8}{15} ln\left(270\frac{(d/D)}{(135+256(d/D))}\right)\right) \\ \alpha = 0.15\left(1 - exp\left(-\sqrt{(d/D)}\right)\right) \\ \beta = 0.687 + 0.313 exp\left(-2\sqrt{(d/D)}\right) \end{gathered} \quad (4)$$

As the authors of [15] stated, the above expression reduces to correct limits for small and large d. Figure 3 shows a normalized drag coefficient (the characteristic value for the normalization is the drag coefficient for an unbounded flow) obtained in the numerical simulation for different dimensionless distances (d/D) from the sphere. Approximations (3) and (4) are also plotted in this figure.

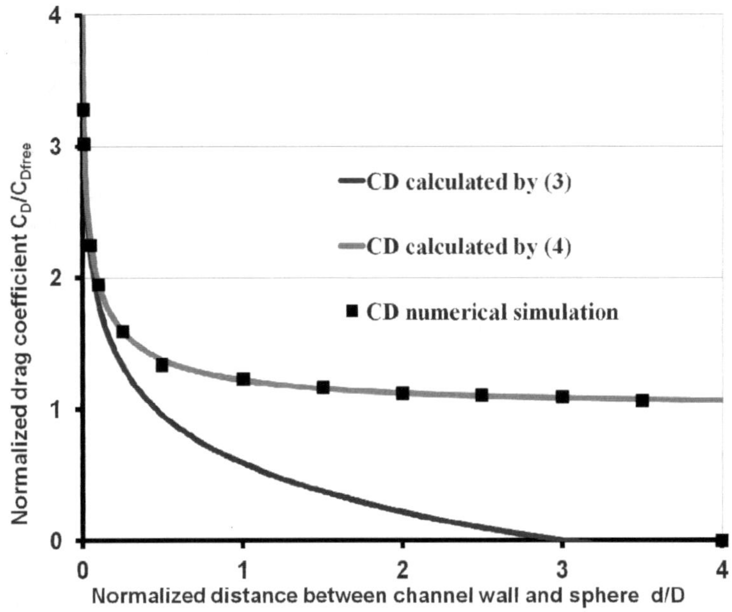

Figure 3. Variation in normalized drag coefficient for different normalized distances from the wall.

It is clearly seen that approximation (3) gives wrong results for large d/D values, but it is in a very good agreement for $d \to 0$ as expected. Approximation (4) shows reasonable values in the entire region of simulation. The data obtained in the numerical simulation are in very good agreement with approximation (4), which confirms the correctness of the choice of the computational zone and mesh sizes.

3. Results and discussion

The main goal of our study was to perform numerical simulation for a solid sphere in a two-wall channel configuration, i.e. between parallel plates. In the course of numerical simulation the channel height was varied from 1.1 to 10 sphere diameters. To compare the results obtained in our numerical simulation with the experiment, we used the experimental data for the centre line of the two-wall channel [16]. In [16], the experimental data were approximated by a 14th degree polynomial. So it is possible to calculate the drag coefficient at the centre line for channels of different heights.

Figure 4 shows the data for the normalized drag coefficient for the channels with the heights equal to 1.1, 2, 5, and 10 sphere diameters. In these figures approximation (4) that presents the drag coefficient for a single-wall configuration is also plotted.

It is clearly seen that the computed results are higher than those approximated for the channel with H/D <5 in the entire range of distances between the wall and sphere surface. The computed results for the channel with H/D=10 are in good agreement with the single-wall approximation. The simulation results for the channel with H/D=5 are slightly higher than those approximated for the single-wall case. There is also an excellent agreement between the results of our numerical simulation and approximation of the experimental data reported in [16] for the center line of the channel.

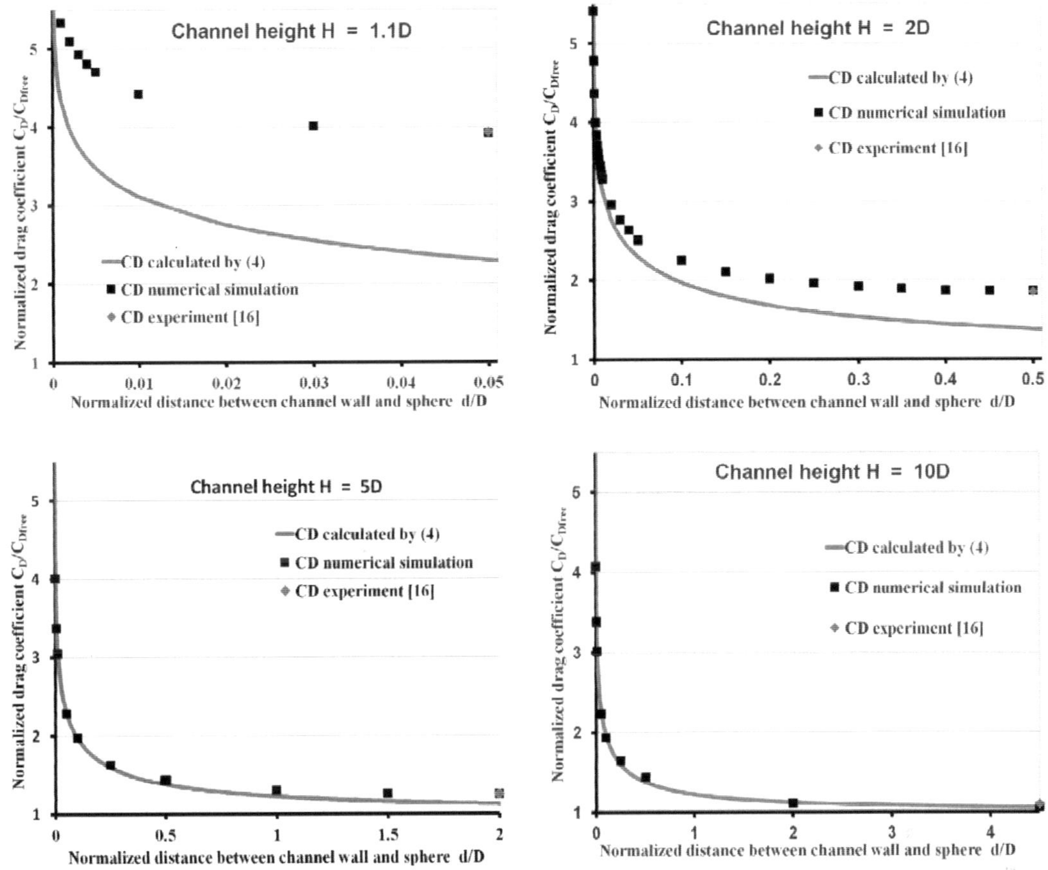

Figure 4. Normalized drag coefficient vs normalized distance between the sphere and wall for different channel heights.

Our results clearly show that the sphere drag coefficient in the two-wall channel configuration can be significantly higher than in an unbounded flow. In the case of two-wall channels with the heights equal to several sphere diameters the drag coefficient is significantly higher than that in a single-wall channel for the same distances between the sphere and the wall. This effect diminishes with increasing channel height.

4. Conclusions

In this paper a numerical simulation of the drag coefficient for the sphere in the two-wall channel is performed. The flow conditions correspond to the micro-particle motion with the Reynolds number of 10^{-3}. The validation of the model is carried out in the conditions of an unbounded flow and a single-wall channel. The simulation data of the validation cases were found to be in good agreement with the experiments and approximations given in the literature. The numerical simulation of the drag coefficient of a sphere in a two-wall channel shows a strong dependence of the sphere drag coefficient on the distance between the sphere and the wall. In the case of a channel with the height of the order of the particle diameter the drag coefficient is much higher than that in an unbounded flow and higher than in a single-wall channel with the same distance between the sphere and the wall.

Acknowledgements
The research of G.S. Sokolovskii was partially supported by the Program 111-7 of the Physical Sciences Division of the Russian Academy of Sciences.

References
[1] A Ashkin and J M Dziedzic 1977 Observation of resonances in the radiation pressure on dielectric spheres *Physical Review Letters* **38** no. 23 p. 1351.
[2] K C Neuman and S M Block 2004 "Optical trapping," *Review of Scientific Instruments* **75**, no. 9, pp. 2787–2809.
[3] K Svoboda and S M Block 1994 Biological Applications of Optical Forces *Annu Rev Biophys Biomol Struct.*
[4] K Dholakia and T Cizmar 2011 Shaping the future of manipulation *Nature Photon.* **5**, pp. 335-342.
[5] M Padgett and R Bowman 2011 Tweezers with a twist *Nature Photon.* **5**, pp. 343-348.
[6] G S Sokolovskii, V V Dudelev, S N Losev, K K Soboleva, A G Deryagin, K A Fedorova, V I Kuchinskii, W Sibbett, E U Rafailov 2014 Bessel beams from semiconductor light sources *Progress in Quantum Electron.* **38**, no.4, pp. 157–188,
[7] G S Sokolovskii, V V Dudelev, S N Losev, A G Deryagin, V I Kuchinskii, W Sibbett, E U Rafailov 2012 Superfocusing of mutimode semiconductor lasers and light-emitting diodes, *Tech. Phys. Lett.* **38**, pp. 402-404.
[8] G S Sokolovskii, V V Dudelev, S N Losev, M Butkus, K K Soboleva, A I Sobolev, A G Deryagin, V I Kuchinskii, W Sibbett, E U Rafailov 2013 Influence of the axicon characteristics and beam quality parameter M^2 on the formation of Bessel beams from semiconductor lasers, *Quantum Electron.* **43**, pp. 423-427.
[9] G S Sokolovskii, S A Zolotovskaya, S N Losev, V V Dudelev, A G Deryagin, V I Kuchinskii, W Sibbett, E U Rafailov 2011 High power Bessel beams from EP-VECSELs, *Proc. SPIE* **7919**, p. 79190J.
[10] G S Sokolovskii, V V Dudelev, S N Losev, A G Deryagin, D A Vinokurov, A V Lyutetskii, N A Pikhtin, S O Slipchenko, I S Tarasov, S A Zolotovskaya, E U Rafailov, V I Kuchinskii, W Sibbett 2010 Study of non diffracting light beams from broad stripe edge emitting semiconductor lasers *Tech. Phys. Lett.* **36**, pp. 9-12.
[11] D A Yanson, E U Rafailov, G S Sokolovskii, V I Kuchinskii, A C Bryce, J H Marsh, W Sibbett 2004 Self-focussed distributed Bragg reflector laser diodes *J. of Appl. Phys.* **95**, pp. 1502-1509.
[12] R G Cox and S K Hsu 1977 The lateral migration of solid particles in a laminar flow near a plane *Int. J. Multiphase Flow* **3**, p. 201.
[13] P Vasseur and R G Cox 1977 The lateral migration of spherical particles sedimenting in a stagnant bounded fluid *Journal of Fluid Mechanics* **80**, no. 3, pp. 561–591
[14] L Zeng, S Balachandar, and P Fischer 2005 Wall-induced forces on a rigid sphere at finite Reynolds number *Journal of Fluid Mechanics* **536**, pp. 1–25
[15] L Zeng, F Najjar, S Balachandar, and P Fischer 2009 Forces on a finite-sized particle located close to a wall in a linear shear flow *Physics of Fluids* **21**, no. 3, p. 033302
[16] A Miyamura, S Iwasaki, and T Ishii 1981 Experimental wall correction factors of single solid spheres in triangular and square cylinders, and parallel plates *International Journal of Multiphase Flow* **7**, no. 1, pp. 41–46
[17] H G Weller, G Tabor, H Jasak, C Fureby 1998 A tensorial approach to computational continuum mechanics using object-oriented techniques *Computers in Physics*, **12**, no. 6
[18] R Clift, J R Grace, and M E Weber 1978 Bubbles, Drops, and Particles, Academic, New York
[19] A J Goldman, R G Cox, and H Brenner 1967 Slow viscous motion of a sphere parallel to a plane wall. I. Motion through a quiescent fluid *Chem. Eng. Sci.* **22**, p. 637

Journal of Physics: Conference Series **769** (2016) 012085 doi:10.1088/1742-6596/769/1/012085

Experimental investigation of the interaction of supersonic airflow with transverse discharge

S V Bobashev[1,2], V P Belik[1], N A Monakhov [1], M V Petrenko[1,2], P A Popov [1,2], and V A Sakharov[1]

[1]Ioffe Physical Technical Institute, Russian Academy of Sciences, St. Petersburg, 194021 Russia
[2]Peter the Great St. Petersburg State Polytechnic University, St. Petersburg, 195251 Russia

e-mail: pavel.popov@mail.ioffe.ru

Abstract. We have experimentally studied the interaction of a supersonic air flow with transverse electric gas discharge. Spectra of radiation emitted from various regions of the flow in the 600–700 nm wavelength range have been measured. Composition of the supersonic air flow is determined and peculiarities of its interaction with discharge have been studied. In particular, significant transfer of the electrode material with gas flow was observed.

1. Introduction

The interaction of high-velocity gas flows with low-temperature plasma is a topical area of research of magneto-plasma aerodynamics [1, 2]. Physical processes accompanying the interaction of plasma with supersonic flow are very complicated and have yet to be sufficiently studied. In particular, the process of gas ionization at high velocities of the flow has some specific features related to the transport of charged particles away from the region of a strong electric field.

This work is a continuation of our investigations devoted to peculiarities of the interaction of supersonic flows with low-temperature plasma. Previously, we have studied the capabilities of a flow control in a supersonic diffuser via the MHD method [3], the influence of gas-discharge plasma on the structure of flow past a body and on the heat flux toward its surface [4]. It was established that a rotating discharge significantly modifies the shock-wave configuration near a body and, hence, heat fluxes toward its surface. It was also established [5] that local near-electrode processes significantly influence the entire discharge dynamics. In the case of subsonic gas flows, similar results were obtained in [6]. Recent investigation [7] of a short-time interaction of a supersonic gas flow with low-temperature laser-generated plasma showed evidence of significant changes in the flow structure and the formation of shock waves. The present investigation is devoted to elucidating peculiarities of the interaction of a supersonic flow with gas discharge plasma.

2. Experimental setup

The experiments were carried out in a pulsed gasdynamic setup based on a Ludwig tube (Figure 1). Conical nozzle 1 with a 60° opening angle had critical and output cross section with diameters of 10 and 60 mm, respectively. The output cross section of the nozzle was placed inside cylindrical vacuum chamber 2 with a 0.3 m inner diameter and height of 0.4 m equipped with a roughing pump. The subsonic part of the nozzle was separated from atmosphere by a plastic diaphragm. The side wall of the chamber was provided with windows 3 for spectroscopic diagnostics. The optical emission spectra were measured using monochromator 4 with a CCD array linked to a computer. A pair of aluminum electrodes was situated 20 mm away from the nozzle output cross section. Central electrode 5 with a diameter of 4 mm was arranged on the nozzle axis, and side electrode 6 was mounted 20 mm away from the nozzle edge The electrodes were connected to pulsed current source 7 representing an LC line consisting of 14 unit circuits, each with capacitance C = 300 μF and inductance L = 3.5 μH. This circuit generated nearly rectangular current pulses with a duration of about 1 ms. Serially connected low-inductance high-voltage transformer formed a voltage pulse of $t \sim 10^{-6}$ s duration and amplitude up to U ~ 30 kV that initiated electric gas discharge.

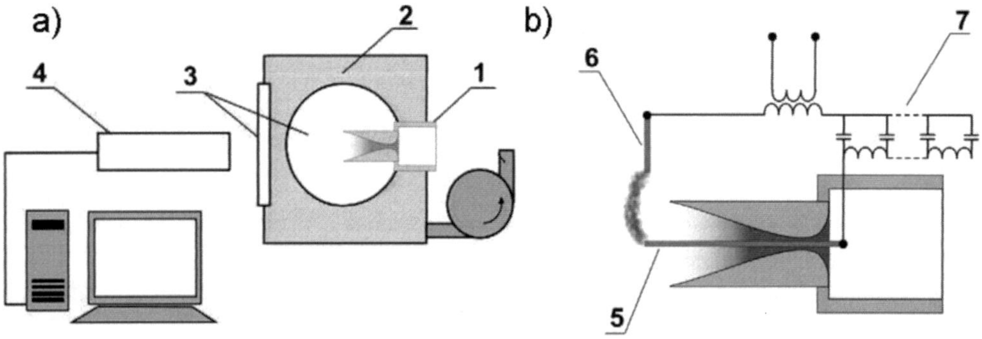

Figure 1. (a) Block scheme of the experimental setup and (b) scheme of the supersonic nozzle with electrodes and pulsed current source (see text for explanations).

Before to the experiment, the chamber was evacuated to a residual gas pressure of P = 0.1 Torr, which corresponded to pressure in supersonic flow at the nozzle output edge. After destruction of the diaphragm, the atmospheric air was ejected into the chamber via the nozzle. The current source was switched on at t =200 μs after destruction of the diaphragm, when airflow in the supersonic nozzle could be considered steady-state. At this moment, an initiating high-voltage pulse was applied between the central and side electrodes, which formed the initial region of discharge. Then, the main storage bank exhibited discharge during t ≈ 1 ms via the initiated channel. In these experiments, the output voltage was varied within U = 300–600 V and the discharge current reached I = 1.7 kA. The total energy deposited in discharge during this pulse varied within Q = 250–700 J depending on the initial voltage of the source.

Parameters of the supersonic air flow in the nozzle output section were determined by simulating non-viscous gas flow in the nozzle, where the pressure was P ≈0.1 Torr and the flow Mach number was M ≈ 5.5. According to calculations, the gas pressure growth in the vacuum chamber during discharge (t ≈ 1 ms) was no more than twofold. At this ratio of pressures, there is no supersonic gas flow separation in the nozzle and the flow structure is close to the calculated pattern.

In these experiments, peculiarities of the interaction of supersonic flow with electric discharge were studied by means of spectroscopic diagnostics in the visible spectral range. For this purpose, we measured the spectra of emission from regions I–IV spaced by 30, 50, 70, and 90 mm, respectively, from the nozzle output along the line passing via the middle of the interelectrode gap.

The emission spectra were measured using a monochromator based on the Seya–Namioka scheme with a CCD array TCD1304DG (Toshiba) detector. The operation of CCD arrays of this type in spectral instruments for diagnostics of pulsed radiation sources was studied in [8]. The minimum time of charge accumulation in these CCD arrays is t ≈ 20 ms. In the given case, this circumstance allows measuring only the integrated emission spectra. In one run, the spectra were measured within Δλ ≈ 100 nm wavelength interval. The inverse linear dispersion of the instrument employed with a

1200 cm^{-1} grating was 1.9 nm/mm. Calibration of the spectrometer with respect to the wavelength scale was performed using Ne I lines from the NIST Atomic Spectra Database [9]. The error of wavelength determination in the entire spectral range studied was $\Delta\lambda < 0.01$ nm.

3. Results and discussion

Here we present the results of experiments performed with supersonic gas flows emitted from the nozzle at pressures P = 0.1 and 10 Torr. The emission spectra were measured in a wavelength range of λ =600–700 nm. This interval was selected with allowance for specific features of the spectrometer employed and the presence of a small number of emission lines of the material of electrodes (Al) and main atmospheric gases (N and O).

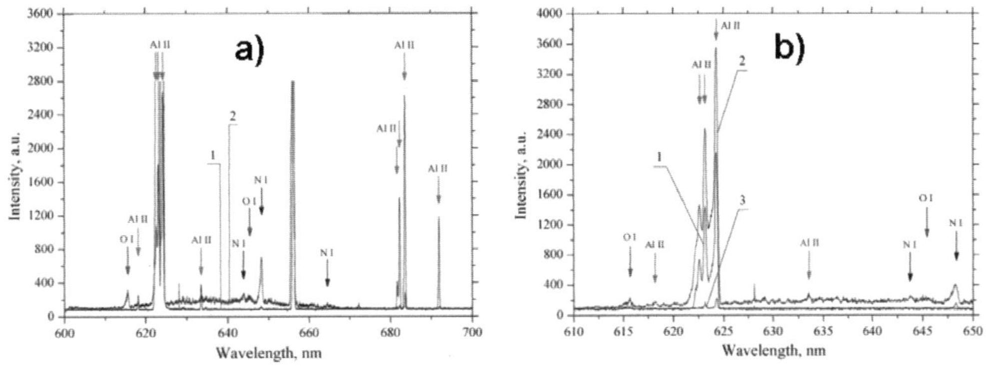

Figure 2. Emission spectra measured in (a) regions I (curve 1) and II (curve 2) for a supersonic flow and (b) region III for stagnant gas at pressures (1, 3) P = 0.1 and 10 Torr, respectively, and (2) for the supersonic flow.

Figure 2 shows the emission spectra measured in various probed regions under conditions of stagnant gas and supersonic flow. The presence of strong emission lines of Al^+ ions (Al II, λ = 622.62, 623.17, and 624.34 nm) is indicative of a noticeable supply of electrode material to the region of discharge. The intensity of these lines in the supersonic flow slightly decreases with distance from the nozzle output and significantly exceeds the intensity of the same lines measured during discharge in the stagnant gas at the same pressure (P = 0.1 Torr). Note also that increasing distance from the nozzle edge is accompanied by changes in the relative intensities of Al II emission lines, which can be explained by the variation of temperature in the flow.

The obtained emission spectra show emission lines due to excited atoms of oxygen (O I, λ = 615.82 nm) and nitrogen (N I, λ = 648.27 nm). It should be noted that these lines only appear at some distance from the nozzle edge (regions II–IV) and their intensities slightly decrease with increasing distance.

In the case of a supersonic flow, regions II–IV exhibit continuous emission spectrum with a maximum at $\lambda \approx 640$ nm. Assuming that this emission has a thermal origin, the temperature corresponding to this maximum can be estimated at $T \approx 0.4$ eV. Note that, in the stagnant gas at P = 0.1 Torr, no continuous emission spectrum was observed.

We have estimated plasma temperature using relative intensities of a pair of spectral lines belonging to the same ion [10]. The temperature in region II was determined using aluminum ion lines Al II with wavelengths λ = 683.71 and 691.99 nm and the lines of excited nitrogen atoms N I with λ = 664.49 and 672.26 nm. In the former case, the estimated temperature was T = 0.13 eV, while the latter estimation gave T = 0.29 eV. In regions III and IV, the temperature estimated using relative intensities of Al II lines was found to be T = 0.22 and 0.27 eV, respectively. This, the time-averaged temperature at a distance of 50–90 mm from the nozzle edge amounts to T = 0.1–0.3 eV.

The discharge develops in air with a concentration of atomic species on the order of $N \sim 10^{16}$ cm^{-3} and the discharge current density reaches $j \sim 10^3$ A/cm^2 [5]. Under these conditions, the ion component of plasma is represented by the electrode material supplied from small-size local emission centers operating on the cathode surface with a current density up to j ~ 10^7 A/cm^2, while the total current is determined by the thermionic emission [11]. According to estimations, the electron concentration in the discharge channel is $n_e \sim 10^{13}$ cm^{-3} and a large current density is related to a low frequency of electron collisions with molecules in the gaseous medium, which leads to a low degree of ionization in the airflow. This is confirmed by the absence of emission lines due to N^+ (and higher charged) ions in the spectrum of discharge plasma emission from air flow in region I and large intensity of emission in Al^+ ion lines.

4. Conclusion

The obtained spectral pattern of emission leads to the conclusions that (I) the discharge region is driven downstream by the oncoming gas flow and (II) there is significant transfer of the electrode material with gas flow over a distance comparable with the transverse size of the supersonic flow. The absence of emission lines corresponding to nitrogen and oxygen ions as well as the continuous emission with a maximum at $\lambda \approx 640$ nm in the spectra of regions spaced by 50–90 mm from the nozzle edge indicates that heating of the supersonic flow in discharge takes place without significant ionization.

References

[1] Starikovskiy A and Aleksandrov N 2011 Nonequilibrium plasma aerodynamics *Aeronautics and Astronautics*, Ed. by M. Mulder (InTech).
[2] Adamovich I, 2010 *Encyclopedia of Aerospace Engineering* (John Wiley & Sons Ltd.).
[3] Vasil'eva R V, Erofeev A V, Lapushkina T A, et al, 2005 *Tech. Phys.*, **50**, 1132.
[4] Bobashev S V, Mende N P, Popov P A, and Sakharov V A, 2010 *Tech. Phys.* **55**, 1768.
[5] Bobashev S V, Mende N P, Monakhov N A, and Sakharov V A, 2012 *Tech. Phys.* **57**, 1637.
[6] Klementyeva I B, Biturin V A, Tolkunov B N, and Moralev I A, 2011 *High. Temp.* **49**, 788.
[7] Garbaruk A V, Gritskevich M S, Kalmykov S G, Mozharov A M, Petrenko M V, and Sasin M E, 2014 *Tech. Phys. Lett.* **40**, 980.
[8] Belik V P, Petrenko M V, and Savushkin A V, 2011 *Tech. Phys. Lett.* **37**, 1037.
[9] NIST Atomic Spectra Database //http://www.nist.gov/pml/data/asd.cfm
[10] Ochkin V N 2010 *Low-Temperature Plasma Spectroscopy* (Moscow: Fizmatlit) [in Russian].
[11] Raizer Yu P 2009 *Physics of Gas Discharge* (Dolgoprudnyi: ID Intellekt) [in Russian].

Laser plasma simulations of the generation processes of Alfven and collisionless shock waves in space plasma

P A Prokopov[1], Yu P Zakharov[1], V N Tishchenko[1], I F Shaikhislamov[1], E L Boyarintsev[1], A V Melekhov[1], A G Ponomarenko[1], V G Posukh[1] and V A Terekhin[2]

[1]Institute of Laser Physic (ILP) SB RAS, Novosibirsk, Av. Lavrentyeva 13/3, 630090, Russia. E-mail: paprok312@gmail.com
[2]All-Russian Scientific Research Institute for Experimental Physics (VNIIEF), Sarov, Av. Mira 37, 607188, Russia

Abstract. Generation of Alfven waves propagating along external magnetic field B_0 and Collisionless Shock Waves propagating across B_0 are studied in experiments with laser-produced plasma and magnetized background plasma. The collisionless interaction of interpenetrating plasma flows takes place through a so-called Magnetic Laminar Mechanism (MLM) or Larmor Coupling. At the edge of diamagnetic cavity LP-ions produce induction electric field E_φ which accelerates BP-ions while LP-ions rotate in opposite direction. The ions movement generates sheared azimuthal magnetic field B_φ which could launches torsional Alfven wave. In previous experiments at KI-1 large scale facility a generation of strong perturbations propagating across B_0 with magnetosonic speed has been studied at a moderate value of interaction parameter $\delta\sim0.3$. In the present work we report on experiments at conditions of $\delta\sim1\div2$ and large Alfven-Mach number $M_A\sim10$ in which strong transverse perturbations traveling at a scale of ~1 m in background plasma at a density of $\sim3*10^{13}$ cm^{-3} is observed. At the same conditions but smaller $M_A\sim2$ a generation, the structure and dynamic of Alfven wave with wavelength ~0.5 m propagating along fields $B_0\sim100\div500$ G for a distance of ~2.5 m is studied.

1. Introduction

With the aim to simulate processes in space plasmas we consider in this paper two different types of disturbances produced by explosive Laser-produced Plasma (LP) expanding in a magnetized Background Plasma (BP).

The first one is torsional Alfven Waves (AW), transverse MHD plasma perturbations propagating along field lines of the external magnetic field B_0. In laboratory it was generated by injecting laser-produced plasma along (and quasi-parallel) the lines of B_0 in a presence of magnetized background plasma. The plasmas are sufficiently collisionless and interaction takes place due to a so-called magnetic laminar mechanism (MLM) or Larmor Coupling [1, 2, 3]. As a result of this interaction the net azimuthal ion current generates sheared azimuthal magnetic field B_φ [1] which could generate AW, propagating along B_0. In experiment besides B_φ a related parallel current J_Z was directly measured as well. It is believed that such disturbances can propagate in the magnetic flux tubes at the Sun chromosphere and transfer energy to the Corona.

The second type of perturbation is the collisionless shock wave (CSW) [4, 5, 6]. It is generated by the injection of LP in magnetized BP across B_0. CSW propagates across the lines of external magnetic field; the registration is made by jump of magnetic field and of BP concentration. The plasma also is in collisionless mode, generation of disturbances is produced by MLM. Similar processes may occur in the solar Corona during a Coronal Mass Ejection.

2. Magnetic laminar mechanism

Magnetic laminar mechanism (MLM) or Larmor Coupling of LP with magnetized BP effectively operates in collisionless regime at high Alfven-Mach numbers $M_A \geq 5$. LP, propagating with velocity V_0, forms a so-called magnetic cavity (area of displaced external magnetic field), the curl electric field E_φ is generated at its boundary. BP-ions accelerate along to this field, while at the same time ions of LP are decelerated in their Larmor rotation (opposite to E_φ). Conditions at which a strong coupling between BP-ions and LP-ions results due to such mechanism were established by for the first time in [1]. Its effectiveness depends mainly on a so-called MLM-parameter $\delta = R_*^2 / R_L R_{L*} \geq 1$, where $R_* = (3N_e / 4\pi n_*)^{1/3}$ – the radius of diamagnetic cavity (N_e – total number of electrons, n_* – concentration of BP-ions), R_L, R_{L*} – Larmor radii of LP-ions and BP-ions respectively. The effective deceleration of LP takes place at a scale of R^*, which is comparable to gas-dynamic scale of $R_m = (3N_e m_i Z_{i*} / 4\pi n_* m_{i*} Z_i)^{1/3}$, and its kinetic energy transferred to BP proportionally to parameter δ (here m_i, m_{i*} - ion mass of LP and BP respectively, Z_i, Z_{i*} - ion charge of LP and BP respectively).

But indeed, the physical processes of electromagnetic fields, responsible for the MLM-interaction, can act in the range up to $M_A \sim 1$ and moreover, could supply some kind of collisionless interaction along the B_0-field, namely important for the AW experiment – opportunity to launch a torsional Alfven Wave (AW) by azimuthal B_φ magnetic field, generated inside of spherical LP at early stage of its expansion. Such fields of the level $B_\varphi / B_0 \sim 1$ were revealed during the first hybrid simulations [1] of MLM-interaction and later their probable consequences in a form of electron whistler [7] or AW [4, 12] were discovered. Usually, the generation [8] of such B_φ-fields are treated on the base of Hall term $K = [rotB \times B]/n$ in the equation for $\partial B_\varphi / \partial t \propto -(C/4\pi e) rotK$. Thus the value of B_φ can be estimated (for the time $\partial t \sim R_b / V_0$ at the characteristic scale $\sim C/\omega_{pi}$, where R_b – deceleration radius of LP in vacuum magnetic field, definition presented in table 1) as $B_\varphi / B_0 \sim R_b / R_L \sim (\delta \cdot m_i Z_{i*} / Z_i m_{i*})^{1/2}$.

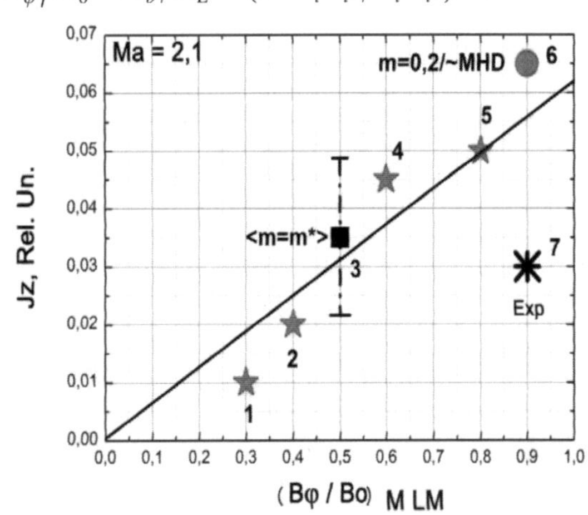

Figure 1. Linear relation between dimensionless J_Z current (generated at the same time as AW) and the initial value of azimuthal B_φ/B_0-field of MLM-model (1) for a set of 3-Fluids simulations:
1 – Run III – δ = 0,08 and $m_i Z_{i*}/m_{i*} Z_i$ = 16;
2 – Run V – δ = 1,3 and $m_i Z_{i*}/m_{i*} Z_i$ = 2;
3 – Run IVa – δ = 4,3 for the special case of equal $m_i Z_{i*}/m_{i*} Z_i$ = 1,4 with a low limit value J_Z;
4 – Run I – δ = 4 and $m_i Z_{i*}/m_{i*} Z_i$ = 1,5;
5 – Run II – δ = 3,5 and $m_i Z_{i*}/m_{i*} Z_i$ = 3;
6 – Run "MHD" with m_i = 0,2 a.m.u.
and experimental point 7 (with maximum level 1000 A, re-calculated to dimensionless J_Z).

The same result could be obtain more exactly with taking into account the main equation of MLM-model, with a calculated maximal velocity $V_{\varphi*} = 0.5\omega_{ci*} R_* \sin\theta$ of BP-ions [9], accelerated by curl electric field during magnetic field exclusion by LP, in dependence upon polar angle θ (where ω_{ci*} – cyclotron frequency of BP-ions). As a result, a frozen to BP initial field B_0 will be torque at most extent near "equator" ($\theta = 90^0$), while a pole, and the incline (or B_φ-field) of its field line could occur. So a resulting B_φ-field in a MLM-model will be:

$$B_\varphi / B_0 = 0.5[\delta_0 (m_i / m_{i*})(Z_{i*} / Z_i)]^{1/2} \sin\theta \cos\theta \quad (1)$$

with the maximal value at $\theta = \pm 45^0$ as was obtained by hybrid simulations [4]. At the figure 1, a successful comparison of such dependence (1) with the results of 3-fluids MHD-model of ILP [10] is presented in a variable values of initial B_φ/B_0 (estimated by MLM-model) and the calculated total current J_Z (in dimensionless units) of AW along to magnetic field B_0. This 3-fluids MHD model simulates plasma as 3 fluid: LP-fluid, BP-fluid and overall electron fluid.

3. Experiment set up and results on Alfven waves

Laboratory experiments are carried out at KI-1 facility [11]. The scheme of experiment is presented on figure 2. Large scale high-vacuum chamber 5 m long, 1.2 m in diameter kept a vacuum of $\sim 10^{-6}$ Torr can be filled by means of θ-pinch with the background plasma (H^+ or He^+). LP-clouds can be generated by two independent microsecond-pulse CO_2-laser systems with similar parameters of radiation focused on a flat (or convex) target of polyethylene (focal size of laser spot ∅2.5 cm). An external magnetic field up to 500 G is created by quasi-stationary current in a solenoid covering the entire outer surface of the vacuum chamber. Diagnostics of plasma is performed using double Langmuir probes P1 (with magnetic probes Mz, Mf), P0 (with magnetic probes Br, Bz, Bf), and a pair of probes IK1, IK2 with the corresponding three-component magnetic probes RM1 (L, T, H), RM2 (L, T, H). For direct detection of current J_Z, related to AW [12, 13, 14], Rogowski coil (PR) 5 cm in diameter was used. It was electrostatically shielded from the interference with potential of the plasma (as magnetic probes). Miniature collectors of direct ion-flux KB1 and KB2 are mainly used for recording such streams along the axis Z, and KB3 as a symmetrical double electrode is used for evaluation of current J_i and plasma electrons temperature T_e.

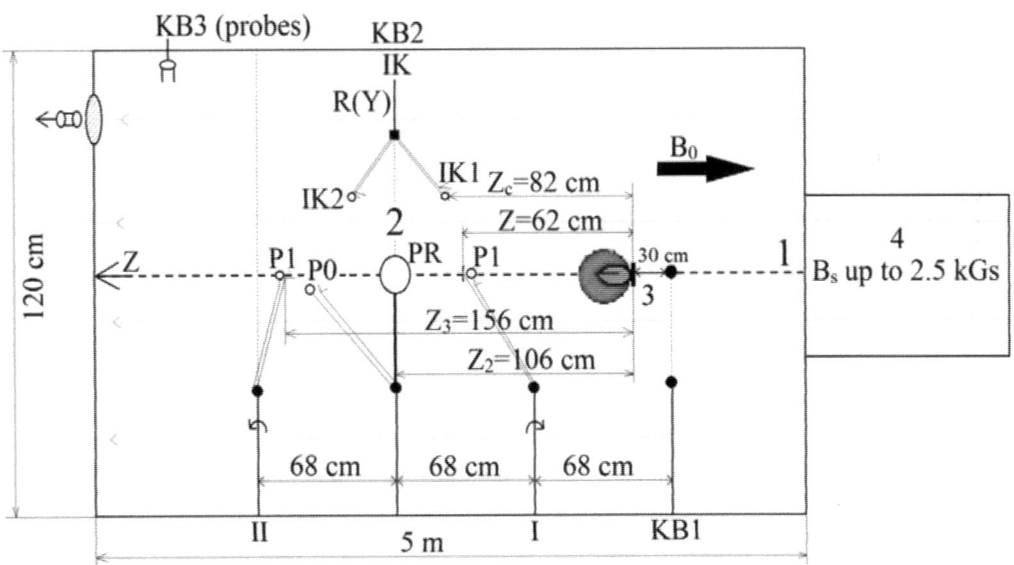

Figure 2. Experimental scheme of Alfven wave experiment at KI-1 facility of ILP [2].

In this experiment a cylindrical coordinate system is used with the Z axis along the axis of symmetry of the vacuum chamber and magnetic field lines. The angular coordinate is φ, and the radial is r. The basic parameters and criteria of similarity of modelling experiment are presented in table 1.
The injection of LP was made along B_0 (Z axis). The propagation of AW is also recorded along Z axis by magnetic probes and Rogowski coil, arranged at different distances from the laser target.

Using BP magnetized by external magnetic field in the range 100-500 Gs we observed propagation of plasma perturbations in a magnetic flux tube. Diameter of this tube is ~15÷20 cm, length is ~2.5 m along the chamber axis.

The propagation velocity of LP in the absence of the external magnetic field and the BP is about 200 km/s, while the speed of BP is order of magnitude smaller ~20÷30 km/s. However, in a vacuum magnetic field LP front was found to move at a speed of about 90 km/s, and in magnetized BP ~60÷80 km/s. LP-ions are slowing down at a time when magnetic cavity forms, so their speed is smaller than in absence of magnetic field. In the last case detailed comparison revealed that we observe in fact perturbation of BP rather than LP ions.

Table 1. List of experimental parameters

Dimensional parameters	AW experiment	CSW experiment
Concentration of H^+/He^+ background, n_*	$0{,}5\div3{,}5*10^{13}$ cm^{-3}	~$3*10^{13}$ cm^{-3}
Ion-skin scale (background)	C/ω_{pi} ~ 5 cm	
External magnetic field (in background)	$B_0 = 100 \div 500$ Gs	$B_0 = 110$ Gs (80 Gs)
Initial velocity of the front of LP-cloud (velocity of n_{max}, maximum of concentration)	$V_0 \approx 200$ km/s ($V_M \approx 150$ km/s)	
Effective energy of LP-cloud $E_0 = E_k *4\pi/\Delta\Omega$ (with E_k – kinetic energy, $\Delta\Omega$ - angle of LP expansion) with N_0 – effective number of electrons	E_0=30 J, with N_0~10^{18}	E_0~1000 J, with N_0~10^{19}
Deceleration radius of LP by magnetic field B_0 (radius of magnetic cavity in vacuum)	$R_b = (3E_0/B_0^2)^{1/3}$ ~ 30 cm with B_0=175 Gs	
Temperature of BP-electrons	T_{e*} ~ 10 eV	$T_{e*} \approx 7\div11$eV
Mean free path of LP-ions in background with initial speed V_0 (or at the lower speed V_R, relative to the background)	λ_{ii*} ~ 200 cm, ion-ion Coulomb collisions (to ≤ 20 cm with $V_R \approx 60$ km/s in H^+ background)	$\lambda_{i\text{-}i*}$ ~ 300 cm, ion-ion Coulomb collisions, $\lambda_{i\text{-}e}$, ion-electron Coulomb collisions
Alfven speed. Alfven wavelength (on Z-axis)	C_A=70 km/s, λ_{AB} ~ 50 cm (B_0=175 Gs, H^+ background, $n^*=3*10^{13}$ cm^{-3})	
Size of background along Z-axis and along r-axis	~ 100÷200 cm and 20 cm	

Dimensionless similarity criteria	AW experiment	CSW experiment
Alfven-Mach number	$M_A = V_M/C_A$ ~ 2	$M_A = V_M/C_A$~7
Larmor radius	$R_L = V_0 m_i C/eZ_i B_0$	
Magnetization of LP-ions with charge-mass composition m_i/Z_i=2,5	$\varepsilon_b = R_L/R_b$=0,7 (≤1), with R_L=21 cm	
Radius of magnetic cavity of LP in background	$R^* \approx (3N_0/4\pi n^*)^{1/3} \approx 20$ cm	$R^* \approx (3N_0/4\pi n^*)^{1/3} \approx 50$ cm
MLM-parameter	$\delta = R_*^2/R_L R_L^*$~3	$\delta = R_*^2/R_L R_L^* \geq 1$
Plasmas beta	$\beta = 8\pi nkT_{e*}/B_0^2$ ~ 0,5	$\beta = 8\pi nkT_{e*}/B_0^2$ ~ 1
Knudsen number	$Kn_i = \lambda_{ii*}/R_B$ ~ 10 (with initial speed V_0, but up to ≤ 1 with real speed $V_R \approx 60$ km/s relative to the background)	

Besides characteristic features of the AW, a high-frequency whistler precursor was registered as well like in our previous work [7]. The measurement of transverse component of the magnetic field has shown that the first disturbance has a right-hand circular polarization that corresponds to electron whistlers, after which the polarization direction is changed with the arrival of AW. The speed of whistler is about 300 km/s. Direct signals from magnetic probes Bf and Br, which corresponds to the derivatives of variations of magnetic field, are presented on figure 3. High-frequency precursor is more clearly visible on this figure. Right-hand circular polarization of transverse component of magnetic field can be seen on figure 4. The polarization changes its direction at the moment 11.5 μs, which corresponds to the arrival of AW.

The main result of this experiment is registration of evidence of AW, namely the transverse component of magnetic field B_φ and longitudinal current J_Z, which are correlated in time. Results of measurement of two magnetic probes and the Rogowski coil arranged at distance 82, 130 and 106 cm from the laser target, respectively, are presented on figure 5.

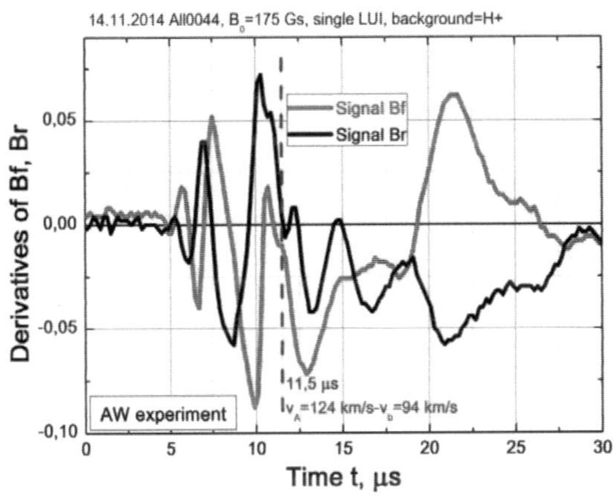

Figure 3. Derivatives of the variations of transverse B_φ, B_r components of magnetic field. Dash line represents the moment of change of polarization and arrival of AW.

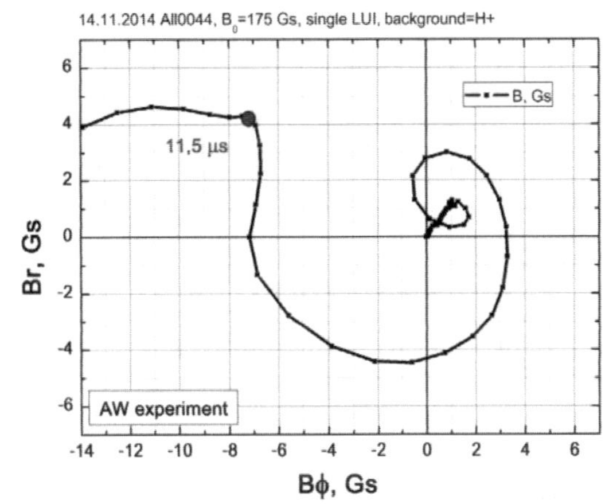

Figure 4. Hodograph of transverse component of the magnetic field variations. The circular polarization changes its direction at the moment 11.5 μs.

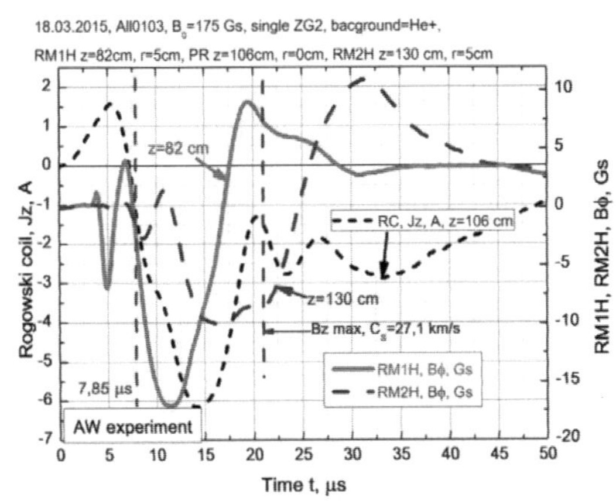

Figure 5. Perturbations of magnetic field B_φ at a distance 82 cm and 130 cm from the target and r=5 cm, detected by probes RM1H and RM2H. Longitudinal current J_Z at a distance 106 cm from the target and r=0, detected by Rogowski coil. First vertical dash line at the moment 7.85 μs represents the arrival of AW. Second one corresponds to maximum of magnetic field displacement, which is caused by slow magnetosonic wave.

It can be seen that the maxima of the perturbations are consistently propagating along the axis of the chamber. Time correlation between peaks of B_φ and J_Z is confirmed by measurement of magnetic probes and Rogowski coil at the same distance from the target. The speed of AW, measured as the ratio of the distance from the target to the probe at the time of arrival of disturbance front, is 105 km/s. Taking into account the own speed of the background (30 km/s), the experimentally measured value of speed of AW is equal 75 km/s. Estimated value of the Alfven speed will be $C_A = B_0(4\pi n_* m_{i*})^{-1/2} = 2.2\cdot 10^{11} B_0 (n_* m_{i*}[a.m.u.])^{-1/2} = 8.6\cdot 10^6 cm/s$, with B_0=175 Gs – external magnetic field, n_*=5·10^{12} cm^{-3} – concentration of BP, m_{i*}=4 a.m.u. – mass of BP-ions, which is close to the experimental value.

4. Experiment set up and results on collisionless shock waves

The scheme of experiment is presented in figure 6 [6]. In this case we use Cartesian coordinate system with X-axis directed along the CSW propagation and Z-axis along the applied magnetic field. The main difference from the previous experiment is that the LP is injected across B_0 =100 Gs. Detailed experimental parameters are given in table 1.

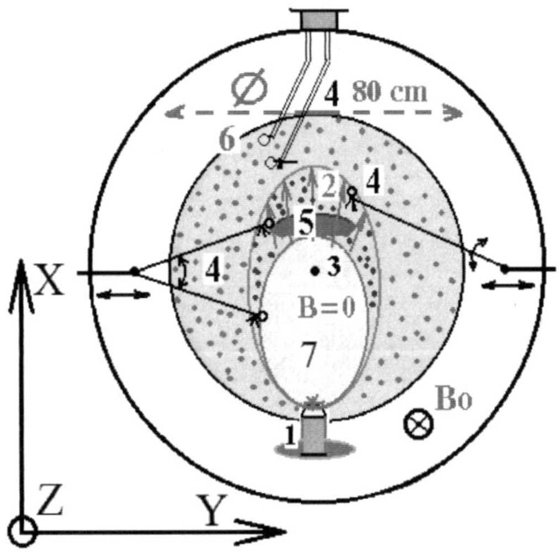

Figure 6. Experimental scheme of stand KI-1 for collisionless shock waves experiment.
1 – Laser target;
2 – Front of LP;
3 – Central axis Z of the chamber;
4 – Diagnostic probes;
5 – Area of interaction between LP and BP;
6 – The boundary of BP;
7 – B=0 – magnetic cavity (area of displaced magnetic field).

The CSW was detected by jump of magnetic field and BP density. Measurements of magnetic probe showed that near (~5÷10 cm) the target displacement of magnetic field is almost complete; the boundary of magnetic cavity is located at the distance of ~50 cm.

To confirm that generation of the CSW is precisely due to MLM, measurements in absence of an external magnetic field were carried out. They show only a small jump of density in vacuum, while LP injection in un-magnetized BP doesn't produce density perturbation at all. When an external magnetic field is on, the strong jump of concentration of the BP caused by the expansion of LP can be observed (figure 7). An external magnetic field of 110 Gs is partially displaced by BP to a level of ~80 Gs. Therefore, at a time of arrival of the disturbance the probe shows a jump of magnetic field with a value of ~35 Gs. Both perturbations have sharp front with duration about ~0.5 μs.

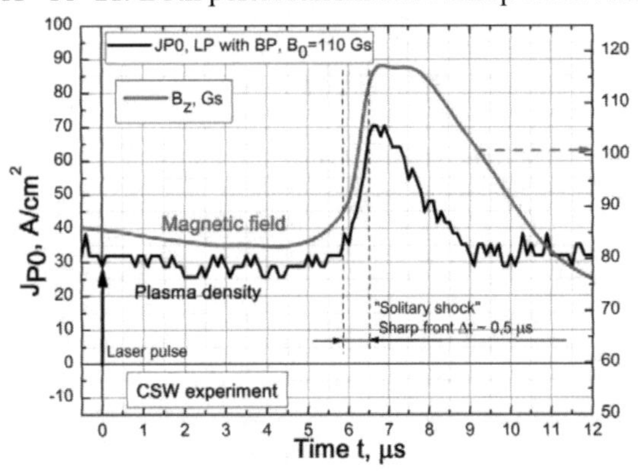

Figure 7. Plasma density and magnetic field jumps, corresponding to the CSW.

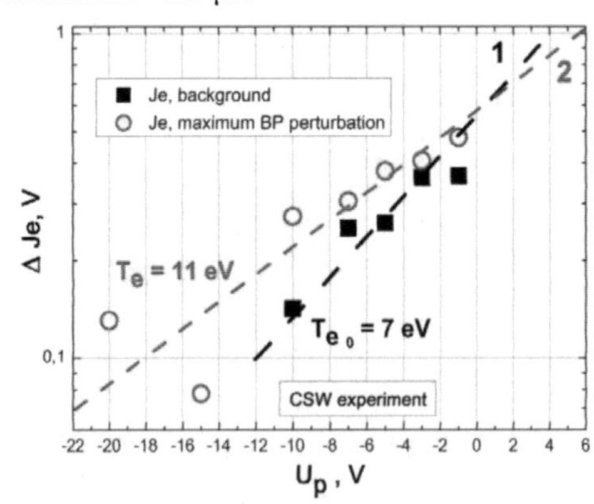

Figure 8. I-V characteristics of P0 probe for unperturbed BP (black line 1, black sqares) and maximum of BP perturbation (red line 2, red circles).

Initial speed of the front of LP-cloud is about ~150 km/s. The front of density and magnetic field jumps move along the X-axis at a speed equal to a half of that value 75 km/s as determined from R-t diagram at a distance of about 50 cm near the expected boundary of the magnetic cavity. This velocity corresponds to their magnetosonic Mach number $M_s \approx 1{,}5 \div 2$ and according to the theory of MHD-shocks [15] we could expect n_*, B_Z-jumps like registered ones (figure 7) and the electron heating in a factor $\approx 1{,}7$ also closed to measured one $\approx 1{,}5$ (figure 8). Temperature measurements based on I-V characteristics of thin cylindrical probe P0 placed at the distance 75 cm from the target are presented on figure 8. Electron temperature of unperturbed BP (black line 1, black squares) is 7 eV; at the maximum of BP perturbation (red dash line 2, red circles) is 11 eV.

5. Conclusion

Conducted modelling experiments on collisionless interaction of laser-produced plasma with magnetized background plasma produced new data on basic plasma processes of wave's generation. The propagation speed of one type of measured perturbations was found to be close to the calculated value of Alfven speed. We believe that they are generated by MLM or Larmor Coupling through formation of azimuthal B_φ fields at the edge of diamagnetic cavity.

High-frequency whistler precursor with right-handed circular polarizations, moving ahead of AW with the speed ~300 km/s, was observed as well. Both of these perturbations are not accompanied by significant variations of plasma concentration.

In another experiment strong perturbations resembling collisionless shock waves propagating across the external magnetic field at a speed of about 75 km/s have been observed. Registration was carried out by jumps of plasma density and magnetic field. Once again, obtained data confirm that these perturbations are generated by the same MLM or Larmor Coupling.

6. Acknowledgments

This work was carried out in the framework of the research ILP SB RAS II.10.1.4 (01201374303); RAS Presidium Program "Fundamentals of breakthrough technologies..."; RAS Presidium Program #21 "Extreme laser radiation: Physics and fundamental applications".

7. References

[1] Terekhin V, Golubev A, Bashurin V 1983 *J. of Appl. Mech. and Tech. Phys.* N**5** 10–17
[2] Zakharov Yu, Antonov V, Bashurin V, Golubev A, Zhmailo V, Ponomarenko A, Posukh V 1985 *J. of Appl. Mech. and Tech. Phys.* N**6** 3
[3] Shaikhislamov I, Zakharov Yu, Posukh V, Melekhov A, Boyarintsev E, Ponomarenko A, Terekhin V 2015 *Plasma Phys.* V**41** 1–9
[4] Winske D, Gary S, 2007 *J. of Geophys. Res.* V**112** A10303
[5] Schaeffer D, Everson E, Bondarenko A, Clark S, Constantin C, Winske D, Gekelman W, Niemann C 2015 *Phys. of Plasmas* V**22** #113101
[6] Zakharov Yu, Ponomarenko A, Terekhin V, Antonov V, Boyarintsev E, Melekhov A, Posukh V, Shaikhislamov I 2014 *Plasma Phys. and Tech.* V**1** 163–5
[7] Dudnikova G et al. 1990 *Plasma Astrophys. ESA* SP-V**311** 191–4
[8] Bergel'son A, Raizer Yu, Surzhikov S 1991 *J. of Appl. Mech. and Tech. Phys* V**3** 22–8
[9] Wright T, 1971 *Phys. of Fluids* V**14** 1905–10
[10] Tishchenko V, Shaikhislamov I 2014 *Quantum Electronics* V**44** 98–101
[11] Zakharov Yu 2003 *Plasma Science, IEEE Transactions on Plasma Science* V**31** 1243–51
[12] Gekelman W et al. 2003 *J. of Geophys. Res.: Space Phys.* V**108** A7
[13] Niemann C, Gekelman W, Constantin C et al. 2013 *Phys. of Plasmas* V**20** #012108
[14] Oraevsky V, Ruzhin Y, Badin V, Deminov M 2002 *Advance in Space Res.* V**29** 1327–34
[15] Balogh A, Treumann R 2013 *Physics of Collisionless Shocks* Springer Science+Buisness Media New York 302–3

Testing of modified curvature-rotation correction for k-ω SST model

A S Stabnikov and A V Garbaruk ,

Peter the Great St.Petersburg Polytechnic University.
Russia, 195251, St.Petersburg, Polytechnicheskaya, 29

an.stabnikov@gmail.com

Abstract. Using eddy viscosity models (EVM) for solving Reynolds Averaged Navier-Stokes (RANS) equations is the most popular and economical way of turbulent flow computations up to date. However this approach, while being very convenient, is not flawless. For example, for accurate prediction of rotating flows and flows with significant streamline curvature (turbine rotors, various vortices etc.) special corrections must be used along with EVM to account for effects of these factors on turbulence. Current work presents the results of testing of the recently suggested modification of the Menter-Smirnov rotation and curvature correction for SST model. A range of internal and external flows is considered: flow in a plane rotating channel, flow in a channel with 180° turn, airfoil tip vortex and 3D boundary layer with the streamwise vortices produced by vortex generators. It is shown that the modified correction results in better prediction than the original version for the three dimensional vortex flows and gives virtually the same results for the 2D boundary layer flows.

1. Introduction

Despite of recent popularity growth of hybrid methods for computing turbulent flows, solving Reynolds Averaged Navier-Stokes (RANS) equations is still the most widely used approach in engineering computations. This approach allows to predict required flow characteristics with decent accuracy using reasonable amount of computer resources. For the closure of the RANS equations eddy viscosity models (EVM) are usually utilized due to their simplicity, stability and low computational cost. Sadly, EVM are not universal, but some of their drawbacks are well known and can be fixed by using ad hoc corrections, like their inability to account for curvature and rotation effects on turbulence. These circumstances motivate creating new and modifying existing curvature and rotation correction terms. In particular, a modification [1] for the Menter-Smirnov [2] correction term for the SST k-ω two equation model turbulence model [3] was proposed based on LES of rotating homogeneous shear flow.

The original Menter-Smirnov correction was developed on the basis of the Spalart-Shur correction [4] for one equation SA model [5] and reads

$$f_{r1,MS} = \max\{\min\{1.25, f_{r1}\}, 0.0\}, \quad (1)$$

where

$$f_{r1} = (1 + c_{r1})\frac{2r^*}{1 + r^*}\left(1 - c_{r3}\tan^{-1}(c_{r2}\tilde{r})\right) - c_{r1}, \quad (2)$$

$r^* = \bar{S}/\bar{\Omega}$, $\tilde{r} = \left[2\Omega_{ik} S_{jk} \left(D\bar{S}/Dt\right)_{ij}\right] / \bar{\Omega} D^3$, $D^2 = \max\{\bar{S}^2, 0.09\omega^2\}$, tensor invariants $\bar{S} = \sqrt{2S_{ij}S_{ij}}$ and $\bar{\Omega} = \sqrt{2\Omega_{ij}\Omega_{ij}}$ of the strain rate S_{ij} and vorticity Ω_{ij}, with constants $c_{r1} = 1$, $c_{r2} = 2$, $c_{r3} = 1$.

The proposed modification [1] is based on matching evolution of turbulent kinetic energy in rotating homogeneous shear flow of LES and SST RANS for different rotation rates and is of form

$$f_{r1,\mathrm{mod}} = \max\{f_{r1}, 0.0\}, \tag{3}$$

where

$$f_{r1} = \left(1 + c_{r1}\right) \frac{1 + c_{r4}}{1 + c_{r4} \cdot |r^* - 2|} \left(1 - c_{r3} \tan^{-1}\left(c_{r2}\tilde{r}\right)\right) - c_{r1}, \tag{4}$$

and $c_{r1} = 0.4$, $c_{r2} = 1.0$, $c_{r3} = 0.6$, $c_{r4} = 0.1$.

In contrast to the original correction, where function (1) is used as a multiplier of the production term for both k and ω transport equations, function (3) multiplies only the production term in k equation.

The purpose of this paper is to test the modified correction on different flows with significant rotation and/or curvature effects. Two-dimensional tasks are represented with plane rotating channel flow and a 180° U-turn duct flow with substantial streamline curvature. Two other test cases are three-dimensional vortical flows: wing tip vortex and co-rotating vortices in the boundary layer downstream of array of vortex generators. Results of three turbulence models were compared: SST model without any correction (SST), and SST with original (SST MS CC) and modified (SST Mod CC) Menter-Smirnov curvature-rotation corrections.

All computations were performed using the NTS code [6], the convective fluxes were approximated using a third order upwind scheme and a central difference second order scheme for viscous fluxes.

2. Two-dimensional boundary flows

2.1 Plane rotating channel

Plane channel flow, rotating at rate Ω around the z axis, normal to the flow plane (fig. 1) is considered at Re $= U_{bulk} \cdot h/\nu = 5.8 \cdot 10^3$ (U_{bulk} – mean flow velocity, h – channel width, and ν – kinematic viscosity) and different rotation rates corresponding to the DNS of Kristoffersen and Andersson [7].

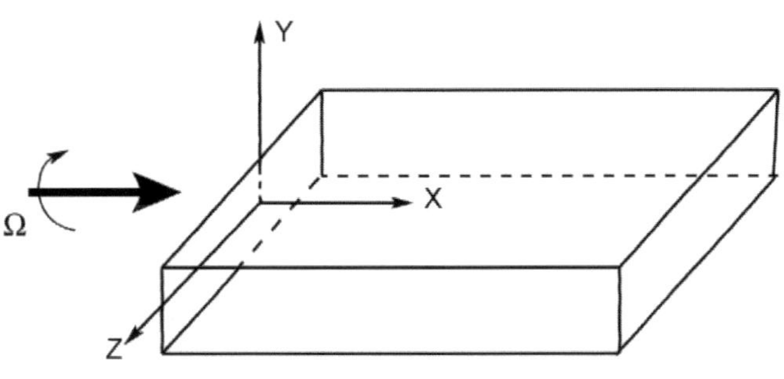

Figure 1. Rotating channel flow scheme

Velocity profiles obtained with both corrections are close to each other for different Rossby numbers Ro=$\Omega \cdot h/U_{bulk}$ (fig. 2a), but results using the modified term are somewhat closer to DNS data. Also, a non-physical eddy viscosity peak at the vicinity of maximum velocity is removed by correction modification (fig. 2b).

2.2 U-turn duct

Flow in U-turn duct at Re=$V_m \cdot H/\nu=10^6$ (V_m – mean bulk velocity, H – channel width) was studied experimentally by Monson et al [8]. Computational domain (fig. 3) contains a *10H* length region upstream of turn, turn region, and a region with length *12H* downstream of turn. The computational grid has dimensions 203×111. The following boundary conditions are applied: developed turbulent velocity profile calculated by SST model at inlet, no slip condition at channel walls and constant pressure at outlet.

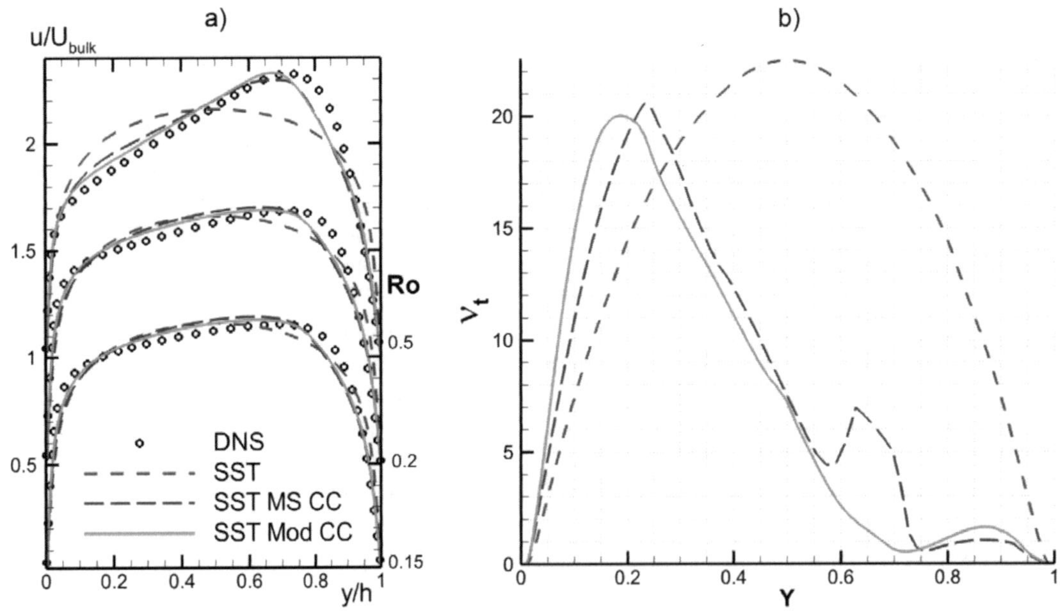

Figure 2. a) – velocity profiles in rotating channel at Ro=0.15, 0.2, 0.5
b) – eddy viscosity profiles at Ro=0.5

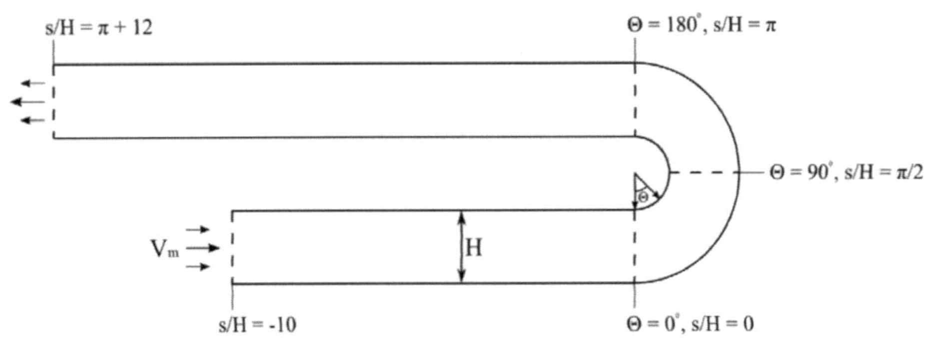

Figure 3. U-duct with 180° turn problem formulation

The results were compared at three channel sections where experiment data is available: start, middle and end of the turn (fig. 3). Profiles, obtained by the two corrections are almost identical (fig. 4) much like in the first test case. So, computations reveal no significant effect of modification for these two-dimensional wall bounded flows.

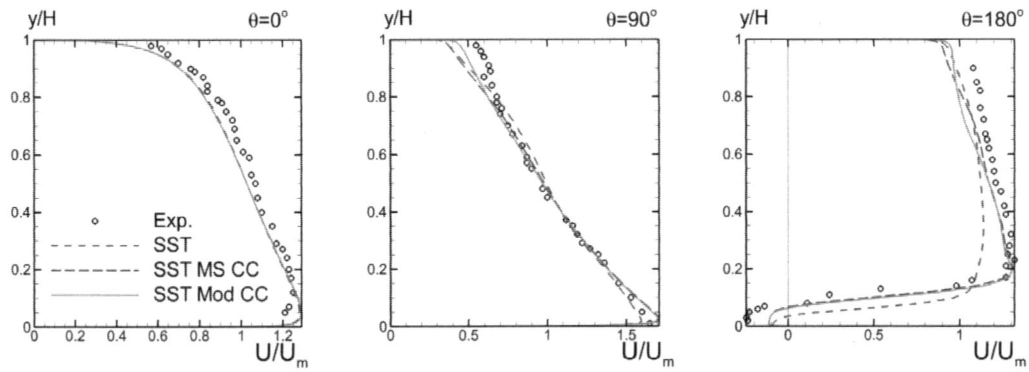

Figure 4. Profiles of streamwise velocity in 180° turn channel flow

3. Three-dimensional vortex flows

3.1 NACA 0012 tip vortex

Finite span wing tip vortex is considered at experimental [9] conditions: a NACA 0012 airfoil with rounded tip is put in a closed wind tunnel (fig. 5). Since experiment inlet Mach number was not high (inlet mean velocity is equal to *51.8* m/s), the incompressible flow at Re=$3.568\cdot 10^6$ is considered. No slip conditions were applied on the airfoil surface while slip conditions are used on the tube walls. Multi-block structured chimera type mesh contains about 5.6 million cells.

Results show (fig. 6) that use of the correction modification leads to significant improvement of prediction of both streamwise and crossflow $U_{crossflow} = \sqrt{v^2 + w^2} / U_{inlet}$ velocity profiles downstream of the wing. It is caused by more effective suppression of eddy viscosity at the vortex center (fig. 7) which leads to delayed decay of the vortex.

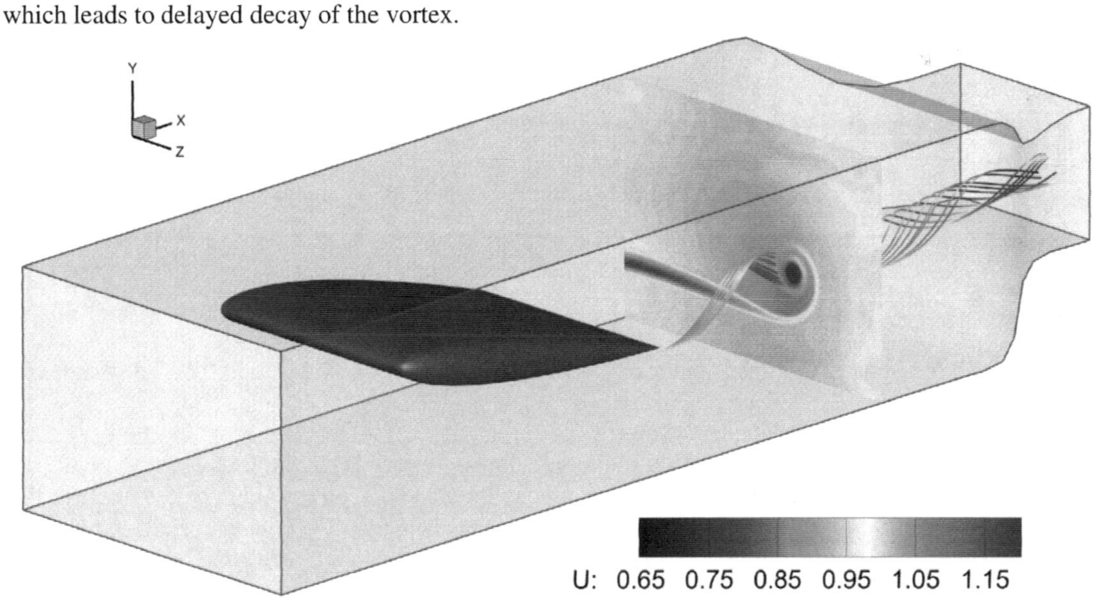

Figure 5. NACA 0012 tip vortex flow scheme

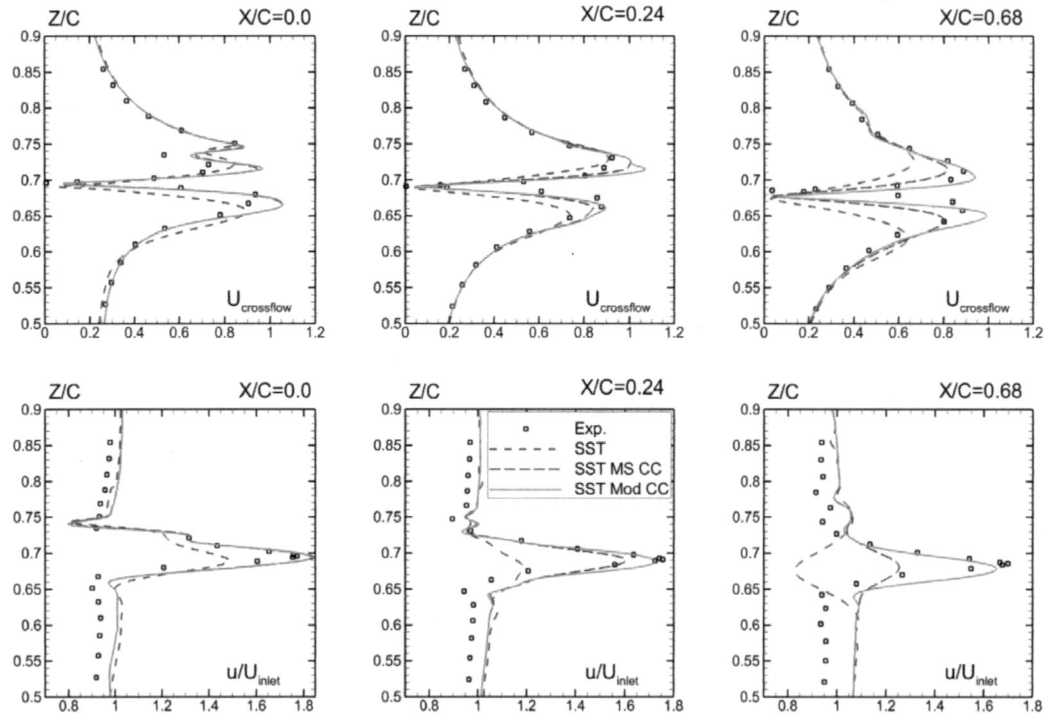

Figure 6. Streamwise and crossflow velocity profiles at the back NACA 0012 wing tip and 0.24C and 0.68C downstream (C is the airfoil chord)

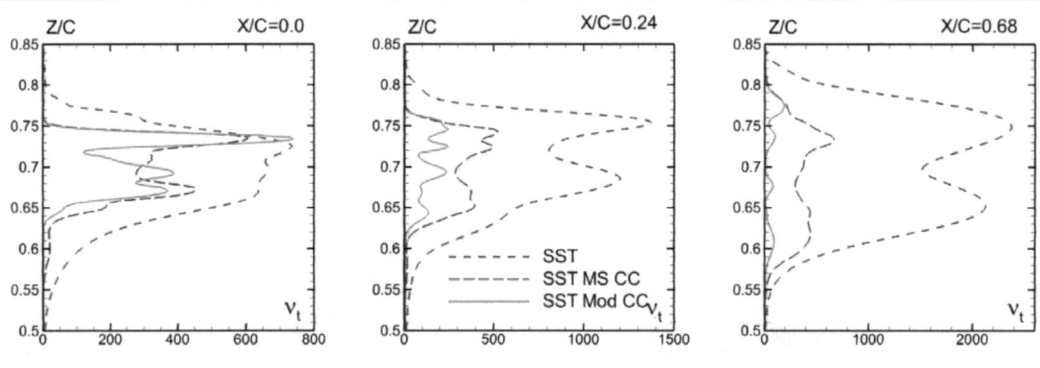

Figure 7. Eddy viscosity profiles at the back NACA 0012 wing tip and 0.24C and 0.68C downstream

3.2 Streamwise vortices in the boundary layer

System of uniform vortex generators (VG) in the flat plate boundary layer with supersonic free stream at M_∞=1.4 is considered. VG height and length are equal to *h* and *4h* correspondingly, the angle to the incoming flow is equal to 18° (fig. 8) and the distance between VG's is *5h*. Flow conditions correspond to DNS [10]: the Reynolds number based on VG's height and free-stream velocity is equal to 10^4. The domain is *10.5h* upstream and *50h* downstream of the front edge of the VG. Width of the domain is equal to the distance between VGs (*5h*), so periodic boundary conditions are specified. The inlet profiles are taken from SST RANS computation of flat plate boundary layer so that boundary layer thickness at VG front edge has thickness equal to *h*. The mesh contains about 7.4 million cells. Comparison of circulation and maximum vorticity downstream of vortex generators shows that vortex decay is much faster in RANS simulations compared to DNS (fig. 9). Usage of curvature corrections somewhat improves the decay rate, using modified correction lessens the mismatch between RANS and DNS even more. This improvement, much like in the case of the NACA 0012 tip vortex problem, is due to decreased eddy viscosity in the vortex center (fig. 10), leading to reduced decay rate of vorticity and circulation.

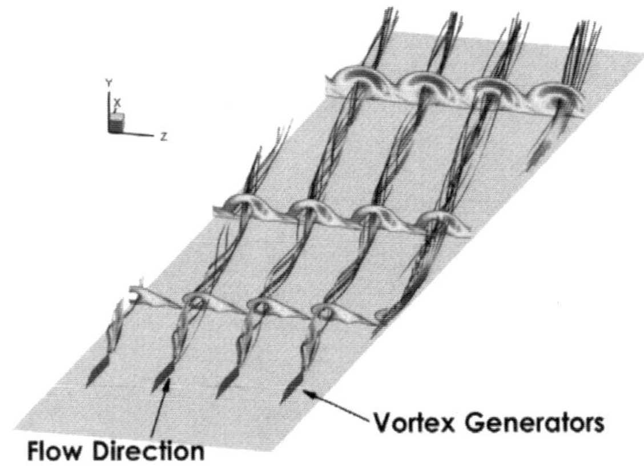

Figure 8. Vortex generators flow scheme

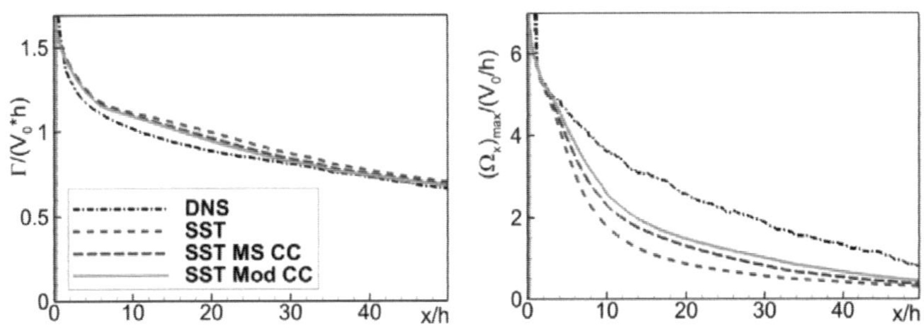

Figure 9. Circulation (left) and maximum vorticity (right) downstream of vortex generators array

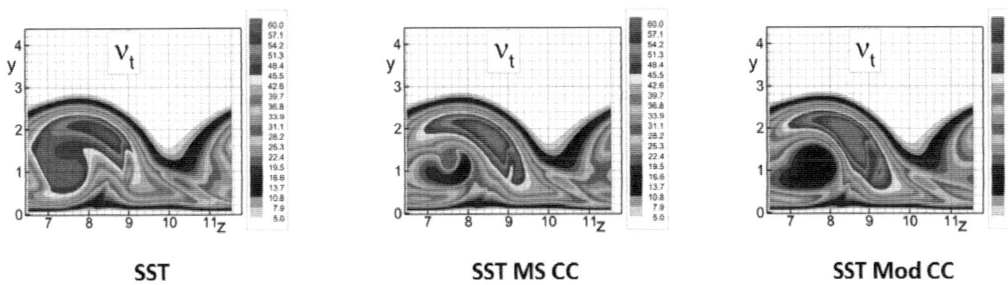

Figure 10. Comparison of the eddy viscosity contours at the section x/h=27.6

4. Conclusions

Testing shows that the effect of correction modification [1] depends on flow type. For two-dimensional wall bounded flows only little improvement is achieved while in three-dimensional vortex flows the modified correction shows significantly better results than the original. Note that the results with modified correction are not inferior to the original version in any of the considered flows. Therefore, the modified Menter-Smirnov correction can be recommended for usage in wide range of flows with curvature and/or rotation effects.

5. References

[1] Stabnikov A and Garbaruk A 2015 Modification of curvature and rotation correction term for SST model based on LES of rotating shear layer. *XX School-seminar for young scientists led by academitian of Russian Academy of Science Leontiev A I "Gas dynamics and heat transfer problems in power settings", Report Theseses,* pp 71-72 (in Russian)

[2] Smirnov P and Menter F 2009 Sensitization of the SST turbulence model to rotation and curvature by applying the Spalart–Shur correction term. *Journal of Turbomachinery*, **131**, 041010.

[3] Menter F 1993 Zonal two equation k-ω turbulence models for aerodynamic flows. *AIAA Paper* **93**-2906

[4] Spalart P and Shur M 1997 On the sensitization of turbulence models to rotation and curvature. *Aerospace Science and Technology*, **1(5)**, 297–302

[5] Spalart P and Allmaras S 1992 A one-equation turbulence model for aerodynamic flows. *AIAA Paper* **92**-0439

[6] Strelets M 2001 Detached eddy simulation of massively separated flows. *AIAA Paper* **2001-**0879

[7] Kristoffersen R and Andersson H 1993 Direct simulation of low-Reynolds-number turbulent flow in a rotating channel. *Journal of Fluid Mechanics*, **Vol. 256**, pp. 163–197

[8] Monson D, Seegmiller H, Mc Connaughey P, and Chen Y 1990 Comparison of experiment with calculations using curvature-corrected zero and two equation turbulence models for a two-dimensional Uduct. *AIAA Paper* **90**-1484

[9] Chow J, Zilliac G and Bradshaw P 1997 Mean and turbulence measurements in the near field of a wingtip vortex. *AIAA J.* **35, No. 10**, pp. 1561-1567

[10] Spalart P, Shur M, Strelets M and Travin A 2015 Direct simulation and RANS modelling of a vortex generator flow. *Flow, Turbulence and Combustion,* **Volume 95, Issue 2**, pp 335-350

Spin relaxation length for medium-energy electrons in palladium thin films

A V Pavlov, A B Ustinov, V N Petrov

Peter the Great St.Petersburg Polytechnic University, 29 Polytechnicheskaya, St.Petersburg 195251, Russia.

E-mail: a.pavlov@physics.spbstu.ru

Abstract. The development of spintronics (the field of science that examines the behavior of systems, based not only on charge transfer, but on magnetic properties of nano-objects as well) requires study of various magnetic and electrical properties of nanoscale systems. To create a new device it is often important to know a length of spin relaxation not only for electrons with energies close to the Fermi surface, but also with energy of several hundred eV. In this study we measured such lengths in palladium at room temperature. Results obtained by secondary electron spectroscopy show that spin relaxation length for energy 100 eV is 11,9±1,6Å and for energy 500 eV is 27,1±10,5Å.

1. Introduction

When electron current goes through a solid, relaxation processes are observed that lead to changes in current properties. Electron energy loss is the simplest effect which is associated with inelastic collisions of electrons with their environment. Number of electrons which passed a certain distance without inelastic scattering is characterized by inelastic mean free path and is well studied and for most of substances can be represented by a universal curve [1].

However, the spin relaxation length, in other words length at which the polarization of the beam decreases *e* times, is still studied poorly. One way to determine this length is to measure conductivity dependence on magnetization for the ferromagnetic material layers. This method allows determination of mean free path with respect to the spin relaxation for electrons with energies close to Fermi level [2]. Nevertheless, for developing new devices it is often important to determine this parameter for electrons with energies of several hundreds eV. For this task suitable method is secondary electron spectroscopy. Study of transport losses for polarized electrons with different energy can be used for determination of spin relaxation length in material, and it is the method we used in this study.

Palladium was chosen as a material for experiments. It is light enough to have significant spin-orbit interaction and possesses a simple electronic structure with filled electron shells. It is vacuum material, can be easily deposited and etched afterwards.

2. Experimental setup

The experimental setup was based on an ultra-high vacuum chamber equipped with facilities for thin-film deposition, argon bombardment and electronic spectrometry [3] including Mott detector for spin polarized measurements [4, 5]. Schematic representation of the setup is shown on figure 1. The deposition source was constructed out of ceramic crucible that was put inside tungsten wire coil. Later was used to heat ceramic and evaporate palladium from inside of it. Main part of argon etching system was ion gun with thermionic cathode and electron gas ionization. Spectrometry setup consisted of dispersive energy analyzer (two pass cylindrical mirror, specially designed for collimation into polarimeter) and classical Mott polarimeter for electron spin polarization measurements.

Basic pressure in the system was $8*10^{-10}$ Torr. The temperature remained equal to room temperature except for sample and chamber annealing processes.

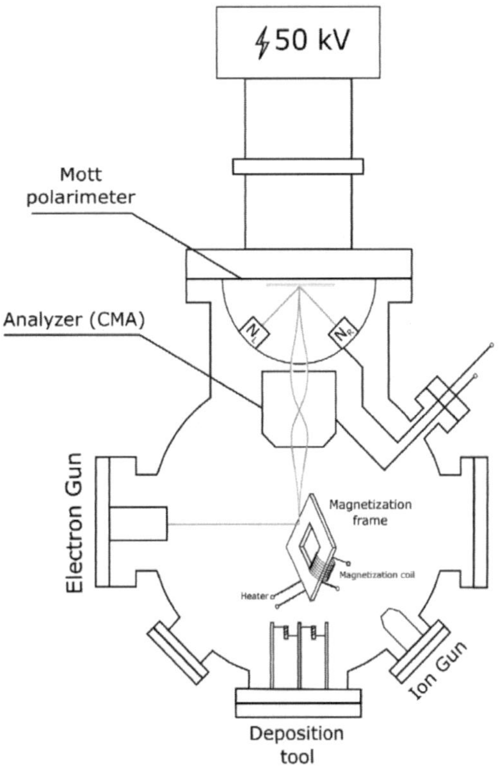

Figure 1. Sketch of experimental setup

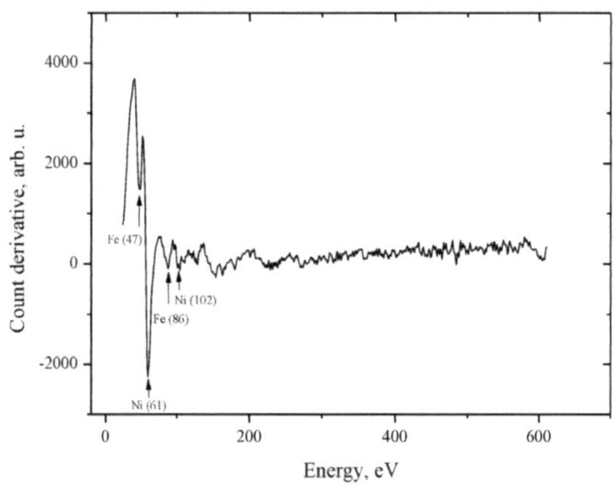

Figure 2. Auger spectrum of clean $FeNi_3$ sample

The main purpose of the experiment was to measure length of spin relaxation in palladium thin films. The basic idea that we used was to record electron polarization hysteresis loops and thus make conclusions about polarization loss rate. This required us to perform the following steps: preparation of single crystal's surface (annealing and ion bombardment), calibration of evaporation source, deposition of thin films (Section 3), recording the hysteresis loops for various film thicknesses and different energies of secondary electrons (Section 4).

In the experiment the frame-shaped $FeNi_3$ (110) single crystal was used. Before the experiment, the crystal was annealed in ultrahigh vacuum. Selected single crystal was used as a robust source of polarized electrons for ongoing spectroscopic measurements. For cleaning off contaminants argon bombardment was applied. Argon ions had an energy of 500 eV, the pressure in the system during the etching procedure was $5*10^{-5}$ Torr of argon. Control of surface's purity was made by means of Auger electron spectroscopy. The cleaning process continued until peaks of typical surface contaminants (carbon, oxygen) disappeared. An example of spectrum after bombardment is shown on Figure 2. Clean Auger spectrum allowed us to state that the crystal's surface was atomically clean.

The magnetization of the sample was carried out as follows. Around one of the sides of the frame coil was wound. Current passing through the coil produces a magnetic field in a closed frame. After magnetization ferromagnetic crystal becomes a source for spin polarized secondary electrons.

3. Film growth

For deposition of palladium onto the $FeNi_3$ surface thermal evaporation method was used. For success of this work it was essential to have precise data on film deposition rates. More than that, growth of films is the key factor, as for our spin relaxation calculations to be correct it is necessary to have all outgoing electrons to pass equal distances in the studied medium.

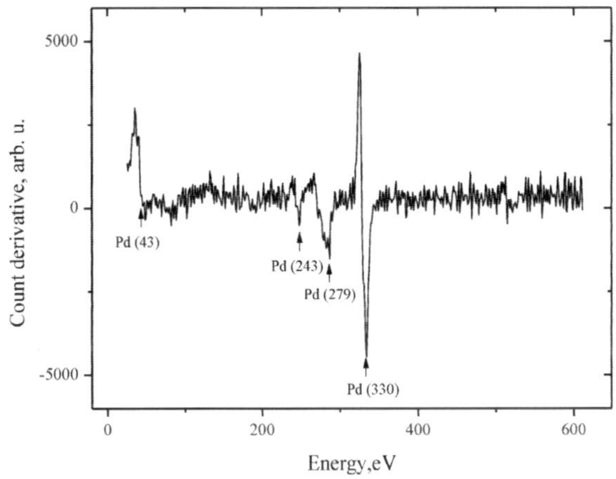

Figure 3. Auger spectra of Pd thin film

Example of thick palladium film spectrum is shown in figure 3. In thermal evaporation deposition rate depends on the current flowing through the heating coil. To calibrate the speed of Pd deposition we needed to choose a heater current, and then deposit Pd from the source onto the crystal surface with equal time intervals to measure amplitude of peaks of Auger transitions in the film and in the substrate. Such measurements in the case of layered growth must lie on an exponential curve. In this case, $I \sim exp\,(-d\,/\,l)$, where I is the signal amplitude, d is the film thickness, l mean free path for electrons in the film is Pd. Knowing the mean free path in the Pd [1], it is possible to link the deposition rate and the thickness of the film on the surface of substrate. The dependence of intensity for substrate peak (Ni 61 eV) and film-material peak (Pd 330 eV) for different thicknesses of film are shown on figure 4. The deposition time is converted into a thickness for convenience of further comparison.

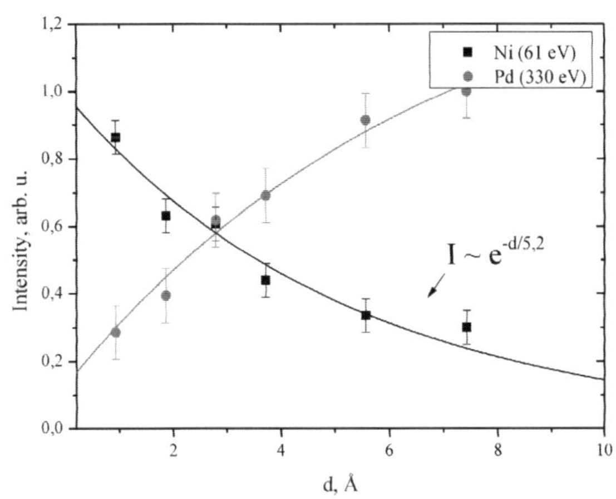

Figure 4. Calibration of Pd evaporation source by Auger peaks amplitude

4. Magnetic measurements

The base of magnetic measurements lay on the measurement of amplitude for electron polarization hysteresis loops for different film thicknesses and different energies of secondary electrons. In our experiment we choose 100 and 500 eV as energies for analysis of secondary electrons, but in following work we will make energy range more complete. Under the hysteresis loop in the case of secondary electrons one should understand result of the following measurement: at fixed analyser's energy we scan through different values of magnetic field inside the frame and measure electron polarization. Result of such measurement gives, really, a hysteresis loop, because electron polarization is tightly connected with sample magnetization and measurements of polarizations were made for different coil currents and thus different applied magnetic field.

Examples of the hysteresis loops for clean crystal and various films deposited onto the crystal surface for 100 eV electrons are shown in figure 5.

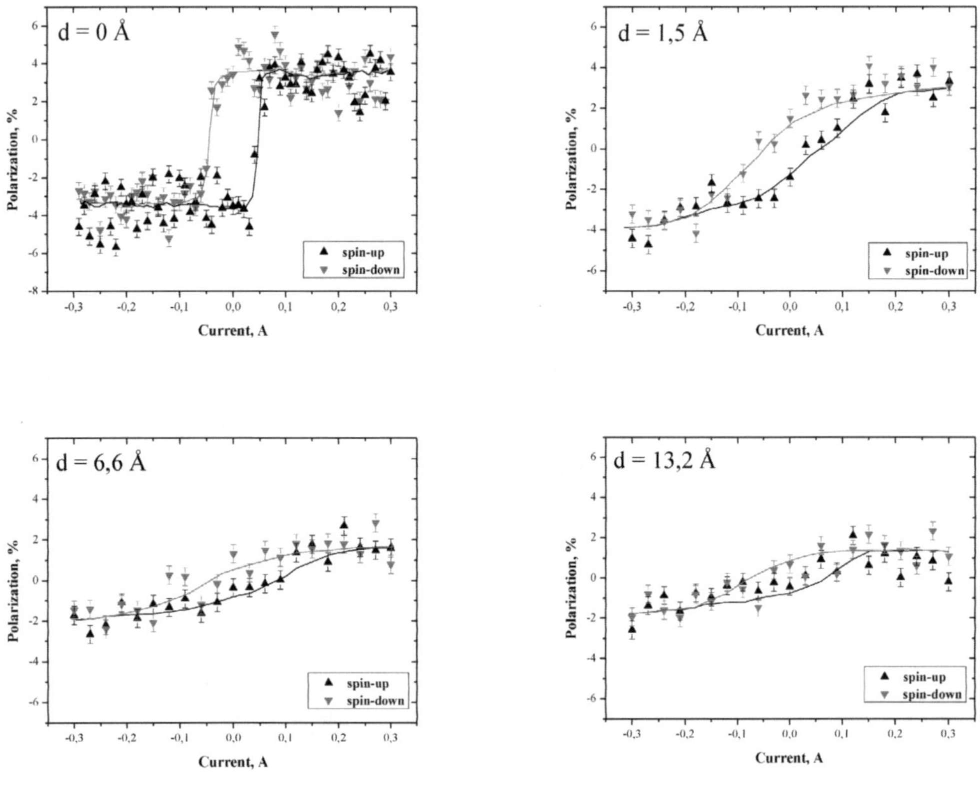

Figure 4. Samples of the hysteresis loops for different film thicknesses (d)

By amplitude of the hysteresis loop we can determine polarization of the electrons that passed through Pd thin film and exited into vacuum towards the detection. From experimental results we derived a dependencies of spin relaxation length in palladium for two energies and various thicknesses. In undergoing work we plan to make those dependencies into the polarization map for better characterization of studied material. Measured electron polarization decreases exponentially (figure 5). This allows us to assume that the polarization is lost during the electron-electron interaction and scattering on the defects [6]. The value in the denominator of exponent can be called the length of spin relaxation of electrons for a given energy in the film of Pd. When we compare the values of the inelastic mean free path with spin relaxation length (figure 4 and figure 5), we see that the spin relaxation length is approximately two times greater than the mean free path. That characterizes Pd as a promising material for the polarization transfer in a spintronic devices at room temperature.

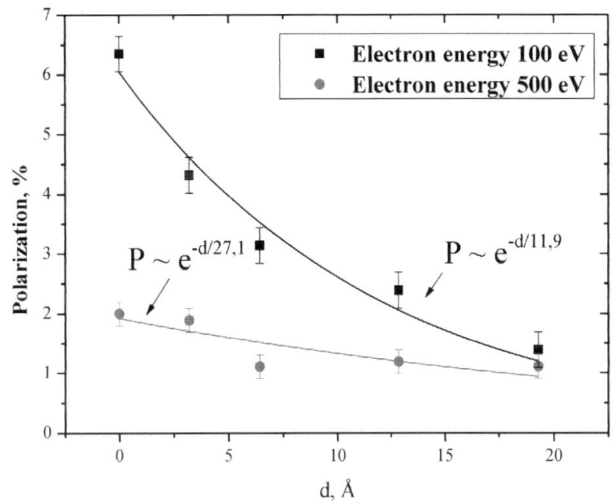

Figure 5. Decrease of secondary electron polarization for different palladium films

The results of this work can be used to create a detector of spin polarization [7]. The main idea of that is to create optical detector of free-electron spin. For this task authors needed a substance that will cover the surface of the detector. In addition to the protective function coating is used as a moderator before electron spin optical analyser. For the application it is very important that the length of spin relaxation for used material to exceed the mean free path at least few times. So, palladium can be used as such a material reducing losses of polarization in the coating.

Another interesting application for palladium film can be the creation of highly polarized positrons sources [8]. Often radioactive isotopes of sodium are used as a positron source. Resulting positron beam is polarized, but has very high energy (few MeV). For applications in spectroscopy, the beam should be slowed, and for that W crystal is often used. After passing through the tungsten crystal positrons lose most of the energy, but also lose their polarization. Replacing W with Pd can increase the polarization of the positron beam. Because, despite of the different charge, positrons and electrons, have similar mechanisms for spin loses.

5. Summary

In this study we obtained spin relaxation length in Pd thin films at room temperature. Measured relaxation length for energy 100 eV is 11,9±1,6Å and for energy 500 eV is 27,1±10,5Å.

References

[1] Seah M. P. & Dench W. A. Quantitative electron spectroscopy of surfaces 1979 *Surface and interface analysis* *1*(1) 2-11.

[2] Gurney B. A., Speriosu V. S., Nozieres J. P., Lefakis H., Wilhoit D. R., & Need O. U. Direct measurement of spin-dependent conduction-electron mean free paths in ferromagnetic metals 1993 *Physical review letters* *71*(24) 4023.

[3] Petrov V. N., & Kamochkin A. S. Energy analyzer for spin polarized Auger electron spectroscopy 2004 *Review of scientific instruments* *75*(5) 1274-1279.

[4] Petrov V. N., Grebenshikov V. V., Grachev B. D., & Kamochkin A. S. New compact classical 40 kV Mott polarimeter 2003 *Review of scientific instruments* *74*(3) 1278-1281.

[5] Petrov V. N., Galaktionov M. S., & Kamochkin A. S. Comparative tests of conventional and

retarding-potential Mott polarimeters 2001 *Review of Scientific Instruments 72*(9) 3728-3730.

[6] Žutić I., Fabian J., & Sarma S. D. Spintronics: Fundamentals and applications 2004 *Reviews of modern physics 76*(2) 323.

[7] Li, X., Tereshchenko O. E., Majee S., Lampel G., Lassailly Y., Paget D., & Peretti J. Optical detection of spin-filter effect for electron spin polarimetry 2014 *Applied Physics Letters 105*(5) 052402.

[8] Petrov V. N., Samarin S. N., Sudarshan K., Pravica L., Guagliardo P., & Williams J. F. Spin polarized low-energy positron source 2015 *Journal of Physics: Conference Series* Vol. 618 No. 1 p. 012043.

Analysis of the broadband chaotic spin-wave excitations in an active ring oscillator based on a metalized ferrite film

A V Kondrashov[1], A B Ustinov[1], B A Kalinikos[1] and S O Demokritov[2]

[1] Department of Physical Electronics and Technology, St.Petersburg Electrotechnical University, 197376 St.Petersburg, Russia
[2] Institute for Applied Physics and Center for Nonlinear Science, University of Münster, 48149 Münster, Germany

E-mail: kondrashov_av@yahoo.com

Abstract. This paper reports the first experimental study of broadband chaotic nonlinear spin-wave excitations which is formed through development of four-wave parametric processes in active ring oscillator based on metallized ferrite film. We find that an increase in the oscillation power leads to Hopf bifurcations sequence. Monochromatic, periodic quasi-periodic and chaotic excitations are observed. Spectra of the chaotic excitations consist of series of chaotic bands separated well in frequency. Parameters of the chaotic attractors are discussed.

1. Introduction

The broadband chaotic phenomena are of great interest for modern technologies. Among them are the chaos-based communication systems [1], fast random-number generation (see [2] and references therein), and radars [3]. Therefore special research interest is given to the broadband nonlinear chaos oscillators. For instance, broadband dynamic chaos in optomechanical system [4], sub-terahertz electromagnetic wave dynamical chaos in semiconductor super-lattices [5], and ultra-wideband chaos in microwave photonic ring systems [6] were recently observed.

In the past, the spin-wave dynamic chaos was observed in different nonlinear ferromagnetic systems such as spheres [7] and thin films [8]. Utilization of such ferromagnetic systems in circuits with an amplifier in the feedback loop allows to fabricate an active ring oscillator. Such delay-feedback ring oscillators have proven to be useful in fundamental studies of nonlinear wave dynamics [9,10]. But the band frequency of the wave dynamics observed in previous papers was limited to several hundreds of kilohertz or megahertz. Obviously, the limitation was due to narrowband excitation and detection system. However in the paper [11] it was shown that spectrum of chaotic spin waves could be broadened up to 1 GHz through the use of the appropriate waveguiding media as well as broadband excitation and detection method.

In this work, broadband chaotic microwave signal excitations were realized with an active ring system based on metalized yttrium iron garnet (YIG) thin film. The film served as the nonlinear dispersive medium for the propagation of magnetostatic spin waves [12]. The metallization of the YIG film, allowed us to increase significantly the spin-wave excitation bandwidth so as to obtain broadband chaotic oscillations.

2. Experimental details

Fig. 1 shows the experimental structure of the active resonant ring. It consisted of a wideband linear microwave amplifier, an adjustable attenuator, a directional coupler and a magnetic film. The microwave amplifier compensated losses in the ring. The adjustable attenuator was used to control the active ring losses [10, 13]. Microwave signal circulating in the ring was detected by the directional coupler and transmitted to an oscilloscope and a spectrum analyzer for temporal and frequency analysis. The ferrite film defined frequencies of self-generated signals. Nonlinear effects occurring in the active ring was caused only by ferrite film nonlinearity.

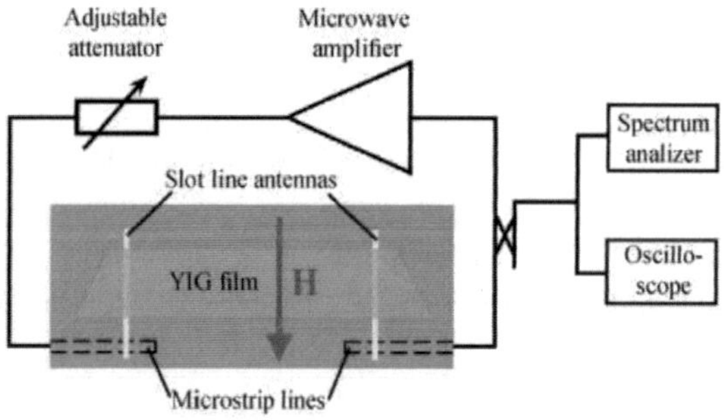

Figure 1. Schematic of the experimental device.

The active ring excitation bandwidth was defined by the YIG film bandwidth. Ferrite film bandwidth is limited by the maximum wave number of spin waves k_{max}, which could be exited and detected by spin-wave antennas. This maximum wave number is defined by antenna electric current localization. For traditionally used microstrip antennas it determined as $k_{max} = 2\pi/w$, where w is the width of the antennas. An increase of excitation bandwidth through a decrease of the antenna width is limited by spin-wave manifold. Application of metalized film allows to increase the spin-wave manifold [14]. It is appropriate to use the slot line or coplanar antennas for excitation of spin waves in metalized films. The slot line antennas are more benefit due to broadband excitation. It was shown recently, that delay line fabricated with metalized yttrium iron garnet film had frequency band in the range of 1 GHz [15].

Magnetic yttrium iron garnet (YIG) film of 65μm-thick, 40 mm-long and 2 mm-width was used for experiments. Its saturation magnetization $4\pi M_s$ was 1750 Gs. Slot-line antennas of 8 mm-length and 50 μm-widths were used for excitation and detection of spin-waves. Distance between antennas was 10 mm.

The antennas were fabricated by etching of slots in a wide metalized dielectric substrate. The film was placed on the antennas. Therefore the film appeared to be metalized. Microwave signal was transmitted to input antenna and fed out from output antenna by microwave microstrip lines on the bottom side of the dielectric substrate.

The film was magnetized to saturation by static magnetic field. The field was directed in a film plane perpendicular to spin wave propagation direction. Such field configuration allows surface spin wave (SSW) excitation. Experimental research was done for several values of bias magnetic field, H_1 = 1462 Oe, H_2 = 1642 Oe.

3. Experimental results and discussion

Preliminary characterization of the film and resonant ring was done. Figure 2 shows dispersion characteristics and amplitude vs. frequency characteristics of the metalized YIG film and active ring. Utilization of the metalized ferrite film increased spin wave group velocity. Thus frequency range corresponding to available wave numbers was broad. One can see that despite the wave number limit of 120 cm^{-1} frequency range of exciting and detecting SSW broadened up to 1 GHz (see Fig. 2(b)).

Increase of bias magnetic field leads to shift of the dispersion characteristics to higher frequencies. But antenna efficiency for spin wave excitation and detection is decreased for higher frequencies. It resulted in increase of transmission losses from 12 dB up to 19 dB (see Fig. 2(a)). It is worth mentioning that dispersion law change leads also to change in nonlinear properties of YIG film.

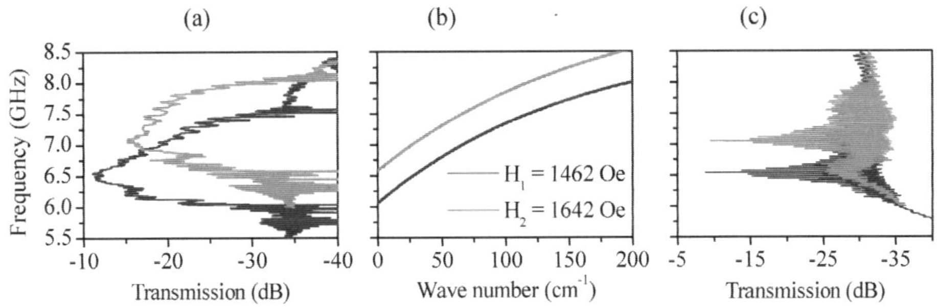

Figure 2. (a) Transmission loss vs. frequency characteristics of spin wave delay line. (b) Dispersion characteristics of the spin waves in metalized ferrite film. (c) Transmission loss vs. frequency characteristics of the active ring.

Characterization of active resonant ring was carried out below self-generation threshold. Below the threshold system behaves as a ring resonator. Frequency of the eigenmodes are defined by the phase condition:

$$k(\omega)\cdot l + \varphi_0 = 2\pi n, \quad (1)$$

where $k(\omega)$ is wave number of the spin waves, l is a distance between spin-wave antennas, φ_0 is a phaseshift of the feedback circuit. The decay rate and quantity of the active ring resonator eigen-modes depends on the spin-wave transmission losses and excitation and detection antennas efficiency. Transmission loss vs. frequency characteristic of active ring is shown in Fig 2(c) for different values of bias magnetic fields. Due to the metallization of the YIG film resonant eigen-modes of the active ring occured in the frequency band of 1 GHz.

As the ring gain coefficient increased up to a threshold value, set here as G = 0 dB, self-generation of monochromatic microwave signal appeared. Frequency of this signal corresponds to the lowest damped eigenmode. Further increase of gain coefficient led to transition from stationary to chaotic self-generation. The scenarios of the transmission to chaos for all values of magnetic bias field were the same. Below we consider in details transition to chaotic dynamics observed for H_2 = 1642 Oe.

At G = 0 dB we observed self-generation of the ring eigenmode with frequency f_0 = 7.032 GHz. At G = 0.1 dB microwave signal power increased to certain threshold level and nonlinear parametric processes appeared in YIG film. All the experimental values of the bias magnetic field were high enough to exclude three wave parametric processes and nonlinearity of the spin system was caused only by four-wave parametric processes. It leads to modulation instability of main eigenmode. The energy of self-generated eigenmode transmitted through nonlinear processes to other active ring eigenmodes. Thus we observed frequency comb spectrum which corresponded to periodically modulated microwave signal. Repulsive nonlinearity of SSW led to formation of the dark envelope soliton trains [16]. Period of solitons was T_s = 30 ns, width on the half of the amplitude corresponded to τ_s = 11 ns. While further increase of the gain coefficient up to G = 0.2 dB secondary modulation instability appeared. Each of harmonics produced its own additional close placed harmonics. In time domain we observed periodical modulation of the pulse amplitude. It corresponded to the generation of the so-called quasiperiodic microwave signal. Increase of the ring gain coefficient up to the value of 0.3 dB led to bifurcation which caused simplification of the active ring dynamic. It manifested itself as a transition back to periodic signal generation. In this regime in contrast to theoretical predictions we observed excitation of bright envelop solitons. Period of bright pulses was 32 ns, width was 4.5 ns.

Change of the pulse period could be explained as follows. A nonlinear phase shift occured in the ferrite film an increase of microwave signal intensity. It led to frequency shift of self-generated harmonics. But each harmonic shifted on its certain frequency. Therefore the bandwidth of self-generated multi comb spectrum decreased and period of pulses slightly increased. The width of the spectrum was broadened and the pulse duration was decreased.

Further, an increase in G led to chaos. First, secondary modulation instability appeared at G = 1.1 dB. Next at G = 2.5 dB close placed sideband harmonics broadened and broadband pedestals appeared around the initial harmonics. Thus the whole spectra of signal consisted of a number of broaden harmonics. Fig. 3 shows such spectra for different values of bias magnetic field. One can see that width of whole spectra is in a range of 1 GHz in the case of H = 1462 Oe and 1.5 GHz in the case of H = 1642 Oe. The widths of each broadened harmonics were approximately 5 MHz.

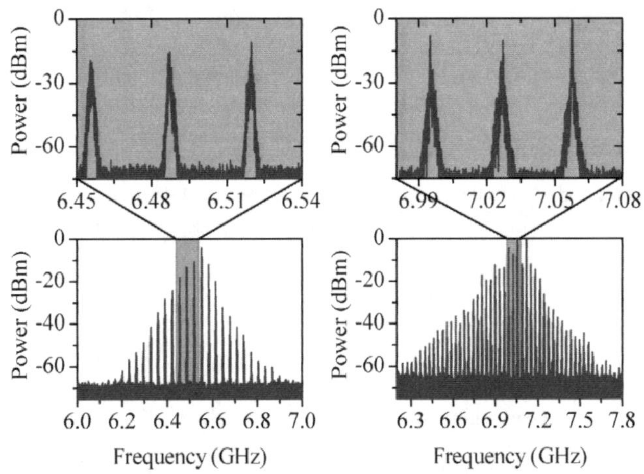

Figure 3. Spin-wave dynamic chaos spectra (a) - H = 1462 Oe, G = 2.3 dB, (b) - H = 1642 Oe, G = 3.2 dB.

For numerical characterization of observed signals we reconstructed multidimensional phase-portraits. We used conventional time-delay method [17]. Attractor in the form of stable focus appeared in the phase space in monochromatic regime. Stable focus became unstable and the dynamic of the active ring was defined by the limit cycle after transition to periodic nonlinear self-generation. 2D-torus was observed after appearance of the secondary modulation instability. Strange attractor corresponded to the chaotic regime. Increase of attractor dimension could be explained by irrational ratio of appeared harmonic frequencies. Described changes of dynamic regimes and attractor forms are called Hopf bifurcations [17]. Transition from regular to chaotic dynamic through sequence of Hopf bifurcation corresponds to Rulle-Takens route.

According to the paper [18] phase portraits reconstructed in compliance with time-delay method could be used for calculation of different attractor parameters like Lyapunov exponent, fractal dimension, Kolmogorov entropy and etc. Dependences of fractal dimension on active ring gain coefficient were calculated for different bias magnetic fields. We used Grassberger-Procaccia algorithm [19]. This method consists of three steps. On the first step one should calculate correlation sum:

$$C(l) = \lim_{N \to \infty} \frac{1}{N^2} \sum_{i,j} \theta\left(l - \left|x_i - x_j\right|\right), \tag{2}$$

where l - characterization distance, N - phase portrait points quantity, $\left|x_i - x_j\right|$ - distance between i-th and j-th points. This parameter characterizes a probability that two points would be divided by distance less than l. In certain range of l dependence of correlation sum connected with characterization parameter could be approximated by the follows proportion:

$$C(l) \sim l^{D_c}, \tag{3}$$

where D_c - correlation dimension. Thus a slope of $\log_{10}(C)$ vs. $\log_{10}(l)$ dependence equal to correlation dimension. Correlation dimension characterize topology of the attractor projection into the phase space with chosen dimension d. Increase of phase space dimension leads to change of the point coordinates quantity and increase of the correlation dimension. It means that projection of attractor is characterized more precisely in higher dimension space. At some value of phase space dimension D_c reaches the saturation. Saturation level is the fractal dimension D_f. It characterizes topology of the whole attractor. Topology of the attractor is defined by a complexity of the investigated dynamic regime and its mathematical model. Thus fractal dimension of the stable point, limit cycle and 2D-torus equal 0, 1, and 2, respectively, which describes attractor topology as a point, a looped line, and a plane. Noisy signal is characterized by the infinite value of D_f, that means no structural dynamic in the phase space and there are no attractor. Phase space dimension corresponds to saturation is called embedding dimension and characterizes the minimum value of phase space dimension that contents the whole attractor. The value of d is proportional to the number of degrees of freedom and defined by a number of necessary equation for numerical simulation of the nonlinear dynamics.

Fig. 5 shows obtained dependences of fractal dimension on active ring gain coefficient for H_1 = 1462 Oe, H_2 = 1642 Oe. One can see that both curves have the same form. For G < 2 dB values of fractal dimension is integer. It confirms that in the phase space regular attractors consequently appeared. Increase of magnetic field value led to change of gain coefficient threshold values corresponding to dynamical regime change. In chaotic regime fractal dimension increased up to certain level and then went to saturation. In simplified model of spin wave dynamical chaos increase of fractal dimension could be explained by increase of present nonlinearity influence or appearance of higher order nonlinear processes.

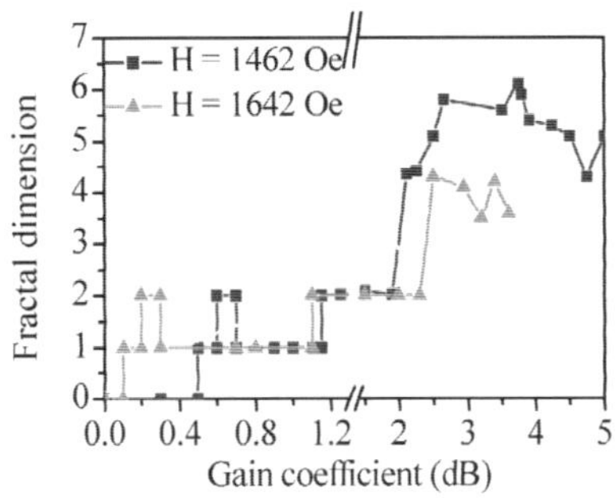

Figure 4. Fractal dimension vs. active ring gain coefficient.

In conclusion we have experimentally investigated broadband self-generation of chaotic signals in active resonant ring based on metalized ferrite film. Frequency spectra of obtained signals were more than 1 GHz. Variation of gain coefficient and magnetic field magnitude allowed to control parameters of self-generated dynamical chaos.

Acknowledgments

The work was supported in part by the Russian Foundation for Basic Research, by Ministry of Education and Science of Russian Federation (Project "Goszadanie"), and by the Deutsche Forschungsgemeinschaft.

References

[1] 2013 *Chaotic Signals in Digital Communications* ed M Eisencraft, R Attux, R Suyama (Boca Raton: CRC Press)
[2] Li W *et al* 2013 *Phys. Rev. Lett.* **111** 044102
[3] 2013*Chaotic Signal Processing* ed. by H. Leung (Philadelphia: SIAM)
[4] Bakemeier L *et al* 2015 *Phys. Rev. Lett.* **114** 013601
[5] Hramov A E *et al* 2014 *Phys. Rev. Lett.* **112** 116603
[6] Wang L X *et al* 2012 *Appl. Opt.* **51** 2935-2940
[7] de Aguiar F M and Rezende S M 1986 *Phys. Rev. Lett.* **56** 1070
[8] Carroll T L 1991 *J. Appl. Phys.* **70** 3197
[9] Kalinikos B A *et al* 1998 *Phys. Rev. Lett.* **80** 4301
[10] Hagerstrom A *et al* 2011 *Phys. Rev. B* **83** 104402
[11] Wu M *et al* 2005 *Phys. Rev. Lett.* **95** 237202
[12] Stancil D D and Prabhakar A 2009 *Spin waves: Theory and Applications* (New York: Springer)
[13] Kondrashov A V *et al* 2010 *Tech. Phys. Lett.* **36** 224
[14] Gurevich A G and Melkov G A 1996 *Magnetization Oscillations and Waves* (Boca Raton : CRC Press Inc.)
[15] Etko Yu L *et al* 2011 *Tech. Phys. Lett.* **37** 1015
[16] Kalinikos B A *et al* 2000 *Phys. Rev. Lett.* **84** 4697
[17] Kantz H and Schreiber T 2004 *Nonlinear Time Series Analysis* (Cambridge: University press)
[18] Takens F 1981 *Lecture Notes in Mathematics* **898** 366
[19] Grassberger P and Procaccia I 1983 *Phys. Rev. Lett.* **50** 346

Journal of Physics: Conference Series **769** (2016) 012090 doi:10.1088/1742-6596/769/1/012090

Magnetic field simulation of wiggler on LUCX accelerator facility using Radia

Y N Sutygina[1*], A E Harisova[1], D A Shkitov[1]

[1] Tomsk Polytechnic University, 30, Lenin St., Tomsk, 634050, Russia

E-mail: yana.sutygina@mail.ru

Abstract. A flat wiggler consisting of NdFeB permanent magnets was installed on a compact linear electron accelerator LUCX (KEK) in Japan. After installing the wiggler on LUCX, the experiments on the generation of undulator radiation (UR) in the terahertz wavelength range is planned. To perform the detailed calculations and optimization of UR characteristics, it is necessary to know the parameters of the magnetic field generated in the wiggler. In this paper extended simulation results of wiggler magnetic field over the entire volume between the poles are presented. The obtained in the Radia simulation magnetic field is compared with the field calculated by another code, which is based on the finite element method.

1. Introduction

On a compact linear electron accelerator LUCX (KEK) in Japan in the end of 2015 was installed a compact planar wiggler [1] consisting of NdFeB permanent magnets. After the installation of this wiggler on LUCX, it is planned to conduct the experiments on the generation of undulator radiation in the terahertz (0.1 - 10 THz) wavelength range [2] on the basis of pre-bunched electron beam with a short distance between bunches which is in order of the bunch length. The electron energy after the electron gun is equal to 8.25 MeV and after the accelerating section is 30 MeV [3]. THz radiation is of a great interest due to the fact that this type of radiation does not ionize organic substances, for example, as X-rays, and therefore it can be used to study biological systems [4].

To perform the detailed calculations and optimize the characteristics of the undulator radiation, it is necessary to know the magnetic field parameters generated in the wiggler. One of the well-known programs that allow us to simulate the undulator radiation characteristics emitted by the electron beam based on freely definable data on a magnetic field is Spectra [5]. Moreover, knowledge of the magnetic field in the whole volume between the wiggler poles is required in order to make the evaluations of the changes and the displacement of the charged particles beam that moves in the wiggler. Those evaluations are necessary to adjust the magnetic system of the accelerator.

This paper presents the magnetic field calculations in the wiggler central plane and the field calculations along the electron beam trajectory (the central axis of the wiggler) for the several values of the gap width between the poles. Also paper presents an analysis of the wiggler rotation relative to it center around all three axes which is influenced to the magnetic field. The magnetic field is calculated on the basis of Radia code [6] which is an integrated module of Mathematica of Wolfram Research [7]. These simulations of magnetic field are compared to experimental data [1] and the data obtained in the simulation based on the finite element method.

2. Device geometry

The wiggler contains five periods consisting of permanent magnets, where the length of each period is equal to 60 mm, the gap width between the poles can be varied from 30 to 60 mm. Three-dimensional model of the wiggler, built in Radia, is shown in figure 1.

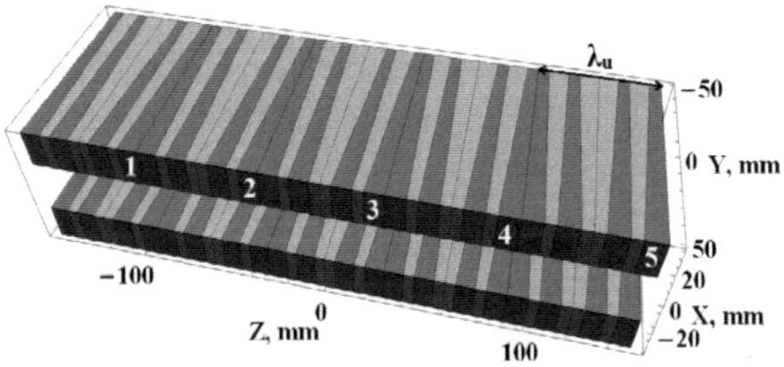

Figure 1. Three-dimensional model of the wiggler.

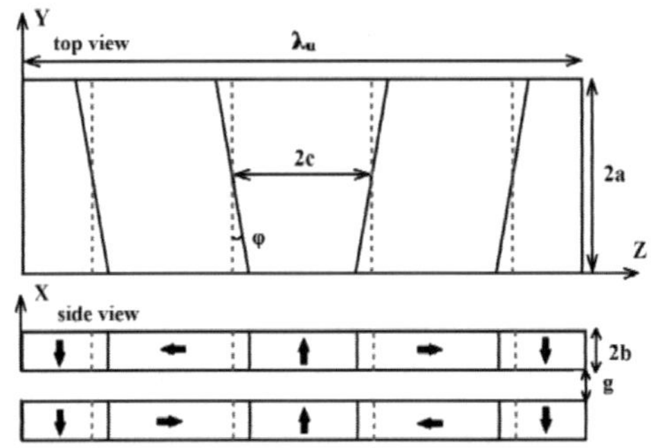

Figure 2. Schemes of one period.

The magnets are in the shape of trapezoidal prism with the same edge angle φ (see figure 2). The magnetic blocks in each period have a horizontal or vertical magnetization direction that is shown in figure 2. The main parameters of the magnetic system are listed in table 1.

Table 1. Main parameters of the wiggler.

Parameters	Value
Period length, λ_u	60 mm
Number of periods	5
Magnet gap, *g*	30 – 60 mm
Edge angle, φ	2.0 deg
Permanent magnet	Nd-Fe-B
Nominal block size, $2b \times 2a \times 2c$	20×100×15 mm
Residual induction of magnet	1.32 T

3. Radia description

The simulation was performed using three-dimensional magnetostatic code – Radia [8]. This software was developed at the ESRF for the magnetic solutions of problems occurring in the design of magnetic lenses, undulators and wigglers. This Radia code is optimized for the tasks of developing namely the undulators and wigglers consisting of permanent magnets and electromagnets. The basis for the calculations in Radia is the boundary element method [9]. During the simulation, the consequent algorithm is to be followed:

1 Construction of the element geometry of the magnetic system.
2 Setting the magnetic properties of the element materials created in step 1.
3 Grouping the elements into containers.
4 If necessary, the geometry segmentation of the system elements into smaller blocks.
5 If necessary, the implementation of an additional transformation to the established system (shifts and/or rotations).
6 The magnetic field calculation generated by the magnet system as a whole or it individual elements.

After constructing the geometry, it is necessary to make a decision on segmenting or non-segmenting of the magnet system. The calculation method used in Radia assumes a uniform magnetization in each element. For the permanent magnet, the magnetic permeability is close to 1 and the magnetization is homogeneous enough for the element volume. However, for materials with high magnetic permeability, magnetization is distributed unevenly. To solve this problem it is necessary to make the partition of the elements into smaller blocks, i.e. segmentation. As a result, the element will be divided into a number of smaller blocks, in which the magnetization will be considered as a homogeneous one. The current version of Radia code does not provide a mechanism for automatic segmentation. The users must specify the segmentation parameters by themselves. The procedure should be stopped when the magnetic field in a given area becomes constant at the desired level of accuracy. The developers recommend firstly to find a sufficient number of blocks in the horizontal direction with no or minimal partition in the longitudinal and vertical directions, and secondly to find a sufficient number of blocks in the longitudinal direction with the previously found value of blocks for the horizontal direction. Segmentation can be applied directly to the whole geometry (i.e. to the container, which combines all the elements) or to different parts of the geometry separately. The simulation has showed that the segmentation is not required.

4. Simulation results

Figure 3a shows a three-dimensional magnetic field distribution in the X axis projection (B_x) in the central plane of the wiggler. As it can be clearly seen from figure 3b, the magnetic field is not symmetrical with respect to the YZ plane. This asymmetry in the magnetic field distribution is caused by the trapezoidal geometry of the magnets, which makes up the wiggler, and their magnetization directions.

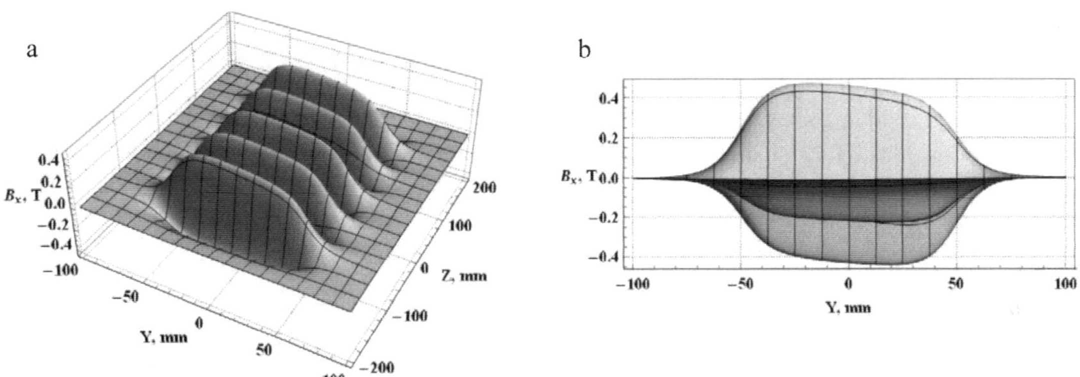

Figure 3. 3D wiggler fields in the central plane with the gap 30 mm width (a), side view (b).

The distribution maxima and minima correspond to the period centers and edges respectively. These extrema of the magnetic field in the YZ plane have a practically constant value and the maximum value spread in the range from -10 to 10 mm along Y-axis is equal to about 4%.

Figure 4 demonstrates the magnetic induction projection on the X axis (B_x) calculated for the gap width values between the wiggler poles equal to 30, 40, 50 and 60 mm. The maximum value of the magnetic induction obtained in the simulation in the wiggler center for the 30 mm gap width is 0.432 T, and the measured value is 0.421 T [1]. The difference between these values is 2.6%. The reason for this discrepancy could be the systematic error of an edge angle in the magnetic block manufacture and not complete accounting of the magnet material properties, in particular, in the simulation, the permeability was assumed to be 1 though the actual value is 1.05 in the direction of the magnetization axis.

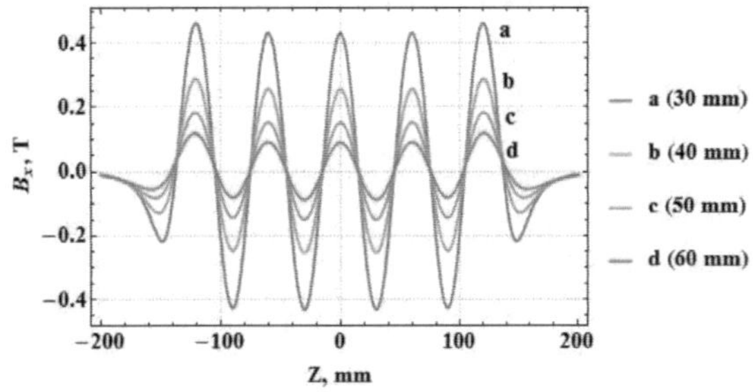

Figure 4. Wiggler magnetic field along the axis for different gap widths between the poles.

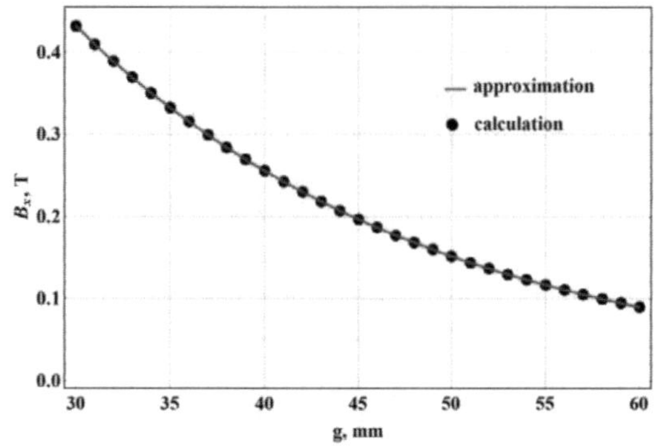

Figure 5. Dependence of the magnetic induction projection on the X axis from the gap width.

Figure 5 shows the magnetic induction projection on the X axis at the wiggler center (0, 0, 0) as a function of the gap width. With the increase of the gap between the poles, the field value decreases exponentially according to the following expression: $a \cdot e^{-b \cdot g}$, where $a = 2.079 \pm 4 \cdot 10^{-4}$ T and $b = 0.052392 \pm 5 \cdot 10^{-6}$ 1/mm. The maximum calculated value of the magnetic field in the wiggler center varies from 0.09 to 0.432 T for the gap width from 60 to 30 mm respectively.

Figure 6 demonstrates the magnetic induction projection (B_x). It was calculated for wiggler with 30 mm gap width along the charged particle beam trajectory. The solid curve corresponds to the case when the beam path coincides with the wiggler axis and the dashed curve corresponds to the magnetic induction calculated for the wiggler rotated at $\alpha = 1/\gamma = 3.548$ degrees angle around the Y axis relative to its center (0, 0, 0) in a counterclockwise direction (see figure 1), where $\gamma = 16.14$ is Lorentz factor of the electron. For comparison of the magnetic field along the beam trajectory at different wiggler positions, the regions O, A and B indicated in figure 6 were chosen which correspond to the wiggler center, the center and the edge of the last period.

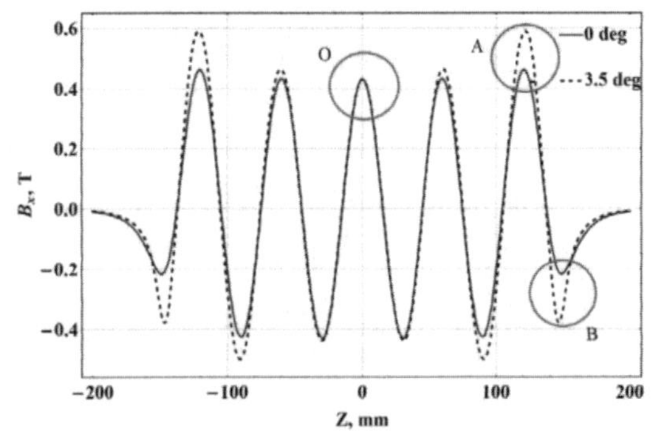

Figure 6. Dependence of the projection magnetic induction (B_x) along the Z axis for the two wiggler positions.

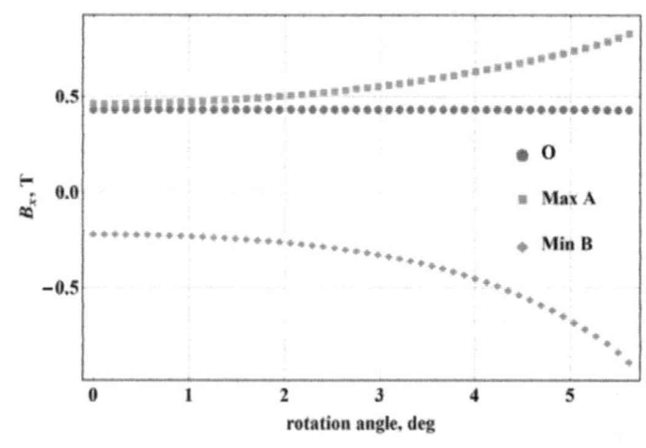

Figure 7. Dependence of the magnetic induction projection (B_x) from the wiggler rotation angle for three locations on the wiggler axis.

Figure 7 shows the dependences of the B_x component, when wiggler turning around the Y axis in the angle range from 0 to 2γ in radians, where 2γ is the maximum angle at which the wiggler does not touch the beam trajectory.

Table 2 to presents the relation (r) of wiggler magnetic fields as a function of the rotation angle (α) relative to the X, Y and Z axes in the regions O, A and B respectively. The parameter r is the ratio of the magnetic induction B_x for the wiggler rotated around an axis at an angle α to the magnetic induction B_x for the wiggler at $\alpha = 0$.

Table 2. Relation of wiggler magnetic fields as a function of the rotation angle.

Value	Rotation axis								
	Y			X				Z	
α,°	1	1.5	2	-4	-2	2	4	-2	2
r (0)	1	1	1	1	1	1	1	0.994	0.994
r (max A)	1.02	1.046	1.087	1.01	1.09	0.988	0.978	0.994	0.994
r (min B)	1.048	1.1	1.2	0.969	0.985	1.013	1.026	0.994	0.994

The simulation has shown that parameter r does not depend on the rotation direction around the Y axis. The parameter r varies depending on the rotation direction around the X-axis due to the internal structure of the wiggler that is the geometry of its magnetic elements. The parameter r varies insignificantly and does not depend on the rotation direction around the Z axis.

5. Method comparison

Figure 8 shows the graphs of the magnetic induction projection on the X axis calculated for 30 mm gap width between the wiggler poles by Radia and Femm. Femm is a program for simulation of the magnetic field in the 2D geometry, which implements the finite element method. It is quite easy to learn, has the graphic user interface and does not require knowledge of programming languages [4].

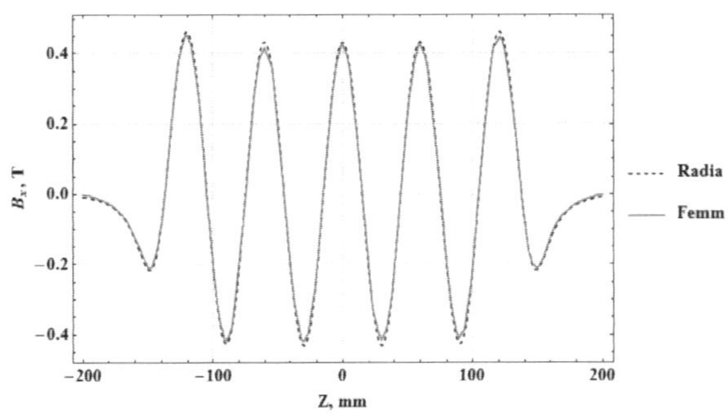

Figure 8. Comparison of the B_x component of the magnetic field calculated in Radia (solid line) and FEMM (dashed line).

The maximum value of the magnetic field in the wiggler center obtained by means of Radia is 0.432 T and Femm is 0.423 T. The difference between these values and the measured value is 2.6 and 0.5% respectively. Apparently, this discrepancy can not only be attributed to the various numerical methods implemented in these programs but also how to set the material magnetic properties, which are different from each other in these applications.

6. Conclusions

Thus, we can conclude that a wiggler model of the LUCX accelerator is developed. This model allows us to calculate the characteristics of the wiggler magnetic field depending on the following parameters of wiggler geometry: the gap width and the rotation angles relative to the beam trajectory. Moreover, on the basis of Radia functional, it is quite easy to move wiggler relative to the beam trajectory, which can be interpreted as a mistake of the wiggler installation or the beam shift relative to wiggler central axis.

The magnetic field is not symmetrical with respect to YZ plane due to the magnet geometry of the wiggler. The magnetic field varies exponentially with changing the gap width. Changing the magnetic field (B_x) in the range from 0.09 to 0.432 T undulator radiation fundamental frequency is changed from 14.5 to 4 THz.

The inaccuracy in the wiggler position adjustment for the rotation around the Y axis mostly affects the magnetic field value along the beam trajectory. The inaccuracies in the adjustment for the rotation around the X and Z axes do not noticeably affect the magnetic field along the beam trajectory. Wiggler rotation about axes X, Y and Z at the angle less than 0.5 degrees leads to the change in the magnetic field (B_x) not more than 1%.

The magnetic field (B_x) obtained by Radia based on the boundary element method, consistent with an accuracy of about 2% with the results obtained by Femm based on the finite element method. However, calculations carried out in Radia are preferred due to it wider range of options.

In the future we plan to simulate the trajectory of the electrons [11] passing through the wiggler at different state of the initial electrons with the help of Radia functionality.

7. References

[1] Kashiwagi S, Kato R, Mihara A, Noda T and Isoyama G 2009 *PRST AB* **12** 120703
[2] Harisova A E, Sutygina Y N, Shkitov D A 2015 *Physical and technical problems in science, industry and medicine: abst. book of the VII Int. sci. conf.* 132
[3] Fukuda M, Araki S, Deshpande A, Higashi Y, Honda Y and Sakaue K 2011 *NIM A* **637** S67
[4] Tonouchi M 2007 *Nature Photon.* **1** 97
[5] Tanaka T, Kitamura K 2001 *J. Synchrotron Rad.* **8** 1221
[6] Chubar O, Elleaume P, Chavanne J 1998 *J. Synchrotron Rad.* **5** 481
[7] https://www.wolfram.com/mathematica/
[8] http://www.esrf.eu/Accelerators/Groups/InsertionDevices/Software/Radia
[9] Gernot B, Smith I, Duenser K 2008 *The Boundary Element Method with Programming: For Engineers and Scientists* (Wien, New-York: Springer)
[10] www.femm.info
[11] Knyazik A, Tikhoplav R, Rosenzweig J B 2009 *Proc. of PAC* **09** 2441

Millimetre-wave isolator based on Al substituted Ba ferrite

A B Ustinov[1], D A Vinnik[2], P Yu Belyavskii[1], S A Gudkova[2,3], I Zakharchuk[4] and E. Lähderanta[4]

[1]Department of Physical Electronics and Technology, St. Petersburg Electrotechnical University, St. Petersburg, 197376 Russia
[2]Laboratory of single crystal growth, South Ural State University, Chelyabinsk, 454080 Russia
[3]Research and educational center of Nanotechnologies, Moscow Institute of Physics and Technology, Dolgoprudny, 141700 Russia
[4]Department of Mathematics and Physics, Lappeenranta University of Technology, Lappeenranta, 53850 Finland

E-mail: ustinov-rus@mail.ru

Abstract. A mm-wave isolator is fabricated and studied. The operating frequency of the devices is 78.5 GHz. A bandwidth at the level of -3 dB equals 1.6 GHz. The device used a flux grown single-crystal aluminum substituted barium ferrite.

1. Introduction

During the past two decades or so the an increase in operating frequencies provides great advantages for the radar and communication systems. One of the approaches to increase frequency of ferrite devices is to use ferrite materials with large magnetocrystalline anisotropy. Magnetic properties of ferrite materials can be widely tuned by substitution. Modification of barium hexaferrite ($BaFe_{12}O_{19}$) by substitution of iron atoms gives significant changes in physical and chemical characteristics such as magnetocrystalline anisotropy, coercive force or Curie temperature [1-17]. Hexagonal ferrites with large magnetocrystalline anisotropy are promising for mm-wave applications as well as for use in sub-terahertz range. In particular, tunable resonators [17-21], isolators [22] and radio-absorbing coatings [13] were developed with the use of films or bulk hexaferrite materials. In this context the elaboration of a robust technology for production of hexaferrite single crystals became important in recent years [23-25].

Purpose of the present work is to grow the aluminum substituted barium ferrite with flux technique and to demonstrate its possible application for mm-wave isolators.

2. Experimental details

Iron oxide (γ-Fe_2O_3, 99.9 %), aluminum oxide (γ-Al_2O_3), barium carbonate ($BaCO_3$) and sodium carbonate (Na_2CO_3) were used for crystal growth [16]. The resistive furnace construction used for single crystal growth was earlier described elsewhere [26]. We used the following composition of initial batch for ferrite crystal growth: Al_2O_3 – 15.63 wt.%, Fe_2O_3 – 51.13 wt.%, $BaCO_3$ – 18.00 wt.%, Na_2CO_3 – 24.75 wt.%.

The raw materials were ground in an agate mortar and set into a platinum crucible. The crucible was placed in a resistance furnace equipped with a type S thermocouple. To homogenize the melt, the furnace was maintained at 1260 °C for 3 h followed by cooling at a rate of 4.5 K/h to 900 °C. After that the system was naturally cooled to room temperature. The spontaneously obtained crystals with a size of up to 8 mm were separated from the solidified melt by leaching in hot nitric acid to remove side phases (e.g. $NaFeO_2$).

Using carbonate flux technique high-quality Al substituted Ba-ferrite crystals of composition $BaFe_{12-x}Al_xO_{19}$ with $x = 0.5$ were grown. Applying the flux technique the crystal growth temperature can be reduced to below 1300 °C, compared with much higher temperatures necessary for flux free crystal growth methods [27, 28]. Fig. 1 presents crystals with edge-length up to 6 mm and the typical shape obtained in our study.

Figure 1. Photograph of $BaFe_{12-x}Al_xO_{19}$ single crystals obtained from flux with mm-scale. One segment of the ruler corresponds to one millimetre.

Using the grown single crystals we made mm-wave waveguide isolator by conventional way [29]. The single crystals were split into narrow slabs. Typical thickness of the slabs was about 100 μm. In-plane sizes of the slabs were of approximately 2 mm^2. The easy axis of the crystal magnetization (c-axis) was orientated perpendicular to the slab plane. To produce an isolator a sample was positioned inside the WR-10 waveguide section on the narrow wall as shown in Fig. 2. The vector network analyzer Rohde&Schwarz ZVA40 with millimeter-wave converters ZVA-Z110 was used. Measurements of S-parameters were carried out for forward and backward propagation direction of test electromagnetic signal through the waveguide section. A part of the measurement cell containing the magnetic sample was magnetized with uniform magnetic field directed perpendicular to the sample plane. The bias field was varied in the range of 0-6 kOe.

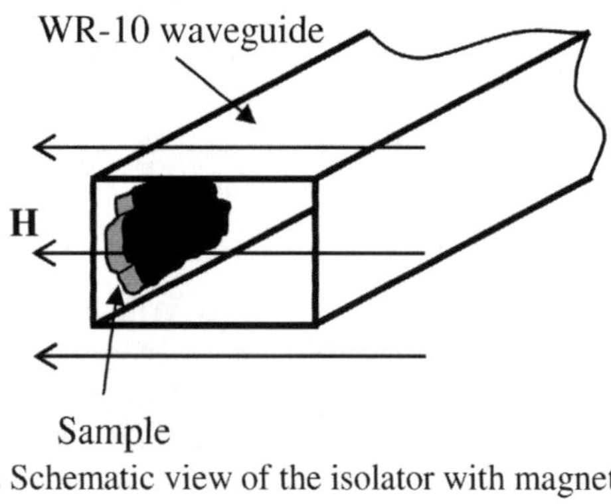

Figure 2. Schematic view of the isolator with magnetic hexaferrite sample.

3. Results and discussion

In accordance with magnetic characterization and previously reported data [16-17], the ferromagnetic resonance frequency for the $BaAl_{0.5}Fe_{11.5}O_{19}$ is in the range of 50-70 GHz for external magnetic fields of H = 0-6 kOe. Therefore, electromagnetic resonances only were observed in range of 75-100 GHz. Fig. 3 shows typical insertion losses in forward and backward direction, i.e. $S_{21}(f)$ and $S_{12}(f)$ characteristics, respectively. The electromagnetic resonances were observed at frequencies of about 78.6 GHz and 96.3 GHz. The measurement device demonstrated reciprocity of the characteristics in zero magnetic field (Fig. 3(a)). The resonance absorption is very similar in both directions of the electromagnetic wave propagation. Application of the bias magnetic field resulted in a non-reciprocal propagation of the test electromagnetic signals in forward and backward directions typical for resonant isolators based on ferrite loaded waveguides (see Fig. 3(b)).

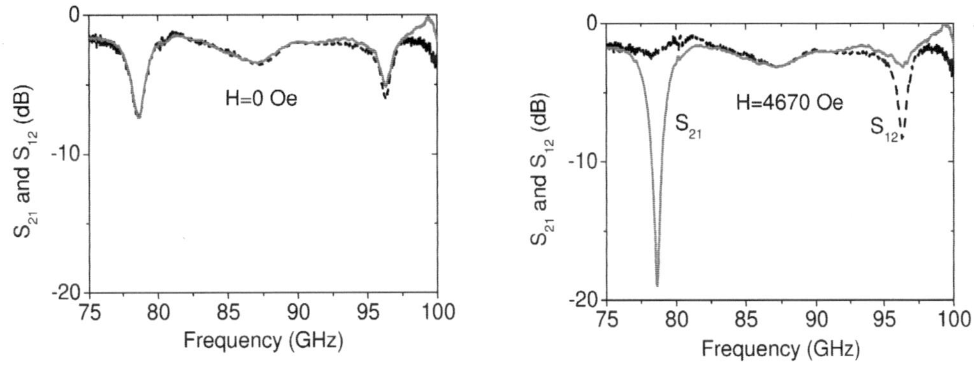

Figure 3. Typical $S_{21}(f)$ and $S_{12}(f)$ characteristics measured for zero magnetic field (left) and for H = 4670 Oe (right).

Considering the resonant peak observed at the frequency of 78.6 GHz it should be noted that for H=0 Oe the resonant frequency f_{res} was 78.743 GHz. An increase of H up to 2820 Oe led to complete diminishing of resonant absorption in backward direction (see $S_{12}(f)$ characteristic) and to the highest absorption in forward direction (see $S_{21}(f)$ characteristic). The resonant frequency of the peak on $S_{21}(f)$ characteristic was slightly decreased down to 78.274 GHz. Further increase of H up to 3265 Oe caused no visible changes in the characteristics. For fields exceeding 3265 Oe an increase of H leads to an increase in f_{res} and the value of the resonance absorption depth is reduced. The results obtained from the measurements of $S_{21}(f)$ and $S_{21}(f)$ characteristics are summarized in Fig. 4.

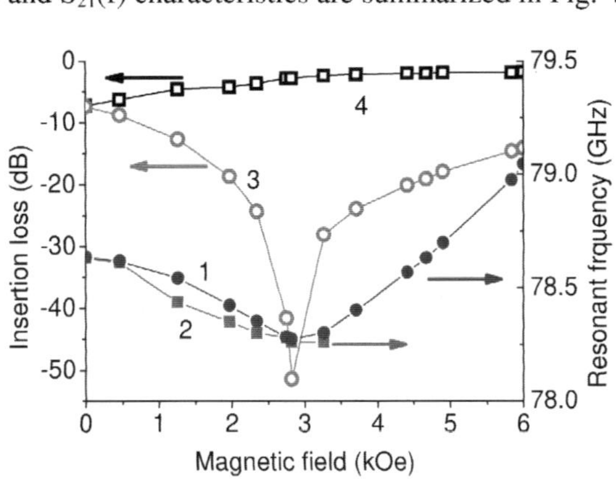

Figure 4. Resonance frequency for S_{21} (curve 1) and S_{12} (curve 2) and insertion loss at the resonance frequency for S_{21} (curve 3) and S_{12} (curve 4) as a function of bias magnetic field.

It is well known that the ferromagnetic resonance is coupled with some electromagnetic modes. Considering this coupling one can conclude that the value of magnetic field H_S for which f_{res} begins to increase corresponds to the transition of the magnetic sample to the saturation state. From this point of view the saturation magnetization $4\pi M_S$ of the investigated $BaAl_{0.5}Fe_{11.5}O_{19}$ hexaferrite can be estimated as $4\pi M_S = H_S = 3265$ G. This value obtained from the microwave measurements is reasonably close to that measured with the SQUID magnetometer (see the Fig. 5).

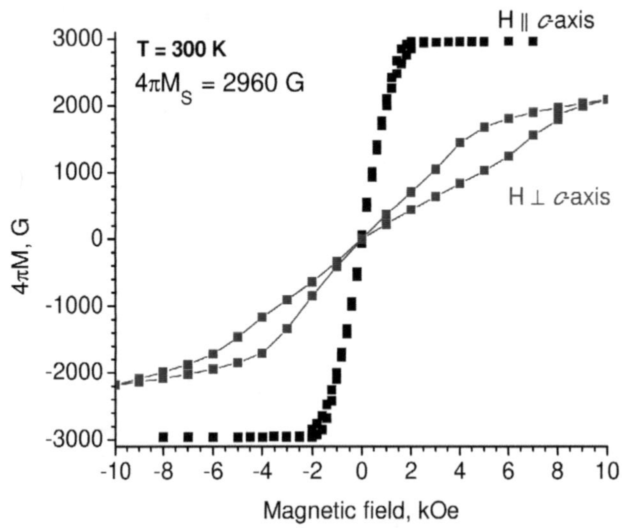

Figure 5. Magnetization as a function of magnetic field at T = 300 K.

The absorption curve (Fig. 3) has a shape close to a Lorentzian profile. The resulting Q-factor of 47 is typical for dielectric modes in the W-band [19, 20]. The obtained characteristics indicate that the single crystals grown by flux technique have good quality and can be used for the development of another millimeter-wave devices. Among them are magnetic field controllable attenuators or notch filters. The operating frequency of the devices would be around 78.5 GHz. A bandwidth at the level of -3 dB from minimum value of the insertion loss equals 1.6 GHz.

4. Conclusions

A mm-wave isolator is fabricated and studied. The operating frequency of the device is 78.5 GHz. A bandwidth at the level of -3 dB from minimum value of the insertion loss equals 1.6 GHz. The use of hexagonal ferrites having a large uniaxial magnetocrystalline anisotropy is a powerful tool to increase significantly an operating frequency of mm-wave devices.

Acknowledgments

The work at SPbETU was supported by the Russian Foundation for Basic Research (grant #16-32-50106 mol_nr), Ministry of Education and Science of Russian Federation (Project "Goszadanie") and by a grant of the President of the Russian Federation for young scientists and PhDs MK-6229.2015.8. The research at SUrSU has been supported by the Russian Foundation for Basic Research, grant № 15-32-51192. This work was partially performed using equipment of MIPT Centers of Collective Usage and with financial support from the Ministry of Education and Science of the Russian Federation (Grant No. RFMEFI59414X0009).

References

[1] Chen D, Liu Y, Li Y, Yang K, Zhang H 2013 *J. Magn. Magn. Mater.* **337-338** 65.
[2] Dhage V N, Mane M L, Keche A P, Birajdar C T, Jadhav K M 2011 *Physica B* **406** 789.
[3] Sl´ama J, Gruskov´a A, Pap´anov´a M, Kevick´a D, Janˇc´arik V, Dosoudil R, Mendoza-Su´arez G, Gonz´alez-Angeles A 2005 *Journal of electrical engeneering* **56 1-2** 21.
[4] Harward I, Nie Y, Chen D, Baptist J, Shaw M J, Liskova E J, Visnovsky S, Siroky P, Lesnak M, Pistora J, Celinski Z 2013 *J. Appl. Phys.* **113** 043903.
[5] Pieper M W, Morel A, Kools F, 2002 *J. Magn. Magn. Mater.* **242-245** 1408.
[6] Marino-Castellanos P A, Anglada-Rivera J, Cruz-Fuentes A, Lora-Serrano R 2004 J. Magn. Magn. Mater. **280** 214.
[7] Vinnik D A, Mashkovtseva L S, Zherebtsov D A, Dyachuk V V, Mikhailov G G 2011 *Bull. S. Ural State Univ. Metall. Ser.* **36 (253)** 41.
[8] Vinnik D A, Tarasova A, Zherebtsov D A, Mashkovtseva L S, Gudkova S A, Nemrava S, Yakushechkina A K, Semisalova A S, Isaenko L I, Niewa R 2015 *Ceram. Int.* **41** 9172.
[9] Shlyk L, Vinnik D A, Zherebtsov D A, Hu Z, Kuo C-Y, Chang C-F, Lin H-J, Yang L-Y, Semisalova A S, Perov N S, Langer T, Pöttgen R, Nemrava S, Niewa R 2015 *Solid State Sciences* **50** 23.
[10] Vinnik D A, Semisalova A S, Mashkovtseva L S, Yakushechkina A K, Nemrava S, Gudkova S A, Zherebtsov D A, Perov N S, Isaenko L I, Niewa R 2015 *Materials Chemistry and Physics* **163** 416.
[11] Vinnik D A, Zherebtsov D A, Mashkovtseva L S 2013 *Doklady physical chemistry* **449** 39.
[12] Vinnik D A, Zherebtsov D A, Mashkovtseva LS, Nemrava S, Perov N S, Semisalova A S, Krivtsov I V, Isaenko L I, Mikhailov G G, Niewa R 2014 *Cryst. Growth Design.* **14** 5834.
[13] Vinnik D A, Zherebtsov D A, Mashkovtseva L S, Nemrava S, Semisalova A S, Galimov D M, Isaenko L I, Niewa R 2015 *J. Alloys Compd.* **628** 480.
[14] Vinnik D A, Zherebtsov D A, Mashkovtseva L S, Nemrava S, Yakushechkina A K, Semisalova A S, Gudkova S A, Anikeev A N, Perov N S, Isaenko L I, Niewa R 2015 *Materials Chemistry and Physics* **155** 99.
[15] Rane M V, Bahadur D, Kulkarni S D, Date S K 1999 *J. Magn. Magn. Mater.* **195** 256.
[16] Vinnik D A, Zherebtsov D A, Mashkovtseva L S, Nemrava S, Bischoff M, Perov N S, Semisalova A S, Krivtsov I V, Isayenko L I, Mikhailov G G, Niewa R 2014 J. Alloy. Compd. **615** 1043.
[17] Ustinov A B, Tatarenko A S, Srinivasan G, Balbashov A M 2009 *J. Appl. Phys.* **105** 023908.
[18] Ustinov A B, Srinivasan G 2008 *Appl. Phys. Lett. 93* 142503.
[19] Popov M, Zavislyak I, Ustinov A, Srinivasan G 2011 *IEEE Trans. Mag.* **47** 289.
[20] Popov M A, Zavislyak I V, Srinivasan G 2011 *J. Appl. Phys.* **110** 024112.
[21] Song Y-Y, Ordóñez-Romero C L, Wu M 2009 *Appl. Phys. Lett.* **95** 142506.
[22] Harris V G 2012 *IEEE Trans. Mag.* **48** 1075.
[23] Harris V G, Chen Z, Chen Y, Yoon S, Sakai T, Gieler A, Yang A, He Y, Ziemer K S, Sun N X, Vittoria C 2006 *J. Appl. Phys.* **99** 08M911.
[24] Harris V G, Geiler A, Chen Y, Yoon S D, Wu M, Yang A, Chen Z, He P, Parimi P V, Zuo X, Patton C E, Abe M, Acher O, Vittoria C 2009 *J. Magn. Magn. Mater.* **321** 2035.
[25] Pullar R C 2012 *Progress in Materials Science* **57** 1191.
[26] Vinnik D A 2014 *Butlerov Communications* **39** 153.
[27] Balbashov A M, Egorov S K 1981 *J. Crystal Growth* **52** 498.
[28] Bugaris D E, zur Loye H-C 2012 *Angew. Chem., Int. Ed.* **51** 3780.
[29] "Advances in Ceramics - Electric and Magnetic Ceramics, Bioceramics, Ceramics and Environment". Edited by Costas Sikalidis (InTech, 2011).

Characterization of delta-doped diamond samples with a planar capacitor

V V Vitko[1,2], A V Kondrashov[1,2], A A Nikitin[1,2], P Y Belyavskiy[1], A B Ustinov[1] and J E Butler[3]

[1]St. Petersburg Electrotechnical University, St. Petersburg, 197376 Russia
[2]Institute of Applied Physics of the Russian Academy of Science, Nizhny Novgorod, 603950 Russia
[3]Cubic Carbon Ceramics, Huntingtown, MD 20639, USA

E-mail: vitaliy.vitko@gmail.com

Abstract. The method for characterization of delta-doped diamonds is developed based on the measurement of a planar capacitor. Theoretical model for calculation of the product of carrier density and carrier mobility is obtained. Theoretical model verified by the experiment. The experimental value of the product of carrier density and carrier mobility was $6{\bullet}10^{14}$ $1/V\cdot s\cdot m$.

One of the most common way to a non-destructive measurement of dielectric properties of artificial diamond films with delta-doped layer is measurement of impedance of a planar capacitor based on this film. RLC-meters are devices that measure the admittance and impedance of a capacitor with dielectric or semiconductor material filling. The planar capacitor based on the delta-doped diamond sample consists of three layers (Fig. 1): a High Temperature High Pressure (HTHP) diamond substrate, a chemical vapor deposited (CVD) diamond film with a delta-doped layer and planar electrodes.

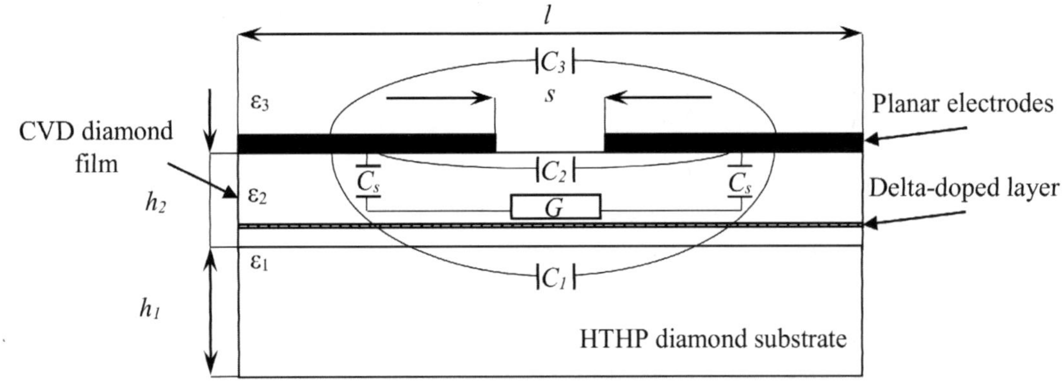

Figure 1. The planar capacitor based on a delta-doped diamond film.

The impedance of the planar capacitor shown in Fig. 1 is formed by (1) the capacitance C_3 of the stray fields in an ambient space with dielectric constant $\varepsilon_3=1$, (2) the capacitance C_2 of the CVD diamond film with the delta-doped layer of thickness h_2 with dielectric constant ε_2, (3) the capacitance C_1 of the HTHP substrate of thickness h_1 with dielectric constant ε_1, (4) the conductivity G of the delta-doped layer and (5) two shunt capacitances C_s due to delta-doped layer conductivity. The equivalent circuit of the structure is shown in Fig. 2.

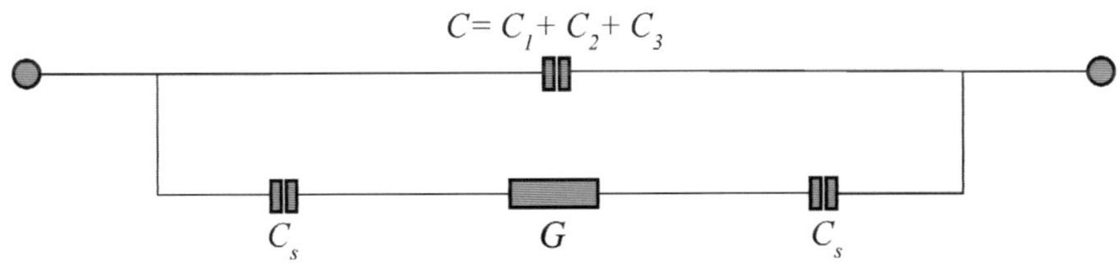

Figure 2. The equivalent circuit of the planar capacitor based on delta-doped diamond film.

The RLC-meter measures resistance R_Σ and capacitance C_Σ of delta-doped diamond structure. One can see from Fig. 2 that the impedance of the investigated structure is determined by the sum of connected in parallel planar capacitances C, as well as connected in series shunt capacitances C_s and conductivity G:

$$Z_\Sigma = R_\Sigma + \frac{1}{i\omega C_\Sigma} = \frac{\frac{2}{i\omega C} + \frac{1}{G}}{1 + i\omega C_S(\frac{2}{i\omega C} + \frac{1}{G})} = \frac{2 + \frac{i\omega C_S}{G}}{i\omega(C_S + 2C + \frac{Ci\omega C_S}{G})}. \quad (1)$$

Therefore, the conductivity is equal to

$$G = \frac{i\omega C_S + C\omega^2 C_S Z_\Sigma}{i\omega C_S Z_\Sigma + 2i\omega C Z_\Sigma - 2}. \quad (2)$$

The shunt capacitance is

$$C_s = \frac{\varepsilon_0 \varepsilon_r S}{h}. \quad (3)$$

The calculation of planar capacitances was based on the partial capacitance method [1, 2]. We represent the composite layered capacitor (Fig. 1) as three simple planar capacitors with homogeneous filler (Fig. 3) connected in parallel, and seek its capacitance as the sum of the three partial capacitances

$$C = C_1 + C_2 + C_3, \quad (4)$$

where C_1, C_2, and C_3 are the capacitances of the component parts of the planar capacitor, which are the fringing field in air, the CVD diamond film with the delta-doped layer, and the HTHP substrate.

Journal of Physics: Conference Series **769** (2016) 012092 doi:10.1088/1742-6596/769/1/012092

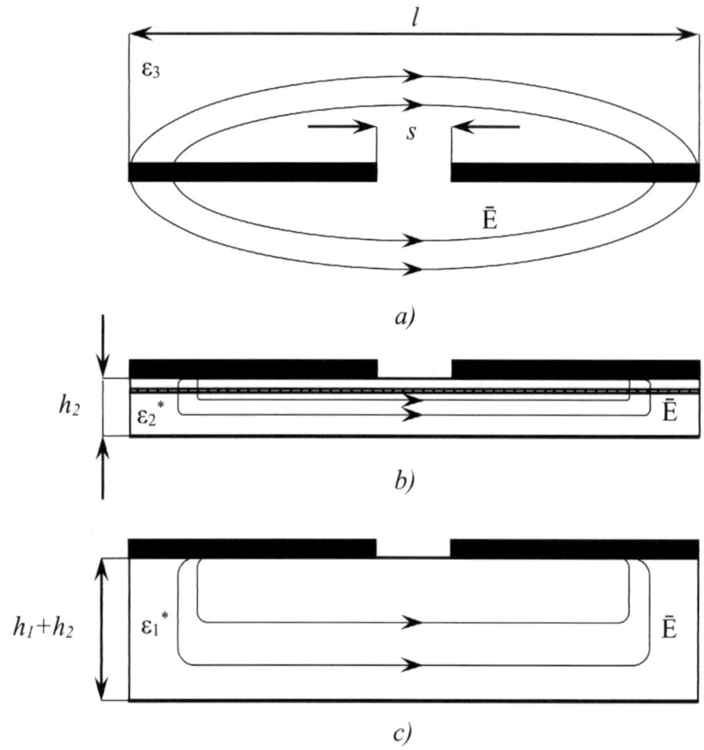

Figure 3. The separation of a planar capacitor shown in Fig. 1 into partial capacitances: a — electrodes in ''air''; b — thin dielectric layer, c — thick dielectric layer.

Here the dielectric constants of the media are decreased in accordance with the equalities

$$\varepsilon_2^* = \varepsilon_2 - \varepsilon_1, \ \varepsilon_1^* = \varepsilon_1 - \varepsilon_3. \tag{5}$$

These changes in the values of the dielectric constants of the layers constitute the main idea of the partial capacitance method, in which each layer is screened from the other layers by a ''magnetic wall'' which does not transmit the normal components of the electric field vectors. The mutual influence of the various layers on the field distribution in each of them is taken into account by the changes in the values of the dielectric constant of each of the layers represented in the case under consideration by relations (5).

The distribution of the electric field in planar capacitors is nonuniform. To calculate the capacitance of a planar capacitor, it is customary to use conformal mapping based on the Christoffel–Schwarz transformation [3]. The capacitance of a sandwich capacitor without fringing fields is easily calculated:

$$C = \frac{\varepsilon_0 \varepsilon_r}{2} \frac{K(k)}{K(k')}, \ k' = \sqrt{1-k^2}, \tag{6}$$

where $K(k)$ is the total elliptic integral of the first kind, k is the modulus of the elliptic integral, ε_0 is the permittivity of free space, and ε_r is the dielectric constant of the medium in which the electric field is concentrated.

For the structures shown in Fig. 3 (a, b, c) we have

$$k_3 = \frac{s}{l}, \tag{7}$$

$$k_2 = th\left(\frac{\pi s}{4h_2}\right) / th\left(\frac{\pi l}{4h_2}\right), \tag{8}$$

$$k_1 = th\left(\frac{\pi s}{4(h_1 + h_2)}\right) / th\left(\frac{\pi l}{4(h_1 + h_2)}\right). \tag{9}$$

The ratio $K(k')/K(k)$ can be represented by the following approximate form [4]:

$$F(k) = \frac{K(k')}{K(k)} = \begin{cases} \pi^{-1} \ln(2\frac{1+(1-k^2)^{0.25}}{1-(1-k^2)^{0.25}}) & \text{for } k^2 \le 0.5, \\ \pi / (\ln(2\frac{1+k^{0.5}}{1-k^{0.5}})) & \text{for } k^2 \ge 0.5. \end{cases} \tag{10}$$

The error of this approximate form does not exceed 10^{-5} [4]. Note that the above formulas give the capacitance of planar structures per unit length in the direction perpendicular to the plane of the figure. Taking into account (6) each capacitance is calculated separately. The capacitance of the air gap on Fig. 3(a) is

$$C_1 = \varepsilon_0 w F(k_1). \tag{11}$$

The capacitance of the CVD diamond with delta-doped layer bounded by the "magnetic wall" on Fig.3(b) is

$$C_2 = \frac{\varepsilon_0 w(\varepsilon_2 - \varepsilon_1)}{2} F(k_2). \tag{12}$$

The capacitance of the HTHP substrate bounded from below by the "magnetic wall" on Fig.3(c) is

$$C_3 = \frac{\varepsilon_0 w(\varepsilon_1 - 1)}{2} F(k_3). \tag{13}$$

Substituting these capacitance in (2) allows to calculate the conductivity of delta-doped diamond layer.

The described method does not allow to investigate the carrier density and carrier mobility of the sample separately. However the CVD diamond fabrication process assumed that the doping rate level is desired value. In this case the carrier mobility can be easily estimated.

As reference we provide an information about desirable mobility for delta-doped CVD diamond samples. It achieves 4500 $cm^2/(V\ s)$ for electrons, and 3800 $cm^2/(V\ s)$ for holes [5]. The carrier density n inside the delta-doped layer can be find as $n = \sigma / e\mu$, where e is electron charge, μ is carrier mobility.

The impedance of investigated structures was measured by RLC-meter. The capacitance of delta-doped diamond structure C_Σ was 0.64 pF and resistance R_Σ was 7783 Ohm. The thickness of HTHP diamond substrate h_1 was 500 μm, the dielectric permittivity ε_1 was 5.7, the thickness of the CVD diamond film was 100 μm with 10 nm delta-doped layer inside and dielectric permittivity ε_2 was 3.5. The conductivity of delta-doped layer calculated by above described algorithm was 96 μS. Thus the product of carrier density and carrier mobility was $2 \cdot 10^{15}$ $1/(V \cdot s \cdot cm)$. Therefore the carrier density was $5.3 \cdot 10^{11}$ $1/cm^3$ assuming that the mobility was 3800 $cm^2/(V\ s)$. Obtained conductivity is two orders lower than reported in the paper [6]. In the same time the value of carrier density is also two orders lower than in another paper [7]. Due to the difference by the same amount of calculated values and results reported in papers, one can assume that the model allows to estimate properties of delta-doped diamond samples. However, the investigated diamond samples had delta layer with low doping rate level.

In summary, this paper reports the non-destructive method of investigation of the dielectric properties of delta-doped diamond. Experimental measurements allow to obtain the product of carrier density and carrier mobility trough the conductivity in the delta-doped layer. Thus, in the present sample, it was $2 \cdot 10^{15}\ 1/(V \cdot s \cdot cm)$. Therefore in case of known doping rate the described method allows to estimate the carrier mobility.

The work was supported in part by the Russian Foundation for Basic Research, Ministry of Education and Science of Russian Federation (Project "Goszadanie"), a grant of the President of the Russian Federation, and by Act 220 of the Russian Government (Agreement No. 14.B25.31.0021 with host institution IAP RAS).

References

[1] Vendik O G, Zubko S P and Nikol'skii M A 1999 *Technical Physics* **44** No 4 349-355
[2] Gevorgian S S, Martinsson T, Linner P L and Kollberg E L 1996 *Microwave Theory and Techniques, IEEE Transactions on* **44** No 6 896-904
[3] Lavrik V I, Savel'ev V N 1970 Handbook on Conformal Transformations *Naukova Dumka, Kiev*
[4] Gupta K C, Garg R and Chadha R 1981 *NASA STI/Recon Technical Report A* **82** 39449
[5] Isberg J, Hammersberg J, Johansson E et al 2002 *Science* **297** No 5587 1670–1672
[6] Kunze M, Vescan A, Dollinger G et al 1999 *Carbon* **37** 787–791
[7] Denisenko A, Kohn E 2005 *Diamond and related materials* **14** No 3 491-498

Journal of Physics: Conference Series **769** (2016) 012093 doi:10.1088/1742-6596/769/1/012093

Theoretical Model of the Stripline Measurement Cell for Doped Diamond Films

V V Vitko[1,2], A V Kondrashov[1,2], A A Nikitin[1,2], P Y Belyavskiy[1], M A Cherkasskii[1], A B Ustinov[1] and J E Butler[3]

[1]St. Petersburg Electrotechnical University, St. Petersburg, 197376 Russia
[2]Institute of Applied Physics of the Russian Academy of Science, Nizhny Novgorod, 603950 Russia
[3]Cubic Carbon Ceramics, Huntingtown, MD 20639, USA

E-mail: vitaliy.vitko@gmail.com

Abstract. A method for measurement of electrophysical properties of delta-doped diamond samples loaded into the measurement cell based on symmetrical stripline is demonstrated. The simulation of S-parameters of the measurement cell with doped diamond samples is carried out. The influence of the carrier density on the transmission characteristics of the symmetric stripline with doped diamond film is investigated. An increase of the carrier density in the doped diamond film from $5\cdot10^{19}$ m^{-3} to 10^{22} m^{-3} leads to frequency shift of transmission characteristic minimum from 15.85 GHz to 16 GHz and increasing of incretion losses from -15 dB to -25 dB.

Recent years an interest to study microwave properties of artificial diamond films with doped layer is observed [1]. In order to use the diamond structures for microwave applications it is necessary to measure the carrier density and mobility. One of the non-destructive investigation technique is based on planar transmission lines. Main advantage of these lines is an absence of cut-off frequency that provides a wide frequency range of measurement [2, 3]. This paper is devoted to design and modeling of the measurement cell based on the symmetric stripline for study the microwave properties of the doped diamond samples.

The insertion losses introduced by the sample in a waveguide path are determined by the imaginary part of the complex dielectric permittivity and magnetic permeability. The method of calculation of the complex permittivity and permeability with the measured transmission coefficient S_{21} and reflection coefficient S_{11} is described in works [4, 5]. In this method, the complex dielectric permittivity can be determined using measured S-parameters of the structure as follows:

$$\varepsilon = \frac{i\left[c/(\omega L)\right]\ln(1/z)}{(1+\Gamma)/(1-\Gamma)}, \tag{1}$$

where c is the velocity of light and ω is an angular frequency, L is a sample length, z is a parameter expressed as $z = \dfrac{S_{11}+S_{21}-\Gamma}{1-(S_{11}+S_{21})\Gamma}$, and Γ is a reflection coefficient expressed as

$$\Gamma=\frac{1-(S_{21}+S_{11})(S_{21}-S_{11})}{(S_{11}+S_{21})-(S_{21}-S_{11})}\pm\sqrt{\left[\frac{1-(S_{21}+S_{11})(S_{21}-S_{11})}{(S_{11}+S_{21})-(S_{21}-S_{11})}\right]^2-1}\,. \quad (2)$$

The sign of the second term in equation (1) is defined by $|\Gamma|<1$. A further specification of the complex dielectric permittivity is carried out in accordance with the Nicholson-Ross algorithm [5].

One of the Maxwell's equation has the following form for the layer with free carriers:

$$\mathrm{rot}\,\mathbf{H}=i\omega\varepsilon_0\left\{\varepsilon'-i\left[\sigma/(\omega\varepsilon_0)\right]\mathbf{E}\right\}=i\omega\varepsilon_0\varepsilon_k\mathbf{E},$$

where **H**, **E** are vectors of the magnetic and electric fields, respectively, and ε_0 is a dielectric permittivity of vacuum, ε' is a real part of the complex dielectric permittivity

$$\varepsilon_k=\varepsilon'-i\left[\sigma/(\omega\varepsilon_0)\right]=\varepsilon'-i\varepsilon'', \quad (3)$$

and $\varepsilon''=\sigma/(\omega\varepsilon_0)$ is its imaginary part.

The conductivity of the doped diamond sample was calculated as

$$\sigma=\omega\varepsilon_0\varepsilon''. \quad (4)$$

Conductivity measurement allows to estimate the doping level of the doped layer in the diamond sample. It is well known that carrier mobility in the diamond achieves 4500 $cm^2/(V\ s)$ for electrons, and 3800 $cm^2/(V\ s)$ for holes [6]. The carrier density n inside the doped layer can be find as

$$n=\sigma/e\mu, \quad (5)$$

where e is electron charge, μ is carrier mobility.

In order to increase functionality of the measurement cell a symmetrical stripline is selected due to features of the field distribution. The symmetrical stripline shown in Fig. 1 consists of a metal strip placed inside a dielectric medium between two grounded metal plates. Stripline is usually filled with a homogeneous dielectric, but it also can be partial filling by different dielectrics. The lowest mode in such a line is a quasi-TEM wave, which is characterized by the absence of the cut-off frequency.

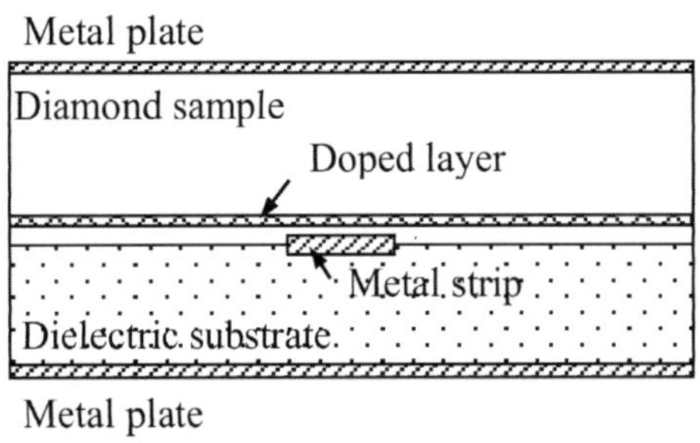

Figure 1. Symmetric stripline with a delta-doped diamond sample.

In order to estimate influence of the carrier density on the transmission characteristics a numerical simulation of S-parameters was carried out. The simulation had following conditions. Ideal metal plates were parallel to the diamond sample and were placed equidistant from the strip. The distance between the metal plates and the strip was equal to the thickness of the diamond sample. As soon as operating frequency range of symmetrical stripline is determined by its dimensions such as the metal strip width and the distance between metal plates [7]. In order to provide operating frequency range from 15.5 GHz to 16.5 GHz following parameters were chosen. A distance between metal plate and strip was about 500 μm. A characteristic impedance of the stripline was of 50 Ohm. The transmission characteristic of investigated measurement cell was calculated with varying carrier density in the doped layer.

Results of the transmission characteristic simulation are shown in Fig. 2. An increase in the carrier density of delta layer leads to a shift of frequency f_{min} corresponding to a characteristic dip in the transmission characteristic. It is also leads to an increase in microwave losses in the sample at the frequency f_{min}. The dependences of frequency f_{min} and transmission coefficient S_{21} at this frequency from the carrier density of the doped layer are shown in Fig. 3 (a, b).

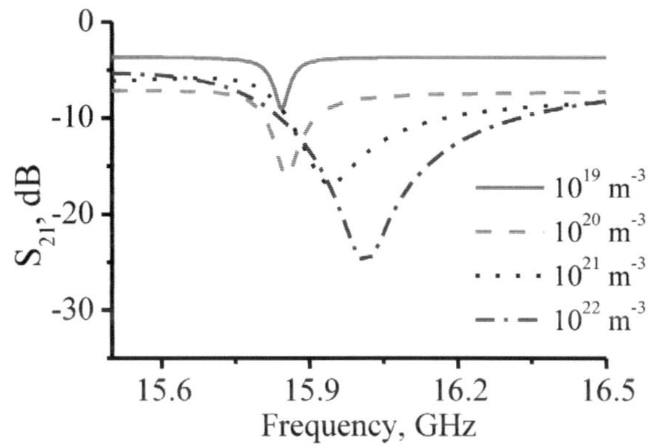

Figure 2. Transmission characteristic of symmetric stripline with a doped diamond sample.

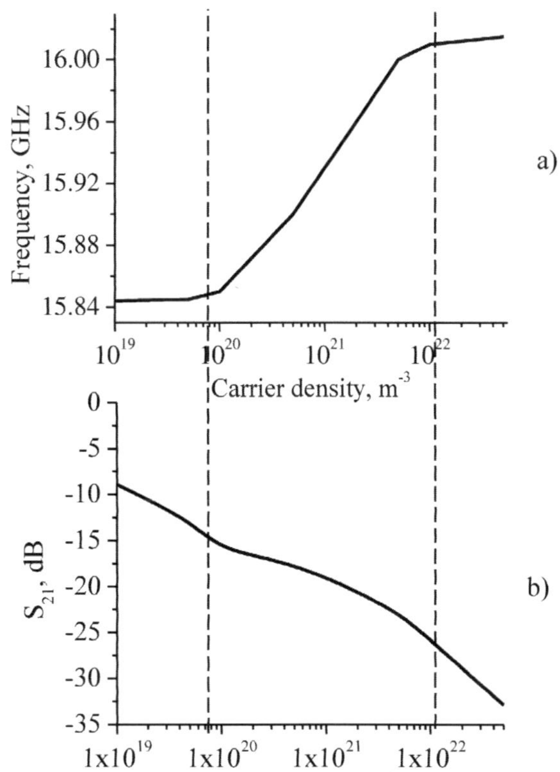

Figure 3. The dependences of the frequency f_{min} (a) and transmission coefficient S_{21} at this frequency (b) from the carrier density in delta-doped layer.

One can see from Fig. 3(a, b) that f_{min} is changed from 15.85 GHz to 16 GHz and insertion losses increase from -15 dB to -25 dB in the carrier density range about $5 \cdot 10^{19}$–10^{22} m^{-3}. Thus, the proposed method allows to determine the carrier density in the doped layer with the measurement the microwave characteristics of the diamond samples.

In conclusion, it was shown that the proposed method based on symmetrical stripline is useful for measurements of microwave parameters of doped diamond samples in a wide frequency range. The simulation of the measurement cell is carried out. It is found that the design of the cell fabricated with a symmetrical stripline has a high sensitivity to microwave parameters in the carrier density range about $5 \cdot 10^{19}$–10^{22} m^{-3}.

The work was supported in part by the Russian Foundation for Basic Research, Ministry of Education and Science of Russian Federation (Project "Goszadanie"), a grant of the President of the Russian Federation for young scientists and PhDs MK-6229.2015.8 and by Act 220 of the Russian Government (Agreement No. 14.B25.31.0021 with host institution IAP RAS).

References

[1] Gurbuz Y, Esame O, Tekin I et al 2005 *Solid-state electronics* **49** No 7 1005-1070
[2] Blackham D V, Pollard R D 1997 *IEEE Trans. instrum. and meas.* **IM-46** No 5 1093-1099
[3] Baker-Jarvis J 1990 . Transmission/reflection and short-circuit line permittivity measurements *Colorado: National institute of standards and technology* 151
[4] Domich P D, Baker-Jarvis J, Geyer R G 1991 *J. res. nation. inst. stand. technol.* **96** No 5 565–575
[5] Nicolson A M, Ross G F 1970 *IEEE Trans. instrum. and meas.* **IM-19** No 4 377–382
[6] Isberg J, Hammersberg J, Johansson E et al 2002 *Science* **297** No 5587 1670–1672
[7] Cohn S B 1995 *IRE Trans.* **3** No 4 16–21

Institute of Physics
Dirac House, Temple Back
Bristol BS1 6BE UK

ISSN: 1742-6588
ISBN 978-1-5108-3385-2

9 781510 833852

CURRAN ASSOCIATES INC.
proceedings
.com